Lecture Notes in Electrical Engineering

Volume 459

The book series *Lecture Notes in Electrical Engineering* (LNEE) publishes the latest developments in Electrical Engineering - quickly, informally and in high quality. While original research reported in proceedings and monographs has traditionally formed the core of LNEE, we also encourage authors to submit books devoted to supporting student education and professional training in the various fields and applications areas of electrical engineering. The series cover classical and emerging topics concerning:

- Communication Engineering, Information Theory and Networks
- Electronics Engineering and Microelectronics
- Signal, Image and Speech Processing
- Wireless and Mobile Communication
- Circuits and Systems
- Energy Systems, Power Electronics and Electrical Machines
- Electro-optical Engineering
- Instrumentation Engineering
- Avionics Engineering
- Control Systems
- Internet-of-Things and Cybersecurity
- Biomedical Devices, MEMS and NEMS

For general information about this book series, comments or suggestions, please contact leontina. dicecco@springer.com.

To submit a proposal or request further information, please contact the Publishing Editor in your country:

China

Jasmine Dou, Associate Editor (jasmine.dou@springer.com)

India

Swati Meherishi, Executive Editor (swati.meherishi@springer.com)
Aninda Bose, Senior Editor (aninda.bose@springer.com)

Japan

Takeyuki Yonezawa, Editorial Director (takeyuki.yonezawa@springer.com)

South Korea

Smith (Ahram) Chae, Editor (smith.chae@springer.com)

Southeast Asia

Ramesh Nath Premnath, Editor (ramesh.premnath@springer.com)

USA, Canada:

Michael Luby, Senior Editor (michael.luby@springer.com)

All other Countries:

Leontina Di Cecco, Senior Editor (leontina.dicecco@springer.com)
Christoph Baumann, Executive Editor (christoph.baumann@springer.com)

** Indexing: The books of this series are submitted to ISI Proceedings, EI-Compendex, SCOPUS, MetaPress, Web of Science and Springerlink **

More information about this series at http://www.springer.com/series/7818

Xinguo Zhang

Editor

The Proceedings of the 2018 Asia-Pacific International Symposium on Aerospace Technology (APISAT 2018)

Volume 2

Asia-Pacific International Symposium
on Aerospace Technology

 Springer

Editor
Xinguo Zhang
Chinese Society of Aeronautics
and Astronautics
Beijing, Beijing, China

ISSN 1876-1100   ISSN 1876-1119 (electronic)
Lecture Notes in Electrical Engineering
ISBN 978-981-13-3304-0   ISBN 978-981-13-3305-7 (eBook)
https://doi.org/10.1007/978-981-13-3305-7

Library of Congress Control Number: 2019935810

This Springer imprint is published by the registered company Springer Nature Singapore Pte Ltd.
The registered company address is: 152 Beach Road, #21-01/04 Gateway East, Singapore 189721, Singapore

Preface

The 2018 Asia-Pacific International Symposium on Aerospace Technology (APISAT 2018) was held in Chengdu City, Sichuan Province, China, during October 16 to 18, 2018. Nearly 400 delegates from China, Australia, Korea, Japan, Canada, Vietnam, and India were present.

Dr. Zuoming Lin, Opening Address

Dr. Zuoming Lin, President of Chinese Society of Aeronautics and Astronautics (CSAA), made the opening address, which has been chaired by Dr. Junchen Yao, Secretary General of CSAA. Following the opening ceremony were four plenary lectures. Dr. Xinguo Zhang, Executive Vice President and CIO of Aviation Industry Corporation of China Ltd. (AVIC), talked on Model-Based Systems Engineering Transformation and Innovation. Dr. Joon-Min Choi, Director of Technology R&D

Head Office of Korea Aerospace Research Insitute (KARI) shed a light on the Space Development in Korea. Dr. Toshio Nishizawa, Director of propulsion research unit of Japan Aerospace Exploration Agency (JAXA), shared an Overview of JAXA's advanced Fan Jet Research (aFJR) Project. Prof. Pier Marzocca of RMIT University introduced its various research activities.

Dr. Xinguo Zhang, Plenary Lecture

The plenary lectures were followed by 80 technical sessions with more than 300 oral presentations, covering topics of aerodynamics, aircraft /UAV design, navigation, combustion and propulsion, guidance and control, structure and materials, air traffic management, etc. Two hundred and fifty peer-reviewed and orally presented papers were published in this book.

On October 18, delegates went on for the technical visit to the Chengdu Tianfu International Aerotropolis and the China Civil Aviation Flight University in Guanghan City, Sichuan Province.

Plenary Lecture Session

The APISAT is a common endeavor among the four professional aerospace societies in China, Australia, Korea, and Japan, namely the Chinese Society of Aeronautics and Astronautics (CSAA), Royal Aeronautical Society Australian Division (RAeS Australian Division), Japan Society for Aeronautical and Space Sciences (JSASS), and Korean Society for Aeronautical and Space Sciences (KSAS). It is hosted in the four countries in turn annually.

The next event will be held in Gold Coast, Australia, during December 4–6, 2019. More information is available on www.apisat2019.com.

About APISAT

Organization

The Asia-Pacific International Symposium on Aerospace Technology (APISAT) is a common endeavor among the four national aerospace societies in China, Australia, Korea, and Japan, namely Chinese Society of Aeronautics and Astronautics (CSAA), Royal Aeronautical Society Australian Division (RAeS Australian Division), Korean Society for Aeronautical and Space Sciences (KSAS), and Japan Society for Aeronautical and Space Sciences (JSASS).

Aim and Scope

APISAT is an annual event initiated in 2009. It aims to provide the opportunity to Asia-Pacific nations for the researchers of universities and academic institutes and for the industry engineers to discuss the current and future advanced topics in aeronautical and space engineering. The official language is English.

Topics

Aerodynamics and Design	Structures and Materials
Computational Fluid Dynamics	Structural Analysis
Wind Tunnel Testing	Structural Testing
Flow Visualization	Smart Structures
Unsteady Aerodynamics	Composite Structures
Acoustics/Aircraft/Helicopter and UAV Design	Structural Dynamics
	Aeroelasticity
Dynamics/Control/Avionics	**Combustion and Propulsion**
Flight Simulation	Combustion Analysis
Navigation	Fuel Injection
Guidance and Control	Turbines
ATM/CNS	Engines
Sensors and Actuators	Cooling Systems
Satellite Attitude Control	Spacecraft Propulsion

APISAT2018 Organization Committee

Executive Committee

Zuoming Lin (President)   CSAA
Andrew Neely (President)   RAeS Australian Division
Youdan Kim (President)   KSAS
Shigeru Obayashi (President)   JSASS

International Program Committee

Xinguo Zhang (Chairman)   CSAA
Song Fu (Co-chairman)   CSAA
Cees Bil (Co-chairman)   RAeS Australian Division
Jae Woo Lee (Co-chairman)   KSAS
Koji Miyaji (Co-chairman)   JSASS

Song Wu	CSAA
Sangchul Lee	KSAS
Tatsunori Yuhara	JSASS

National Organizing Committee (CSAA)

Junchen Yao (Chairman)
Ce Yu (Vice Chairman)
Xue Zhang (Secretariat)
Zhenghong Gao
Jun Zhou
Yongling Fu
Yahong Chen
Jianping Wang
Pinqi Xia
Jun Hua
Jinsong Leng
Wenbo Du
Changchun Zhou

National Organizing Committee (RAeS Australian Division)

Cees Bil (Chairman)
Douglas Nancarrow
Hideaki Ogawa
Pier Marzocca
Murray Scott
Vincent John

National Organizing Committee (KSAS)

Jae Woo Lee (Chairman)
Shangchul Lee (Vice Chairman)
Sang Joon Shin (Secretariat)
Chang-Kyung Ryoo
Changjeon Hwang

National Organizing Committee (JSASS)

Koji Miyaji (Chairman)
Tatsunori Yuhara (Secretariat)
Naoto Azusawa
Yoshitaka Kondo
Hirotomo Kimata
Zhong Lei
Yoshinori Matsuno

Contents

Dynamics/Control/Avionics

Contents

xxvii

Acoustics Aircraft Helicopter and UAV Design

Low Boom Supersonic Aircraft Configuration Optimization Using Inverse Design Method

Yidian Zhang[1], Jiangtao Huang[2], and Zhenghong Gao[1(✉)]

[1] School of Aeronautics, Northwestern Polytechnical University,
Xi'an 710072, China
millerjs@126.com, zgao@nwpu.edu.cn
[2] China Aerodynamics Research and Development Center,
Mianyang 621000, China
hjtcyf@163.com

Abstract. Mitigation of sonic boom to an acceptable stage is a key point for the next generation of supersonic transports. Meanwhile, designing a supersonic aircraft with an ideal ground signature is always the focus of research on sonic boom reduction. This paper presents an inverse design approach to optimize the near-field signature of an aircraft making it close to the shaped ideal ground signature after the propagation in the atmosphere. Using the proper orthogonal decomposition (POD) method, a guessed input of augmented Burgers equation is inversely achieved. By multiple POD iterations, the guessed ground signatures successively approach the target ground signature until the convergence criteria is reached. Finally, the corresponding equivalent area distribution is calculated from the optimal near-field signature through the classical Whitham F-function theory. To validate this method, an optimization example of Lockheed Martin 1021 is demonstrated. The modified configuration has a fully shaped ground signature and achieves a drop of perceived loudness by 7.94 PLdB. Finally, a non-physical ground signature is set as the target to test the robustness of this inverse design method.

Keywords: Sonic boom · Low-boom optimization · Inverse design · Supersonic aircraft

1 Introduction

The physical phenomenon of sonic boom attracted attention in 1947 when the trail aircraft Bell X-1 crossed sonic barrier for the first time in human history. However, over almost ten years after this milestone moment, the physical formation of sonic boom was understood as the natural nonlinear evolution of the near-field pressure as it propagated away from the aircraft [1] (see Fig. 1).

The sad failure of the first generation of supersonic aircraft, such as Concorde and Tupolev TU-144, are largely due to the strict restriction on supersonic flight overland. It has long been a constant pursuit for aircraft designers to achieve transports that fly at supersonic speed quietly and economically. To meet the harsh requirement of noise

© Springer Nature Singapore Pte Ltd. 2019
X. Zhang (Ed.): APISAT 2018, LNEE 459, pp. 1023–1041, 2019.
https://doi.org/10.1007/978-981-13-3305-7_82

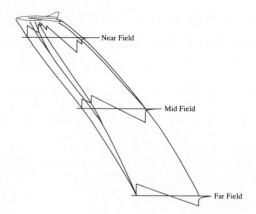

Fig. 1. Sonic boom generation and propagation [2]

level, a large number of studies have been conducted which could be roughly divided into two categories: sonic boom simulation or assessment [3, 4] and sonic boom optimization [5, 6].

The modern sonic simulation approach is normally solved by two steps. First, the near-field overpressure is obtained. Second, the near-field overpressure is extrapolated along the ray path to a concerned altitude (normally the ground). The Whitham F-function theory [7] from linearized supersonic aerodynamics is widely used in the first step to get the near-field signature while computational fluid dynamics (CFD) is gradually adopted in recent years [8] accounting for the nonlinear effects around air-crafts of complex configurations. For the second step, wave parameter method [9] and augmented Burgers equation [10] are the mainstream of today's approach. The main improvement of the latter is that it takes into consideration the attenuation of propagation in real atmosphere.

The prototype flight test in the shaped sonic boom demonstrator (SSBD) [11] verified the effectiveness of aircraft shaping to modify the boom ground signature, which provides a powerful factual basis for sonic boom optimization. Compared with looking for an optimal signature in countless waveforms by forward search, it seems to be more efficient to specify an ideal target signature and achieve it. For that reason, inverse design approaches are mainly used in low-boom designs. Basically, most of the methods for sonic boom inverse design can be categorized into two groups: (1) near-field target matching [12, 13], (2) ground target matching [14, 15]. Besides, Rallab-handi [16] proposed a novel way to match the equivalent area. The near-field matching method designates a target near-field overpressure distribution and achieves it by shaping the aircraft while the far-field matching method specifies a target ground signature. Compared with the second method, the near-field target matching method is easier to operate because the relationship between the near-field signature and aircraft configuration, such as the distribution of equivalent area, is straightforward which is helpful for the shaping of aircraft. However, a well-shaped near-field signature does not necessarily ensure an optimal ground signature after a propagation in the atmosphere. In this work, a desired ground signature is given to match.

The critical point of this inverse design framework is to find the corresponding near-field signature of the given ground signature, or a near-field signature adjacent to it in case it does not exist physically. Based on classical wave parameter method, the inverse design approach has been proposed by George and Seebass [17]. However, there are few counterparts for high-fidelity augmented Burgers equation method. Li [18] presented a novel reversed augmented Burgers to reversely propagate the signature from ground to a near-field position. Despite the straightforwardness of this method, the reverse diffusion equation is inherently ill-posed and hence needs regularization to stabilize the numerical solution. Like the artificial viscosity in CFD, the regularization process contaminates the real solution. Moreover, if the given ground signature is non-physical, some other numerical problems may be caused as it does happen in other inverse problems of partial differential equations [19, 20].

The proper orthogonal decomposition (POD) is widely used in a number of fields, including dimensionality reduction [21], image inpainting [22], pattern recognition [23] and signal processing [24]. In addition, it also yields satisfactory results in aerodynamics, comprising reduced-order model for unsteady aerodynamics [25, 26], inverse design for airfoils [27] and analysis of turbulent flows [28]. This technique extracts a set of empirical modes from samples, which illustrate the dominant behavior of the concerned system. Impressively, the POD based method exerts no restriction on the objective of inverse design, even if it is non-physical, and it will be showed in this paper.

This paper introduces an extended POD method, so-called gappy (incomplete) POD method [29, 30] (GPOD), to the inverse design of sonic boom. In short, the method is stated as follows: given a target ground signature, the optimal near-field signature can be determined by appropriate interpolation of known samples. First, the mathematical models used in this work is introduced in Sect. 2. Then, the optimization framework is constructed in Sect. 3. Section 4 exhibits an optimization example of Lockheed Martin 1021. Section 5 concludes this paper.

2 Mathematical Models

2.1 Proper Orthogonal Decomposition

The fundamental POD method is introduced here in brief. Detailed deduction and discussion are provided by Berkooz [28].

Despite the big success in many nonlinear problems, POD is essentially a linear method. The core thought of it is to find a set of optimal POD modes ($\mathbf{\Phi}$) to maximize the projection of the raw data on it. Mathematically, it can be described as a maximum problem with a constraint:

$$\max_{\mathbf{\Psi}} \left\langle |(\mathbf{U}, \mathbf{\Psi})|^2 \right\rangle = \left\langle |(\mathbf{U}, \mathbf{\Phi})|^2 \right\rangle$$
$$\text{s.t.} \, (\mathbf{\Psi}, \mathbf{\Psi}) = 1 \tag{1}$$

Where $\langle \cdot \rangle$ is an averaging operation, which may be time, space, or ensemble average. $(\cdot)$ means the inner product of two vectors and $|\cdot|$ represents the norm induced by inner product:

$$|\Psi| = (\Psi, \Psi)^{1/2} \tag{2}$$

From (1), it can be shown that, Φ is the eigenfunction of the correlation tensor:

$$\int_{\Omega} \langle |(\mathbf{U}(x), \mathbf{U}^*(x'))| \rangle \Phi(x')dx' = \lambda\Phi(x) \tag{3}$$

Where $\mathbf{U}^*$ is the hermitian of $\mathbf{U}$.

The method of snapshots, firstly introduced by Sorovich [31], is a great improvement for POD. This method transforms the original analytical problem to a sample-based problem which is easy to solve numerically. Most of today's applications of POD are based on the method of snapshots.

In the linear space spanned by a set of linearly independent snapshots (samples), Φ is simply a linear combination of the M centralized snapshots $\{\mathbf{U}^i\}_{i=1}^{M}$:

$$\Phi^{(j)} = \sum_{i=1}^{M} \beta_i^{(j)} \mathbf{U}^i \tag{4}$$

For finite samples, (3) can be converted to the following eigenproblem:

$$\mathbf{R}\beta = \lambda\beta \tag{5}$$

Where $\mathbf{R}$ is the correlation matrix:

$$R_{ij} = \frac{1}{M}(\mathbf{U}^i, \mathbf{U}^j) \tag{6}$$

The correlation matrix has M eigenvectors: $\beta = \left(\beta^{(1)}, \beta^{(1)} \dots \beta^{(M)}\right)$, which are corresponding to M eigenvalues: $\lambda = (\lambda_1, \lambda_2 \cdots \lambda_M)$. Based on (4), the M optimal POD basis vectors (also called empirical basis) are determined. The degree of dominance of each basis is given by the corresponding eigenvalue. Provided the abovementioned eigenvalues are in descending order, the "energy" captured by the first N modes is determined by:

$$C = \frac{\sum_{i=1}^{N} \lambda_i}{\sum_{j=1}^{M} \lambda_j} \times 100\% \tag{7}$$

To approximate the desired space, a linear combination of L selected modes is used:

$$\mathbf{U} \approx \sum_{i=1}^{L} \alpha_i \mathbf{\Phi}^{(i)} \tag{8}$$

Where the selection of modes is determined by (7), typically, an energy percent of 99% is selected.

POD method also offers a good way to fill the incomplete data, which is regarded as gappy POD method. The incomplete data can be given in the following form:

$$\mathbf{I} = \begin{bmatrix} \mathbf{K} \\ \mathbf{X} \end{bmatrix} \tag{9}$$

Where $\mathbf{K}$ represents the known elements while $\mathbf{X}$ represents the unknowns. In this paper, $\mathbf{K}$ stands for the target ground signature and $\mathbf{X}$ means the unknown near-field signatures.

A set of snapshots are sampled and the first L modes $\{\mathbf{\Phi}^{(i)}\}_{i=1}^{L}$ are extracted via POD analysis. The selected modes can be reorganized into the following form:

$$\mathbf{\Phi} = \begin{bmatrix} \mathbf{\Phi}_{\mathbf{K}} \\ \mathbf{\Phi}_{\mathbf{X}} \end{bmatrix} \tag{10}$$

The subscripts $\mathbf{K}$ and $\mathbf{X}$ represent the components of the basis corresponding to knowns and unknowns respectively.

Then, the known components in the incomplete data is reconstructed by the corresponding POD basis using the least square method:

$$\mathbf{\Phi}_{\mathbf{K}}^{\mathrm{T}} \mathbf{\Phi}_{\mathbf{K}} \mathbf{\Gamma} = \mathbf{\Phi}_{\mathbf{K}}^{\mathrm{T}} \mathbf{K} \tag{11}$$

Where $\mathbf{\Gamma} = (\gamma_1, \gamma_2 \ldots \gamma_L)^{\mathrm{T}}$ is the new coordinates of the known data in the coordinates system defined by the L eigenvectors.

Finally, the repaired vector $\tilde{\mathbf{X}}$ of GPOD is given by:

$$\tilde{\mathbf{X}} = \mathbf{\Phi}_{\mathbf{X}} \mathbf{\Gamma} \tag{12}$$

It should be noted that, the POD makes no special assumption about the analyzed data. In addition, no restriction is exerted on the incomplete data which ensure the robustness of this method.

2.2   Augmented Burgers Equation

The augmented Burgers equation is an accurate physical model to simulate the propagation of sonic boom in a stratified real atmosphere with low computational cost compared with direct CFD simulation. Therefore, this method is widely adopted by sonic boom researchers. In the second sonic boom prediction workshop (2nd SBPW) [32] ten of the eleven participants selected Burgers equation based method as the tool for the propagation cases in that workshop. The form of it is stated here briefly. A comprehensive introduction is provided by Cleveland [10]. Equation (13) is the non-dimensional form of it:

$$\frac{\partial P}{\partial \sigma} = P\frac{\partial P}{\partial \tau} + \frac{1}{\Gamma}\frac{\partial^2 P}{\partial \tau^2} + \frac{1}{2\rho_0 c_0}\frac{\partial \rho_0 c_0}{\partial \sigma}P - \frac{1}{2S}\frac{\partial S}{\partial \sigma}P + \sum_v \frac{C_v \frac{\partial^2}{\partial \tau^2}}{1 + \theta_v \frac{\partial}{\partial \tau}}P \qquad (13)$$

Where the fives terms in (13) represents nonlinear distortion, classical attenuation, atmospheric stratification, geometry spreading and molecular relaxation respectively. The dimensionless pressure is $P = p'/p_0$, where p_0 is the reference pressure. σ is the non-dimensional distance normalized by shock formation distance $x_{sf} = \rho_0 c_0^3/(\beta\omega_0 p_0)$. Where ρ_0, c_0 are the ambient density and sound speed. β is a constant define by specific heat ratio $\gamma = 1.4$, where $\beta = 1 + (\gamma - 1)/2$. The dimensionless time is $\tau = \omega_0 t'$ where $1/\omega_0$ is the reference time and t' is the reduced time. The classical dimensionless attenuation parameter is defined by $\Gamma = b\omega_0/(2\beta p_0)$, where b is the viscosity coefficient. The dimensionless relaxation time is $\theta_v = \omega_0 \tau_v$. The dimensionless time is given by $C_v = m_v \tau_v \omega_0^2 x_{sf}/(2c_0)$, where the dimensionless sound speed disturbance is $m_v = 2(\Delta c)_v/(c_0)$.

Equation (13) can be solved by splitting method. The initial equation is split into five equations:

$$\frac{\partial P}{\partial \sigma} = P\frac{\partial P}{\partial \tau} \qquad (14)$$

$$\frac{\partial P}{\partial \sigma} = \frac{1}{\Gamma}\frac{\partial^2 P}{\partial \tau^2} \qquad (15)$$

$$\frac{\partial P}{\partial \sigma} = \sum_v \frac{C_v \frac{\partial^2}{\partial \tau^2}}{1 + \theta_v \frac{\partial}{\partial \tau}}P \qquad (16)$$

$$\frac{\partial P}{\partial \sigma} = \frac{1}{2\rho_0 c_0}\frac{\partial \rho_0 c_0}{\partial \sigma}P \qquad (17)$$

$$\frac{\partial P}{\partial \sigma} = -\frac{1}{2S}\frac{\partial S}{\partial \sigma}P \qquad (18)$$

(14)–(18) are solved sequentially and periodically from the near-field to a desired altitude. Lee [33] showed that, the error of splitting is small if the time step is small enough.

3 Inverse Design Framework

This paper focuses on the development of an efficient methodology to determine an optimal near-field signature for a given ground signature using GPOD. In essence, GPOD of snapshots is a sample-based method. To enhance the optimizer, the sample library is dynamically adjusted during the design process by means of the following two ways. (1) The guessed near-field signature that fail to approach the given ground signature is added to the sample library as a calibration sample. (2) The sampling range shrinks as the optimizer advances for fine search at later stages. In each GPOD process, using the dominant modes extracted from the samples, the target ground signature is fitted. Then, the same fitting coefficients are applied to the samples to get the predicted

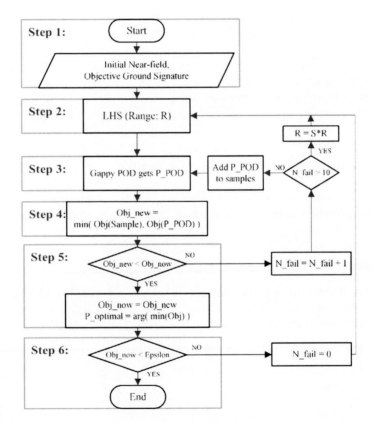

Fig. 2. Sonic boom inverse design framework (The objective function will be given in Sect. 4.2)

near-field signature as is introduced in Sect. 2.1. A flow chart of the whole framework is depicted in Fig. 2. The operations of each step are listed as follows:

- **Step 1** – *Initialization*: Define the initial near-field signature to optimize; generate an ideal ground signature as the target for the following steps.
- **Step 2** – *Sampling*: A set of samples are generated based on Latin hypercube sampling design, the sampling range is: $R = (r_{lb}, r_{ub})$.
- **Step 3** – *Gappy POD reconstruction*: The guessed near-field signature is got by GPOD method. The true ground signature for the guessed near-field signature is calculated.
- **Step 4** – *Comparison and selection*: Select the best near-field signature from the samples and the guessed near-field signature as the design result for current design loop.
- **Step 5** – *Enhancement judgement*: Judge whether the cost drops. If it drops update the current optimum and go to stop 6. If not, then add the current guessed near-field signature to sample library as the calibration sample and go to step 3. If it fails to drop more than 10 times, then shrink the sampling range: $R = S \times R$ and go to step 2. S is the constriction factor. $(0 < S < 1)$
- **Step 6** – *Convergence judgement*: Judge whether the convergence criteria is reached, if it converges then end the optimization, if not go to step 2.

4   Validation and an Optimization Example

In this section, the code for the simulation of sonic boom propagation is validated firstly. Then an optimization example is set and demonstrated. Finally, the results of the optimizer are discussed.

4.1   Validation of Code for Sonic Boom Propagation

The accurate physical model is the basis for optimization. To validate the code used in this paper, two cases in 2nd SBPW [32] are utilized. The calculated ground signature are compared against existing validated code sBOOM [34] developed by NASA.

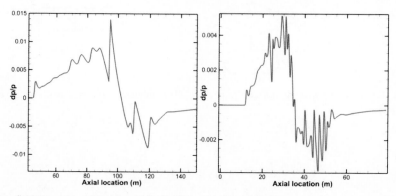

(a) Near-field signatures of Lockheed Martin 1021   (b) Near-field signatures of axis-symmetric body

Fig. 3.   Near-field signatures of two cases in 2nd SBPW (rolling angle: $0°$)

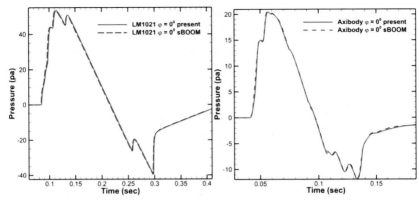

(a) Ground signatures of Lockheed Martin 1021   (b) Ground signatures of axis-symmetric body

Fig. 4. Comparisons of ground signatures in 2nd SBPW against sBOOM (atmospheric conditions: standard atmosphere profile, constant relative humidity of 70%, rolling angle: $0°$)

The two cases of Lockheed Martin 1021 (LM 1021) and axis-symmetric body in 2nd SBPW are selected which are the cases most of the participants have close results. The atmospheric conditions for the selected cases are standard atmosphere profile and constant relative humidity of 70% across the profile. Figure 3 shows the near-field signatures of them which are extracted from several body lengths away from the aircraft. Figure 4 gives the comparison of the ground signatures of both two cases against sBOOM. The signatures in both cases conform very well and validate that the results obtained from the present code are reliable for the following optimization.

4.2   Setting of the Optimization Example

For the sake of fidelity and convenience, the same case of LM 1021 is selected as the initial configuration for the optimizer. The configuration of it is illustrated in Fig. 5.

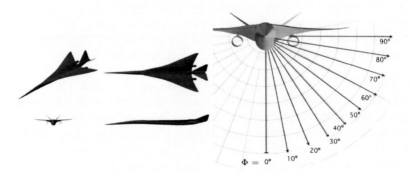

Fig. 5. The configuration of LM 1021(32)

Before starting the optimization process in Sect. 3, an interesting nature of the signature of sonic boom is introduced first.

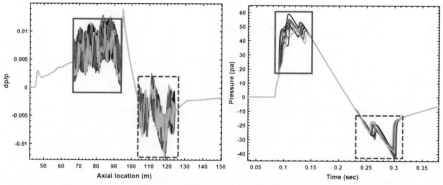

(a) A set of disturbed near-field of LM 1021

(b) The corresponding ground signatures after propagation

Fig. 6. A set of disturbed near-field signatures and the corresponding ground signatures of LM 1021 (The signatures in red solid boxes and blue dashed boxed represent the corresponding signature sections before and after propagation.)

As is illustrated in Fig. 6, if we impose perturbation to two isolated regions in the near-field signature, the corresponding ground signature is also disturbed in two isolated regions. In Fig. 6, two pairs of red solid boxes and blue dashed boxes represent the corresponding signature sections before and after propagation. It shows that the local disturbance only exert influence on a finite region rather than the whole waveform. From the macroscopic aspect, these two pairs of regions in signature can be regarded as two independent physical based input-output systems. The mechanism behind this phenomenon is that the influence of the nonlinear effect and attenuation is limited and only affects the local characteristics of the whole waveform. This nature can be exploited in the following optimization. The inverse design framework only considers the disturbed regions in Fig. 6. Besides, these two pair of regions are treated as two independent subsystems. For instance, the guessed near-field signature of the first disturbed region only depends on the signatures in the two red solid boxes in Fig. 6.

Depends on the low-boom principles summarized by Plotkin [35], the original ground signature is shaped into a sine-like boom which would remove the audible high frequency energy. The target ground signature is shown in Fig. 7.

The perceived noise level of the ground boom is evaluated by the method proposed by Stevens [36] and adapted by Shepherd [37]. This metric is widely used in today's sonic boom research which correlates with human perception of sonic booms very well. The perceived noise level of the shaped target gains a reduction of 8.7 PLdB from 91.93 PLdB to 83.23 PLdB.

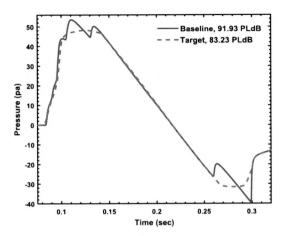

Fig. 7. The baseline and target ground signature for LM 1021 (a reduction of 8.7 PLdB after shaping)

To describe the degree of proximity of the intermediate design ground signature to the target, the Euclidean metric is naturally selected as the objective function:

$$Obj(D) = \sqrt{\sum_{i=1}^{N} \left(P^i_{Design} - P^i_{Target}\right)^2} \tag{19}$$

Where D is the design variables, P_{Design} is the ground signature of the design and P_{Target} is the target ground signature. The objective function has the dimension of Pascal. N is the number of discrete ground signature points. Although this objective function is ubiquitous in sonic boom design, some questions are still pending and a discussion on it will be given in Sect. 4.4.

The design variables in this problem is the discrete points in the near-field pressure distribution. The number of them is 200 in total. In this case, only the points inside the two disturbed regions are active. The settings of the optimization example are as follows. For the first disturbed region (region in red solid boxes in Fig. 6(a)), 64 design variables are active. And for the second disturbed region (region in blue dashed boxes in Fig. 6(a)), 42 design variables are active. For both of these two regions, the initial sampling range is: R = $(-4 \times 10^{-3}, 4 \times 10^{-3})$ and the constriction factor in this case is $S = 0.8$. In each sampling, 30 samples are generated and calculated. 99% of energy is chosen as the threshold for POD modes selection. In each optimization, a total of 50 times of sub-iterations are conducted. Since The main purpose of this paper is to demonstrate the effectiveness of the method, a relatively sparse grid is used in this paper. It requires about 10 s for a call of sonic boom simulation routine on a 64-bit Inter Quad Core 3.8 GHz processor with 8 GB RAM.

4.3   Results of the Optimization Example

To eliminate the impact of the uncertainty of the samples, the optimizer is conducted 200 times in total. By checking the intermediate results of the optimization, it can be found that the designs of the lowest cost do not necessarily have the lowest noise level. Hence both of the design of the lowest cost and the design of the lowest noise level are stored during the optimization loops. The statistical results of them are listed in Table 1. As reported in Table 1, the mean noise level of the designs is 87.01 PLdB which has an enhancement of 4.94 PLdB compared with the baseline by almost 600 calls of sonic boom propagation model. For the optimal design, a reduction of 7.34 PLdB is achieved, which is regarded as a huge improvement in sonic boom optimization because the perceived loudness of sonic boom is in logarithmic scale.

Table 1. The statistical results of 200 times optimization

	Baseline	Minimum	Mean	Standard deviation
Cost	218.49 (pa)	60.61 (pa)	72.58 (pa)	5.32 (pa)
Noise level	91.93 (PLdB)	84.59 (PLdB)	87.01 (PLdB)	0.78 (PLdB)

The ground signature of the design with the lowest noise level is plotted in Fig. 8. As can be seen in Fig. 8, the optimized ground signature is fully shaped. The sharp shocks are shaped into smooth wiggles across the target signature.

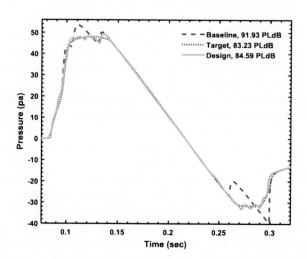

Fig. 8. The comparison of the designed ground signature against the target

The corresponding designed near-field and the iteration history is presented in Fig. 9. The former illustrates that the original smooth near-field signature is optimized into repeated oscillations. These sharp shocks will be damped by atmospheric attenuation and finally become comparatively smooth waves near the ground. The latter of Fig. 9 reveals a typical phenomenon in the inverse design of sonic boom. Since the optimizer is driven by objective function (19), undoubtedly, the cost of it declines strictly. However, the perceived noise level does not drop monotonously, although the downward trend of it is apparent. At the last step, a drop of the cost results in a small increase on the noise level.

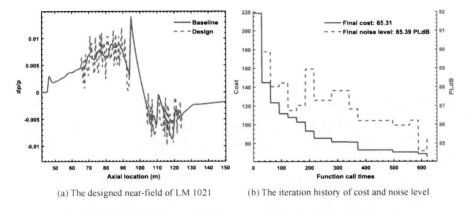

(a) The designed near-field of LM 1021   (b) The iteration history of cost and noise level

Fig. 9. The design result and iteration history of the optimizer

Once the near-field signature is determined, the equivalent area could be obtained by means of the inverse Abel transform [18] based on the assumption for slender bodies [38]:

$$A_e(x) = \frac{\sqrt{2R} \times \sqrt{M^2 - 1}}{\gamma M^2} \int_0^x \left(\frac{dp}{p}\right)\sqrt{x - y}\,dy \qquad (20)$$

Based on (20), the distribution of the equivalent area is got by numerical integration and is plotted against the initial equivalent area distribution in Fig. 10. The equivalent area consists of two parts accounting for the contributions of volume and lift respectively.

Figure 10 shows that, the equivalent area of the design is obviously tightened at the tail section, which means a volume loss for the aircraft if the lift distribution is assumed to be unchanged. In addition, the optimized equivalent area distribution is smoother than the baseline.

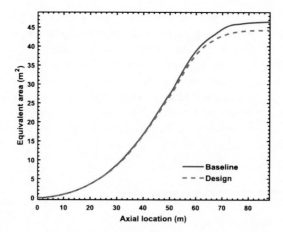

Fig. 10. The equivalent area of the designed near-field signature

4.4   Discussions on the Optimization Example

In this section, some discussions about the intermediate and final results of the optimization is given. First the optimized ground signature is inspected in frequency domain. Then the 'energy distribution' of POD modes and the rationality of the current objective function is investigated. Finally, a nonphysical target ground signature is utilized to test the robustness of the presented method.

To check the result of the optimal design, the shaped ground signature is inspected in the frequency domain by one-third octave analysis [36]. The frequency spectra of sound pressure level (SPL) and the loudness in the scale of sone is presented in Fig. 11 respectively, where one sone is referenced to a noise of 32 dB at 3150 Hz. As we can see clearly in Fig. 11(b), the optimized ground signature reduces the loudness

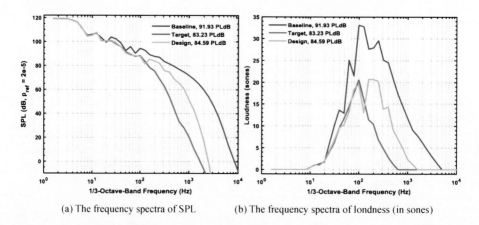

(a) The frequency spectra of SPL   (b) The frequency spectra of londness (in sones)

Fig. 11. The frequency spectra of optimized ground signature against baseline and target (Bands around 100 Hz are the dominant bands of the noise level)

significantly at bands around 100 Hz, which are the dominant bands of the noise level. In the frequency domain, the strength of the ground boom is dominated by the bands of the highest SPL, the contributions from other bands are negligible. To attenuate the overall noise level, special attention should be given to noise components near 100 Hz.

Figure 12 indicates the energy captured by each POD mode when 30 samples are selected. The first 16 modes capture more than 99% of the overall energy which are selected to reconstruct the target ground signature.

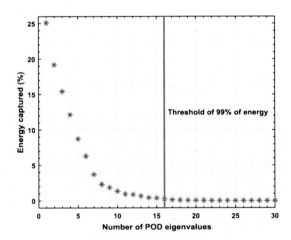

Fig. 12. Energy captured by each POD mode (The first 16 modes capture more than 99% of the overall energy)

A question remained in Sect. 4.3 is that, a drop of the cost of the objective function cannot ensure a drop in perceived noise level. To illustrate this question more thoroughly, the cost and the noise level of all designs at the last step of iteration are plotted in pair in Fig. 13. As is presented in Fig. 13, although there are few final designs of high cost but low noise level and vice versa, the relationship between cost and noise level is not very clear. Despite this uncertainty, the presented target has achieved good results in previous optimization. From the author's point of view, the target guides the direction of the optimizer rather than a requirement must be met. For this reason, the storing of both the design closest to target and the design of the lowest noise level during the optimization process is strongly recommended. In addition, blindly pursuing the proximity to the target is meaningless. Furthermore, some advanced objective functions should be proposed to describe to degree of closeness to the target in term of the frequency domain rather than the stiff Euclidean metric.

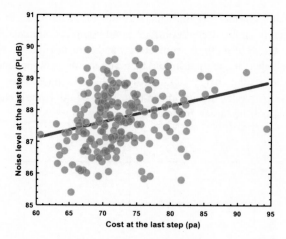

Fig. 13. Cost versus the corresponding noise level when iteration ends (The noise level here is the loudness corresponding to the waveform at the last iteration rather than the optimal noise level of the whole optimization)

In the last part of this section, the robustness of the present method is examined when non-physical solutions are set as the desired ground signature. A zigzag solution is made whose first order derivative is discontinuous at inflection points. This kind of ground signature is physically non-existent because of the occurrence of the atmospheric attenuation. The parameters and convergence criteria are exactly the same as the optimization example in Sect. 4.2 expect for the desired ground signature. Figure 14 gives the optimized ground signature against the baseline and target. As shown in Fig. 14, the optimized ground signature matches the target to a great extent, although the target ground signature is non-physical.

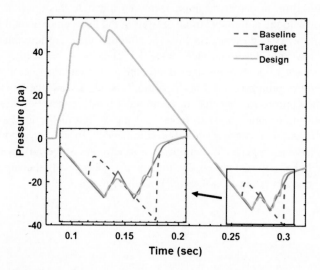

Fig. 14. The optimized ground signature matching a physically non-existent target

5 Conclusion

An inverse design framework using gappy proper orthogonal decomposition (GPOD) method combined with augmented Burgers equation was developed and investigated, aiming to shape the ground signature into a desired target. The developed framework was verified based on an optimization example to inversely design the distribution of the equivalent area of Lockheed Martin 1021. A non-physical target was generated to test the robustness of the developed method. Some conclusions can be drawn as follows:

- The proposed inverse design frame work provides a promising and efficient avenue to find the corresponding near-field signature for a given target ground signature based on augmented Burgers equation of high-fidelity. The optimized ground signature achieves a huge enhancement of 7.34 PLdB in perceived noise level in spite of the big difference between the baseline and target signature.
- This method is robust enough for various inputs. Even if the target ground signature is non-physical, the present method could still offer a solution which has a ground signature very adjacent to the unreal target. This nature is very designer-friendly which does not require rich experience in sonic boom engineering for aircraft designers.
- The Euclidean metric of the designed signature to the target is not always consistent to the proximity in frequency domain, although it does guide the direction of the optimization. Some advanced cost function should be proposed to describe to degree of closeness to the target in term of noise perceived level.

Until now, the proposed design methodology focuses on the field of purely deterministic approach. Since the atmospheric properties are of natural variability, the application of uncertainty quantification and robust design [39, 40] is very demanding.

Acknowledgements. The author thanks Miss Siyi Li for her early work on POD and enlightening discussions with her.

References

1. Alonso JJ, Colonno MR (2012) Multidisciplinary optimization with applications to sonic-boom minimization. Ann Rev Fluid Mech 44(44):505–526
2. Pilon AR (2007) Spectrally accurate prediction of sonic boom signals. AIAA J 45(9):2149–2156
3. George AR, Plotkin KJ (1969) Sonic boom waveforms and amplitudes in a real atmosphere. AIAA J 7(10):1978–1981
4. Plotkin KJ (2002) State of the art of sonic boom modeling. J Acoust Soc Am 111(2):530–536
5. George AR, Seebass R (1971) Sonic boom minimization including both front and rear shocks. AIAA J 9(10)
6. Makino Y, Aoyama T, Iwamiya T, Watanuki T, Kubota H (2015) Numerical optimization of fuselage geometry to modify sonic-boom signature. J Aircr 36(4):668–674

7. Whitham GB (1952) The flow pattern of a supersonic projectile. Commun Pure Appl Math 5(3):301–348
8. Cheung SH, Edwards TA, Lawrence SL (1992) Application of computational fluid dynamics to sonic boom near- and mid-field prediction. J Aircr 29(5)
9. Thomas CL (1972) Extrapolation of sonic boom pressure signatures by the waveform parameter method. NASA TN D-6832
10. Cleveland RO (1995) Propagation of sonic booms through a real, stratified atmosphere. University of Texas at Austin, Austin (1995)
11. Pawlowski JW, Graham DH et al (2005) Origins and overview of the shaped sonic boom demonstration program. In: 43rd AIAA aerospace sciences meeting and exhibit - Meeting Papers 2005
12. Aftosmis M, Nemec M, Cliff S (eds) (2013) Adjoint-based low-boom design with Cart3D (Invited)
13. Nadarajah S, Jameson A, Alonso J (2002) Sonic boom reduction using an adjoint method for wing-body configurations in supersonic flow. AIAA Paper 40(10):1954–1960
14. Rallabhandi S (ed) (2013) Sonic boom adjoint methodology and its applications. In: AIAA applied aerodynamics conference
15. Rallabhandi SK, Nielsen EJ, Diskin B (2012) Sonic-boom mitigation through aircraft design and adjoint methodology. J Aircr 51(2):502–510
16. Rallabhandi SK (2013) Application of adjoint methodology to supersonic aircraft design using reversed equivalent areas. J Aircr 51(6)
17. Seebass R, George AR (1972) Sonic-boom minimization. J Acoust Soc Am 51(49):72
18. Li W, Rallabhandi S (2014) Inverse design of low-boom supersonic concepts using reversed equivalent-area targets. J Aircr 51(1):29–36
19. Kirsch A (2011) An introduction to the mathematical theory of inverse problems. Springer, New York, pp 585–586
20. Tarantola A (2005) Inverse problem theory and methods for model parameter estimation. Society for Industrial & Applied Mathematics, Philadelphia PA xii, p 342
21. Jain A (1981) Image data compression: a review. Proc IEEE 69(3):349–389
22. Sirovich L, Kirby M (1987) Low-dimensional procedure for the characterization of human faces. Joptama 4(3):519
23. Fukunaga K (1972) Introduction to statistical pattern recognition, 2nd ed. Academic Press, pp 2133–2143
24. Han S, Feeny B (2003) Application of proper orthogonal decomposition to structural vibration analysis. Mech Syst Sig Process 17(5):989–1001
25. Willcox K, Peraire J (2002) Balanced model reduction via the proper orthogonal decomposition. AIAA J 40(11):2323–2330
26. Willcox K (2004) Unsteady flow sensing and estimation via the gappy proper orthogonal decomposition. Comput Fluids 35(2):208–226
27. Tan BT, Damodaran M, Willcox KE (2004) Aerodynamic data reconstruction and inverse design using proper orthogonal decomposition. AIAA J 42(8):1505–1516
28. Berkooz G, Holmes P, Lumley JL (1993) The proper orthogonal decomposition in the analysis of turbulent flows. Ann Rev Fluid Mech 25(1):539–575
29. Stein ML (2006) Interpolation of spatial data, vol 45(36). Springer, pp 238–240
30. Venturi D, Karniadakis GE (2004) Gappy data and reconstruction procedures for flow past a cylinder. J Fluid Mech 519(519):315–336
31. Sirovich L (1987) Turbulence and the dynamics of coherent structures. 1 - coherent structures. 2 - symmetries and transformations. 3 - dynamics and scaling. Q Appl Math 45(3):561–571

32. Rallabhandi SK, Loubeau A (2017) Propagation summary of the second AIAA sonic boom prediction workshop. In: AIAA applied aerodynamics conference, Grapevine, USA, 9–13 January 2017
33. Lee YS, Hamilton MF (1995) Time-domain modeling of pulsed finite-amplitude sound beams. J Acoust Soc Am 97(2):906–917
34. Rallabhandi S (ed) (2013) Advanced sonic boom prediction using augmented burger's equation. In: AIAA aerospace sciences meeting including the new horizons forum and aerospace exposition
35. Plotkin K, Sizov N, Morgenstern J (eds) (2006) Examination of sonic boom minimization experienced indoors. In: AIAA aerospace sciences meeting and exhibit
36. Stevens SS (1972) Perceived level of noise by Mark VII and decibels (E). Jacoustsocam 51 (2B):575–601
37. Shepherd KP, Sullivan BM (1991) A loudness calculation procedure applied to shaped sonic booms: NASA langley technical report server
38. Plotkin K (ed) (1989) Review of sonic boom theory. In: American institute of aeronautics and astronautics conference
39. Huan Z, Zhenghong G, Fang X, Yidian Z (2018) Review of robust aerodynamic design optimization for air vehicles. Arch Comput Methods Eng
40. Zhao H, Gao Z, Gao Y, Wang C (2017) Effective robust design of high lift NLF airfoil under multi-parameter uncertainty. Aerosp Sci Technol 68:530–542

Multi-disciplinary Optimization of Large Civil Aircraft Using a Coupled Aero-Structural Adjoint Approach

Jiangtao Huang[1(✉)], Jing Yu[1], Zhenghong Gao[2], Zhu Zhou[1], and Biaosong Chen[3]

[1] China Aerodynamics Research and Development Center,
Computational Aerodynamics Institute, Mianyang, China
`hjtcyf@163.com`
[2] Northwestern Polytechnical University, Xi'an, China
[3] Dalian University of Technology, Dalian, China

Abstract. The coupled aero-structural adjoint-based approach for large civil aircraft multi-disciplinary optimization is studied in this paper. Firstly, computational techniques are introduced. Then the coupled aero-structural adjoint system (CASA) is developed and constructed, solving the structural adjoint equations derived from the structural static equations, with the help of the parallel flow adjoint code PADJ3D and lagged coupled adjoint method. Afterwards, the multi-disciplinary optimization model for the large civil aircraft is established, while the sequential quadratic programming algorithm is used for optimization, freeform deformation method is employed for geometric parameterization and parallel RBF-TFI grid reconstruction technology is used for mesh perturbation. On accounting of the coupled aerodynamic and structural disciplines, the multi-disciplinary optimization for large civil aircraft wing is carried out based on the developed CASA system. The simulation results demonstrate the effectiveness of the CASA system. Under the stress constraints, the aerodynamic drag and the structural weight can be effectively optimized.

Keywords: Coupled aero-structural adjoint · Structural stress · Aeroelastic · Multi-disciplinary optimization · Coupled sensitivity

1 Introduction

In recent decades, more attentions are paid to flexible wing design that considers aeroelastic deformation and structural weight reduction. New advances in composite materials with relatively light weight have made possible the idea of flexible and morphing wing design. The successful application of composite materials greatly promotes the development of flexible wing design technology, but it also issues aerodynamic and structural coupling problems.

The conventional differential-based gradient optimization and evolutionary algorithms have extremely low design efficiency in multidisciplinary and multi-objective optimization, and the computational complexity is large and unbearable. The coupled

© Springer Nature Singapore Pte Ltd. 2019
X. Zhang (Ed.): APISAT 2018, LNEE 459, pp. 1042–1054, 2019.
https://doi.org/10.1007/978-981-13-3305-7_83

sensitivity analysis based on the adjoint system will play a significant and superior role in dealing with the flexible and morphing wing design, because of its efficient in coupled sensitivity calculation.

So far, a series studies on coupled aero-structrual adjoint (CASA) problems have been done by researchers. Professor Martins's MDO team [1–7] has constructed the aerodynamic structure coupling adjoint optimization system based on a high-confidence analysis method. The Jameson team carried out the multidisciplinary aerodynamic-structural optimization of the wing plane shape based on the CASA system [8]. Mohammad Abu-Zurayk and Meryem Marcelet et al. also studied and established the CASA optimization method, based on the CFD solver, TAU [9], and elsA [10, 11] respectively.

In this paper, the CASA system for Aircraft Multi-Disciplinary Optimization is studied and constructed. Combined with the lagged coupled method, the coupled adjoint equations are solved together to realize the assembly of the CASA system. Section 2 describes the computational techniques. Then the CASA system is constructed in Sect. 3, solving the structural adjoint equations derived from the structural static equations, with the help of the parallel flow adjoint code PADJ3D. Finally, the multi-disciplinary optimization design for large civil aircraft is carried out to verify the feasibility and effectiveness of the CASA optimization system.

2   Computational Techniques

2.1   Parallel CFD Solver PMB3D

PMB3D is a large-scale parallel CFD code developed by CARDC (China Aerodynamics Research and Development Center), which solves the N-S equations for arbitrary curvilinear coordinates and performing numerical simulation of flow field:

$$\frac{\partial \widehat{Q}}{\partial t} + \frac{\partial \widehat{E}}{\partial \xi} + \frac{\partial \widehat{F}}{\partial \eta} + \frac{\partial \widehat{G}}{\partial \varsigma} = \frac{\partial \widehat{E}_v}{\partial \xi} + \frac{\partial \widehat{F}_v}{\partial \eta} + \frac{\partial \widehat{G}_v}{\partial \varsigma} \tag{1}$$

PMB3D code [12] provides options for variety of turbulence models, such as SA, SST turbulence models, and Langtry-Menter transition prediction models; and options for variety of spatial discretization schemes, such as JST, Roe, Vanleer, AUSM, and HLLC. LUSGS implicit method is employed for time progression. PMB3D supports 1-to-1 block, patched, and overlapped grid, as well as MPI-based massively parallel computation.

2.2   Parallel RBF-TFI Mesh Deformation Technology

After the CFD surface grid is updated, the radial basis function (RBF) method combined with the transfinite interpolation (TFI) technique is used to perform the spatial grid deformation. The RBF technique is used to build an exact interpolation model based on the deformation of the vertex of the object grid. Then the established model is used to calculate the deformation of block vertex. This method is easy to maintain the consistency of the grid topology. Furthermore, edge, face, and interior point in ξ, η, γ direction can be updated by the TFI method for each block [13]:

$$\overset{\to}{dx}^1 (\xi, \eta, \gamma) = (1 - S_{\xi, \eta, \gamma}) \overset{\to}{dx}(0, \eta, \gamma) + S_{\xi, \eta, \gamma} \overset{\to}{dx}(NI, \eta, \gamma) \tag{2}$$

$$\begin{aligned} \overset{\to}{dx}^2 (\xi, \eta, \gamma) &= \overset{\to}{dx}^1 (\xi, \eta, \gamma) + (1 - t_{\xi, \eta, \gamma})(\overset{\to}{dx}(\xi, 0, \gamma) \\ &- \overset{\to}{dx}^1 (\xi, 0, \gamma)) + t_{\xi, \eta, \gamma}(\overset{\to}{dx}(\xi, NJ, \gamma) - \overset{\to}{dx}^1 (\xi, NJ, \gamma)) \end{aligned} \tag{3}$$

$$\begin{aligned} \overset{\to}{dx}^3 (\xi, \eta, \gamma) &= \overset{\to}{dx}^2 (\xi, \eta, \gamma) + (1 - u_{\xi, \eta, \gamma})(\overset{\to}{dx}(\xi, \eta, 0) \\ &- \overset{\to}{dx}^2 (\xi, \eta, 0)) + t_{\xi, \eta, \gamma}(\overset{\to}{dx}(\xi, \eta, NK) - \overset{\to}{dx}^2 (\xi, \eta, NK)) \end{aligned} \tag{4}$$

In Eqs. (2)–(4), $\overset{\to}{dx}$ denotes the grid point displacement, while S, t, u describe the logical coordinate value in three directions respectively. Parallelization of grid deformation is carried out by MPI. Firstly, the main process carries out the RBF operations, performing the deformation of the spatial grid block vertices, and broadcasts results to the other processes; Then, TFI interpolation is used to update the displacement of edges, faces, and volumes of each partitioned grid; Finally, the grid deformation results of all processes are collected by the main process. The parallel code framework is shown in Fig. 1.

```
if(my_id==1)then
   call New_wallSurf
   call RBF(nnode, mat)
endif
call MPI_Bcast(alphaxyz, nnode, mpi_double_precision,0,comm,err_code)
call Edge_TFI(newx,newy,newz)
call Face_TFI(newx,newy,newz)
call Vol_TFI(newx,newy,newz)
call writegrid_proc(newx,newy,newz,filename_proc)
if(my_id==1)then
   call gathergrid(x_grid, y_grid, z_grid,filename)
endif
```

Fig. 1. The parallel code framework of RBF_TFI technology

2.3 Static Aeroelasticity Numerical Simulation

In this paper, the LDLT [14] method is used to solve the structural static equations:

$$\boldsymbol{Kd} = \boldsymbol{F} \tag{5}$$

The stiffness matrix can be converted to the following form by LDLT decomposition:

$$\frac{\partial R_s}{\partial d} = A = LDL^T$$

$$L = \begin{bmatrix} 1 & & & \\ L_{21} & 1 & & \\ \vdots & & \ddots & \\ L_{6N_s,1} & L_{6N_s,2} & \cdots & 1 \end{bmatrix} \quad D = \begin{bmatrix} d_{11} & & & \\ & d_{22} & & \\ & & \ddots & \\ & & & d_{6N_s,6N_s} \end{bmatrix} \quad (6)$$

After L_{ij}, d_{ii} element is decomposed, the structural static equation can be quickly solved by the following three steps:

$$\begin{aligned} L\varphi &= -\left(\frac{\partial I}{\partial d}\right)_{js} - \sum_{i=1}^{N_a} \left(\frac{\partial R_{a,i}}{\partial d_{js}}\right)^T \tilde{\psi}_{a,i} \\ D\vartheta &= \varphi \\ L^T \psi_s &= \vartheta \end{aligned} \qquad (7)$$

Where ϑ, φ are intermediate variables, K is a stiffness matrix, d is the deformation displacement of the structural node and F is the aerodynamic load.

2.4 Parallel Adjoint Code PADJ3D

The discrete adjoint equations derived from the Navier-Stokes equations is manually constructed. PADJ3D code [15] is programmed for solving these adjoint equations, making use of the second-order JST scheme, artificial viscosity and SST two-equation turbulence model. The adjoint equations are as follows.

$$\begin{aligned} R_c(\lambda)_{j,k,l} - R_D(\lambda)_{j,k,l} - R_v(\lambda)_{j,k,l} &= 0 \\ V_{j,k,l}\frac{\partial \lambda}{\partial t} + R_c(\lambda)_{j,k,l} - R_D(\lambda)_{j,k,l} - R_v(\lambda)_{j,k,l} &= 0 \end{aligned} \qquad (8)$$

In Eq. (8), $R_c(\lambda), R_D(\lambda), R_v(\lambda)_{j,k,l}$ represents the convective term, the artificial viscosity term, and the physical viscosity term of the adjoint equation respectively. The corresponding physical boundary conditions of the discrete adjoint equations are in the form of matrix, and the turbulent viscosity coefficient is frozen. Equation (2) is solved by the LU-SGS method, and the implicit boundary conditions used for time advancement are as follows:

$$\Delta\lambda^* = 0, \quad \Delta\lambda = 0 \qquad (9)$$

When solving the discrete adjoint equations, during the process of parallelization, the load balance is measured by the number of units, and the message is communicated by MPI. The CFD code employed in this paper uses a 1-to-1 block grid, therefore, the information translated by the MPI are the adjoint variables information on the two-level virtual grids in the partitions of each process.

3 Coupled Aero-Structural Adjoint System

3.1 Coupled Aero-Structural Adjoint Equations

Considering the flow field convergence residual constraints and the aeroelastic equilibrium constraints, the minimization problem can be expressed as:

$$\min_{w.r.t.\,D} I(W, X, D) \tag{10}$$

$$R_a(W, X, D) = 0 \tag{11}$$

$$R_S(W, X, D) = Kd - F = 0 \tag{12}$$

In formulas (10)–(12), R_a represents the residual of the flow field and R_s denotes the residual of the structure. K, d, F are the structural stiffness matrix, structure displacement and the load distribution respectively. By introducing Lagrangian operator ψ to Eq. (3), the objective function L can be constructed.

$$L = I + \psi^T R \tag{13}$$

Where $\psi = [\psi_A, \ \psi_S]^T$ are the flow field adjoint variables and structural adjoint variables respectively. By deriving the expression (13), the following expression can be obtained:

$$\frac{dI}{dX} = \frac{d}{dX}\left(I(W, X) + \psi^T R(W, X)\right)$$
$$= \left\{\frac{\partial I}{\partial W} + \psi^T \frac{\partial R}{\partial W}\right\}\frac{dW}{dX} + \left\{\frac{\partial I}{\partial X} + \psi^T \frac{\partial R}{\partial X}\right\}$$

Let the first term of the above formula equal to zero, and the aerodynamic/structural multidisciplinary coupling equation can be obtained

$$\frac{\partial I}{\partial W} + \psi^T \frac{\partial R}{\partial W} = 0 \tag{14}$$

It can be expressed as a matrix [1],

$$
\begin{bmatrix}
\dfrac{\partial R_a}{\partial w_i} & \dfrac{\partial R_a}{\partial d_j} \\[2mm]
\dfrac{\partial R_s}{\partial w_i} & \dfrac{\partial R_s}{\partial d_j}
\end{bmatrix}^T
\begin{bmatrix}
\psi_a \\[6mm]
\psi_s
\end{bmatrix}
=
\begin{bmatrix}
\dfrac{\partial I}{\partial w_i} \\[2mm]
\dfrac{\partial I}{\partial d_j}
\end{bmatrix}
\tag{15}
$$

3.2 Lagged Coupled Adjoint Equations

The system described in Eq. (15) is a complete representation of the adjoint equations for the coupled aerodynamic/structural multidisciplinary system. The resulting system of equations would be extremely large and as a result extremely costly to solve. Even if the iterative method is used to address the system, there are difficulties in memory storage, such as the storage of non-diagonal cross-derivative term $\left(\frac{\partial R_s}{\partial w_i}\right)^T \psi_s$. In order to reduce the memory requirements, Martins [1] introduced lagged adjoint variables $\tilde{\psi} = [\tilde{\psi}_a \quad \tilde{\psi}_s]^T$, which delayed the calculation of the off-diagonal term. It convert the off-diagonal term to forced terms of the in Eq. (15), then the off-diagonal term is used as the source term for flow adjoint and structural adjoint equations respectively. The lagged coupled adjoint method reduces the difficulty of solving the coupled adjoint equations, and the initial flow adjoint system does not need to be drastically modified.

$$
\left(\frac{\partial R_a}{\partial w}\right)^T \psi_a = -\frac{\partial I}{\partial w} - \left(\frac{\partial R_s}{\partial w}\right)^T \tilde{\psi}_s
$$
$$
\left(\frac{\partial R_s}{\partial d}\right)^T \psi_s = -\frac{\partial I}{\partial d} - \left(\frac{\partial R_a}{\partial d}\right)^T \tilde{\psi}_a
\tag{16}
$$

In the FSI3D code, the radial base interpolation method is adopted for the conversion of aerodynamic loads to structural loads [16]. In the X direction, the relationship between structural loads and aerodynamic forces can be expressed in matrix and vector form (the other directions are exactly the same form).

$$
\begin{bmatrix}
F_{sx,1} \\
F_{sx,2} \\
F_{sx,3} \\
\vdots \\
\vdots \\
F_{sx,N_s}
\end{bmatrix}
=
\begin{bmatrix}
A_{11} & A_{12} & A_{13} & \cdots & \cdots & A_{1,N_a} \\
\vdots & & & & & \\
\vdots & & & \ddots & & \\
\vdots & & & & & \\
\vdots & & & & & \\
A_{N_s,1} & A_{N_s,2} & A_{N_s,3} & \cdots & \cdots & A_{N_s,N_a}
\end{bmatrix}
\begin{bmatrix}
F_{ax,1} \\
F_{ax,2} \\
F_{ax,3} \\
\vdots \\
\vdots \\
F_{ax,N_a}
\end{bmatrix}
\tag{17}
$$

F_{sx}, F_{ax} represent the structural load and aerodynamic force respectively. $A_{i,j}$ is the interpolation matrix, and N_s, N_a represent the number of structural nodes and the number of CFD surface grid respectively. Using the expressions above, $\frac{\partial R_s}{\partial w}$, the derivative of the structural residual with respect to the flow field variable can be derived. Since the stiffness matrix and the displacement vector are independent of the flow field variable, the corresponding derivative term is zero.

3.3   The Derivative of the Objective Function to the State Variable

To build aerodynamic and structural coupling adjoint optimization system, the derivative of aerodynamic forces and structural stress with respect to state variables are required. The state variable is $\tilde{W} = [W, d]^T$, where W, d are the flow field conservation variable and the structure displacements, respectively. The freedom degree of structural displacement depends on the type of the finite element (e.g. shells, solids, rod beams, etc.)

The drag coefficient derivative with respect to the state variable can be expressed as:

$$\frac{\partial C_D}{\partial w} = \frac{\partial C_D}{\partial Q}\frac{\partial Q}{\partial w}, \quad \frac{\partial C_D}{\partial d} = \frac{\partial C_D}{\partial X_{Surf}}\frac{\partial X_{Surf}}{\partial d} == \frac{\partial C_D}{\partial X_{Surf}}A_{FSI} \tag{18}$$

Where, X_{Surf} represents the coordinate of wall surface grid and A_{FSI} denotes the fluid-solid coupling operation matrix. Taking Von Mises stress as independent variable, the KS function [1] and its derivative with respect to the state variable are as follows

$$KS = \frac{1}{\beta}\ln(\sum_n \exp(\beta \frac{\sigma_n - \sigma_0}{\sigma_0}))$$
$$\frac{\partial KS}{\partial w} = \frac{\partial KS}{\partial Q}\frac{\partial Q}{\partial w} = 0, \quad \frac{\partial KS}{\partial d} = \frac{\partial KS}{\partial \sigma}\frac{\partial \sigma_m}{\partial d_j} = \frac{\partial KS}{\partial \sigma}S_{mj} \tag{19}$$

In Eq. (19), σ_n, σ_0 are Von Mises stress and material yield stress, respectively. The calculation of each derivative in Eq. (19) is implemented by the CASA system, while the calculation of S_{mj} is performed by the finite element analysis software SiPESC [17], which is developed by Dalian University of Technology.

After solving the coupled adjoint equations, the gradient information can be solved by (20).

$$\frac{dI}{dX} = \frac{\partial I}{\partial X} + [\tilde{\psi}_a^T, \tilde{\psi}_s^T]\frac{\partial}{\partial X}\begin{bmatrix} R_a \\ R_s \end{bmatrix} \tag{20}$$

Based on the chain derivation, the terms of (20) can be further expressed as

$$\frac{\partial I}{\partial X} = \frac{\partial I}{\partial S_{wall}}\frac{\partial S_{wall}}{\partial X} \tag{21}$$

$$\tilde{\psi}^T\frac{\partial R}{\partial X} = \tilde{\psi}^T\frac{\partial R}{\partial G}\frac{\partial G}{\partial S_{wall}}\frac{\partial S_{wall}}{\partial X} \tag{22}$$

S_{wall} and G represent the wall surface and the space grid coordinates respectively. It can be seen that the expression form of $\frac{\partial G}{\partial S_{wall}}$ depends on the grid deformation method, and that of $\frac{\partial S_{wall}}{\partial X}$ depends on the parameterization method. Based on the modules above, the CASA system can be established, as shown in Fig. 2.

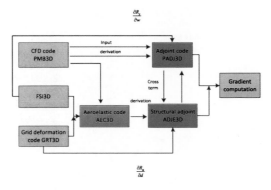

Fig. 2. The CASA system [18]

4   Validation of CFD Solver PMB3D

CFD solver plays an important role in aerodynamic optimization. In this section, the self-developed CFD solver PMB3D will be studied and validated. The DPW-4 CRM (Common Research Model) wing-body-tail model is employed for simulation example [19, 20]. The wind tunnel testing data of this model is of a Mach number that is similar to the design state of the aircraft which would be optimized in Sect. 5. The computational grid consists of 15 million cells, and y^+ is close to 0.5. The computational state is $M = 0.85$, $CL = 0.5$, $Re = 5.0 \times 10^6$. Figures 3, 4, 5 and 6 gives the pressure distribution comparison between CFD results and experimental data at typical spanwise stations. It can be seen that the CFD calculation results are in good agreement with the wind tunnel test data. So, it can be concluded that the PMB3D code can provide reliable calculation results for aerodynamic design.

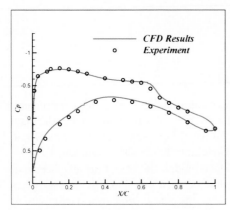

Fig. 3. Pressure distribution of 13.1% spanwise

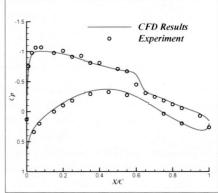

Fig. 4. Pressure distribution of 28.39% spanwise

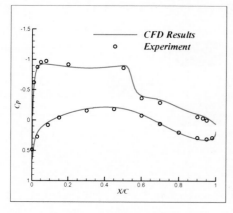

Fig. 5. Pressure distribution of 50.18% spanwise

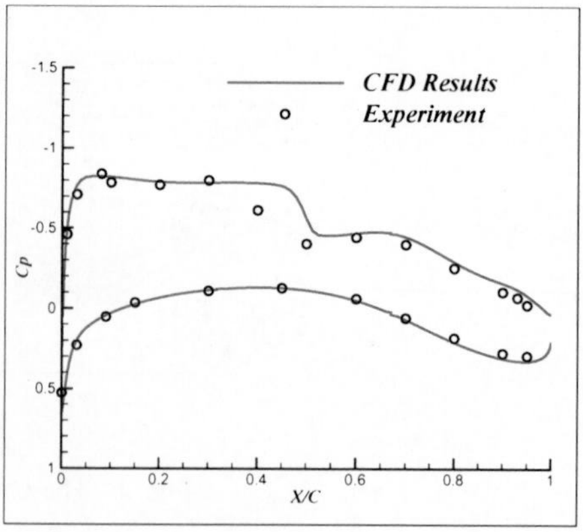

Fig. 6. Pressure distribution of 72.9% spanwise

5 Aero-Structural Optimization by Coupled Adjoint Approach

The research on optimization design was carried out based on Common Research Model (CRM), in order to facilitate other aspects of research, we added a vertical tail to the baseline CRM. And the new model is employed for aerodynamic and structural multidisciplinary optimization. The optimization objective is to minimize the drag coefficient and the weight, considering the lift coefficient and the stress constraints. The mathematical optimization model used in this paper is as follows.

$$
\begin{aligned}
&\min I = \alpha C_D + \beta W \\
&subject\ to: \\
&C_L = 0.5 \\
&KS \geq 0 \\
&t/c \geq (t/c)_{initial}
\end{aligned}
\tag{23}
$$

In Eq. (23), α and β separately represent the weight coefficients of drag and structural weight. C_D describes the drag coefficient, while W gives the structural weight. The main components of the configuration are wing, fuselage, horizontal tail and vertical tail. The CFD grid contains 12.6 million cells, divided into 290 blocks, as shown in Fig. 7. The JST spatial discrete scheme, SST turbulence model, LU-SGS time advancement, and multi-grid acceleration convergence techniques are put into use and 64 cores are employed for calculation. Figure 8 shows the FFD (Freeform Deformation) parameterization lattice. 200 control points distributed on the FFD lattice are used as the design variables. Simplified shell element is used as the structural finite element model, mainly including beams, ribs, etc., as shown in Fig. 9. For structural parameterization, FFD technology is used to control the deformation of the finite element grid.

Figures 10 and 11 show the convergence process of the coupled adjoint equation when the aerodynamic force and KS function is used as the objective function. The convergence process of the structural adjoint equation residual is showed in Fig. 12. It can be seen that after 6 coupling calculations, the adjoint variables of the structural adjoint equation tend to converge.

Figures 13 and 14 show the design results of different optimization methods, where (a) represents the initial configuration, (b) represents the aerodynamic optimization results without structural stress constraints, and (c) represents CASA-based aero-structural optimization results. It can be seen that for both cases with and without structural consideration, the shock is sharply reduced, and it tends to be shock-free in case (b) in

From Fig. 14, it can be seen that, after optimization, the structural stress of both configurations are increased. But the stress gain of the case with stress constraint is less than that of the case without stress constraint. As a matter of fact, the weight coefficient has considerable influences on the design results, i.e. the drag reduction and stress design results. In the future work, considering the specific design requirements, the weight function impact on the optimization result needs to be studied. Table 1 shows the optimization results of different methods. It can be seen that the CASA system can realize the integrated design of drag reduction and weight reduction while ensuring the constraint of yield stress.

Fig. 7. CFD grid

Fig. 8. FFD parameterization

Fig. 9. Distribution of finite element grid of skin structure [18]

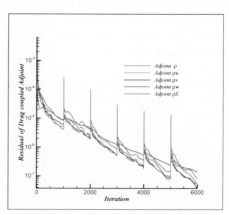

Fig. 10. Process of residual convergence of coupled adjoint system (Drag)

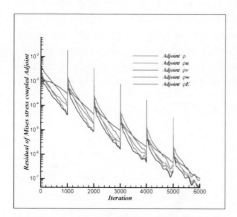

Fig. 11. Process of residual convergence of coupled adjoint system (Von Mises, KS function)

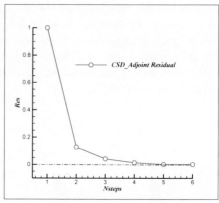

Fig. 12. Process of residual convergence of structural adjoint equation

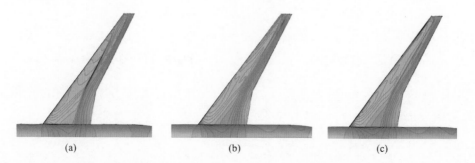

Fig. 13. Pressure contour of different configuration

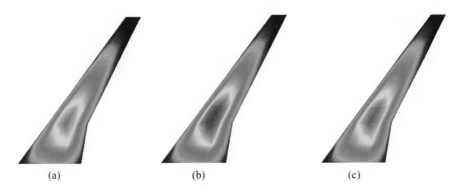

(a)   (b)   (c)

Fig. 14. Stress contour of different configuration

Table 1. Drag and weight of initial and optimized wing

Different configuration	Drag	Weight
Initial	0.02976	2.61324783e+004 kg
Aerodynamic optimization	0.02863	—
Aero-structural optimization	0.02907	2.51865619e+004 kg

6   Conclusion

A coupled aero-structural adjoint approach was studied in this paper. Firstly, each computational technique is introduced, as shown in the derivation of the coupled equations, the expression of the derivative of each objective function depending on the grid deformation and the parameterization method. The derivative of each objective can be explicitly obtained by chain derivation. With the help of the parallel flow adjoint code PADJ3D and lagged coupled adjoint method, the coupled aero-structural adjoint system (CASA) is constructed. The lagged coupled adjoint method could largely reduce the memory requirements, as well as making easy for programming and modularity. After several coupling calculations, the coupled adjoint equations residuals and the structural residuals tend to converge. It shows that the solutions of the lagged coupled adjoint approach are consistent with the initial coupled system. Then the multi-disciplinary optimization for large civil aircraft was studied by using CASA system. The multidisciplinary optimization results of the supercritical wing show that the aerodynamic optimization without stress constraint can significantly reduce the drag, but brings about the problem of increasing the stress. Considering the structural stress constraints, the coupled adjoint system can effectively reduce the drag and the weight, providing powerful technical support for aero-structural integrated design. In the future work, the impact of the weight function on the optimization result will be studied for the specific design requirements.

Acknowledgements. This research was supported by the National Key R&D Program of China (2016YFB0200704).

References

1. Joaquim RRA, Martins A (2002) Coupled-adjoint method for high-fidelity. aero-structural optimization. The department of aeronautics and astronautics and the committee of stanford university 2002
2. Mader CA, Kenway GKW, Martins JRRA (2008) Towards high-fidelity aerostructural optimization using a coupled adjoint approach. In: 12th AIAA/ISSMO multidisciplinary analysis and optimization conference, 10–12 September 2008, Victoria, British Columbia, Canada
3. Kenway GKW, Kennedy GJ, Martins JRRA (2014) Scalable parallel approach for high-fidelity steady-state aeroelastic analysis and adjoint derivative computations. AIAA J 52(5):935–951
4. Martins JRRA, Hwang JT (2013) Review and unification of methods for computing derivatives of multidisciplinary computational models. AIAA J 51(11):2582–2599
5. Kenway GKW, Martins JRRA (2014) Multipoint high-fidelity aerostructural optimization of a transport aircraft configuration. J Aircr 51(1):144–160
6. Kennedy G, Martins J (2010) Parallel solution methods for aerostructural analysis and design optimization. In: 13th AIAA/ISSMO multidisciplinary analysis optimization conference, p 9308
7. Martins JRRA, Alonso JJ, Reuther JJ (2004) High-fidelity aerostructural design optimization of a supersonic business jet. J Aircr 41(3):523–530
8. Leoviriyakit K, Jameson A (2002) Case studies in aero-structural wing planform and section optimization. In: AIAA, p 5372
9. Abu-Zurayk M, Brezillon J (2011) Shape optimization using the aerostructural coupled adjoint approach for viscous flows. In: Evolutionary and deterministic methods for design, optimization and control.© CIRA, Capua, Italy
10. Ghazlane I, Carrier G, Dumont A (2012) Aerostructural adjoint method for flexible wing optimization. In: AIAA, p 1924
11. Marcelet M, Peter J, Carrier G (2008) Sensitivity analysis of a strongly coupled aero-structural system using direct and adjoint methods. In: AIAA, p 5863
12. Gang L, Zhongyun X, Jiantao W, Fan L (2015) Numerical simulation of missile air-launching process under rail slideway constraints. ACTA Aerodynamica Sinica 33(02):192–197
13. Spekreijse SP, Boerstoel JW (1998) An algorithm to check the topological validity of multi-block domain decompositions. In: Proceedings 6th international conference on numerical grid generation in computational field simulations, NLR-TP-98, Greenwich, p 198
14. Hua QIN, Yanzi XU (2011) Improving SVM's learning efficiency by using matrix LDLT parallel decomposition. Comput Eng Appl 47(12):200–212
15. Huang JT, Liu G, Zhou Z et al (2017) Investigation of gradient computation based on discrete adjoint method. Acta Aerodynamica Sinica 35(4):554–562
16. Allen CB, Rendall TCS (2007) Unified approach to CFD-CSD interpolation and mesh motion using radial basis functions. In: AIAA, p. 3804
17. Zhang H, Chen B, Li Y et al (2011) Advancement of SiPESC for development of integrated CAE software systems. Comput. Aided Eng. 20(2):39–49
18. Huang J, Zhou Z, Liu G, Gao Z, Huang Y, Wang Y (2018) Numerical study of aero-structural multidisciplinary lagged coupled adjoint system for aircraft. ACTA Aeronauticaet Astronautica Sinica 39(5):121731
19. Wang YT, Zhang SJ, Meng DH (2013) Numerical simulation and study for DPW4 wing/body/tail. ACTA Aerodynamic Sinica 31(6):739–744
20. Vassberg JC, Dehaan MA, Rivers SM, et al (2008) Development of a common research model for applied CFD validation studies. In: AIAA, p 6919

Efficiency Estimation of Formation Flight Types

Yang Tao[(⊠)], Zhiyong Liu, Neng Xiong, Yan Sun, and Jun Lin

China Aerodynamics Research and Development Center,
High Speed Aerodynamic Institute, Mianyang 621000, Sichuan, China
50323222@qq.com

Abstract. Formation flight is usually seen in birds' migration and military operations, and will be more often in unmanned aircrafts applications in future. It is known that drag reduction and fuel saving can be achieved in formation flight and positional parameters are important to this effect. But few literatures about efficiency of formation flight types on this effect are published. Wake vortexes of the leading aircraft in formation flight are assumed to be a pair of reversed vortexes. Using potential flow techniques, models of the trailing aircraft's lift and drag coefficients are constructed which are functions of lateral spacing and vertical spacing. An estimation of energy saving is conducted on three formation flight types-V mode, Λ mode and echelon mode. The formation studied consists of three the same aircrafts. The total drag of three aircrafts in a formation is regarded as an index for energy saving efficiency under a precondition that all aircraft's lift equals to their own weights. Each type is under its best status. Results show that echelon mode has the highest energy saving efficiency, Λ mode is a little low and V mode is markedly lower than the two others. This result accords well with formation flight types used in birds' migration.

Keywords: Formation flight · Surfing wake vortex ·
Optimization · Computational fluid dynamics · Stability

1 Introduction

With the rapid development of drones, the use of drones to perform various tasks has become more and more frequent. Research shows that drone formation flight can improve the success rate of mission execution and the ability to resist accidents. It can also expand the scope of investigation and provide more abundant regional information [1]. And Multiple drone formation flights can obtain the aerodynamic performance equivalent to a large display string ratio aircraft, while structural strength does not need to be strengthened, can increase the voyage and Airtime [2].

The research on formation flight focuses on the analysis of drag reduction mechanism and the study of formation flight control technology [3–7]. Based on these goals, theoretical modeling, wind tunnel experiments, flight tests and numerical calculations have been carried out, and abundant research results have been obtained. It is generally believed that the tail eddy currents of the front aircraft in the formation can increase the effective angle of attack of the rear aircraft. To increase the lift the aircraft behind the

© Springer Nature Singapore Pte Ltd. 2019
X. Zhang (Ed.): APISAT 2018, LNEE 459, pp. 1055–1064, 2019.
https://doi.org/10.1007/978-981-13-3305-7_84

cruise can fly at an angle smaller than the cruising angle of attack, thereby reducing the flight drag. Further studies have shown that the drag reduction effect is related to the formation parameters, especially the lateral spacing [4–8].

Bangash et al. tested the drag reduction effect of the three formation modes (see Fig. 1) through wind tunnel experiments, pointing out that both the trapezoidal mode and the V-mode can achieve a significant increase in the lift-to-drag ratio of wingmen, and the front and rear modes lead to a reduction in the lift-to-drag ratio of wingmen. V-mode has the largest lift-to-drag ratio increment [9]. Other wind tunnel experiments and flight experiments are mostly formations using trapezoidal modes. They mainly pay attention to the aerodynamic performance of wingmen affected by the formation parameters. At present, few people have assessed the energy saving efficiency of the formation mode. In this paper, a mathematical model based on Eddy current model is proposed. The model is validated first, and the energy-saving efficiency evaluation of the three-aircraft formation flight with three modes of V-shape, B-shape and trapezoidal shape (see Fig. 2) is carried out. The lift of the three aircraft is limited to the cruising lift, that is, the lift is equal to the gravity of the aircraft, and the total drag of the three aircraft is used as an energy saving indicator to reduce drag. The total drag is directly related to the total fuel consumption, which can reflect the energy saving effect of the formation.

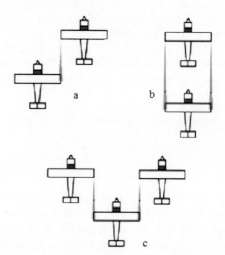

Fig. 1. Types of aircraft formations investigated: (a) echelon, (b) lead-trail, and (c) V.

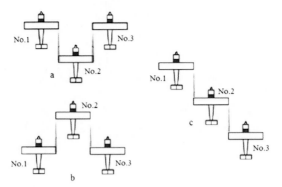

Fig. 2. Types of formation flight estimated: (a) V mode, (b) Λ mode, and (c) echelon mode.

2   Mathematical Model

2.1   Eddy Current Model

Ayumu Inasawa et al. studied the velocity distribution of the wingtip vortex in the flow section at low speeds through flow field display and PIV measurement method [10]. Using more than 100 PIV data fitting, it is found that the velocity distribution with the Osei vortex model is in good agreement.

$$v_\theta = \frac{\Gamma_0}{2\pi r}(1 - \exp(-\frac{r^2}{r_c^2})) \tag{1}$$

Among them, v_θ is the tangent induced speed, r is the distance from the vortex center, r_c is the vortex core radius, Γ_0 is the vortex strength. According to the Kutta-Joukowski theorem, vortex strength can be obtained by the following formula.

$$\Gamma_0 = \frac{L}{\rho v_\infty b'} \tag{2}$$

Here L is lift, ρ is aero density, v_∞ is the inflow flow speed, b' is Effective wingspan length. For an oval-shaped wing $b' = \pi b/4$, b is wingspan length.

2.2   Upper Washing Effect

In formation flight, the tail vortex of the front aircraft produces induced velocity at the rear aircraft. The average effect of the induced speed changes the effective angle of attack of the rear aircraft [11–14] (See Fig. 3). The average effect is that the lift drag ratio can be improved, the aerodynamic performance can be improved, and the drag reduction and energy conservation can be achieved.

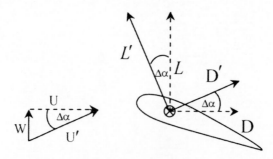

Fig. 3. Up wash effect

The appearance of the upper washing speed W deflects the wind axis system $\Delta\alpha$, resulting in changes in the lift and drag in the vertical coordinate system.

$$\Delta L = L' \cos(\Delta\alpha) + D' \sin(\Delta\alpha) - L \qquad (3)$$

$$\Delta D = D' \cos(\Delta\alpha) - L' \sin(\Delta\alpha) - D \qquad (4)$$

Normally the drone's cruising angle of attack is less than 10. So there are:

$$C_L = C_{L0} + C_L^\alpha \alpha \qquad (5)$$

$$C_D = C_{D0} + C_D^{\alpha^2} \alpha^2 \qquad (6)$$

$$\Delta\alpha \approx \tan(\Delta\alpha) = \frac{W}{U} \qquad (7)$$

In the formula, respectively, the single-machine zero angle of attack lift coefficient and the drag coefficient, the superscript is the derivative. According to the relationship between the rise, the drag, and the angle of attack at the small angle of attack, and the approximate relationship of the trigonometric function, the second-order small quantities are omitted:

$$\Delta C_L = (C_L^\alpha + C_{D0} + C_D^{\alpha^2} \alpha^2)\Delta\alpha \qquad (8)$$

$$\Delta C_D = (2\alpha C_D^{\alpha^2} - C_{L0} - C_L^\alpha \alpha)\Delta\alpha \qquad (9)$$

It can be seen that the change of lift and drag is related to its own angle of attack, and is related to the basic aerodynamic performance of the aircraft, and is related to the washing speed.

2.3   Up Wash Speed

The tail vortex is reduced to a pair of vortices that rotate inwards and the induced velocity at the rear aircraft is calculated using the vortex model (see Fig. 4).

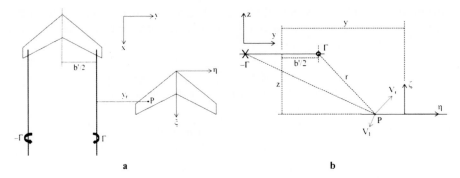

Fig. 4. Calculation of upwash velocity (a: top view; b: back view)

In Fig. 4, the XYZ coordinate system is a long machine coordinate system, in which the relative position of the wingman is given. The $\xi\eta\zeta$ coordinate system is a wingman coordinate system, in which the position of P point on the wing of the aircraft is given. The upper washing speed induced at point P is:

$$V_\zeta = \frac{L_1}{2\pi\rho V_{1\infty} b'_1}[\frac{y - \frac{b'}{2} - \eta}{(y - \frac{b'}{2} - \eta)^2 + z^2}(1 - e^{-\frac{(y - \frac{b'}{2} - \eta)^2 + z^2}{r_c^2}})$$
$$- \frac{y + \frac{b'}{2} - \eta}{(y + \frac{b'}{2} - \eta)^2 + z^2}(1 - e^{-\frac{(y + \frac{b'}{2} - \eta)^2 + z^2}{r_c^2}})]$$

(10)

The subscript 1 in formula (10) indicates that the parameter belongs to the long machine. The average induction speed can be obtained by extending the formula (10) along the machine. Substituting the average induction velocity into the formula (7)–(9), the mathematical model of wingman rise and drag coefficient increment is obtained.

$$\Delta C_L = f(y, z)$$

(11)

$$\Delta C_D = g(y, z)$$

(12)

For multi-aircraft formations, the induced speed is the result of the joint action of the tail Eddy currents of all aircraft in front of the object aircraft.

3   Calculation Results and Analysis

3.1   Model Validation

William et al. Experimental data based on experimental data published in the article of the Journal of Aircraft in 2004. The experimental flow speed is 19.8171 m/s. The model used is a tailless triangular wing configuration with a leading edge of 65° swept angle and a trailing edge zigzag of 25° swept angle. The two models used in the experiment are slightly different. The data of B model is used in this paper. In the experiment, the B model was used as a wingman. Some of the parameters used in the calculations are shown in Table 1.

Table 1. Parameters used in calculation

b	0.8796 m	C_{D0}	0.013091
$C_D^{\alpha^2}$	0.0006289	C_{L0}	0.062369
C_L^{α}	0.0365	s	0.4448 m^2
α	8°	α_1	8°

Through programming, the change of wingman's lift coefficient and drag coefficient increment in the range of $y = 0.5b \sim 1.5b, z = 0$ is calculated. The calculation of vortex core radius uses Atilla et al. The calculation method is compared with the results of their model predictions, as shown in Fig. 5. It can be seen that the prediction results of the mathematical model presented in this paper are in good agreement with the experimental results, and the incremental prediction results of drag coefficient are obviously better than those of Atila et al. It is worth noting that the parameters in Table 1 are given according to the angle and are converted to radians in the actual operation.

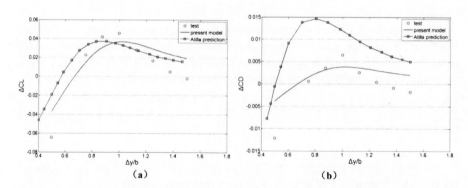

Fig. 5. Results comparison of prediction and experiments (a: lift coefficient increments; b: drag coefficient increments)

3.2 Energy Conservation Assessment

Using the aircraft model in 2.1 and the incoming velocity, 8° is assumed to be the cruising angle. For the V-shaped model, only the middle aircraft can obtain a drag reduction effect. It is also washed by the tail vortex of the two aircraft in the front and right. When calculating the washing speed, the formula (10) must be multiplied by 2 times.

For the shape model, both of the latter two aircraft can achieve a drag reduction effect and can be calculated directly using the 2.1 section model.

For the trapezoidal model, the two aircraft behind can enjoy the drag reduction effect, but there are differences. For the middle aircraft, it can still be calculated using a 2.1 section model. For the last aircraft, the formula (2) should also be changed, except that the formula (10) needs to be modified. Because the tail vortex strength generated by the middle aircraft is less than the value calculated using the cruise lift, the lift calculated at the flight angle of attack should be used.

Using the mathematical model established by the above method, the energy-saving efficiency evaluation of three formation modes is carried out. The best efficiency of each mode is obtained by using the following methods 1. The fixed long angle of attack is the cruising angle of attack, and the position parameters Y and Z are optimized at a certain angle of attack of the wingman to find the maximum lift and drag ratio position. The lift coefficient and drag coefficient of the wingman at this time are recorded; 2 Change wingman angle of attack, repeat the previous process, obtain a series of lift coefficient and drag coefficient; 3 By interpolation, the lift of the machine is equal to the angle of cruising lift, and corresponding drag coefficient is obtained. 4 Repeat the first three steps to find the drag coefficient of all wingmen. The total drag coefficient of three aircraft is used as the evaluation index for energy conservation. One of the prerequisites for this assessment is that the impact of the aircraft behind it on the aircraft in front of it is negligible.

Based on B, the position parameters Y and Z are dimensionalized. In$(Y, Z) = [0.7, 1.5; -0.3, 0]$ The optimal state is found in space, and the lift coefficient of the wingman changes with the angle of attack curve under the shape-shaped mode. When the angle of attack $\alpha = 6.4°$, the lift coefficient curve jumps. After analysis, it is found that when $\alpha \leq 6.4°$, the maximum lift drag ratio is dominated by the drag coefficient when $\alpha > 6.4°$. The maximum lift drag ratio at 6.4° is dominated by the lift coefficient. The position of the maximum lift-to-drag ratio before and after $\alpha = 6.4°$ changes from $(Y, Z) = (0.7, -0.3)$ to $(1.025, 0)$. The existence of this kind of jump causes the interpolation part to be unable to carry out and needs to improve the way to find the best efficiency (Fig. 6).

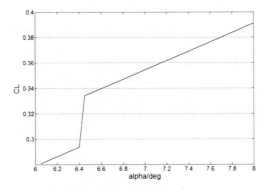

Fig. 6. Optimization results of the trailing aircraft's lift coefficient (in Λ mode formation)

In order to avoid the situation where the lift ratio is insufficient in the maximum position and the lift ratio is not minimal in other positions and the maximum position drag is not minimal, give up the restriction of maximum lift ratio and find the position where the minimum drag under the lift constraint is satisfied. The results of the re-optimization are shown in Fig. 7.

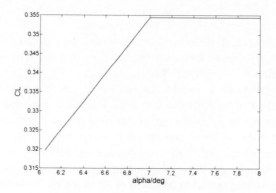

Fig. 7. New optimization results of the trailing aircraft's lift coefficient (in Λ mode formation)

According to the new method of finding the best efficiency, the distribution of the drag coefficient of the three formations is shown in Table 2, and the corresponding formation position is shown in Table 3. The aircraft number is shown in Fig. 2. It can be seen that the total drag of the trapezoidal model is the smallest, and the energy conservation efficiency is the highest, which is about 6.6%; The shape model is slightly less efficient than the trapezoidal model, which is approximately 6.3%; Although the V-shaped No. 2 aircraft has the smallest drag, its total drag is significantly greater than the first two and its efficiency is the lowest, at about 4.6%. In the flight experiments conducted by Beukenberg and Hummel, the wingmen in the two trapezoidal formations obtained an average of 10% of fuel savings during the 150 s time interval [16]. The total energy saving efficiency converted to the three-machine model is about 6.7%, which is equivalent to the predicted value of the model.

Table 2. Drag coefficients

Mode	No. 1	No. 2	No. 3	Total
V	0.0533	0.0459	0.05334	0.1526
Λ	0.0483	0.0533	0.0483	0.1500
Echelon	0.0533	0.0483	0.0478	0.1495

Table 3. Positional parameters

Mode	No. 1	No. 2	No. 3
V	0,0	1.5,0	3.0,0
Λ	0,0	1.025,0	2.05,0
Echelon	0,0	1.025,0	2.025,0

4   Conclusion

In this paper, the incremental mathematical model of rising and drag coefficient of wingman in formation flight is established through the technique of potential flow. The predicted results of the model agree well with the experimental results. This model is used to predict and evaluate the energy saving efficiency of formation flight with three modes: V-shaped, trapezoidal and trapezoidal. The minimum total drag of the three aircraft that meet the lift constraints is an energy-saving evaluation index. The results show that the trapezoidal mode is the most efficient, the trapezoidal mode is slightly less efficient than the trapezoidal mode, and the V-shaped mode efficiency is significantly lower than the former two. The results were consistent with the formation pattern used in bird migration.

In addition, it was also found that when the formation flight is the most energy-efficient, the wingman is not necessarily in the space with the largest lift-to-drag ratio, and is affected by the wingman's angle of attack. The maximum lift and drag ratio of the wingman is dominated by drag when the angle of attack is small, and is dominated by lift when the angle of attack is large.

Acknowledgments. This research work was funded by the National Natural Science Foundation of China under Grant No.11372337.

References

1. Bearc R, Lawton J, Hadaegh F (2001) A coordination Architecture for spacecraft formation control. IEEE Trans Control Syst Technol 9(6):777–790
2. Anderson MR, Robbins CR (1998) Formation flight as a cooperative game. In: AIAA guidance navigation and control conference, Reston, VA
3. Maskew B (1997) Formation flying benefits based on vortex lattice calculations. NASA CR-154974, May 1977
4. Hummel D (1983) Aerodynamic aspect of formation flight in birds. J Theor Biol 104:321–347
5. Beukenberg M, Hummel D (1990) Aerodynamics, performance and control of airplanes in formation flight. In: International council of the aeronautical sciences, Paper 90-5.9.3, September 1990
6. Binetti P, Kartik BA, Krstic M et al (2003) Formation flight optimization using extremum seeking feedback. J Guidance Control Dyn 26(1):132–142
7. Zou Y, Pagilla PR, Ratliff RT (2009) Distributed formation flight control using constraint forces. J Guidance Control Dyn 32(1):112–120

8. Dogan A, Venkataramanan S, Blake W (2005) Modeling of aerodynamic coupling between aircraft in close proximity. J Aircr 42(4):941–955
9. Bangash ZA, Sanchez RP, Ahmed A et al (2006) Aerodynamics of formation flight. J Aircr 43(4):907–912
10. Inasawa A, Mori F, Asai M (2012) Detailed observations of interactions of wingtip vortices in close-formation flight. J Aircr 49(1):206–213
11. Yuan Y, Yan J-G, Qu Y-H (2013) The aerodynamic coupling simulation in formation flight of multi-UAV. Flight Dyn 31(1):29–32
12. Fan Q.-j (2008) Key techniques research of cooperative formation biomimetic flight control for multi-UAV. Nanjing University of Aeronautics and Astronautics, April 2008
13. Liu C.-g (2008) Research on biomimetic close formation flight control of UAVs. Nanjing University of Aeronautics and Astronautics, January 2008
14. Longhui E (2009) Research on UAV coordinated formation flight based on bionic vision technology. Nanjing University of Aeronautics and Astronautics, January 2009
15. Blake WB, Gingras DR (2004) Comparison of predicted and measured formation flight interference effects. J Aircr 41(2):201–207
16. Beukenberg M, Hummel D (1990) Aerodynamics, performance and control of airplanes in formation flight. In: International council of the aeronautical sciences, Paper 90-5.9.3

Drag Reduction of Transonic Wings with Surrogate-Based Optimization

Jichao Li[⊠], Jinsheng Cai, and Kun Qu

National Key Laboratory of Aerodynamic Design and Research,
Department of Fluid Mechanics, School of Aeronautics,
Northwestern Polytechnical University, Xi'an, China
cfdljc@gmail.com

Abstract. Preliminary design of aircraft wings requires multiple cycles of optimization to compromise the influences from different disciplines and constraints. Surrogate-based optimization is a popular choice in this circumstance because surrogate models can be reused in different cycles once constructed, while gradient-based optimization might be less efficient despite its fast convergence in each cycle. However, surrogate-based optimization suffers from the curse of dimension. The selection of design variables has a big influence on the optimization efficiency and performance. We investigate two different approaches to drag minimization in the design of transonic wings: (1) optimize all the section shapes and their twists independently (2) scale all the wing sections together and optimize them with their twists. We find that surrogate-based optimization by the former approach cannot ensure a better solution in spite of its larger design space. The latter approach is generally more efficient, which is more practical in preliminary design of wings.

Keywords: Drag reduction · Aerodynamic shape optimization ·
Surrogate-based optimization

1   Introduction

As important components of aircraft, wings determine the aerodynamic performance and flight efficiency to a large extent, so the design of wings is vital in aircraft design. From the view of aerodynamics, aircraft wing design includes determining the planform, section shapes, section twists, dihedrals, and so on. In recent years, high-fidelity computational fluid dynamics (CFD) like Reynolds-averaged Navier–Stokes (RANS) is playing an increasingly important role in aircraft wing design. Drag reduction by optimizing the wing shape of commercial aircraft has brought numerous benefits to airlines.

Aerodynamic shape optimization with CFD is mainly based on two different approaches—gradient-based optimization (GBO) and gradient-free optimization. GBO uses gradient-based algorithms to optimize the aerodynamic function. Gradient can be obtained by different approaches [1, 2], and the adjoint method [3–7] is the most efficient one for high-dimensional aerodynamic shape optimization problems since it provides gradient values to all parameters almost simultaneously. The computational

© Springer Nature Singapore Pte Ltd. 2019
X. Zhang (Ed.): APISAT 2018, LNEE 459, pp. 1065–1080, 2019.
https://doi.org/10.1007/978-981-13-3305-7_85

cost of adjoint method is approximately equal to the CFD simulation. Although GBO is criticized due to local convergence in multimodal functions, it is efficient in aerodynamic shape optimization [8–10] and has been successfully applied in aircraft design [11, 12]. Gradient-free optimization is generally coupled with surrogate models, so this approach is also known as surrogate-based optimization (SBO) [13–18]. SBO is more likely to converge to the global optimum because the search can be easily conducted more widely with a gradient-free algorithm or gradient-based algorithm with a multi-start execution [19].

In the early stage of aircraft design, designers need to quickly evaluate the best performance subject to different constraints like wing area or the lift coefficient at cruise. In this circumstance, multiple optimizations are required, so GBO may be not an efficient choice. On the contrary, SBO could be more efficient because the surrogate model trained in one optimization can be reused in other optimizations. Actually, optimizations subject to different constraints could share one common surrogate and then the computation is reduced. So SBO is more widely adopted in preliminary design of aircraft wings.

For an aerodynamic problem with 10–20 parameters, SBO works efficiently [14, 15, 17]. With the increase of the input dimension, curse of dimensionality is an issue for SBO where an accurate surrogate is more difficult to train and the optimization becomes less efficient or effective. Applying SBO in a larger design space is not always beneficial even though better designs may exist. Li et al. [20] adopted the active subspace method to reduce the input dimension in aerodynamic shape optimization and achieved better solution than the optimization in the original design space. Han et al. [13] applied SBO in the design of a transonic wing with 40 and 80 shape parameters, and they found that the optimized drag with 40 parameters was smaller than the optimized drag with 80 parameters. Therefore, a proper selection of design variables is important for the design. As the primary parameters, section shapes and their twist angles are vital to the performance of wings. On the one hand, one can allow sections to change their shapes independently in order to give more freedoms to the wing design, but many control parameters are required. On the other hand, one can scale all the sections to change their shapes together, and in this way, about 20 parameters are enough to control the wing shape. The latter approach gives less freedoms while the former might be influenced by the curse of dimensionality. A comparison of SBO in the two approaches using the same case is of great interest.

In this paper, we minimize the drag of two baseline wings subject to different lift constraints to investigate the performance of the two approaches. 75 and 19 parameters are used in the two approaches, respectively, and we note them as Approach I and Approach II. We use the Kriging model [21] to surrogate the aerodynamic functions and use the EI criterion [22–24] to infill new training points during the optimization. We find that the aerodynamic profiles are not that flat or multimodal because gradient-based optimization starting from different points always converge to the same solution. So, in order to reduce the computational cost, we minimize the drag subject the lift constraint with a gradient-based algorithm associated with multi-start execution in each optimization cycle. We use multi-start gradient-based optimizations and a gradient-free algorithm to maximize the EI function considering the multimodality and computational efficiency. The optimization results show that using Approach I is not necessarily better, while Approach II is more efficient.

The rest of this paper is organized as follows: We first introduce the Kriging model and the optimization infilling strategy in Sect. 2. We compare the performance of the gradient-free algorithm and the gradient-based algorithm in the inner loop of SBO and propose our optimization strategies in Sect. 3. We describe the optimization problem in Sect. 4.1 and compare the SBO results of both approaches in Sect. 4.2. We end this paper with a summary of conclusions in Sect. 5.

2 Surrogate-Based Optimization

Surrogate models provide a low computational cost alternative to time-consuming high-fidelity models. Kriging is a famous surrogate model proposed by Krige [21], and it provides not only the prediction of a function but also the mean square error of the prediction. Kriging models have been widely applied in SBO for aerospace design problems [13, 16, 25].

2.1 Kriging Model

For a system with input $\mathbf{x} \in \mathbb{R}^n$ and a scatter output y, $\mathbf{y}_s = (y^{(1)}, y^{(2)}, \dots, y^{(n_s)})^{\mathrm{T}}$ is the vector of observed outputs on input points $\mathbf{x}^{(1)}, \mathbf{x}^{(2)}, \dots, \mathbf{x}^{(n_s)}$. We follow the derivation of previous researches [25, 26], and the Kriging prediction $\hat{y}$ of system output function with respect to input $\mathbf{x}$ can be expressed as

$$\hat{y}(\mathbf{x}) = \widehat{\beta}_0 + \mathbf{r}^{\mathrm{T}} \mathbf{R}^{-1} \left(\mathbf{y}_s - \mathbf{1} \widehat{\beta}_0 \right), \tag{1}$$

where $\widehat{\beta}_0 = \left(\mathbf{1} \mathbf{R}^{-1} \mathbf{1} \right)^{-1} \mathbf{R}^{-1} \mathbf{y}_s$. $\mathbf{1} \in \mathbb{R}^{n_s}$ is a column vector whose elements are all filled with ones. $\mathbf{R} := [R(\mathbf{x}^{(i)}, \mathbf{x}^{(j)})]_{ij} \in \mathbb{R}^{n_s \times n_s}$, and $\mathbf{r}(\mathbf{x}) := [R(\mathbf{x}^{(i)}, \mathbf{x})]_i \in \mathbb{R}^{n_s}$. $R(\mathbf{x}, \mathbf{x}')$ is the kernel of Kriging model, which is a spatial function of $\mathbf{x}$ and $\mathbf{x}'$ that depends on the Euclidean distance between them. In this paper, we use the Gaussian exponential correlation function as the kernel, that is,

$$R(\mathbf{x}, \mathbf{x}') = \prod_{i=1}^{n} \left(-\theta_i \left(x_i - x_i' \right)^2 \right), \forall \theta_i \in \mathbb{R}^+. \tag{2}$$

Generally, numerical optimization is adopted to find best hyper-parameters θ_i $(i = 1, \dots, n)$ that maximize the likelihood function $\tilde{L}$ expressed in Eq. (3).

$$\tilde{L} = \frac{1}{\sqrt{(2\pi\hat{\sigma}^2)^{n_s} |R|}} \exp \left(-\frac{1}{2} \frac{(\mathbf{y}_s - \widehat{\beta}_0 \mathbf{1})^{\mathrm{T}} \mathbf{R}^{-1} \left(y_s - \widehat{\beta}_0 \mathbf{1} \right)}{\hat{\sigma}^2} \right), \tag{3}$$

where $\widehat{\sigma}^2 = \left(\mathbf{y}_s - \mathbf{1}\widehat{\beta}_0\right)^{\mathrm{T}} \mathbf{R}^{-1}\left(\mathbf{y}_s - \mathbf{1}\widehat{\beta}_0\right)/n_s$. In practice, we maximize the natural logarithm of the likelihood function in Eq. (4) for simplification.

$$\ln\tilde{L} = -\frac{n_s}{2}\ln\sigma^2 - \frac{1}{2}\ln|R|. \tag{4}$$

This function is highly multimodal [27], so we use a gradient-free algorithm—the genetic algorithm to maximize it.

Then Eq. (1) is used to predict the output of any given input $\mathbf{x}$. In addition, Kriging provides the mean squared error $s^2(\mathbf{x})$ of the prediction, which indicates the uncertainty of the prediction.

$$s^2(\mathbf{x}) = \widehat{\sigma}^2\left[1 - \mathbf{r}^{\mathrm{T}}\mathbf{R}^{-1}\mathbf{r} + \left(1 - \mathbf{1}\mathbf{R}^{-1}\mathbf{r}\right)^2/\left(\mathbf{1}^{\mathrm{T}}\mathbf{R}\mathbf{1}\right)\right]. \tag{5}$$

2.2   Aerodynamic Optimization with Kriging

We use the Kriging model introduced above to surrogate the aerodynamics in SBO. The flowchart of SBO used in this paper is shown in Fig. 1. First, samples infilling the design space are generated in the design of experiments (DoE) step using a sampling method—Latin hypercube sampling (LHS) method. They are evaluated by the high-fidelity aerodynamic model—RANS. Then, surrogate models are trained by these initial samples. We only consider drag minimization with a lift constraint in this paper, so we train two surrogates for C_D and C_L, respectively. Initial surrogates are usually not accurate enough, so they are refined cycle by cycle before the optimum is eventually obtained. We refer to it as the outer loop in Fig. 1. New samples used to refine surrogates are obtained through optimizations in the inner loop. Many infilling criteria [25] have been proposed to improve the efficiency and global searching capability. We adopt a robust criterion introduced in the Efficient Global Optimization (EGO) method [22]. That is, in addition to the solution of minimizing C_D. subject to the C_L constraint, we add another point by maximizing the expected improvement (EI) in each cycle.

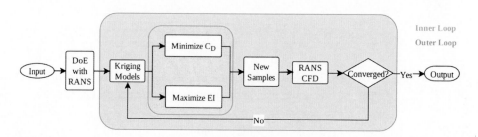

Fig. 1. The flowchart of SBO.

The EI function of an unconstrained problem is given in Eq. (6).

$$E[I(\mathbf{x})] = \begin{cases} (y_{min} - \widehat{y}(\mathbf{x}))\Phi\left(\frac{y_{min}-\widehat{y}(\mathbf{x})}{s(\mathbf{x})}\right) + s(\mathbf{x})\phi\left(\frac{y_{min}-\widehat{y}(\mathbf{x})}{s(\mathbf{x})}\right) & \text{if } s > 0 \\ 0 & \text{if } s = 0 \end{cases} \quad (6)$$

We use $g(\mathbf{x}) = C_L^{(con)} - C_L(\mathbf{x}) \leq 0.0$ represent the constraint for C_L in this paper. $C_L(\mathbf{x})$ is predicted by the Kriging model as $\widehat{C}_L(\mathbf{x})$. We assume $\widehat{g}(\mathbf{x}) = C_L^{(con)} - \widehat{C}_L(\mathbf{x})$ to be a random variable $G(\mathbf{x}) \in N\left[\widehat{g}(\mathbf{x}), s_g^2(\mathbf{x})\right]$, where $\widehat{g}(\mathbf{x})$ and $s_g^2(\mathbf{x})$ are the prediction and its mean square error, respectively. For the C_L constrained problems in this paper, Eq. (6) is extended to a constrained form (Eq. (7)) based on the derivation in [23, 24].

$$E[I(\mathbf{x})]^{(con)} = E[I(\mathbf{x})] \times P[G(\mathbf{x}) \leq 0.0],$$
$$\text{where } P[G(\mathbf{x}) \leq 0.0] = \Phi\left(-\frac{\widehat{g}(\mathbf{x})}{s_g^2(\mathbf{x})}\right) \quad (7)$$

In each inner loop, we minimize the drag coefficient subject to the lift constraint and maximize the EI in Eq. (7) to obtain two new points. We discuss the optimization algorithms for these two inner-loop optimizations in Sect. 3.

3 Optimization Algorithms in SBO

When conducting optimizations in the inner loop of SBO, we consider both the computational efficiency and the global searching capability. Although gradient-free algorithms improve the global searching capability, it is much more costly than gradient-based algorithms. Unless the number of local optima is very high, multi-start execution of gradient-based algorithms should be more efficient [28]. Drag minimization subject to a lift constraint might be multimodal, but previous researches have shown that the multimodality can be insignificant [8, 29, 30].

Table 1. Description of aerodynamic cases in the comparison of gradient-based and gradient-free algorithms in the inner loop of SBO.

	Case 1 Airfoil shape optimization	Case 2 Wing shape optimization
Number of design variables	19	75
Number of training points	50	80
Objective function	C_d	C_d
Lift constraint	$C_l \geq 0.82$	$C_L \geq 0.5$
Mach number	0.73	0.85

We compare the performance of both kinds of algorithms to investigate which approach is more suitable for aerodynamic optimization problems of wings. For gradient-based optimization, we use the sequential least squares programming algorithm (SLSQP) implemented in PyOptSparse [31] as the optimizer. LHS method is adopted to generate starting points in gradient-based optimization to reduce the possibility of being trapped to local optima. For gradient-free optimization, we use the particle swarm optimization (PSO) method in pyswarm[1].

Table 2. Description of parameters in four PSO combinations.

	PSO 1	PSO 2	PSO 3	PSO 4
Number of particles in the swarm	100	100	500	500
Maximum number of searching iterations	100	1000	100	1000
Particle velocity scaling factor	0.5	0.5	0.5	0.5

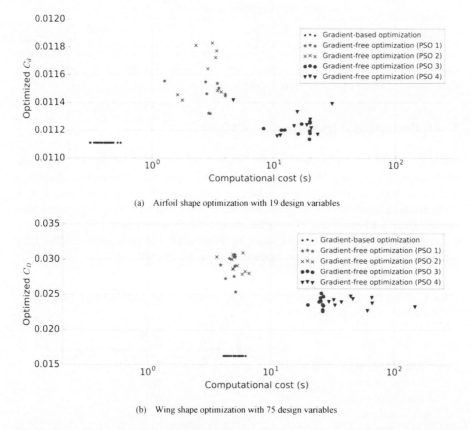

(a)   Airfoil shape optimization with 19 design variables

(b)   Wing shape optimization with 75 design variables

Fig. 2. Comparison of gradient-based optimization and gradient-free optimization in drag minimization problems subject to a lift constraint.

---

[1] https://pythonhosted.org/pyswarm.

As shown in Table 1, two drag minimization cases are involved in this investigation. In gradient-based optimization, we use the finite difference to compute gradient because predictions by surrogates are fast. In gradient-free optimization, parameters of PSO have an influence on the optimization results and different executions of PSO could find different solutions even with the same PSO parameters. So we use four combinations of PSO parameters which are noted as PSO 1–4 in Table 2, and we run ten optimizations with each combination.

The drag minimization results and corresponding computational costs are shown in Fig. 2. Gradient-based optimizations find the best results in both cases. Modification of PSO parameters improves the results to some extent. For example, in Fig. 2(a), some tests in PSO 3 and PSO 4 have comparable results with those in gradient-based optimization. However, this modification dramatically increases the computational cost. Even worse, as shown in Fig. 2(b), the improvement by this modification becomes insignificant in the case with a larger input dimension. A further increase in numbers of particles and searching iterations might make PSO converge to better results, but it is even more costly. So, in order to obtain a better convergence and reduce computational cost, we use multistart gradient-based optimizations for drag minimization problems in this paper.

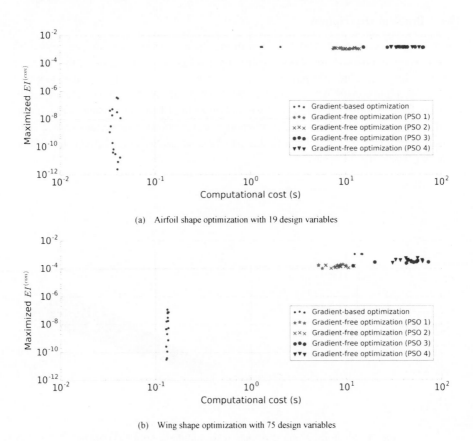

(a)   Airfoil shape optimization with 19 design variables

(b)   Wing shape optimization with 75 design variables

Fig. 3. Comparison of gradient-based optimization and gradient-free optimization in maximizing the EI function of drag minimization problems with lift constraints.

EI functions in Eqs. (6) and (7) are highly multimodal [25], and finding the global optimum is expensive [32]. In order to find a suitable approach to the maximization, we use the cases in Table 1 as well to compare the performance of gradient-based optimization and gradient-free optimization. The results are shown in Fig. 3. Many gradient-based optimizations are stuck in small values. This reinforces the multi-modality of the EI function. Nevertheless, there are still some gradient-based optimizations that converge to much larger values, which are even better than those optimized by PSO, especially in the wing case where the input dimension is higher. Modifications of PSO parameters influences the results only to a small extent. Considering the advantages and disadvantages of both approaches, we adopt a combination of them for EI maximization in this paper. That is, we use a multi-start execution of gradient-based optimizations and one gradient-free optimization with PSO 1. To be noticed, gradient-based optimizations that are stuck to small local minima only cost a little time due to fast convergences, so the total cost of multi-start execution is low.

4   Drag Minimization of Transonic Wings

4.1   Problem Description

In this paper, we use a RANS solver with the Spalart–Allmaras turbulence model to simulate aerodynamics. Free-Form Deformation (FFD) is used to control the shape of wings, which includes manipulations of section shapes and twists. In inner loops of SBO, we minimize the drag coefficient using gradient-based optimizations (SLSQP) starting from 10 different points, and we maximize the EI function using both a multi-start execution of SLSQP with 30 different starting points and a gradient-free optimization (PSO 1). The problems in this paper are all subject to a lift constraint considering the demand in practice.

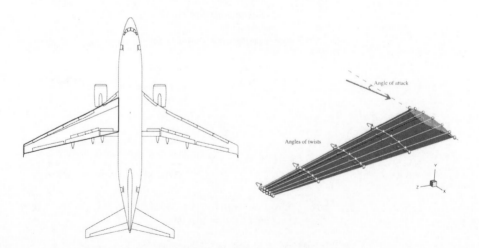

Fig. 4. Planform of the B737-like wing and the control FFD points.

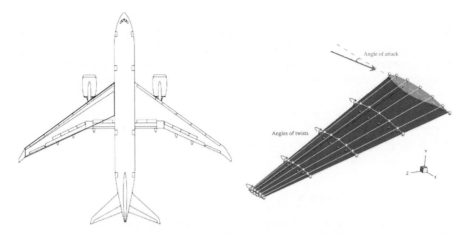

Fig. 5. Planform of the B787-like wing and the control FFD points.

We consider shape optimizations of two wings with typical planforms in transonic commercial aircraft. As shown in Figs. 4(a) and 5(a), we define their planforms based on Boeing 737-700 and 787-800 jets, respectively. We use the RAE2822 airfoil as the initial section shape in both cases, and FFD boxes with $7 \times 2 \times 2$ control points, shown in Figs. 4(b) and 5(b), are used to manipulate wing shapes. The mean chords of the two wings are 1.35 m, and their sweep angles at 1/4 chord lines are $25°$ and $32°$, respectively. The flow conditions, objective functions, constraints, and bounds of design variables in two cases are described in Table 3.

Table 3. Description of aerodynamic optimization problems for both wings.

	Case A B737-like wing	Case B B787-like Wing
Objective function	Minimizing C_D	Minimizing C_D
Lift constraint	$C_L \geq 0.6$	$C_L \geq 0.5$
Mach number	0.78	0.85
Reynolds number	5×10^6	5×10^6
with respect to:	$\alpha \in \left[1.0°, 5.0°\right]$	$\alpha \in \left[2.0°, 6.0°\right]$
	twists $\in [-10.0°, 10.0°]$	twists $\in [-10.0°, 10.0°]$
	$\Delta Y_{\text{FFD}} \in [-0.05, 0.05] \times Y_{\text{FFD}}^{initial}$	$\Delta Y_{\text{FFD}} \in [-0.05, 0.05] \times Y_{\text{FFD}}^{initial}$

Table 4. Numbers of design variables in two approaches.

	Approach I	Approach II
Section shape	70	14
Section twist	4	4
Angle of attack	1	1
Total	75	19

For each case, we use two approaches to solve the optimization. In Approach I, any FFD points can change independently along the Y-axis to manipulate the wing shape. We allow the change of four sections' twists based on 1/4 chord points, and the angle of attack is a parameter to satisfy the lift constraint. There are 75 design variables in total in Approach I, which is challenging for surrogates to handle. Approach II involves fewer design variables, where the five section shapes are manipulated together. This decrease of geometry freedoms is helpful for surrogates, but it may exclude some good designs outside the design space. Investigation of this compromise in SBO is of great interest for preliminary design of wings. The details of design variables in the two approaches are shown in Table 4.

4.2   Optimization Comparison

We use 75 and 20 samples in DoE to train initial Kriging surrogate models for Approach I and Approach II, respectively. This choice is to make the number of DoE samples approximately equal to the input dimension which is suggested in [25]. We use a computational budget of 300 CFD simulations for each optimization, and after that the optimization is terminated. Convergence histories are shown in Figs. 6 and 7. Feasible and infeasible points are marked by considering if the lift constraint is satisfied (within a tolerance of 2 lift counts).

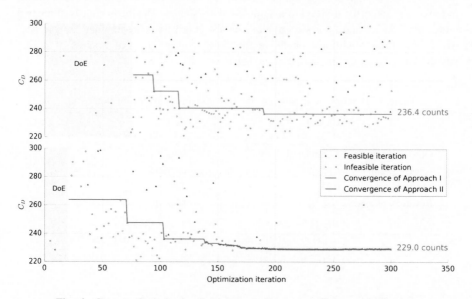

Fig. 6. Case A: Optimization by Approach II is more efficient and effective.

As shown in Fig. 6, Approach II is both more effective and efficient than Approach I in Case A. Optimization in Approach I converges to $C_D = 237.4$ from the baseline $C_D = 263.82$, which is a 10.4% reduction, while Approach II converges to $C_D = 229.0$ with a reduction of 13.2%. In addition, considering the distribution of the

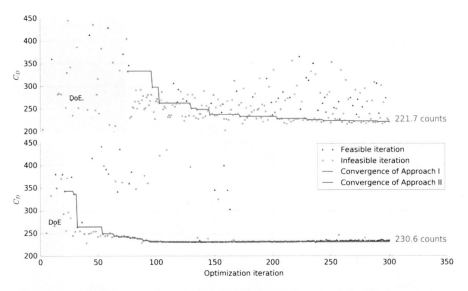

Fig. 7. Case B: Convergence of Approach II is much faster.

new added samples, Approach II conveys a clear convergence after about 200 CFD simulations. However, we cannot see any signs of convergence even after 300 CFD simulations in Approach I. From another point of view, this means optimization in Approach I might converge to a better solution if a larger computational budget is allowed, but it will further reduce its efficiency. Geometries and C_P counters of the baseline and optimized wings are shown in Fig. 8. Approach I reduces the drag mainly by manipulating section shapes, while the drag reduction in Approach II primarily comes from the changes of angles of attack and twists.

For Case B shown in Fig. 7, optimizations in Approach I and Approach II obtain 43.8% and 41.5% drag reductions, respectively, which is comparable. Similar with Case A, Approach II converges quickly to $C_D \approx 230$ counts within 100 CFD simulations, while, at this time, the best solution in Approach I is $C_D \approx 340$ counts. Approach I catches up with Approach II after about another 100 CFD simulations, and it finally converges to a slightly better result at the end of optimization. As shown in Fig. 9, both approaches satisfy the lift constraint by increasing the angle of attack and reduce the drag by increasing negative angles of twist, especially for sections near the wingtip. This makes the lift along the wing span closer to the elliptical distribution and then reduces the induced drag.

The results in these two cases show that SBOs by both approaches reduce the drag coefficient, and optimization results are generally similar with the given computational budget (300 CFD simulations). If a smaller computational budget is allowed, Approach I might be less effective due to its slow convergence. The convergence by Approach II is much faster. So if higher optimization efficiency is pursued, Approach II is a good choice.

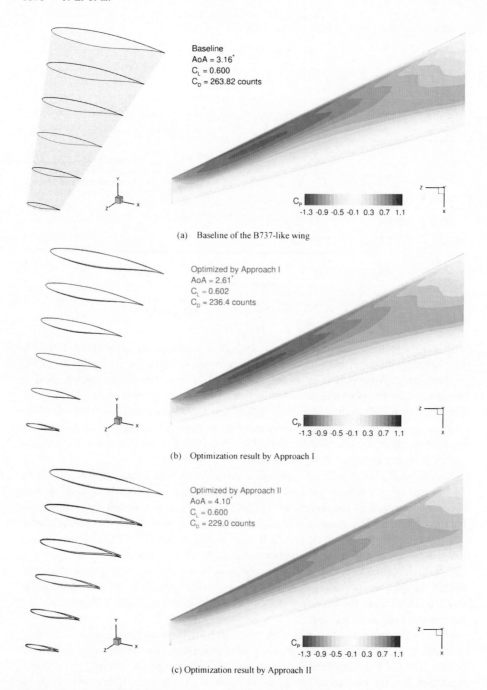

Baseline
AoA = 3.16°
C_L = 0.600
C_D = 263.82 counts

C_P
-1.3 -0.9 -0.5 -0.1 0.3 0.7 1.1

(a)   Baseline of the B737-like wing

Optimized by Approach I
AoA = 2.61°
C_L = 0.602
C_D = 236.4 counts

C_P
-1.3 -0.9 -0.5 -0.1 0.3 0.7 1.1

(b)   Optimization result by Approach I

Optimized by Approach II
AoA = 4.10°
C_L = 0.600
C_D = 229.0 counts

C_P
-1.3 -0.9 -0.5 -0.1 0.3 0.7 1.1

(c) Optimization result by Approach II

Fig. 8. Case A: Approach II converges to a better solution by changing the angle of attack and twists more effectively.

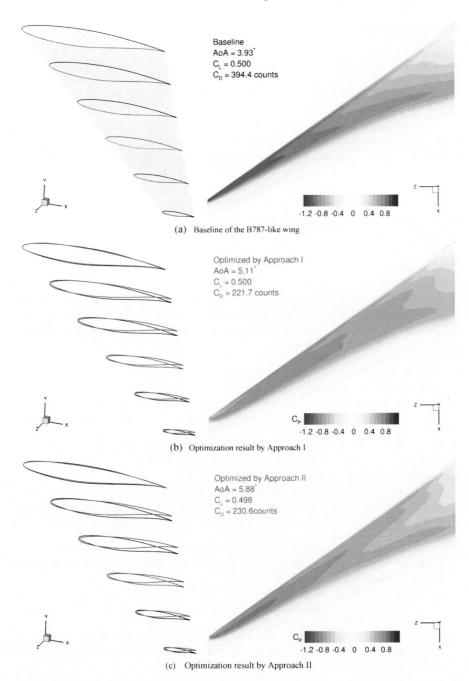

Baseline
AoA = 3.93°
C_L = 0.500
C_D = 394.4 counts

-1.2 -0.8 -0.4 0 0.4 0.8

(a) Baseline of the B787-like wing

Optimized by Approach I
AoA = 5.11°
C_L = 0.500
C_D = 221.7 counts

C_P
-1.2 -0.8 -0.4 0 0.4 0.8

(b) Optimization result by Approach I

Optimized by Approach II
AoA = 5.88°
C_L = 0.498
C_D = 230.6counts

C_P
-1.2 -0.8 -0.4 0 0.4 0.8

(c) Optimization result by Approach II

Fig. 9. Case B: Final results by two approaches are comparable. Approach I and Approach II reduce the drag coefficient by 43.8% and 41.5%, respectively.

5 Conclusions

This paper focuses on the investigation of two selections of design variables in SBO of transonic wings—scale section shapes together or not. Multi-start gradient-based optimizations are adopted in inner loops of SBO to improve the convergence and reduce the computational cost.

Although more freedoms are involved by manipulating section shapes independently, better results are not necessarily obtained via SBO in this approach. Its performance is easily influenced by an early termination of the optimization when the computational budget is limited. However, the reduction of design variables by scaling all section shapes together makes SBO converge much faster, and the solution is comparable with the other approach. This conclusion may provide some useful insights for preliminary design of wings in the transonic regime.

References

1. Martins JRRA, Sturdza P, Alonso JJ (2003) The complex-step derivative approximation. ACM Trans Math Softw 29(3):245–262. https://doi.org/10.1145/838250.838251
2. Peter JEV, Dwight RP (2010) Numerical sensitivity analysis for aerodynamic optimization: a survey of approaches. Comput Fluids 39(3):373–391. https://doi.org/10.1016/j.compfluid.2009.09.013
3. Jameson A (1988) Aerodynamic design via control theory. J Sci Comput 3(3):233–260. https://doi.org/10.1007/bf01061285
4. Mader CA, Martins JRRA, Alonso JJ, van der Weide E (2008) ADjoint: an approach for the rapid development of discrete adjoint solvers. AIAA J 46(4):863–873. https://doi.org/10.2514/1.29123
5. Lyu Z, Martins JRRA (2013) RANS-based aerodynamic shape optimization of a blended-wing-body aircraft. In: 21st AIAA computational fluid dynamics conference, San Diego, CA. http://doi.org/10.2514/6.2013-2586
6. Li J, Qu K, Cai J, Cao C (2016) Adjoint approach based on reduced-order model for steady PDE systems. In: 17th AIAA/ISSMO multidisciplinary analysis and optimization conference. American Institute of Aeronautics and Astronautics. http://doi.org/10.2514/6.2016-3668
7. He P, Mader CA, Martins JRRA, Maki KJ (2018) An aerodynamic design optimization framework using a discrete adjoint approach with OpenFOAM. Comput Fluids. http://doi.org/10.1016/j.comp uid.2018.04.012
8. Lyu Z, Kenway GKW, Martins JRRA (2015) Aerodynamic shape optimization investigations of the common research model wing benchmark. AIAA J 53(4):968–985. https://doi.org/10.2514/1.J053318
9. Kenway GKW, Martins JRRA (2016) Multipoint aerodynamic shape optimization investigations of the common research model wing. AIAA J 54(1):113–128. https://doi.org/10.2514/1.J054154
10. Li J, Cai J, Qu K (2018) Adjoint-based two-step optimization method using proper orthogonal decomposition and domain decomposition. AIAA J 56(3):1133–1145. https://doi.org/10.2514/1.j055773

11. Chen S, Lyu Z, Kenway GKW, Martins JRRA (2016) Aerodynamic shape optimization of the common research model wing-body-tail configuration. J Aircr 53(1):276–293. https://doi.org/10.2514/1.C033328
12. Kenway GKW, Martins JRRA (2017) Buffet onset constraint formulation for aerodynamic shape optimization. AIAA J 55(6):1930–1947. https://doi.org/10.2514/1.J055172
13. Han ZH, Abu-Zurayk M, Gortz S, Ilic C (2018) Surrogate-based aerodynamic shape optimization of a wing-body transport aircraft configuration. In: Notes on numerical fluid mechanics and multidisciplinary design. Springer International Publishing, pp 257–282. http://doi.org/10.1007/978-3-319-72020-316
14. Leifsson L, Koziel S (2010) Multi-fidelity design optimization of transonic airfoils using physics-based surrogate modeling and shape-preserving response prediction. J Comput Sci 1 (2):98–106. https://doi.org/10.1016/j.jocs.2010.03.007
15. Andres E, Salcedo-Sanz S, Monge F, Pérez-Bellido A (2012) Efficient aerodynamic design through evolutionary programming and support vector regression algorithms. Expert Syst Appl 39(12):10700–10708. https://doi.org/10.1016/j.eswa.2012.02.197
16. Zhang Y, Han ZH, Leifsson LT (2017) Surrogate-based optimization applied to benchmark aerodynamic design problems. In: 35th AIAA applied aerodynamics conference. American Institute of Aeronautics and Astronautics. http://doi.org/10.2514/6.2017-4367
17. Wu X, Zhang W, Song S (2017) Robust aerodynamic shape design based on an adaptive stochastic optimization framework. Struct Multi Optim 57(2):639–651. https://doi.org/10.1007/s00158-017-1766-5
18. Allen CB, Poole DJ, Rendall TCS (2018) Wing aerodynamic optimization using efficient mathematically-extracted modal design variables. Optim Eng 19(2):453–477. https://doi.org/10.1007/s11081-018-9376-7
19. Chernukhin O, Zingg DW (2013) Multimodality and global optimization in aerodynamic design. AIAA J 51(6):1342–1354. https://doi.org/10.2514/1.j051835
20. Li J, Cai J, Qu K (2018) Surrogate-based aerodynamic shape optimization with the active subspace method. Struct Multi Optim. http://doi.org/10.1007/s00158-018-2073-5
21. Krige DG (1951) A statistical approach to some basic mine valuation problems on the witwatersrand. J Chem Metall Min Soc 52:119–139
22. Jones DR, Schonlau M, Welch WJ (1998) Efficient global optimization of expensive black-box functions. J Global Optim 13(4):455–492. https://doi.org/10.1023/A:1008306431147
23. Sasena MJ, Papalambros P, Goovaerts P (2002) Exploration of metamodeling sampling criteria for constrained global optimization. Eng Optim 34(3):263–278. https://doi.org/10.1080/03052150211751
24. Parr JM, Keane AJ, Forrester AI, Holden CM (2012) Infill sampling criteria for surrogate-based optimization with constraint handling. Eng Optim 44(10):1147–1166. https://doi.org/10.1080/0305215x.2011.637556
25. Liu J, Song WP, Han ZH, Zhang Y (2016) Efficient aerodynamic shape optimization of transonic wings using a parallel infilling strategy and surrogate models. Struct Multi Optim 55(3):925–943. https://doi.org/10.1007/s00158-016-1546-7
26. Han Z (2016) Kriging surrogate model and its application to design optimization: a review of recent progress. Acta Aeronautica et Astronautica Sinica 37(11):3197–3225. https://doi.org/10.7527/S1000-6893.2016.0083
27. Mardia KV, Watkins AJ (1989) On multimodality of the likelihood in the spatial linear model. Biometrika 76(2):289–295. https://doi.org/10.1093/biomet/76.2.289
28. Haftka RT (2016) Requirements for papers focusing on new or improved global optimization algorithms. Struct Multi Optim 54(1):1. https://doi.org/10.1007/s00158-016-1491-5

29. Li J, Bouhlel MA, Martins JRRA (2019) Data-based approach for fast airfoil analysis and optimization. AIAA J 57(2):581–596. https://doi.org/10.2514/1.J057129
30. Yu Y, Lyu Z, Xu Z, Martins JRRA (2018) On the influence of optimization algorithm and starting design on wing aerodynamic shape optimization. Aerosp Sci Technol 75:183–199. https://doi.org/10.1016/j.ast.2018.01.016
31. Perez RE, Jansen PW, Martins JRRA (2012) pyOpt: a python-based object-oriented framework for nonlinear constrained optimization. Struct Multi Optim 45(1):101–118. https://doi.org/10.1007/s00158-011-0666-3
32. Jones DR (2001) A taxonomy of global optimization methods based on response surfaces. J Global Optim 21(4):345–383. https://doi.org/10.1023/A:1012771025575

Aero-Structural Optimization of a Supersonic Wing Model Using Adjoint-Based Optimization Algorithm

Jingrui Guo[1], Min Xu[1], and Yi Li[2(✉)]

[1] School of Astronautics, Northwestern Polytechnical University,
Xi'an 710072, Shaanxi, People's Republic of China
[2] School of Aeronautics, Northwestern Polytechnical University,
Xi'an 710072, Shaanxi, People's Republic of China
liyi504@nwpu.edu.cn

Abstract. The paper present a supersonic aero-structural optimization method based on the gradient information obtained by adjoint approach. The adjoint approach works on the acquisition of the partial derivative matrix and the iterative operation of the matrix. A swept wing model is studied to illustrate this methodology, in which the objective function is the lift-to-drag ratio considering the elastic deformation of wing structure, and independent variables are the thickness of the skin and the size of the spar. The results show that the higher lift-to-drag ratio can be achieved, while the weight of wing structure keeps light.

Keywords: Aero-structural optimization · Adjoint approach · Supersonic

1 Introduction

Supersonic aircraft is a complex system whose performance is determined by multiple disciplines such as aerodynamics and structure. The traditional serial design method doesn't consider the relationship between disciplines at every stage, so that the ability of coupled analysis method cannot be fully used and the system is hard to achieve optimal situation. Therefore, we need a supersonic aero-structural optimization method to replace the traditional one. At the same time, the cost of aero-structural analysis decides that we should choose the gradient-based optimization algorithm. However, calculating gradient information by finite-differences is very expensive. For supersonic aero-structural optimization, the cost is proportional to the number of design variables, which is usually large. Therefore, the adjoint method, for which the calculation time is independent of the number of design variables, should be a suitable choice.

Several developments have been done over the last decade to build aero-structural analysis model [4]. For adjoint approach, many valuable researches has been done during the past decade. In 2002, Martins [3] built a coupled aero-structural adjoint solver and used it to solve complete configuration aero-structural optimization problem. Gauger [1] developed a method to solve the aerodynamic shape optimization problem using continuous adjoint approach. Timothee [6] presented two high-fidelity aero-structure gradient computation methods for design variables. Sanchez [5] presented a technique to calculate the gradients in general fluid-structure interaction problems using adjoint approach.

© Springer Nature Singapore Pte Ltd. 2019
X. Zhang (Ed.): APISAT 2018, LNEE 459, pp. 1081–1091, 2019.
https://doi.org/10.1007/978-981-13-3305-7_86

This paper is organized as follows: Subsect. 2.1 gives the flow chart about the aero-structural optimization method. Subsection 2.2 describes the derivation of aero-structural adjoint equation and gives the objective function which is expressed by the adjoint vector. Subsection 2.3 discusses the calculation method and the formula of partial derivative matrix, where the objective function is the lift-to-drag ratio and the independent variable is the thickness of the structural analysis model. Finally, Sect. 3 gives a case of supersonic aero-structural optimization, in which the objective function is lift-to-drag ratio and independent variable is structural size.

2 Optimization Method

This section describes the aero-structural optimization method and it includes 3 sub-sections. Firstly, the flow chart about the optimization process is given. Then, we discuss the specific calculation method of the adjoint approach. The third subsection considers the calculation of the partial derivative matrix.

2.1 Aero-Structural Optimization Process

During the design process, we usually consider the static aeroelastic problem of the aircraft and the general optimization problem can be abstracted as

$$\begin{cases} p = [p_1, p_2, \ldots, p_n]^T \\ s.t. A_k(u(p), w(p), p) = 0(k, 1, 2, \ldots, n_A) \\ S_l(u(p), w(p), p) = 0(l, 1, 2, \ldots, n_S) \\ \max I(u(p), w(p), p) \quad or \quad \min I(u(p), w(p), p) \end{cases} \tag{2.1}$$

where I is the objective function, which can be calculated from the independent variable p as shown in Fig. 1.

A_k and S_l are the governing equations for aerodynamic analysis and structural analysis respectively, and they are equal to 0. n_A and n_S are the numbers of aerodynamic and structural governing equations respectively. w and u represent the fluid conservative variables and the structural displacements respectively and both of them are decided by p.

Because of the reasons stated in introduction, we chose the gradient optimization algorithm and use the adjoint approach to get the gradient information. The optimization process is shown in Fig. 2.

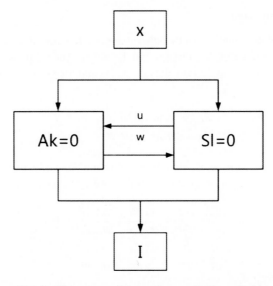

Fig. 1. Flow chart of the static aeroelastic analysis

The adjoint part in the Fig. 2 calculates the gradient information based on the partial derivative matrix and the specific calculation method of the matrix is described in details in the following subsection.

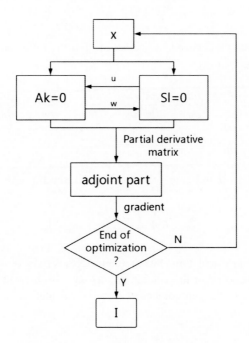

Fig. 2. Flow chart of the optimization process

2.2   Adjoint Approach

The specific calculation method of the adjoint approach is discussed in this subsection. Here, the papers of Joaquim Martins [2] and Timothee Achard [6] are referenced. The optimization problem can be described as Eq. (2.1) and the gradient information can be calculated as

$$\frac{dI}{dp} = \frac{\partial I}{\partial u}\frac{du}{dp} + \frac{\partial I}{\partial w}\frac{dw}{dp} + \frac{\partial I}{\partial p} \tag{2.2}$$

At the same time, the total derivative of A_k and S_l with respect to any design variable must be zero, and the gradient information can be calculated as

$$\begin{aligned}
\frac{dA}{dp} &= \frac{\partial A}{\partial p} + \frac{\partial A}{\partial u}\frac{du}{dp} + \frac{\partial A}{\partial w}\frac{dw}{dp} = 0 \\
\frac{dS}{dp} &= \frac{\partial S}{\partial p} + \frac{\partial S}{\partial u}\frac{du}{dp} + \frac{\partial S}{\partial w}\frac{dw}{dp} = 0
\end{aligned} \tag{2.3}$$

In order to obtain the adjoint equations, Eq. (2.2) is formulated by adding the total derivative of A_k and S_l with respect to p. ϕ and φ are usually called the adjoint vector.

$$\begin{aligned}
\frac{dI}{dp} &= \frac{\partial I}{\partial p} + \frac{\partial I}{\partial u}\frac{du}{dp} + \frac{\partial I}{\partial w}\frac{dw}{dp} + \phi(\frac{\partial S}{\partial p} + \frac{\partial S}{\partial u}\frac{du}{dp} + \frac{\partial S}{\partial w}\frac{dw}{dp}) \\
&+ \varphi(\frac{\partial A}{\partial p} + \frac{\partial A}{\partial u}\frac{du}{dp} + \frac{\partial A}{\partial w}\frac{dw}{dp})
\end{aligned} \tag{2.4}$$

Equation (2.4) can be changed by factorizing out the differential terms $\frac{du}{dp}$ and $\frac{dw}{dp}$, then Eq. (2.5) can be yielded.

$$\begin{aligned}
\frac{dI}{dp} &= (\frac{\partial I}{\partial u} + \phi\frac{\partial S}{\partial u} + \varphi\frac{\partial A}{\partial u})\frac{\partial u}{\partial p} + (\frac{\partial I}{\partial w} + \phi\frac{\partial S}{\partial w} + \varphi\frac{\partial A}{\partial w})\frac{\partial w}{\partial p} \\
&+ \frac{\partial I}{\partial p} + \phi(\frac{\partial S}{\partial p}) + \varphi(\frac{\partial A}{\partial p})
\end{aligned} \tag{2.5}$$

The adjoint vectors ϕ and φ are chosen such that the Eq. (2.6) can be satisfied.

$$\begin{pmatrix} \frac{\partial S}{\partial u} & \frac{\partial A}{\partial u} \\ \frac{\partial S}{\partial w} & \frac{\partial A}{\partial w} \end{pmatrix}^T \begin{pmatrix} \phi \\ \varphi \end{pmatrix} = \begin{pmatrix} -\frac{\partial I}{\partial u} \\ -\frac{\partial I}{\partial w} \end{pmatrix} \tag{2.6}$$

In order to improve calculation efficiency, Eq. (2.6) is transformed into the form of the Eq. (2.7) and the value of the adjoint vectors can be obtained by iterative calculation. In Eq. (2.7), t is the step number of iterative calculation.

$$\frac{\partial S^T}{\partial u} \phi^{t+1} = -\frac{\partial I}{\partial u} - \frac{\partial A^T}{\partial u} \varphi^t$$

$$\frac{\partial A^T}{\partial w} \varphi^{t+1} = -\frac{\partial I}{\partial w} - \frac{\partial S^T}{\partial w} \phi^t$$

(2.7)

According to Eq. (2.8), the gradient information of the coupled model can be calculated.

$$\frac{dI}{dp} = \frac{\partial I}{\partial p} + \phi(\frac{\partial S}{\partial p}) + \varphi(\frac{\partial A}{\partial p})$$

(2.8)

2.3   Partial Derivative Matrix Detail

In this subsection, the calculation of the partial derivative matrix in Eq. (2.7, 2.8) are discussed. Equation (2.8) is the total sensitivity equation, which is made up of three partial derivative terms: $\frac{\partial I}{\partial p}$, $\frac{\partial A}{\partial p}$ and $\frac{\partial S}{\partial p}$. When using linear finite-element models for structural analysis, the governing equations for structural analysis can be written as Eq. (2.9) [2]. K is the global stiffness matrix of the structure and f_s is the vector of applied structural nodal forces.

$$S_l = Ku - f_s = 0$$

(2.9)

$\frac{\partial S}{\partial p}$ can be written as Eq. (2.10).

$$\frac{\partial S}{\partial p} = \frac{\partial K}{\partial p} u$$

(2.10)

$\frac{\partial I}{\partial p}$ and $\frac{\partial A}{\partial p}$ can be calculated by automatic differentiation method. If the objective function is the lift-to-drag ratio and the independent variable is the structure size, $\frac{\partial I}{\partial p}$ and $\frac{\partial A}{\partial p}$ are equal to zero [2], and Eq. (2.8) can be written as [6]

$$\frac{dI}{dp} = \phi(\frac{\partial S}{\partial p})$$

(2.11)

Equation (2.7) is the aero-structural adjoint equation, which is made up of six partial derivative terms. When using linear finite-element models for structural analysis, $\frac{\partial S}{\partial u}$ and $\frac{\partial S}{\partial w}$ can be calculated by Eq. (2.12) [2] and Eq. (2.13).

$$\frac{\partial S}{\partial u} = K$$

(2.12)

$$\frac{\partial S}{\partial w} = -\frac{\partial f_s}{\partial w} = -\frac{\partial f_s}{\partial f_a} \frac{\partial f_a}{\partial w}$$

(2.13)

f_a is the vector of applied aerodynamic nodal forces and $\frac{\partial f_s}{\partial f_a}$ is determined by the interpolation algorithm which is used to transmit data between aerodynamic analysis model and structural analysis model. $\frac{\partial f_a}{\partial w}$ can be calculated by automatic differentiation method. $\frac{\partial A}{\partial u}$ can be obtained by Eq. (2.14).

$$\frac{\partial A}{\partial u} = \frac{\partial A}{\partial x_a} \frac{\partial x_a}{\partial x_s} \frac{\partial x_s}{\partial u} \tag{2.14}$$

In Eq. (2.14), x_a and x_s are the grid node coordinates of aerodynamic and structural grid, respectively. $\frac{\partial x_a}{\partial x_s}$ is determined by the interpolation algorithm, which is the similar to $\frac{\partial f_s}{\partial f_a}$. $\frac{\partial A}{\partial x_a}$ can be calculated by automatic differentiation method.

In general, $\frac{\partial l}{\partial u}$, $\frac{\partial l}{\partial w}$ and $\frac{\partial A}{\partial w}$ can be calculated by automatic differentiation method. However, the specific content of the objective function must be taken into account when calculating the first two items.

3 Optimization Case

This section presents the results of the aero-structural optimization of a wing's lift-to-drag ratio after aerodynamic deformation using the approach described in Sect. 2, and the independent variables are the thickness of skin and the size of the spar. The flight speed is 1.5 Mach and the angle of attack is 8°. The optimization problem can be written as Eq. (3.1). Cl and Cd represent the lift coefficient and the drag coefficient respectively.

$$\begin{cases} p = [p_1, p_2, \ldots, p_{98}]^T \\ s.t. A_k(u(p), w(p), p) = 0 (k, 1, 2, \ldots, n_a) \\ S_l(u(p), w(p), p) = 0 (l, 1, 2, \ldots, n_s) \\ \max I = \frac{Cl}{Cd} \end{cases} \tag{3.1}$$

The wing shown in the Fig. 3 is made up by skins, ribs and spars. The unit of Fig. 3 is millimeter.

The structural analysis model and the aerodynamic analysis model used in this case is shown in Fig. 4, and the material of structural analysis model is aluminum alloy. The rid line in Fig. 4(a) indicates the position of spar, the green one is rib and the black one is skin.

The original sizes of the upper and lower skins are same, which are shown in Fig. 5.

After optimization, the thickness information of the upper skin is shown in Fig. 6. The thickness information of the lower skin is shown in Fig. 7.

The sizes of spar are divided into 7 parts which are represented as No. 1–7 from root to tip, as shown in the Table 1. W and H are the optimized sizes, and the original ones are represented by W_o and H_o.

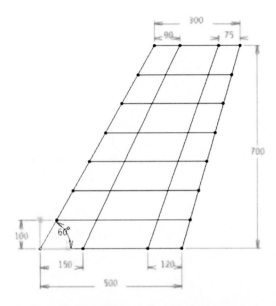

Fig. 3. Diagram of wing structure

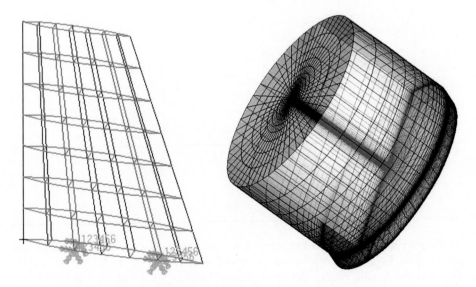

Fig. 4. (a) The structural analysis model; (b) The aerodynamic analysis model

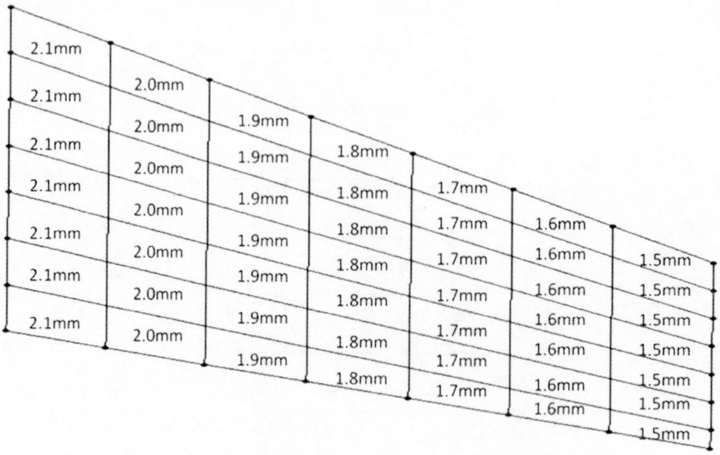

Fig. 5. The original thickness information of skin

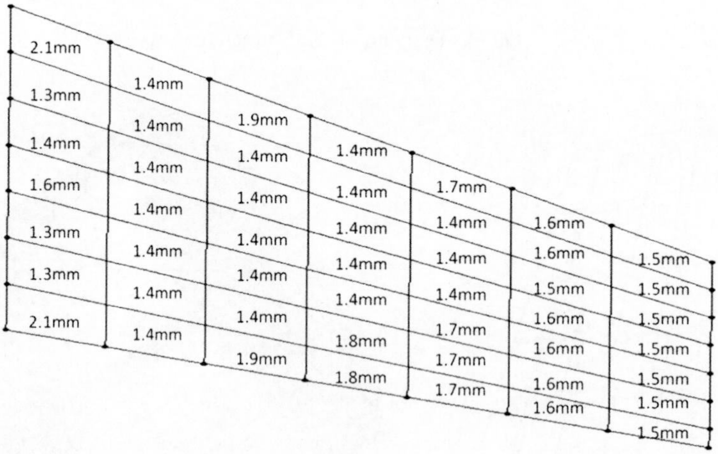

Fig. 6. The thickness information of the upper skin

As we can see, the structural sizes of the wing root parts are reduced during the optimization. Figure 8 shows the optimization history of the drag coefficient (in count), the lift coefficient (in count), the wing structural weight (in kilogram) and the lift-to-drag ratio.

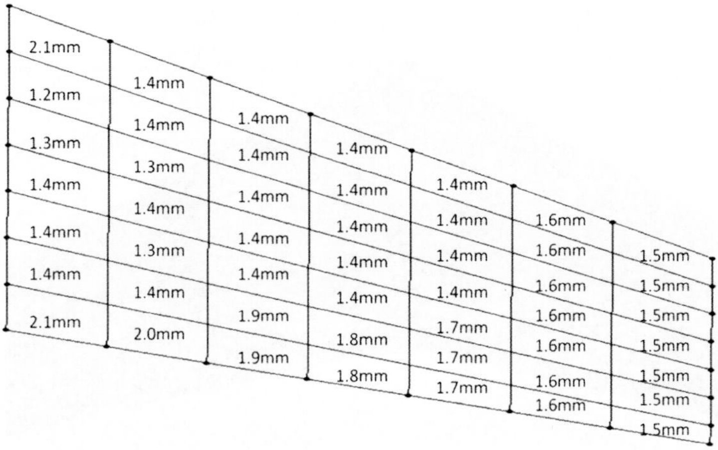

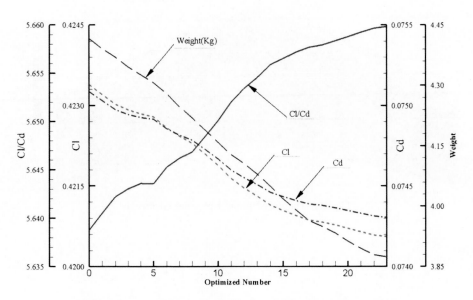

Fig. 7. The thickness information of the lower skin

Table 1. The sizes of spar

Part	W (mm)	W_o (mm)	H (mm)	H_o (mm)
No.1	16.3	22.0	1.4	2.1
No.2	19.7	20.0	1.4	2.0
No.3	18.0	18.0	1.4	1.9
No.4	16.0	16.0	1.4	1.8
No.5	14.0	14.0	1.4	1.7
No.6	12.0	12.0	1.6	1.6
No.7	10.0	10.0	1.5	1.5

Fig. 8. The change of aerodynamic parameters and structural weight

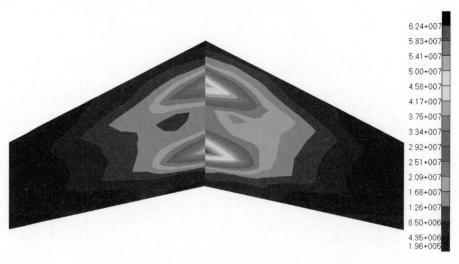

Fig. 9. The stress distribution of optimized structure

The weight, *Cl* and *Cd* are reduced during the optimization, but the lift-to-drag ratio increase from 5.6388 to 5.6598. Figure 9 indicates the stress distribution of optimized structure and the distribution of pressure coefficient is shown in Fig. 10. Left parts of Figs. 9 and 10 are the distributions without optimization and right parts are the optimized ones. After optimization, the value of stressed areas which are located in the wing root increased, and the change of *Cp* is small because of the comparatively small deformation of the wing.

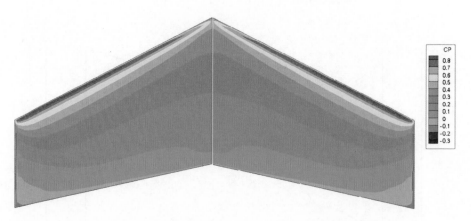

Fig. 10. The distribution of pressure coefficient

4 Conclusions

Aero-structural optimization is an important way to improve the overall performance of the aircraft. This paper presents a gradient-based aero-structural optimization method which is illustrated by a swept wing model in the state of Mach 1.5. The gradient information is obtained by adjoint approach which includes the acquisition of the partial derivative matrix and the iterative operation of the matrix. The objective function is the lift-to-drag ratio considering the elastic deformation of wing structure. There are total number of 112 design variables including the thickness of the skin and the size of the spar. Results show that the lift-to-drag ratio increased from 5.6388 to 5.6598 while the weight decreased from 4.4165 kg to 3.87485 kg.

References

1. Gauger N (2006) Adjoint approaches in aerodynamic shape optimization and MDO context. In: ECCOMAS CFD
2. Martins JRRA (2003) A coupled-adjoint method for high-fidelity aero-structural optimization. Dissertation abstracts international, vol 64–03, section: B, p 1349. Adviser: Juan J. Alons
3. Martins J, Alonso J, Reuther J (2002) Complete configuration aero-structural optimization using a coupled sensitivity analysis method 28(10):2000–4754
4. Mcnamara JJ, Friedmann PP (2011) Aeroelastic and aerothermoelastic analysis in hypersonic flow: past, present, and future. AIAA J 49(6):1089–1122
5. Sanchez R, Albring T, Palacios R, Gauger NR, Economon TD, Alonso JJ (2017) Coupled adjoint-based sensitivities in large-displacement fluid-structure interaction using algorithmic differentiation. Int J Numer Meth Eng
6. Timothée A, Christophe B, Roger O (2017) Comparison of high-fidelity aero-structure gradient computation techniques. Application on the CRM wing design. In: IFASD

The Optimization Design of Lift Distribution and Propeller Performance for Rotor/Wing Compound Helicopter

Xiaoxin Liu[1]([✉]), Lili Lin[2], Minghua Peng[2], and Jianbo Li[2]

[1] Nanjing University of Aeronautics and Astronautics, Nanjing, China
601345269@qq.com
[2] National Laboratory of Science and Technology on Rotorcraft Aeromechanics, Nanjing University of Aeronautics and Astronautics, Nanjing 210016, China

Abstract. The future development trend of helicopter will possess the function of high speed, far distance and high ceiling. Now, the main development trend of compound helicopters at home and abroad are the conventional configuration with auxiliary propulsion unit and the wing. But for this configuration, the force on the parts are very different from the ordinary. To solve this problem, it need to ensure the allocation of the lift between rotor and wing in the stage of general design. In this paper, for Rotor/wing compound helicopter, build a calculation model of required power and put up a kind of lift distribution strategy between the rotor and the wing and analyze it in different flight status. Through the research, the project of lift distribution has been confirmed. At the same time, to improve the working efficiency of the propellers, build an optimization model to modify propellers geometric parameters in I-sight. According to the optimized parameters, the working efficiency of the propellers have been verified to promote a lot.

Keywords: Compound helicopter · Lift distribution · Propeller performance

Nomenclature

b	Blade chord length
C_D	Drag coefficient of propeller airfoil
C_L	Lift coefficient of propeller airfoil
C_{Lw}	Wing lift coefficient
$C_{L\alpha w}$	Wing lift line coefficient
C_{x7}	Drag coefficient of rotor airfoil
C_{y7-90}	Lift coefficient of rotor airfoil in $90°$
J	Induced power correction factor
J_0	Induced power correction factor in power
k_c	Weight coefficient in cruise
k_h	Weight coefficient in hover
k_m	Weight coefficient in maximum forward flight speed
k_p	Coefficient of correction of profile drag inhomogeneity
k_{p0}	Coefficient of correction of profile drag inhomogeneity in hover
k_t	The coefficient of power transmission loss

© Springer Nature Singapore Pte Ltd. 2019
X. Zhang (Ed.): APISAT 2018, LNEE 459, pp. 1092–1107, 2019.
https://doi.org/10.1007/978-981-13-3305-7_87

k_T	Thrust factor
ΔM	The difference value of tip Mach number
M_1	The tip Mach number in hover
M_{1-90}	The tip Mach number for advancing blade in $90°$
M_{ljx}	Mach number caused by sudden increase in profile drag
M_{pl}	The moment of left propeller
M_{pr}	The moment of right propeller
$\bar{N}_{pb}$	Unit wave drag power
$\bar{N}_{pi}$	Unit induced power
$\bar{N}_{pr}$	Unit profile power
N_B	Blade number of propeller
P	Disk loading
p_{pr}	The power of right propeller
p_{pl}	The power of left propeller
q_w	Wing flowing pressure
S_w	Wing area
σ	Disc solidity
κ	The coefficient of tip loss
Δ	Density ratio
ρ_0	Sea level atmospheric density
μ	Advance ratio

1 Introduction

The configuration of conventional helicopter has advantages in hovering, vertical take-off and landing, but have no ability to catch up with the fixed-wing plane in some performance such as velocity, range and ceiling.

The compound helicopter which based on the conventional rotor, adding auxiliary propulsion unit and wing, which have become mainstream in both domestic and overseas including Bell 533, X-49A in America and X-3 high speed compound helicopter produced by the European helicopter company. These types of compound helicopter are different from the coaxial rigid rotor compound helicopter. Meanwhile, it not only the flight velocity has improved but also lift-to-drag ratio.

In this paper, the researcher beginning to study the lift distribution between rotor and wing in different flight status for a kind of Rotor/Wing compound helicopter. The configuration of helicopter is shown in the Fig. 1.

The rotor which undertake most of lift in hovering, vertical take-off and landing is regard as attitude control surface. As the forward speed increasing, the wing produce lift gradually to unload the rotor. In this way, the pilot can reduce collective pitch and disc angle of attack. Until the helicopter meet specific velocity, reducing the rotor revolution down and keep it till reach the maximum velocity in high speed flight model.

Fig. 1. The diagrammatic drawing of rotor/wing compound helicopter

The helicopter has two lift system and its relevant control mechanism. On one hand, it can use collective pitch and periodic pitch to control it in different flight status. On the other hand, it can use aileron, elevator and the collective pitch of two propellers to control the helicopter. By this means, can it have advantages in aerodynamic characteristics in hovering, vertical take-off and landing. And it also has high lift-to-drag ratio which just like fixed-wing plane in high speed forward flight model.

Because of the specific configuration of the helicopter, its rotor and wing can produce lift in different flight status. So it need to think about the lift distribution between rotor and wing in general design stage.

In the current research situation at home and abroad, Cao Fei has studied the effect on flight performance between lift distribution relationship for a kind of single main rotor compound helicopter. According to the research, when the rotor maintains a moderate speed, only keep the lift increasing linearly with forward velocity raising can meet the need for high speed flight state.

An American scholar called Johnson, who put up a kind of Rotor/Wing compound configuration. He calculated the results of Rotor/Wing lift distribution in cruise flight status and get a conclusion that the optimum rotor revolution will lower than the revolution of autogyro.

In this paper, the researcher set the minimum required power as the optimization target in different flight status, by this means, it can confirm the Rotor/Wing lift distribution ratio in this state. Meanwhile they also have optimized the design of propeller's aerodynamic parameter to enhance its working efficiency.

2   The Model of Required Power

In the required power calculation model of whole aircraft, it need to calculate the required power of rotor and propellers in hover, low speed and forward flight model.

In the hover state, it only need to consider the balance between the gravity of helicopter and the lift which is produced by rotor. In addition, the moment caused by propellers can offset the anti-torque of rotor.

In the forward flight state, the lateral force need not be considered temporarily. The lift which is produced by rotor and wing can offset the gravity of helicopter in direction Y. The thrust which is produced by the propellers can offset the drag from fuselage, tail and wing in direction X. The analysis of moment balance is the same as that in hover state (Fig. 2).

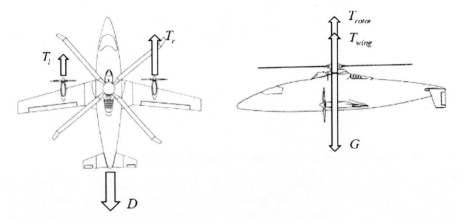

Fig. 2. The diagrammatic drawing of force analysis for the helicopter in forward flight state

2.1   The Required Power Calculation Model of Rotor

The required power calculation model of rotor is given below.
    The unit required power of the rotor is:

$$\bar{N}_p = \bar{N}_{pr} + \bar{N}_{pi} + \bar{N}_{pb}$$
$$= \frac{3}{4 \times 1000} \frac{K_p C_{x7}}{K_T \kappa C_{y7}} \Omega R + \frac{1}{2450} \frac{J}{\kappa V_0} \frac{P}{\Delta} + \left(\frac{m_{Kb}}{\sigma}\right) \sigma/G \tag{2.1}$$

The wave power:

$$M_{1\cdot 90} \approx (1 + \overline{V_0})M_1 \tag{2.2}$$

$$C_{y7\cdot 90} \approx (1 - 2.7\mu)C_{y7} \tag{2.3}$$

$$(\Delta M) = M_{1\cdot 90} - M_{1jx} \tag{2.4}$$

The solidity of the rotor:

$$\sigma = \frac{k}{\pi R^2} \int b dr \tag{2.5}$$

In the formula 1.1:

$$K_p = (1 + 3\mu^2)K_{p0} \tag{2.6}$$

$$J = (1 + 3\mu^2)J_0 \tag{2.7}$$

Therefore, the required power of rotor is:

$$P_{rotor} = \overline{N}_p * G \tag{2.8}$$

The anti-torque of rotor is:

$$M_r = \frac{P_r}{n} \tag{2.9}$$

2.2   The Required Power Calculation Model of Propeller

Compared with the conventional helicopter, the configuration of compound helicopter is more novelty, two propellers are arranged on both sides of the wing. The configuration of propellers is tractor propeller because it can face to the flow direction and have high efficiency and good heat dissipation.

This type of propellers has two functions, one is to set as auxiliary propulsion plant of the compound helicopter, the other is to balance the rotor anti-torque and realize heading control which depends on the thrust of differential motion from the both sides of propellers.

The calculation methods of the thrust of propellers is shown as:

First the torque of rotor can be calculated:

$$M_r = \frac{P_r}{\omega} \tag{2.10}$$

$$M_r = M_p \tag{2.11}$$

The thrust from both sides of the propellers can balance the anti-torque of rotor:

$$\left| T_{pr} - T_{pl} \right| = \frac{M_p}{l} \tag{2.12}$$

Then the thrust of propellers in hover state:

$$T_{pA} = \left| T_{pB} \right| \tag{2.13}$$

The thrust of propellers in forward flight state:

$$D = D_{wing} + D_{fuselage} \tag{2.14}$$

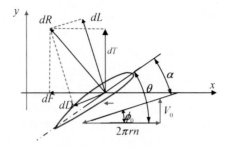

Fig. 3. The diagrammatic drawing of the aerodynamic calculation of propeller's airfoil

$$T_{pA} + T_{pB} = D \tag{2.15}$$

According to the formula, the drag of fuselage can be get from the method of CFD, and the drag of wing can be calculated which consult from the aircraft design manual.

Next, calculating the collective pitch of propellers on both sides on the basis of forward thrust of propellers and obtain the torque of propellers.

The calculation of aerodynamic force of propellers is based on the Blade Element-Momentum theory (Fig. 3).

The lift of blade element:

$$dL = \frac{1}{2}\rho W_0^2 C_L b dr \tag{2.16}$$

The drag of blade element:

$$dD = \frac{1}{2}\rho W_0^2 C_D b dr \tag{2.17}$$

The thrust of blade element:

$$dT = dR \cos(\phi + \gamma) \tag{2.18}$$

The torque of blade element:

$$dM = \frac{1}{2}\rho V^2 Q_c dr \tag{2.19}$$

The thrust after integration:

$$T = \frac{1}{2}\rho V^2 N_B \int_{r_0}^{R} T_c dr \qquad (2.20)$$

The torque after integration:

$$M = \frac{1}{2}\rho V^2 N_B \int_{r_0}^{R} Q_c dr \qquad (2.21)$$

Now, the required power of propellers can be calculated:

The torque of propellers on both sides can be calculated on basis of thrust of propellers and then get the corresponding power of propellers.

$$P_{pr} = M_{pr} \cdot \omega_r \qquad (2.22)$$

$$P_{pr} = M_{pr} \cdot \omega_l \qquad (2.23)$$

$$P_p = P_{pr} + P_{pl} \qquad (2.24)$$

2.3   The Required Power Calculation Model of Whole Aircraft

In conclusion, the required of the whole aircraft is:

$$P_{com} = \frac{P_{rotor} + P_{propeller}}{1 - K_t} \qquad (2.25)$$

3   The Optimum Strategy of Rotor/Wing Lift Distribution

3.1   The Strategy of Reducing Rotor Revolution

It will suffer from three parts of restrictions if the maximum velocity is wanted to improve. First there is the shock wave restrictions on the advancing blade, partial shock wave will reduce the aerodynamic efficiency on the rotor to a great extent. Then, retreating blade will improve the angle of attack to balance the lift which produced by advancing blade. In this way, the airflow on the blade will separate from the surface of the blade, and it's also a way to lower the aerodynamic efficiency. At last, the helicopter of conventional configuration always improve the forward thrust of rotor by increasing the disc angle of attack. But due to the restriction from the whole configuration, the forward thrust of rotor is limited by the angle.

In this paper, the compound configuration can reduce the rotor revolution in forward flight state because the wing will undertake some part of lift to unload the loading from the main rotor. And the propellers can also provide enough thrust to maintain the forward speed. So the problems which mentioned before have been basically solved.

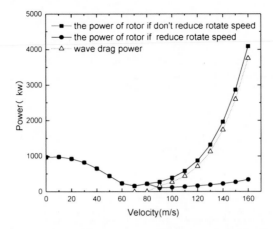

Fig. 4. The variation tendency of power with the speed increasing for reduce rotor revolution before and after

Looking at the Fig. 4, the rotor produces major wave drag power and it's required power rising sharply with the forward speed increasing. Because, at that time, rotor is in the high advance ratio state. While it can reduce the required power of the helicopter effectively if the rotor revolution is reduced at appropriate point.

When the forward flight speed lower than 80 m/s, the wave drag power is relatively small. It's insignificant for rotor to lower the rotor revolution at that time. Otherwise, reducing rotor revolution in small velocity scale will make the rotor undertake larger lift. Resulting in the collective pitch of rotor will get large and come up with airflow separation on the retreating blades.

When the forward flight speed surpasses to 80 m/s, the wave drag power rise sharply, and it will make more negative effect on the performance of high flight speed state for Rotor/Wing compound helicopter.

Therefore, the rotor revolution will be decided to reduce at the point of 80 m/s. In this way, can it not only reduce wave drag effectively but also avoid the airflow separation on the retreating blade.

3.2   The Optimization Strategy of Lift Distribution

In this section, the components of lift distribution are selected as optimization variable and set the lowest required power as the optimization target. And it can get the final objective laws of lift distribution between the rotor and wing in the lowest required power consuming state (Fig. 5).

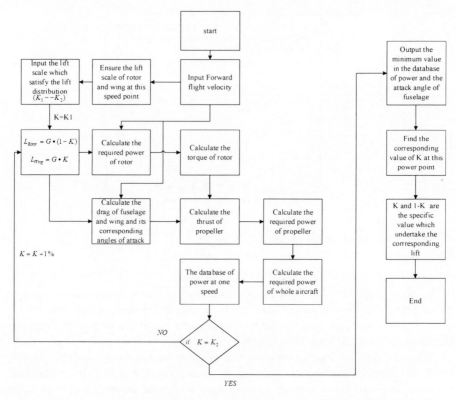

Fig. 5. The optimization flow path of lift distribution for the Rotor/Wing compound helicopter

3.3   Analysis of Examples

It's time to continue calculating the required power of the example helicopter by the means of the method and get the lift distribution between the rotor and wing in different flight status. Some parts of general design parameters are shown in Table 1.

Table 1. Some general parameters of the example helicopter

Item/unit	Value
Take-off weight/kg	5000
Wing span/m	8.8
Wing area/m^2	8.82
Rotor diameter/m	10.4
Blade number	4
Rotor solidity	0.082
Rotor revolution/rpm	367.5/257.2[*]
Propeller diameter/m	2.0
Propeller revolution/rpm	2200

*Two rotor revolution represent the rotor revolution which reduce before and after respectively.

The optimized results are given below.

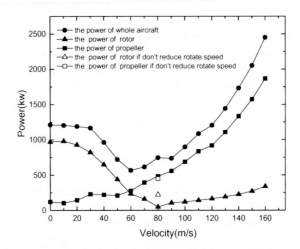

Fig. 6. Compare the power of rotor/propeller in different forward flight speed

As is shown in the Fig. 6, the highest line on the figure is the curve of the whole aircraft power, and the square and triangle of hollow dots represents the power value of the rotor and propellers before reducing revolution. Compared with the power after reduce revolution, the power of rotor decrease significantly. On the contrary, the power of propellers changes a little.

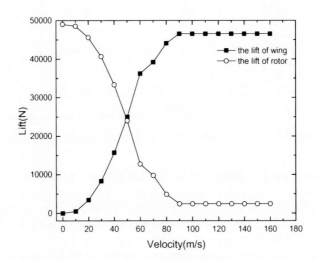

Fig. 7. The lift distribution tendency for rotor and wing in different forward flight speed

As is shown in Fig. 7, the lift which is undertake by the wing increases constantly as the forward speed gets more. While the lift from the rotor is decreasing constantly. Meanwhile, the lift of rotor will be restricted and require it not lower than 5%. So, the results of lift distribution of rotor are 5%.

The reason of restricting lift distribution ratio is that, the rotor need to keep power inputting and keep it in not autorotation state but in steerable state. In this way, can it reduce much drag from the rotor and keep smaller of attack angle of rotor.

At last the attack angle of wing can be calculated which relys on the lift from the wing.

$$L_w = q_w S_w C_{lw} \qquad (3.1)$$

$$C_{lw} = C_{l\alpha w} \bullet \alpha \qquad (3.2)$$

The angle of attack will be output as the restriction condition to satisfy the corresponding lift distribution of the rotor and wing in different flight states. The target angle of attack in different flight velocity is shown in Fig. 8.

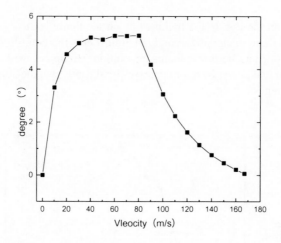

Fig. 8. The target angle of wing in different forward flight speed

4   Aerodynamic Parameter Optimization of Propeller

The function of propellers in Rotor/Wing compound helicopter are improving forward flight thrust, keeping the balance of anti-torque of the main rotor and controlling the heading movement by changing its collective pitch.

The characteristic of this type of propellers is not only having larger range of thrust change but also needing to satisfy the needs of different working condition. So, it's necessary for us to optimum the aerodynamic parameters of the propellers.

4.1   The Optimization Model

(1) The optimization model

The aerodynamic design optimization variable including revolution, thread pitch, solidity and the taper ratio.

(2) The optimization target

The highest working efficiency of the propellers is set as the optimization target.

At first, the highest working efficiency of the propellers is set as the optimization target in hover state, cruise state and maximum forward flight speed state as the objective function, and distribute weight coefficient to every flight state.

First of all, the efficiency of the cruise state determines the flight performance of the whole aircraft, and the efficiency of the cruise state should take a large proportion in the three states. Then, an important function that the compound helicopter needs to meet is to realize the hovering, while the hovering state consumes a large amount of power, so improving the efficiency of the hovering state can also reduce the required power of the whole aircraft. Therefore, the weight value of the hover state's efficiency is greater than the weight value of the maximum flying speed state (Table 2).

$$\eta = k_h \eta_h + k_c \eta_c + k_m \eta_m \tag{4.1}$$

Table 2.   The weight coefficient of components

Item	Weight coefficient
k_h	0.3
k_c	0.5
k_m	0.2

(3) The constraint condition

① The propellers on both sides of the fuselage need to satisfy the needs of forward thrust.

② The resultant moment which produced by the propellers needs to satisfy the needs of the rotor anti-torque.

(4) The optimization methods

The aerodynamic parameter optimization of the propellers is based on the I-sight parameter optimization platform and the calculation model of required power for propellers. For the variables such as revolution, thread pitch, solidity, and taper ratio will take the optimization design. The specific block diagram is shown in Fig. 9.

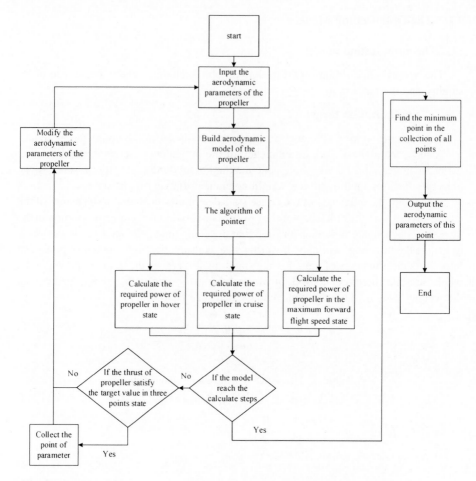

Fig. 9. The block diagram of the aerodynamic parameter optimization design for propellers

Table 3. The initial value and data scale of the parameter for left propeller

Variable	Lower limit	Upper limit	Initial value
Solidity	0.1	0.25	0.19
Thread pitch	0	100	20
Taper of blade	0.5	1	1
Revolution	1800	2600	2200

4.2   The Optimization Result

By numerical test, it begins to ensure the initial value and data range of the optimization problem of propeller's aerodynamic parameters. And the initial value and data range from the both sides parameters of propellers are shown in Tables 3 and 4.

Table 4. The initial value and data scale of the parameter for right propeller

Variable	Lower limit	Upper limit	Initial value
Solidity	0.1	0.25	0.19
Thread pitch	0	100	60
Taper of blade	0.5	1	1
Revolution	1800	2600	2200

Table 5. Comparison of the aerodynamic parameter for propellers before and after optimization

	Right propeller		Left propeller	
	Before optimization	After optimization	Before optimization	After optimization
Solidity	0.191	0.182	0.191	0.176
Thread pitch	60	51	20	91
Taper of blade	1	0.999	1	0.994
Revolution	2200	2130	2200	2580

According to the data in the Table 5, it can be saw the solidity, revolution for the two propellers are different which depend on their different required thrust. Compared the solidity and revolution before and after optimization, it can be see the taper ratio change a little but thread pitch changes a lot. Specifically, the thread pitch of right propeller reduces slightly and the thread pitch of left propeller grow greatly.

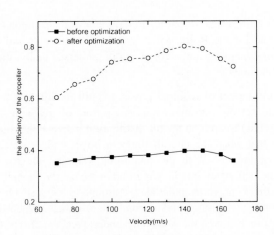

Fig. 10. The working efficiency of the left propeller in different forward flight speed

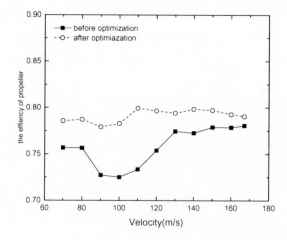

Fig. 11. The working efficiency of the right propeller in different forward flight speed

The efficiency of the both sides of propellers before and after optimization in different flight state are shown in Figs. 10 and 11. For the right propeller, its working efficiency stay in the high before the optimization, and it get further enhance to nearly 80%. On the contrary, for the left propeller, its efficiency can only reach nearly 40%. After the optimization, it can reach nearly 60%–80%. So it can prove that, the optimal method is efficient.

5   Conclusion

The main content of this article is the research on the lift distribution of rotor and wing in steady flight state for the Rotor/Wing compound helicopter and the optimization problem of the aerodynamic parameters of the propellers. Finally, the researcher gets the following conclusions.

(1) The calculation model of required power is built in different flight state for the helicopter and set the minimum required power of the helicopter as the optimization target to research on the lift distribution between the rotor and wing. The results show that when the aircraft flight in hover or low speed state, the rotor offers most of the lift. With the increasing velocity, the rotor unloads some loadings of the helicopter and the wing undertake nearly the half of lift, after that, the ratio that the wing which undertaken continue to increase. On the contrary, the ratio that the rotor which undertaken continue to reduce. When the forward flight speed exceeds to 100 m/s, the rotor only undertakes 5% of the whole lift and the wing undertake nearly 95% of the lift.

(2) For the reason of that the characteristic of this type of propellers is not only having larger range of thrust change but also needing to satisfy the needs of different working condition, some aerodynamic design optimization variable is set such as revolution, thread pitch, solidity and the taper ratio. The aerodynamic parameter optimization of the propellers is based on the I-sight parameters optimization platform and set the highest working efficiency of the propellers as the optimization target in hover state, cruise state and maximum forward flight speed state as the objective function, and distribute weight coefficient to every flight state. According to the verification result, it find that the working efficiency of the propellers have improved obviously which compare the optimization before and after.

References

1. Ni X (2008) Future development prospect of helicopter technology. Aeronaut Manufact Technol 03:32–37
2. Li J (2016) Progress of compound helicopter technology. J Nanjing Univ Aeronaut Astronaut 48(02):149–158
3. Cao F, Chen M (2016) Performance characteristics of single rotor compound helicopter. J Beijing Univ Aeronaut Astronaut 42(04):772–779
4. Yeo H, Johnson W (2009) Optimum design of a compound helicopter. J Aircr 46(4):19
5. Zhang Y (2003) Research of computing method for helicopter required power. Helicopter Tech 01:1–5
6. Mcdonald RA (2013) Optimal propeller pitch scheduling and propeller–airframe matching for conceptual design. In: AIAA aviation technology, integration, and operations conference
7. Yang X (2006) Aerodynamic performance analysis of aero-model propeller. Shandong University, Jinan
8. Cheng Y, Li G, Nie W (2012) Comparison of unsteady aerodynamic performance of propeller airfoil at low Reynolds number. Helicopter Tech 01:16–19

Research on Optimal Design Method of Tilt-Rotor Electric Propulsion System

Dengyan Duan, Hong Zhao, Minghua Peng, and Jianbo Li[✉]

National Laboratory of Science and Technology on Rotorcraft Aeromechanics, Nanjing University of Aeronautics and Astronautics, Nanjing 210016, China
ljbl01@nuaa.edu.cn

Abstract. Improve the efficiency of propulsion system is an effective mean to archive flight performance improvement of electric-powered tilt-rotor. In this paper, the motor is constructed by combining the motor equivalent circuit model and the positive polynomial loss model. The proprotor is modeled by Goldstein vortex theory and the validity is verified. Based on this, aimed at an effective optimization goal, a comprehensive optimization method for the propulsion system is proposed using optimization algorithm. Then the optimization of the propulsion system was carried out and an effective optimization program was obtained for a small electric-powered tilt quad rotor. The research results show that this optimization method can effectively solve the problem of motor and proprotor matching. By optimizing the design of the propulsion system, the power demand can be effectively reduced and the flight performance can be improved.

Keywords: Electric propulsion system · Tilt-rotor · Optimization · Motor · Proprotor

1 Introduction

Electric-powered multi-rotor UAVs has a good prospect and it can be used to aerial photography, pesticides spraying, power line patrolling and so on. But due to the low battery energy density, endurance of such UAVs is short, which greatly limits the expansion of their application fields [1, 2]. Electric-powered tilt-rotor has a better flight performance by combining the vertical lift capability of multi-rotor UAVs with the speed and range of a conventional fixed-wing aircraft, but it also faces the same problem. While expecting further breakthroughs in battery technology, it is one of the current technological breakthroughs to continuously improve the efficiency of electric propulsion system. Electric propulsion system usually has a motor and a propeller, the problem is how to comprehensively consider the characteristics of both the motor and propeller. But research results in this area are still few, and most of the electric propulsion system has not been systematically optimized. Especially in the design of Electric-powered tilt-rotor propulsion, it is more difficult because it has two different flight modes of hovering and flying. To achieve the optimal efficiency of the two working states is so sophisticated that the research is carried out less.

© Springer Nature Singapore Pte Ltd. 2019
X. Zhang (Ed.): APISAT 2018, LNEE 459, pp. 1108–1119, 2019.
https://doi.org/10.1007/978-981-13-3305-7_88

For the matching between motor and propellers, some research has been carried out. Chen constructed the propulsion system based on the matching relationship between the motor and the propeller, obtained the optimal reduction ratio at different heights, and determined the optimal working area of the propulsion system [3]. But his research is designed for a fixed-size propeller that does not optimize the aerodynamic shape. Besides, the research has a reduction gear, which adds weight and complexity to the system. Muzar and Lanteigne get the motor and propeller efficiency under the corresponding speed and torque by establishing the equivalent circuit model for the DC motor and using the empirical coefficient method to carry out the propeller modelling [4]. The equivalent circuit model is relatively intuitive, but it does not account for mechanical losses, copper losses and iron losses of the motor. McDonald has carried out many works. His research indicates that the motor equivalent circuit model cannot reflect the actual efficiency well, so he proposes positive polynomial loss model based on Larminie and Lowry's calculation method of motor losses. Then he models the propeller same as Dominic Muzar and Eric Lanteigne's method and carries out lots of matching design works between motors and propellers for a certain aircraft in the flight envelope [5–7]. In addition, electric-powered tilt-rotor has two fight modes. While hovering, the inflow ratio is small but the disk loading is large. While flying, the inflow ratio is large but the disk loading is small. So the empirical coefficient method cannot reflect the aerodynamic performance of the proprotor well [4, 8].

This paper establishes a brushless DC motor model by combining the motor equivalent circuit model and the positive polynomial loss model. Considering the two working states of the electric-powered tilt-rotor, the proprotor model is constructed according to the Goldstein vortex theory. On this basis, a comprehensive optimization method for the propulsion system is proposed. Then considering the hovering and cruising states of a small electric-powered quad tilt-rotor, regard power consumption of a typical mission profile which both contains hover and forward flight least as the target, we optimize the Variable-speed and adjustable pitch propulsion system using genetic algorithms based on the characteristics of the motor.

2 DC Motor Modeling

The motor losses are neglected in the equivalent circuit model, although it is quite intuitive. The positive polynomial loss model can calculate the actual motor efficiency, however, some parameters of the model need to be gained by experience or experiments. Consequently, to derive the parameters aforementioned, the calculating data of the equivalent circuit model at maximum efficiency and efficiency of 0.8 is combined with the positive polynomial loss model, and then the motor is modelled.

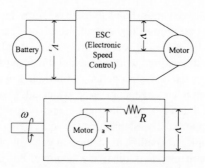

Fig. 1. Equivalent circuit model of the brushless DC motor

Figure 1 represents equivalent circuit model of the brushless DC motor, where V_s (volt) is the power supply voltage as seen by the ESC, v (volt) is equivalent driving voltage as seen by the motor and V_m (volt) is back EMF of the motor corresponding to the motor's shaft speed. The ESC adjusts v by using pulse width modulation technology (PWM), and the motor reaches maximum rev when β equals 100%. The relationship between the battery voltage and the input voltage of the motor is shown in Eq. (2.1).

$$v = \beta V_s \tag{2.1}$$

Motor's speed constant K_v (rpm/volt), winding resistance $R(\Omega)$, and motor's line-peak current at no-load i_0(A) are three characteristic parameters of the motor. Besides, the motor's shaft speed ω (rad/s) and the motor output torque Q (N * m) can be worked out using Eqs. (2.2) and (2.3).

$$\omega = \frac{2\pi(v - iR)K_v}{60} \tag{2.2}$$

$$Q = \frac{60(i - i_0)}{2\pi K_v} \tag{2.3}$$

Next, the motor output power P_{shaft} (W), input power P_{ele} (W) and efficiency η can be derived from above parameters, which are as follows.

$$P_{shaft} = Q\omega = (v - iR)(i - i_0) \tag{2.4}$$

$$P_{ele} = vi \tag{2.5}$$

$$\eta = \frac{P_{shaft}}{P_{ele}} = \frac{(v - iR)(i - i_0)}{vi} \tag{2.6}$$

The motor losses P_L in Eq. (2.7) mainly includes friction loss, windage loss, copper loss and iron loss. Among them, the friction loss and iron loss are proportional to ω, wind resistance loss to ω^3, and copper loss to Q^2.

$$P_L = C_0 + C_1\omega + C_2\omega^3 + C_3Q^2 \tag{2.7}$$

Given the motor loss P_L, the motor efficiency η can be gained according to Eq. (2.8).

$$\eta = \frac{\omega Q}{\omega Q + P_L} \tag{2.8}$$

Let Eq. (2.8) take partial derivations with respect to Q, ω, we get Eqs. (2.9) and (2.10).

$$\frac{\partial \eta}{\partial Q} = \frac{\omega}{\omega Q + P_L} - \frac{\omega Q}{(\omega Q + P_L)^2}\left(\omega + \frac{\partial P_L}{\partial Q}\right) \tag{2.9}$$

$$\frac{\partial \eta}{\partial \omega} = \frac{Q}{\omega Q + P_L} - \frac{\omega Q}{(\omega Q + P_L)^2}\left(Q + \frac{\partial P_L}{\partial \omega}\right) \tag{2.10}$$

To get maximum efficiency, we let the partial derivations equal 0, and the results are expressed in Eqs. (2.11) and (2.12).

$$P_L - Q\frac{\partial P_L}{\partial Q} = 0 \tag{2.11}$$

$$Q\frac{\partial P_L}{\partial Q} - \omega\frac{\partial P_L}{\partial \omega} = 0 \tag{2.12}$$

Then Eq. (2.7) is substituted into Eqs. (2.11) and (2.12), meanwhile, $\bar{\omega}$ and $\bar{Q}$ are the motor's shaft speed and torque at maximum efficiency $\bar{\eta}$. As a result, the relationship of C_1, C_2 and C_3 to C_0 can be seen in Eqs. (2.13) and (2.14).

$$\begin{bmatrix} \bar{\omega} & \bar{\omega}^3 & \bar{Q}^2 \\ 0 & -2\bar{\omega}^3 & \bar{Q}^2 \\ \bar{\omega} & -\bar{\omega}^3 & 0 \end{bmatrix} \begin{Bmatrix} C_1 \\ C_2 \\ C_3 \end{Bmatrix} \begin{Bmatrix} -C_0 + \bar{\omega}\bar{Q}^1\frac{-\bar{\eta}}{\bar{\eta}} \\ -C_0 \\ -2C_0 \end{Bmatrix} \tag{2.13}$$

$$P_{L,0.8} = C_0 + C_1\omega_{0.8} + C_2\omega_{0.8}^3 + C_3Q_{0.8}^2 \tag{2.14}$$

When substituting C_1, C_2 and C_3 with C_0 and applying motor's shaft speed, torque and power at the efficiency of 0.8, namely $\omega_{0.8}$, $Q_{0.8}$ and $P_{L,0.8}$, C_0 can be calculated by Eq. (2.14) and the same to C_1, C_2 and C_3.

Table 1. Characteristic parameters of XM-14.5

Motor	K_v(rpm/volt)	$R(\Omega)$	i_0(A)
XM-14.5	180	0.039	0.7

The characteristic parameters of XM-14.5 are listed in Table 1, and the motor's efficiency map is shown in Fig. 2.

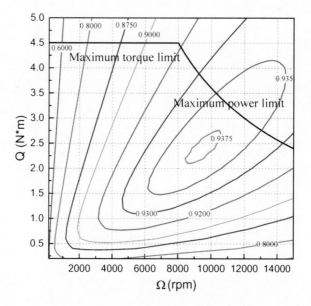

Fig. 2. Efficiency map of motor XM-14.5

3   Proprotor Modeling

While hovering, the inflow ratio is small but the disk loading is large. While flying, the inflow ratio is large but the disk loading is small. The study found that the Goldstein vortex theory is applicable to both light-loaded propellers and heavy-loaded propellers [4].

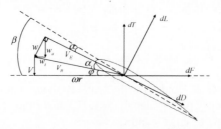

Fig. 3. The velocity and force acting on the blade element

The proprotor is modeled according to the Goldstein vortex theory. Figure 3 shows the velocity and force acting on the blade element. R(m) is the radius and r is the distance from the center of the hub to any point in the blade profile, where x is the dimensionless value of r. σ is the blade solidity, ω (rad/s) is the proprotor speed,

V (m/s) is the inflow velocity, and λ is the inflow ratio. φ is the flow angle of the blade element, φ_T is the flow angle of the blade element at the tip of the blade, and α_i is the interference angle. V_E is the resultant speed, ω_a and ω_t are the axial and circumferential components of the induced velocity.

$$w_t = \frac{B\Gamma}{4\pi r\kappa} \tag{3.1}$$

$$\frac{w_a}{\omega R} = \frac{1}{2}\left(-\lambda + \sqrt{\lambda^2 + 4\frac{w_t}{\omega R}\left(x - \frac{w_t}{\omega R}\right)}\right) \tag{3.2}$$

Γ is the circulation of blade element, and κ is the Goldstein coefficient, which is replaced by the Prandtl tip loss coefficient F, as shown in the following equation.

$$F = \frac{2}{\pi}\cos^{-1}\exp\left(-\frac{B(1-x)}{2\sin\varphi_T}\right) \tag{3.3}$$

The thrust coefficient C_T and the power coefficient C_P can be obtained by the Eqs. (3.4) and (3.5), where C_L is the lift coefficient and C_D is the drag coefficient.

$$C_T = \frac{\pi^3}{8}\int_{x_h}^{x_T}\left(\frac{V_E}{\omega R}\right)^2\sigma$$
$$(C_L\cos(\varphi + \alpha_i) - C_D\sin(\varphi + \alpha_i))dx \tag{3.4}$$

$$C_P = \frac{\pi^4}{8}\int_{x_h}^{x_T}\left(\frac{V_E}{\omega R}\right)^2\sigma x$$
$$(C_L\sin(\varphi + \alpha_i) + C_D\cos(\varphi + \alpha_i))dx \tag{3.5}$$

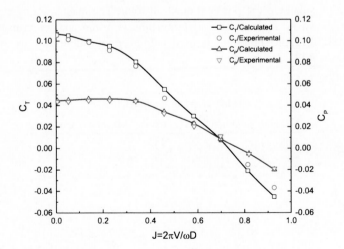

Fig. 4. Comparison between calculation and experimental results

The APC10*7 propeller was calculated by the Goldstein vortex theory, as shown in Fig. 4 comparison between calculation and experimental results. It shows that the calculated values are in good agreement with the experimental values no matter large or small inflow ratios it is, so the Goldstein vortex theory can be used for the optimization design of the propulsion system.

4 Optimization of Propulsion System

The electric-powered tilt-rotor has two working states: hovering and cruising. It is necessary to ensure high efficiency of the propulsion system in both working states, so the double objective optimization is carried out with the goal of minimizing the power consumption of the motor during both hovering and cruising.

The objective function is defined as Eq. (4.1), where $En(W \cdot h)$ is the total power consumption, P_{hover} is the hovering power, P_{cruise} is the cruising power, η_{hover} is the motor efficiency of hovering, η_{cruise} is the motor efficiency of cruising, t_{hover} (min) is hovering time, t_{cruise} (min) is the cruising time.

$$En = \frac{t_{hover}P_{hover}}{60\eta_{hover}} + \frac{t_{cruise}P_{cruise}}{60\eta_{cruise}} \qquad (4.1)$$

Proprotor diameter, pitch, solidity, and taper ratio are optimized with the goal that the objective function should be minimum. In order to reduce the structure weight and complexity, no reduction gear is added. In order to avoid stall, the angle of attack at the proprotor characteristic profile (0.7R) of hovering, cruising, and maximum speed should be less than the stall angle. Besides, the power of the proprotor and current during hovering and cruising should be less than the maximum sustainable power and current of the motor.

The genetic algorithm with strong robustness and global optimization is used for optimization design. The following is a proprotor optimization based on motor characteristics for a small electric-powered quad tilt-rotor. The main working states of the aircraft are shown as Table 2. The hovering time is 5 min and the cruising time is 30 min. The 20*12APC proprotor is used before optimization, and the total power consumption is 899.2 $W \cdot h$.

Table 2. Main working state

Working condition	Hover	Cruise	Maximum speed
Velocity/m/s	0	35	50
Force/N	50	5	8

The range of each optimization variable is as shown in Eq. (4.2).

$$\left.\begin{array}{l} 18 \leq D \leq 35 \\ 8 \leq Pitch \leq 20 \\ 0.03 \leq \sigma \leq 0.07 \\ 0.2 \leq \gamma \leq 1 \end{array}\right\} \tag{4.2}$$

4.1 Variable-Speed Propulsion System Optimization

The optimization result of the Variable-speed propulsion system is $En = 744.8\ W \cdot h$, which is 17.2% lower than that before optimization, as shown in Table 3.

Table 3. Variable-speed propulsion system optimization results

Designed value	Diameter/inch	26.8
	Pitch/inch	15.3
	Solidity	0.04
	Taper ratio	0.2
Calculated value	Rotational speed of hover/rpm	4101.8
	Rotational speed of cruise/rpm	4202.1
	Hovering power/W	526.1
	Cruising power/W	252.2
	Motor efficiency of hover	0.93
	Motor efficiency of cruise	0.91
	Total power consumption/ $W \cdot h$	744.8

After the optimal scheme is determined, we use the interval factor method to analyze the sensitivity of the parameters. In order to more intuitively see the factors affecting the total power consumption of the propulsion system, the influence of each design variable on the total power consumption, proprotor cruising power, hovering power, motor cruising efficiency and hovering efficiency is analyzed.

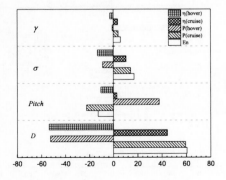

Fig. 5. Variable-speed propulsion system sensitivity analysis

1116   D. Duan et al.

Figure 5 shows the sensitivity analysis results of the Variable-speed propulsion system. As can be seen from the figure, the diameter and pitch are important factors affecting the proprotor hovering and cruising power. Increasing the diameter or reducing the pitch will reduce the hovering power and increase the cruise power. Motor efficiency is primarily related to the diameter of the proprotor.

4.2   Adjustable Pitch Propulsion System Optimization

The optimization result of the adjustable pitch propulsion system is $En = 551.9\ W \cdot h$, which is 38.6% lower than that before optimization, as shown in Table 4.

Table 4. Adjustable pitch propulsion system optimization results

Designed value	Diameter/inch	30.7
	Pitch/inch	12.0
	Solidity	0.03
	Taper ratio	0.2
	Rotational speed of hover/rpm	3723.2
	Rotational speed of cruise/rpm	1572.2
Calculated value	Blade pitch of hover/°	0
	Blade pitch of cruise/°	19.4
	Hovering power/W	413.9
	Cruising power/W	192.9
	Motor efficiency of hover	0.92
	Motor efficiency of cruise	0.88
	Total power consumption/ $W \cdot h$	587.4

Figure 6 shows the sensitivity analysis results of the adjustable pitch propulsion system. As can be seen from the figure, the diameter and rotation speed are important factors affecting the proprotor hovering and cruising power. Increasing the hovering or

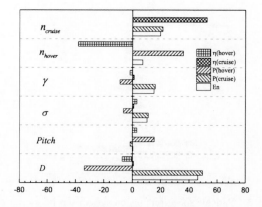

Fig. 6. Adjustable pitch propulsion system sensitivity analysis

the cruising rotation speed will increase the hovering and the cruising power. Increasing diameter will reduce the hovering power and increase the cruising power. Motor efficiency is primarily related to the rotation speed.

In addition, as can be seen from Figs. 5 and 6, whether it is a Variable-speed or adjustable pitch propulsion system, the influence of variable changes on hovering and cruising is mostly reversed. This contradiction increases the difficulty in designing the propulsion system of the electric-powered tilt-rotor.

4.3   Comparative Analysis

Figure 7 shows the lift drag ratio distributions of Variable-speed and adjustable pitch proprotor. It can be seen from the figure that the lift drag ratio of the adjustable pitch proprotor is better than Variable-speed proprotor whether hovering or cruising. This is because the adjustable pitch proprotor achieves a favorable angle of attack by increasing the pitch, while the variable speed proprotor can only achieve a favorable angle of attack by increasing the speed to reduce the inflow angle. But the increase of rotational speed causes an increase in proprotor profile power, which indirectly lead an increase in total power consumption.

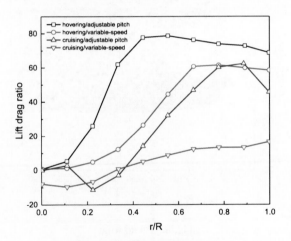

Fig. 7. Lift drag ratio distribution

Aim at the minimum of hovering power and cruising power, draw the pareto front of Variable-speed propulsion system as shown in Fig. 8(a) and adjustable pitch propulsion system as shown in Fig. 8(b). As we can see, the adjustable pitch propulsion system is easier to balance the two working states of hovering and flying.

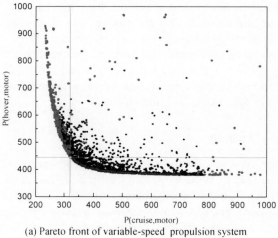

(a) Pareto front of variable-speed propulsion system

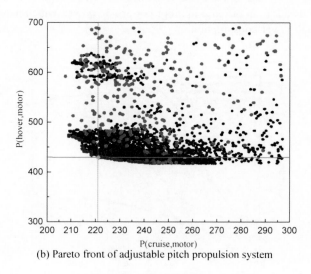

(b) Pareto front of adjustable pitch propulsion system

Fig. 8. Pareto front

5 Conclusions

(1) The electric-powered tilt-rotor has two flight modes: hovering and cruising. But the influence of variable changes on hovering and cruising performance is almost reversed, which makes the design of the electric-powered tilt-rotor propulsion system more difficult. Simple motor and proprotor matching methods severely limit flight performance improvements.

(2) The electric propulsion system optimization method aiming at the minimum of total power consumption and considering hovering and forward flight is effective. The optimization can make the propulsion system in high efficiency working state, eliminate the reduction gear and reduce weight and complexity.

(3) The adjustable pitch propulsion system is more suitable for the electric-powered tilt-rotor, which can archive high efficiency both in hovering and forward flight. And it can effectively improve the efficiency of the propulsion system, reduce the power demand and improve the flight performance.

References

1. Liang B, Wang H, Yuan C (2010) Design and optimization of power system for electric small unmanned aerial vehicles. Aeronaut Comput Tech 06:78–80 (in Chinese)
2. Chen J, Yang S, Mo L (2009) Modeling and experimental analysis of UAV electric propulsion system. J Aerosp Power 06:1339–1344 (in Chinese)
3. Chen S, Song B, Wang H (2013) Simulating parameter matching for propulsion system of high altitude airship. J Northwest Polytechnical Univ 04:530–534 (in Chinese)
4. Muzar D, Lanteigne E (2016) Experimental characterization of brushless DC motors and propellers for flight application
5. Mcdonald RA (2013) Electric propulsion modeling for conceptual aircraft design
6. Mcdonald RA (2015) Modeling of electric motor driven variable pitch propellers for conceptual aircraft design
7. Mcdonald RA (2013) Optimal propeller pitch scheduling and propeller–airframe matching for conceptual design
8. Tjhai C (2013) Developing stochastic model of thrust and flight dynamics for small UAVs. Dissertations & Theses-Gradworks

Aerodynamic/Stealthy Integrated Design Optimization of Airfoil for Supersonic Fighter

ZhongYuan Liu[✉], BinQian Zhang, WenTing Gu, MingHui Zhang, and ZhenLi Chen

School of Aeronautics, Northwestern Polytechnical University,
Xi'an 710072, China
liuzy@mail.nwpu.edu.cn

Abstract. Airfoil is very important in the aircraft performance. To figure out the conflicts between the requirements of the aerodynamic and stealthy design of supersonic fighter airfoil, we established a multi-objective optimization platform, which is based on high fidelity method, Parsec parameterization method, artificial neural network and Pareto genetic algorithm, to reduce the supersonic drag and radar cross section (RCS) of airfoils. Three optimized airfoils are obtained based on the optimization design. Compared with the initial airfoil, the geometric shapes of optimized airfoils have smaller leading-edge radius, later maximal thickness location and diamond-shaped thickness distribution. Reductions of the supersonic drag coefficients, the values of pitch moment and RCSs in the key azimuth of optimized airfoils are included in the design condition. Meanwhile, the lift coefficients are improved. Results also indicate that the multi-objective optimization framework could be implemented.

Keywords: Supersonic fighter · Airfoil design · Stealthy design ·
Multi-objective optimization method

1 Instruction

During the development of fighters, researchers always focus on the airplanes' stealthy and supersonic aerodynamic performance. However, these two properties may be contradicted with each other in the design procedure. To ensure favourable stealthy and aerodynamic performance, small-aspect swept configuration is adapted in the design of the most advanced fighters. Though the main flow of the configuration is three-dimensional, the two-dimensional flow investigation in the analysis of airfoils and the results of the airfoil analysis are still valuable to improve fighters' performance. On the one hand, the supersonic drag characteristics of the airfoil have a direct impact on the fighters' supersonic cruise capacity. On the other hand, the shape of the airfoils' leading edge has great influence on the stealthy and aerodynamic integrated design. The aerodynamic design aims to weak the wave shock to reduce the wave drag, while the stealthy design aims to minimize the radar cross section (RCS) of the airfoils. Considering that the airfoils' shapes based on these two aspects are conflicted, it is necessary to develop an aerodynamic/stealthy integrated design of the airfoil for supersonic fighters.

© Springer Nature Singapore Pte Ltd. 2019
X. Zhang (Ed.): APISAT 2018, LNEE 459, pp. 1120–1130, 2019.
https://doi.org/10.1007/978-981-13-3305-7_89

There are a lot of investigations on the stealthy and aerodynamic design of airfoils. Vinh and van Dam [1] adopted the finite difference time domain Maxwell equation combined with the potential equation to carry out the transonic aerodynamic and stealthy optimization design of the airfoils. Zhu and Islam et al. [2] employed the vector flux decomposition to calculate the transonic aerodynamic data and RCS of an airfoil and utilized a numerical optimization method to get a high-aerodynamic-efficiency and low-observability airfoil. Yu and Zhu [3] created a new fitness function and adopted the determinacy and genetic algorithm to proceed bi-disciplinary/ bi-objective optimization of airfoil's transonic aerodynamic and stealthy performance. Besides, Pittman [4] utilized the numerical optimizer combined with nonlinear full potential solver to optimize a supersonic airfoil. Kroo, Willcox and March et al. [5] adopted the linear method for plate supersonic flow, shock expansion wave method, Euler solver combined with Gradient-Free algorithm to design a supersonic airfoil. Lattarulo, Seshadri and Parks [6] employed the multi-objective alliance algorithm to optimize the aerodynamic performance of a supersonic airfoil. None of information mentioned above focused on both the supersonic aerodynamic design and stealthy design.

In this paper, a multi-objective optimization platform is constructed to optimize supersonic aerodynamic and the stealthy performance of the chosen airfoil aiming at one certain fighter.

2   Analysis and Calculation Method

In this section, the methods to analyse the aerodynamic and stealthy performance are introduced.

2.1   Drag Analysis

For supersonic airfoil, the aerodynamic performance is related to the drag of the airfoil. Friction drag and wave drag are the main compositions of the total drag of a supersonic airfoil. Therefore, the main work of improving aerodynamic performance is to reduce the friction drag and wave drag. According to the reference [7], to reduce the friction drag, we should increase the region of laminar flow and decrease the wetted area. However, it is difficult to increase the region of laminar flow in the design of the supersonic natural-laminar-flow airfoil [8]; and decreasing the wetted area is available in the three-dimensional design. As to the wave drag, there are two popular methods to reduce the wave drag: the supersonic area design method for three-dimensional design [9], and the direct design method, which is used in this paper to optimize the airfoil based on the geometric parameters influence study. Meanwhile, drag coefficient C_d is chosen as the aerodynamic objective of the airfoil.

2.2   Aerodynamic Analysis Method

Aerodynamic numerical method is based on the upwind finite volume method to solve the two-dimensional steady NS equation. Second order accuracy is obtained by MUSCL interpolation combined with a stable limitator. NS equations with the assumption of thin layer are solved via the Runge-Kutta time discretization and the FDS-Roe scheme. One-equation Spalart-Allmaras turbulence model is utilized because of the high calculation stability and accuracy. The method has been proved to be capable of dealing with the supersonic flow of airfoil in reference [10].

The airfoil grid has a C-type topology, which has 201 nodes on the tangential direction of the surface and 169 nodes in the normal direction of the surface, as shown in Fig. 1.

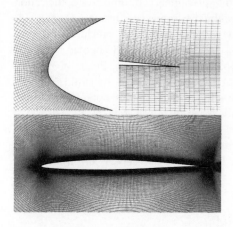

Fig. 1. Airfoil grid   **Fig. 2.** Different method results comparison

2.3 Stealthy Analysis Method

The radar cross section (RCS) is a key factor in designing stealthy airfoil. According to the reference [19], the airfoil is transformed into a three-dimensional wing, which is 1 m in width and in chord direction respectively when computing the RCS of airfoil.

There are mainly two categories for the RCS calculation: the approximation algorithm and the accurate algorithm. The methods of approximation algorithm mainly include the Geometric Optics (GO), the Physical Optics (PO), the Uniform Geometric Theory of Diffraction (UTD), the Geometric Theory of Diffraction (GTD) and the equivalent current method. Some methods of accurate algorithm are as follows: the Method of Moment (MOM), the Multilevel Fast Multipole Method (MLFMM), the Finite Difference Time-Domain Method (FETD), and the Finite Element Time-Domain Method (FDTD) [11]. PO or its hybrid method is the most popular method used in the aircraft stealthy design due to its efficiency. MOM and PO methods are employed to calculate the RCS of NACA0012 at the frequency 1 GHz, and the results are compared with the numerical result based on Crank-Nicolson scheme [12], as shown in Fig. 2. It indicates that the MOM result agrees with the reference numerical result better than the result of PO. Thus, MOM will be chosen in the study.

3 Multi-objective Optimization Platform

The optimization platform will be introduced in this section, which includes airfoil parameterization method, surrogate model construction method, optimization method and the road map of the platform.

3.1 Parameterization Method of Airfoil

The Parsec method is used to parameterize the airfoil geometry. Parsec method is a polynomial method focusing on the main geometric parameters of the airfoil, which has specific and independent geometric significances.

In this paper, thirteen parameters are used to describe an airfoil. They are the upper/lower leading edge radius (R_{le-up}, R_{le-low}), the upper/lower peak position (X_{up}, Y_{up}, X_{low}, Y_{low}), the upper/lower peak curvature (X_{up-xx}, X_{low-xx}), the upper/lower trailing edge position (X_{te}, Y_{up-te}, Y_{low-te}), the trailing edge angle (β) and the trailing oriented angle (α), as is shown in Fig. 3.

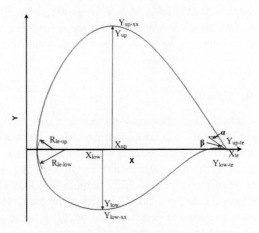

Fig. 3. Parameterization sketches

Parsec method expresses the upper and lower surface of the airfoil by two polynomials [13, 14]:

$$y_{up} = \sum_{n=1}^{6} a_n x^{n-1/2} \tag{3.1}$$

$$y_{low} = \sum_{n=1}^{6} b_n x^{n-1/2} \tag{3.2}$$

Then, two systems of equations could be obtained by imposing the related geometric parameters. The values of a_n and b_n can therefore be found by solving the two systems of equations.

3.2   Surrogate Model Construction

The high-fidelity aerodynamic and stealthy calculations of each airfoil require so much computation time and resources that the surrogate model is introduced in the optimization platform to improve the efficiency. The surrogate model has the characteristics of simpler structure, less calculated quantity. It can also satisfy the engineering requirements and has a good non-sample-point prediction capability. The surrogate model in this paper is based on the Radial-based Function (RBF) neural network [15, 16]. RBF neural network not only has the neural network's natural characteristics of nonlinearity, universality and non-convexity, but also has the advantages of powerful prediction capability, fast learning ability and strong consistency approximation, which is suitable to deal with the intense non-liner problem between aerodynamics and electromagnetism.

Before constructing surrogate model, samples are chosen in the design space by the Latin Hypercube Sampling (LHS) method first to get the sampling airfoils. Then the aerodynamic and stealthy data are calculated by the method mentioned in Sect. 2. Finally, the surrogate model could be constructed by RBF neural network based on the sample data.

3.3   Pareto Genetic Algorithm

The aerodynamic/stealthy design optimization method is used to deal with the multi-objective optimization problems in the design of the airfoil. The traditional way to deal with the problem is to apply weight coefficients to transform multi-objective problems into single-objective ones, which in fact is still a single-objective optimization. Pareto method [17] can get all of the corresponding optimum solutions of different weight-allocation, which reflects the nature of multi-objective problem. Pareto genetic algorithm consists of Pareto method and genetic algorithm [18]. It can get multiple Pareto optimum solutions of multi-objective optimization by a single search, and offer the candidate solutions in the form of Pareto fronts, which makes the optimization design more applicable.

3.4   Optimization Platform Construction

Based on the above methods, an aerodynamic/stealthy design optimization platform of airfoils can be constructed as follows:

1. Chose proper design space and obtain initial samples by LHS;
2. Obtain the aerodynamic/stealthy data of initial samples by solvers, and build RBF neutral network;

3. Regard the initial samples as population, proceed Pareto genetic algorithm, work out the objection function with surrogate model, sort the population, and acquire the fitness value;
4. Revise the optimized Pareto fronts with the solvers, and adopt the revised data to update the samples;
5. Check the accuracy. If the accuracy satisfies the requirement, the process will be stopped, if not, turn to Step 2, optimize until the accuracy reaches the requirement;
6. Obtain the Pareto front from Pareto solution sets, get the optimization results.

The flow chart of the optimization process is showed in Fig. 4.

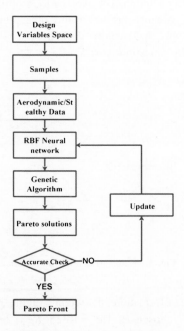

Fig. 4. Optimization road map

4   Optimization Design

According to the multi-objective optimization design platform, the airfoil will be optimized in this section.

4.1   Initial Airfoil

The US fighters in service mainly adopt NACA64A or NACA65A series as basic airfoils of the wings [19]. The geometric characteristics of these foils have small leading radius, rear maximum thickness, which indicates a good drag-divergence performance. Therefore, NACA 64A206 is chosen as the initial airfoil for optimization.

4.2   Design Condition

The RCS changes according to the frequency of irradiation. Regularly, an aircraft has three regions according to the ratio of characteristic length d to wave length λ: rayleigh region, resonance region and high-frequency region. Most of the radars in service are at S, C, X, Ku wave bands, which belong to high-frequency scattering region when comparing the wave length with the typical length of fighter. According to stealthy method mentioned above, the characteristic length of the airfoil is 1 m. Chose 1 GHz as the frequency of irradiation. The value of d/λ is about 3.3, which belongs to high-frequency region according to reference [20].

According to the reference [21], the research on stealthy performance of airfoil should focus on key pitching azimuth. The average RCS of the airfoil with the azimuth changing from –30° to 30° is chosen as the stealthy objection function of the airfoil, as shown in Fig. 5. Therefore, the stealthy design condition is that the frequency $f = 1$ GHz, and the stealthy objection is $\overline{RCS}_{-30° \sim 30°}$.

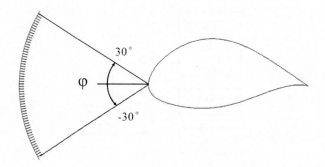

Fig. 5. Radar irradiation on airfoil

Besides, based on the information of a certain fighter, the aerodynamic design condition of the airfoil is chosen as $Ma = 1.325$, $Re = 1 \times 10^7$, $\alpha = 2.7°$, and the aerodynamic objective is C_d.

4.3   Airfoil Optimization Design

Based on the optimization design platform, the airfoil optimization design can be carried out. Objective function and constraints of the airfoil optimization design are as follows:

Objection function: minimized C_d, $\overline{RCS}_{-30° \sim 30°}$
Design condition: $Ma = 1.325$, $Re = 1 \times 10^7$, $\alpha = 2.7°$, $f = 1$ GHz
Constraints: $C_l \geq C_{l0}$, $t \geq t_0$, $|C_m| \leq |C_{m0}|$

In which, C_{l0} t_0 and C_{m0} represent the lift coefficient, thickness and pitching moment coefficient of the initial airfoil respectively. The reference point of piching moment is at the quarter chord.

The optimized results is shown in Fig. 6. On the Pareto front, three airfoils are chosen to discuss, named Foil 1, Foil 2, Foil 3, respectively. Obviously, the RCSs of the airfoils decrease while the drag coefficients C_d increase from Foil 1 to Foil 3.

Figure 7 shows the geometric comparison between initial airfoil and the optimized ones. The Table 1 shows the parameters of the airfoils, in which X_{te}, $Y_{up\text{-}te}$, $Y_{low\text{-}te}$ are constants while the other parameters are design variables. When contrasting the optimized parameters with the initial ones in the Table 1, the value of X_{low} changes a little, the values of Y_{up}, $R_{le\text{-}up}$, $R_{le\text{-}low}$, α and $X_{low\text{-}xx}$ decrease and the values of X_{up}, Y_{low}, β increases. Besides, the values of $X_{up\text{-}xx}$ of Foil 1 and Foil 2 increase while that of Foil 3 decreases. It is found that the optimized airfoils have thinner leading edge radius, later maximal thickness location and diamond-shaped thickness distribution than the initial one. Meanwhile, the aft cambers of optimized airfoil are smaller than that of the initial one. These geometric characteristics unload the pressure of aft airfoil resulting in the reductions of node-down pitching moment, the wave drag and RCS, which can improve the aerodynamic and stealthy performance.

The Table 2 shows the objective and constraint values of the initial airfoil and the optimized ones. At the design condition, the drag coefficients C_d of the optimized airfoils reduce by 8.65%, 7.16% and 3.2%, the RCSs of the optimized airfoils decrease by 12.17%, 15.45% and 17.20%, respectively. At the same time, the pitching performance of the optimized airfoils are obviously improved with 37.83%, 38.87% and 36.15% reductions of pitch moment coefficients C_m, which is beneficial for tailless configuration when considering trim in the future. Besides, the lift coefficients C_l increase by 5.78%, 5.04%, and 4.25%, while the thickness show a slight increase by 0.16%, 0.33% and 1.31%.

Figure 8 shows the pressure distribution of the airfoils. The leading edge suction peaks of the optimized airfoils increases due to the smaller leading edge radius, while the pressure differences decrease around the 30% of the chord and the aft of the airfoils Therefore, the pitching moments decrease while the lifts remain almost the same. Figure 9 shows the RCSs distributions of the airfoils, the up-surface RCSs of the optimized airfoils have evident diminution with the low-surface RCSs remaining almost the same. Besides, the azimuths corresponding to the maximum RCSs of the low-surface move backward.

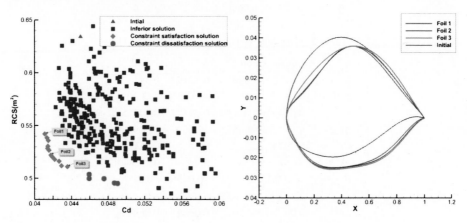

Fig. 6. Pareto front of the optimization design   **Fig. 7.** Geometric comparison

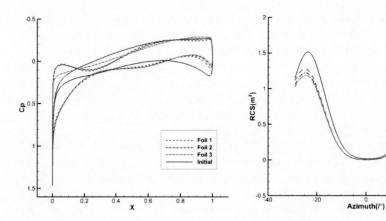

Fig. 8. Pressure distribution comparison   **Fig. 9.** RCS distribution comparison

Table 1. Geometric parameters of the initial and optimized airfoils

	Initial	Foil 1	Foil 2	Foil 3
X_{te}	1	1	1	1
Y_{up-te}	0	0	0	0
Y_{low-te}	0	0	0	0
X_{up}	0.3983	0.4779	0.4900	0.4923
Y_{up}	0.0403	0.0359	0.0359	0.0359
X_{up-xx}	−0.3376	−0.5323	−0.4773	−0.3230
X_{low}	0.3354	0.3486	0.3319	0.3334
Y_{low}	−0.0196	−0.0249	−0.0252	−0.0258
X_{low-xx}	0.1886	0.1504	0.1508	0.1536
R_{le-up}	0.0029	0.00075	0.00074	0.00075
R_{le-low}	0.0015	0.0004	0.0005	0.0005
α	−0.0607	−0.0489	−0.0471	−0.0395
β	0.0343	0.0496	0.0607	0.0620

Table 2. The values of objection and constraint

	Initial	Foil 1	Foil 2	Foil 3
RCS	0.6172	0.5421	0.5219	0.5111
$\Delta RCS\%$	–	−12.17%	−15.45%	−17.20%
C_d	0.04499	0.04110	0.04177	0.04355
$\Delta C_d\%$	–	−8.65%	−7.16%	−3.19%
C_l	0.2024	0.2141	0.2126	0.2110
$\Delta C_l\%$	–	5.78%	5.04%	4.25%
C_m	−0.0762	−0.0474	−0.0466	−0.0486
$\Delta C_m\%$	–	−37.83%	−38.87%	−36.15%
t	0.06	0.0601	0.0602	0.0608
$\Delta t\%$	–	0.16%	0.33%	1.31%

The three optimized airfoils have different performance in stealth and aerodynamics. Foil 3 has the best stealthy performance because the trailing edge angle of Foil 3 is the biggest among them, which can reduce the radar wave diffraction and creeping to reduce the RCS. Foil 1 has smallest drag because of the smaller leading edge radius and trailing edge angle, bigger upper-surface curvature and trailing oriented angle, which can reduce the wave drag. Foil 2 is a compromise of Foil1 and Foil 3.

5   Conclusion

In this paper, an airfoil for supersonic fighter is designed based on aerodynamic/ stealthy integrated design optimization. The airfoil design is conducted by a multi-objective optimization platform consisting of high-accurate stealthy and aerodynamic analysis, parameterization research, surrogate model construction and Pareto genetic algorithm study. It is proved that the optimization results can improve the stealthy and aerodynamic performance. Therefore, the platform is turned out to be reliable for aerodynamic/stealthy integrated design optimization of the airfoil for supersonic fighters. Future work will focus on the three-dimensional design of the fighter, including stealthy design, supersonic cruise performance design, and transonic cruise and maneuvering design.

References

1. Vinh H, van Dam CP (1993) Shape optimization for aerodynamic efficiency and low observability. In: AIAA 24th fluid dynamics conference, AIAA 93, p 3115
2. Zhu ZQ, Islam Z et al (1998) Airfoil shape optimization for high aerodynamic efficiency/low observability. Acta Aeronautica et Astronautica Sinica 19(6):641–654 (In Chinese) 朱自强, Zubair Islam等(1998) 翼型外形高气动效率/低可探测性的优化. 航空学报 19(6):641–654
3. Yu RX, Zhu ZQ (2002) Computation of the bi-objective/bi-disciplinary numerical optimization. Acta Aeronautica et Astronautica Sinica. 23(4):330–333 (In Chinese) 吁日新, 朱自强 (2002) 双目标双学科数值优化计算. 航空学报 23(4):330–333
4. Pittman JL (1986) Supersonic airfoil optimization. NASA report 86-1818, NASA Langley research center
5. Kroo I, Willcox K, March A et al (2010) Multifidelity analysis and optimization for supersonic design. NASA/CR-2010-216874
6. Lattarulo V, Seshadri P, Parks GT (2013) Optimization of a supersonic airfoil using the multi-objective alliance algorithm. In: GECCO 2013 Proceeding of the 15th annual conference on genetic and evolutionary computation, Netherlands
7. Sturdza P (2003) An aerodynamic design method for supersonic natural laminar flow aircraft. Dissertation of Stanford University
8. Sturdza P (2007) Extensive supersonic natural laminar flow on the aerion business jet. In: 45th AIAA aerospace sciences meeting and exhibit, Reno, Nevada 2007, p 685
9. Zhenli C, Binqian Z, Jing S (2006) A more efficient design method for supersonic drag reduction. J Northwest Polytechnical Univ 24(4). (In Chinese) 陈真利, 张彬乾 (2006) 基于面积律概念的超音速减阻设计方法. 西北工业大学学报, 24卷, 4期

10. Md Hossain S, Raiyan, MF et al (2014) A comparative flow analysis of NACA 6409 And NACA 4412 aerofoil. Int J Res Eng Technol. eISSN: 2319-1163

11. Luo L (2014) Aerodynamic and stealth integrated optimization design of airfoils for blended-wing-body tailless configuration. Master thesis. Northwestern Polytechnical University.(In Chinese) 罗烈(2014) 翼身融合无尾布局翼型气动隐身设计.西北工业大学硕士论文

12. Shankar V, Hall WF, Mohammaian AH (1989) A time-domain differential solver for electromagnetic scattering problems. In: Stone WR (ed) Radar cross section of complex objects. IEEE Press, pp 127–139

13. Jiao ZH (2012) Investigation on the effects of geometric parameters on airfoils' stealth characteristics. Mech Sci Technol Aerosp Eng 31(12):1980–1987. (In Chinese) 焦子涵 (2012) 翼型几何参数对隐身特性的影响研究. 机械科学与技术 31(12):1980–1987

14. Castonguay P, Nadarajah SK (2007) Effect of shape parameterization on aerodynamic shape optimzation. In: 45th AIAA aerospace sciences meeting and exhibit 8–11 January, Reno, Nevada

15. Broomhead DS, Lowe D (1988) Multi-variable functional interpolation and adaptive networks. Complex Syst 2:321–355

16. Jackson IRH (1988) Convergence properties of radial basis function. Control Approximation 4(2):243–264

17. Koopmans TC (1951) Analysis of production as an efficient combination of activities. In: Activity analysis of production and allocation, cowies commission monograph, vol. 13. Wiley, New York

18. Cheng FY, Li D (1998) Genetic algorithm development for multiobjective optimization of structures. AIAA J 36(6)

19. Baorui F (1997) Airplane aerodynamic configuration design. Aviation Industry Press. (In Chinese) 方宝瑞(1997)飞机气动布局设计. 航空工业出版社, 北京

20. Liu ZH, Huang PL, Wu Z (2009) Frequency response scattering characteristic of aircraft. Acta Aeronautica et Astronautica Sinica 30(4):643–648. (In Chinese) 刘战合, 黄沛霖, 武哲 (2009) 飞行器目标频率响应特性 航空学报 30(4):643–648

21. Zhang BQ, Luo l, Chen ZL et al (2014) On stealth airfoil optimization design for flying wing configuration. Acta Aeronautica et Astronautica Sinica 35(4):957–967. (In Chinese) 张彬乾, 罗烈,陈真利,等 (2014) 飞翼布局隐身翼型优化设计. 航空学报 35(4):957–967

Aerodynamic Shape Optimization of the Common Research Model Based on Improved SQP Algorithm

Jing Yu[1], Jiangtao Huang[1], Dong Hao[2($\boxtimes$)], and Zhu Zhou[1]

[1] Computational Aerodynamics Institute,
China Aerodynamics Research and Development Center, Mianyang, China
yujinghd@hotmail.com
[2] High Speed Aerodynamics Institute,
China Aerodynamics Research and Development Center, Mianyang, China
haodongcardc@outlook.com

Abstract. Aircraft Aerodynamic Shape Optimization is a complex, large-scale and expensive optimization problem. Adjoint-based gradient optimization method plays a significant role in aerodynamic shape design. In this paper, we mainly focus on the optimization algorithm applied to aircraft aerodynamic design field. Sequential Quadratic Programming (SQP) is employed here, and some experience-based improvements are made in line search to accelerate the convergence of the algorithm. After improvements, the physical change of variables in each iteration can be evaluated, set and control. The optimization model with the drag coefficient minimization objective and wing thickness constraints for Wing-Body-Tail Common Research Model (CRM) is established. The optimization strategy and the improved SQP algorithm are detailed and verified afterwards. Aerodynamic shape design for CRM with Wing-Body-Tail Configuration is carried out. The optimization results are compared and discussed. Our optimization procedure reduced the drag from 167.9 counts to 149.4 counts (an 11% reduction) within 40 iterations, calling CFD solver 49 times. The optimization results demonstrate the effectiveness of the improved SQP method proposed in this paper.

Keywords: Sequential Quadratic Programming ·
Aerodynamic shape optimization · Common Research Model

1 Introduction

Aircraft Aerodynamic Shape Optimization is a complex, large-scale and expensive optimization problem [1]. Benefited from the advancement of high performance computing, and the development of computational fluid dynamics (CFD) models, numerical simulation and optimization of aircraft aerodynamic design has become possible. The use of numerical optimization reduces much of the wind-tunnel testing, shortens the design cycle, and makes it easy to automatically change and further improve the aerodynamic design.

© Springer Nature Singapore Pte Ltd. 2019
X. Zhang (Ed.): APISAT 2018, LNEE 459, pp. 1131–1149, 2019.
https://doi.org/10.1007/978-981-13-3305-7_90

The optimization can be performed with gradient-free or gradient-based methods [1]. The gradient-free methods, such as Particle Swarm Optimization (PSO), Genetic Algorithm (GA) et al., are easy to implement and claimed to converge to the global optimum theoretically. But when dealing with the large-scale variable problems, their computational cost is extremely high. The gradient-based methods, such as quasi-Newton method, BFGS, and Sequential Quadratic Programming (SQP) et al., relatively converge fast and are easy to deal with large-scale variables, but they require the gradient information of the objective function and the constraints which is hard to obtain in certain problems, and the gradient-based methods usually converge to a local optimum, not the global one. In Aircraft Aerodynamic Shape Optimization, slight local change in wing shape would issue large impact on aerodynamic performance. To finely improve the aerodynamic design, large-scale variables should be taken into account, and gradient-based methods are preferred options for optimization.

Much work has been done for gradient-based aerodynamic shape optimization. Some of them focused on the study of adjoint method [2–8], which assisted in calculating the aerodynamic shape derivatives, and enabling the computation of gradients at a cost independent of the number of design variables. Some other focused on the aerodynamic performance analysis. Based on the Common Research Model (CRM), the Martines' team addressed a series of aerodynamic shape optimization problems with single or multipoint design [9–14]. The self-developed multilevel optimization acceleration technique is put into use, and several optimization algorithms, including gradient-based and gradient-free methods are compared and analyzed. It is concluded that the gradient-free methods require two to three orders of magnitude more computational effort when compared to the gradient-based [1]. In literatures [15–20], adjoint-based aerodynamic optimization methods are used for various models, such as Blended-Wing-Body Aircraft, Busemann-type supersonic biplanes, airfoils, 1303 UCAV and so on. During the optimization process, most of the researches employed commercial optimization software, e.g. SNOPT, IPOPT, and PSQP et al. Few of them paid attention to the studies on optimization algorithm. The mature commercial software is of good versatility and suitability. Their algorithms have been used and verified in various studies. But they still have potential to make improvements, especially for certain specific application area. In this paper, we mainly focus on the optimization algorithm applied to aircraft aerodynamic design field. Sequential Quadratic Programming (SQP) is employed here, and some experience-based improvements are made to accelerate the convergence of the algorithm.

The organization of this paper is as follows. Section 2 describes the optimization problem, including the design variables, the constraints the objectives and the optimization model. Section 3 presents the optimization strategy and the corresponding tools. The SQP algorithm and its improvements are detailed and verified. Section 4 gives the simulations and discusses the optimization results. Section 5 makes conclusions.

2   Problem Formulation

Aerodynamic Design Optimization Discussion Group (ADODG) issued a set of benchmark cases for aerodynamic design optimization, ranging from the optimization of a two-dimensional airfoil using the Euler equations to three-dimensional shape optimization using the RANS equations. In this paper, the NASA Common Research Model (CRM) with Wing-Body-Tail Configuration is employed to optimize. This paper aims at performing lift-constrained drag minimization of the NASA CRM Wing-Body-Tail Configuration using SQP. SQP is a gradient-based optimization method. A complete problem description will be provided in this section.

2.1   Baseline Geometry

CRM Wing-Body-Tail Configuration is used to be the baseline geometry. The main reference parameters for this baseline configuration are listed in Table 1.

Table 1. Baseline geometry configuration

Parameters	Value
Reference area	191.8 m^2
Reference chord	6.9 m
Moment reference	(0.0; 0.0; 0.0) m
Reynolds number (Mach = 0.85)	5×10^6

Moment of force is not considered either as a constraint or an objective in this paper, so the moment reference is simply set as zero, the original point of the coordinate.

2.2   Mesh Generation

The structured mesh of the CRM Wing-Body-Tail is generated. The mesh we generated for the test case optimization contains 6,620,160 cells (see Fig. 1). The mesh is marched out from the surface mesh using an O-grid topology to a farfield located at a distance of 30 times the span.

Fig. 1. Mesh generation for CRM Wing-Body-Tail

2.3   Design Variables

Free-form deformation (FFD) approach is employed to parameterize the CRM geometry in this paper. Ten airfoil sections are extracted from the CRM wing along with the span-wise direction. The shape of each airfoil is defined by 16 control points, where half of these on the top and the other half on the bottom. So there are 160 design variables in total (see Fig. 2).

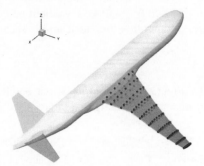

Fig. 2. The shape design variables are the z-displacements of 160 FFD control points

2.4   Constraints

Only two constraints are considered in this paper, the first one is the constant lift, and the second one is the thickness of the wing.

(1) Lift constraint: the lift coefficient is set to be 0.5.
(2) Thickness constraint: The relative maximum thickness constraints of three airfoils are taken into account (see Fig. 3). One airfoil is located at the wing root; the second one is located at the kink; and the third one is at the wing tip. The original relative maximum thicknesses of these airfoils are 13.69%, 10.52%, and 9.51%, respectively. The relative thickness (Rthick) is calculated by Eq. (1), where *thickness* represents the actual thickness and *chord* denotes the chord length of the corresponding airfoil.

$$Rthick = thickness/chord \qquad (1)$$

In this paper, the relative maximum thicknesses of the three airfoils are constrained to be equal or greater than 13.5%, 10.5% and 9.5%, respectively.

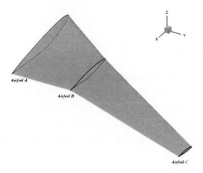

Fig. 3. The constrained airfoil sections

2.5   Optimization Model

With the objective of minimizing the drag coefficient of the CRM Wing-Body-Tail geometry in nominal cruising state, the shape design variables are set to be the z-coordinate movements of 160 control points on the FFD volume. All of these points are distributed in ten airfoil sections (see Fig. 2). The lift constraint and the maximum thickness constraint of airfoil sections A, B and C (see Fig. 3) are taken into account. The complete optimization problem is described in Table 2.

Table 2. Aerodynamic shape optimization problem

Function/Variable	Description	Quantity
Minimize C_d	Drag coefficient	1
With respect to z	FFD control point z-coordinates	160
	Total design variables	160
Subject to $C_L = 0.5$	Lift constraint	1
$\max(Rthick_A) - 13.5\% \geq 0$	Maximum thickness constraint of airfoil section 1	1
$\max(Rthick_B) - 10.5\% \geq 0$	Maximum thickness constraint of airfoil section 2	1
$\max(Rthick_C) - 9.5\% \geq 0$	Maximum thickness constraint of airfoil section 3	1
	Total constraints	4

To maintain the lift constraint, the attack angle should also be optimized. The attack angle would be optimized implicitly during Flow Simulation. And this variable would not appear here.

3   Methodology

3.1   The Main Framework

The main framework used for aerodynamic shape optimization in this paper is illustrated in Fig. 4. There are five key numerical tools and methods that are used for shape optimization, i.e. Geometry Parameterization, Mesh Perturbation, CFD solver for Flow Simulation, Gradient Solver, and Optimizer. The optimization algorithm employed by Optimizer would be detailed in Sect. 3.2. The other four components are described in this section briefly.

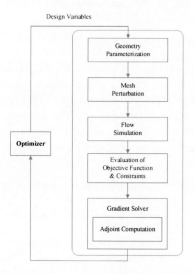

Fig. 4. The main framework for aerodynamic shape optimization

In Geometric Parametrization, the free-form deformation (FFD) volume approach is employed. Instead of parameterizing the geometry itself, the FFD parameterizes the geometry changes, making it easier to manipulate complex geometries.

Mesh Perturbation: When the design variables (i.e. the control points on FFD volume) change, the geometry would consequently change. Then it is necessary to perturb the mesh for CFD simulation solving the modified geometry. Based on the radial basis functions, a mesh deformation technique is developed and employed for hybrid structured grid. Firstly, make an interpolation according to the displacements of boundary grid points, as well as constructing a series of radial basis functions. Then based on the constructed radial basis function series, modify the displacements of the whole grid points in the calculation domain [21].

In Flow Simulation, a self-developed CFD solver PMB3D is employed here. PMB3D is a large-scale parallel solver for RANS equations, employing finite-volume discrete and implicit method. It provides options for variety of turbulence models with algebra, one or two equations. The steady flow is simulated by time dependent method, while the unsteady flow is done by dual-time method. Multi grid, residual mean, and local time step methods are used for accelerating the convergence. There are various options for convection term discretization, including Jameson Schmidt Turkel (JST) scheme, Van Leer upstream scheme, Roe upstream scheme for conservation variables, and AUSMPW + scheme et al. MUSCL et al. is used for interpolation. It also provides options for variety limiters with Van Albada, Vanleer, Condiff and Venkatakrishnan et al. Central discrete scheme is employed for viscous dissipation term. PMB3D could deal with variety grids, including facing-connected grid, surface lapped grid, multi structural grid, and unstructured grid.

In Gradient Solver, another self-developed adjoint solver Adjoint3D is employed for gradient computation [22].

3.2   SQP Algorithm and Its Improvements

Sequential Quadratic Programming (SQP) method is employed here for nonlinear equality and in-equality constraints. It is capable of solving large-scale nonlinear optimization problems with thousands of constraints and design variables. In this section, we will detail the method and its improvements.

(1)   The generic optimization problem

The nonlinear programming problem under consideration is stated as

$$\begin{aligned} \min \quad & f(x) \\ \text{s.t.} \quad & h_i(x) = 0, i \in E \\ & g_i(x) \geq 0, i \in I \end{aligned} \tag{2}$$

Where $f : \mathbb{R}^n \to \mathbb{R}$, $h : \mathbb{R}^n \to \mathbb{R}^l$, and $g : \mathbb{R}^n \to \mathbb{R}^m$ are smooth functions. We write the Lagrangian of this problem as

$$L(x, \mu, \lambda) = f(x) - \mu^T h(x) - \lambda^T g(x) \tag{3}$$

The KT (Kuhn-Tucker) condition of Eqs. (2) and (3) is

$$\begin{cases} \nabla_x L(x^*, \mu^*, \lambda^*) = 0 \\ h_i(x^*) = 0, i \in E \\ g_i(x^*) \geq 0, i \in I \\ \lambda_i^* \geq 0, i \in I \\ \lambda_i^* g_i(x^*) = 0, i \in I \end{cases} \tag{4}$$

(2) The QP sub-problem

To solve the nonlinear programming problem as stated in Eq. (2), we should first solve its QP sub-problem. The QP sub-problem of Eq. (2) at point (x_k, μ_k, λ_k) can be built as

$$
\begin{cases}
\min & F(d_k) = \nabla f(x_k)d_k + \frac{1}{2}d_k^T B_k d_k \\
s.t. & H_i(d_k) = h_i(x_k) + A_{i,k}^E d_k = 0, i \in E \\
& G_i(d_k) = g_i(x_k) + A_{i,k}^I d_k \geq 0, i \in I
\end{cases}
\tag{5}
$$

Where B_k is a positive definite approximation to the Hessian Matrix of the Lagrangian formulation (3), $A_k^E = \nabla h(x_k)$, and $A_k^I = \nabla g(x_k)$.

The KT (Kuhn-Tucker) condition of Eq. (5) is

$$
\begin{cases}
\nabla_d L(d^*, \mu^*, \lambda^*) = 0 \\
H_i(d^*) = 0, i \in E \\
G_i(d^*) \geq 0, i \in I \\
\lambda_i^* \geq 0, i \in I \\
\lambda_i^* G_i(d^*) = 0, i \in I
\end{cases}
\tag{6}
$$

Making use of the KT condition, Eq. (5) is equivalent to

$$
\begin{cases}
M_1(d, \mu, \lambda) = B_k d - (A_k^E)^T \mu - (A_k^I)^T \lambda + \nabla f(x_k) = 0 \\
M_2(d, \mu, \lambda) = h(x_k) + A_k^E d = 0 \\
\lambda \geq 0 \\
g(x_k) + A_k^I d \geq 0 \\
\lambda^T(g(x_k) + A_k^I d) = 0
\end{cases}
\tag{7}
$$

The last three equations in Eq. (7) can be seen as a linear complementarity problem. Define a smoothing FB-function as

$$
\varphi(\varepsilon, a, b) = a + b - \sqrt{a^2 + b^2 + 2\varepsilon^2}
\tag{8}
$$

Where $\varepsilon > 0$ is a smooth parameter. Let

$$
\Phi(\varepsilon, d, \lambda) = (\varphi_1(\varepsilon, d, \lambda), \varphi_2(\varepsilon, d, \lambda), \ldots, \varphi_m(\varepsilon, d, \lambda))^T
\tag{9}
$$

$$
\varphi_i(\varepsilon, d, \lambda) = \lambda_i + [g_i(x_k) + (A_k^I)_i d] - \sqrt{\lambda_i^2 + [g_i(x_k) + (A_k^I)_i d]^2 + 2\varepsilon^2}
\tag{10}
$$

Where $(A_k^l)_i$ describe the $i - th$ row of the matrix A_k^l. Denoting $z = (\varepsilon, d, \mu, \lambda)$, then Eq. (7) can be rewritten as

$$M(z) := M(\varepsilon, d, \mu, \lambda) = \begin{bmatrix} \varepsilon \\ M_1(d, \mu, \lambda) \\ M_2(d, \mu, \lambda) \\ \Phi(\varepsilon, d, \lambda) \end{bmatrix} = 0 \tag{11}$$

The Jacobian matrix of $M(z)$ is

$$M'(z) = \begin{bmatrix} 1 & 0 & 0 & 0 \\ 0 & B_k & -(A_k^E)^T & -(A_k^l)^T \\ 0 & A_k^E & 0 & 0 \\ \upsilon & D_2(z)A_k^l & 0 & D_1(z) \end{bmatrix} \tag{12}$$

Where

$$\upsilon = \nabla_\varepsilon \Phi(\varepsilon, d, \lambda) = (v_1, v_2, \ldots, v_m)^T \tag{13}$$

$$v_i = -\frac{2\varepsilon}{\sqrt{\lambda_i^2 + [g_i(x_k) + (A_k^l)_i d]^2 + 2\varepsilon^2}} \tag{14}$$

$$D_1(z) = diag(a_1(z), \ldots, a_m(z)) \tag{15}$$

$$D_2(z) = diag(b_1(z), \ldots, b_m(z)) \tag{16}$$

$$a_i(z) = 1 - \frac{\lambda_i}{\sqrt{\lambda_i^2 + [g_i(x_k) + (A_k^l)_i d]^2 + 2\varepsilon^2}} \tag{17}$$

$$b_i(z) = 1 - \frac{g_i(x_k) + (A_k^l)_i d}{\sqrt{\lambda_i^2 + [g_i(x_k) + (A_k^l)_i d]^2 + 2\varepsilon^2}} \tag{18}$$

Solving Eq. (7), the approximated solutions of the descending direction d_k, the corresponding λ and μ can be obtained.

(3) Merit function

The SQP algorithm convergence is promoted by imposing decrease in the ℓ_1 merit function:

$$\phi(x, \sigma) = f(x) + \sigma^{-1}(\|h(x)\|_1 + \|g(x)_-\|_1) \tag{19}$$

where $g(x) \triangleq \max\{0, -g_i(x)\}$, and $\sigma > 0$ is a penalty parameter.

(4) The framework of SQP algorithm

The SQP algorithm is specified as follows.

Step 0: Give the initial data (x_0, μ_0, λ_0), and the symmetric positive definite matrix B_0. Compute $f(x_0)$, $\nabla f(x_0)$, $A_0^E = \nabla h(x_0)^T$, $A_0^I = \nabla g(x_0)^T$, $A_0 = \begin{bmatrix} A_0^E \\ A_0^I \end{bmatrix}$. Let $k = 0$.

Step 1: Check if the solution satisfy the termination condition. If satisfy, go to step 7, else go to step 2.

Step 2: Solve the QP sub-problem at x_k, and then get d_k, μ_k and λ_k.

Step 3: Chose the proper parameter σ_k by the following method and then calculate the merit function.

Let $\tau = \max(\|\mu_k\|, \|\lambda_k\|)$, then $\sigma_k = \begin{cases} \sigma_{k-1}, & \text{if } \sigma_{k-1}^{-1} \geq \tau + \delta \\ (\tau + 2\delta)^{-1}, & \text{if } \sigma_{k-1}^{-1} < \tau + \delta \end{cases}$, where $\delta > 0$ is a random parameter.

Step 4: Line search. Compute the steplength α_k, satisfying the condition:

$$\phi(x_k + \alpha_k d_k, \sigma_k) - \phi(x_k, \sigma_k) \leq c\alpha_k \phi'(x_k, \sigma_k)$$

The constant c is usually chosen to be small, and this condition asks for little more descent in f.

Step 5: Let $x_{k+1} = x_k + \alpha_k d_k$. Compute $\nabla f(x_{k+1})$, A_{k+1}^E, and A_{k+1}^I. Compute B_{k+1} using quasi-Newton method.

Step 6: Let $k = k + 1$, and then go to step 1.

Step 7: End.

(4) Line search algorithm

Line search, also called one-dimensional search, is a local search in the SQP framework, aiming at finding one step length to satisfy the descending condition. Lots of line search methods have been proposed, such as golden section search, interpolation, Wolfe principle, and Armijo principle etc. No matter what method is put into use, the key problem in practice is to define the initial step length. A good initial step length could sharply reduce the local iteration number, accelerating the convergence of the whole algorithm. In addition, due to the expensive computation cost in CFD, in the application of aircraft aerodynamic design, we need minimum number of calls to CFD solver and the gradient solver. In this paper, quadratic interpolation is employed to search the step length. In order to define a reasonable initial guess of the step length, we normalize the descent direction d_k first, and then define an initial guess by engineering experience. When d_k is normalized, the maximum number in d_k vector is 1. And the step length then represents the maximum change in the variables, i.e. the step length has a physical meaning.

Let $\phi(\alpha_k) = \phi(x_k + \alpha_k \bar{d}_k, \sigma_k)$, and $\bar{d}_k$ is the normalized d_k. The line search decrease condition is $\phi(\alpha_k) \leq \phi(0) + c\alpha_k \phi'(0)$. Suppose that the initial guess α_0 is given. If we have $\phi(\alpha_0) \leq \phi(0) + c\alpha_0 \phi'(0)$, then we terminate the search. Otherwise, we know that the interval $[0, \alpha_0]$ contains acceptable step lengths. We form a quadratic approximation $\phi_q(\alpha)$ based on three pieces of information available $\phi(0)$, $\phi'(0)$, and $\phi(\alpha_0)$.

$$\phi_q(\alpha) = \left(\frac{\phi(\alpha_0) - \phi(0) - \alpha_0 \phi'(0)}{\alpha_0^2} \right) \alpha^2 + \phi'(0)\alpha + \phi(0)$$

The new trial value α_1 is defined as the minimizer of this quadratic:

$$\alpha_1 = - \frac{\phi'(0)\alpha_0^2}{2[\phi(\alpha_0) - \phi(0) - \phi'(0)\alpha_0]}$$

The whole optimization process of the improved SQP method is depicted in Fig. 5.

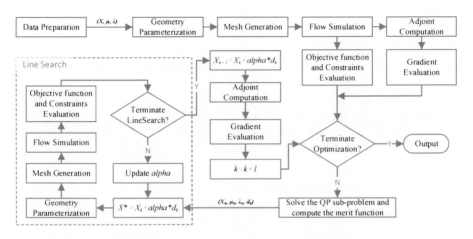

Fig. 5. The main framework of optimization process

3.3   SQP Algorithm Verification

This section is aiming at testing the efficiency of the SQP method. Three simple mathematical function cases are employed and the line search improvement is not put into use, since the improvements are only suitable for aerodynamic design applications. The optimization results for each test case are showed in Tables 3, 4 and 5. In each table, "Iter" means the number of iterations, "Eval" presents the number of objective function evaluations, while "Con" gives the sum or the detail value of the constraints.

(1) Case 1 with three equality constraints

$$
\begin{aligned}
\min \quad & f(x) = e^{x_1 x_2 x_3 x_4 x_5} - 0.5(x_1^3 + x_2^3 + 1)^2 \\
s.t. \quad & x_1^2 + x_2^2 + x_3^2 + x_4^2 + x_5^2 - 10 = 0 \\
& x_2 x_3 - 5 x_4 x_5 = 0 \\
& x_1^3 + x_2^3 + 1 = 0
\end{aligned}
$$

Table 3. Optimization results for case 1

The initial guess	Iter	Eval	Optimal points	$f(x)$	$\|d_k\|$	Con
$(-1.7,1.5,1.8, -0.6, -0.6)$	8	9	$(-1.72,1.60,1.83,-0.764,-0.764)$	0.0539	8.6e-8	0.00
$(-1.7,1.6,1.8, -0.7, -0.7)$	16	17	$(-1.72,1.60,1.83,-0.764,-0.764)$	0.0539	1.5e-6	0.00
$(-1.8,1.7,1.9, -0.8, -0.8)$	5	6	$(-1.72,1.60,1.83,-0.764,-0.764)$	0.0539	3.3e-5	0.00
$(-2,1.5,2, -1, -1)$	7	8	$(-1.72,1.60,1.83,-0.764,-0.764)$	0.0539	8.5e-6	0.00
$(-3,2,3, -2, -2)$	9	10	$(-1.72,1.60,1.83,-0.764,-0.764)$	0.0539	1.8e-5	0.00

(2) Case 2 with one equality constraint and two inequality constraints

$$\begin{aligned}
\min \quad & f(x) = -\pi x_1^2 x_2 \\
s.t. \quad & \pi x_1 x_2 + \pi x_1^2 - 150 = 0 \\
& x_1 \geq 0 \\
& x_2 \geq 0
\end{aligned}$$

Table 4. Optimization results for case 2

The initial guess	Iter	Eval	Optimal points	$f(x)$	$\|d_k\|$	Con
(3,3)	7	8	(3.99,7.98)	−398.94	8.3e-5	(0.00,3.99,7.98)
(4,4)	11	12	(3.99,7.98)	−398.94	7.3e-6	(0.00,3.99,7.98)
(1,1)	14	15	(3.99,7.98)	−398.94	1.1e-5	(0.00,3.99,7.98)

(3) Case 3 with four inequality constraints

$$\begin{aligned}
\min \quad & f(x) = x_1^2 + x_2^2 - 16x_1 - 10x_2 \\
s.t. \quad & -x_1^2 + 6x_1 - 4x_2 + 11 \geq 0 \\
& x_1 x_2 - 3x_2 - e^{x_1 - 3} + 1 \geq 0 \\
& x_1 \geq 0 \\
& x_2 \geq 0
\end{aligned}$$

Table 5. Optimization results for case 3

The initial guess	Iter	Eval	Optimal points	$f(x)$	$\|d_k\|$	Con
(4,4)	7	8	(5.24,3.75)	−79.81	1.0e-7	(−0.00,-0.00,5.24,3.75)
(1,2)	6	7	(5.24,3.75)	−79.81	3.3e-5	(−0.00,-0.00,5.24,3.75)
(6,7)	4	5	(5.24,3.75)	−79.81	1.4e-8	(−0.00,-0.00,5.24,3.75)

It can be concluded from Tables 3, 4 and 5 that the SQP method employed in this paper could solve the nonlinear optimization problems with equality and inequality constraints effectively.

4 Aerodynamic Shape Design Simulations

The aerodynamic design optimization results for the CRM wing-body-tail benchmark problem described in Table 2 under the nominal flight condition (Mach 0.85, Re = 5 e6) are presented in this section. The grid of 6,620,160 cells is used for the optimization. Based on the FFD method, 160 design variables described in Fig. 2 are optimized. Constant lift constraint and the airfoil thickness constraint are considered. The initial step length is 0.01, which means that the maximum change of the aircraft shape at each optimizing iteration is 1 cm. Our optimization procedure reduced the drag from 167.9 counts to 149.4 counts (an 11% reduction).

4.1 Variable Sensitivity Analysis

Before optimization, the variable sensitivity is drawn and analyzed first, examining the potential improvements of the baseline geometry. The surface sensitivity of the drag with respect to the airfoil shape is illustrated in Fig. 6, as a contour plot of the derivatives of the drag with respect to shape variations in the z derection. The comparison between the initial sensitivity and descent direction of the design variables are depicted in Fig. 7. In each figure, "NGradient" means the normalized gradient value, and "Ndk(IE)" means the normalized descent direction in inequality constraint.

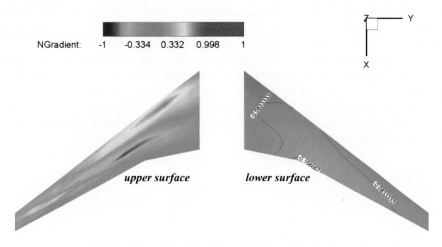

Fig. 6. The surface sensitivity of the baseline geometry

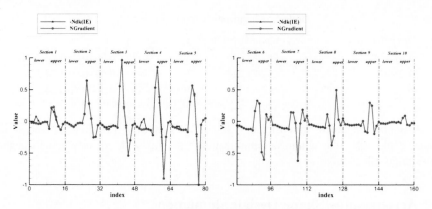

Fig. 7. The initial sensitivity and descent direction of the design variables

It can be seen from Fig. 6 that the regions with the highest drag gradient are close to the shock on the upper surface and the wing leading edge (the initial pressure distribution of the baseline geometry can be seen in Fig. 8). This indicates that local shape changes on these regions would drive the major drag reduction at the beginning of the optimization. Since these sensitivity plots are a linearization about the current design point, they provide no information about the constraints. Nonetheless, these sensitivity plots indicate what drives the design at this design point.

It can be seen from Fig. 7 that the drag gradient direction and the negative descent direction are almost the same at the beginning of the optimization. The slight differences are caused by the information of the constraints.

The experience-based step length for the SQP line search is set to be 0.01 on account of the initial descent direction, i.e. the maximum change of the aircraft shape at each iteration is 1 cm.

4.2   Single-Point Aerodynamic Shape Optimization

Our optimization procedure reduced the drag from 167.9 counts to 149.4 counts (an 11% reduction) within 40 iterations, calling CFD solver 49 times, and the whole optimization process is terminated by the failure of line search. The pressure distribution comparisons between the initial model and the optimization results are illustrated in Fig. 8. Eight airfoil sections are extracted from the CRM wing along with the span-wise direction. Starting from the wing root to the wing tip, each airfoil section is numbered from section 1 to section 8 in order. Section 1 and section 4 are separately the constrained airfoil A and airfoil B depicted in Fig. 3. Figure 9 gives a detailed comparison on pressure distribution and geometry between the baseline airfoil sections and the optimized ones.

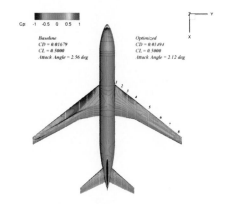

Fig. 8. Comparison of the optimization results with the baseline configuration

In Figs. 8 and 9, the baseline results are depicted in blue and the optimized results are shown in red. At the optimal point, the lift coefficient constraint is satisfied, and the corresponding attack angle is changed from 2.56° to 2.12°.

It can be seen from Fig. 9 that the thickness of the optimized sections are generally thinner than the original ones. Airfoils A, and B are only allowed to be slightly thinner than the original ones, so the maximum thickness change of sections 1 and 4 are not strong. But the thickness of other sections decreases much, since there is a strong incentive to reduce the airfoil thicknesses in order to reduce wave drag.

In Fig. 8, there are closely spaced pressure contour lines exhibited in the baseline wing, spanning a significant portion of the wing. And the intensive pressure contour lines indicate a shock. While in the optimized wing, the pressure contour lines are near uniform spacing, indicating shock elimination under the nominal flight condition. The airfoil Cp distributions of different position are depicted in Fig. 9. After optimization, the sharp increase in local pressure due to the shock becomes a gradual change from the leading edge to the trailing edge, indicating a shock-free state.

Figure 10 shows the convergence process along each iteration, while Fig. 11 describes the constraints value change along optimization procedure. Our optimization procedure reduced the drag from 167.9 counts to 149.4 counts (an 11% reduction) within 40 iterations, calling CFD solver 49 times. During the first 40 iterations, at each line search, the first local search could always find a better location. And at iteration 41, the line search tries 8 times to find a proper steplength but fail. The optimization process is terminated as the step length is too small to continue. This, to some degree, demonstrates that the improvements in line search are efficient.

Figure 11 shows that during the optimization, the maximum thickness of airfoil A decreases gradually, that of airfoil B changes little, while that of airfoil C increases a lot. Different from airfoils A and B, airfoil C is not completely located on the X-Z plane, that is to say the thickness increment for airfoil C does not indicate the thickness increment for the airfoils extracted by the X-Z plane in the wing tip.

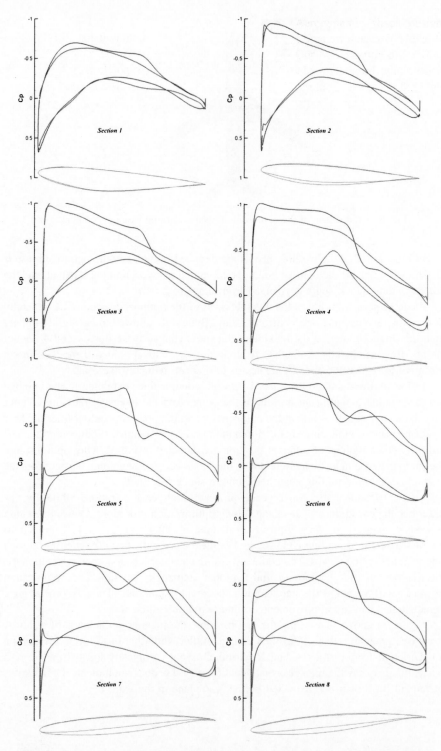

Fig. 9. Comparison of the pressure distribution and geometry for each airfoil section

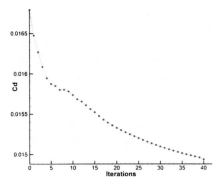

Fig. 10. Iteration process

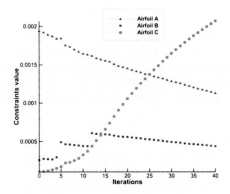

Fig. 11. Constraints variation

The whole optimization demonstrate that the improved SQP method proposed in this paper can successfully and efficiently optimize and design the aerodynamic shape, and the line search improvements can enhance the optimization convergence, i.e. reducing the drag by calling the CFD solver as less as possible.

5 Conclusion

The Sequential Quadratic Programming (SQP) method, which is capable of solving nonlinear optimization problems with equality and inequality constraints and large-scale design variables, is studied and improved in this paper. Based on its application to aircraft aerodynamic design field, some experience-based improvements are made in line search to accelerate the convergence of the algorithm. Based on the improvements, the physical change of variables in each iteration can be evaluated, set and control.

Aerodynamic shape design with the drag coefficient minimization objective for Common Research Model (CRM) with Wing-Body-Tail Configuration is carried out. Our optimization procedure reduced the drag from 167.9 counts to 149.4 counts (an 11% reduction) within 40 iterations, calling CFD solver 49 times. The optimization results demonstrate the effectiveness of the improved SQP method proposed in this paper.

Acknowledgement. This work was supported by National Key R & D Program of China under Grant No. 2016YFB0200704, and Equipment Pre-Research Project of China under Grant No. 41406030301.

References

1. Lyu Z, Xu Z, Martins JRRA (2014) Benchmarking optimization algorithms for wing aerodynamic design optimization. In: The eighth international conference on computational fluid dynamics (ICCFD8), Chengdu, Sichuan, China, 14–18 July
2. Reuther JJ, Jameson A, Alonso JJ, Rimlinger MJ, Saunders D (1999) Constrained multipoint aerodynamic shape optimization using an adjoint formulation and parallel computers, part 1. J Aircr 36(1):51–60. https://doi.org/10.2514/2.2413
3. Reuther JJ, Jameson A, Alonso JJ, Rimlinger MJ, Saunders D (1999) Constrained multipoint aerodynamic shape optimization using an adjoint formulation and parallel computers, part 2. J Aircr 36(1):61–74. https://doi.org/10.2514/2.2414
4. Hicken JE, Zingg DW (2010) Aerodynamic optimization algorithm with integrated geometry parameterization and mesh movement. AIAA J 48(2):400–413. https://doi.org/10.2514/1. 44033
5. Hicken JE, Zingg DW (2010) Induced-drag minimization of nonplanar geometries based on the Euler equations. AIAA J 48(11):2564–2575. https://doi.org/10.2514/1.J050379
6. Lyu Z, Kenway G, Paige C, Martins JRRA (2013) Automatic differentiation adjoint of the Reynolds-averaged Navier-Stokes equations with a turbulence model. In: 43rd AIAA fluid dynamics conference and exhibit, June
7. Jameson A, Martinelli L, Pierce N (1998) Optimum aerodynamic design using the Navier-Stokes equations. Theor Comput Fluid Dyn 10(1–4):213–237. https://doi.org/10.1007/s001620050060
8. Martins JRRA, Hwang JT (2013) Review and unification of methods for computing derivatives of multidisciplinary computational models. AIAA J 51(11):2582–2599. https://doi.org/10.2514/1.J052184
9. Liem RP, Martins JRRA, Kenway GKW (2017) Expected drag minimization for aerodynamic design optimization based on aircraft operational data. Aerosp Sci Technol
10. Kenway GKW, Martins JRRA (2016) Multipoint aerodynamic shape optimization investigations of the common research model wing. AIAA J 54(1):113–128
11. Lyu Z, Martins JRRA (2014) Aerodynamic shape optimization investigations of the common research model wing benchmark. AIAA J
12. Chen S, Lyu Z, Kenway GKW, Martins JRRA (2015) Aerodynamic shape optimization of the Common Research Model wing-body-tail configuration. J Aircr. https://doi.org/10.2514/1.C033328

13. Kenway GKW, Martins JRRA (2016) Aerodynamic shape optimization of the CRM configuration including buffet-onset conditions. In: 54th AIAA aerospace sciences meeting. American Institute of Aeronautics and Astronautics, January. https://doi.org/10.2514/6. 2016-1294

14. Martins JRA (2015) Wing design via numerical optimization. SIAG/OPT Views News 23 (1):2–7

15. Lyu Z, Martins JRRA (2014) Aerodynamic design optimization studies of a blended-wing-body aircraft. J Aircr. https://doi.org/10.2514/1.C032491

16. Petrone G, Hill DC (2014) Single-objective and multi-objective robust optimization of airfoils using adjoint solutions. In: 32nd AIAA applied aerodynamics conference, 16–20 June 2014, Atlanta, GA

17. Hu R, Jameson A, Wang Q (2012) Adjoint-based aerodynamic optimization of supersonic biplane airfoils. J Aircr 49(3):802–814

18. Coppin J, Qin N, Birch T (2013) Adjoint based aerodynamic optimisation of a UCAV. In: 51st AIAA aerospace sciences meeting including the new horizons forum and aerospace exposition, 07–10 January 2013, Grapevine (Dallas/Ft. Worth Region), Texas

19. Bisson F, Nadarajah S, Shi-Dong D (2014) Adjoint-based aerodynamic optimization of benchmark problems AIAA-2014-0412. In: 52nd aerospace sciences meeting, 13–17 January 2014, National Harbor, Maryland

20. Bisson F, Nadarajah S (2015) Adjoint-based aerodynamic optimization of benchmark problems AIAA-2015-1948. In: 53rd AIAA aerospace sciences meeting, 5–9 January 2015, Kissimmee, Florida

21. Wang G, Lei BQ, Ye ZY (2011) An efficient deformation technique for hybrid unstructured grid using radial basis functions. J Northwest Polytechnical Univ 29(5):783–788 (in Chinese)

22. Huang J, Liu G, Zhou Z, Gao Z, Huang Y (2017) Investigation of gradient computation based on discrete adjoint method. Acta Aerodynamica Sinica 35(4):554–562

Productivity Analysis and Optimization of Aircraft Assembly Line Based on Delmia-Quest

Heng Zhong[✉], Xiaojun Zhang, Jun Hu, Shuntao Liu,
and Xinyun Shao

Chengdu Aircraft Industrial (Group) Co., Ltd., Chengdu 610091, China
fenghuo1423@163.com

Abstract. To analyze the influence of different methods on the productivity of aircraft assembly line more quickly and flexibly, this paper proposes a virtual simulation method combined with industrial engineering theory instead of traditional empirical formula and Gantt chart method. On the basis of the process flow, key data and layout of the aircraft assembly line, the simulation model is built. The productivity, productive tempo and utilization of equipment of the assembly line are studied, and the bottleneck of the assembly line is analyzed. At the end of this paper, the method of productivity upgrading is studied. The optimization of process flow, equipment, processing technology, and processing time are summarized. Taking the cost as the constraint, the simulation experiment method is proposed to analyze different schemes, and the optimal scheme to meet the capability requirements is obtained.

Keywords: Productivity · Aircraft assembly line · Delmia-Quest · Virtual simulation

1 Introduction

At present, traditional methods such as Gantt chart, personal experience and formula are commonly used in the design and optimization of aircraft assembly lines. They can play a reference role to some extent, but they are unpredictable for more elaborate operation situations of assembly line such as equipment utilization, resource cross-influence, and process blockage. At the same time, they cannot compare different layout schemes and optimization strategies quickly and flexibly. Using traditional methods is time consuming and laborious in the design and optimization of assembly lines, and it is difficult to get optimal results.

The simulation software for production system, such as Delmia-Quest, Witness, Flexsim and Em-plant, can use virtual simulation technology to map the actual assembly line to the virtual simulation model. Through the simulation, all kinds of data of the assembly line can be outputted, and the different implementation schemes of the assembly line can be analyzed and compared quickly and flexibly according to the constraints, so as to provide the data support for the designers. At present, a lot of scholars have carried out research and application of virtual simulation technology in

© Springer Nature Singapore Pte Ltd. 2019
X. Zhang (Ed.): APISAT 2018, LNEE 459, pp. 1150–1159, 2019.
https://doi.org/10.1007/978-981-13-3305-7_91

logistics management, productivity optimization, and cost control. Delmia-Quest is used to analyze the production process and logistics system with the aim to reduce the workload task, improve the efficiency of the production line, and optimize the inventory level and the site space utilization [1–3]. Flexsim is used to simulate the production line and logistics centre with the aim to verify the system optimization scheme and achieve cost control [4–7]. Virtual simulation technology has been applied in aircraft assembly lines, mainly focused on assembly line modelling and parameter analysis, but there are still shortcomings in the analysis and selection of productivity simulation optimization methods [8–11].

This paper studies the productivity analysis and optimization methods of the aircraft assembly line with the virtual simulation technology. On the basis of analyzing the characteristics of aircraft assembly line, the simulation modelling and analysis methods are discussed. The parameters such as production capacity and equipment utilization are obtained, and the bottleneck is analyzed. Then, the optimization of aircraft assembly line is studied in the aspects of process flow, equipment, processing technology, and processing time, and the optimal optimization scheme is obtained by using the virtual simulation experiment method with the cost as the constraint.

2 Features of Aircraft Assembly Line

The aircraft assembly line has many unique characteristics. The unique characteristics which have the large impact on productivity and need special attention in the modelling process are summarized as follows:

- Discrete type production. The production units are mostly spatially distributed, and there is no assembly line between upstream and downstream processes. The logistics and semi-finished products between processes need to be transported to the next process through special transit links.
- The process route is complex, and the assembly process is numerous. The parallel operation and sequential operation are cross-correlated between different processes, and a fixed and in-variable technological sequence is not formed.
- The flexibility of production equipment is poor, and most of them are special equipment. In order to ensure high-precision positioning and shape requirements of aircraft products, the equipment is designed to be rigidly positioned. Therefore, the product can only rely on the corresponding equipment to produce, and the stability of the equipment directly affects the productivity of the assembly line.
- The blockage caused by the equipment occupying in the assembly process has a serious impact on the productivity. Due to the large size of aircraft components, there are few areas for product buffer storage near the assembly line. If there is a problem in the follow-up process, the product cannot flow normally, it can only continue to occupy the equipment, which often leads to the stop of the production in all the preceding processes and affects the productivity.

- There is a big gap between the assembly processing cycle and the logistics time between processes. The processing cycle of each station in the aircraft assembly is calculated in days or weeks, while the logistics time is generally tens of minutes or a few hours. Therefore, the logistics time has less influence on the entire assembly cycle. In order to reduce the complexity of the simulation model, the logistics activities between processes can be omitted.

3   Modeling and Analysis of Aircraft Assembly Line

The construction of assembly line simulation model needs to go through the process shown in Fig. 1. First of all, it is necessary to determine the process flow, the number and layout of the equipment according to the actual assembly line, so that the 3D model of the assembly line can be built. Secondly, on the basis of the 3D model, the key data parameters of the assembly line, such as process cycle, equipment failure rate, maintenance time, working class, and the logical relationship between the assembly processes are needed. The actual key parameters are assigned to the 3D model, and the actual assembly line can be mapped to the virtual model. Finally, run the simulation model and output the corresponding data to analyze whether the target requirements can be achieved.

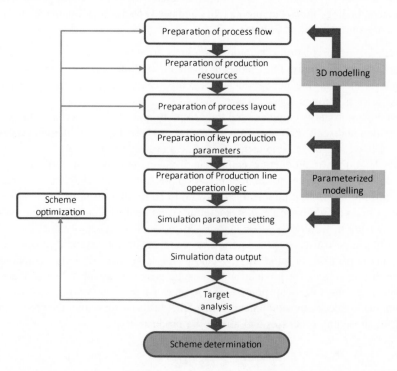

Fig. 1. Simulation flow chart

3.1 Preparation of Simulation Modelling Data

Data preparation is the most important step in the process of assembly line simulation. Adequate data preparation is the key to success in simulation. Data preparation is to collect information that can accurately describe the assembly line, including the physical description of the production process, the description of the logical relation, the parametric description. It usually requires collection of process flow, key data and layout of the assembly line.

3.1.1 Assembly Process Flow

Assembly process flow is the foundation of simulation model and the key to determine the operation logic of assembly line. Before the simulation model is built, the craft process of the aircraft assembly should be combed first to determine the input and output relation of the material or product. The refinement degree of the process flow can be set up according to the requirement of the assembly line analysis and optimization. Figure 2 shows the process flow of an aircraft assembly line. The assembly line will be used for the analysis of the case later.

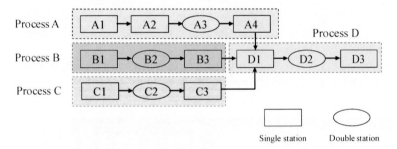

Fig. 2. Process flow chart of the assembly line

From Fig. 2, we can see that the assembly line is made up of four parts assembly lines, namely A, B, C and D. The components of A, B and C are assembled separately, and finally assembled into parts D. Each figure (rectangle or ellipse) in Fig. 2 represents an assembly process. The rectangle indicates that the assembly process has only one assembly station, and the ellipse indicates that the assembly process has two interchangeable assembly stations and can be processed at the same time for two products.

3.1.2   Key Data of Assembly Line

The key data for building assembly line simulation model are process cycle, quantity of equipments, number of input and output parts of every process, equipment failure data, equipment maintenance data, operation rate of logistics equipment, supply cycle of parts, and initial inventory of parts, as shown in the following Table 1:

Table 1. Key input parameters

Variables	Attribute
Quantity of equipments	Constant
Process cycle	Constant
Number of input parts in process	Constant
Number of output parts in process	Constant
Equipment maintenance cycle	Constant
Equipment maintenance time	Constant
Operation rate of key logistics equipment	Constant
Supply cycle of parts	Constant
Initial inventory of parts	Constant

3.1.3   Layout of Assembly Line

Layout of assembly line is the basis of constructing assembly line physical model. The following is the layout of the assembly line in the case (Fig. 3).

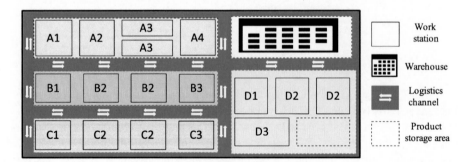

Fig. 3. Layout of the assembly line

3.2   Construction of Simulation Model

This paper uses Delmia-Quest to build the model. The process of model construction can be divided into two steps. The first step is to build the 3D geometric model. The models of the equipments and products involved in the process are built up, and arranged according to the layout of the assembly line. The second step is to map the actual situation of the assembly line to the 3D model. The key data, process flow and logic of the assembly line should be assigned to the 3D model to determine the inflow and outflow of the material, the transmission route of the production information flow, the process cycle and the logic of the operation of the process. Table 2 shows the key data of the assembly line in the case, and the simulation model is shown in Fig. 4.

Table 2. Key parameter table of the assembly line

No.	Working procedure	Process cycle (day)	Equipment	Input parts	Number of input parts	Output product	Number of output parts
1	A1_p	2	A1_m	A0	1	A1	1
2	A2_p	3	A2_m	A1	1	A2	1
3	A3_p	8	A3_m	A2	1	A3	1
4	A4_p	5	A4_m	A3	1	A4	1
5	B1_p	1	B1_m	B0	1	B1	1
6	B2_p	6	B2_m	B1	1	B2	1
7	B3_p	3	B3_m	B2	1	B3	1
8	C1_p	4	C1_m	C0	1	C1	1
9	C2_p	2	C2_m	C1	1	C2	1
10	C3_p	7	C3_m	C2	1	C3	1
11	D1_p	3	D1_m	A4, B3, C3	1	D1	1
12	D2_p	6	D2_m	D1	1	D2	1
13	D3_p	8	D3_m	D2	1	D3	1

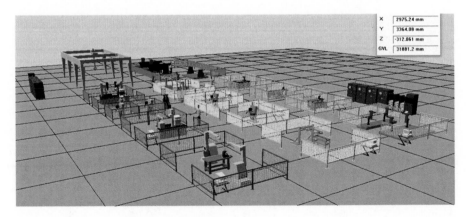

Fig. 4. Simulation model of the assembly line

3.3   Simulation Analysis

After building the simulation model, the simulation system time needs to be determined according to the actual working time. In the case, according to the actual production situation of the assembly line, the productivity of the assembly line is evaluated on a yearly basis, and the key production data unit is days. In order to unify the time unit, the working time of one year of the production system operation is defined as 300 days. Since all products on the assembly line are not available when the simulation model starts running, in order to get closer to the stable operation of the assembly line, the simulation time needs to be adjusted. The specific formula is as follows:

$$T = T1 + T2 \tag{1}$$

T is the actual simulation time in the formula, $T1$ is the number of days worked in normal year (defined as 300 days), and $T2$ is the time from starting to stable operation of assembly line. Simulation analysis shows that $T2$ of the case is 35 days, so the simulation time T is set to 335 days.

By running the simulation model, the operation status, output, and equipment utilization of each equipment in the assembly line can be obtained, as shown in Table 3, and the utilization of equipments is shown in Fig. 5.

Table 3. Simulation results of the case

Equipment	State times			Utilization (%)	No. of products
	Idle	Busy - Processing	Blocked - Wait block		
A1_m	0	94	241	28.06	47
A2_m	2	138	195	41.194	46
A3_m1	5	180	150	53.731	22
A3_m2	8	176	151	52.537	22
A4_m	13	214	108	63.881	42
B1_m	0	46	289	13.731	46
B2_m1	1	136	198	40.597	22
B2_m2	2	132	201	39.403	22
B3_m	7	129	199	38.507	43
C1_m	0	184	151	54.925	45
C2_m1	9	46	280	13.731	23
C2_m2	12	44	279	13.134	22
C3_m	6	298	31	88.955	42
D1_m	48	126	161	37.612	42
D2_m1	27	124	184	37.015	20
D2_m2	30	120	185	35.821	20
D3_m	27	308	0	91.94	38

From the simulation results, it can be seen that the annual output of final product D3 is 38, and the utilizations of equipment C3_m and D3_m are 88.955% and 91.94% respectively. Generally, bottlenecks in assembly lines often occur in relatively high loading rates. Therefore, in the process of the assembly line optimization or capacity upgrading, process C3_p and D3_p should be considered.

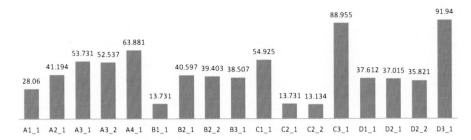

Fig. 5. Utilization of equipments (%)

4   Productivity Optimization

4.1   Optimization Method of Assembly Line

The methods used to increase the productivity of aircraft assembly lines can be summarized as follows: optimization of process flow, optimization of equipment, optimization of processing technology.

- Optimization of process flow usually involves splitting, adjusting, and merging process flows. For processes with a long assembly cycle, a small number of equipment, and a fixed processing sequence, the splitting of the processes flows can be carried out, that is, the process completed at a certain station can be decomposed into multiple stations to complete. Conversely, for other processes with a short assembly cycles and a large number of equipment, adjusting or merging of the processes flows may be considered to reduce resource waste and excess production capacity.
- The optimization of production equipment involves adding and reducing equipment. The equipment with a larger load should be added to increase the productivity in the process, and the equipment with a small load should be reduced to save resources.
- The optimization of processing technology is usually through technical means to optimize the process or equipment, in order to improve process efficiency, reduce process cycle, and reduce equipment failure or maintenance time.

Usually, while optimizing the assembly line and increasing productivity, it is also necessary to consider input constraints such as cost and time. At present, there are four optimization schemes in this case named M1, M2, M3, and M4. The optimization method and cost of the four schemes are shown in Table 4. Now we need to analyze the different combinations of schemes and choose one that can produce more than 45 final products and needs the least cost.

Table 4. Optimization schemes

Process	Scheme	Optimization method	Cost (ten thousand)
C3	M1	Add a equipment of C3	50
	M2	The processing cycle of C3 is reduced by 20%	20
D3	M3	Add a equipment of D3	60
	M4	The processing cycle of D3 is reduced by 20%	15

4.2   Simulation Optimization Experiment and Analysis

According to the different schemes, the corresponding parameters are changed in the simulation model, and the contrast experiment is set up in the Delmia-Quest. The experimental results can be obtained after running the simulation models, as shown in the Table 5.

Table 5. Experimental results

Scheme	Quantity (pieces)	Cost (ten thousand)	Schemes	Quantity (pieces)	Cost (ten thousand)
M1	38	50	M1&M2	38	70
M2	38	20	M1&M3	60	110
M3	44	60	M1&M4	48	75
M4	44	15	M2&M3	55	80
			M2&M4	48	35
			M3&M4	44	75

It can be seen from Table 5 that optimizing the C3 process cannot increase the productivity, because the bottleneck is in the D3 process. Therefore, optimizing the D3 process can achieve greater productivity. However, optimizing them separately cannot meet the demand for productivity. By superimposing experiments on different schemes, it can be seen that adding one equipment to C3 and D3 separately can obtain the maximum productivity, but the cost is more than other schemes. Reducing the processing cycle of 20% of C3 and D3 processes respectively can meet the demand, at the same time, the cost is the least. So the combination of M2 and M4 is the optimal scheme.

5   Conclusions

This paper analyzes the operation characteristics of aircraft assembly line, and expounds the key parts of virtual simulation technology in the simulation research of aircraft assembly line. By using a hypothetical case, the simulation model is built, the data of the assembly line and the utilizations of the equipments are obtained, and the bottleneck process is analyzed. Combined with the common optimization methods of aircraft assembly line, optimization schemes of the assembly line are presented. With the constraints considered, the production capacities of different optimization schemes are quickly obtained by using the simulation comparison experiment method, and the optimal scheme is obtained. The simulation analysis method provided in this paper can provide a reference for the rapid optimization of aircraft assembly line and bring great benefits to the improvement of aircraft assembly efficiency.

References

1. Salleh NAM, Kasolang S, Mustakim MA et al (2017) The study on optimization of streamlined process flow based on Delmia quest simulation in an automotive production system. Procedia Comput Sci 105:191–196
2. Halim NHA, Yusuf N, Jaafar R et al (2015) Effective material handling system for JIT automotive production line. Procedia Manuf 2:251–257
3. Salleh NAM, Kasolang S, Jaffar A (2012) Simulation of integrated Total Quality Management (TQM) with Lean Manufacturing (LM) practices in forming process using Delmia quest. Procedia Eng 41:1702–1707
4. Meng Z, Wang H-j (2015) Simulation and optimization of the mixed assembly line based on Flexsim software. Modular Mach Tool Autom Manuf Tech 01:142–145
5. Zhu X, Zhang R, Chu F et al (2014) A Flexsim-based optimization for the operation process of cold-chain logistics distribution centre. J Appl Res Technol 12(2):270–278
6. Qiu Y-j, Tu H-n (2015) Research on simulation and optimization of the mixed-model production line based on Flexsim and genetic algorithm. Modular Mach Tool Autom Manuf Tech 08:119–123
7. Zhang G-h, Zhang L-j (2016) Mixed assembly line balancing and optimization based on Flexsim simulation technology. Modular Mach Tool Autom Manuf Tech (06):131–133+137
8. Lu H, Liu X, Pang W et al (2012) Modeling and simulation of aircraft assembly line based on quest. Adv Mater Res 569:666–669
9. Caggiano A, Marzano A, Teti R (2016) Resource efficient configuration of an aircraft assembly line. Procedia CIRP 41:236–241
10. Du J, He Q, Fan X (2013) Automating generation of the assembly line models in aircraft manufacturing simulation. In: IEEE international symposium on assembly and manufacturing. IEEE, pp 155–159
11. Ziarnetzky T, Biele A (2014) Simulation of low-volume mixed model assembly lines: modeling aspects and case study. In: Simulation Conference. IEEE, pp 2101–2112

Experimental Study on Aerodynamic Performance of Flapping Wing with One-Way Holes/Gaps

Wenqing Yang[1(✉)], Bifeng Song[1], Guanglin Gao[2], and Kun Zhang[3]

[1] School of Aeronautics,
Northwestern Polytechnical University, Xi'an 710072, China
yangwenqing@nwpu.edu.cn
[2] Beijing Institute of Aerospace Technology, Beijing 100074, China
[3] Beijing Institute of Mechanical and Electrical Engineering,
Beijing 100074, China

Abstract. Gaps between feathers can be observed when some birds are flying. They can control the opening and closing of gaps to achieve the goal of good aerodynamic performance. The effects of the wing gaps on aerodynamic performance is investigated by experimental study. Firstly, a kind of flapping wing with one-way holes is designed and tested. The holes are connected by hinge, which is open during the up-stroke process and close during the down-stroke process. One-way holes can increase the lift significantly. However, the hinge connection makes the movement process less continuous. And there's more resistance because the windward area is increased by hinged holes. Further on, the method was improved by using one-way deforming characteristics of bird feathers. The goose feathers are used as a covering skin of flapping wing, which can have continuous opening and closing gaps, making the flapping process more smoothly and getting smooth aerodynamic characteristics. The results show that the one-way holes/gaps flapping wing can effectively increase the lift and reduce the power consumption. The using of natural feathers as flapping wing covering skin is a conducive way to improve the aerodynamic performance.

Keywords: Flapping wing · Experimental study · Aerodynamic performance · One-way · Holes · Gaps

1 Introduction

Natural birds have excellent flying ability [1]. Many researchers have tried to tell why bird can fly and want to adapt these flying mechanism on the manmade air vehicle design [2–4]. During the flight of birds, gaps between the feathers can be observed, as shown in Fig. 1. Inspired by the feather gaps, if we open some holes on the surface of the flapping wing, and control the opening and closing of the holes in the up and down stroking processes, what will be happened? Therefore, the research ideas of this paper are obtained. Opening holes on surface of the wing is mainly based on observation of the wings of birds, by controlling the opening and closing of holes in the wing surface may achieve the goal of improving aerodynamic performance.

© Springer Nature Singapore Pte Ltd. 2019
X. Zhang (Ed.): APISAT 2018, LNEE 459, pp. 1160–1168, 2019.
https://doi.org/10.1007/978-981-13-3305-7_92

Fig. 1. Gaps between feathers on birds wings

One important difference between the flapping wing and the fixed wing flight air vehicles is the aerodynamic force produced by flapping. The flapping induced lift are peculiar to the flapping wing flight vehicle. The factors include the frequency and amplitude of flapping, which describe the movement parameters. Early studies showed that planar flapping wing can generate positive lift in the down-stroke process and generate negative lift in the up-stroke process. There are few research concentrated on the flapping lift. With the development of the flapping wing flight vehicle and the high performance requirements, the existing design methods have been unable to meet the requirements.

The exploration of flapping lift is part of the essential characteristics of flapping air vehicles, which is not only beneficial to reveal the mystery of flapping flight, but also important to improve flight performance. The flapping lift can be obtained by opening one way holes on the flapping surface.

Bird wings are a typical example of flapping lift. First of all, the airfoil of birds' wings has positive camber, with geometrical shape of convex on upper surface and concave on lower surface. Therefore, the vertical direction flapping will produce a net upward flapping lift, and the lift force increased along with the intense of flapping.

Secondly, birds can fold the wing in the process of the upstroke to reduce the negative lift, and to increase the cycle average lift. At the same time it can also reduce the consumption of energy, improve the efficiency of the flight.

Finally, the special structure of a bird's feathers also gives flapping lift. Park et al. [5] found that the permeability of birds' primary flying feathers to air flow in different directions is different. The resistance of the down-stroking stage is about 30% greater than that of the up-stroking stage, resulting in flapping lift. In addition, in the process of the up-stroking process, the slots between primary feathers can let air through, which can reduce the negative lift. And under the down-stroking process, the primary feathers can touch each other and close the gaps between the feathers, increased the positive lift. The method of one way slotted flapping wing is inspired by the characteristics of bird wings, which will be studied by experiments.

2 Experimental Model

As shown in Fig. 2, there are several one way holes on the wing surface, and the air flow is controlled by the opening and closing of the holes. The airflow pressure on the lower surface is large during the down-stroke process, and the holes are pressed to close. The airflow pressure on the upper surface is large during the up-stroke process, and the holes are pressed to open, the airflow can through the wing surface to reduce the pressure and reduce the negative lift.

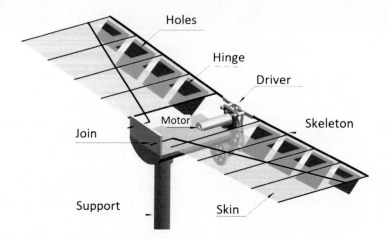

Fig. 2. The experimental model of the wing with one-way holes

3 Experimental Setup

The experiments are carried out in the wind tunnel for MAV, shown in Fig. 3. The main technical parameters are shown in Table 1. The Nano17 series Force/Torque sensor with six components of ATI companies in the United States is mainly used for the test. The experiment used Nano17 SI-50-0.5 sensor, its performance parameters are shown in Table 2.

Fig. 3. Experimental system of wind tunnel

Table 1. Main parameters of experimental wind tunnel

Parameters	Values
Wind tunnel length	6.25 m
Test section size (length, width and height)	0.7 × 0.5 × 0.5 m
Experimental wind speed range	3–21 m/s
Dynamic stability	0.004–0.008
Air turbulence	<0.22%
Mean airflow deflection	<0.2°
Range of incidence	−4 ~ 22°
Angle of attack accuracy	±6′

Table 2. Parameters of Nano17 SI-50-0.5 force/torque sensor

	X,Y force	Z force	X,Y torque	Z torque
Range	±50 N	±70 N	±500 N · mm	±500 N · mm
Resolution	1/160 N	1/130 N	1/64 N · mm	1/64 N · mm
Maximum load	±350 N	±750 N	±2400 N · mm	±3100 N · mm
Natural frequency of vibration	7.2 kHz			
Volume	3.1 cm^3			
Weight	9.4 g			
Strain material	Silicon semiconductor material			

4   Test and Results

The wings with and without holes in the same planform were tested. The angle of attack is 5°, flapping frequency range 3 Hz–7 Hz, wind speed change from 4 m/s to 7 m/s, the force/torque sensor can show instantaneous lift, thrust, lateral force, three-axis torque and instantaneous power consumption, reading sampling rate of 2000 Hz.

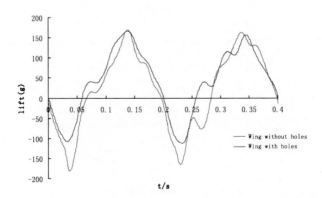

Fig. 4. Instantaneous lift comparison between the wings with and without holes

Figure 4 shows the comparison of instantaneous lift between the wings with and without holes in the flapping frequency of 5 Hz. The lift of wing with holes is bigger than the wing without holes. The peak differ about 50 g, which means the wings with one way holes can generate much bigger lift than the wings without holes.

Figure 5 shows that the lift of wings without holes increase less than the wings with holes with the increase of flapping frequency. For this hinge holes method, when the wind speed is high, the process of opening and closing holes become more difficulty, which makes the lift increase less effective.

Figure 6 shows that the thrust of the wing with holes is generally smaller than that of the wing without holes. One reason is that the opening and closing of the hinge plate generates additional resistance, resulting in the reduction of total thrust. Because the hinge plate used in this experiment is around the leading edge of the wing, the opening of the hinge plate increases the wing's windward area and leads to increased drag. Secondly, the opening of the hinge plate in the up-stroke process unloads the aero-dynamic force of the wing surface, reduces the elastic deformation of the up-stroke process and reduces the thrust.

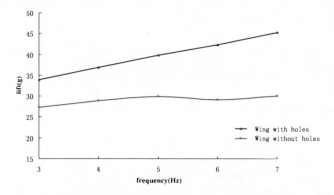

Fig. 5. Lift performance between wings with and without holes

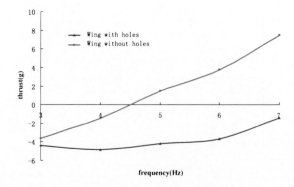

Fig. 6. Thrust performance between wings with and without holes

Figure 7 shows the power consumption between the wings with and without holes. The wings with holes can generate less negative lift during the up-stroke process, so the power to overcome negative lift is less, which reduces the total power consumption, and improve the energy utilization rate.

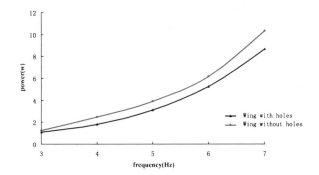

Fig. 7. Power consumption between wings with and without holes

5   Improved Experimental Model

Through the above experiment, it is found that there is a discontinuous problem in the movement of hinge wing during flapping. The hinge opening and closing is rapid, and there is a slight vibration. At the same time, due to the existence of hinge, resistance increased. Due to the low continuity of the opening and closing process of the experimental wings with hinge holes, the aerodynamic change is quite severe. We have improved the experimental of the manner of opening holes.

To take advantages of the one-way deformation of natural feathers. A feathered gapped flapping wing was made, shown in Fig. 8. The aerodynamic characteristics of the feather covered flapping wing were tested. To compare, we test two kinds of flapping wings. One is made of carbon fiber and polyester membrane. The other kind of flapping wing has feather cover, which has same shape and area with the basal wing for comparison. The two kinds of wings have same size, with half span of 26 cm and root chord of 10 cm. The main structure skeleton are also the same compose of a leading spar and a slant spar. The membrane wing has a mass of 6.9 g, and the feather wing has a mass of 7.1 g, they are similar in mass. The order of the feathers on the flapping wing is like the arrangement of the birds' wing. In the two adjacent feathers, the one near the wing tip is below, and the one near the wing root is above.

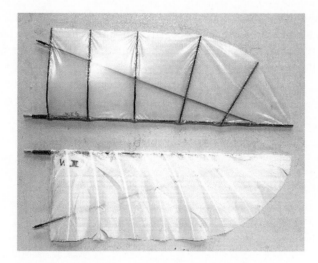

Fig. 8. Feather covered flapping wing and membrane wing in the same planform

6   Results and Discussion

The aerodynamic performance and corresponding power of two kinds of flapping wings are tested. The averaged lift force, thrust force, and consumed power of the two wings varies with the flapping frequency is shown in Fig. 9.

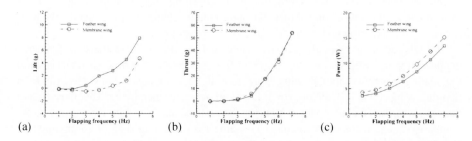

Fig. 9. Averaged forces and power comparison between feather wing and membrane wing

Figure 9 shows that the feather wing produced more lift than the membrane wing especially when the flapping frequency bigger. The thrust generated by two wings are almost the same, which increase with the increase of flapping frequency. At the same time, the consumed power of feather wing is less than that of the membrane wing.

The periodic curve of lift, thrust, and consumed power in a flapping cycle at 7 Hz frequency are shown in Fig. 10. The sample rate of the experimental system is 2000 points per second, so the results appear to be continuous.

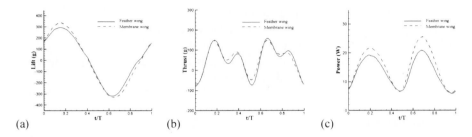

(a)   (b)   (c)

Fig. 10. The periodical forces and power comparison between feather wing and membrane wing

For the cycle lift curve, the lift force of the feather wing is less than the membrane wing during down stroke process, but opposite in the up stroke process. In the whole flapping period, the amount of increase is greater than the decrease, so the total lift increases for the feather wing. The fluctuating range of feather wing is small than that of the membrane wing.

For the thrust curve, the thrust of the feather wing is less than the membrane wing during down stroke process, but opposite in the up stroke process. However, in the whole period, the amount of increase and decrease is almost the same.

For the consumed power curve, the power of the feather wing is less than the membrane wing during the whole flapping cycle, which is mainly because the lift fluctuating range of feather wing is small to reduce the power consumption.

The whole process was filmed with a high-speed camera. For both wings, the apparent deformation can be observed in whole flapping process. The maximum deformation and the amount of time occurs similar. The main difference between the two wings are happen in the up-stroke, there is a noticeable gap happens between each adjacent feathers of feather wings shown in Fig. 11. This behavior significantly reduces the negative lift from the up stroke process, this is also the secret of the feather wing can generate more lift.

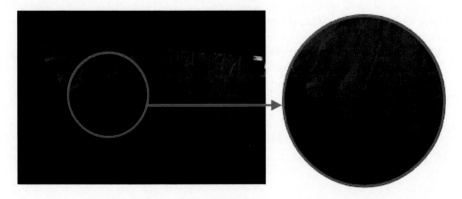

Fig. 11. Clear gap between feathers during up-stroke process

In the photos, the trailing edge deformation of feather wing is a little greater than the membrane wing, because the trailing edge of feather is very thin, its stiffness changes continuous. This feature makes the range of lift fluctuate less, which in one of the things that man-made materials can't do. It is the decrease in lift amplitude that leads to the reduction of consumed power.

7 Conclusions

The aerodynamics of flapping wings with one-way holes/gaps are designed and tested experimentally. In the down-stroke process, the one-way holes/gaps makes the wing as a whole surface. However, in the up-stroke process, there will be holes/gaps appearing, which let some air passed and accordingly decrease the negative lift generated during up-stroke, results in the averaged lift increase in a whole flapping cycle. At the same time, the consumed power also decrease.

According to the results of experiments, the thrust generated by hinge holes is less than the wing without holes. However the thrust generated by feather gapped wing is almost same with the membrane wing. This means the feathered wing have continuous deforming process, which is good for aerodynamic performance.

The wings with holes/gaps can produce a greater lift and consume less power, it means the aircraft can fly more easily and has a longer endurance. Because of the greater lift, the angle of attack can be reduced in cruise stage, and the drag is reduced accordingly, the power consumption is reduced further.

Acknowledgments. This work was supported in part by the National Key Research and Development Program of China under Grant 2017YFB1300102, and was supported by the National Natural Science Foundation of China under grant U1613227.

References

1. Pennycuick CJ (2008) Modelling the flying bird. Academic Press, London
2. Shyy W, Berg M, Ljungqvist D (1999) Flapping and flexible wings for biological and micro air vehicles. Prog Aerosp Sci 35:455–505
3. Tanaka H, Okada H, Shimasue Y, Liu H (2015) Flexible flapping wings with self-organized microwrinkles. Bioinspiration Biomimetics 10(4). https://doi.org/10.1088/1748-3190/10/4/046005
4. Heathcote S, Wang Z, Gursul I (2008) Effect of spanwise flexibility on flapping wing propulsion. J Fluids Struct 24:183–199
5. Park HC, Lee SY, Sang ML et al (2004) Design and demonstration of flapping wing device powered by LIPCA. Smart Struct Mater 5390:212–216

Measurement of Propeller Characteristics at a Negative Advance Ratio Using a Whirling Arm Facility

Yuto Itoh[1(✉)] and Atsushi Satoh[2(✉)]

[1] Graduate School of Engineering,
Iwate University, 4-3-5 Ueda, Morioka, Japan
t5718002@iwate-u.ac.jp
[2] Faculty of Science and Engineering,
Iwate University, 4-3-5 Ueda, Morioka, Japan
satsushi@iwate-u.ac.jp

Abstract. Although wind tunnels are the most popular aerodynamic measurement tool of today, whirling arms are another type of tool which is especially useful for the measurement at very low airspeed, including zero. The authors developed a modern whirling arm facility for the measurement of the characteristics of small-scale propellers. In this work, an experiment to measure the characteristics of APC SF 8x6 propeller at a negative advance ratio (from 0.0 to –0.8) is conducted. The rotation of the arm is controlled by a servo motor to maintain the steady rotational speed (i.e. axial airspeed of the propeller) against the thrust fluctuation of the propeller attached at the end of the arm. The very small standard error and standard deviation of the thrust and torque measurement demonstrate the developed system's ability for precise aerodynamic measurement. In a certain range of advance ratio (from –0.4 to –0.8), remarkable fluctuation of thrust and torque was observed, which suggests the propeller was in a non-steady working condition such as vortex ring state.

Keywords: Whirling arm · Propeller characteristics · Low advance ratio · Vortex ring state

Nomenclature

a	Lift curve slope, rad^{-1}
B	Number of blades
c	Blade chord, m
C_Q	Torque coefficient
C_T	Thrust coefficient
dD	Differential aerodynamic drag generated in blade elements, N
D	Propeller diameter, m
J	Advance ratio
dL	Differential aerodynamic lift generated in blade elements, N
m	Mass of the measurement device, kg
n	Rotational speed, s^{-1}

© Springer Nature Singapore Pte Ltd. 2019
X. Zhang (Ed.): APISAT 2018, LNEE 459, pp. 1169–1188, 2019.
https://doi.org/10.1007/978-981-13-3305-7_93

O Center of rotation
Q Propeller torque, N · m
r Span from axis of rotation, m
R Radius, m
s Standard deviation, N, N · m
T Thrust, N
U Resultant airspeed, m/s
v Induced velocity, m/s
V Airspeed, m/s
α Angle of attack, rad
δ Blade drag coefficient in the cross sectional area of the propeller, rad^{-1}
η Efficiency
θ Pitch angle, rad
λ Airspeed ratio
ρ Air density, kg/m^3
ϕ Inflow angle, rad
ψ Rotational angle, rad
ω Angular velocity, rad/s

Subscript

c Centrifugal parameters
h Parameters in hover
i Parameters induced by the propeller
p Parameters of the propeller
r Representative parameters
s Parameters in the steady state
wa Parameters of the whirling arm

1 Introduction

Most of the small multirotor UAVs (Unmanned Aerial Vehicles) of today are using fixed-pitch propellers for the production of lift (or thrust). Nondimensionalized thrust and torque of the propeller are called "thrust coefficient" and "torque coefficient", respectively. These coefficients heavily depend on the advance ratio which is the parameter to indicate the working state of the propeller.

The advance ratio is defined as a ratio between the axial speed and the rotational speed at the tip of the propeller. Suppose that the axial speed is positive when the rotating propeller is advancing like screw in still air. Then, the "positive advance ratio" is a normal working state and it suggests a vertical climb of a multirotor UAV. On the other hand, "negative advance ratio" suggests vertical descent. In this case, the direction of the airflow flowing into the disc plane of the propeller is opposite to the direction of the propeller downwash flow.

When a propeller is working in negative advance ratio, thrust and torque may fluctuate irregularly, and the stability and controllability of the flight are greatly affected. Furthermore, it is difficult to return to the normal working state, once the propeller falls into a vortex ring state (VRS). Therefore, it is important to measure the propeller characteristics at a negative advance ratio because this data is useful for preventing multirotor UAVs from falling into hazardous conditions in steep descent or descent in an up-current of air.

As far as the authors' knowledge, there is not so much experimental data of small-scale propellers in descent. At present, wind tunnel tests are mainly used for experimental analysis of propeller characteristics at a negative advance ratio [1, 7–9]. However, the speed of the steady airflow created by wind tunnel has a lower limit, and the slower the airflow, the greater the turbulence. Therefore, it is difficult to precisely measure the propeller characteristics at the low advance ratio (in both positive and negative sides) by the wind tunnel testing. Moreover, the aerodynamic characteristics obtained by the wind tunnel experiment need to be corrected in consideration of the tunnel wall, especially when the downwash hits the wall.

To overcome the difficulties described above, the use of a whirling arm is another option. Whirling arm facility was originally devised by Robins [3] to measure the aerodynamic characteristics of objects before the development of wind tunnel. The object to be measured is attached at the end of the arm and the arm is rotated to make a desired airspeed. The aerodynamic force acting on the attached object is measured by the sensors on the rotating arm. In recent researches, the whirling arm facility is applied to measure lateral and directional stability derivatives [2, 5] and analyze the aerodynamic performance of cornering automobiles [4]. In measurement of propeller characteristics, the whirling arm has a clear advantage that it has no limitation on minimum airspeed which can be created. This means steady airflow can be created even when the desired airspeed is very low (i.e. almost zero).

The purpose of this paper is to measure the aerodynamic characteristics of the propeller at negative and low advance ratio, by using a whirling arm facility. The use of a whirling arm enables us to measure the propeller characteristics at very low advance ratio, including zero.

For the measurement of propeller characteristics, a whirling arm whose rotation speed is controlled by a servo motor has been developed. The rotation speed of the test propeller is also controlled by PI feedback controller. Therefore, a developed facility is capable of measuring propeller characteristics for any desired advance ratios, namely, any combination of the rotational speed of the propeller and the arm, including the region of negative advance ratio.

In this paper, propeller characteristics at the advance ratio over a range of 0.0 to −0.80 were measured. The ability of the system for precise aerodynamic measurement was verified by the very small standard deviation in thrust and torque measurement over a range of advance ratio from 0.0 to −0.40.

In Sect. 2.1, the whirling arm facility for the measurement of propeller characteristics is introduced. In Sect. 2.2, the detail of the measurement using whirling arm is explained. The data correction method is discussed in Sect. 2.3. The measurement was carried out for the advance ratio from approximately 0 to −0.80. The measurement results and statistical evaluation are presented in Sects. 3.2 and 3.3.

2   Measurement Methodology

2.1   Whirling Arm Facility

The whirling arm facility developed by the authors is shown in Figs. 1, 2 and 3. The primary specification of this equipment is shown in Table 1. At the end of the whirling arm, force/moment sensor (WACOH-TECH WEF-6A200-4-RC5) is attached to measure the torque and thrust produced by a test propeller. The test propeller and a brushless motor are supported by a motor supporting rod and the rod is attached to the force/moment sensor. Note that the arm is not perpendicular to the supporting rod (see Fig. 3) so that the direction of the relative wind produced by the whirling arm at test propeller is orthogonal to the propeller rotating plane.

The arm is supported by a main shaft and rotates in the horizontal plane. The rotation of the arm is controlled by a 200 W servo motor (Oriental Motor BXS6200AM motor) equipped in the base of the facility. Therefore the steady rotational speed of the arm is maintained no matter how the thrust of the test propeller is changed by throttle setting or unsteady aerodynamics.

The brushless motor to drive the test propeller is powered by a lithium polymer battery (rated voltage is 11.1 V). To prevent the change of rotational speed of the test propeller caused by battery voltage drop or change of propeller working condition, PI controller is implemented on the mbed microcontroller on the table of the arm. For the measurement of the rotational speed required for feedback control, Melexis Technology US1881LSE magnetic sensor and the permanent magnets attached on the motor body are used.

Fig. 1. The whirling arm facility

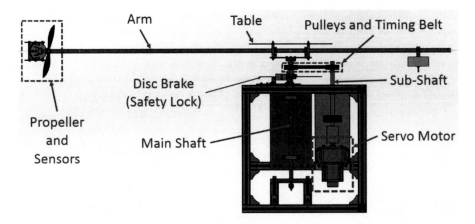

Fig. 2. The side-view of the whiling arm facility

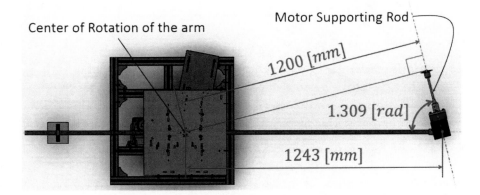

Fig. 3. The top-view of the whirling arm facility

Table 1. The primary specification of the whirling arm facility

Item	Value, unit
Length of the whirling arm	1.243, m
Turning radius of the propeller	1.200, m
Angle between the whirling arm and thruster's rod	1.309, rad

Figure 4 shows the control and measurement system of the whirling arm facility. The desired rotational speed of the arm is controlled by the motor driver set on the ground (Oriental motor BXSD200) and measured by an attached rotary encoder (Mutoh UN100). The torque and thrust measurement data from the force/moment sensor are received by a battery-powered small PC on the table on the arm ("Data collection PC" in Fig. 4). The power consumption of the brushless motor driving the test propeller is measured by a current/voltage sensor (Texas Instruments INA226) connected at the battery side of the ESC (Electric Speed Controller) and the measurement data is also received by "Data collection PC". Using a remote desktop application, "Data collection PC" is remotely controlled from ground PC ("Command transmission PC" in Fig. 4) via WiFi network. The desired rotational speed of the test propeller can be set from "Data collection PC" using remote desktop.

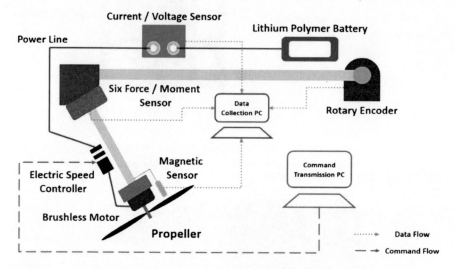

Fig. 4. Measurement system of the whirling arm facility

During the measurement in negative advance ratio, the arm is forced to rotate against the torque generated by the thrust of the test propeller, by a servo motor. The advance ratio of the propeller can be arbitrarily set for the measurement because the rotational speed of the arm and the propeller can be set individually. Moreover, the rotational speed of the arm is maintained by the servo motor mentioned above, even when the thrust fluctuation of the propeller generated by a non-steady working condition, such as the vortex ring state, exists.

2.2   Propeller Characteristics Model

Propeller characteristics are derived from the measured airspeed, thrust, and torque of the propeller. Note that the airspeed produced by the rotation of the arm varies with respect to the radial distance from the rotary shaft. Therefore, the thrust and the torque generated by a propeller blade also vary with respect to rotary position of the blade because the angle of attack of the blade periodically varies as the propeller rotates.

In this section, the aerodynamic characteristics of propellers and the representative models used to analyze propeller characteristics are formulated.

Radial Distribution of Airspeed on the Rotor Disc

Figure 5 shows the whirling arm facility is in operation. The parameters related to the rotation of the arm are also shown in Fig. 5. Let us assume ω_{wa} and ω_p are constant throughout this paper.

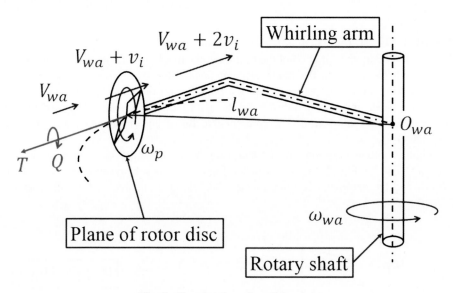

Fig. 5. Rotational motion of the arm

As shown in Fig. 6, the airspeed produced by the rotation of the whirling arm depends on r_{wa}. V_{wa} and the airspeed linearly changes with respect to r_{wa} on a propeller disc.

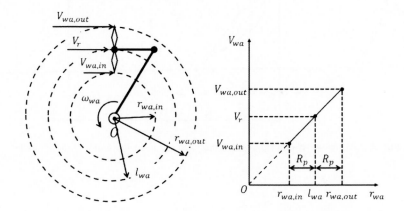

Fig. 6. Relation between turning radius of propeller and airspeed

Let us define the representative airspeed, V_r as the averaged V_{wa} (distributed air-speed over the propeller disc) in Eq. (1). We can see V_r is equal to the airspeed at the center of the propeller by calculation.

$$V_r = \frac{1}{r_{wa,out} - r_{wa,in}} \int_{r_{wa,in}}^{r_{wa,out}} \left(\frac{1}{2\pi} \int_0^{2\pi} V_{wa} d\psi_p \right) dr_{wa} = l_{wa}\omega_{wa}. \tag{1}$$

Propeller Thrust and Torque

In this subsection, the variation of the thrust and the torque of a rotating single blade are formulated, by applying the blade element theory. From this result, representative thrust and the representative torque for the propeller on the whirling arm are derived.

In this paper, the averaged thrust and the averaged torque over a single rotation of the propeller, are adopted as the representative thrust and the representative torque, respectively.

Figure 7 shows the geometrical relationship between the axis of rotation of the arm ("Rotary shaft" in Fig. 7) and the propeller. Figure 8 shows the resultant velocity of the airflow and the aerodynamic force of the blade element at a radius from the rotation axis of r_p.

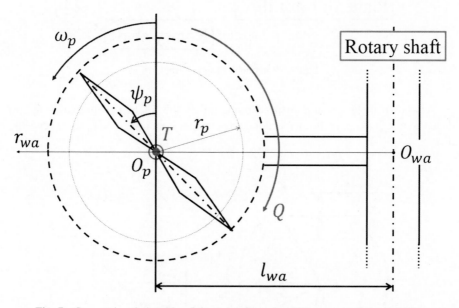

Fig. 7. Geometric relationship of the axis of rotation of the arm and the propeller

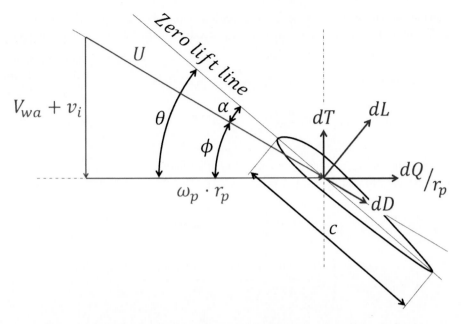

Fig. 8. Velocity components entering the blade element and aerodynamic forces

When the rotary position of a blade is at $\psi_p = 0$ or π, the steady thrust T_s and the steady torque Q_s of the blade are expressed respectively by Eqs. (2) and (3),

$$T_s = \frac{1}{2}\rho a \bar{c} R_p^2 \cdot \omega_p \left(\frac{\bar{\theta}\omega_p R_p}{3} - \frac{l_{wa}\omega_{wa} + v_i}{2} \right), \tag{2}$$

$$Q_s = \frac{1}{2}\rho \bar{c} R_p^2 \cdot \left\{ \frac{2\bar{\delta}\omega_p^2 - a\omega_{wa}^2}{8} R_p^2 + a(l_{wa}\omega_{wa} + v_i)\left(\frac{\bar{\theta}\omega_p R_p}{3} - \frac{l_{wa}\omega_{wa} + v_i}{2} \right) \right\}. \tag{3}$$

In Eqs. (2) and (3), $\bar{c}$ is the average chord length of a blade, $\bar{\delta}$ is the average blade drag coefficient, and a is the lift curve slope of the blade element, namely,

$$a = \frac{\partial C_l}{\partial \alpha}. \tag{4}$$

a is assumed to be constant.

Let $\bar{\theta}$ be an averaged pitch angle, α is expressed by the following Eq. (5),

$$\alpha = \bar{\theta} - \phi \simeq \bar{\theta} - \frac{\left(l_{wa} + r_p \sin \psi_p \right)\omega_{wa}}{r_p \omega_p}. \tag{5}$$

Moreover, the periodic component of thrust and torque (T_v and Q_v) are expressed as the function of ψ_p, if we ignore the unsteady aerodynamics of the blade for simplicity.

$$T_v = -\frac{1}{6}\rho a \bar{c} R_p^3 \cdot \omega_{wa}\omega_p \sin\psi_p, \tag{6}$$

$$Q_v = \frac{1}{2}\rho a \bar{c} R_p^3 \cdot \omega_{wa}\left\{\frac{2\bar{\theta}\omega_p \sin\psi_p + \omega_{wa}\cos 2\psi_p}{8}R_p - \frac{2}{3}(l_{wa}\omega_{wa} + v_i)\sin\psi_p\right\}. \tag{7}$$

Accordingly, the total thrust T and the total torque Q are expressed by the sum of equations (2), (6) and the sum of equations (3), (7), and shown as Eqs. (8) and (9),

$$T\left(\psi_p\right) = \frac{1}{2}\rho a \bar{c} R_p^2 \cdot \omega_p\left\{\frac{R_p}{3}\left(\bar{\theta}\omega_p - \omega_{wa}\sin\psi_p\right) - \frac{l_{wa}\omega_{wa} + v_i}{2}\right\}, \tag{8}$$

$$Q\left(\psi_p\right) = \frac{\rho \bar{c} R_p^2}{2} \cdot \left\{\frac{2\bar{\delta}\omega_p^2 - a\omega_{wa}^2 + a\omega_{wa}\left(2\bar{\theta}\omega_p \sin\psi_p + \omega_{wa}\cos 2\psi_p\right)}{8}R_p^2 \right.$$
$$\left. + a(l_{wa}\omega_{wa} + v_i)\left(\frac{\bar{\theta}\omega_p - 2\omega_{wa}\sin\psi_p}{3}R_p - \frac{l_{wa}\omega_{wa} + v_i}{2}\right)\right\}. \tag{9}$$

Therefore, let us define the representative thrust T_r and the representative torque Q_r as the average of $T\left(\psi_p\right)$ and $Q\left(\psi_p\right)$ over a single rotation. These are equivalent to thrust T_s and torque Q_s.

$$T_r = \frac{1}{2\pi}\int_0^{2\pi} T\left(\psi_p\right)d\psi_p = T_s \tag{10}$$

$$Q_r = \frac{1}{2\pi}\int_0^{2\pi} Q\left(\psi_p\right)d\psi_p = Q_s \tag{11}$$

Propeller Characteristics
By using representative values, $V_r, T_r,$ and Q_r, the propeller characteristics for the propeller attached to the whirling arm are defined as follows,

$$J = \frac{V_r}{n_p \cdot D_p}, \tag{12}$$

$$C_T = \frac{B \cdot T_r}{\rho \cdot n_p^2 \cdot D_p^4}, \tag{13}$$

$$C_Q = \frac{B \cdot Q_r}{\rho \cdot n_p^2 \cdot D_p^5}, \tag{14}$$

$$\eta = \frac{1}{2\pi} \cdot J \cdot \frac{C_T}{C_Q}. \tag{15}$$

Induced Velocity Characteristic [6]

Analysis of the induced velocity characteristics is one way to evaluate the working state of the propeller. This evaluation is useful to know the working state of the propeller which varies with respect to the induced velocity. The induced velocity characteristic is expressed by the non-dimensional induced velocity which is a function of the non-dimensional descent velocity.

The induced velocity characteristics of the propeller in descent flight are analyzed by λ_d and λ_i shown in Eqs. (16) and (17),

$$\lambda_d = \frac{V_r}{v_h}, \tag{16}$$

$$\lambda_i = \frac{-V_r + \sqrt{V_r^2 + \frac{2T}{\pi \rho R_p^2}}}{v_h}. \tag{17}$$

Here v_h is expressed by Eq. (18). In Eq. (18), T_h is the static thrust during hovering and is equivalent to the thrust when the progress rate is zero.

$$v_h = \sqrt{\frac{T_h}{2\pi \rho R_p^2}} \tag{18}$$

From the simple momentum theory [6] and the measurement results of Wayne [10], Shetty [7] and others, it is known that the value of λ_i rises to more than about twice the hover value (when $\lambda_d = 0$, $\lambda_i = 1$) in the vortex ring state, then falls steeply to about the hover value at entry to the windmill brake state.

The induced velocity characteristics of the propeller are calculated from the above three equations, by using the parameters set in the measurement experiment and the measured thrust.

2.3 Measurement Correction

Influence of the Rotation of the Arm

When the arm is rotating, the air around the arm rotates at a constant angular velocity with the rigid motion of the arm, thereby forming a swirling flow around the rotation axis of the arm. The swirling flow can be regarded as a compound vortex, consisting of a forced vortex caused by the rotation of the arm and a free vortex caused by the potential flow outside the arm's circle. Since the area in which the propeller performs exists inside the arm's circle, the swirling flow is dominant for the forced vortex.

In this work, it is assumed that the swirl does not significantly affect the measurement because the arm rotates at a low speed. Therefore, circumferential velocity and pressure distribution can be ignored by adopting airspeed at the rotational center of the propeller.

Correction for Centrifugal Effects
When the arm is rotating, the centrifugal force and the torque due to the rotation of the arm act on the end of the arm, which influences the measurement of the thrust and the torque of the test propeller. Therefore, these centrifugal effects on the measurement data are eliminated using the calculated centrifugal force from the arm geometry.

Figure 9 shows the simplified geometry of the arm. Assume that the arm is rigid and the component parts of the arm are symmetrical with respect to the rotating surface of the arm. Let the position of the force/moment sensor, the end of the arm, and the center of gravity of the measuring device be S, A and C, respectively.

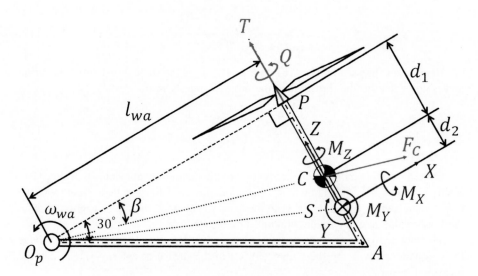

Fig. 9. Geometric model of the arm and centrifugal force component

Then, the centrifugal force F_c which acts on C (the center of gravity of the measuring device) is expressed as the Eq. (19),

$$F_c = m\|\omega_{wa}\|^2 \cdot l_{wa} \sec \beta. \tag{19}$$

In Eq. (19), β is expressed by l_{wa} and d_1, as the Eq. (20),

$$\beta = \tan^{-1}\left(\frac{d_1}{l_{wa}}\right). \tag{20}$$

On the other hand, F_c can be decomposed into the X axis component $F_{c,X}$ and the Z axis component $F_{c,Z}$ and

$$F_{c,X} = \|\boldsymbol{F_c}\| \cdot \cos\beta = ml_{wa} \cdot \|\boldsymbol{\omega_{wa}}\|^2, \tag{21}$$

$$F_{c,Z} = \|\boldsymbol{F_c}\| \cdot (-\sin\beta) = -ml_{wa} \cdot \|\omega_{wa}^2\| \cdot \tan\beta. \tag{22}$$

Then, moments $M_{c,Y}$ and $M_{c,Z}$, due to the centrifugal force, are expressed as

$$M_{c,Y} = ml_{wa} \cdot \|\boldsymbol{\omega_{wa}}\|^2 \cdot d_2, M_{c,Z} = 0. \tag{23}$$

The parameters required for calculating centrifugal components are shown in Table 2.

Table 2. Geometric parameters of a whirling arm

Parameter	Value, unit
Length l_{wa}	1.200, m
Mass of the measuring device m	1.200, kg
Distance d_1	0.237, m
Distance d_2	0.014, m
Angle β	0.212, rad

2.4   Measurement Conditions

In the measurement experiment performed in this work, n_p have been set to the following values,

$$n_p = 46, 60, 75, 89.$$

For each n_p, ω_{wa} are appropriately set to achieve these advance ratios.

$$J = -0.8, -0.6, -0.5, -0.4, -0.2, -0.1, 0.$$

2.5   Data Processing

The measured thrust and the torque become oscillatory due to the centrifugal force and the natural vibration of a whirling arm facility. In order to analyze propeller characteristics, it is essential to remove these oscillatory components from the obtained data.

To overcome this problem, an appropriate time interval in which the arm angular velocity is steady have been extracted. The low-pass filter designed with the natural frequency of a facility also has been applied to these extracted data.

3   Result and Discussion

3.1   Experiment Environment

The electric propulsion system to be measured consists of components shown in Table 3.

Table 3. Components of the electric propulsion system

Component	Parts' Name
Propeller	APC Slow Flyer 8x6SF
Brushless DC motor	Hacker A30-24M-UAV
Electric speed controller	Hacker JETI model HI Copter 40A Opto

Measurement experiments have been performed at the Iwate university gymnasium 1. Since a whirling arm facility has been put in a sufficiently large space, there is no airflow from the outside, and aerodynamic interference with the wall surface and other structures.

3.2   Measured Propeller Characteristics

Figures 10, 11, 12, and 13 show measured propeller characteristics. The vertical error bars show the standard error at each measurement point.

In the range where J is more than -0.40, C_T increase gradually or remain almost constant, as J decreases gradually. In the same range, C_Q becomes almost constant in all conditions. Therefore, η gently decreases as J decreases.

In the range where J is less than -0.50, thrust characteristics and torque characteristics drop sharply. It can be considered that thrust becomes negative in lower advance ratio because the angle of attack becomes less than 0 and the direction of lift is reversed. When C_Q are in the vicinity of 0, the efficiency sharply drops, and the standard error becomes very large. Moreover, when C_Q is less than 0, it is considered that the propeller is operating in the area of windmill and torque occurs in the direction of rotation of the rotor. On the other hand, in the region where J is less than -0.70, the efficiency decrease sharply and may be greater than 0 in some cases, with the large standard error. Therefore obtained characteristic values in such a condition have accuracy required to be represented.

In both thrust characteristics and torque characteristics, since standard errors at most measurement points are sufficiently small, obtained characteristics have a high precision.

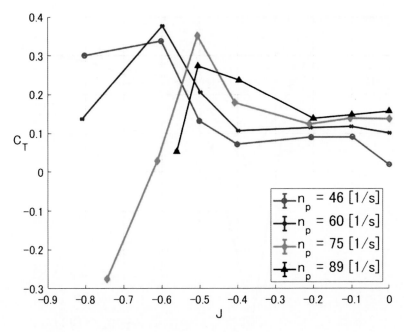

Fig. 10. The thrust characteristic curve related to the advance ratio

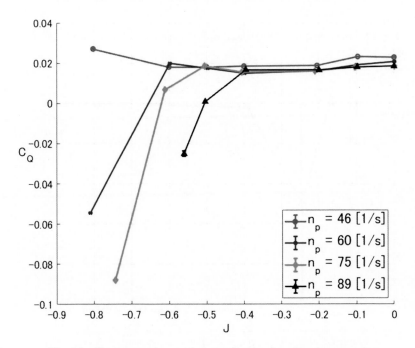

Fig. 11. The torque characteristic curve related to the advance ratio

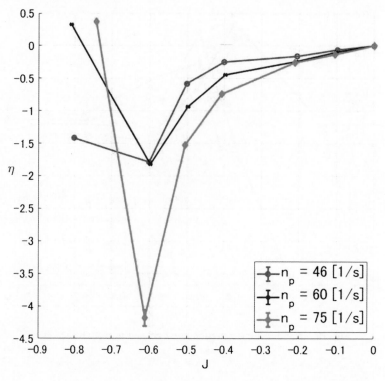

Fig. 12. The efficiency curve related to the advance ratio ($n_p = 46, 60, 75[1/s]$)

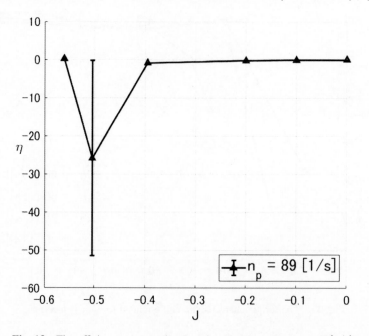

Fig. 13. The efficiency curve related to the advance ratio ($n_p = 89[1/s]$)

Figure 14 shows the induced velocity characteristics calculated from the measured data. The induced velocity characteristic at the lowest rotational speed greatly increases as the descend velocity increases, compared with other characteristic curves. The other three induced velocity characteristics gently increase to approximate 3.0 of λ_i.

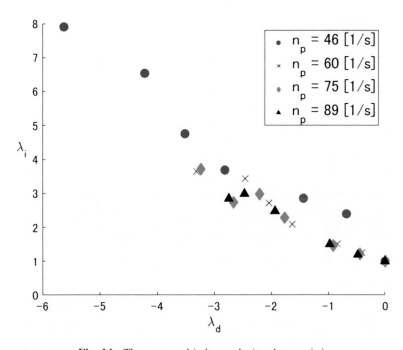

Fig. 14. The measured induce velocity characteristic

3.3   Statistical Evaluation

Figure 15 shows the standard deviations of thrust data s_T for each J. Figures 16 and 17 show the standard deviations of torque data s_Q for each J. The magnitude of s_T tends to increase as J decreases. Moreover, the magnitude of s_Q becomes higher at the condition where C_Q becomes less than 0.

In some operating state where the advance ratio is less than -0.5 and C_Q is less than 0, the aerodynamic forces produced by the propeller become non-steady. At the same time, it can be said that these large fluctuations suggest the propeller is in the non-steady working condition such as vortex ring state.

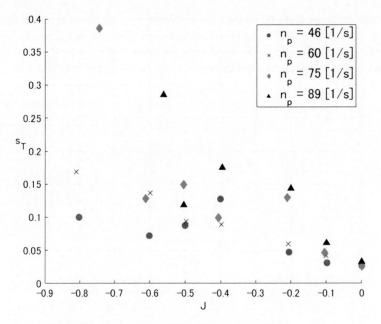

Fig. 15. Standard deviations about measured thrust data

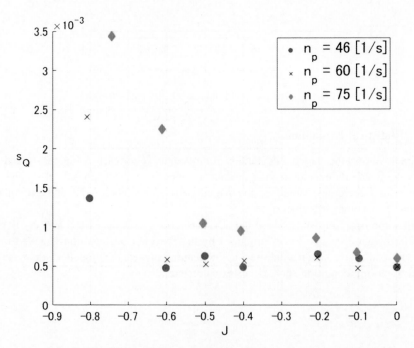

Fig. 16. Standard deviations about measured torque data ($n_p = 46, 60, 75[1/s]$)

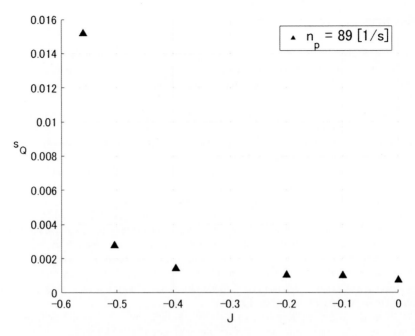

Fig. 17. Standard deviations about measured torque data ($n_p = 89[1/s]$)

4 Conclusion

In this paper, propeller characteristics of a small-scale propeller at the negative advance ratio were measured, by using a whirling arm facility. The very small standard error and standard deviation of the measurement data (i.e. thrust and torque) were observed in the range of 0.0 to −0.4. This result proves the ability of the developed whirling arm to maintain the very low and steady airspeed, which is important to measure the aerodynamic characteristics of the propeller. In the range of −0.4 to −0.8, a relatively large standard deviation of the data was observed. These large fluctuations suggest the propeller is in the non-steady working condition such as vortex ring state.

The standard deviations of thrust characteristics and torque characteristics increase as the advance ratio decreases. Also, When the thrust and torque characteristics drop sharply, their standard deviation increase greatly compared with other measurement conditions.

In the future work, the authors will apply propeller characteristics measured by a whirling arm to the modeling of multirotor UAVs' flight dynamics.

Acknowledgement. The authors would like to thank Mamoru Kikuchi for the technical assistance in developing the whirling arm facility. The authors also thank the technical advisory group of Japan Aerospace Technology Foundation (JAST) for the assistance in measuring propeller characteristics.

This work is partially supported by the discretionary budget of the Dean of the Faculty of Science and Engineering, Iwate University.

References

1. Betzina M (2001) Tiltrotor descent aerodynamics: small-scale experimental investigation of vortex ring state. In: Proceedings of the 57th annual forum of the American Helicopter Society, 9–11 May 2001, Washington, DC, Fairfax, American Helicopter Society International
2. Gill AP, Battipede M (2001) Experimental validation of the wing dihedral effect using a whirling arm equipment. J Aircr. https://doi.org/10.2514/2.2874
3. Gina H (2013) Modeling ships and space craft: the science and art of mastering the oceans and sky. Springer, New York, pp 160–163
4. Keogh J, Barber T, Diasinos S, Doig G (2015) Techniques for aerodynamic analysis of cornering vehicles. SAE technical paper. https://doi.org/10.4271/2015-01-0022
5. Mulkens JMM, Ormerod OO (1993) Measurements of aerodynamics rotary stability derivatives using a whirling arm facility. J Aircr. https://doi.org/10.2514/3.48263
6. Seddon J, Newman N (2011) Rotor in vertical flight: momentum theory and wake analysis. In: Seddon J (ed) Basic helicopter aerodynamics, 3rd edn. Wiley, New York, pp 23–61
7. Shetty RO, Selig SM (2011) Small-scale propellers operating in the vortex ring state. In: Proceedings of the 49th American institute of aeronautics and astronautics aerospace sciences meeting, 4–7 January 2011, Orlando, USA, Sunrise Valley Drive, American Institute of Aeronautics and Astronautics
8. Washizu K, Azuma A, Koo J, Oka T (1966) Experiments on a model helicopter rotor operating in the vortex ring state. J Aircr. https://doi.org/10.2514/3.43729
9. Yaggy P, Mort K (1963) Wind-tunnel tests of two VTOL propellers in descent. NACA TN D-1766
10. Wayne J (2005) Model for vortex ring state influence on rotorcraft flight dynamics. NASA/TP-2005-213477

Evaluating the Combat Effectiveness
of Anti-ship Missile in Cooperative Operation

Qijia Yun[1(✉)], Bifeng Song[1], Huayu Gao[2], Chaojie Liang[2],
and Yang Pei[1]

[1] School of Aeronautics, Northwestern Polytechnical University,
Xi'an 710072, China
605105764@qq.com
[2] China Academy of Launch Vehicle Technology, Beijing 100076, China

Abstract. An analysis scheme and a physics-based mission system model are
applied to evaluate the parameter influence on combat effectiveness of anti-ship
missiles in cooperative operations. The cooperative operation is divided into 3
steps: cooperative penetration, cooperative detection and cooperative strike. The
technical models which include radar detection model, hit probability model and
vulnerability model are adapted to a suitable level of detail to combat effectiveness evaluation. The synthesis evaluation method of cooperative combat
effectiveness is analyzed according to operation steps. The case studies are
specifically applied to the problem of linking the design parameters to mission
effectiveness. The simulation results indicate that increasing the number of anti-
ship missiles is the major way of improving combat effectiveness especially in
cooperative operations. Besides, increasing of speed, armor thickness and
maneuver overload as well as decreasing the RCS of anti-ship missiles are also
effective approaches to improve combat effectiveness. Although the models are
at basic level, the complexity of the cooperative operation is demonstrated. The
mission system model applied is feasible for assessing the combat effectiveness
of cooperative operations.

Keywords: Cooperative operation · Combat effectiveness · Anti-ship missile ·
Detection · Vulnerability

1 Introduction

The design of an anti-ship missile is categorized into three phases: conceptual design,
preliminary design and detailed design [1]. The conceptual design phase focuses on the
determination of the quantifiable design goals which considers the customer requirements. The preliminary design phase is used to analyze the design space and select the
most promising design scheme. The detailed design phase then analyze and optimize
every parameter of the selected design scheme in iteration. The anti-ship missile design
process has experienced a great development during the last decades due to the blos-
som of computing power and computer-aided design software. However, although 85%
of the life-cycle costs are determined in conceptual design phase, the conceptual design
phase did not progress that much like the last two phases [2].

© Springer Nature Singapore Pte Ltd. 2019
X. Zhang (Ed.): APISAT 2018, LNEE 459, pp. 1189–1201, 2019.
https://doi.org/10.1007/978-981-13-3305-7_94

In the past, the design requirements are determined by reference to similar missile capabilities of foreign countries. This method can achieve usable missile designs, but far from well. Designers must understand how these design requirements are determined. While the anti-ship missile is designed to be used in combat field, the design requirements should obviously be determined according to the operations to be performed. Therefore, the combat effectiveness of these operations needs to be considered during the design parameter selection and evaluation [3].

To maximize the combat effectiveness of an anti-ship missile, it is vital to balance the combat capability, susceptibility and vulnerability through-out the design process [4]. But these attributes are always limited by things such as speed, radar cross section signature level, operating environment, material performance, technical maturity etc., makes the balancing process extremely difficult. Enhancing one capability may lead to penalties for others. Extensive researches on combat effectiveness evaluation have been carried out, among which the typical works include ADC (Availability, Dependability, Capability) method [5], SEA(System Effectiveness Analysis) method [6], Lanchester equation method [7], index method [8], fuzzy evaluation method [9], information entropy method [10] etc. Most of these researches are focused on the combat effectiveness of single missile in independently combat situation, little research addresses the cooperation of missiles.

However, war has transformed from linear wars where battles are fight between weapon platforms independently to multidimensional wars where battles are fight between integrative joint operation systems which is composed of high-tech weapons. Take the aircraft carrier strike group as an example, it has a multilayer defense system which can intercept any target entering its defense area with waves of air-defense missiles. Considering the complicated battle field and crucial survival environment of future warfare, anti-ship missiles could no longer fight on their own, they must cooperate tightly as a team to win the battle [11, 12].

In this study, we intend to improve anti-ship missile design requirements by introducing combat effectiveness analysis of cooperative operations into the early design phases.

The rest of this paper is structured as follows. Section 2 introduces the concept of cooperative operation of anti-ship missile. Section 3 proposes the technical models to evaluate the combat effectiveness of anti-ship missile in cooperative engagement. Section 4 describes a quantitative model which is used in the synthesis and measures of effectiveness. Section 5 presents two case studies based on real scenarios. The paper ends with a conclusion in Sect. 6.

2   Cooperative Combat Method of Anti-ship Missile

Because of the extremely elaborate defense system of modern aircraft carrier strike group, anti-ship missiles need to cooperate with each other to maximize the utilization of different missiles. An understanding of how anti-ship missiles fight cooperatively is required to understand how to calculate the combat effectiveness according to real-world operations. The cooperative operation method consists of 3 parts: cooperative penetration, cooperative detection and cooperative strike.

2.1 Cooperative Penetration

The defense system structure of modern aircraft carrier strike group is multiple layers, which is made up of aircraft and sea-based air-defense missiles of different ranges. The radius of their defense area could be as far as 400 km. It is a serious problem to get through this defense area. The typical methods used to increase the penetration probability of anti-ship missile are divided into two parts: on one hand, anti-ship missiles can reduce their RCS or through sea-skimming flight to delay the time of being detected; on the other hand, anti-ship missiles can increase their armor thickness or improve their maneuverability to avoid being killed. But these methods are either too expensive or have penalties to other features. Cooperative penetration is an efficient way to increase the penetration probability of anti-ship missiles.

One typical cooperative penetration method is to fire missiles that cooperate in both time dimension and spatial dimension which exceed the enemy's intercepting capability. If the defense system can't intercept all the anti-ship missiles in the short time window, part of the anti-ship missiles will penetrate the defense area without being intercepted. Another cooperative penetration method is formation flight that can fly a relatively safe route by cooperative control and guidance [13–15].

2.2 Cooperative Detection

Due to the weight and volume limit of the anti-ship missile, it can't afford to take high power radar to search the target. Though radar system is becoming much smaller thanks to technology development, the increase of detection capability of the anti-ship missile is still limited. Cooperative detection is a good way to solve this problem. The anti-ship missiles can detect a target from different angles and share the detected target information to increase the detection probability through data fusion. Another cooperative detection method is called distributed search. If the targets spread in a large area, the anti-ship missiles can work together to divide the area into small parts and search each part with one anti-ship missile to increase the search efficiency. Cooperative detection can decrease the cost through equipping part of the missiles with high power radars and the rest with simple guidance receiver devices [16].

2.3 Cooperative Strike

The ultimate purpose of an operation is to destroy the enemy with minimum cost, to achieve this goal, we will not allow to waste an anti-ship missile on a depot ship instead of an aircraft carrier. To make the most of every anti-ship missile, an allocation scheme needs to be made so that the high value targets are assigned with more missiles. In this way, the total combat effectiveness is increased because more high value targets are destroyed. In addition, cooperative control and guidance can increase the precision of the anti-ship missile. Cooperative target assignment depends on high resolution radar system to recognize warship types and robust data link to exchange target information. If the radar resolution is high enough, the anti-ship missile will be able to recognize and strike the critical components of the warship precisely. So that the combat effectiveness is increased dramatically [17].

3 Technical Models

In this paper, we propose a new methodology for cooperative combat effectiveness analysis of anti-ship missile by combat simulation. The technical models used in the combat scenario are deliberately kept at basic level. The aim of this paper is to demonstrate the methodology's feasibility, not to optimize the performance of anti-ship missile through high resolution simulation. Following assumptions are made for simplifying the calculation:

(1) The radar model only considers the RCS and range of the target, no electronic countermeasures are considered.
(2) The RCS of the anti-ship missile is angle-independent values.
(3) Anti-ship missile and intercept missile has constant speeds.
(4) The range of intercept missiles is enough to reach anti-ship missiles detected by warship.
(5) The speed of the ship is negligible with respect to the speed of missiles.
(6) The anti-ship missile is detected and locked the minute it enters the detection area of the warship.
(7) The fragments of the intercept missile all have the same speed, mass, volume and the hit angle are all vertical to the anti-ship missile surface.

3.1 Missile Detection Model

Though anti-ship missiles have experienced great development in the last few decades, its guidance system is still radar seeker since GPS is not suitable for tracking moving targets and the detection range of IR seeker is limited.

Because of the type of warship RCS signature is Swerling III, IV [18], the detection probability of the anti-ship missile radar is:

$$P_d = \left[1 + \frac{2S_{Nm}}{(2 + S_{Nm})^2} \ln\left(P_f^{-1}\right) \right] P_f^{\frac{2}{2 + S_{Nm}}} \tag{1}$$

Where, P_{fa} is the false alarm probability of the anti-ship missile, S_{Nm} is the signal to noise ratio of the anti-ship missile.

The signal to noise ratio of a target with an RCS of σ_t at a distance of R_t can be obtained through the following equation:

$$S_N = \frac{\sigma_t R_0^4}{\sigma_0 R_t^4} S_{N0} \tag{2}$$

Where, σ_0 is the RCS of the reference target, R_0 is the distance of the reference target, S_{N0} is the signal to noise ratio of the reference target, R_t is the distance to target of the anti-ship missile when it turns on its radar.

3.2   Warship Detection Model

Because of the earth curvature and the electromagnetic wave of the warship radar travels in straight lines, the warship radar can't see anti-ship missiles below the sea level. So that a sea-skimming flight anti-ship missile is very easy to sneak attack a warship. To solve this problem, modern air craft carrier strike groups are equipped with AWACS (Airborne Warning and Control System) in most situations. AWACS can detect targets far beyond the warship's optical range and guide the intercept missiles to engage the anti-ship missiles. For the convenience of calculation, it is assumed that the AWACS patrols just above the aircraft carrier.

To an AWACS radar with a minimum signal to noise ratio $S_{N_{min}}$, if the signal to noise ratio of a given reference target with a RCS of σ_0 in a distance of R_0 is S_{N0}, then the maximum radar detection range of a target with a RCS of σ_t could be calculated with the following function:

$$R_F = \left(\frac{\sigma_t}{\sigma_0} \frac{S_{N0}}{S_{N_{min}}} R_0^4 \right)^{1/4} \tag{3}$$

3.3   Hit Probability Model

To a missile with a miss distance of σ_m, the probability of hitting a target with an exposure area of A_P is:

$$P_H = \frac{A_p}{2\pi\sigma_{miss}^2 + A_p} \tag{4}$$

Assume the miss distance σ_{miss} is influenced by the maneuverability of its target. The maneuverability of warship is negligible. The miss distance of intercept missile [19] can be obtained as follows:

$$\sigma_{miss} = \sqrt{\sigma_s^2 + \sigma_m^2} \tag{5}$$

$$\sigma_m = \left(\phi_{MVR} K_M \tau^5 \right)^{1/2} \tag{6}$$

Where, σ_s is the miss distance of the missile control & guidance system error in meters, σ_m is the maneuver induced miss distance, ϕ_{MVR} is the power spectral density of target maneuver noise, K_M is the adjoint coefficient which vary with the effective navigation ratio N, τ is the sampling time.

The power spectral density of target maneuver noise ϕ_{MVR} can be obtained by:

$$\phi_{MVR} = N_T^2 / T_M \tag{7}$$

Where, N_T is the target maneuver overload, T_M is the target maneuver time.

3.4   Vulnerability Model of Anti-ship Missile

Anti-ship missile has a simple and compact structure, critical components such as seeker, warhead and engine fill its inner space. The anti-ship missile will be considered destroyed if its shell is damaged. Assume the damage mode of the intercept missile is fragment damage that explodes nearby the anti-ship missile and destroy the anti-ship missile with fragments.

The probability of the anti-ship missile being killed by one piece of fragment [20] is:

$$P_0 = \begin{cases} 1 + 2.65e^{-0.347 \times 10^{-8}e_b} - 2.96e^{-0.143 \times 10^{-8}e_b}, e_b > 4.5 \times 10^8 \\ 0, e_b < 4.5 \times 10^8 \end{cases} \tag{8}$$

Where the kinetic energy received per unit volume e_b is:

$$e_b = \frac{m_s v_d^2}{2Sh_d} \tag{9}$$

Where, m_s is the mass of the fragment, S is the area of the fragment, h_d is the thickness of the anti-ship missile's armor, v_d is the speed when the fragments hitting the anti-ship missile.

$$v_d = v_0 \cdot \exp\left(-\frac{C_D S \rho_a R_s}{2m_s}\right) \tag{10}$$

Where, C_D is the drag coefficient of the intercept missile fragment, ρ_a is the air density, R_S is the miss distance of the intercept missile, v_0 is the initial velocity of the intercept missile fragment.

$$v_0 = v_c \sqrt{\frac{\beta}{1 + 0.5\beta}} \tag{11}$$

Where, v_c is Gurney constant, β is the explosive load parameter.
The number of the fragments that hit the anti-ship missile is:

$$n_s = A_k \frac{M}{2\pi R_s^2 (\cos \varphi_1 - \cos \varphi_2)} \tag{12}$$

Where, M is the total fragments number, A_k is the exposure area of the anti-ship missile, R_S is the miss distance of the intercept missile, φ_1 and φ_2 are the splashing angles of the fragment.

The probability of the anti-ship missile being killed is:

$$P_k = 1 - (1 - P_0)^{n_s} \tag{13}$$

3.5 Vulnerability Model of Warship

Warships are huge in size with components distributed in different cabins. It is not easy to evaluate the vulnerability of a warship hit by one anti-ship missile. Since hitting different part of the warship can cause different casualties. The vulnerability calculation of the warship is influenced by many factors, but this is not the main purpose of this paper. To simplify the calculation process, we choose 3 major factors here: the gross tonnage of the warship, the equivalent TNT explosive load of the anti-ship missile warhead and the vulnerability of the warship. The probability of being destroyed of a warship by an anti-ship missile can be calculated as follows:

$$P_K = \frac{1}{1 + \alpha \frac{d}{G}} \tag{14}$$

Where, d is the gross tonnage of the warship, G is the equivalent TNT explosive load of the anti-ship missile warhead, and α is the vulnerability of the warship.

4 Combat Effectiveness Model of Cooperative Operation

The aim of this part is to acquire the combat effectiveness of cooperative operation which was defined as the sum of each warship's probability of kill multiplied by its importance coefficient. The cooperative operation is divided into 3 steps: cooperative penetration, cooperative detection and cooperative strike.

4.1 The Synthesis Measure of Cooperative Combat Effectiveness

Assume there are m warships, each has an importance coefficient which is determined by the military value and vulnerability of the warship w_i, and is hit by n_i anti-ship missiles. The synthesis measure of cooperative combat effectiveness is defined as the probability of all warships being destroyed multiplied by their importance coefficient:

$$CE = \sum_{i=1}^{m} w_i[1 - (1 - P_{ki})^{n_i}] \tag{15}$$

Where, P_{ki} is the probability of the i-th warship being killed by one anti-ship missile.

The probability of the i-th warship being killed by one anti-ship missile is defined as the product of detection probability, hit probability and kill probability:

$$P_{ki} = P_{Pi}P_{Di}P_{Hi}P_{Ki} \tag{16}$$

Where, P_{Pi} is the probability of the anti-ship missile penetrating the defense area of the warship, P_{Di} is the probability of the anti-ship missile detecting the warship, P_{Hi} is the probability of the anti-ship missile hitting the warship, P_{Ki} is the probability of the anti-ship missile killing the warship.

4.2   Cooperative Penetration

The survival probability of anti-ship missile engaged with k intercept missiles is:

$$P_P = \prod_{i=1}^{k} (1 - P_{ki}) \tag{17}$$

Where, P_{ki} is the probability of getting killed at the i-th engagement with intercept missile of the anti-ship missile, k is the total engagement times.

The cooperative penetration probability of the anti-ship missiles is related to the number of intercept missiles they engaged. If the number of anti-ship missiles exceeds the number of intercept missiles that the warship can afford to fire once, the average penetration probability of anti-ship missile can be calculated as follows:

$$P_P = \prod_{i=1}^{k} \frac{1}{n} [n - z + z(1 - P_{ki})] \tag{18}$$

Where, n is the total number of the anti-ship missiles fired, z is the maximum intercept missiles the warship can fire at one time.

The engagement times k can be obtained through calculating the flight distance of the anti-ship missile during the time from the intercept missile is fired to engagement, until it is less than 0. Then the engagement times k is acquired. The flight distance of the anti-ship missile during the time from the intercept missile is fired to engagement is:

$$d_i = v_a \frac{R_{\max} - \sum_{j=1}^{i-1} (d_j + v_a t_e)}{v_a + v_i} \tag{19}$$

$$d_1 = v_a \frac{R_{\max}}{v_a + v_i} \tag{20}$$

Where, v_a is the speed of the anti-ship missile, v_i is the speed of the intercept missile, $R_{\max}$ is the actual maximum detection range of the warship radar, t_e is the damage evaluation time of the warship.

4.3   Cooperative Detection

If the anti-ship missiles can share the target information, and search the target from different angles, each anti-ship missile has a detection probability according to the RCS of the target. The cooperative detection probability will be any anti-ship missile detects the target:

$$P_D = 1 - \prod_{i=1}^{nP_P} (1 - P_{di}) \tag{21}$$

Where, P_P is the probability of the anti-ship missile penetrating the defense area of the warship, P_{di} is the detection probability of the i-th anti-ship missile, n is the total number of the anti-ship missiles fired.

4.4 Cooperative Strike

If the anti-ship missiles can work cooperatively and distribute the anti-ship missiles among the warships according to the coefficient w_i, then the number of anti-ship missiles hitting the i-th warship is:

$$n_i = w_i n P_P \tag{22}$$

Where, n is the total number of the anti-ship missiles fired, P_P is the probability of the anti-ship missile penetrating the defense area of the warship. Since we can't split one missile to strike two targets, so the number of anti-ship missiles hitting a warship needs to be rounded to the nearest integer, and of course, the sum of n_i must be n.

If the anti-ship missiles work independently, the anti-ship missiles will hit the target they met randomly, so that the number of anti-ship missiles hit the i-th warship n_i will be:

$$n_i = n/m \tag{23}$$

Where, n is the total number of the anti-ship missiles fired, m is the number of the warships.

5   Simulation and Result Analysis

The fictive mission was defined as neutralizing an aircraft carrier strike group with anti-ship missiles. It is assumed that the anti-ship missiles are fired out of the warship defense area. Warships can fire intercept missiles if the anti-ship missiles enter their defense area.

Simulation of the proposed model is performed to evaluate the combat effectiveness of cooperative combat in different cases and compare the results with independent combat. For the sake of convenient calculation, assume the aircraft carrier strike group is composed of one aircraft carrier and two frigates.

The mission model was implemented and simulated using parameter values from Table 1.

Table 1. List of main inputs

Parameter	Value	Parameter	Value
Intercept missile speed v_i	3 Ma	Gross tonnage of the aircraft carrier d_1	60 kt
Damage evaluation time t_e	10 s	Gross tonnage of the frigate d_2	6 kt
Number of aircraft carriers	1	Total fragments number M	1000
Importance coefficient of Aircraft carrier w_1	0.80	Exposure area of anti-ship missile A_F	1 m^2
Number of frigates	2	Leading edge fragment splash angle φ_1	60°
Importance coefficient of frigate w_2	0.10	Trailing edge fragment splash angle φ_2	120°
Anti-ship missile maneuver time T_M	3 s	Gurney constant v_c	2316 m/s
Fragment mass m_s	2 g	Explosive load parameter β	0.8
Fragment area S	1 cm^2	Drag coefficient C_D	0.97
Adjoint coefficient K_M	3.77	Air density ρ	0.364 kg/m^3
Sampling time τ	0.5 s	Anti-ship missile radar turn on distance R_t	30 km
Exposure area of aircraft carrier A_{P1}	20000 m^2	Maximum intercept missiles fired once	6
Exposure area of frigate A_{P2}	2000 m^2	AWACS radar minimum signal to noise ratio $S_{NA_{min}}$	15
RCS of aircraft carrier σ_{t1}	1000 m^2	RCS of frigate σ_{t2}	100 m^2
Vulnerability of aircraft carrier α_1	10	Vulnerability of frigate α_2	5
Anti-ship missile radar reference target RCS σ_0	5 m^2	AWACS radar reference target RCS σ_{A0}	5 m^2
Anti-ship missile radar reference target range R_0	20 km	AWACS radar reference target range R_{A0}	200 km
Anti-ship missile radar reference target signal to noise ratio S_{N0}	20	AWACS radar reference target signal to noise ratio S_{NA0}	20
False alarm probability P_{fa}	10^{-6}	System miss distance of anti-ship missile σ_s	3 m
Armor thickness of anti-ship missile h_d	2 cm	Maneuver overload of anti-ship missile N_T	50 m/s^2

There are a lot of design parameters having influence on the combat effectiveness of anti-ship missiles in cooperative operation. Analyzing all of them needs a huge workload. To verify the feasibility of the proposed analysis model, four parameters are selected. The ranges of anti-ship missile design parameters to analyze are shown in Table 2.

Table 2. Ranges of anti-ship missile design parameters

Parameter	Definition	Unit	Minimum value	Maximum value
n	Number		6	30
σ_t	RCS	m^2	0.01	5
v_a	Speed	Ma	0.8	3
G	Explosive load	kg	100	500

The influence of anti-ship missile design parameters on the combat effectiveness of anti-ship missiles in cooperative operation is shown if Fig. 1. And the combat effectiveness of anti-ship missiles in independent operation with the same design parameter is shown if Fig. 2.

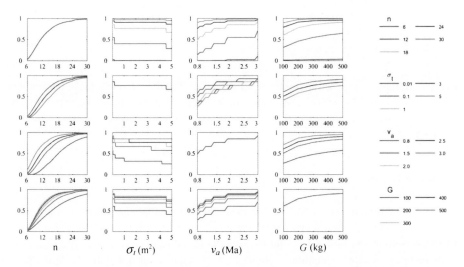

Fig. 1. The influence of design parameters on combat effectiveness in cooperative operation

The result of the simulation demonstrates that with the same initial parameters, huge combat effectiveness increase can be made through cooperative operation. Cooperation is an efficient way of improving the combat effectiveness. The number of anti-ship missiles is the main factor that affects the combat effectiveness. The anti-ship missile number is the main parameter that affects the combat effectiveness. As the number of anti-ship missiles increase, advantage of cooperation becomes more obvious. The increase of combat effectiveness slows down with the number of anti-ship missiles gets larger because fewer valuable targets are left. The RCS, speed and explosive load of anti-ship missile are also important factors for combat effectiveness. Increasing the speed and decreasing the RCS can reduce the time in enemy defense area and increase the penetration probability. Because the penetration process is to engage with waves after waves of intercept missiles, the survival probability decreases

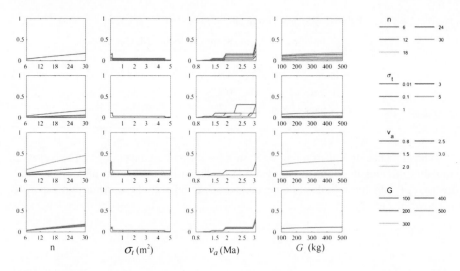

Fig. 2. The influence of design parameters on combat effectiveness in independent operation

every time they engage with an intercept missile, so that the curves of these parameters are discontinuous. Increasing the explosive load can increase the strike capability of anti-ship missile, which lead to an increase on the combat effectiveness. But the influence is limited in independent operations.

6   Conclusion

In this work a quantitative model for the evaluation of combat effectiveness of anti-ship missile is presented. And a simulation example is performed to demonstrate how the model was applied within a relevant mission context during concept design. The result demonstrates that the evaluation methodology is suitable for qualitative evaluation of anti-ship missile at different cooperative operations. The advantage of cooperative operation is remarkable and can be more important than some design parameters. To achieve a reliable design project, the operation type and combat process must be considered. Among the four selected parameters, the number of anti-ship missiles is the main factor that influences the combat effectiveness. The other three factors can also affect the combat effectiveness in cooperative operations, but their influence is limited in independent operations.

Further studies should take other cooperative operation situations into consideration, such as leading missiles equipped with high power radar cooperating with cheap radio silence missiles or mix the anti-ship missile group with electronic jamming missiles. Analyze the optimized combination of different kind of missiles that can achieve the maximum effectiveness-cost ratio.

References

1. Park JH, Seo K-K (2004) Incorporating life-cycle cost into early product development. J Eng Manuf 218(9):1059–1066
2. Frangopol DM, Maute K (2003) Life-cycle reliability-based optimization of civil and aerospace structures. Comput Struct 81(7):397–410
3. Wang HW, Li SL, Tong ZX (2011) Modeling and trade-off of aircraft survivability based on combat capability. J Beijing Univ Aeronaut Astronaut 37(8):933–936
4. Ball RE (2003) The fundamentals of aircraft combat survivability analysis and design, 2nd edn. AIAA
5. Li TJ (2000) The effectiveness of missile weapon system and its analysis. National Defence Industrial Press, Beijing
6. Du Z, Wang CZ, Chen WC (2013) Fast analytical model of cruise missiles penetration ship-based anti-missile system. J Beijing Univ Aeronaut Astronaut 39(11):1449–1454
7. Ling YX (2006) Combat model and simulation. National University of Defence Technology Press, Washington, DC
8. Tan SL, Yan SK, Chen XS (2010) Fighting efficiency evaluation model of cruise missile based on index method. Fire Control Command Control 35(5):883–889
9. Sugeno M, Kang GT (1998) Structure identification of fuzzy model. Fuzzy Sets Syst 28:15–33
10. Fei ZC (2009) Research on entropy weight-analytic hierarchy process and grey-analytic hierarchy process. Tianjin University, Tianjin, pp 5–10
11. Shiyi G (2004) Some discussions about smart missile. Tactical Missile Technol 4:1–7
12. Zhengdong H, Tao L, Shifeng Z (2007) Study on the concept of missiles formation cooperative engagement system. Winged Missile J 10:13–18
13. Shima T (2011) Optimal cooperative pursuit and evasion strategies against a homing missile. J Guid Control Dyn 34(2):414–425
14. Yan L, Wu ST (2008) Research on the combat effectiveness analysis method of formation guidance of aerodynamic missile. Mod Defense Technol 36(5):15–20
15. Shima T, Oshman Y, Shinar J (2002) Efficient multiple model adaptive estimation in ballistic missile interception scenarios. J Guid Control Dyn 25(4):667–675
16. Zhao JB, Yang SX (2017) Review of multi-missile cooperative guidance. Acta Aeronautica et Astronautica Sinica 38(1):020256
17. Lee ZJ, Su SF, Lee CY (2003) Efficiently solving general weapon-target assignment problem by genetic algorithms with greedy eugenics. IEEE Trans Syst Man Cybern Part B 33(1):113–121
18. He Y (1999) Radar automatic detection and CFAR process. Tsinghua University Press, Beijing
19. Ding C, Mao S (1996) Miss distance dynamics in active radar homing missiles. In: CIE international conference of radar, vol 10, pp 735–738
20. Pei Y, Song BF, Han Q (2006) A generic calculation mode for aircraft single-hit vulnerability assessment based on equivalent target. Chin J Aeronaut 19(3):183–189

Research on Flight Dynamic Modeling and Interference of Components for Rotor/Wing Compound Helicopter

Lili Lin[1](✉), Xiaoxin Liu[2], Minghua Peng[2], and Jianbo Li[2]

[1] Nanjing University of Aeronautics and Astronautics, Nanjing 210016, China
Linlilid@126.com

[2] National Laboratory of Science and Technology on Rotorcraft Aeromechanics, Nanjing University of Aeronautics and Astronautics, Nanjing 210016, China

Abstract. In order to study the rotor/wing compound helicopter characteristic of flight dynamics, this paper establishes the helicopter flight dynamics model, including rotor, propeller, wing with aileron, horizontal tail with elevator, vertical tail, fuselage and other aerodynamic models; and then the CFD and momentum source method are used to analysis the aerodynamic interference of the fuselage/wing/flat tail to rotor, the rotor to the fuselage/wing/flat tail, the propeller to the wing/flat tail, and corrected the aerodynamic calculation model of each component in the form of interference coefficient. The target pitch angle is introduced as a constraint to solving the problem of lift distribution between rotor and wing in the configuration aircraft. The manipulation redundancy problem is solved by setting the three stages as hovering low speed, pre-transition flying, and high-speed front flying to assign the manipulations. The helicopter is mainly controlled by collective pitch, horizontal/longitudinal period variable pitch and the propeller pitch in the hovering low-speed stage; In the high speed stage it is controlled by aileron, elevator and propeller pitch; It is controlled by all the manipulations through introducing the weighting factor in the pre-transition flight stage. Finally, get the trimming strategy of the rotor/wing compound helicopter through the trimming analysis of the example.

Keywords: Compound helicopter · Trim · High advanced ratio · Aerodynamic interference

Notation

A_1 lateral period variable pitch of rotor
a_{1s} lateral cycle waving coefficient
A_f maximum cross-sectional area of fuselage
b blade chord length
b_{1s} longitudinal cycle waving coefficient
B_1 longitudinal period variable pitch of rotor
e hinge offset ratio
I_b moment of inertia of blade waving
K_0 blade flapping stiffness
l_f length of the fuselage

© Springer Nature Singapore Pte Ltd. 2019
X. Zhang (Ed.): APISAT 2018, LNEE 459, pp. 1202–1221, 2019.
https://doi.org/10.1007/978-981-13-3305-7_95

M_b	aerodynamic moment about flapping axis
α_h	horizontal tail angle of attack
α_w	wing angle of attack
α_{0h}	horizontal tail zero lift angle of attack
δ_a	aileron deflection angle
δ_e	elevator deflection angle
M_i	hub motion moment of inertia
q_f	dynamic pressure of fuselage
q_h	dynamic pressure of horizontal tail
q_v	dynamic pressure of vertical tail
q_w	dynamic pressure of wing
S_h	horizontal tail area
S_v	vertical tail area
S_w	wing area
T	rotor lift
V	flight speed
φ_l	left propeller pitch
φ_r	right propeller pitch
φ_7	collective pitch
γ	roll angle
θ	pitch angle

1 Introduction

The compound helicopter has many characteristics of flight performance, such as vertical take-off and landing, hovering and low-speed flight, high-speed, long-range and long-haul. It can be applied in a wide range of fields and has received extensive attention from many scholars at home and abroad.

Many scholars have carried out research on flight dynamics models for different types of compound helicopters. Among them, Jean-paul found that after the introduction of the aileron, elevator and propellers, the change of manipulation has a significant impact on trimming results by studying the flight dynamics model of a compound helicopter. He took UH-60A modification which increased a short wing and a tail propeller as an example. It is concluded that the stabilator can reduce the lift of the wing, and the introduction of the aileron can improve the aerodynamic efficiency of the rotor.

In addition, Duan Saiyu has studied the flight dynamics models of compound coaxial helicopter with a propulsion propeller and wing. Based on the characteristics of compound coaxial helicopter, he considered the aerodynamic interference between the upper and lower rotors, the rotor and the wing, built the aerodynamic calculation model of the rotor, wing and other components. Furthermore, he has analyzed the physical principle of compound coaxial helicopter and the flight characteristics over the entire stable flight speed range.

Kong Weihong has studied the flight dynamics model of UH-60L/VTDP compound high-speed helicopter. She proposed that the key to the calculation of compound high-speed helicopter trimming is how to deal with the problem of manipulation redundancy and has got relationship between the manipulation, attitude, lift and thrust of the configuration at different speeds.

Zhang Xiaochi has conducted a flight dynamics modeling research work on a rotor fixed-wing compound aircraft. They have proposed that such a configuration makes the lift-to-drag characteristics of high-speed flight and the propulsion efficiency of the dynamic system higher than that of the conventional type of high-speed helicopter and tilt-rotor, thus having a wider flight envelope, and it can decouple lift and horizontal thrust, get smooth transition flight.

There are many pneumatic components in the compound helicopter, and the aerodynamic interference between the components is obvious. Wanjia has conducted a correlation analysis on the individual rotor/wing model and found that the interference effect is greatly affected by the forward flight speed. As the speed of the forward flight increases, the interference from the wing to the rotor is strengthened, and the interference of the rotor to the wing is weakened. In addition, Zhao Yinyu also analyzed the aerodynamic interference problem of the rotor/propeller model in for rotor/wing compound helicopter.

In this paper, the flight dynamics characteristic of a rotor/wing compound helicopter is studied. At the same time, the aerodynamic interference between components is reflected by introducing the interference coefficient in the flight dynamics model.

The schematic diagram of the helicopter is shown in Fig. 1. It mainly includes rotor, propellers, wing, horizontal tail, vertical tail, fuselage and other components. It has two sets of lift systems with rotor and fixed wing and corresponding operating mechanism. In the vertical take-off and landing, hovering and low-speed flight, through the collective pitch, lateral and longitudinal period variable pitch manipulation. At high speed flight, through ailerons, elevator and propeller pitch maneuvers. It not only exerts the good aerodynamic characteristics of the helicopter in vertical take-off, hovering and low speed, but also has the high lift-to-drag ratio characteristics of the fixed-wing aircraft at high speed.

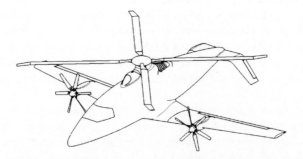

Fig. 1. Schematic of rotor/wing compound thrust helicopter

2   Flight Dynamics Model

Compared with the mature aerodynamic modules of fuselage, tail rotor, horizontal tail and vertical tail in conventional helicopter the main considerations for the rotor/wing compound helicopter are the aerodynamic model of wing and the propellers. In addition, the rudder surface aerodynamic models is added to the wing and horizontal tail.

In the modeling process of this paper, the body coordinate system is adopted, and the machine center of gravity is taken as the coordinate origin. The body structure axis is the X axis, and the vertical X axis is the Y axis. The right hand rule is used to determine the Z axis to the right. Figure 2 shows a schematic diagram of the sample helicopter coordinate system.

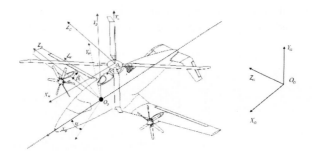

Fig. 2. Schematic diagram of sample helicopter coordinate system

2.1   Flight Dynamics Equation

The mass of the rotor/wing compound helicopter is m, F is the external force acting on the center of gravity. The kinematic equation of the aircraft is obtained by Newton's second law:

$$\begin{cases} \sum F_x = mV'_x - m(V_y\omega_z - V_z\omega_y) + mg\,\sin\theta \\ \sum F_y = mV'_y - m(V_z\omega_x - V_x\omega_z) + mg\,\cos\theta\,\cos\phi \\ \sum F_z = mV'_z - m(V_x\omega_y - V_y\omega_x) + mg\,\cos\theta\,\sin\phi \end{cases} \quad (2.1)$$

The rotation of the aircraft comes from the external moment acting on the aircraft, and using the momentum moment theorem to establish the dynamic equation of the rotation of the aircraft around the center of gravity:

$$\begin{cases} I_x\omega'_x = (I_y - I_z)\omega_y\omega_z + I_{xy}(\omega'_y - \omega_z\omega_x) + M_y \\ I_y\omega'_y = (I_z - I_x)\omega_z\omega_x + I_{zx}(\omega'_z - \omega_x\omega_y) + M_z \\ I_z\omega'_z = (I_x - I_y)\omega_x\omega_y + I_{yz}(\omega'_z - \omega_y\omega_z) + M_z \end{cases} \quad (2.2)$$

In addition, the kinematic relationship between the attitude angle of the aircraft and the angular velocity is supplemented:

$$\begin{cases} \phi' = \omega_x - \tan\theta(\omega_x\cos\phi - \omega_z\sin\phi) \\ \theta' = \omega_y\sin\phi + \omega_z\cos\phi \\ \varphi' = \frac{1}{\cos\theta}(\omega_y\cos\theta - \omega_z\sin\phi) \end{cases} \tag{2.3}$$

2.2   Rotor Aerodynamic Model

The rotor uses the dynamic inflow model of Pitt and Peters, the model takes into account the uneven distribution of the induced velocity on the disc and its hysteresis with aerodynamic changes in the rotor, those put the aerodynamic load of the rotor contact with the transient change in the induced velocity of the rotor.

For a steady flight state, the components of the disc induced velocity do not change over time.

$$\begin{Bmatrix} v_0 \\ v_s \\ v_c \end{Bmatrix} = L \begin{Bmatrix} C_T \\ C_L \\ C_M \end{Bmatrix} \tag{2.4}$$

$$L = \frac{1}{v_m} \begin{bmatrix} \frac{1}{2} & 0 & \frac{15}{64}\sqrt{\frac{1-\sin\alpha_w}{1+\sin\alpha_w}} \\ 0 & -\frac{4}{1+\sin\alpha_w} & 0 \\ \frac{15}{64}\sqrt{\frac{1-\sin\alpha_w}{1+\sin\alpha_w}} & 0 & -\frac{4\sin\alpha_w}{1+\sin\alpha_w} \end{bmatrix} \tag{2.5}$$

Mass flow parameter of inflow:

$$v_m = \frac{V^2 + 3V \cdot v_{eq} \cdot \sin\alpha_H + 2v_{eq}^2}{\sqrt{V^2 + 2V \cdot v_{eq} \cdot \sin\alpha_H + v_{eq}^2}} \tag{2.6}$$

The rigid rotor is selected so that it can be controlled to keep the disc at a small angle of attack to make the drag small when the rotor reduce the revolution (Fig. 3). There is a rigid rotor flapping motion model:

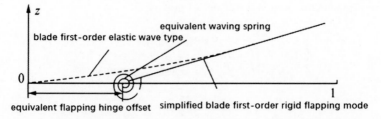

Fig. 3. Schematic diagram of equivalent method of rigid rotor waving

There is a rotor blade flapping motion equation:

$$\ddot{\beta} + \left[\frac{K_0}{I_b} + \left(1 + e\frac{K_s}{I_b} \right)\Omega^2 \right]\beta = \frac{M_b}{I_b} + M_i \tag{2.7}$$

The force and torque acting on the blade element is:

$$
\begin{aligned}
dL &= \frac{1}{2}\rho V^2 a_\infty \alpha b dr \\
dD &= \frac{1}{2}\rho V^2 C_x b dr \\
dT_s &= dT \cos \beta \\
dH_s &= dQ \sin \psi - dT \sin \beta \cos \psi \\
dS_s &= -dQ \cos \psi - dT \sin \beta \sin \psi \\
dM_K &= dQr \cos \beta
\end{aligned}
\tag{2.8}
$$

Hub pitching torque is:

$$MG_z = \frac{k}{2}M_b\Omega^2 e a_{1s} \tag{2.9}$$

Hub rolling torque is:

$$MG_x = \frac{k}{2}M_b\Omega^2 e b_{1s} \tag{2.10}$$

In summary, there are lift, back force, lateral force and torque of the rotor:

$$
\begin{aligned}
T &= \kappa\frac{k}{2\pi}\int_0^{2\pi}\int_{r0}^{R} dT_s dr d\psi \\
H &= \kappa\frac{k}{2\pi}\int_0^{2\pi}\int_{r0}^{R} dH_s dr d\psi \\
S &= \kappa\frac{k}{2\pi}\int_0^{2\pi}\int_{r0}^{R} dS_s dr d\psi \\
M_k &= \frac{k}{2\pi}\int_0^{2\pi}\int_{r0}^{R} dM_k dr d\psi
\end{aligned}
\tag{2.11}
$$

Finally, there are the aerodynamics forces and torques of the rotor to center of gravity of the machine:

$$
\left\{ \begin{array}{c} F_x \\ F_y \\ F_z \end{array} \right\} = \left[\begin{array}{ccc} \cos \delta_s & \sin \delta_s & 0 \\ -\sin \delta_s & \cos \delta_s & 0 \\ 0 & 0 & 1 \end{array} \right] \left\{ \begin{array}{c} -H \\ T \bullet K_{lr} \\ S \end{array} \right\} \tag{2.12}
$$

$$\begin{Bmatrix} M_x \\ M_y \\ M_z \end{Bmatrix}_{MR} = \begin{bmatrix} \cos \delta_s & \sin \delta_s & 0 \\ -\sin \delta_s & \cos \delta_s & 0 \\ 0 & 0 & 1 \end{bmatrix} \begin{Bmatrix} MG_x \\ -M_k \\ MG_z \end{Bmatrix} + \begin{bmatrix} 0 & z_{MR} & -y_{MR} \\ -z_{MR} & 0 & x_{MR} \\ y_{MR} & -x_{MR} & 0 \end{bmatrix} \begin{Bmatrix} F_x \\ F_y \\ F_z \end{Bmatrix}_{MR}$$

(2.13)

2.3   Propeller Aerodynamic Model

The momentum-blade element combination theory is used to calculate the propeller aerodynamic. The aerodynamic model is established separately from the propellers on both sides. The following is only one side (Fig. 4).

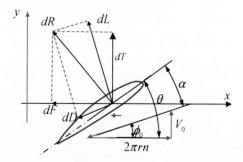

Fig. 4. Schematic diagram of propeller airfoil aerodynamic force calculation

The elementary force and torque of propeller is:

$$dL = \frac{1}{2} \rho W_0^2 C_L b dr$$

$$dD = \frac{1}{2} \rho W_0^2 C_D b dr$$

$$dT = dR \cos(\phi + \gamma)$$

$$dM = \frac{1}{2} \rho V^2 Q_c dr$$

(2.14)

Giving for the total thrust and torque:

$$T = \frac{1}{2} \rho V^2 N_B \int_{r_0}^{R} T_c dr$$

$$Q_{PR} = \frac{1}{2} \rho V^2 N_B \int_{r_0}^{R} Q_c dr$$

(2.15)

2.4   Wing Aerodynamics

Because the ailerons on the wing are differentially deflected, the aerodynamic models should be established separately for the left and right wings.

Among them, the lift of wing is:

$$L_w = q_w S_w (C_{lw} + C_{lf}) \tag{2.16}$$

The lift coefficient of wing is:

$$C_{lw} = C_{l\alpha w} \bullet \alpha_w + C_{l0} \tag{2.17}$$

The lift coefficient of aileron is:

$$C_{lf} = (C_{lf}^{\delta})_{pm} \delta \tag{2.18}$$

The drag of wing is:

$$D_w = q_w S_w (C_{dw} + C_{df}) \tag{2.19}$$

The drag coefficient of wing is:

$$C_{dw} = C_{d0w} + C_{diw} \tag{2.20}$$

The drag coefficient of aileron is:

$$C_{df} = 0.008 - 0.003 C_{lf} + 0.01 C_{lf}^2 \tag{2.21}$$

Convert those force to the body coordinate system:

$$\left\{ \begin{array}{c} F_x \\ F_y \\ F_z \end{array} \right\}_w = B_w \left\{ \begin{array}{c} -D_w \bullet K_{dw} \\ L_w \bullet K_{lw} \\ 0 \end{array} \right\} \tag{2.22}$$

2.5   Aerodynamics of Horizontal Tail

The lift of horizontal tail is:

$$L_h = q_h S_h (C_{lh} + C_{le}) \tag{2.23}$$

The lift coefficient of horizontal tail is:

$$C_{lh} = C_{l\alpha h} (\alpha_h - \alpha_{0h}) \tag{2.24}$$

The lift coefficient of elevator is:

$$C_{le} = (C_{le}^{\delta})_{pm}\delta \tag{2.25}$$

The drag of horizontal tail is:

$$D_h = q_h S_h (C_{dh} + C_{de}) \tag{2.26}$$

The drag coefficient of horizontal tail is:

$$C_{dh} = C_{d0h} + C_{dih} \tag{2.27}$$

The drag coefficient of elevator is:

$$C_{de} = 0.008 - 0.003C_{le} + 0.01C_{le}^2 \tag{2.28}$$

Convert those force to the body coordinate system:

$$\begin{Bmatrix} F_x \\ F_y \\ F_z \end{Bmatrix}_H = B_H \begin{Bmatrix} -D_h \bullet K_{dh} \\ L_h \bullet K_{lh} \\ 0 \end{Bmatrix} \tag{2.29}$$

2.6   Aerodynamics of Vertical Tail

The yaw operation of the configuration is differentially operated by the propellers, so the rudder is not considered, and only the resistance of the vertical tail need be calculated.

$$D_v = q_v S_v C_{dv}$$
$$C_{dv} = C_{d0v} \tag{2.30}$$

2.7   Aerodynamics of Fuselage

According to the CFD method, the drag coefficient, lift coefficient, and pitching moment coefficient of the fuselage at different angles of attack are calculated, and the point of the pitching moment is taken at the center of gravity of the whole machine.

$$D_f = C_{df}q_f A_f$$
$$L_f = C_{lf}q_f A_f$$
$$M_{yf} = C_{Myf}q_f A_f l_f \tag{2.31}$$

Among them, the lift and drag of the fuselage at the aerodynamic center and the pitching moment curve is as shown at the cruising speed (75 m/s) (Fig. 5).

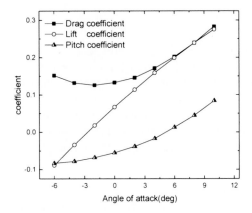

Fig. 5. The lift/drag/pitching moment coefficient curve change with angle of attack

Convert those force to the body coordinate system:

$$\begin{bmatrix} F_{x,Fg} \\ F_{y,Fg} \\ F_{z,Fg} \\ M_{x,Fg} \\ M_{y,Fg} \\ M_{z,Fg} \end{bmatrix} = \begin{bmatrix} -D_f \bullet K_{df} \\ L_f \bullet K_{lf} \\ 0 \\ 0 \\ 0 \\ M_{zf} \end{bmatrix} \tag{2.32}$$

3   Analysis of the Aerodynamic Interference Between Components

The aerodynamic interference between the pneumatic components of the rotor/wing compound helicopter is significant. In order to consider the aerodynamic interference of components in the flight dynamics model, CFD and momentum source method are used. According to the characteristics of the rotor/wing compound helicopter, the interference characteristics between components are analyzed, and various interference coefficients are defined and calculated.

3.1   The Aerodynamic Interference of Fuselage/Wing/Horizontal Tail to the Rotor

The aerodynamic characteristics of the model are analyzed by CFD method, and get L_{wr-r} which is the thrust of the rotor at different speeds under the wing and the horizontal tail, compare with L_r which is the thrust of the rotor under individual rotor and get L_{wr-r} as thrust interference factor

$$K_{lr} = L_{wr-r}/L_r \qquad\qquad (3.1)$$

It can be seen from Fig. 6 that in hovering and low-speed flight, the degree of interference is slightly larger, and the actual thrust is greater than the thrust generated by the individual rotor. However, the thrust of the rotor under the blockage of components is slightly less than the thrust when the interference is not considered at high speeds. Before the forward flying speed reaches 80 m/s, the interference level become weak, and then the interference continues to increase in the range of 80–100 m/s. As the forward flying speed continues to increase, the degree of thrust interference is weakened again.

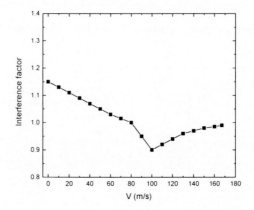

Fig. 6. Chart of interference factor of fuselage/wing/horizontal tail to rotor thrust changes with speed change

3.2   The Aerodynamic Interference of the Rotor to Fuselage/Wing/Horizontal Tail

In the hovering and low-speed forward flight phase, most lift is assumed by rotor. At this time, the drag of the fuselage wing horizontal tail is small, so the interference factor of the rotor's lift and drag to the corresponding components is not considered. The interference factor is considered after 50 m/s.

By using the CFD and momentum source method to simulate the rotor's under-washing interference, it will be found that under the interference of the rotor, the lift and drag of the fuselage changes.

The sinking of the rotor wake will cause the wing load to increase, and the effective lift of the wing is reduced. The horizontal tail is in the rotor wake behind the body. The lower wash of the rotor directly impacts the horizontal tail surface, resulting in a large downward load, which affects the aerodynamic characteristics of the horizontal tail directly (Fig. 7).

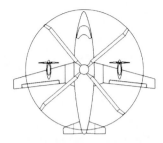

Fig. 7. Schematic diagram of the influence range of the rotor on the fuselage/wing/horizontal tail

Rotor-to-fuselage lift interference factor is:

$$K_{lf} = L_{fr-r}/L_f \tag{3.2}$$

Rotor-to-fuselage drag interference factor is:

$$K_{df} = D_{fr-r}/D_f \tag{3.3}$$

Rotor-to-wing lift interference factor is:

$$K_{lw} = L_{wr-r}/L_w \tag{3.4}$$

Rotor-to-wing drag interference factor is:

$$K_{dw} = D_{wr-r}/D_w \tag{3.5}$$

Rotor-to- horizontal tail lift interference factor is:

$$K_{lt} = L_{tr-r}/L_t \tag{3.6}$$

Rotor-to- horizontal tail drag interference factor is:

$$K_{dt} = D_{tr-r}/D_t \tag{3.7}$$

It can be seen from Fig. 8 that as the speed of the forward flight increases, the degree of interference of the rotor to the lift of the fuselage/wing/horizontal tail becomes smaller and smaller. This is because the higher the speed of the forward flight, the less the interference effect of the rotor flow field washing.

In terms of drag, as it can be seen from Fig. 9, the drag interference of the rotor to fuselage and wing decreases as the forward speed increases, while the drag interference of the horizontal tail increases first and then decreases. The rotor wake produces a downwash flow at horizontal tail, which changes the actual pitch angle of horizontal tail. Especially in the high-speed forward flight state, the effect is more obvious.

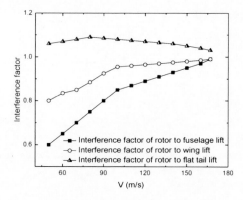

Fig. 8. Chart of interference factor of rotor to fuselage/wing/horizontal tail thrust changes with speed change

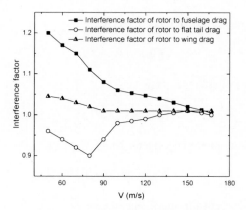

Fig. 9. Chart of interference factor of rotor to fuselage/wing/horizontal tail drag changes with speed change

3.3   Propeller Interference on the Wing/Horizontal Tail

The wing behind the propeller is affected by the forward propeller induced speed. Increased flow velocity leads to increased dynamic pressure. Compared with the non-propeller, the lift and drag will have a large change.

As shown in Fig. 10, the shaded portion is the slip flow region, and the incoming flow velocity is the sum of forward flight speed and propeller induced speed. There are slipstream areas on the wing surface and the horizontal tail. The non-shaded area is a free-flow area, and the incoming flow speed is consistent with the forward flying speed.

Propeller-to-wing lift interference factor is:

$$K_{lwp} = L_{wp-p}/L_w \tag{3.8}$$

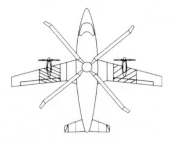

Fig. 10. Schematic diagram of the interference area of the propeller slipstream to the wing/horizontal tail

Propeller-to-wing drag interference factor is:

$$K_{dwp} = D_{wp-p}/D_w \qquad (3.9)$$

Propeller-to-horizontal tail lift interference factor is:

$$K_{ltp} = L_{tp-p}/L_t \qquad (3.10)$$

Propeller-to-horizontal tail drag interference factor is:

$$K_{dtp} = D_{tp-p}/D_t \qquad (3.11)$$

Figure 11 shows the interference lift and drag factors of propeller to wing and the tail.

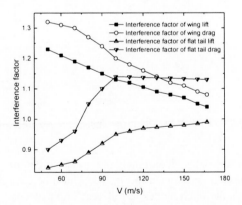

Fig. 11. Chart of interference factor of propeller to wing/horizontal tail lift and drag changes with speed change

4   Example Trim Analysis

4.1   Trim Method

In order to solve the problem of maneuver redundancy, the configuration helicopter is divided into three stages, namely, the hovering low speed stage, the pre-transition flight stage and the high-speed forward flight stage. The manipulations and attitude angles required for the sample helicopter trimming are $[\varphi_7, A_1, B_1, \varphi_l, \varphi_r, \theta, \gamma]$ in the hovering low speed stage, $[\varphi_7, A_1, B_1, \varphi_l, \varphi_r, \delta_a, \delta_e, \theta, \gamma]$ in the pre-transition flight stage, and $[\varphi_7, \varphi_l, \varphi_r, \delta_a, \delta_e, \theta, \gamma]$ in the high-speed forward flight stage.

When it is flying stably, the linear acceleration, angular acceleration and angular velocity of the body are zero, and the forces and moments acting on the components must be balanced. The corresponding rigid body Euler equation is reduced to:

$$\sum F_x = 0 \quad \sum F_y = 0 \quad \sum F_z = 0$$
$$\sum M_x = 0 \quad \sum M_y = 0 \quad \sum M_z = 0 \tag{4.1}$$

Since the state variables that need to be trimmed are more than the number of equilibrium equations, constraints are introduced to meet the needs of solving the equation.

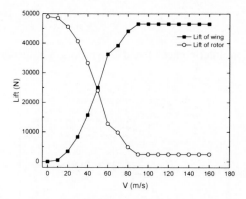

Fig. 12.   Diagram of lift distribution rule under different forward flight speeds

The compound helicopter determines the pitch angle of the aircraft by the rotor/wing lift distribution determined in Fig. 12. And it be used as a constraint of the trim. Figure 13 shows the pitch angles at different flight speeds.

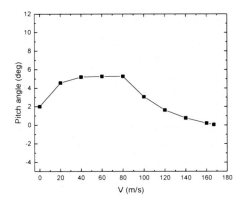

Fig. 13. Target pitch angle changes with forward flight speed

In addition, since the trimming constraint is insufficient in the pre-transition flight mode, a weighting factor $f_{h1}, f_{z1}, f_{h2}, f_{z2}$ is set for each manipulation. Their range is between 0 and 1 where f_{h1}, f_{z1} decreases with increasing speed until it reaches 0, and f_{h2}, f_{z2} increases with speed and ends with 1.

The actual operating range of each manipulation under pre-transition flight mode is:

$$\begin{cases} A_1 = f_{h1}(v)\delta_h \\ \delta_a = f_{h2}(v)\delta_h \end{cases} \\ \begin{cases} B_1 = f_{z1}(v)\delta_z \\ \delta_e = f_{z2}(v)\delta_z \end{cases} \tag{4.2}$$

The flight dynamics equation can be solved using the Sequential Quadratic Programming (SQP) algorithm based on the lift distribution constraints and the weight factor assignment.

4.2 Example Analysis

In order to verify the correctness and effectiveness of the rotor/wing compound helicopter flight dynamics model, this paper takes a rotor/wing compound helicopter as an example to trim and analysis the various states of sample machine from hovering to high-speed flight. Table 1 shows the overall parameters related to the sample machine.

Figures 14, 15 and 16 shows the trends of collective pitch, lateral period variation, longitudinal period variation, left propeller pitch, right propeller pitch, aileron deflection angle, and elevator deflection angle at different forward flight speeds.

Table 1. Sample conceptual helicopter part of the general parameters

Item/unit	Value
Take-off weight/*kg*	5000
Wing span/*m*	8.8
Wing area/*m²*	8.82
Horizontal tail area/*m²*	2.36
Vertical tail area/*m²*	1.17
Blade diameter/*m*	10.4
Blade number	4
Rotor revolution/*rpm*	367.5/257.2[*]
Propeller diameter/*m*	2.0
Rotor solidity	0.082
Propeller revolution/*rpm*	2200
Propeller blade number	6
Left propeller pitch/*inch*	40
Right propeller pitch/*inch*	60

[*]The two rotor revolutions represent the rotor revolution which reduce before and after respectively.

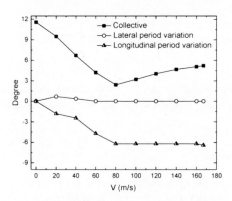

Fig. 14. Collective pitch, lateral/longitudinal period variation changes with the forward flying speed change

In the hovering low speed stage, the rotor takes a larger part of the lift, and the collective pitch is larger. As the forward flight speed increases, the sample helicopter enters the pre-transition flight stage, and the horizontal and vertical maneuvers are converted into ailerons, elevator from the lateral and longitudinal period variable pitch operations. At this time, in order to keep the angle of attack of the rotor disc in a small range, the longitudinal period variable pitch gradually increases. After the current flying speed exceeds 80 m/s, the rotor still needs to maintain 5% lift, and the collective

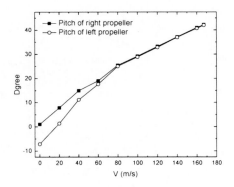

Fig. 15. Pitch of both side propellers change with the forward flying speed change

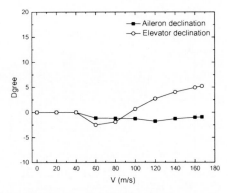

Fig. 16. Deflection angle of aileron/elevator change with the forward flying speed change

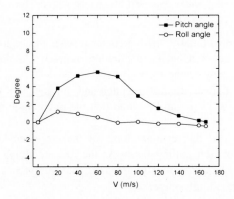

Fig. 17. Pitch angle and roll angle change with the forward flying speed change

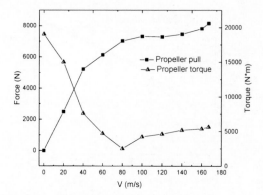

Fig. 18. Composition of forces/torques of propellers change with the forward flying speed change

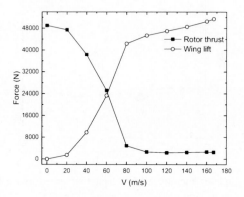

Fig. 19. Rotor/wing actual generated lift change with the forward flying speed

pitch gradually increases. Finally, the helicopter reaches the high-speed forward flight stage, and the body needs a balanced pitching moment. At this time, the elevator increases with the increase of the forward flight speed.

Figure 17 reflects the variation of the pitch and roll angles of the body with speed. By comparing with Fig. 13, it can be found that the pitch angle of trimming can well satisfy the constraint.

As shown in Fig. 18, the resultant force of the propellers on both sides of the fuselage provides a forward pulling force. As the forward flying speed increases, the pulling force needs to be continuously increased, and the pitches of the propellers on both sides are correspondingly increased.

Their combined torque is responsible for balancing the counter-torque of the rotor. As the forward speed increases, the rotor is continuously unloaded, the counter torque is reduced, and the combined torque of the propeller is correspondingly reduced.

Comparing Figs. 12 with 19, it can be seen that the lifts assigned by the rotor and wing of the flight dynamics model satisfy the target lift, which proves that the trimming operation satisfies the corresponding constraints.

5 Conclusion

In this paper, research the flight dynamics trimming characteristics of the rotor/wing compound helicopter, and the following conclusions are obtained:

(1) Establish a flight dynamics model for a rotor/wing compound helicopter, including aerodynamic models such as rotor, propeller, fuselage, wing, horizontal tail, and vertical tail. Among them, the rotor aerodynamic model considers the dynamic inflow model with non-uniform induced velocity. And it considers the equivalent flapping effect of rigid rotor too.
(2) Researched the effects of rotor and propellers on the aerodynamic interference of the whole machine components and the interference of the fuselage/wing/horizontal tail to the rotor thrust. Those are expressed by the interference factor.
(3) Get the variation rule of the manipulations over the entire speed range by trimming. With the increase of the forward flight speed, the collective pitch decreases first and then increase, and the variation of the lateral period variation is not large. The longitudinal period variable first increases and then gradually stabilize, and the pitch of the propellers on both sides increases continuously. The aileron deflection angle does not change much and the elevator deflection angle gradually increases.

References

1. Reddinger JP, Gandhi F (2014) Physics-based trim optimization of an articulated slowed-rotor compound helicopter in high-speed flight. J Aircr 52(6):1–11
2. Duan S-Y, Chen M (2011) Investigation of compound coaxial helicopter trim. Aircr Des 31 (03):13–17 + 36
3. Kong W (2011) Research on some key technical issues of compound high speed helicopter, Nanjing University of Aeronautics and Astronautics
4. Zhang X, Wan Z (2016) Conceptual design of rotary wing and fixed wing compound VTOL aircraft. Acta Aeronautica et Astronautica Sinaca 37(01):179–192
5. (2009) Influence of wing location on rotor_wing interaction of compound helicopter. J Beijing Univ Aeronaut Astronaut 35(05):519–522
6. Zhao Y, Lin X (2017) Analysis on rotor-propellers interaction flowfield for compound double-thrust-propeller high-speed helicopter. J Nanjing Univ Aeronaut Astronaut 49 (02):154–164

Research on a Modeling Method of Ducted Propulsion System for Vertical Take-Off and Landing Aircraft

Min Chang[1], Weixiang Zhou[2], Bo Peng[3], and Junqiang Bai[1(✉)]

[1] Unmanned System Research Institute,
Northwestern Polytechnical University, Xi'an, China
junqiang@nwpu.edu.cn
[2] School of Aeronautics, Northwestern Polytechnical University, Xi'an, China
[3] School of Aeronautic Science and Engineering,
Beihang University, Beijing, China

Abstract. This paper carries out a modeling method of ducted propulsion system of vertical take-off and landing aircraft. On axial flow condition, the model of ducted propulsion system, based momentum theory, is established and verified. Based on the non-dimensional definition of open propeller, the non-dimensional method, applied to the ducted blades, is proposed. The research shows that the ducted propulsion system model and the ducted blade non-dimensional method are reasonable and effective, for high performance aircraft configuration design with tilting duct-fan propulsion system.

Keywords: Vertical take-off and landing aircraft · Ducted propulsion system ·
Modeling method · Non-dimensional method · Momentum theory

1 Introduction

Vertical take-off and landing (VTOL) aircraft include tilting rotor type, tilting culvert type, vector thrust type, lift fan type and so on. In recent years, more and more attention has been paid to the VTOL aircraft with tilting ducts, as AD-150 [1] and Trifan 600 [2]. The usual ducted simulation analysis methods include blade element theory, momentum source method [3], vortex theory and so on. Momentum source method simplifies the complex blade as an actuator disk [4], it can quickly analyze macro performance of the duct. However, this method ignores the detailed flow around the blade and transfers the periodic flow to the quasi steady flow by time average [5]. Acheson used this method to study the distributed ducted fan [6]. The results show that the momentum source method is not accurate for lateral inflow simulation of ducted blade. The lip stall will seriously affect simulation accuracy of ducted blade. The advance ratio J is an important non-dimensional parameter to describe open propeller. It is expected to improve analysis accuracy by applying advance ratio to ducted blade on the non-axial flow condition.

© Springer Nature Singapore Pte Ltd. 2019
X. Zhang (Ed.): APISAT 2018, LNEE 459, pp. 1222–1230, 2019.
https://doi.org/10.1007/978-981-13-3305-7_96

Based on dimensionless parameters, Brandt conducted a large number of wind tunnel tests on various open propellers [7]. Liu [8] proposed that motion of blades is similar under the same advance ratio. This concept can also be extended to ducted blade. Naipei [9] analyzed the variation of thrust and efficiency with advance ratio of ducted propulsion system at different angles of attack. Due to the same advance ratio definition as open blade, as increasing of advance ratio, the ducted propulsion system behaves inconsistent at different angles of attack. Therefore, it is necessary to modify advance ratio of open propeller to suitable for ducted blade. It can be applied to determine design parameters of VTOL aircraft.

In this paper, the classical ducted momentum theory is applied and verified to model ducted propulsion system. And then, based on definition of open propeller advance ratio, modify and propose a non-dimensional method of ducted blade.

2   The Model of the Ducted Momentum Theory and Its Verification

The model of ducted momentum theory on axial flow conditions is shown as Fig. 1. The model adopts to 4 assumptions: (1) air passes through the propeller uniformly, (2) the effect of impeller only increases static pressure uniformly, but not increase the circumferential velocity, (3) addition power of propeller is not enough to change air density, (4) expansion/contraction in nozzle and ejection from nozzle uniform velocity are ideal. In Fig. 1, P_0, V_0, P_1 and V_c is static pressure of free stream, velocity of free stream, static pressure before lip-inlet of duct and velocity of before and after blade respectively. According to the third assumption and mass conservation law, and cross-sectional area before and after paddle plate does not have a sharp change, the velocity V_c passing propeller disk is same. According to the second assumption, static pressure increased to P_2 through oar plate. After the uniform expansion or further compression by nozzle, static pressure changes from P_2 to P_e and velocity changes uniformly into V_e. And static pressure P_e has returned to normal atmospheric pressure through the nozzle P_2.

Based on above description, Bernoulli equation and momentum theory, we can derive:

The blade thrust force is: $T_{rotor} = \frac{1}{2}\rho(V_e^2 - V_0^2)A_c$

The total thrust of ducted propulsion system is: $T_{all} = \rho A_e V_e \cdot (V_e - V_0)$

The ideal power consumption is: $P_{\dot{m}} = \frac{1}{2}\rho A_e V_e(V_e^2 - V_0^2) = T_{rotor} \cdot V_c$

The propulsion efficiency of ducted propulsion system is: $\eta = \frac{2 \cdot V_0}{V_e + V_0}$

In order to verify the accuracy of model of ducted propulsion system, we carry out verification of NASA test model [10], which has been carried out wind tunnel test for a ducted propulsion system from -10 to 110 degrees of attack angle. The test equipment and model are shown in Fig. 2.

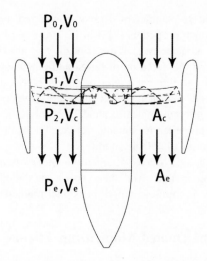

Fig. 1. The model of ducted momentum theory on axial flow condition

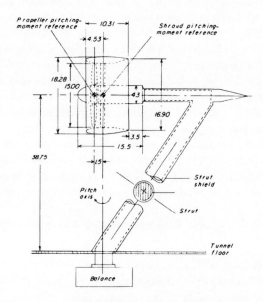

Fig. 2. NASA test model

In this paper, the experimental results of 8000 RPM duct blades under hovering condition are selected. The ducts produce a pull force of 120 Ns and a pull force of 80 Ns, so the pressure on the excitation disk changes to a uniform mutation, the value of which is 730 Pa.

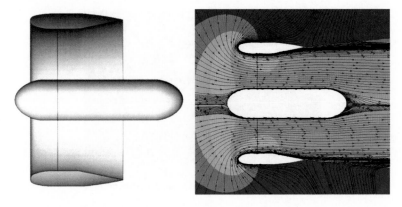

Fig. 3. (left) Geometric model (right) Space flow field

In view of the shape data given in the report [10], the catia 3D geometric modeling is carried out as shown in the left of Fig. 3. The model is divided by ANSYS ICEM mesh division tools and the ANSYS CFX is used to analyze the aerodynamic analysis of the model based on the RANS equation, and the spatial flow field of the experimental model under the hover state is obtained, such as Fig. 3. As shown on the right.

The calculated ducts pull force is 113 N and the error from the experimental results is 5.8%, so the accuracy of CFD analysis using momentum source calculation method is proved. On the basis of the calculated results, the mass flow rate of the ducts is 4.71577 kg/s. In order to verify the momentum theory, the calculated ducted thrust, the pull force of the equivalent propeller disk and the diffuser ratio of the ducted nozzle are back pushed by the momentum theory. The mass flow rate is 4.89537 kg/s, which is similar to the CFD analysis, and the error is only 4.7%. It can be seen that the accuracy of momentum theory in ducted axial flow is very well.

From the analysis of the flow field, it can be seen that the diffuser of the nozzle does not completely expand the air flow. Because of the existence of the inverse pressure gradient, the thickness of the surface layer of the nozzle is thicker, and the actual nozzle area is less than the area of the flow tube, as shown in Fig. 4. When only the total thrust and mass flow and the geometric expansion ratio are known, the mass flow calculated by the momentum theory is calculated on the premise of the default geometric expansion ratio is the real diffuser ratio, which will result in the larger mass flow compared to the real mass flow. However, in general, considering the performance of ducts at high and low speeds, the diffuser ratio of ducts is designed to be 1. The true diffuser ratio is almost identical with geometric diffuser ratio, and the accuracy of momentum theory is further guaranteed.

Fig. 4. Actual sliding flow tube area

3 Dimensionless Method of Blade Performance for Ducted Fan

3.1 The Definitions of Dimensionless Advance Ratio for Duct Blade

Naipei [9] has analyzed the variation of tensile force and efficiency with advance ratio at different angles of attack for ducted propulsion systems. From the calculated results, it can be seen that at different angles of attack, the same ducted propulsion system exhibits completely inconsistent behavior with the increase of the advance ratio. It is have to analyze a performance curve at each duct inclination angle, so it is unreasonable to apply the dimensionless advance ratio definition of the open propeller directly to the duct blade. Because of the rectifying effect of the chute lip, it can be considered that the airflow in the blade inside the duct is always an axial flow under different external working conditions of the duct, and the theory of motion similarity under different working conditions is also true. Based on this, the dimensionless advance ratio of duct blade can be deduced as follows:

$$J_c = \frac{V_c}{nD}$$

V_c is the axial velocity of the inner flow passing through the blade, and D denotes the ducted blade diameter, n for the speed of the duct blade, and the unit is rev/sec.

3.2 Verification of Dimensionless Definition of Duct Blade Performance

The blade performance under pure internal flow conditions is calculated by using the model shown in Fig. 5.

As shown in the figure, the inlet and outlet of the whole flow tube are set as mass flow boundary condition, and the blade speed is set, as shown in Fig. 6, is the flow line of the inner flow under a certain internal flow condition. By setting the mass flow rate of the inlet and outlet at 0.47698 kg/s, the rotating speed of the blade varies from

12000RPM to 32000RPM. Figure 7 shows the variation of blade performance curve with dimensionless advance ratio. It can be seen from the right figure of Fig. 7 that the ideal efficiency value of the blade appears in the working condition of the advance ratio $J_c = 1.78$ and the efficiency value is 79%.

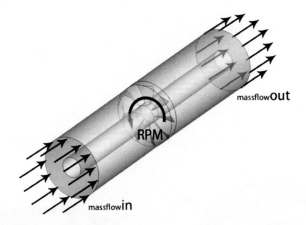

Fig. 5. Calculation model of blade under pure internal flow

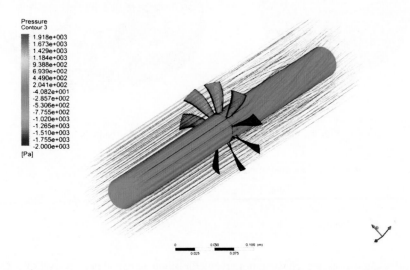

Fig. 6. Flow line

In order to verify the feasibility of the dimensionless method, it is also necessary to calculate the performance of the actual ducted propulsion system under different working conditions to determine that the dimensionless performance of the ducted blades is consistent with the calculated performance under pure inflow conditions. The surface mesh of the ducts and blades used in the calculation is shown in Fig. 8.

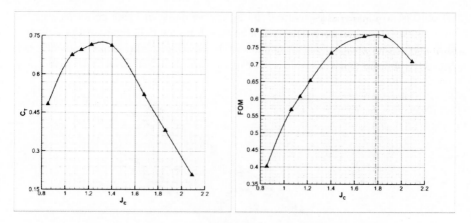

Fig. 7. (left) Force coefficient, (right) Blade efficiency value

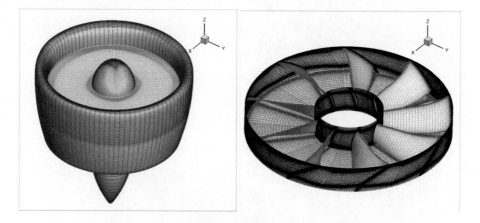

Fig. 8. The surface meshes of the culvert and paddles

Table 1. Calculation results of two typical working conditions

	RPM	V	α	J	C_T	C_P	η
Condition 1	12000	8	0	1.4708	0.6736	1.3232	0.749
Condition 2	24000	8	45	1.47	0.687	1.332	0.758

The calculation uses the quasi-stationary method and is calculated by ANSYS CFX based on the RANS equation. The calculation flow field is divided into a rotating domain and a static domain. The static domain includes a ducted body and a far field, and the rotating domain includes a rotor blade and a hub of the rotating portion, and data is transmitted between the rotating domain and the stationary domain through the interpolation plane. The geometric parameters of the calculated ducted fan are: the blade diameter is 120 mm, the geometrical expansion ratio of the duct is 1, and the gap between the blade tip and the duct is 0.5 mm.

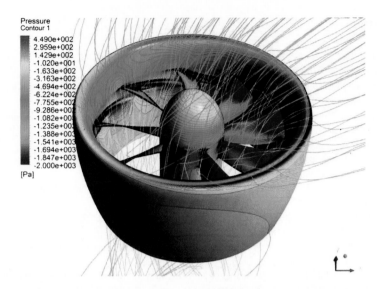

Fig. 9. Flow pashlines and contours of AOA 45 deg of a duct fan

The calculation results are shown in Table 1. V is flow speed; α is angle of attack; J is advance ratio; C_T is force coefficient; C_P is power coefficient; η is Blade efficiency. It can be seen that the two different speeds of different angles of attack have basically similar dimensionless advance ratios, and the blade performance parameters are basically the same at similar forward ratios. Compared with the pure in-flow condition in Fig. 7, the performance exhibited by the blade under the conditions of the ducted condition is also in agreement with the performance curve under the pure in-flow condition. It can be verified that the non-dimensionalization method of this paper is reasonably available (Fig. 9).

4   Conclusion

In this paper, based on momentum theory, a model of duct propulsion system of ducted-fan VTOL aircraft is established and verified. On this basis, a non-dimensional method of ducted blade is developed, and is useful to understand the performance of the ducted fan propulsion. It is convenient and effective to employ the ducted fan nondimensional method to carry out integrated designing considering hovering mode and level flight mode for ducted-fan configuration aircraft.

References

1. Cyrus A (2010) Aerodynamic Analysis and Simulation of a Twin-Tail Tilt-Duct Unmanned Aerial Vehicle. University of Maryland
2. Dennis DO (2016) Conceptual design of XTI Aircraft TriFan 600. In: 16th AIAA Aviation Technology, Integration, and Operations Conference. Washington, D.C
3. Rajagopalan RG (1991) Laminar flow analysis of a rotor in hover. J Am Helicopter Soc
4. Rajagopalan RG, Mathur SR, Three Dimensional Analysis of a Rotor in Forward Flight. AIAA-1989-1815
5. Zori LAJ, Rajagopalan RG (1995) Navier-stokes calculations of rotor-airframe interaction in forward flight. J Am Helicopter Soc, April 1995
6. Acheson KE (2011) Fan interaction study for distributed ducted fan systems in hover and transition. In: 29th AIAA Applied Aerodynamics Conference
7. Brandt JB (2011) Propeller performance data at low reynolds numbers. In: 49th AIAA Aerospace Sciences Meeting. Orlando, FL
8. Liu P (2005) Theory and Application of Air Propellers. Beijing University of Aeronautics and Astronautics Press
9. Naipei P, Bi KRKD (2009) Performance investigation of ducted aerodynamic propulsors. In: First International Symposium on Marine propulsors. Trondheim, Norway
10. Kalman J, Grunwald KWG (1962) Aerodynamic loads on an isolated shrouded-propeller configuration for angles of attack from -10 to 110, in NASA TECHNICAL NOTE. NASA

OrbitPlus Open CubeSat Platform Feasibility Study and Preliminary System Design

Hamed Ahmadloo[1,2], Alireza Mazinani[1,3], Sara Pourdaraei[1,3],
MohammadReza Bayat[1,3], Mahyar Naderi[1,3],
and Ehsan Sherkatghanad[1,3(✉)]

[1] OrbitPlus Technologies Group, Haidian District, Beijing, China
[2] Beijing Institute of Technology,
5 South Zhongguancun Street, Haidian District, Beijing, China
[3] Beihang University, No. 37 Xueyuan Road, Haidian District, Beijing, China
ehsan_shgh@yahoo.com

Abstract. Following the commercial trend of space sector and taking advantage of smaller satellites to be launched especially in LEO; OrbitPlus Technology Group began the feasibility study and preliminary system design for a 3U CubeSat which can support a general designed payload in one unit and two other units which filled by essential subsystems. This paper contains the main subsystems initial conditions to support longer orbital lifetime with altitudes higher than 500 (Km) for different types of payload. To have a clear idea on required subsystems, Structural, Command and Data Handling, Telecommunication, Power Generation, Attitude Control, Propulsion and End-of-Life Disposal subsystems have been investigated. This study can be first step of initial detail system design of OrbitPlus CubeSat open platform with respect to specific requirements of different payloads.

Keywords: Space system design · CubeSat · Satellite subsystems · LEO

1 Introduction

Current number of new space missions with small payloads is dramatically increasing with a large percentage of lunched CubeSats from different space agencies, private companies and educational sectors like universities. Technology developments are leading the space exploration programs to take advantage of more complex systems in small size platforms such as CubeSats with one to 12 unit of cubes with fixed 10 (cm) edge size and maximum mass of 1.33 (Kg) per unit. This new horizon of small payloads is giving the importance to reduction of mission cost and less risk of failure in each mission. CubeSats are the most common standardization of small payloads; able to perform a space service for an educational, commercial and even a scientific missions.

In this paper, the open platform for supporting a payload with longer orbital lifetime with respect to available CubeSat standards has been studied. The proposed system of 3 unit CubeSat which can support less than 5 years of in-orbit lifetime for a general designed payload with a maximum size of one unit is investigated. Two units out of three are containing the required subsystems that can support the payload and the

© Springer Nature Singapore Pte Ltd. 2019
X. Zhang (Ed.): APISAT 2018, LNEE 459, pp. 1231–1246, 2019.
https://doi.org/10.1007/978-981-13-3305-7_97

platform is adoptable based on payload and mission requirements. To keep the flexibility of this 3U satellite for different type of payloads, the considered altitude can even be higher than 500 (Km).

For above explained 3U CubeSat platform, all required subsystems for supporting the main payload has been studied. For main structural frame new lightweight materials is considered, new method for command data handling subsystem which is adaptable for simultaneous control of main subsystems and payload is proposed, wide range and reliable telecommunication link implementation is suggested, passive and active attitude control system by taking advantage of GPS signals for telemetry propose is investigated, power generation subsystem to support the orbital performance is explored, propulsion subsystem specifics from available technologies is selected and finally, the de-orbiting subsystem for End-of-Life disposal phase is presented. This paper will address to the first step preliminary system design of a 3U CubeSat platform, which can support payloads with higher altitude and longer orbital lifetime. The aim of this study is designing a commercial open platform of CubeSat to facilitate space environment access of any payload with no need to design other required subsystems.

2 Structural Subsystem

Generally, the Structural Subsystem of CubeSat is made of a lightweight material that provides appropriate connection to the other subsystems to make a safe passage through all phases of the mission. The simplicity of production and montage, having lightweight structure, free space for sensors, battery, circuitry, and payloads are key features in the design of the CubeSat structure. In addition, the CubeSat's structure can insert multiple payload sensors integrated into the subsystem easily. In fact, the aim of the CubeSat structure is to supply a simple, firm structure that will pass launch loads successfully, while guaranteeing an easily accessible data and power bus in order to debug and assembling of components. Due to the limitations of geometrical size and small financial budget of the CubeSat, this must be made based on maximizing usable interior space, while minimizing the intricacy and cost of the design. The CubeSat configuration is basically a cube, with dimensions of $10 \times 10 \times 10$ cm in outer sides and 3.0 mm spacing above each side of the cube for the installation of outer components such as the antenna, data link, and power charger inlet port. The mass center of the CubeSat must be within about ± 2 cm of the geometrical one. The maximum allowable mass of CubeSat is 1.33 kg, and it is expected that the mass structure should not be more than 30% of the total CubeSat mass approximately [5]. The structure should also satisfy harmonic and random vibration requirements in related tests.

Lightweight materials such as aluminum alloys and composite material present considerable feasibility for usage in aerospace structures. After the introduction of advanced laminated hybrid materials like fiber metal laminates (FMLs), the tendency to produce modern aerospace components with composite materials is also growing manifolds [9, 17, 21]. FMLs are a kind of hybrid materials consisting of alternating layers of metal sheets and fiber-reinforced epoxy prepreg.

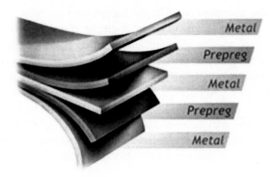

Fig. 1. Fiber metal laminate structures

Aluminum is the most conventionally used metal and Kevlar, glass or carbon can be used as fibers. CARAL is the trade name of FML using carbon fibers in the middle layers and GLARE is the trade name of FML using glass fibers in the middle layer. These materials offer a pioneer combination of characteristics such as distinguished fatigue resistance, superior specific static properties, significant impact resistance, flame resistance and corrosion properties. In addition, for normal components, manufacturing and repair are not difficult. Also, some solutions have been presented to solve the production of these materials. Zafar et al. modified a new method named as "3A method" and has been offered to produce complex shaped, as well as monolithic and metallic sheet materials. Their methodology is generating of metallic blanks in the desired form simultaneously, by using hydroforming process [24]. Sherkatghanad et al. investigated an innovative methodology applying semi-cured FML blanks instead of a solidified blank to form complex shaped parts using hydroforming method [22]. Anyway, several materials are considered in this research selecting the best choice for material selection. The criteria for selection are based on characteristics as follows: Strength, Weight and Machinability (Table 1).

Table 1. Characteristics of suggested lightweight material

Material	Elasticity modules (Gpa)	Density (gr/cc)	Ultimate tensile strength (Mpa)	Machinability
Aluminum 2024-O	73	2.78	425	Good
Carbon Epoxy-woven (50%–50%)	75	1.6	800	Weak
Glass Epoxy-woven (50%–50%)	31	1.8	600	Weak
GLARE	60	2.5	627	Normal
CARAL	72	2.4	Wide range	Normal

According to the above-mentioned parameters, carbon epoxy is considered as a nominal choice for structural subsystem of this CubeSat platform. In generally conditions, due to corrosion, environmental effects, and outgassing issues utilizing this material needs to be investigated in details. So, using metallic layers on the outer sides can solve all these problems and FML can be a really good option. As the outer sides of these materials covered by Aluminum, utilizing this material has both advantages of a metallic part as well as composites. In this condition, there are two appropriate options. Although Caral is lighter and superior than Glare the issue of Galvanic corrosion may happen in Caral material. Altogether Glare can be a new choice to make the CubeSat structure with advantage of decreasing the structure mass by about 10%.

3   Command and Data Handling

There are some restrictions to design of commercial CubeSat. These restrictions can be considered in two categories. The first category of restrictions is due to mass and size constraints that it can be explained as:

- C&DH boards have to small enough to be adjusted in CubeSat dimensions.
- C&DH components must to be light enough to be adjusted in CubeSat requirement.
- Power providing has limitation due to size constraints of battery and solar cells so power consumption of C&DH should be as low as possible.

Another hand, second category of restrictions is due to complexity and flexible development of subsystems. Because of its commercial application, basis of each mission and related payload, other subsystems need to develop and modify independently. According to [4, 6, 18]. There are three kinds of structure for C&DH, that it has shown at Table 2.

Table 2. Categorization of C&DH structures in CubeSat

1	Centralized structure:	All of subsystems are connected to a central processor in C&DH
2	Duplicated structure:	Two (or more) processors can operate simultaneously or only one of them is operating at a time according operation plan or schedule
3	Distributed structure:	Every subsystem has its own processor

It should mention that processor word describe processor board that it may referred to microcontroller unit or processor unit include microprocessor, Read Only Memory (ROM), Random Access Memory (RAM) and bus interfaces. Each kind of structures has comparative advantages related to others. Despite this, there are significant disadvantages that should be considered for designing which briefly has been shown in Table 3:

Table 3. Advantages and disadvantages of C&DH structures in CubeSat

	Advantages	Disadvantages
Centralized structure:	1. Simplicity in its implementation 2. Reduction in components cost 3. Depending on the type of processor, Power consumption can be low	1. Development has low flexibility 2. Single point failure 3. Assign and manage the interface for all subsystems
Duplicate structure:	1. Solve single point failure 2. Power consumption can be low by cold redundancy method	1. Development has low flexibility 2. Components cost 3. Bug detection problem
Distributed structure:	1. Each module can be developed separately 2. Speeds up development time 3. Reducing the workload of C&DH 4. Reducing the codes' complexity	1. Components cost 2. High power consumption 3. More space

Figure 2 shows our proposed structure that C&DH can operate under three processes of modules simultaneously or only with one master process unit or with all of processes of modules and units at the same time. Indeed it can work in two modes that developers can adjust it base on different payloads, for example main mode can operate under three modules simultaneously and if any problem accrues in any of them, master processor will work as alternative and the workload will be swapped on it by using watchdog method, So, in ordinary situation, master processor works in standby mode to reduce power consumption. Also proposed structure provides a good mechanism in preventing single point failure and experts can develop payload and other subsystems separately and then integrate them.

To reach an optimal design for proposed structure, it is necessary to select main components optimally. In each section experts can develop related module independently. So the critical factor to be considered is the processor unit's specifications. Parameters that needed to be considered are processor selection, ROM size, RAM size, bus interfaces and power consumption. These parameters are definitely depending on the payload and mission.

Three factors must be considered in components selection of commercial CubeSat: (a) cost-effective, (b) convenient to use and (c) reliable for mission. So with Considering mission and payload, most of components should be the best chose for instance if we require to fast data rate we should be chose SPI as interface and if we consider restriction in number of wires connection we should be chose I2C as interface so basis of proposed structure, the SPI is suitable because it used for communication between multiple devices with master device over short distances at high speed. According to

interface, Arduino Uno boards with ATmega328 can be used for modules and
LPC3250 SOM with ARM926EJ-S CPU (Johl et al. 2014) can be used as master
process unit because of it is designed for low-power, high-performance applications,
which is ideal for CubeSat flight computers.

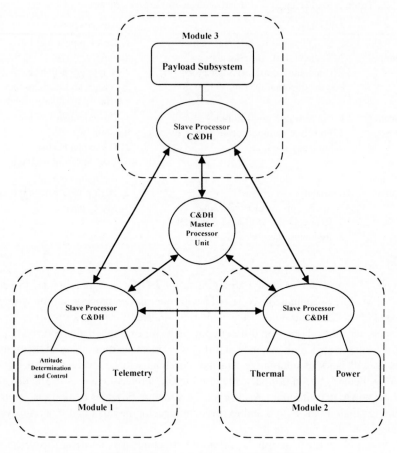

Fig. 2. Proposed structure of C&DH

4   Telecommunication

Function of communication module in CubeSat is receiving commands from the
communication facility and transmitting to the command and Data Handling system in
uplink path and transmit payload and engineering data to the ground facility in
downlink path.

In order to enable affordable and reliable link implementations, the use of commercial off-the-shelf (COTS) radios is preferred, which are modified for use in space.
A significant consideration in selecting the transceiver is implied by the parameters of
the digital modulation techniques which are determined by the requirements of the

payload. Specifically, knowing the amount of information I to be transmitted to the ground station during one trajectory pass along with the visible time T of cubesat from ground station simply data rate R on the CubeSat-to-ground station radio link as $R = I/T\,bits/s$. According to the design parameters, the height of 550 km of Low Earth Orbit (LEO), total orbit period time is 1 h 35 min and 39 s and viability time of the satellite is about 12.2 min. Also in this scenario 2 ground stations had been defined, so the total sending and receiving time in a day would be 24.4 min equal to about 1464 s. Most of the traditional components are proposed maximum 1 Mbps data rate in VHF, UHF and L bands. Another approach in CubeSat can be usage of Software Design Radios (SDR) which has become a proper alternative in recent years for the implementation of multi-purpose receivers. SDRs easily perform most of the processing in software without devoting special hardware to modify radio devices and are based on the use of field programmable gate array (FPGA) technology. FPGAs can be reprogrammed to suit different needs and to enable versatile transmitters and receivers for CubeSat telecommunication subsystems. This technology has the potential to reduce cost and development time while providing significant flexibility in terms of the modulation and demodulation techniques that are available for use. Multiple choices to implement transceivers for both the CubeSat and the ground stations are available in this scenario rather than traditional and customized choice. Although each configuration can have its own pose and cons, based on our design criteria it would be the best possibility. Reference [2] provides a fully integrated, single board radio platform with continuous frequency coverage from 70 MHz to 6 GHz and up to 56 MHz of real-time bandwidth. The main claim about designing a multi-purpose CubeSat has also been met the telecommunications subsystem needs.

5   Power Generation

Three basic parts for any CubeSat power generation Subsystem can be represented as: power sources (Solar cells), energy storage (battery), and power management and distribution (PMAD).

For the case of near earth satellite like CubeSats, the common source of energy is provided by Sun and solar panel arrays are required for generating sufficient power of all subsystem during the mission. The two phases of Sun illumination and eclipse of Earth shadow needed to be considered carefully. The sizing process for required solar panel arrays should support the subsystems workload and also charging the battery for eclipse phase. The orbital elements of the satellite is defining the initial parameters for power generation subsystem design, since, orbital altitude and inclination of the orbit defines the eclipse time of the satellite. Typically, multi-junction cells are used in space applications because they are consisted of several materials and multi-layer, which allow them to have a high efficiency over a wide range of the spectrum related to single junction cells. One of famous multi-junction solar cells is Clyde Space from EMCORE and Spectrolab [3].

The main responsibility of the satellite battery is providing enough power for satellite during the eclipse, the capacity of the batteries are selected based on the eclipse time and the required power of the subsystems' of the CubeSat platform. The range of accessible power from GaAs solar panels for 3U CubeSat is between 7.9 (W) to 20 (W) with four deployed panels [10].

By having the solar panel arrays and batteries, PMADs provide an appropriate allocation of power to subsystems and prevent damage to electronic devices, and also it manages the usage of the CubeSat battery, voltage amplification, controlled capacitor charging/discharging and voltage signal conditioning. Most PMADs are designed to meet payload specific criteria and the final selection from commercial systems will be addressed based on objectives of the payload.

6 Attitude Control

Due to mission objectives, implementing attitude control can be vital to gives the ability to the payload for successful service during the mission. It can be challenging to reach sufficient accuracy and precision based on mission requirements. The existing attitude control subsystems for other satellites are not suitable for our platform since the mass, required volume and cost are higher out defined margins. Considered Attitude Control System (ACS) should firstly contains the accurate sensors to predict the orientation of the CubeSat and secondly, mechanism for adjusting the satellite to the desired orientation. The available sensors for typical CubeSats can be divided into two types of Digital Sun Sensors and Digital Compass which both of them can deliver sufficient accuracy for typical payloads. Selection criteria can be adjusted to platform final cost, since the mass and volume is relatively low (mass is less than 20 (mg) and $60 \times 30 \times 10$(mm)). Sun Sensors are more accurate in comparison to the Digital Compass, with accuracy of $\pm 0.5°$ and $\pm 2°$, respectively [8].

The general approach for combined magnetorquer and reaction wheels are the same for this platform. Since, by continues usage of reaction wheels they can get saturated, magnetorquers are necessary for avoidance of this situation. From different types of magnetorquers, the rod shapes with tight wound coils are preferred. Due to limitation of total subsystems mass and power, the acceptable range can be mass of less than 80 (g) and maximum power of less than 0.5 (W). For case of reaction/momentum wheels, first option is manufacturing the system in our group after accurate simulation of the whole platform; second option can be usage of existing commercial ones. In commercial existing systems, power consumption range is 45 (mW), maximum speed of 6500 (RPM), maximum torque is 3.2 (mNm), with mass of 300 (g). The ACS design and selection will be addressed in future publications, since, the selection of the payload and its requirements are necessary for further steps.

6.1 Navigation

There are different methods for orbit determination of a CubeSat in Low Earth Orbits (LEO) missions, including Global Navigation Satellite System receiver, Sun Sensor, Magnetometer and Gyroscope. Among these methods the most cost-effective one is Global Navigation Satellite System (GNSS) which is widely used in LEO missions, as a navigation system or even as a payload for remote sensing goals. Besides orbit determination GNSS can also be used for onboard positioning and timing, attitude determination, formation flying and remote sensing. These are the reasons that GNSS receiver is assumed to be installed on this CubeSat platform.

Table 4. GNSS receivers characteristics [14, 16, 20]

	GPSRM 1	piNAV-L1	Venus838FLPx-L
Size(mm^3)	96 x 90 x 18	75 x 35 x 12.5	10 x 10 x 1.3
Weight (g)	106	47	0.3
Power (w)	1.3	0.12	0.18
Power supply (v)		2.7 to 3.6 V	3.3
Supported constellation/ frequency	L1/L2/L2C GPS L1/L2 GLONASS	L1 GPS	L1 GPS
Dynamic sensitivity (km/s)	...	9	7.8

In order to use GNSS receiver in space different requirements should be considered. As an example in High Earth Orbits (HEO), GNSS receiver should have a good performance in low SNR signal environments, but in LEO the high dynamic performance of GNSS receiver is more important [11]. In this paper the requirements of a GNSS receiver in a CubeSat for the lifetime of 5 years with orbit altitude of 550 km is considered and some COTS receivers which are best-fitted ones are listed. Although GNSS receiver in LEO uses the benefit of more visible GNSS satellites and less ionosphere errors than terrestrial receiver, it faces to some challenges, such as high Doppler and Doppler rate, rapidly changing geometry, radiation, and onboard processing limitations. If the GNSS receiver is aimed for a CubeSat or nanosatellites other limitations of size and weight should be considered, too.

The most challenging part of a GNSS receiver that make it distinguished for LEO missions is related to the satellite high dynamic motion as one of the most important requirements. The orbital velocity in the altitude of 550 km is about 7.58 km/S, that it leads to a high Doppler effect on the GNSS signals. The Doppler frequency is estimated through the Eq. (1):

$$f_d = f_r - f_s = -f_s \cdot \frac{v_R}{c} \tag{1}$$

The receivers in LEO should be able to work in Doppler shift of ±50 kHz. In the Table 4, some COTS receivers that support the orbital velocity of 7.58 km/s are listed with their characteristics. Since this GNSS receiver will be installed on a commercial CubeSat platform the most cost-efficient selection can be Venus838FLPx-L receiver.

7 Propulsion

The propulsion subsystem is one of the primary devices for satellite that can enhance the capability for orbit modifications and attitude control. Due to the limitations, until now CubeSats mostly have been used for near earth missions, this fact magnifies the importance of using the propulsion system on higher orbit mission CubeSat. Correct

Table 5. Available propulsion systems for CubeSat (1)

	Product/propellant	Thrust	Isp	Dry mass	Power	Size	Flight heritage
Chemical propulsion systems and thrusters	Hydros/water	0.25–0.6 N	258 s	0.5–1 kg	–	0.5–1 U	–
	CDM-1/AP/HTPB	76 N	226 s	0.46 kg	–	153 cm³	–
Cold gas	NANOPS/SF6	35 mN	46 s	480 g	–	5 × 5 × 10 cm	CanX-2
	CNAPS/SF6	12.5–50 mN	45 s	240 g	–	7 × 12.5 × 18 cm	CanX-5
	MEMS/xenon	100 mN	30 s	188 g	–	9.1 × 9.1 × 2.2 cm	MEPSI-3
	T3-μ PS/nitrogen	6 mN	69 s	<10 g	–	100 mL	Delfi-n3xt
	MEPSIMiPS/isobutane	53 mN	65 s	456 g	–	9.1 × 9.1 × 2.5 cm	–
	Custom/R236fa	110 mN	64 s	290 g	–	10 × 9 × 4.4 cm	Bevo-2
	Microprop system/N2,Xe, He	0.1–1 mN	50–75 s	115 g	–	4.4 × 4 × 5.1 cm	–
Electrostatic systems	SiEPS	74 μN	1150 s	100 g	1.5 W	0.2u	Aero-Cube-8 IMPACT
	MiXI/xenon	0.4–1.5 mN	1760–3100 s	200 g	14–50 W	–	–
	μ NRIT-2.5		360–2850 s	210 g	35 W	–	–
	BIT-1/xenon, iodine	100–180 μN	2150–3200 s	53 g	10–55 W	–	–
Electrodynamic systems	μ PPT/PTFE	10–20 μ N-s	1000 s	25 g	0.5–4 W	–	–
	PPTCUP/PTFE	40 μ N-s	650 s	280 g	2 W	9 × 9 × 2.5	–
	PPT/PTFE	2.5–10 μ N-s	900 s	300 g	1 W	9 × 9 × 3 cm	–
	μ PPT/PTFE	0.5 mN	700 s	550 g	2 W	0.5U	Falcon-Sat 3
	BmP-220/PTFE	0.02 mN-s	536 s	500 g	1.5–7.5 W	370 cm³	–
	μ CAT/metal	1–20 μN	3000 s	200 g	<10 W	200 cm³	BRICSat-P
	μ BLT/Al	54 μN		250 g	4 W	4 × 4 × 4 cm	–
Electrothermal systems	MEMS system/butane	10 μ N-1mN	60–92 s	295 g	2 W	0.44 U	–

selection of the Propulsion System is among one of the most important steps of the conceptual design phase that can help designer and project managers to make effective decisions. This will help to assess the impact of new or alternative concepts, refine the requirements, predict and improve the system configuration in the early design or enhancement phases before requiring significant funding to be expensed [12, 13] (Fig. 3).

The propulsion subsystem is one of the primary devices for satellite that can enhance the capability for orbit modifications and attitude control. Due to the limitations, until now CubeSats mostly have been used for near earth missions, this fact magnifies the importance of using the propulsion system on higher orbit mission CubeSat. Considering the propulsion systems available for CubeSat, the following solutions can be proposed: Cold gas, chemical propulsion systems and electrical propulsion systems (including electrostatic, electrodynamic, electro thermal). Regarding the published data, the information of propulsion systems for CubeSat with a dry mass lower than 0.6 (kg) is presented in Table 5. Based on the predicted thrust and the requirement for the propulsion system to be less than one Unit of a 3U CubeSat, the Cold gas propulsion system with a dry mass less than 600 g is proposed and the related performance is depicted in Fig. 1.

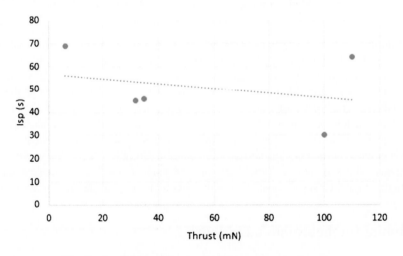

Fig. 3. Available cold gas propulsion system performance

8   Disposal

Regarding to higher altitude required for some payloads to be operated by OrbitPlus CubeSat, it is necessary to comply with mitigation guidelines of InterSpace Agency Debris Mitigation Committee (IADC). Since, the designed orbital altitude is higher than 500 (km) and lifetime is considered to be 5 years; the disposal mechanism of the satellite at the end of the mission must guarantee 25 years of de-orbiting. Based on limited mass and volume of 3 unit CubeSat, the active de-orbiting mechanism is not suitable for this type of satellites. On the other hand, low earth orbits with altitude less than 1000 (km) have the advantage of using atmospheric drag for spacecraft de-orbiting. Proposed disposal mechanism for this satellite is mainly based on lightweight drag sail which can be deployed at the End-of-life moment of the CubeSat [1, 19, 23] (Fig. 4).

Fig. 4. Compact drag sail front and back side

With this disposal subsystem, there is no need of active attitude control subsystem during the de-orbiting phase and more importantly, the sail can be adjustable for exact de-orbiting time as the mission requirements. Drag sail subsystem takes advantage of hollow telescopic moon shape booms with stages which can be extended to the final required size. The booms are in diagonal format to minimize the required length of the booms and each boom can be extended symmetrically from both sides to reach the final size. The inner sail part is basically four rectangular shape stretchable panels with reinforcements on the both sides to avoid losing the sail in case of collision with small fragments. The sail will be located the on end side of the 3 unit CubeSat and the deployment mechanism is considered with pressurized gas inside the hollow booms.

9   Initial Configuration and Model

Although in this step the preliminary design of this platform is presented, the first draft designed configuration of the system is presented as well. The main structure of the system for providing two units for the supporting subsystems and one unit for the payload is presented in Fig. 5. The access for the payload for Earth observation payloads is provided and the main structure is adaptable for specific payload requirements.

Fig. 5. OrbitPlus open platform structure

For power generation subsystem as it mentioned before four solar panel arrays are considered initially, this is common approach for CubeSats in the same format. The four panels will be deployed in on direction and from on end side of the CubeSat. Figure 6 is illustrating the solar panels and their deployment mechanism.

Fig. 6. Solar panel arrays configuration after deployment

The total system in general case with not considering specific requirement of the payload is presented in Fig. 7. Although this design would not fulfill all the requirements of different payloads, it can be modified easily regarding to the mission requirements and objectives.

Fig. 7. Initial structural and solar panel arrays configuration

10 Conclusion

In this paper, the primary system design criteria for each subsystem of OrbitPlus open platform of 3U CubeSat has been investigated. The feasibility of 3U CubeSat which can support a general payload with 1U size is illustrated. Based on each subsystem study, structural frame can be manufactured with 5 percent lower mass, C&DH subsystem is proposed with new approach with higher redundancy, telecommunication subsystem is sized for 2 accessible ground stations, power generation subsystem is considered with general requirements of any mission, attitude control subsystem with commercial existing GPS signal revivers is selected, from different propulsion methods cold gas is chosen in this step as the main propulsion unit and disposal mechanism for 25 years of de-orbiting time is designed. Considering the system design for OrbitPlus Open Platform, all the subsystems are feasible to be fitted in this satellite with expected performance for different type of payloads.

References

1. Ahmadloo H, Zhang J (2017) De-orbiting collision risk assessment and detailed orbital simulation of LEO space debris removal drag sail. In: The 9th asian-pacific conference on aerospace technology and science& the 2nd asian joint symposium on aerospace. Beijing, China, pp 201–215
2. Alvarez JL, Rice M, Samson JR, Koets MA (2016) Increasing the capability of CubeSat-based software-defined radio applications. In: 2016 IEEE aerospace conference, pp 1–10
3. Cote K, Gabriel J, Patel B, Ridley N, Taillafer Z, Tetreault S (2011) Mechanical, power and propulsion subsystem design for A Cubesat. Worcester Polytechnic Institute
4. Fiala P, Vobornik A (2013) Embedded microcontroller system for PilsenCUBE picosatellite. In: IEEE 16th international symposium on design and diagnostics of electronic circuits & systems (DDECS), pp 131–134
5. James Wells G, Stras L, Jeans TG, Foiisy D (2003) Canada's smallest satellite: the Canadian advanced nano space experiment (CanX-1). In: Proceedings of the 16th annual AIAA/USU conference on small satellites, Logan, UT, USA, August 2002
6. Johl S et al (2014) A reusable command and data handling system for university CubeSat missions. In: Department of aerospace engineering & engineering mechanical University of Texas at Austin, Austin TX USA Aerospace Conference. IEEE
7. Lemmer K (2017) Propulsion for CubeSats. Acta Astronaut 134:231–243
8. Li J et al (2013) Design of attitude control systems for CubeSat-Class Nanosatellite. J Control Sci Eng 2013:15
9. Marom G (1996) Book review: advanced composite materials. In: Pilato LA, Michno MJ. Springer Verlag (1994), p 208 + xvii. Price: DM 138.00. ISBN 3-540-57563-4. Heterogeneous Chemistry Reviews 3(1):75–75
10. Melone CW (2009) Preliminary design, simulation, and test of the electrical power subsystem of the TINYSCOPE Nanosatellite. M.S. Thesis, Naval Postgraduate School, Monterey, CA, USA, December 2009
11. Pourdaraei S, Qin H, Jin T, Naderi M (2017) High earth orbit navigation by vector tracking. In: 68th International Astronautical Congress (IAC2017), Adelaide, SA, Australia: International Astronautical Federation, IAF
12. Naderi M, Liang G, Karimi H (2017) Modular simulation software development for liquid propellant rocket engines based on MATLAB Simulink. In: International conference on mechanical, material and aerospace engineering (2MAE). Hong Kong society of mechanical engineers (HKSME), Beijing, China, 12–14 May, 2017
13. Naderi M, Liang G, Karimi H (2017) Modeling and simulation of staged combustion cycle LPRE. In: 8th international conference on mechanical and aerospace engineering (ICMAE). IEEE Xplore, Prague, Czech Republic
14. NanoAvionika: Cubesat GPS receiver [internet]. GPS receiver piNAV-L1 [about 2 screens]. Accessed 29 June 2018. https://n-avionics.com/cubesat-components/navigation-systems/cubesat-gps-receiver/
15. Poghosyan A, Golkar A (2017) CubeSat evolution: analyzing CubeSat capabilities for conducting science missions. Prog Aerosp Sci 88:59–83
16. Pumpkin: CubeSat Kit [internet].Datasheet: GPSRM 1 GPS Receiver Module (Hardware Revision: C), p 19, July 2014. Accessed 1 July 2018. http://www.cubesatkit.com/docs/datasheet/DS_CSK_GPSRM_1_710-00908-C.pdf
17. Reyes G, Kang H (2007) Mechanical behavior of lightweight thermoplastic fiber–metal laminates. J Mater Process Technol 186(1–3):284–290

18. Sabri SF, Yuhaniz SS, Kamardin K (2016) Designing a low-cost CubeSat's command and data handling subsystem kit. ARPN J Eng Appl Sci 11(10):6259–6264
19. Sangregorio M et al (2016) Orbital simulation of low-cost drag sail sub-system designed for LEO space debris removal. J Beijing Inst Technol 25(2):15–20
20. SkyTraq Technology: Skytraq [internet]. Datasheet: Venus838LPx GPS Module, p 24. Accessed 13 July 2018. http://www.skytraq.com.tw/products/Venus838FLPx_PB_v1.pdf
21. Sokolova O, Carradò A, Palkowski H (2011) Metal-polymer–metal sandwiches with local metal reinforcements: a study on formability by deep drawing and bending. Compos Struct 94(1):1–7
22. Sherkatghanad E, Lang L, Liu S, Wang Y (2017) Innovative approach to mass production of fiber metal laminate sheets. Mater Manufact Process 33(5):552–563
23. Visagie L, Lappas V, Erb S (2015) Drag sails for space debris mitigation. Acta Astronautica 109:65–75. https://doi.org/10.1016/j.actaastro.2014.12.013
24. Zafar R, Lang L, Zhang R (2014) Experimental and numerical evaluation of multilayer sheet forming process parameters for lightweight structures using innovative methodology. Int J Mater Form 9(1):35–47

Waveriders Designed for Given Planform Leading Edge Curves

Xiaoyan Wang[1,2(✉)], Jun Liu[1,2], and Shaohua Chen[1,2]

[1] College of Aerospace Science and Engineering, National University
of Defense Technology, Changsha 410073, Hunan, People's Republic of China
wangxiaoyan_seven@163.com
[2] Science and Technology on Screamjet Laboratory, National University
of Defense Technology, Changsha 410073, Hunan, People's Republic of China

Abstract. Planform leading edge curve (PLEC) is one of the waverider's characteristic curves. As sweep angles of the waveriders designed for given the curve can be controlled directly, the design method for given PLEC is worth being studied. In this paper, a parameterization design method for a more practical waverider given PLEC by use of the osculating cone theory was introduced in detail, and the key of the design process is to parameterize the PLEC and compute the leading points and the flowfield on osculating plane. Constraints of the method were also given in this study, when the leading points are existed. As the cubic spline curve could be controlled parametrically more conveniently, this kind of curves were used to describe PLEC. A series of waveriders were generated for given different PLECs. Furthermore, the numerical simulation method was used to verify the proposed design method and to analyze the effects of the key parameters of PLEC on the waverider configurations. Conclusions were got: different special parts of waveriders could be controlled. In the conditions this paper set up, waveriders with double body could be generated when controlling two design points, and waveriders with finlets and with chined back were generated when controlling three design points, waveriders with wings were got when controlling four design points. Some aerodynamic performances of the typical waveriders were also calculated.

Keywords: Waverider · Planform leading edge curve · Cubic spline curve · Design method · Special parts

1 Introduction

A waverider [1, 2] is one kind of hypersonic aerodynamic configurations with good properties, named by "wave-riding" because of shock wave attaching to the leading edge [2]. Waveriders [3–5] are applied to hypersonic vehicles widely relying on high lift-to-drag ratio and conveniently integrating design with inlets. By now, the cone-derived method [6–9], put forward by Jones [10], is often used to design waveriders. The basic flowfield of the cone-derived waverider is a conical supersonic at zero angle of attack, but it limits the shape of the shock wave to the arc. Subsequently, Sobieczky [11] came up with osculating cone theory [11, 12], which expanded Jone's design

© Springer Nature Singapore Pte Ltd. 2019
X. Zhang (Ed.): APISAT 2018, LNEE 459, pp. 1247–1260, 2019.
https://doi.org/10.1007/978-981-13-3305-7_98

theory. It was used in waveriders design successfully [13–17] and made the base plane of waveriders could be designed reasonably according to inlet lips' shapes, which made the design space and applied range of waveriders more widely [18–21].

The waverider design based on osculating cone theory [22] is derived by two curve, one is shock wave profile curve (SWPC) to control the shock wave shape, and another is a curve from which the shape of the leading edge on the shock wave can be determined, including upper surface profile curve (USPC), lower surface profile curve (LSPC) and planform leading edge curve (PLEC). A given conical flowfield solution is applied to osculating planes that are perpendicular to the SWPC, while being scaled to match the curvature of the local shock wave. The ensemble of osculating plane describes a three-dimensional flowfield in which streamlines can be traced [23–26] to generate the "wave-riding" lower surface of the waverider configuration.

Waverider design method for given USPC [22] is the most common, the method can specify the shape of leading edge on the base plane, and its design thought is very direct: lower surface of waverider was obtained through calculating leading edge and tracing streamline from that, but it is should be noticed that USPC must locate between shock wave profile and osculating cone. Method for given LSPC controls the shape of lower surface on the base plane, which is benefit for integrating design of waverider forebody with inlet, but inverse streamline tracing is required and there is constraint for location of lower surface profile, too. Method for given PLEC can control the parameters of waverider's shape, length and width, and sweep angles directly, so research of this method has important value not only in theoretical aspect but also in practice.

2   Waverider Design Method for Given PLEC

The design method is an inverse process. Flowfield can be calculated when knowing SWPC and parameters of flowfield. Lower surface is got by tracing streamlines from the leading edge, which is obtained by PLEC and the flowfield, and upper surface is got by use of free-streamline method. Then, a waverider configuration has been generated.

2.1   Design Process

2.1.1   Known Conditions

Parameters of flowfield including Mach number M_a and shock wave angle β, and curves including SWPC and PLEC.

2.1.2   Specific Steps

Firstly shock wave profile curve is dispersed to execute the following steps for the dispersing point:

A. Computing flowfield on osculating plane

Computing curvature circle, which is also conical shock, going though the dispersing point P_1, exhibited in Fig. 1. According to β and M_a, osculating cone can be computed. O_1 is the center point of Conical shock. The location of osculating plane is determined by P_1

and O_1, which is shown in Fig. 2. Flowfield on the osculating plane and the cone-half angle δ are obtained through computing Taylor-Maccoll equation [27].

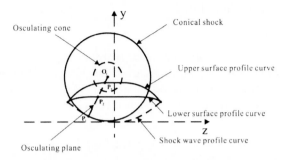

Fig. 1. Osculating plane features on the base plane (zy plane)

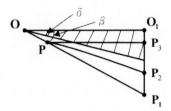

Fig. 2. Osculating plane

B. Computing local leading point

In Fig. 3, O is vertex of conical shock, on osculating plane, leading point P is the intersection of conical shock OP_1 and PLEC.

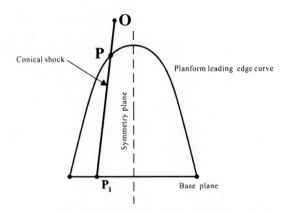

Fig. 3. Conical shock and PLEC from a plan view (zx plane)

C. Tracing streamlines

From leading point P, lower surface streamline PP_2 was calculated by tracing streamline approach [25], and the upper surface streamline PP_3 was calculated by free-streamline approach, shown in Fig. 2.

D. Generating configurations

Repeating all of the above operations to every dispersing point, streamlines on each osculating plane were got. Lofting each streamline into surface respectively could generate upper surface and lower surface, and then closing surface got the configuration. By now the whole waverider was designed completely.

2.2   Constraint Conditions

It is of course not arbitrary curves can be specified as PLEC, there is request out of the proposed design method that the following equation set must have solution when $z \in (0, \frac{1}{2}W)$:

$$\begin{cases} f(x, z) = 0 \\ g(x, y, z) = 0 \end{cases} \tag{1}$$

With $f(x, z) = 0$ the PLEC function, and $g(x, y, z) = 0$ the shock linear function. Because the equation set does not have solution, it means the leading point does not exist.

3   Cases of Waveriders

3.1   Parameterization Design of PLEC

Based on waverider design method for given PLEC, cubic spline curve is used as planform profile here. The parameterization approach of cubic spline curve is shown in Fig. 4.

In the figure, M and N are certain while the length and width of waverider are given, points can be inserted at proper locations (point 1...point N − 1), the n + 1 points divide PLEC into n parts. Control parameters cover: the coordinates of inserts, the angle between axis z and tangent line at point N θ_N (the angle at M is zero by default). The PLEC can be calculated by given parameters, and conditions that the curve has first and second order continuity. So the form of PLEC function is:

$$x = \begin{cases} a_1 z^3 + b_1 z^2 + c_1 z + d_1 \\ a_2 z^3 + b_2 z^2 + c_2 z + d_2 \\ \cdots \\ a_n z^3 + b_n z^2 + c_n z + d_n \end{cases} \tag{2}$$

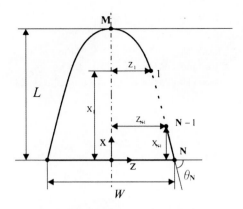

Fig. 4. Parameters of cubic spline curve

3.2   Generation of Waverider Configurations

In order to study the influence of PLEC on waverider configurations, some design parameters were fixed: Mach number $Ma = 10$, design altitude $H = 0$ km, shock wave angel $\beta = 10°$, the length of waverider $L = 4$ and width $W = 3$, shock wave profile is:

$$y = (8/9) \tan 10° \times z^2 - 4 \tan 10° \qquad (3)$$

3.2.1   Basic Two Points Control (no Insert)

In this condition, the shape of planform profile is only decided by the angle θ_N. Values of θ_N are presented in Table 1:

Table 1. Values of θ_N when no insert

Case	a1	a2	a3	a4	a5	a6
$\theta_N/°$	5	25	45	75	80	85

Waveriders related are shown in Fig. 5.

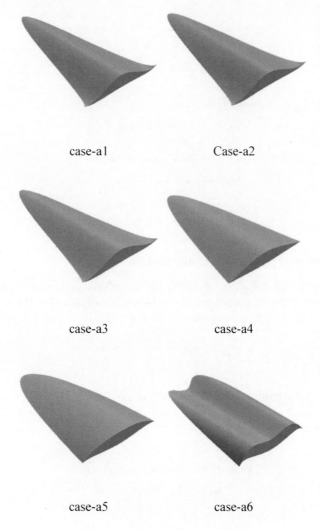

case-a1   Case-a2

case-a3   case-a4

case-a5   case-a6

Fig. 5. Waveriders when zero insert

By analyzing the configurations, it was found that waveriders looking like case-a6 would be generated until θ_N came to a value. This kind of waverider configurations was named "double-body" waverider here because of its bulges at both sides. In the condition of satisfying the constraints by method itself, the value range of θ_N to generate "double-body" waverider is related to length and width of waverider:

$$\theta_N > \arctan(6L/W) \tag{4}$$

This "double-body" waverider is special, it may be more proper to apply in integrating design with two inlets, also it is a new approach to increase volumetric ratio somehow.

3.2.2   Three Points Control

Three points control means there is one insert. Shapes of PLEC depend on not only θ_N's value but also location of the insert. $\theta_N = 45°$, and setting control parameters in accordance with Table 2.

Table 2. Values of control parameters when one insert

Case	b1	b2	b3	b4	b5	b6	b7	b8
X_1/L	0.9	0.7	0.5	0.3	0.1	0.7	0.7	0.7
$Z_1/\frac{1}{2}W$	0.3	0.3	0.3	0.3	0.3	0.1	0.2	0.4

Generated waveriders are in Fig. 6.

The one more insert enriched the shapes of PLEC, and waverider configurations were more various: besides "double-body" waverider (case-b1/b8), waverider with finlets (case-b5) and waverider with chined back (case-b6) appeared.

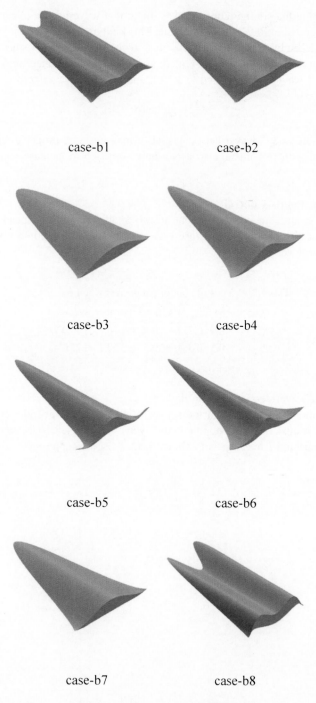

case-b1   case-b2

case-b3   case-b4

case-b5   case-b6

case-b7   case-b8

Fig. 6. Waveriders when one insert

3.2.3   Four Points Control

$\theta_N = 45°$, there are two inserts, values of parameters were shown in Table 3:

Table 3. Values of control parameters when two inserts

Case	c1	c2	c3	c4	c5	c6	c7
X_1/L	0.98	0.48	0.78	0.78	0.78	0.78	0.78
$Z_1/\frac{1}{2}W$	0.14	0.14	0.04	0.08	0.14	0.14	0.14
X_2/L	0.4	0.4	0.4	0.4	0.3	0.7	0.4
$Z_2/\frac{1}{2}W$	0.25	0.25	0.25	0.25	0.25	0.25	0.4

Waveriders are in Fig. 7.

It is of course that configurations became more and more complicated. There were waveriders with wings, which did not appeared in former two conditions. The waverider (case-c2/c3) looks more like a vehicle.

3.3   Numerical Simulation Method

The Euler equations are employed to numerically simulate the inviscid flow fields in the aerodynamic performances of the waveriders by using the commercial software ANSYS Fluent [28]. Specifically, the Euler equations are solved using the density-based (coupled) implicit solver. The second-order spatially accurate upwind scheme that applies the AUSM method to the flux vector is utilized, and the least-squares cell-based method is used to compute the gradients. The Courant-Friedrichs-Lewy (CFL) number is set at 0.5. As indicated previously, the air is assumed to be a thermally and calorically perfect gas.

The iterative process and its solutions can be considered to converge if the residuals reach minimum values, that is, values less than four orders of magnitude of their original value, and when the difference between the computed inflow and outflow mass flux falls below 0.1%.

Because of the symmetry of the geometric configurations, only half of the flow field is required to perform the numerical simulation. To ensure accuracy in the solution of the turbulent flow, a value of y+ less than 1 is used for the main portion of the wall flow region. The numbers of structured grid cells employed in case-b1, case-b5, case-b6, and case-c3 are 3,152,370, 3,391,356, 4,064,418, 3,196,589, respectively. Furthermore, Fig. 8 shows the grid employed in case-b5.

In this study, the design condition of the cases is as follows: a flight Mach number of 10.0 and a cruising height of 0 km with a 0° angle of attack.

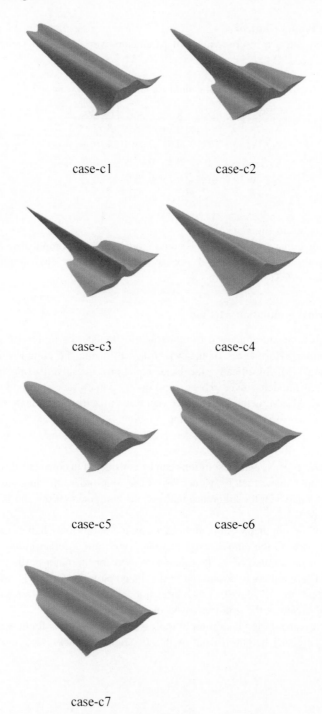

case-c1

case-c2

case-c3

case-c4

case-c5

case-c6

case-c7

Fig. 7. Shapes of waveriders when two inserts

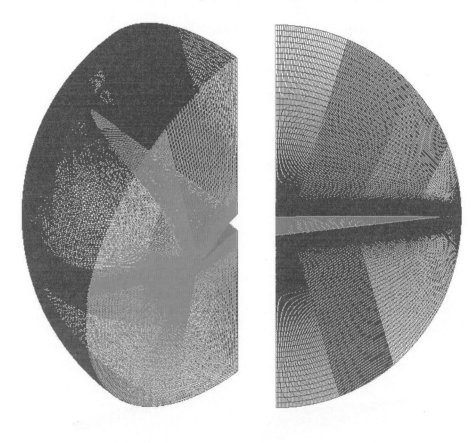

a)   Three dimensional grid   b)   Grid at symmetry plane

Fig. 8. Grid for case-b5

4   Results and Discussion

A part of aerodynamic performances of waveriders at design point are shown in Table 4, where C_L is lift coefficient, C_D is drag coefficient, L/D is lift-to-drag ratio, Cmz is pitching moment coefficient, r is volumetric ratio which is calculated as the following.

$$r = V^{2/3}/S_p \tag{5}$$

with V the volume of waverider and S_p the projected planform area.

Table 4. Aerodynamic performances of waverider cases

Case	C_L	C_D	L/D	Cmz	r
case-b1	0.1686	0.0172	9.8185	−0.1888	0.1438
case-b5	0.0786	0.0078	10.0275	−0.1255	0.1443
case-b6	0.0773	0.0077	10.0192	−0.1371	0.1319
case-c3	0.0779	0.0078	9.9829	−0.1562	0.1076

The non-dimensional pressure contours of selected waveriders are shown in Fig. 9. In the inviscid flow field, there is almost no flow spillage from the lower surface to the upper surface. This indicates that the four cases can ride on the shock wave under the design flight condition when the flow is inviscid, and the shock wave remains attached to the leading edge. In other words, these contours display true waverider characteristics and validate that each special part is "wave-riding" so the proposed design method is correct.

And pressure contours on symmetry plane are shown in Fig. 10.

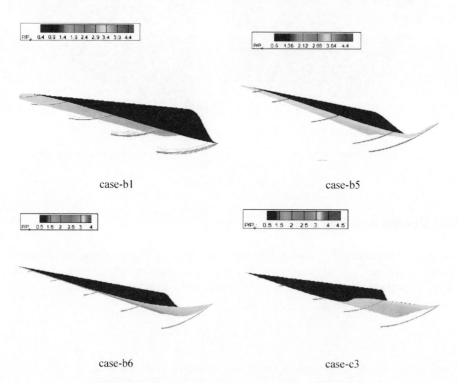

case-b1   case-b5

case-b6   case-c3

Fig. 9. Non-dimensional pressure contours of waveriders

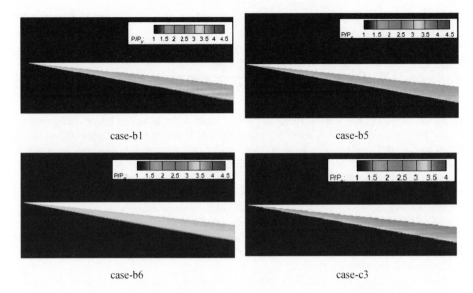

case-b1   case-b5

case-b6   case-c3

Fig. 10. Non-dimensional pressure contours at the symmetry plane

5   Conclusions

(1) Based on osculating cone theory, a waverider design method for given planform leading edge curve was introduced detailedly in this study, and the constraint of method is that the leading point exists;

(2) By setting parameters, a series of waveriders were generated. The more parameters, the more complicated configurations were generated, such as "double-body" waverider, waverider with finlets, waverider with chined back and waverider with wings.

(3) The method was validated to be correct by numerical simulation, every kind of configurations had a good "wave-riding" effect, including at the special parts such as bulges, finlets and wings.

References

1. Nonweiler TRF (1959) Aerodynamic problems of manned space vehicles. J Roy Aeronaut Soc 63:521–528
2. Lunan D (2015) Waverider, a revised chronology: AIAA Paper
3. Haney JW, Beaulieu WD (1994) Waverider inlet integration issues. In: Proceedings of the AIAA Paper 94-0383
4. You YC, Cai DWLK (2018) Numerical research of three-dimensional sections controllable internal waverider hypersonic inlet
5. You YC, Zhu CX, Guo JL (2009) Dual waverider concept for the integration of hypersonic inward-turning inlet and airframe forebody

6. He XH (1992) Computational analysis of hypersonic flows past generalized cone-derived waveriders. University of Oklahoma, Norman
7. Lin SC, Rasmussen ML (1988) Cone-derived waveriders with combined transverse and longitudinal curvature: AIAA Paper
8. Rasmussen ML, Clement LW (1986) Cone-derived waveriders with longitudinal curvature. J Spacecraft 23(5):461–469
9. He XZ, Le JL, Wu YC (2009) Design of a curved cone derived waverider forebody
10. Jones JG, Moore KC, Pike J, Roe PL (1968) A method for designing lifting configurations for high supersonic speeds, using axisymmetric flow fields. Ingenieur-Archiv 37:56–72
11. Sobieczky H, Dougherty FC, Jones K (1990) Hypersonic waverider design from given shock waves. In: Proceedings of the First International Waverider Symposium, University of Maryland. College Park (1990)
12. Konstantinos Kontogiannis AS (2015) On the conceptual design of waverider forebody geometries. In: Proceedings of the 53rd AIAA aerospace sciences meeting, Florida. American Institute of Aeronautics and Astronautics (2015)
13. Ding F, Liu J, Shen C-B, Huang W (2015) Novel approach for design of a waverider vehicle generated from axisymmetric supersonic flows past a pointed von Karman ogive. Aerosp Sci Technol 42:297–308
14. Ding F, Liu J, Shen C-B, Huang W (2015) Novel inlet-airframe integration methodology for hypersonic waverider vehicles. Acta Astronautica 11:178–197
15. Ding F, Shen C-B, Liu J, Huang W (2015) Influence of surface pressure distribution of basic flow field on shape and performance of waverider. Acta Astronautica 108:62–78
16. Ding F, Shen C-B, Liu J, Huang W (2015) Comparison between novel waverider generated from flow past a pointed von Karman ogive and conventional cone-derived waverider. Proc Inst Mech Eng Part G J Aerosp Eng 229(14):2620–2633
17. Ding F, Liu, J, Shen, C-B, Liu, Z, Chen, S-H, Fu, X (2017) An overview of research on waverider design methodology. Acta Astronautica 140:190–205
18. Graves RE (1999) Aerodynamic performance of osculating-cones waveriders at high altitudes. University of Oklahoma, Boulder
19. Miller RW, Argrow BM, Center KB, Brauckmann GJ, Rhode MN (1998) Experimental verification of the osculating cones method for two waverider forebodies, 4–6 March 1998
20. Strohmeyer D (1998) Lateral stability derivatives for osculating cones waveriders in sub- and transonic flow
21. Ding F, Liu J, Shen C-B, Huang W (2015) Simplified osculating cone method for design of a waverider
22. Sóbester KKA (2017) Efficient parameterization of waverider geometries. J Aircr 54(3):12
23. Kothari AP, Tarpley C, Mclaughlin TA, Babu BS, Livingston JW (1996) Hypersonic vehicle design using inward turning flow fields: AIAA Paper
24. Billig FS, Baurle RA, Tam C-J, Wornom SF (1999) Design and analysis of streamline traced hypersonic inlets: AIAA Paper
25. Billig FS, Kothari AP (2000) Streamline tracing: technique for designing hypersonic vehicles. J Propul Power 16(3):465–471
26. Slater JW (2014) Methodology for the design of streamline-traced external-compression supersonic inlets: AIAA Paper
27. Taylor GI, Maccoll JW (1933) The air pressure on a cone moving at high speed. Proc Roy Soc London A: Math Phys Eng Sci 139(838):278–311
28. ANSYS, Inc. ANSYS FLUENT 13.0 Theory Guide (2010)

On Aircraft Design Under the Consideration of Hybrid-Electric Propulsion Systems

D. Felix Finger[1,2(✉)], F. Götten[1,2], C. Braun[1], and C. Bil[2]

[1] FH Aachen University of Applied Sciences,
Hohenstaufenallee 6, 52064 Aachen, Germany
f.finger@fh-aachen.de
[2] RMIT University, GPO Box 2476, Melbourne, VIC 3001, Australia

Abstract. A hybrid-electric propulsion system combines the advantages of fuel-based systems and battery powered systems and offers new design freedom. To take full advantage of this technology, aircraft designers must be aware of its key differences, compared to conventional, carbon-fuel based, propulsion systems. This paper gives an overview of the challenges and potential benefits associated with the design of aircraft that use hybrid-electric propulsion systems. It offers an introduction of the most popular hybrid-electric propulsion architectures and critically assess them against the conventional and fully electric propulsion configurations. The effects on operational aspects and design aspects are covered. Special consideration is given to the application of hybrid-electric propulsion technology to both unmanned and vertical take-off and landing aircraft. The authors conclude that electric propulsion technology has the potential to revolutionize aircraft design. However, new and innovative methods must be researched, to realize the full benefit of the technology.

Keywords: Hybrid-electric aircraft · Aircraft design · Design rules · Green aircraft

Abbreviations

EM	= electric motor
EO/IR	= electro-optical and infrared
ICE	= internal combustion engine
L/D	= lift-to-drag ratio
MTOM	= maximum take-off mass
P/W	= power-to-weight ratio
rpm	= revolutions per minute
T/W	= thrust-to-weight ratio
UAV	= unmanned aerial vehicle
VTOL	= vertical take-off and landing
W/S	= wing loading

© Springer Nature Singapore Pte Ltd. 2019
X. Zhang (Ed.): APISAT 2018, LNEE 459, pp. 1261–1272, 2019.
https://doi.org/10.1007/978-981-13-3305-7_99

1   Introduction

Interest in cleaner propulsion has increased in the past decades due to the environ-mental effects of fossil fuel and its limited resources, leading to a rising number of emission regulations. Future emission targets will be more ambitious, as shown by Flightpath 2050 [1], imposed by the European Commission, or NASA's N+3 and N+4 program goals [2].

Fully electric propulsion systems are currently not able to replace conventional systems consisting of only fuel burning engines, for the same performance. This is caused mainly by high battery weight resulting from a low specific energy compared to fossil fuel, which does not allow fully electric powered, long-range flights. The advantages that electric propulsion entail, though, are a huge opportunity to make aviation more efficient and eco-friendly. Electric powertrains offer high energy con-version efficiencies in comparison to Carnot-efficiency restricted fuel burning engines, as well as low emissions and low noise pollution.

A hybrid-electric propulsion system design combines the advantages of fuel-based systems and battery powered systems and can contribute to the solution of the thrust matching problem by preventing inefficient operations of gasoline engines at reduced power settings. Additionally, hybrid designs offer new degrees of freedom for overall aircraft design like distributed propulsion (e.g. [3], [4] or [5]). This is possible since power can be transmitted to light-weight electric motors using wires instead of multiple heavy engines or complex mechanical driveshafts.

Hybrid-electric propulsion systems may offer significant reductions in fuel burn and emissions over conventional aircraft. Yet, aircraft design under consideration of these new propulsion configurations is not treated in the classical design textbooks (e.g. [6–8]). Therefore, this paper gives an overview of the challenges and potential benefits associated with the design of aircraft that use hybrid-electric propulsion systems.

This paper is structured the following way: Following the introduction, the most popular architectures of hybrid-electric powertrains will be introduced. Then, the implications of the new technology on the aircraft design process will be reviewed. Finally, special consideration is given to unmanned aircraft and vertical take-off and landing vehicles.

2   Hybrid-Electric Propulsion Configurations

To establish the nomenclature of hybrid-electric propulsion, an overview of the five most significant configuration choices for power specific propulsion systems will be presented:

1. Fully electric
2. Conventional combustion engine
3. Serial-hybrid
4. Turbo-electric
5. Parallel-hybrid

As this paper focuses on general aviation airplane applications, the internal combustion engine (ICE) is the machine of choice to convert carbon-based fuels to shaft power. Obviously, different sources of power could be employed in the place of ICE, like for example fuel cells or turbo-machines.

2.1 Fully Electric

An electric motor (EM) is supplied with power from batteries and drives the propeller (Fig. 1). Depending on the application, it might be highly desirable to integrate a gearbox between propeller and

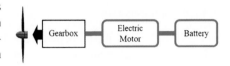

Fig. 1. Fully electric

EM. The first electric-powered flight was made as early as 1883, when an EM was fitted to an airship. Reports [9] indicate that the heavy batteries limited its range, a problem that still troubles electric aviation today.

Aircraft designers tend to assume a specific power per unit of mass to estimate the weight of powertrain components. While weight indeed scales with power for combustion engines, the EM's weight scales with torque. Consequently, because power and torque are connected by the rotational speed, it is very beneficial to run electric motors at high rpm and low torque for a given power requirement, as such machines are substantially lighter. Propellers, on the other hand, are most efficient at very low rpm, when compared relative to EMs. They are limited to a helix tip-speed of about Mach 0.8 [6], which limits rpm to below 3000 rpm for most general aviation aircraft. At smaller scales (e.g. UAVs), a higher rpm is possible. If both propeller and EM are optimized together, a gearbox can significantly improve performance of the system. A detailed trade study, which considers both the weight of the EM and the weight of the gearbox, must be conducted to choose the right gear ratio. A first order weight estimation method for gearboxes is presented in [10]. Since the physical motor size is also driven by torque, a geared solution can be much more compact. Again, a trade study is necessary, to assess this metric at a system level.

Preliminary studies indicate that gear ratios between two and four are sufficient to significantly reduce system weight.

2.2 Conventional Internal Combustion Engine

An internal combustion engine (ICE) drives a propeller (Fig. 2), sometimes with a gearbox in between. It is the counterpart to the fully electric system and known as the conventional powertrain.

Combustion engines cannot be designed as freely as electric motors. The tradeoff between torque, rotational speed and power, as described for electric machines, is not possible. While reduction

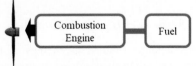

Fig. 2. Conventional ICE

drives have been used for high-power piston engines, the propellers of most general aviation aircraft are most commonly directly connected to the engine crankshaft. To match the maximum propeller speed, aircraft engines develop peak power at about 3000 rpm.

Between both the fully electric and the conventional powertrain which represent the extremes of the powertrain spectrum, there are the partially electric hybrid powertrains.

2.3   Serial-Hybrid Configuration

In a serial-hybrid system (Fig. 3), the propeller(s) are driven by the EM(s) only and an ICE or gas turbine is used to generate electricity for the electric system. The electric system is supplemented by a battery. Sometimes, the unit of ICE and generator are referred to as a range extender.

For the serial hybrid powertrain, the propeller is solely driven by an EM, which is sized to the maximum performance requirements. A multi-propeller layout, where each propeller has its own EM, is also possible. Depending on the layout of the system, the ICE can run at a constant speed and power setting, which allows it to be optimized for one specific design point. It drives a mechanically-coupled generator that produces electric power, which is then fed to the power management system. In this system, power is either routed directly to the EM which drives the prop, or the power is split between the battery (to recharge it) and the EM. The power management system also allows the battery to provide additional power to the EM, if the required power is higher than what the ICE/generator system can provide. Thus, if the ICE is sized for efficient cruise, ICE and battery can still deliver high power for take-off and climb if used together.

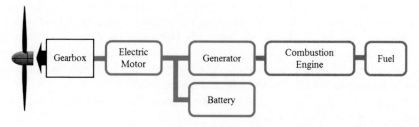

Fig. 3. Serial-hybrid

If the battery is sized accordingly, an aircraft with a serial-hybrid powertrain can also operate in an all-electric mode. This might be desirable from a noise perspective, even for take-off and initial climb.

2.4   Turbo-Electric Configuration

In this layout, combustion engines or turbo machines are used to drive electrical generators, which then power electric motors to driven propellers or fans. The turbo-electric configuration can be considered a serial hybrid, without batteries (Fig. 4).

It avoids the battery weight penalty. However, a generator system and the electric motor are added to a conventional powertrain. Those components infer a weight and efficiency penalty. For special vehicles (e.g. VTOL aircraft – see Sect. 4) the turbo-electric configuration can be of benefit.

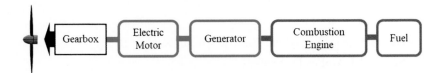

Fig. 4. Turbo-electric

2.5   Parallel-Hybrid Configuration

The parallel hybrid is a configuration in which ICE and EM work in conjunction. For the parallel hybrid powertrain, an electric motor and an ICE are mechanically connected to a propeller shaft, often via a gearbox when both are connected to one propeller shaft (Fig. 5).

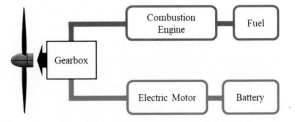

Fig. 5. Parallel-hybrid – Variant 1

An implementation where an electric motor drives one propeller and an ICE drives another propeller is also considered a parallel hybrid (Fig. 6). Just as for the serial hybrid case, the design with multiple propulsion units for one aircraft is possible.

The total required power is split between ICE and EM by a certain degree

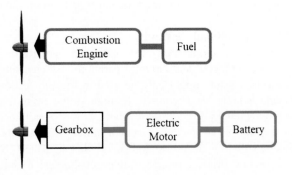

Fig. 6. Parallel-hybrid – Variant 2

of hybridization. This allows for a smaller combustion engine, as well as a small electric motor. Compared to the EM of a serial-hybrid configuration, the EM in a parallel-hybrid powertrain can be smaller and lighter, since it will only provide a certain fraction of the total power.

Because power from the ICE is mechanically transmitted to the propeller, a dedicated generator is not present in this set-up. This reduces complexity and weight and has a favorable effect on the efficiency chain of the powertrain.

3   Influence of Hybrid-Electric Propulsion on Aircraft Design

In the following section, the benefits and challenges on aircraft design associated with the integration of hybrid-electric propulsion systems will be discussed. General aspects are presented first and then the specifics of certain hybrid-electric propulsion layouts are addressed.

3.1   General Aspects of Hybrid-Electric Propulsion

3.1.1   Scale-Free Motor Technology

Electric motors offer high specific power, power density and efficiency, independent of their scale [11]. In contrast, combustion engines and jets lose efficiency with a reduction in size. This is a tribute to smaller tolerance levels of larger engines and a better ratio between internal surface area and swept volume, a factor greatly affecting specific consumption. Additionally, further system penalties like cooling drag apply and degrade system performance further. This behavior allows to employ many small electric motors instead of one big electric motor without a significant reduction in efficiency – even though propeller efficiency does drop slightly due to Reynolds number effects [12]. This opens up the design space for distributed propulsion architectures. Those are covered in Sect. 3.2.

Because electric motors have few moving parts, they can obtain higher reliability and better maintainability than ICEs [13]. They also create much less vibration [14]. Because they have a comparatively low operating temperature range, they do not require a warm-up phase. Rather, they can instantaneously supply full power.

If the electric system is designed accordingly, an electric motor can also act as a generator. This allows to recuperate energy from the flow.

3.1.2   Operational Flexibility

Hybrid-electric propulsion offer significant benefits regarding the flexibility of the configuration layout and aircraft operations. By choosing a certain operation strategy the electric system might only be used in specific sections of the flight, when energy demand is high, or, if required, throughout the entire flight. It is possible to vary the operation strategy from flight to flight, as it does not need to be prescribed. However, any specific design can only vary that strategy in certain boundaries, since the operation strategy will drive the power and energy that the battery system needs to provide. Thus, the operation strategy strongly influences the design and vice versa.

3.1.3   Noise

A significant amount of noise is produced by reciprocating engines. If quiet operation of the aircraft during a certain flight phase is required, the system can be designed to be operated on the electrical system only. The configuration designer can also use the flexibility in placement of electric motors for noise reduction: propulsors can be shielded from the ground by using the aircraft as a barrier. This can be combined with boundary-layer ingestion technology. That way, hybrid-electric propulsion systems might improve both noise and efficiency.

3.2   Specific Aspects of Hybrid-Electric Configurations

3.2.1   Parallel-Hybrid Configuration

The direct connection of the combustion engine to the propeller shaft of a parallel hybrid powertrain offers major efficiency benefits in comparison to a serial hybrid configuration. This leads to the investigation of the parallel hybrid design being an inevitable step to accomplish the aspiring emission targets.

To make full use of all advantages that parallel hybrid-electric powered aircraft can deliver, an optimum design of the propulsion system regarding the proportion of electric power and power from fuel-based propulsion devices is as important as the optimum design of the whole aircraft itself [15].

3.2.2   Serial-Hybrid Configuration

Serial-hybrid configurations offer great flexibility and geometric design freedom. The electric motor(s) can be installed independently of the location of the combustion engine(s). Power can be produced anywhere on the aircraft and routed to small electric motors with cables, wherever desired. This is very desirable for the configuration designer, as small EMs can be integrated with ease. Bulky ICEs can be moved to places where volume is available, without having to connect them directly to a propeller or a fan. Thus, new configurations, which can take advantage of favorable aero-propulsion interaction are possible. Often times, such configurations are unachievable with traditional propulsion systems. This includes propellers placed at the tip of lifting surfaces (wing, horizontal tail, vertical tail [16]) or even the implementation of a distributed propulsion configuration (see e.g. [17], [18] or [3]). Thereby, the weight increase of the propulsion system is traded against improved aerodynamic efficiency.

One of the benefits of distributed propulsion is the ability to put propulsion where the drag is. As an example, a propeller behind a blunt body can significantly reduce flow separation and result in superior performance. Propulsors can fill in wakes or tailor downwash distributions, thereby improving aerodynamic efficiency [18]. Distributed propulsion can also be used as a high-lift device. By blowing over the wing surface, the section lift coefficient is considerably increased [3]. While it might be possible to design a distributed propulsion layout using mechanical linkages, such as shafts, to distribute power, these systems are typically very complex and heavy. Electric power transfer, in contrast, is much simpler.

The Aurora Flight Sciences' XV-24A "LightningStrike" VTOL design used this configuration and carried the distributed propulsion concept to the extreme. It employed a distributed embedded electric fan system that consists of 24 separate motors. Power was provided by a turboshaft engine driving triple redundant generators [19]. The setup would not have been possible by the means of mechanical power transfer.

Parallel- and serial-hybrid propulsion systems should not be compared on a 1:1 basis, because this would not be a fair comparison. As shown in [20], the serial-hybrid will always perform worse than the parallel-hybrid for a similar set of parameters. This is caused by the additional mass that the generator system will add to the aircraft, the reduction of the total powertrain efficiency, and the fact that the electric motor mass is higher, because the motors must be sized to the maximum P/W. Instead, the serial hybrid configuration must be used to improve the overall efficiency of the aircraft. If an increase in aerodynamic efficiency is able to offset the added weight and the reduced powertrain efficiency, than this system pays off.

3.3   Challenges Introduced by Hybrid-Electric Propulsion

3.3.1   Weight

Aircraft are highly sensitive to weight. Weight drives cost, energy efficiency and emissions. Lighter aircraft perform better. This is a key difference that distinguishes the electric aircraft from the electric car, where this influence is much less significant.

While electric motors are much lighter on a power per weight ratio than combustion engines, batteries offer far less energy density compared to carbon based fuels. This weight penalty of the battery must be offset by the resulting gains in propulsive and aerodynamic efficiency to bring a net benefit for electric and hybrid-electric designs [15]. If the weight is not carefully traded off, any new design with this technology is doomed to fail.

3.3.2   Heat

Even though electric systems are much more efficient then combustion engines, they still produce waste heat. This heat is not trivial, especially when high-power electrical systems are used. Electric motors can be damaged by excessive heat, as can motor controllers, and batteries will even fail catastrophically and react with thermal runaway.

Therefore, an effective cooling solution must be designed for these systems. Batteries are special in this regard because they do not only require thermal management when they are discharged, but also while they are charged – at least when the charging process involves high levels of power.

When designing the cooling system, the designer must consider different cases than for ICE propelled aircraft: ICE power lapses with altitude, as does the cooling mass flow. Electric motors can produce full power independently of their operational altitude. This means that the critical conditions for cooling might be high power flight at high altitude – a case that is typically not critical for conventional aircraft.

3.4   Design Rules

The full benefit of hybrid-electric propulsion is realized, when a new aircraft incorporates such a propulsion system from the beginning of the design process [11]. The full benefits can only be obtained by a clean sheet design. Potential performance and efficiency are left on the table if an existing design is simply modified to use hybrid-electric propulsion. The overall design needs to change.

The design point in terms of wing loading (W/S) and power-to-weight ratio (P/W) cannot be chosen as for aircraft with conventional propulsion systems. Instead, the lowest weight design is achieved by using smaller wings and higher installed power [20]. In hybrid-electric aircraft, the electric system can boost the overall power for short period peaks in the power profile, such as take-off and climb phases. This allows the combustion engine to be sized to a continuous power constraint. This might be cruise for a manned aircraft or loiter for a UAV. Meanwhile, the electric system is sized with respect to the remaining delta in maximum required power. This is highly beneficial, as electric motors offer a 3–5 times higher specific power than combustion engines, this reducing the propulsion system's mass. The downside is that the energy for the electric motors is carried in batteries which are heavy. Battery weight prohibits continuous use of electric motors during sustained periods of time. While batteries with higher mass specific energy naturally result in lighter and better performing designs, they do not change the design point dramatically. This is, of course, fundamentally different for fully-electric aircraft.

Typically, the design point is shifted to higher wing loadings and to a higher power-to-weight ratio. This is true for both parallel- and serial-hybrid designs, regardless of mission length, cruising airspeed, altitude variations, aerodynamic performance or battery technology level [21].

Moving the design point this way has several consequences: Overall installed power is increased and take-off distance or climb performance become the limiting constraints. The maximum wing loading is of course limited by the stall speed constraint. This is usually set by certification specification or by landing field length requirements.

Higher wing loadings reduces the impact of turbulence on an aircraft. This is very beneficial for general aviation aircraft, as these are often highly sensitive to gusts. While the ride in turbulence is rarely a primary consideration of the aircraft owner, the passengers' comfort in rough air should always be considered. This is also a consideration for unmanned vehicles: These often carry optical payloads and much less image stabilization is required for aircraft that provide a smooth ride in all meteorological conditions.

4   Special Considerations

Hybrid-electric propulsion systems are widely applicable. In this section, some special considerations for niche applications – unmanned vehicles and vertical take-off and landing (VTOL) designs – are discussed.

4.1   Unmanned Aerial Vehicles

Unmanned aerial vehicles (UAVs) are often used in surveillance and reconnaissance roles. Long time on station is a key performance metric for these vehicles. Accordingly, their operating point is usually minimal required power. However, total required propulsive power can be significantly higher than that. This might be the case, if certain time-to-climb or dash speed requirement need to be met. Take-off performance can also be a driving factor for larger UAVs that are runway based. For smaller UAVs (typically up to a MTOM of 100 kg) the take-off is usually assisted by a catapult system. Such aircraft will naturally not be subject to take-off performance requirements.

A large difference between maximum and loiter power might also be a consequence of the high fuel fraction (often on the order of 40% of the MTOM), which is typical for UAVs. As fuel is burned off, power required for steady flight is reduced.

At the same time, it might be desirable for UAVs to achieve high speed, if moving targets are observed from above or if a quick location change is required. However, typical UAVs (see Fig. 7) carry their optical payloads externally and suffer from high drag [22]. For this reason, a top speed constraint might be a driving factor in a design.

A hybrid-electric propulsion system can thus be beneficial to unmanned aircraft, since it allows ICEs to be sized for a minimum power operating point. The electrical

Fig. 7. Exemplary, high drag UAV configurations

system can then be used to provide boosts in power when needed. Therefore, the parallel-hybrid layout is particularly suited for such operations.

Additionally, the ability to fly only on electric power for a specific time is an interesting proposition for UAVs. By shutting down the ICE, the noise signature can be substantially reduced, provided the propeller noise is not the dominant noise factor. Such means could also be used to alter the infrared (IR) signature of the vehicle, even though this is typically not a primary consideration for reconnaissance UAVs.

4.2   Vertical Take-off and Landing (VTOL) Aircraft

Finding a propulsion system that is efficient in cruise and in hover is a huge challenge for VTOL aircraft. VTOL aircraft require a thrust-to-weight ratio greater than unity (usually, a T/W of at least 1.3 is assumed) for the vertical part of their mission, however, once transition to forward flight is commenced, typically not more than a thrust-to-weight ratio of 0.1 (depending on the achieved L/D – a L/D of 10 is assumed here) is needed to sustain steady flight. The ratio between hover thrust and cruise thrust is 13, or inversely, the engine thrust required for cruise is just 7.7% of the installed thrust. Due to this huge gap in required power, a single propulsion system for both hover flight and cruise flight suffers from reduced efficiency, as its primary operation points are very far from each other. Moreover, it is highly questionable that a combustion engine or turbo-machine could be throttled to such a low power setting. Electric motors do not have a throttling limit like combustion engines and permit a very wide variation in torque and rotational speed at acceptable efficiency. While this alone does not completely solve this thrust matching problem, at least it alleviates it [23].

Recently, a new 'hybrid' configuration [24] has gained a lot of popularity amongst designers of unmanned transitioning VTOL aircraft: The fusion between a multicopter and a conventional aircraft [25]. Recent improvements in electrical motor technology and battery systems allow to add an electric hover-propulsion system to an otherwise conventional UAV, giving it VTOL capability (Fig. 8). According to the propulsion configurations showcased before, this can be considered a separated parallel-hybrid system.

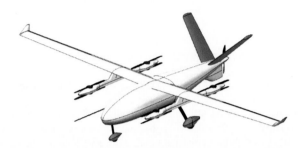

Fig. 8. Transitioning VTOL UAV – Octocopter configuration

5 Conclusion

Hybrid-electric and fully electric aircraft have the potential to revolutionize aircraft propulsion and aircraft design. However, new and innovative methods must be researched, to realize the full benefit of the technology. Best performance is only obtained by approaching the aircraft design problem holistically from the top down. This is especially true for aircraft that employ tight aero-propulsive coupling.

This paper gave an overview of the most significant challenges and potential benefits that the new technology introduces. Key points that drive the design of such aircraft were highlighted. The applicability of the methods extends from small UAV applications to man-carrying aircraft.

References

1. European Commission (2011) Flightpath 2050: Europe's Vision for Aviation. Publications Office of the European, Luxembourg
2. Suder KL (2012) Overview of the NASA Environmentally Responsible Aviation Project's Propulsion Technology Portfolio, NASA, Washington, D.C
3. Borer NK, Patterson MD, Viken JK, Moore MD, Bevirt J, Stoll AM, Gibson AR (2016) Design and performance of the NASA SCEPTOR distributed electric propulsion flight demonstrator. In: 16th AIAA aviation technology, integration, and operations conference, Washington, D.C

4. Stoll AM, Mikic GV (2016) Design studies of thin-haul commuter aircraft with distributed electric propulsion. In: 16th AIAA aviation technology, integration, and operations conference, Washington, D.C

5. Gohardani AS, Doulgeris G, Singh R (2011) Challenges of future aircraft propulsion: a review of distributed propulsion technology and its potential application for the all electric commercial aircraft. Progress Aerosp Sci 47:369–391

6. Raymer DP (2012) Aircraft design: a conceptual approach, 5th edn. AIAA, Virginia

7. Mattingly JD, Heiser WH, Pratt DT (2002) Aircraft engine design, 2nd edn. AIAA, Virginia

8. Finger DF (2016) Comparative performance and benefit assessment of VTOL and CTOL UAVs. In: 65 Deutscher Luft- und Raumfahrtkongress DLRK 2016, Braunschweig

9. Tissandier G (1886) La Navigation aérienne, L'aviation et la direction des aérostats, Hachette, Paris

10. Brown GV, Kascak AF, Ebihara B, Johnson D, Choi B, Siebert M, Buccieri C (2005) NASA glenn research center program in high power density motors for aeropropulsion. NASA, Cleveland

11. Moore MD, Fredericks B (2014) Misconceptions of electric propulsion aircraft and their emergent aviation markets. AIAA, Virgina

12. Deters RW, Ananda GK, Selig MS (2014) Reynolds number effects on the performance of small-scale propellers. In: 32nd AIAA applied aerodynamics conference, Atlanta

13. Gundlach J (2014) Designing unmanned aircraft systems: a comprehensive approach, 2nd edn. AIAA, Virginia

14. Gudmundsson S (2014) General aviation aircraft design: applied methods and procedures. Butterworth-Heinemann, Oxford

15. Pornet C (2015) Electric drives for propulsion system of transport aircraft. In: New Applications of Electric Drives. InTech, 115–141

16. Geiß I, Voit-Nitschmann R (2016) Sizing of the energy storage system of hybrid-electric aircraft in general aviation. CEAS Aeronaut J 8:53–65

17. Finger DF, Braun C, Bil C (2017) A review of configuration design for distributed propulsion transitioning VTOL aircraft. In: Asia-Pacific international symposium on aerospace technology - APISAT2017, Seoul, Korea

18. Ko A, Schetz JA, Mason WH (2003) Assessment of the potential advantages of distributed-propulsion for aircraft. In: XVI international symposium on air breathing engines (ISABE), Cleveland

19. Schaefer CG (2016) LightningStrike VTOL X-Plane program. In: 72nd American helicopter society international annual forum, West Palm Beach

20. Finger DF, Braun C, Bil C (2018) An initial sizing methodology for hybrid-electric light aircraft. In: AIAA Aviation Forum, Atlanta

21. Finger DF, Braun C, Bil C (2018) Case studies in initial sizing for hybrid-electric general aviation aircraft. In: AIAA/IEEE electric aircraft technologies symposium (EATS), Cincinnati

22. Goetten F, Havermann M, Braun C, Bil C, Gomez F (2018) On the applicability of empirical drag estimation methods for unmanned air vehicle design. In: AIAA Aviation Forum, Atlanta

23. Colucci F (2016) Lift Where You Need It, Vertiflite, November-December 2016

24. Finger DF, Braun C, Bil C (2017) The impact of electric propulsion on the performance of VTOL UAVs, in 66. Deutscher Luft- und Raumfahrtkongress DLRK 2017, Munich

25. Warwick G (2017) Aviation Week & Space Technology - Uber Unveils 2020 Plans For Electric VTOL Air-Taxis Demos, 28 April 2017. http://aviationweek.com/aviation-week-space-technology/uber-unveils-2020-plans-electric-vtol-air-taxis-demos. Accessed 1 July 2018

Research on Civil Aircraft Design Based on MBSE

Yunong Wang[1], An Zhang[1(✉)], Delin Li[1], and Haomin Li[2]

[1] School of Aeronautics, Northwestern Polytechnical University,
Xi'an 710072, China
wangyn@mail.nwpu.edu.cn
[2] Shanghai Aircraft Design and Research Institute, Shanghai, China

Abstract. Civil aircraft is a complex system. With the development of aviation industry, scale and complexity of the systems are getting larger and larger. Hence conventional design process can no longer satisfy the following requirements: integrality and consistence of information, capability of describing different activities and flexibility of requirements changes. However, MBSE (Model based System Engineering) has shown its potential of handling the challenges. Instead of natural language, MBSE adopts different models as the basic elements to storage and transfer data. Hence the relation between requirements of different design levels will be more intuitive and a faster response to requirements modification become possible. In this paper, from the top requirements of civil aircraft, we introduce a V&V activity model to the existing Harmony-SE to construct a both efficient and effective design framework. Comparing with conventional V design process model, our method enables the incremental and iterative developing method as well as a validation step after each design stage. These will produce better-quality aircraft within shorter development period.

Keywords: Civil aircraft design · MBSE · V&V activity model · Harmony-SE

1 Introduction

The main purpose of system engineering is to construct a system which can satisfy all the complex requirements. The system engineering in conventional civil aircraft design process belongs to TBSE (Text based System Engineering). The main outputs of TBSE is a series of "natural language" based texts [1]. Although the text is very expressive and can convey a lot of information, it might be ambiguous under some circumstance. Moreover, natural language is not capable to deal with specific cases in some complicated activities.

Civil aircrafts are very complex systems. They are non-linear combinations of more than one hundred subsystems. With the development of aviation industry, this number is still growing. These subsystems can accomplish their specific task, they also can make up a civil aircraft to accomplish more complex tasks [2]. During the conventional civil aircraft design process, reports, drawings, experimental data and financial information generated during the engineering process are mostly recorded in the form of text

© Springer Nature Singapore Pte Ltd. 2019
X. Zhang (Ed.): APISAT 2018, LNEE 459, pp. 1273–1283, 2019.
https://doi.org/10.1007/978-981-13-3305-7_100

documents. Therefore, maintenance and managements of the information could be a big challenge. Also, there is no guarantee that during the transformation, the integrality and consistence of information can be well-remained.

The drawbacks of text drove people to consider different basic element to storage engineering information. And then MBSE came into being. Instead of natural language texts, MBSE adopts different models as the basic element of systems. Comparing to TBSE, it has many irreplaceable advantages, and there are major changes in modelling language, approaches and tools [3].

Besides the type of basic element, the design process itself also met some challenges. The whole design process of large civil aircraft is a complex and long-term work. Engineers usually adopt a traditional V model life cycle (Fig. 1) to organize their work. It can be roughly divided into the following steps: firstly, conceptual design will be determined according to the design requirements; the second is the preliminary design and detailed design; the final is the prototype test and experiment, flight test and finalization until mass production and put into use.

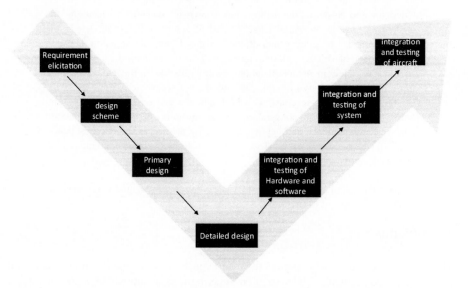

Fig. 1. Traditional V life circle model in civil aircraft design.

Although this model is quite straight forward, it cannot match the complex requirements of modern airplanes. The procedure is very classic, however, it's a one-way process. Hence an incremental iteration cannot be realized under this framework. Also, in this model, we expect that each work product created in every activity is complete, accurate, and correct at that stage.

Unfortunately, with the development of aviation industry, the function and complexity of civil aircraft are multiplying since the multi-disciplinary and multi-disciplinary technologies; however, the product development cycle is constantly shortening under market requirements. These ask for a both efficient and accurate design process. The key problems of current TBSE V circle model are as followed:

1. The quality of product. The large amount of engineering data cannot be described precisely by texts. Hence during the design process, the designing scheme cannot satisfy the stakeholder needs.
2. Long designing time. As we mentioned above, V circle doesn't allow incremental iteration. Therefore, regardless of where the modification occurs, the process will start again from the requirements analysis phase, increasing the design cycle and highly reducing work efficiency [4].
3. Large work quantity and poor efficient. During the conventional civil aircraft design process, the validation of engineering data will not begin until the whole design procedure is finished. These lead to defects in the design that cannot be detected in time, reducing work efficiency and affecting product quality.
4. Slow response to requirement modification. There is no traceability link between design requirements of different levels. If any of these requirements have been changed during the design process, researchers cannot find whole influenced requirements and adjust the requirements in time.

We think Traditional Systems Engineering can no longer handle the complex design process and a revolution is needed. Many engineers think a MBSE based framework can cope with most of these challenges. Although MBSE has been widely used in many areas, including aeronautics and aerospace industries, existing works mainly focus on small subsystems. Kaslow et al. verified how to realize the task of RAX CubeSat (Radio Aurora Explorer Cube Satellite) in outer space by establish executable MBSE model [5]. Pessa et al. use MBSE method to integrate CMs (Control Maintenance system) and Fs (Fule system) and emulate the integrated system [6]. Ferreiral and Gorlach exploited the controller of AGVs (Automated Guided Vehicles) based MBSE in Visual Paradigm software and realized the control in Microsoft Visual Studio environment. They utilized designed AVG realized the traction of tow truck according to the predetermined path [7].

The existing wildly-used variations of MBSE include Harmony-SE, OOSEM (Object-Oriented Systems Engineering Method) and RUP-SE (Rational Unified Process-Systems Engineering). Among all these, Harmony-SE is most suitable to solve some of our problems we stated above. In fact, this method has been widely used in aviation industry. Liu and Cao et al. researched the design methods of concept aircraft based SysML and Simulink. They taking the unmanned flight control system-Predator as an example, designed and emulated the concept aircraft [8]. Han et al. combining traits of spacecraft, put forward the MBSE method which regard IDS (Interface Data Sheet) as unified data source [9]. Xue et al. realized the design of avionic communication system by Harmony-SE method and finished the establishment of models of requirements, functions and architecture [10]. However, all these applications are limited to small subsystems.

But our purpose is to construct a framework for large civil aircraft design. In this large system, a validation step of each stage is needed to solve the problem we mentioned in the third problem of V circle model. Therefore, in this work, we combine the V&V active model with an existing method Harmony-SE, which will be discussed in next section, to construct a particle design process.

The paper is organized as followed: a brief introduction of MBSE and its variation Harmony-SE is placed in Sect. 2. Then, a detailed discussion of our novel framework will be given in Sect. 3. And we will conclude our work in Sect. 4.

2   MBSE Methodology

MBSE is a methodology which will lead to a "model-based" or "model-driven" system engineering. It's a summary of methods, working-flow and tools used in corresponding systems. In the framework of MBSE, we build demand models, functional models, and architectural models through UML (Unified Modeling Language) to realize the decomposition and allocation of requirements, functions and architectures. MBSE is a formal application of modeling methods that enable modeling methods to support system requirements, design, analysis, validation, and validation activities, starting with the conceptual design phase and continuing through design development and all subsequent life cycle stages [11]. As a better approach than text specification, models are introduced to create, manage, and validate engineering data. The model used by MBSE, including dynamic information of requirements, structure, behavior, and parameters. The model enables a more intuitive understanding and expression system for all types of professional engineering and technical personnel throughout the organization, ensuring that the entire model is delivered and used based on the same model.

Also, we can implement "validation" and "confirmation" of system requirements and functional logic through model. And it can drive co-simulation, product design, implementation, testing, synthesis, verification and validation.

The INCOSE (the International Council on Systems Engineering) Joint OMG (Object Management Group) developed a SysML (System Modeling Language) suitable for describing engineering systems based on the UML (Unified Modeling Language). The company also developed the corresponding tools to support SysML, and integrated the SysML modeling tools with existing professional analysis software such as FEA, CAD, etc. And they also proposed the overall solution of MBSE, which has the basis of the actual development engineering system [12].

Under the framework of MBSE, all parties work on demand analysis, system design, simulation, etc. around the system model to facilitate the collaborative work of the engineering team, so that the entire design team can make better use of the advanced results of models and software tools of various professional disciplines.

The Harmony-SE method extends the life cycle of the traditional V model and transforms it into an incremental iterative periodic activity stream, as shown in Fig. 2.

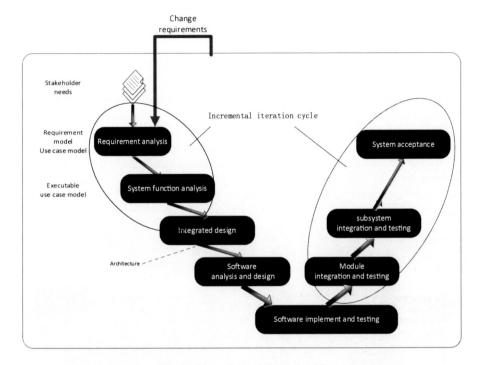

Fig. 2. Flow-chart of Harmony-SE.

This extension is performed though introducing an extra incremental iterative periodic activity stream. Traditional V model life cycle use this life cycle for only one time. For example, only when all requirements were analyzed, the work of system function analysis can begin. However, in the Harmony-SE life cycle, the activities can be executed for several times, and the engineering data can be transmitted to the downstream engineering incrementally. During the design process, researchers divide use cases into several parts, and every part of requirements begin to work solely. When requirements be changed, researchers only need to analyze the changed requirements.

Harmony-SE method has its specific process norm and tools of designing. IBM specializes in the Harmony-SE toolkit, leveraging Rational Rhapsody's execution capabilities to perfectly implement the Harmony-SE process. Like we wrote in Sect. 1, Harmony-SE has been adopted by aviation induction for years. For instance, Harmony-SE-based Rhapsody series tools are currently widely used in the industry, and its users include COMAC (Commercial Aircraft Corporation of China Ltd) and some subsidiaries of AVIC (Aviation Industry Corporation of China) [13].

Nevertheless, Harmony-SE is designed for small systems. For a complex project like civil aircraft design. It lacks a validation step of each design phase. Based on this idea, we are going to design a model database which can perform real-time inspection of engineering data to improve the efficiency and accuracy of civil aircraft development.

3   Civil Aircraft Design Process Based MBSE

To handle all the challenges, we propose the civil aircraft design process as shown in Fig. 3.

Firstly, model-based engineer data transformation method can ensure the veracity and integrality of engineer data, reduce possibility of losing and inaccurate describing data during the transfer process. Meantime, researchers can manage the requirements by establishing the traceability link. Once any of the requirements have been changed, researchers will make a risk assessment in time and change the rest influenced requirements.

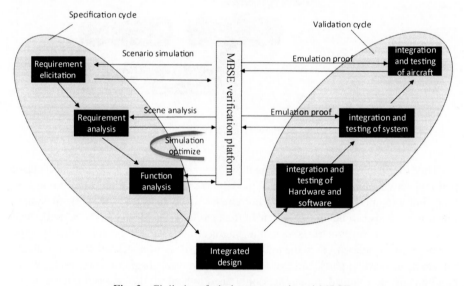

Fig. 3. Civil aircraft design process based MBSE.

Aiming to the problem of slow response to requirement modification, we adopt the flow path based on Harmony-SE, using incremental activities to reduce response time and lead time. To solve the problem that cannot find the inaccurate design in time leading a mass of rework, we adopt V&V model life cycle, establish the MBSE verification platform. This will verify the data during the design process, and insure the validity of engineer data. Researchers will find the inaccurate design and amend them in time, avoiding the rework and reducing the workload.

Based on the V model, we establish a V&V model and introduce a verification work at each stage. In the left half of the large "V" process, the data is post-tested by a small "V" to ensure the correctness of the design requirements. In the right half of the large "V" process, a small "V" is used for the priori process, looking for problems in the simulation and avoiding unnecessary losses. A virtual simulation test that puts the model into a defined simulation scenario for inspection can more clearly verify that the design meets the requirements. After each stage of work is completed, the model data is placed in the MBSE verification platform for inspection to ensure that the engineering data meets the requirements before proceeding to the next stage of work.

The process is divided into the following stages:

1. Requirements elicitation: Requirements are the first analysis and definition in civil aircraft design. At this stage, researchers should identify the stakeholders and capture stakeholder needs from stakeholders. Stakeholders are any living or non-living things that are affected by the system. The designer divides the stakeholders into deterministic stakeholders, prospective stakeholders, and potential stakeholders based on the number of stakeholder ownership of the company's legality, rights, and urgency. For civil aircraft, the identified stakeholders are mainly aircraft developers, civil aviation companies, civil aviation bureaus, airports, government departments, pilots and various flight attendants. The prospective stakeholders mainly include people who have boarded the aircraft and competitors etc. Potential stakeholders are all the people. After identifying stakeholders in systems engineering projects, researchers need to obtain their original needs from stakeholders through a variety of survey research methods. Typical survey research methods include field survey, questionnaires, interviews, and joint requirements planning.

2. Requirements analysis: By refining and formalizing the captured requirements, the model is built and simulated in the virtual scene to generate detailed development requirements. Due to the more functional requirements of the civil aircraft, DODAF (Department of Defense Architecture Framework) was introduced. According to DODAF, the requirements analysis and definition of the civil aircraft were carried out, and the requirements were captured in the scene to obtain the design requirements of the civil aircraft system. The design requirements use an engineering-oriented technical language representation that describes the behavioral state of the system in detail.

3. Function analysis: Detailed design requirements are obtained through function modeling, function analysis and other methods. Function analysis is based on use cases, creating use cases based on the development requirements of the requirements analysis stage output. The use case is an intermediate point that connects requirements and functions. To ensure that all functional requirements and related performance requirements are covered by these use cases, it is necessary to establish a connection and provide traceability. The workflow to the use case is shown in Fig. 4.

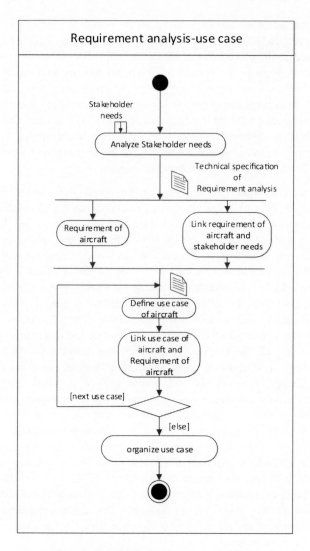

Fig. 4. Requirements-use cases workflow

Function analysis creates executable models to describe continuous and concurrent behaviours and events that occur over time in each use case. Through the logical relationship of these behaviours, the specific activities and resource/data flow and activity interface information are determined to generate specific design requirements. The function analysis workflow is shown in Fig. 5.

4. Synthesize design: Comprehensive design according to design requirements. In the design process, the detailed design requirements and interface information determined in the function analysis stage should be designed to ensure consistency with the design requirements.

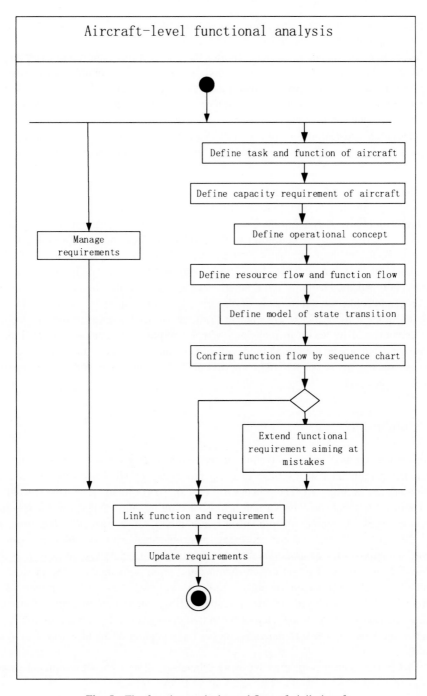

Fig. 5. The function analysis workflow of civil aircraft

5. Integration and testing of hardware and software: Test the software and hardware of the design, build models and save them.
6. Integration and testing of system: Before the physical integration and experiment, the model established in the previous stage is putted into the simulation scenario for verification. After confirming that the system is correct and meets the design requirements, the physical system is integrated and tested to ensure that the system does meet the requirements in the physical world.
7. Integration and testing of civil aircraft: Similar to the previous stage, the aircraft model is first simulated in a virtual scene and then related physical experiments are performed.

The civil aircraft development process can realize the verification and confirmation of system requirements and function logic through core technology links such as requirements analysis, system function analysis and synthesize design in the early stage of civil aircraft R&D. The definition of the product is clear and accurate early in the entire project, and the correctness of these data and models is continuously verified throughout the design process. In addition, due to the increasingly fierce market competition and the changing requirements, we require that the function architecture, logical architecture and interfaces of the civil aircraft system be flexible and can respond quickly to changes. Through the process of civil aircraft development, the requirements can be traced in real time. When a requirement changes, it can immediately find and amend the remaining levels of demand associated with it and conduct a risk assessment to assess the risk, improving design efficiency and shortening the development cycle.

4   Conclusion

In this paper, we analyse the conventional civil aircraft design process and summarizes the existing problems. In response to these problems, an MBSE-based civil aircraft design process is proposed. The process is based on the Harmony-se process. Establish traceability links between the various levels of requirements. When one of requirements changes, the remaining influenced requirements can be discovered in time, so that the researchers can conduct risk assessment and deal with the changed requirements. To improve the design efficiency and robustness of whole system as well as the quality of aircraft, we first introduce incremental iterative methods to the process to enable a simultaneous development of multiple activities. Moreover, a novel V&V activity model is employed to construct the MBSE integration platform. And this makes engineering data in each step of the activity process can be verified by real-time simulation, which will ensure the correctness of engineering data and improve product quality.

This process can greatly improve work efficiency, shorten the development cycle, and provide an effective method for civil aircraft development.

Acknowledgements. The research presented in this paper is supported by Ministry of Industry and Information Technology of the People's Republic of China, under Grant MJ-2016-F-02.

References

1. Zhu J, Yang H, Gao Y et al (2016) Summarize of model-base system engineering. Aeroengine 42(4):12–16
2. Li W.-j (2005) Aircraft configuration design. Northwestern Polytechnical University Press
3. Chen H-t, Deng Y-c, Yuan J-h et al (2016) Fundamental of model-base system engineering. Aerosp Chin 32(3):18–23
4. Hoffmann HP (2011) Systems engineering best practices with the rational solution for systems and software engineering, version 4.1. IBM Corporation, Somers, NY (US)
5. Kaslow D, Anderson L, Asundi S, et al (2015) Developing a CubeSat model-based system engineering (MBSE) reference model-interim status. In: IEEE aerospace conference, pp 1–16. IEEE
6. Pessa C, Cifaldi M, Brusa E, et al. (2016) Integration of different MBSE approaches within the design of a control maintenance system applied to the aircraft fuel system. In: 2016 IEEE international symposium on systems engineering (ISSE), pp 1–8. IEEE
7. Ferreira T, Gorlach IA (2016) Development of an automated guided vehicle controller using a model-based systems engineering approach. S Afr J Industr Eng 27(2):206–217
8. Liu X-h, Cao Y-f, et al (2011) Flight control system conceptual prototype design based on SysML and Simulink. J Electron Sci Technol 40(6):887–891
9. Han F-y, Lin Y-m, Fan H-t (2014) Research and practice of model-based systems engineering in spacecraft development. Spacecr Eng 23(3):119–125
10. Xue W, Jia C-q et al (2016) Application of MBSE in avionics communication systems. Electron Sci Technol 29(5):45–48
11. Friedenthal S, Griego R, Sampson M (2007) INCOSE model based systems engineering (MBSE) initiative. In: INCOSE 2007 symposium
12. Friedenthal S, Moore A, Steiner R (2006) OMG systems modeling language (OMG SysML™) Tutorial. In: INCOSE international symposium
13. Zhang Y, Ni Z-j, Xiao Z-b et al (2012) Research and application of harmony-based system modeling method in integrated data management system. Avionics Technol 1:42–47

Design of Wave Rider Based on Shock Fitting Method

Guoliang Li[(✉)], Anlong Gong, Qiang Liu, Chuqun Ji, Yunjun Yang,
and Weijiang Zhou

China Academy of Aerospace Aerodynamics, No. 17, West Road Yun Gang,
Beijing, China
liboyang0929@sina.com

Abstract. Design of wave rider is attracting more and more attention right now and is of special interest for hypersonic applications. The article imports Shock Fitting Methods to calculate the three dimensional hypersonic flow to capture shock wave precisely. The choice of shock generating body is very important, because it affects the shock flow field. It can be single-stage or multiply-stage. In this paper, the double-cone shock generating body is introduced in order to modify lift/drag and the center of pressure of wave rider. According to the restriction of ratio of length to width, the shock surface is cut and then the outer line of wave-riding surface is decided. Relying on engineering application, the leeward surface is designed. The results show that lift/drag of engineering wave rider is relatively higher than ordinary vehicle.

Keywords: Shock generating body · Wave rider · Shock-fitting

1 Introduction

The wave rider is a typical configuration of hypersonic vehicle. Shock is fully attached to the front edge of vehicle and the windward airflow is limited by attached shock and doesn't leak into the leeward surface. As a result, the ratio of L/D (lift/drag) is higher than vehicle with traditional configuration.

The traditional design methods of wave rider have two categories. The forward method such as conical-wave rider-like methods in which the shock wave is fixed and the limits to vehicle L/D ratio exists, although the flow field is simulated accurately and get easily. As a result, the method is not used widely. The reverse method such as osculation cones method in which the flow field is simulated not as exactly as the former method because of neglecting the cross flow. Therefore, it is difficult to the ensure the performance of wave riding. At present, the design of wave rider is focused worldwide and some new methods spring up.

Some researchers suggest a design method of wave rider with controlled the leading and trailing edge, based on the supersonic aerodynamic principle combined with the streamline tracing and geometric reconstruction technique. Some researchers put forward an aerodynamic configuration by increasing sweepback angle to generating leading edge vortex so as to increase the lift of leeward surface, which can improve the low-speed aerodynamic performance.

© Springer Nature Singapore Pte Ltd. 2019
X. Zhang (Ed.): APISAT 2018, LNEE 459, pp. 1284–1290, 2019.
https://doi.org/10.1007/978-981-13-3305-7_101

In order to design the wave rider with better aerodynamic performance. The method for ordinary 3d flow field is needed to be set and then the hypersonic flow field for ordinary geometrical configuration such as elliptic cone or cone with attack angle is computed in priority. The shock wave flow field is get by both shock capturing methods and shock fitting methods. The deficiency of the former method for wave rider design is the location of shock wave strides over several meshes, which does not apply for wave rider design. However, the latter method is preferred for design of wave rider due to its high resolution.

In this paper, the procedure of design based on shock-fitting method contains as follows: the decision of shock generating body; the computation of shock flow field; the choice of windward surface. The decision of shock generating body is underlined because it affects the shock flow field. Therefore, the suitable shock generating body is needed and it can be flexible including single-stage, second-stage, and even multiply-stage.

2   Shock Fitting Method

The computational process of shock wave flow field is usually consist of two stages. The shock capturing method is utilized during the first half of total steps. The shock fitting method is used for the second half of total steps. The kind of arrangement can accelerate the convergence. The following pictures shows the moveable process of shock in Fig. 1.

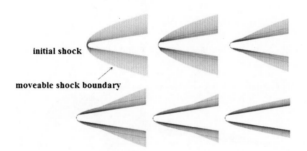

initial shock

moveable shock boundary

Fig. 1. Moveable process of shock

After the shock flow field is computed certain steps by the capturing shock method, usually 3000 steps or even more. The Rankine-Hugoniot shock formula is used to compute the flow parameters behind shock, as shown in Eq. 2.1:

$$\left\{ \begin{array}{c} \rho_1(V_1 \cdot n - z) = \rho_2(V_2 \cdot n - z) \\ p_1 n + \rho_1 V_1(V_1 \cdot n - z) = p_2 n + \rho_2 V_2(V_2 \cdot n - z) \\ \rho_1 V_1 \cdot n + \rho_1 E_1(V_1 \cdot n - z) = p_2 V_2 \cdot n + \rho_2 E_2(V_2 \cdot n - z) \end{array} \right. \tag{2.1}$$

Where p is the pressure; ρ is the density; V is the velocity vector; E is the energy. z is the velocity of shock wave. n is the normal vector of shock wave. The subscript 1 means the parameter before shock. The subscript 2 means the parameter behind shock.

In addition, the feature compatibility formula is needed to close the system. Hence, the flow parameter can be computed by the system. The mesh system is dynamic because the shock wave is moveable during computation. The movement velocity of shock wave converges to zero. Unsteady shock wave converges to steady shock wave. The outer surface of shock wave can apply for design of wave rider.

3   Generation of Windward Surface of Wave Rider

As mentioned above, some steps are needed in constructing wave rider. An simple example will illustrate these steps as follows. The L/W (ratio of length to width) of wave rider is set 2. Flight parameter: Ma = 6.0, AOA (angle of attack) = 0°. The elliptic cone is regarded as shock generating body. As shown in Fig. 2, the outer boundary of shock is get by shock fitting method. The distribution of Uy/Ux behind shock is shown in Fig. 3.

Fig. 2.   Outer boundary of shock and shock generating body

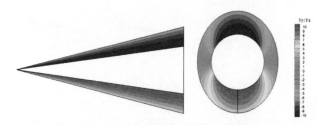

Fig. 3.   Distribution of Uy/Ux behind shock

According to limitation of L/B = 2, the outer boundary is cut. The cut line is the front edge of wave rider. The streamline tracking method is utilized for getting windward surface of wave rider, as shown in Fig. 4.

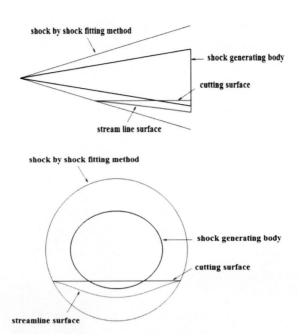

Fig. 4. Generation of windward surface of wave rider

4 Aerodynamic Analysis of Wave Rider

Here the wave rider is designed for engineering. In principle, double-cone is usually chosen in Fig. 5, which can modify both L/B and Xcp (center of pressure). In order to attain relative higher L/D, the shock had better attach to the wall of wave generating body. The little cone angle is preferred. In order to lessen Xcp, rear cone angle should be smaller than front cone angle. As a result, the flow field as a whole can be divided into the front compressible wave and the rear expansive wave. Therefore, the distribution of higher pressure lies in front part of wind ward surface of wave rider. The configuration parameter is shown in Table 1.

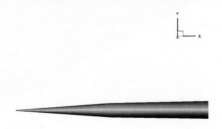

Fig. 5. Double-cone wave generating body

Table 1. Shock generating body configuration

Front stage: rear stage	Front cone angle	Rear cone angle
1:1	4°	1°

Designing parameter: H = 35 km, a = 8.0, AOA = 8°. The accurate boundary of shock is obtained by shock fitting method in Fig. 6.

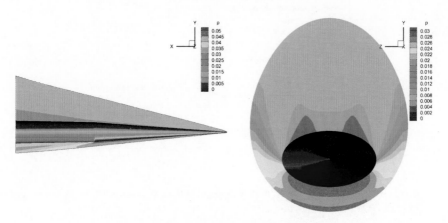

Fig. 6. Pressure contour of shock flowfield

According to L/B = 2.5, the suitable location along the flow direction is chosen to cut the outer boundary of shock. After the front edge is get, the streamline tracking method is used to define windward surface, as can be seen in Fig. 7.

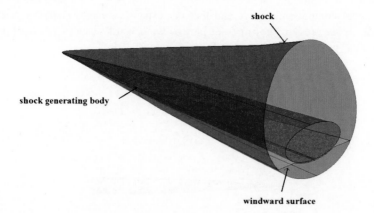

Fig. 7. Windward surface of wave rider

The aerodynamic performance of theoretical configuration is shown in Table 2. Leeward surface is free stream surface and windward surface is wave-riding surface. As shown in Fig. 8, the wave-riding surface is kept and the design of leeward surface mainly applies for engineering application.

Table 2. Aerodynamic performance of theoretical configuration

AOA	Xcp	L/D	Cmz
6°	5.9123E−01	3.4405E+00	9.8679E−03
8°	5.9927E−01	4.0998E+00	1.2088E−02
10°	6.0388E−01	4.0389E+00	1.3671E−02

Fig. 8. Engineering design of wave rider

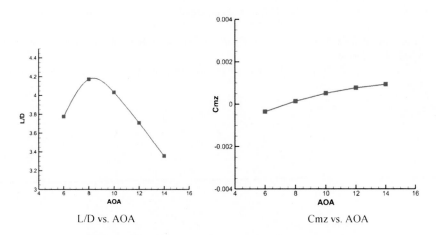

L/D vs. AOA   Cmz vs. AOA

Fig. 9. Aerodynamic properties of engineering wave rider

As shown in Fig. 9, the max L/D attains almost 4.2 with AOA valued at 8°. The pitching moments change nearby 0, which has advantage of rudder control.

5   Conclusion

The design of wave rider is constructed by introducing shock fitting method, which can apply for ordinary three dimensional flowfield. The L/D can be changed by modifying the configuration of shock generating body. The results for engineering configuration show that the wave rider has higher L/D than ordinary vehicle. The pitching moments change nearby 0 and the wave rider has the advantage of rudder control.

References

1. Nonweiler TR (1959) Aerodynamic problem of manned space vehicles. J R Aeronaut Soc 63:521–530
2. Center K, Sobiecaky H, Doughtery F (1991) Interactive design of hypersonic waverider geometries. In: AIAA, 1991–1697
3. Liu CZ, Bai P, Chen BY et al (2016) Rapid design and optimization of waverider from 3D flow. J Astronaut 37(5):535–543
4. Rasmussen ML (1980) Waverider configuration derived from inclined circular and elliptic cones. J Spacecr Rocket 17(6):537–545
5. Yongzhi Li, Guangxi Li, Kunyuan Zhang et al (2018) Aerodynamics modification design and analysis of waverider with controllable leading and trailing edges. J Rocket Propul 44(2):1–10
6. Song F, Yan C, Ma B, et al (2018) Uncertainty analysis of aerodynamic characteristic for cone-derived waverider configuration. Acta Aeronautica et Astronautica Sinica 39(2): 121519-1–121519-10
7. Cui K, Xu Y, Xiao Y et al (2017) Effect of compression surface deformation on aerodynamic performances of waveriders. Chin J Theor Appl Mech 49(1):75–83
8. Rasmussen ML, Clement LW (1984) Cone-derived waveriders with longitudinal curvature. In: AIAA Paper 84-2100
9. Song F, Yan C, Ma B (2017) Design and aerodynamic analysis of a wide speed waverider. Phys Gases 2(5):25–36

A Design Method of Civil Commercial Aircraft Cabin Integration Based on System Engineering Thought

Zhaoliang Zou[(✉)], Xu Zhang, and Dayong Dong

Shanghai Aircraft Design and Research Institute (SADRI),
Jinke Road #5188, Shanghai, China
zouzhaoliang@comac.cc

Abstract. The cabin is a part of a civil commercial aircraft that directly provides service to the end user, including necessary safety facilities for passengers and cabin crews, the physical environment such as sound, light and temperature, essential service facilities and service facilities which enrich the travel experience. Specifically, it includes cabin lighting system, environment control system, passenger oxygen system, passenger address and cabin interphone system, in-flight entertainment system, cabin external communication system, cabin decoration and necessary cabin service equipment such as the seat, the kitchen and the bathroom. Ultimately, the cabin management system implements integrated control and display operation.

The cabin of civil aircraft is not only a platform which provides the most direct service to passengers and cabin crews, but also an embodiment of the civil aircraft's brand value. All systems and components in the cabin need to be considered as a whole in order to ensure their consistent design style and the unique design concept. The application of system engineering thought to carry out the Top-down integrated design guarantees the high value of cabin brand, also it reduces the project risk and avoid the cabin potential safety hazard. This paper will study and summarize the method of cabin design based on system engineering thought.

Keywords: Cabin integration · System engineering ·
Requirement verification · Requirement validation

1 Civil Aircraft Cabin Overview

From CRT monitors, tape/CD players to digital audio and video system with LED display screen, from the flying Information isolated Island to the air-ground connection, from the traditional cabin lighting to the scene lighting system, from the electric fan to the air-conditioning, from the basic safety equipment to the emergency equipment system. Civil aircraft cabin has experienced a long course of development, from meeting the minimum travelling needs of "can ride" to pursuing higher level of passenger vision, listening, sense of the full range of "comfort". The cabin gradually becomes a whole system, and naturally becomes the front line of applying these new technology and complex system applications.

© Springer Nature Singapore Pte Ltd. 2019
X. Zhang (Ed.): APISAT 2018, LNEE 459, pp. 1291–1297, 2019.
https://doi.org/10.1007/978-981-13-3305-7_102

As shown in the Fig. 1 above, typical civil aircraft is consist of cabin decoration, emergency equipment, cabin lighting system, environment control system, passenger oxygen system, passenger address and cabin interphone system, in-flight entertainment system, cabin external communication system, and cabin service facilities such as seats, kitchens and lavatories. All these systems and facilities can communicate and share information through cabin core network, ultimately, the cabin management system implements integrated control and display operation. The systems described above is the main part of cabin system whose development tendency is digitization, modularization and informationization. Since the majority of cabin system has no great impact on flying safety, it greatly reduces the difficulty of applying the highly demanded new technology compared with the flying essential systems.

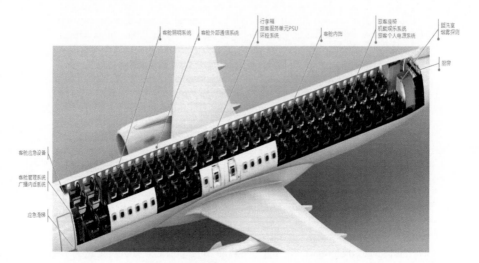

Fig. 1. Typical civil aircraft configuration

2　The Challenges to Cabin Design

While the flexible architecture has brought unlimited possibilities for application development, the increasingly complex and highly integrated cabin also brings great challenges to engineering design simultaneously. The core difficulty and value embodiment of civil aircraft cabin is no longer limited in the engineering realization level, it is the ingenious function that not only can bring the most practical value to the customers but also express the manufacturer's brand value.

After delivery of a certain model, users report service equipment inconvenience which mainly includes the kitchen, seat and even the door handrails. In addition, some design problems such as lacking of emergency call and broadcast squeal are revealed. In view of the above problems, this paper summarizes the reasons for the process and method of cabin development.

- *Lack of Customer-oriented Requirements Capture Process and Method*

Customer demand is the most important way of requirement capturing for civil aircraft cabin design. Insufficient users' requirement capturing in the initial stage of project design will lead to the deviation of cabin design in the most basic market target. Therefore, we should define the standard user requirement capture process and make scientific capture method to ensure the comprehensiveness of the requirement.

- *Lack of Cabin Integration Level Requirements Definition and Validation System*

The function design of the cabin should be carried out according to the typical operation scene. It is an essential section in the cabin design to fully analyze the service function demand of the cabin crew and the passengers in each typical scene, and to confirm the function demand through design review, static text evaluation, simulation evaluation, prototype evaluation and original rationality test.

- *Lack of Completed Integrated Design and Validation Activity at Cabin Integration Level*

In order to meet the requirement of cabin integration and verify the requirement realization, in addition to the integration test of system level, we should plan the customer-oriented assessment as the verification means and carry out the comprehensive evaluation among the airline users and passenger representatives based on a specified method, process and criterion.

- *Imperfect Project Structure and Unclear Division of Responsibilities*

Since cabin design involves a number of complex systems whose responsibility should not be independent of each other, a main body which will be responsible for the cabin integration, coordinating the system design team, and carrying out cabin design work from top to bottom.

- *The assessment and validation environment in the development process is not planned or qualified to support the corresponding assessment and validation work.*
- *While the traditional design process plans a product decomposition structure PBS based on the ATA Chapter, an up-to-date civil aircraft cabin with top design concept and brand characteristics should be considered as a complete product in the design process.*

On the whole, it is an effective way to solve the cabin design problem by using the system engineering design method to carry out the development process of demand capturing, function definition, requirement analysis, design synthesis, demand confirmation and verification.

3   The Application of System Engineering Thought in Cabin

With the increasingly complex and integrated integration of civil aircraft, high level of cross-linking and coupling between the systems, such as the cabin, poses greater challenges to the design. In recent years, application of systems engineering methods get more attention and promotion in civil aircraft design field. As a censor, National Civil Aviation Bureau, also recommended the use of systems engineering design methods to carry out civil aircraft complex system design. The system engineering method is an interdisciplinary design method, which includes both product integration and process integration, and it emphasizes the tradeoff analysis of the components in the whole life cycle in order to achieve the global optimum and satisfy the stakeholders. The "validation" and "verification" activities are the core of the system engineering design method.

3.1 Design Object Definition

For civil aircraft cabin, the application of system engineering design method begins with the accurate and reasonable definition of design object. As the most close to the user and the market aircraft products, the cabin product should be analyzed from multiple dimensions at the beginning of development process.

- *User and Function Dimension*

The most essential design consideration in the cabin design of civil aviation is that the cabin provides service to end users directly. In the definition of civil aircraft cabin design process, first of all, we need to consider how the cabin can meet the use of each service object needs during various stages of aircraft operation. All of the use requirements together should be able to form a complete operation of the task coverage, and ultimately produce a cabin functional dimension definition.

- *Physical Dimension*

The physical definition of the cabin is different from other aircraft products, in addition to the basic design, civil aircraft cabin needs a complete system of industrial design, which is competent to express the unique requirements in aesthetics and comfort. Industrial design can bring a unified design concept and design language for the cabin, furthermore, it brings the brand characteristics and brand value display for the aircraft. Industrial design provides the necessary design input and constraints for each system or equipment in the cabin, thus a core consideration in defining the cabin from a physical dimension.

3.2 Development Process and Activity Planning

According to the engineering idea of "confirming" and "verifying" based on the requirement of the SAE ARP 4754A, the design process of civil aircraft cabin is summed up as the activity set shown in the following Fig. 2.

As shown in the figure, according to the typical product decomposition structure, the civil aircraft cabin design activities can be divided into aircraft level, cabin integration level, system integration level, product implementation level. In the process of engineering implementation, levels can be merged or cropped according to the complexity of market target positioning. However, the top-bottom requirement decomposition and the bottom-up system synthesis process must be followed.

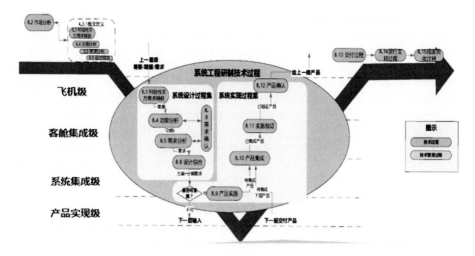

Fig. 2. Typical civil aircraft cabin design process based on system engineering

Specifically, at the upper level of requirements, it is necessary to define the objects, targets and criterion, then we should decompose the requirements and trace them back to the objects of the next level through the architecture design of this level in order to define the requirements that each object performs. The system synthesis begins with product realization, will be verified step by step, in order to check whether the demand is realized completely and accurately. Typical cabin design activities at all levels are defined as follows:

- *Aircraft Level*

As the top design requirement of cabin system, aircraft level requirement synthesizes the market requirements, user requirements, product serialization requirements, security requirements, and so on, then try to achieve optimum to satisfy stakeholders. During this process, top level requirements such as cabin seats layout and users' characteristics will be produced.

Aircraft level requirement will be verified by carrying out the assessment and inspection among the actual end users before the aircraft is delivered.

The most concerned design elements in aircraft level are the basic space parameters and seats numbers which are based on overall design of the aircraft. Cabin's stack position, door position and cabin's dimensions are important input and restriction of the next level's design, which need to take the influence of aircraft's shape, layout, structure and commercial load design into consideration. The cabin-level is the most direct embodiment of overall design and market demand, also it has the most direct impact on the cabin layout design and cabin safety design.

- *Cabin Integration Level*

Cabin integration level is aim at capturing end user's requirement for the cabin based on defined typical operation scenario. It mainly focus on balancing all the advantages and disadvantages of different architecture design which emphasize on engineering technology and project management, and requirement will then be decomposed to be systems' and facilities' requirement. Validation process will be carried out by industrial design department through simulation, prototype and principle test.

Cabin functional requirements, cabin safety-related design requirements, cabin industrial design concept, industrial design language and other non-functional requirements such as interface requirements between systems, cabin core network design specifications, cabin layout plan, cabin preliminary layout plan, cabin industrial design plan and so on will be produce at this level. Among them, the cabin functional requirements should be based on analysis of the typical operation scenario, the resulting functional requirements should be able to cover all the scene needs, then whether the advanced scene needs will be provided or not can be weighed according to the aircraft's market objectives. Cabin safety-related design requirements are designed to address the requirements of airworthiness regulations such as emergency evacuation, all these requirements will be decomposed into each system through the architectural design of this level during the design activities at the cabin integration level. For the cabin non-functional requirements, it is important to consider the requirement of the cabin's aesthetics, cabin human factors, and the derived interface requirements which will implement the functional requirement.

- *System Integration Level*

Taking each system in cabin system as design objective, system integration level is aim at carrying out system-level requirements decomposition and validation. System requirements are graded to hardware and software level through optimization of architecture design.

System design requirements, system scheme, mechanical and electrical interface control files, human-machine interface and decoration specifications will be carried out at this level.

- *Product Implementation Level*

Product Implementation level is aim at conducting hardware and software requirements definition and development, also hardware and software level test and validation will be carried out.

4   Conclusion

Based on the research of the development trend of civil aircraft, this paper analyzes the potential challenges in the field of civil aircraft cabin design at present. The design process of civil aircraft cabin design based on system engineering thought is proposed, and the design activities of each level are defined. It can be used as civil aircraft cabin development and project implementation reference.

References

1. 《飞机设计手册》总编委会 (2005) 飞机设计手册.第5册，民用飞机总体设计. 航空工业出版社
2. Specification for manufacturers' technical data (1956) ATA specification no. 100. Air Transport Association of America
3. Li Z, Li M, Huang D, et al (2015) Aviation maintenance publication translation: based on ATA-2200 specification. J Beijing Univ Aeronaut Astronaut
4. Landi A, Nicholson M (2011) ARP4754A/ ED-79A - guidelines for development of civil aircraft and systems - enhancements, novelties and key topics. SAE Int J Aerosp 4(2):871–879

Design and Test of Plasma Control Surface on Unmanned Aerial Vehicle

Jiageng Cai, Chang Li, Huaxing Li[(⊠)], and Xuanshi Meng

Department of Fluid Mechanics, School of Aeronautic,
Northwestern Polytechnical University, Xi'an, China
hxli@nwpu.edu.cn

Abstract. A flying wing layout is an aircraft layout with no tail only huge wings. It has the advantages of lightweight, low flight resistance and good stealth, but it also has shortcomings such as poor maneuverability. The use of the active flow control technology to replace or enhance the control surfaces, therefore improve the lateral maneuverability of flying wings has attracted the interest of various research teams. The plasma flow control technology can change flow near the actuator to achieve the effect of controlling the local pressure of wing, thereby completing the lateral manipulation of the unmanned aerial vehicle (UAV). Many teams have also carried out relevant experiments, which laid the theoretical foundation for this paper. This paper mainly designs a sensor board system of flying wing UAV, which is used to collect the local pressure on the wing of UAV and control the switch of actuator. An experiment was carried out on the ground. First, this board can collect and monitor the pressure data at a certain point. Second, the circuit system can also react to specific situations (such as local pressure values) and automatically control the actuator's switches. Third, the plasma actuator can also be actively controlled by remote control. This experiment lays the theory and practice foundation for the UAV flight experiments in the future.

Keywords: Flying wing UAV · Plasma actuator · Control system design · Ground test

1 Introduction

A flying wing is an aircraft that has no tail only huge wings. It is developed from a tailless aircraft. Compared with the traditional layout of the aircraft, the aircraft with a flying wing layout has the advantages of high aerodynamic efficiency, high lift, large range and low fuel consumption. Its stealth performance is very good, so it has value of military application. However, the aircraft in the flying wing has reduced stability in all directions due to the elimination of the tail and the fusion of the fuselage and the wing. Moreover, his longitudinal movement is often combined with the movement of the lateral and heading, which makes it more difficult to maneuver. Moreover, it has the disadvantage of poor take-off and landing performance [1].

© Springer Nature Singapore Pte Ltd. 2019
X. Zhang (Ed.): APISAT 2018, LNEE 459, pp. 1298–1311, 2019.
https://doi.org/10.1007/978-981-13-3305-7_103

Plasma flow control technology now provides a solution to this problem. Plasma flow control technology has the advantages of simple structure, lightweight, small thickness and low power consumption. It can provide more possibilities for application on flying wing. In aviation, many academic teams have extensively studied plasma flow control method—an important method of flow control. For nearly 20 years, more than 100 research teams committed to this direction. They placed one or several pairs of electrode on the airfoil surface. Under high voltage actuation, plasma generate between the electrode groups, thereby changing the temperature field and velocity field near the electrodes, thus forming an influential trend [2]. Under the disturbance induced by the plasma actuator, the turning or separation of the wing boundary layer can be effectively controlled to achieve the effect of increasing lift and reducing drag [3]. The plasma discharge modes of plasma generation are: Corona Discharge (CD), Dielectric Barrier Discharge (DBD), Spark Discharge (SD), and Arc Discharge (AD). Dielectric Barrier Discharge Plasma Actuator has the advantages of stable working condition, fast response time, flexible control position, simple structure, no need for additional air supply devices, etc. [4]. So it has great advantages in the application of the flying wing (Figs. 1, 2, 3, 4, 5, 6 and 7).

Fig. 1. A flying wing

Moeller and Rediniotis used a series of surface-mounted aerodynamic eddy current control actuators to control the pitching moment of a 60-degree sweeping delta wing model at high angles of attack. The controlling is achieved by changing the phenomenon of vortex damage, which affects the distribution of the chord lift on the wing and ultimately causes the pitching moment. Nelson et al. placed a plasma actuator on the 1303 UAV to generate the same rolling torque as the conventional control rudder. Lopera et al. studied the plasma aerodynamic control of the Unmanned Aerial Vehicle (UAV), Patel et al. performed a wind tunnel test of the plasma pneumatic actuator on the 1303 flying wing UAV. The results show that the plasma actuator can significantly increase lift and reduce drag of the aircraft by controlling the flow field in a large angle of attack. This conclusion is proposing the concept of flight control based on the "plasma wing" [5–6]. Grundmann et al. [7] used a plasma actuator to control the UAV flow separation.

Du et al. [8] did a plasma actuator to test the aerodynamic torque control of the micro-aircraft's transverse direction. This experiment used a method of adjusting the actuation parameters to achieve proportional control of the aerodynamic torque of the aircraft. It began an attempt to control the aircraft using Plasma flow control technology. The results of the wind tunnel test of the flying wing using the effects of plasma actuation indicate that the operation of the actuators placed at different positions of the aircraft can control the roll, yaw and pitching moments of the aircraft. This experiment also proves that the aerodynamic torque obtained by the plasma actuator can achieve the same control effect as the conventional rudder surface with a certain deflection angle. The current results show that different actuators can be arranged at different positions of the aircraft to control the roll, yaw and pitching moments of the aircraft. Moreover, the use of plasma deployed on the flying wing to achieve the effect of increased altitude and aerodynamic torque control is a major research hotspot currently [9–11].

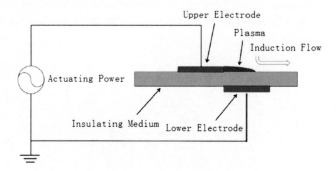

Fig. 2. Simple plasma actuator structure

To this end, this paper mainly design a more mature control system, firstly to achieve a good front-end control of the experimental UAV model, which could be used to design the flying wing UAV using plasma-actuation aerodynamic rudder surface in the future. Therefore, this paper will design a complete experimental control system, which has the functions of collecting data, real-time monitoring, active response, remote control, etc., and then simply tests its working effect on the ground. Through this test, this paper evaluates the control effect of the plasma virtual aerodynamic rudder on the UAV model.

2 Experiment Setup

2.1 Experimental Material Parameters

See Tables 1 and 2.

Table 1. Flying wing model parameters

No.	Material name	Remarks
1	Body length	About 800 mm
2	Wing length	About 2100 mm
3	Wing area	About 80 dm^2
4	Flying weight	1500 g–7000 g
5	Flying center of gravity	430–450 mm backwards on leading edge the wing
6	Battery	6S10000 mah
7	Motor	Have
8	Propeller	Chord length is about 200 mm

Fig. 3. Experimental flying wing UAV

Table 2. Materials required for experiment

	Name and model the device or consumable	Quantity	Name and model the device or consumable	Quantity
Equipment and consumables requirements	Flying wing UAV model	1	Plasma actuators	1
	Plasma actuators manufacturing material (dielectric layer, electrode copper sheet)	Several	Flight test equipment (sensors, etc.)	Several

Fig. 4. Consumables used in the test experiment

2.2   Data Collection System

In order to observe and evaluate the working effect of the actuator, this experiment needs to collect the local pressure value on the wing. The effect of the actuator's operation on the roll and yaw moment of the UAV is evaluated by collecting the local pressure of the surface of the model before and after control.

Therefore, this paper has produced the following data collection system, which is based on the basic functions of the Arduino UNO board, and adds some functional modules to obtain the expected functions. This circuit board system consists of pressure sensors, relays with different trigger modes, and serial port access terminals. This board is powered by a standard USB data cable.

Fig. 5. Arduino UNO main broad

The specific functions of this collection system are as follows: I. Real-time collection of pressure data and display on the screen, II. Control of actuator switches according to a specific trigger condition, III. Active control of the main switch of the experiment and actuator switch.

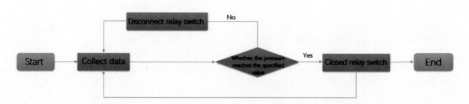

Fig. 6. Logic diagram of the control circuit

The pressure sensor LQ-062-5D is a differential pressure sensor, which is mainly used to measure the difference between the static pressure of the model surface and the standard pressure in the cabin. Considering about the estimated phenomena of this experiment and the requirements of the experimental results, the experiment chooses this model of sensor. This sensor has a range of 5 Psi, an accuracy of 0.01 Psi, and an collection frequency of 20 Hz.

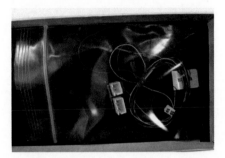

Fig. 7. LQ-062-5D pressure sensor

The Arduino UNO main board has digital signal input/output and analog signal input/output ports. This experiment can use the functions of these ports to collect real-time pressure data. A schematic diagram of the data collection circuit portion is shown in Fig. 8. In this step, the analog signal input port, sensor, and relay are connected together by a plurality of wires, and then the code having the function of allowing the analog signal input port to start collecting data at a frequency of 20 Hz is burned into the Arduino UNO main board. Finally, the collection circuit part was successfully produced.

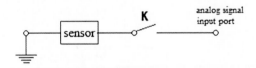

Fig. 8. Data collection circuit diagram

This experiment looking forward to have the effect that the plasma actuator can be automatically turned on in a particular scene—such as when a local pressure is above a certain pressure value—to achieve automatic adjust the effect of local pressure on the surface of the wing. So this paper also designed and made a part of the circuit of the closed-loop control relay, its circuit diagram is shown in Fig. 9. A unit can be controlled by the combination of the Arduino UNO board and the low-level trigger relay. Because the relay is suitable for driving high-power electrical components, such as cranes. Therefore, using the Arduino UNO main board connecting the relay can achieve the function of low voltage controlling high voltage. This relay has three output ports, which represent the normally opening, the public port and the normally closed port. The public port and the normally open end form a normally open switch, the public port and the normally closed end form a normally closed switch. When inputting a high level to the receiving end of the relay, the public port is closed with the normally open end. This experiment continues to write a code that outputs a high level when the acquired value is greater than a specific value we set. Burn this code into the Arduino board and connect the circuit. At this point, the production of the circuit of the closed-loop control is over.

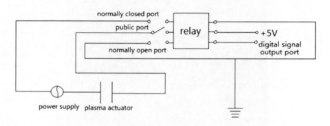

Fig. 9. Circuit diagram of automatic control circuit

This experiment also hopes to actively control the start or stop of the collection, as well as directly control the switching of the plasma actuator, so this paper also designed the following part of the circuit, as shown in Fig. 10. This part uses a remote control relay with a working voltage of 12 V, a working frequency of 315 MHz and an ideal working distance of 1000 m, as shown in Fig. 11, and is connected as shown in the figure. This relay has three modes of interlocking, self-locking and jog control. The relays K1 and K2 are switches for actively controlling the plasma actuator. When K1 is closed, the actuator works, K1 and K2 are both disconnected, the actuator is not working. In addition, the entire system is also equipped with a safety switch. When the safety switch is disconnected, the entire system won't work.

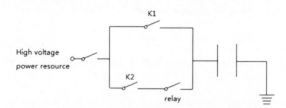

Fig. 10. Remote active control circuit diagram

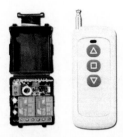

Fig. 11. The remote control relay

Fig. 12. Plasma actuator paste position

The plasma actuation system includes actuators and an actuating power source. This experiment used a symmetrically arranged actuator that used copper foil as the upper and lower electrodes. The lower electrode has a width of 10 mm and the upper electrode has a width of 4 mm. The leading edge of the upper electrode is approximately 4 mm from the leading edge of the wing. Both the upper and lower electrodes are 0.05 mm thick copper foil tape. The actuator uses 4 layers of Kapton tape as the insulating medium, and the tape has a thickness of 0.056 mm.

This experiment will paste the plasma actuator on the left wing, which is the one to be tested, as shown in the Fig. 12.

3 Experiment

3.1 Power Supply Test

This experiment uses the Minipuls 2.1 integrated power supply, which consists of a full bridge converter and an interpole transformer, as shown in Fig. 13. It converts a DC voltage signal input from a lithium battery into a high voltage sine wave signal. It also has a high voltage divider and current detector for monitoring the operation of the equipment.

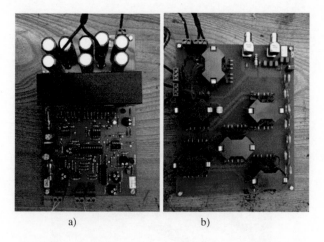

a)   b)

Fig. 13. (a) Full-bridge converter (b) Inter-pole transformer

When a 22-volt DC lithium battery powers the device at the DC input port, the current through the interpole transformer can be changed to 8 kV, 10 kHz high voltage sine wave by adjusting the knob on the full bridge converter. At this time, the power supply problem of the actuator is solved. After proper adjustment of the potentiometer, the power supply has been successfully debugged according to the result displayed by the oscilloscope. The output voltage is 8 kV and the frequency is 10 kHz. Experiments can be performed (Figs. 14, 15, 16 and Table 3).

Fig. 14. Power supply testing

Table 3. Minipuls 2.1 supply information

Voltage peak range	0–12kVp-p
Excitation frequency range	5–20 kHz
Duty cycle	0–100%
Pulse frequency	0–250 Hz
Power supply	30 W
Weight	340 g

3.2   Remote Control Function Test

This experiment used a cpc6000 pressure controller from the Mensor Company from America to provide pressure values to the acquisition board. It consists of two independent precision pressure control adjustment channels, each channel can be configured with up to two standard pressure modules, and each standard pressure module contains two separate standard range segments. This means that each CPC6000 can be configured with up to 8 standard pressure ranges. By connecting the pressure sensor to the cpc6000 pressure controller, it can set the pressure value of the output to verify and calibrate the function of the acquisition board.

Fig. 15. CPC6000 pressure controller from the Mensor company

First, the remote control circuit should be tested to work. In the first step, when the relay is short-circuited, both K1 and K2 are disconnected by remote control, and the actuator does not work. Closing K2, the actuator still does not work. Disconnect K2, close K1, the actuator works and emits a purple glow. In the second step, when the relay is disconnected, if K1 is disconnected, the actuator will not work regardless of whether K2 is open or closed. If K1 is closed, the actuator can work and emit a purple glow, whether K2 is open or closed. This means that the remote control circuit can work successfully.

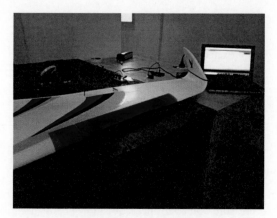

Fig. 16. Remote control function testing

3.3   Data Collection and Automatic Control Function Test

Second, the test of the acquisition circuit and the control circuit will be performed together. After connecting the entire board to the cpc6000, enter the pressure signal as shown in Fig. 17. The trigger condition is set to: When the pressure value is greater than 0.10 Psi, the actuator will be opened. Then the whole process will go on, the following phenomenon can be observed: the serial monitor interface displays the pressure data collected by the sensor in real time, and the collected pressure curve can also be seen in real time in the serial plotter interface. The plasma actuator is off when the trigger condition is not met (i.e. the red area). After the trigger condition is met (i.e. the blue area), the plasma actuator will be turned on, which emits a purple glow. At this time, the switch of the actuator is actively controlled by the above-mentioned remote control circuit, and it is found to be still feasible. This means that the collection point and control circuits are working properly, and that the entire board can work together successfully without interference (Figs. 18, 19, 20 and 21).

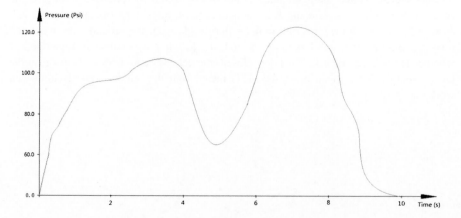

Fig. 17. Pressure curve output from pressure controller

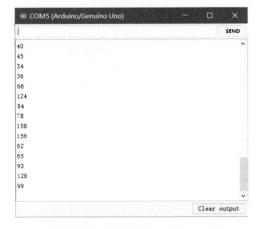

Fig. 18. The interface of serial monitor

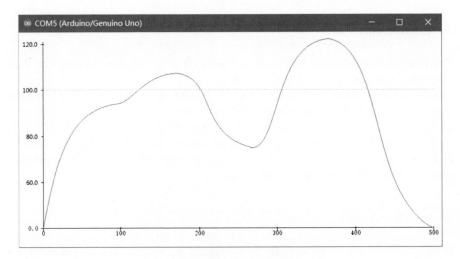

Fig. 19. The interface of serial plotter

Fig. 20. The plasma actuator emits a purple glow when it is working

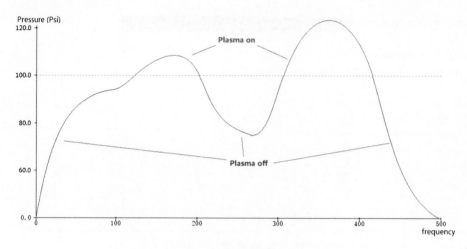

Fig. 21. State of operation of the plasma actuator

4 Conclusion

This paper mainly designed a circuit board, which has the functions of collecting real-time data, monitoring the state of the wing surface, automatically controlling the actuator, and controlling the actuator remotely. The board was tested with a flying wing model and a standard pressure source. The results show that the design of this board is successful: it can monitor the local pressure on the surface of the wing in real time. And it also can react to a specific situation to control the switch of the actuator. It also allows people to control the actuator's switch actively and remotely. This board is operating successfully and this experiment is success.

5 Further Work

In the future, a wind tunnel experiment to measure the control pressure threshold will be designed and implemented, and a code that can intelligently analyze the UAV's real-time flight status and calculate the instantaneous ideal threshold will be written. Finally, when these preparations are completed, the UAV flight test will also be carried out. This flight experiment will give a variety of data on the plasma virtual aerodynamic surface UAV in a series of flight conditions. A more mature design of UAV with virtual plasma pneumatic will be presented.

Acknowledgement. This work was supported by National Natural Science Foundation of China (Grant Nos. 11672245 and 11772263), the Peak Experience Project of Northwestern Polytechnical University, the National Key Laboratory Research Foundation of China (Grant No. 9140C420301110C42), the Fundamental Research Funds for Central Universities (3102018JCC008), and the 111 Project (No. B17037). This paper also thank Qichen Bao, a student from the University of Liverpool, United Kingdom, for giving inspiration and technical guidance to the board design part of this experiment.

References

1. Ma S, Wu C, Chen H (2006) Research on stability and maneuverability of flying wings. China Flight Dyn. 24(3)
2. Sidorenko AA, Budovsky AD, Pushkarev AV et al (2008) Flight testing of DBD plasma separation control system. Institute Theoretical and Applied Mechanics, Novosibirsk
3. Yao J, He H, Zhou D, He C, Shi Z, Du H (2017) Experiment of plasma-driven roll control of flying wing layout aircraft. J Beijing Univ Aeronaut Astronaut 43(4)
4. Zhang Y, Hou Y, Wang F (2016) Overview of plasma flow control technology for dielectric barrier discharge around aircrafts. AVIC First Aircraft Design and Research Institute, Xi'an Aerospace Science and Technology, Shaanxi, 15 June 2016, vol 27, No 06 05-10
5. Kaparos P, Koltsakidis S, Panagiotous P Experimental investigation of DBD plasma actuator on a BWB aerial vehicle model. Laboratory of Fluid Mechanics & Turbomachinery, Department of Mechanical Engineering, Aristotle University thessaloniki, 54124, Greece
6. Zhao C, Pang C, Xu J et al (2016) The effect of plasma in controlling the flow separation flying-wing of B WB Models. In: Proceedings of 2016 IEEE Chinese guidance, navigation and control conference, 12–14 August 2016, Nanjing
7. Grundmann S, Frey M, Tropea C (2012) Unmanned Aerial Vehicle (UAV) with plasma actuators for separation control. Institute of Fluid Mechanics and Aerodynamics, Technical University of Darmstadt, 64287, Germany
8. Du H, Shi Z, Ni F et al (2013) Aerodynamic torque control of flying wing layout based on plasma excitation. Aerosp J 34(9)
9. Chen K, Liang H (2015) Wind tunnel experiments on flow separation control of a UAV by nanosecond discharge plasma aerodynamic actuation. School of Aeronautics, Northwestern Polytechnical University, Xi'an, 710032, China. School of Aeronautics and Astronautics Engineering, Air Force Engineering University, Xi'an 710038, China
10. Nelson RC, Corke TC, He C (2007) Modification flow structure over a UAV wing for roll control. Aerospace Sciences Meeting, University of Notre Dame, 8–11 January 2007, Reno, Nevada
11. Zhang X Study on plasma control of supercritical wing separation flow. PhD thesis Northwestern Polytechnical University Xi'an, Shaanxi

Civil Aircraft Fly Test Frequency-Domain Data Method Research

Bin Gao[1,2(✉)] and Zhengqiang Li[2]

[1] Nanjing University of Aeronautics and Astronautics, Nanjing 210016, China
[2] Shanghai Aircraft Design and Research Institute, Shanghai 201210, China
{gaobin, lizhengqiang}@comac.cc

Abstract. In this paper, the method of converting time-domain data into frequency data processing is studied. Flight test data is recorded according to the time interval of time domain data, test data such as frequency, power spectrum analysis, frequency domain identification characteristics shall be carried out in the frequency domain, this needs to be time domain data transformation processing, converted into frequency domain data.

Keywords: Frequency domain data · Fourier transform · Frequency sweep test

1 Introduction

Flight test data is recorded according to the time interval of time domain data, but the time domain data is not enough for characteristics analysis of the flight, the amplitude-frequency character and the phase-frequency character is important for the flight test data analysis, the frequency character shall be analysed and compared to desktop simulation data, the good conclusion can be made after the analyzing and comparing. This paper give a method that can convert time-domain data into frequency data, and can be used in flight data analysis.

2 Frequency Response

The frequency response will be defined by a simple dynamic system (SISO, linear, stable, time invariant). Suppose the sine wave (periodic) with amplitude A and frequency f $x(t)$ is used to drive the system.

$$x(t) = A \sin(2\pi ft) \tag{2.1}$$

When the response transient process gradually disappears, The system output $y(t)$ will also be a sine wave (periodic) of the same frequency f, but the corresponding amplitude is B and there is a phase shift φ:

$$y(t) = B \sin(2\pi ft + \varphi) \tag{2.2}$$

© Springer Nature Singapore Pte Ltd. 2019
X. Zhang (Ed.): APISAT 2018, LNEE 459, pp. 1312–1321, 2019.
https://doi.org/10.1007/978-981-13-3305-7_104

In cases of the periodic input and output signals, the Fourier series can be in the form of numerical calculate the value of the A, B and φ at this time in Fourier series contains only first-harmonic (sine and cosine).

The frequency response function $H(f)$ is the complex function value, which is generally represented by the amplitude frequency characteristic and phase frequency characteristic data curve of the frequency f as the independent variable:

$$|H(f)| = \frac{B(f)}{A(f)} \tag{2.3}$$

And

$$\angle H(f) = \varphi(f) \tag{2.4}$$

The frequency response can be obtained by the experimental method, which is to use sine input signals of different frequencies to drive the system and measure the corresponding output amplitude and phase shift.

Frequency response $H(f)$ the best linear system input and output behavior description way completely describe the dynamic characteristic of the system, this description without any prior knowledge about the dominant equation of motion system's internal structure.

3 Fourier Transform and Frequency Response

Sinusoidal input can be extended to any non-periodic input, and can also be used in systems with stable or unstable dynamic response characteristics. The Fourier series can be obtained by the Fourier transform, and the Fourier transform can also be used for non-periodic signals.

Non-periodic time domain signal $x(t)$ and $y(t)$ can be converted to frequency domain signal $X(f)$ and $Y(f)$ by Fourier transform, the standard Fourier transform is defined as follow:

$$X(f) = \int_{-\infty}^{\infty} x(t)e^{-j2\pi f t}dt \tag{3.1}$$

and

$$Y(f) = \int_{-\infty}^{\infty} y(t)e^{-j2\pi f t}dt \tag{3.2}$$

$X(f)$ and $Y(f)$ is Fourier coefficient. The frequency domain response $H(f)$ is a function which have complex value, it has a relation with the input Fourier coefficient $X(f)$ and the output Fourier coefficient $Y(f)$:

$$Y(f) = H(f)X(f) \qquad (3.3)$$

It is same with the function that has a constant sinusoidal input (function (2.3) and function (2.4)).

The frequency domain response function $H(f)$ is the ratio of the Fourier output and Fourier input, which can be presented as real part and complex part as follow:

$$H(f) = \frac{Y(f)}{X(f)} = H_R(f) + jH_I(f) \qquad (3.4)$$

The amplitude ratio of the function can be obtained as follow:

$$|H(f)| = \sqrt{H_R^2(S) + H_I^2(S)} \qquad (3.5)$$

The phase shift of the function (2.4) can be obtained by $H(f)$

$$\varphi(f) = \angle H(f) = \tan^{-1}\left[\frac{H_I(f)}{H_R(f)}\right] \qquad (3.6)$$

In practical application, the limitation of the Fourier transform calculated by function (3.1) and (3.2) is that the domain input and output signal time domain are integrated (Dirichlet condition)

$$\int_{-\infty}^{\infty}|x(t)|\,dt < \infty \quad 和 \quad \int_{-\infty}^{\infty}|y(t)|\,dt < \infty \qquad (3.7)$$

Obvious, this condition make the open-loop frequency sweep test of the unstable system can't be done. When the bounded frequency sweep signal passes the unstable system, the output of the signal is unbounded, so the function (3.7) is not fulfilled.

These condition shall be satisfied during the frequency sweep test: (1) start and stop in the equilibrium condition, (2) adjust the plane maneuver to ensure the boundedness of the transient process, and ensure the approximate symmetric against the reference flight state. When doing the frequency sweep test, pilot (control system) shall provide the feedback signal to ensure the boundedness of the closed loop response without considering the steady-state of the plane. So, whether the dynamic character of the plane is steady or non-steady, the Dirichlet condition of the function is meet, the frequency response function can be obtained by frequency sweep test. Frequency sweep test done in the closed loop condition will bring some bias to the estimation of the frequency response, but in practise, this bias is not obvious when the ratio of the noise is so little in the stimulate signal.

The maneuver input of the pilot can't be used in frequency response test because of the boundedness requirement of the function (3.7). For example, when the pilot just make the forward step input of the longitudinal sidestick, the correspond response of the plane is a constant angular velocity. The response of the plane will accumulate gradually (e.g. speed, descent rate) until the maneuver stopped by the pilot. This is not a good input for the frequency response calculation, because of the integration of $x(t)$ in function (3.7) is divergent until the end of the data record, this is the same with the frequency response calculation of the unstable system without the feedback structure.

4 Frequency Response Calculation Example

In this paper, frequency sweep signal is taken as an example to illustrate the calculation process of frequency response. The system to be identified is a typical second-order oscillation link as follows:

$$H(s) = \frac{1}{s^2 + 2\zeta\omega_n + \omega_n^2} \tag{4.1}$$

The damping ration is $\zeta = 0.35$, undamped frequency is $\omega_n = 3$ rad, about 0.48 Hz. The parameter of the input is set:

- sampling frequency: $f_s = 50$ Hz
- sampling time interval (simulation step length): $\Delta t = 0.02$ s
- initial frequency: $f_{start} = 0.1$ Hz (about 0.6832 rad/s)
- stop frequency: $f_{end} = 1.5$ Hz (about 9.4248 rad/s)
- recording time: T = 80 s

The output data also is recorded in 50 Hz, the single record length is 80 s.

4.1 Parameter Estimation

4.1.1 System Frequency Range
The choose maximum frequency is 3 times of the undamped frequency in this second-order system, as $f_{max} = 1.5$ Hz; the choose minimum frequency is $f_{min} = 0.15$ Hz, the maximum frequency and the minimum frequency is 10 times.

4.1.2 Data Requirements
The initial frequency of the frequency sweep test is $f_{start} = 0.1$ Hz, the stop frequency is $f_{end} = 1.5$ Hz, the interest frequency is covered, the system response is stimulated, and the frequency requirement is meet.

According to the criterion described above, the filter frequency is

$$f_f = 5f_{max} = 7.5 \, \text{Hz}$$

Also, sampling frequency is:

$$f_s = 5f_f = 25f_{max} = 37.5 \, \text{Hz}$$

The actual sampling frequency is $f_s = 50 \, \text{Hz}$, which meet the requirement above.

The interest minimum frequency is $f_{min} = 0.15$, so the interest maximum frequency is $T_{max} = 1/0.15 = 6.67 \, \text{s}$. The initial frequency is 0.1 Hz, the initial period is 10 s, the requirement is meet.

The recording signal is 80 s, which is more than 5 times of the interesting maximum period, and also is more than 5 times of the initial signal period. The low frequency character is completed recorded, which meet the requirement.

4.1.3   Window Parameter Estimation

According to the window estimation method, the window shall include 2 oscillation period of the minimum frequency, also is 2 times of the maximum period:

$$T_{win} = 2T_{max} = 13.3 \, \text{s}$$

Then, the maximum window length is no longer than 50% of the single record, so

$$T_{max} \leqslant 0.5 \times 80 = 40$$

The maximum window length is no longer than 40 s, no shorter than 13.3 s.

The minimum window shall provide 10 times frequency range between the minimum effective frequency and the maximum interesting frequency f_{max}, so the minimum window is:

$$T_{min} \geqslant 20 \frac{1}{f_{max}} = 20/1.5 = 13.3$$

The minimum window is no longer than 13.3 s.

The window length is 13.3 s–40 s according to the interesting frequency range.

15 s is choose by the window length of the single window estimation.

14.5 s, 21 s, 27.5 s, 34 s, 40 s can be choose of the multi-window analysis.

As the window length is 15 s, 5 times of the window is 75 s, the single length is 80 s, which meet the requirement of the 5 times of the window.

For the compound divided window, the two record can be connected together as the single record is not enough, which the length is 160 s, meet the requirement.

4.2   The Time Domain Data CZT Transform

4.2.1   Data Evaluation

According to the convention method, the time domain data shall be CZT transformed separated, the frequency content, the data quality shall be evaluated, then the high quality record data is connected to meet the length requirement of the expected data.

The single CZT transform of the input signal is described above, the result is as Fig. 1.

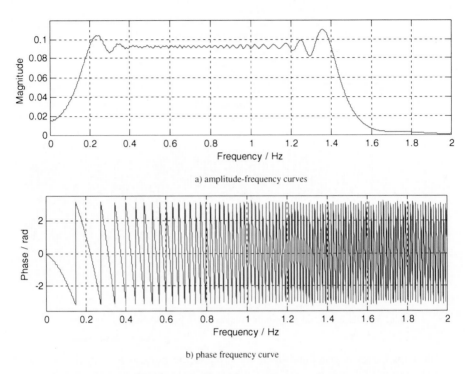

a) amplitude-frequency curves

b) phase frequency curve

Fig. 1. Input signal amplitude-frequency and phase-frequency curves

According to the amplitude-frequency, the amplitude is high quality and less change during 0.15 Hz–1.5 Hz, which can stimulate system during this frequency.

Figure 2 shows the time domain curve of the output record, Fig. 3 shows the CZT transform result of the output record (Fig. 4).

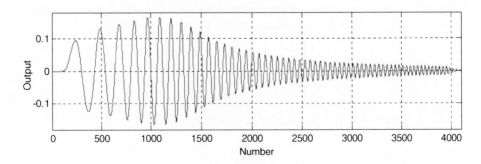

Fig. 2. Output time domain signal

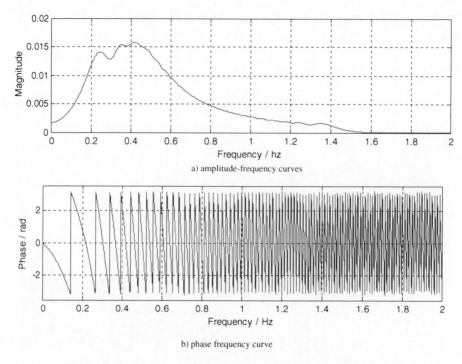

a) amplitude-frequency curves

b) phase frequency curve

Fig. 3. Output signal amplitude-frequency and phase-frequency curves

The output data is generated by simulation, so the noise is too small that can be ignored, then the two test data record can be connected together direct to generate the record signal which can meet the length requirement.

4.2.2   Apply the CZT of the Split Window Method

According to the analysis above, the window size is choose to $T_{win} = 15$ s, the window length is L = 750. The use data is single recorded, the record time is $T_F = 81.92$ s, the record length is N = 4096.

To compare with the result that not using window, the choose frequency number is 4096. The Hanning window is choose refer to CIFER, the window overlap ratio is choose 0.8 refer to CIFER, the overlap number of window is (Fig. 5):

$$n_r = 1 + \frac{T'_{rec}/T_{win} - 1}{1 - x_{frac}} = 1 + \frac{81.92/15 - 1}{1 - 0.8} = 24$$

According to the above set, the CZT transform is invoked, the every transform result is weighted averaged, the result is as above figure.

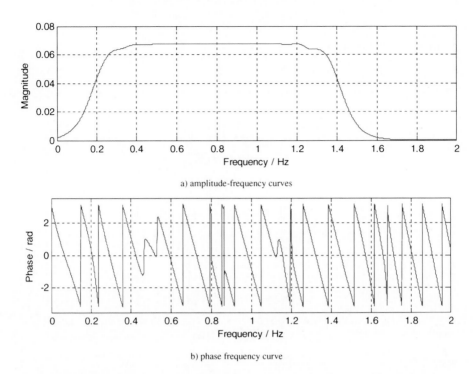

a) amplitude-frequency curves

b) phase frequency curve

Fig. 4. Amplitude-frequency and phase-frequency curves calculated by input signal window (after adjustment)

Obvious, the CZT transform is more smooth using split window. As showed in the Amplitude-frequency curve, compare with single CZT transform result, the curve is the same, but the result is more smooth and the amplitude is different using split window method. The multi-window averaging reduces the wave amplitude value in Amplitude-frequency, and the sidelobe leakage produced by the rectangle window is reduced by using Hanning window.

As showed in the phase-frequency curves, the difference between single CZT transformation and split method is big. That is because of the phase shift reduction caused by the multi-window average. So the phase-frequency result of the single CZT transformation is more accurate.

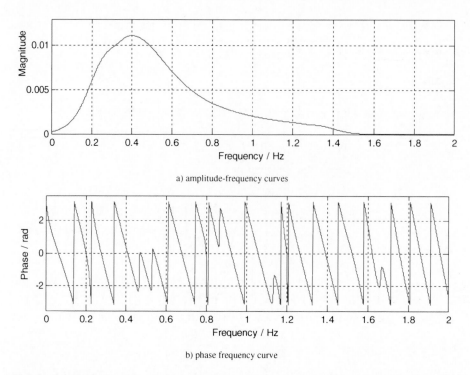

a) amplitude-frequency curves

b) phase frequency curve

Fig. 5. Amplitude-frequency and phase-frequency curves calculated by output signal window (after adjustment)

5 Conclusion

In a word, assume that the stimulate signal (input) and the measured value of the correspond response (output) are given, the frequency response of the system can be determined, the abundant signal of the dynamic system character is provided in the condition that don't male any assumption of the system structure and stability. With the Bode plot, Nichols plot and Nyquist method, the frequency response function shall support the dynamic system analysis, control system design, verification of the simulation model without the parametric model (e.g. transfer function, state space model). In addition, when the dynamic system is nonlinear, the frequency response extracted by Fourier transform is first harmonic describing function, this function is the best linear model describing nonlinear system.

The CZT transform is more smooth using split window. Comparing with the normal signal, the frequency sweep signal is more special, the frequency distribution is changed with time, when catching signal in the different location of the time domain using windows, the different location response of frequency domain is mapped. So this shall be pay more attention when using split window method.

The flight test data can be obtained using this method, and the frequency domain character can be analysed and compared to desktop simulation data, this will give a good conclusion to flight test and is important to flight test.

References

1. Jategaonkar RV (2005) Flight vehicle system identification-engineering utility. J Aircr 42(1):11
2. Morelli EA, Klein V (2005) Application of system identification to aircraft at NASA Langley research center. J Aircr 42(1):12–25
3. 李乃宏, 吴瑶华, 黄文虎 (1995) 改进的非线性连续-离散系统的极大似然参数估计及其应用. 自动化学报 21(4):471–475
4. 张友民, 张洪才, 戴冠中, 贺志斌 (1994) 非线性滤波方法及其在飞行状态及参数估计中的应用. 航空学报 15(5):620–626
5. 刘利生 (2000) 外测数据事后处理. 国防工业出版社, 北京, pp 262–330
6. 周自全 (2010) 飞行试验工程. 航空工业出版社, 北京
7. Morelli EA (1995) Estimating noise characteristics from flight test data using optimal Fourier smoothing. J Aircr 32(4):689–695
8. Lu XJ, Chen Q, Li P et al (2014) Real-time simulation for online identification of aircraft parameters using Fourier transform regression. In: IEEE Chinese Guidance, Navigation and Control Conference (CGNCC). IEEE, pp 1373–1378. (EI:20150700517665)

An Aircraft Level System Test Facility Based on Individual System Test Benches

Chen Wu[(✉)], Guirong Zhou, Guanglei Xu, and Bin Gao

COMAC Shanghai Aircraft Design and Research Institute, No. 5188 Jinke Road,
Pudong New District, Shanghai, People's Republic of China
wuchenl@comac.cc

Abstract. This paper describes the method of creating an aircraft level system integration test facility by connecting distributed individual system test integration benches using a fiber–optic network and signal acquisition and reconstruction. The facility has been used for x type aircraft integration testing to support first flight. The experience proved this aircraft level system integration test facility can satisfy the aircraft function integration test requirements. Connecting existing system test facilities to reduce the integration tasks on the aircraft reduced the time taken to get to first flight. The cost of a new aircraft level system test facility was saved by reusing the distributed test resources.

Keywords: Aircraft level systems integration test ·
Remote signals interconnecting · Remotely collaborative comprehensive test ·
Timeslot · Optic signal switch

1 Introduction

The integration testing in aircraft development has to be carried out at many levels, subsystem, system and aircraft. The aircraft level system integration test is traditionally carried out either on aircraft, in dedicated aircraft level laboratory or by connecting system integration benches concentrated at one site. Testing on aircraft implies that any design changes identified will be at a late stage of the development process and the later a change is introduced, the costlier it becomes. Furthermore, it is difficult to install specific diagnostic devices on an aircraft. A dedicated aircraft level laboratory provides a very good solution but at a high price. The solution of using individual, co-located system benches with representative connections is not practical because of the amount of equipment that would be needed at one site [1].

Aircraft system integration physically assembles all systems at one site. Physical integration means that at first the focus is on hardware integration and related problems (e.g., connectors or plugs do not fit) rather than on the interaction of the systems and the related problems (e.g., incomplete/incompatible interfaces, interface misinterpretation, configuration problems). To overcome some of these drawbacks, testing can be performed in two separate phases: During the first phase, the systems are function integrated by connecting the multi-system integration test rigs which are typically located at different sites, and in the second phase, the systems are physically integrated at one site or on-the-aircraft. For first integration steps, laboratory environments like an

© Springer Nature Singapore Pte Ltd. 2019
X. Zhang (Ed.): APISAT 2018, LNEE 459, pp. 1322–1329, 2019.
https://doi.org/10.1007/978-981-13-3305-7_105

iron bird, avionic system bench or electric system bench are often considered more meaningful and cost-effective than on-the-aircraft integration [2].

The development of digital systems has made communication far easier to establish. Moreover, the use of fiber-optic, with signal acquisition and reconstruction, with its characteristics of low transmission loss, high speed and good electromagnetic interference resistance, allow long distances transmissions [3]. This approach allows individual system test benches to be integrated into an aircraft level laboratory and thus saving the cost and human resources needed to build dedicated aircraft level laboratory.

2   The Goal and the Capability of the System Integration Test Benches

For a recent Aircraft development program, an Avionics System Test Bench, Iron Bird, Reverse Thrust System Test Bench and Electrical System Test Bench were built on the same site but separated by equipment around them up to 100 meters. This made it impossible to directly connect these systems together without signals attenuation which would make their integration unrepresentative. These benches were built with the intension of using them for aircraft level integration, so a solution for this problem had to be found.

In general, these system integration test benches would contain all the equipment making up the system under test connected to simulations of the systems it interfaces with. In this way the equipment making up the system would be integrated. The next step is to replace the simulated systems with real hardware to perform the aircraft level integration testing. The benches also provided the acquisition system, monitoring system, failure injection system, auto-test function, and the physical interface for integration testing.

2.1   The Iron Bird

The Iron Bird included all flight control, empennage, landing gear, brakes and hydraulic systems, all with representative sizing, stiffness and strength. There were platforms for operators and clear zones to accommodate movement of the flight control surfaces and landing gear extension/retraction and three-axis turntable. An electric motor used to simulate the engine driver pump work to pump the hydraulic fluid. The mechanical aspects of these systems could therefore be integrated on the rig, but the Avionics and other systems had to be simulated.

For full aircraft integration testing, the systems on Iron Bird needed signals transferred via the real avionics system to and from the cockpit to verify control and monitoring functions on flight deck. These could only be made available from the Avionics System Bench. Similarly, The Electrical System Test Bench was needed to provide truly representative power conditions and supply distribution control.

2.2   Electrical System Test Bench

The Electrical System Test Bench includes the aircraft generator and electrical control system, along with hardware to simulate the electrical load. As with the Iron Bird, the control system on the Electrical System Test Bench needed signals from the real avionics system and cockpit provided by the Avionics System Test Bench. It would also benefit from the more sophisticated Power Plant Simulator on the Reverse Thrust System Test Bench to drive its generator as well as more representative electrical distribution control using the display control panel on Avionic System Bench.

2.3   Reverse Thrust System Test Bench

The Reverse Thrust System Test Bench was made up of the real reverse thrust system and a simulation of the engine system. During the aircraft level system integration test the Reverse Thrust System Test Bench needed the Avionics System Test Bench, Electrical System Test Bench and Iron Bird to verify landing performance.

2.4   Avionics System Bench

The Avionic System Bench comprise all real avionics equipment, including the data network, IMA, communication, navigation, display and on-boarding maintenance, and the cockpit deck and 80% of non-avionic system electronic control units with simulated sensor inputs. The other system benches will work with Avionic System Bench to verify data forwarding, flight data indication, systems state monitoring, communication, navigation and fault reporting.

3   The Method Connecting the Individual System Test Benches

Figure 1 show the aircraft level system test facility constructed from the individual system test benches. The Signal Processing Devices, which provide aircraft system electrical signal processing as well as simulator signal and clock signal interfaces, was introduced into each of the individual test benches. The signals from the connected test benches now replaced those generated by the individual simulators for stand-alone test. The clock network ensures that all the individual benches work in unison. A Reflective Memory Network transfers the simulation signals between the benches. A Single-mode fiber-optical cable, carrying the aircraft system signals, and two multi-mode fiber-optical cables, carrying the simulator and clock signals, are laid between individual test benches. A central signal Switch is set up for forwarding aircraft system, simulator and clock signals between these benches. The individual test benches are therefore connected into an aircraft level system test facility by these three networks.

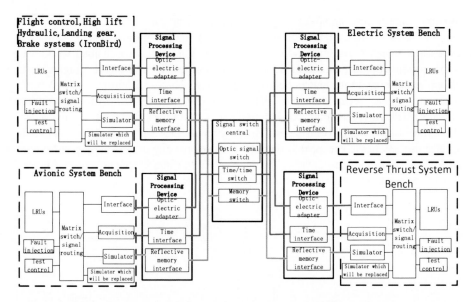

Fig. 1. Block diagram of Aircraft level system test facility test benches

3.1 Fiber-Optic Communication Network of Aircraft System Signals

The fiber-optic communication network transmitting aircraft system signals consists of optic-electronic adapters, optic signal switches, and single mode fiber-optic. The Optic-electronic adapter undertakes the aircraft system electrical signals acquisition and conversion and reconstruction and digital signals communication. Optic switches forward frames from one Signal Processing Device to another.

The time division multiple access communication method is used in the fiber-optic communication network. The available bandwidth of the single mode fiber-optic is 2 Gbps. The consecutive 1000 bit is a frame data that is divided into 125 timeslots. Each timeslot contains 8 bit data. So the available bandwidth of the frame is 16 Mbps (2 Gbps/125 = 16 Mbps). A group of signal channels has fix size timeslots depended on its signal type. The sender packs the signal to their timeslots on a frame and the receiver unpacks the signals from a frame.

4 Optoelectronic Adapter

The optoelectronic adapter consists of signal processing adapter cards (which fulfil the role of signal conditioning, signal acquisition/reconstruction, analog-to-digital/digital-to-analog conversion, timeslot coding/decoding) and communication processing board (which deal with the timeslot packing/unpacking, optical-electrical conversion, optical signals transferring/receiving).

Analogue signals, discrete signals, A664 signals and A429 signals are transmitted between the system test benches by fiber-optic networks. The adapter card type is designed according to these signal types. Each type of adapter is combined with multiple transmitting and receiving channels and can simultaneously transmit and receive signals. The 2 MHz acquisition frequency and 16 bit acquisition accuracy is used for low-speed analogue and discrete signals. The Receiving channel works with signal conditioning, signal acquisition, analogue-to-digital converting, coding, packing, electric-optic converting and optic signal emitting. The Transmitting channel works with optic signal receiving, optic- electric converting, unpacking (including frame preamble detecting, data unpacking, frame parity checking), decoding, digital- analog converting and signal reconstruction. The A429 and A664 signal channels work as the similar process flows except that they do not perform any analog-digital and digital-analog converting. Figure 2 shows the adapter transmitting channel diagram. Figure 3 shows the signal processing diagram of the receiving channel.

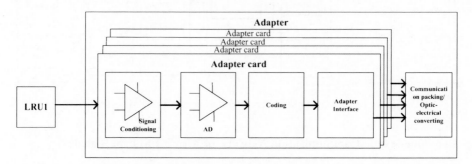

Fig. 2. Signal processing on the adapter transmitting channels

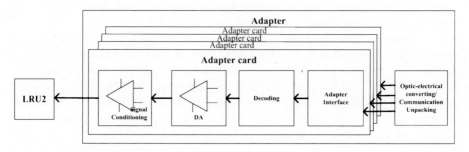

Fig. 3. Signal processing on the adapter receiving channels

One adapter can have four type signal processing adapter cards. The timeslot of every type adapter card is determined by Eq. (1). Every adapter's total timeslots is no longer than 120.

$$N = S * A * n/B \tag{1}$$

Where N is timeslot, S is sampling rate, A is accuracy, n is quantity of channel, B is bandwidth.

The number of adapter cards on one Signal Processing Device is determined by the quantity of the signal transmitted and received on the system bench. Every four adapter cards need an adapter. An adapter connect to a fiber-optic.

Figure 4 shows the adapter diagram. The adapter card is connected to the communication processing module through the adapter connector. Signals are packed or unpacked by the FPGA. The FLASH memory, which stores the load program of PowerPC and FPGA is read/written by the CPLD. The external data storage of the PowerPC and FPGA is fulfilled by DDR. The fiber-optic signal is transmitted and received by the SFP.

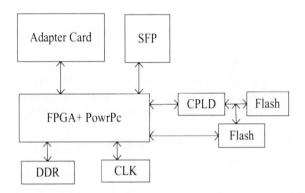

Fig. 4. Block diagram of the adapter communication processing module

5   Optic Signal Switch

One master timeslot switch and many slave timeslot switches forward the optical signals among adapters. An optic signal switch connect eight adapters. The network topology diagram drew on the configuration PC generate the routing table which is loaded on optic signal switch. Optic signal switch forward signal by extracting the adapter and adapter card number from the management timeslot. The main optic signal switch synchronizes clock of all adapters and slave switches by its reference time clock. Figure 5 is the network topology diagram.

5.1   Simulator Signal Reflective Memory Network

The Simulator signal reflective memory network consists of simulator PCs on the distributed system test benches, reflective memory switch and multimode fiber-optic. The simulator PCs share data and work in a closed loop with aircraft LRUs connected by the Fiber-optic communication network.

The simulation models include flight simulation model (providing pressure altitude, Mach number, total temperature of the atmosphere on the Iron Bird), power plant

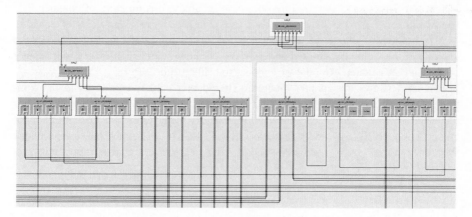

Fig. 5. The network topology configuration

model (providing thrust, N1, N2, fuel consumption rate, engine-driver on Reverse Thrust Test Bench) and the communication and navigation excitation model (which receives aircraft position and attitude from the flight simulation model) located on Avionic System Bench. The power plant model receives indicated air speed, throttle lever angle and bleed air command from LRUs on Avionic System Bench. It also receives pressure altitude, Mach number, and total temperature from the flight simulation model. The flight simulation model receives thrust, N1, N2, and fuel consumption rate from power plant model. The hydraulic engine driving pump simulator and electrical system generator receive N2 from the power plant model.

5.2 Clock Synchronization Network

The clock synchronization network consists of a master clock, clock switch and multimode fiber-optic. The master clock synchronizes the slave clocks on the distributed system test benches. The adapters, timeslot switches, acquisition boards and avionic clock on fiber-optic communication network and simulator PC, reflective memory switch on simulation signal reflective memory network all work with the same time reference.

5.3 Latency of Aircraft System Signals

Latency of Aircraft System Signals on Fiber-Optic Communication Network comes from signal processing adapter cards, communication processing board, Optic signal switch and fiber-optic. Latency can be calculated by Eq. (2).

$$Ts = 2 * T4 + 2 * T2 + T1 + T3 \tag{2}$$

Where T4 is latency of signal processing adapter cards, T2 is latency of communication processing board, T1 is latency of Optic signal switch, T3 is latency of fiber-optic.

Test show A664 signal latency is less than 225 μs, A429 signal latency is less than 12 μs, Analog signal latency is less than 100 μs.

6 The Testing on the Aircraft Level System Test Facility

Distributed flight control system, landing gear system, brake system, hydraulic system, electric system, reverse thrust system and avionic system test with one another or multilateral systems on the Aircraft Level System Test Facility. The following tests have been done:

- The end-to-end function integration testing with load, including interface communication test, logical function test.
- The first flight profile testing for aircraft status checking, system level function, system interface on all flight phase and for normal take off and landing, go-around in air and touchdown on flight leg test. One function and multiple functions inoperative testing.
- OATP assessment with normal and abnormal flight status, function on ultimate condition and harsh environment (high temperature, high humidity, wind shear), aircraft protection (stall, anti-collision, anti-icing, rain removal).
- First flight process assessment for flight test schedule, manipulating adaptability, emergency response, major function failure, EICAS.
- Familiar with the layout of cockpit and flight operating procedures, training for flight test task on flight phase.

These tests help the system designer ahead for flight operating and the coordination mechanisms among systems. Many potential function design conflicts, mismatch in interfaces, cable mistakes are found. Pilot increase their confidences by familiar with first flight process and requisites. The first flight test speed up.

7 Conclusion

The aircraft level system integration testing is divided into functional integration testing and performance integration testing. The functional integration testing is fulfilled on the Aircraft Level System Test Facility based on distributed aircraft system test benches instead of aircraft to accelerate the development progress. The expense of a new aircraft level system test facility is saved. It makes sense for more remote test labs to be connected by the fiber-optic network.

References

1. Lansdaal M, Lewis L (2000) Boeing's 777 system integration lab. IEEE Instrum Meas Mag 9:13–18
2. Von Aliki O (2007) System testing in the avionics domain, p 43
3. Li Y, Wang Y, Shen J, Li Q (2009) Fiber-optic communication network. Xidian University Press, p 2

Design and Experimental Study on a Flapping Wing Micro Air Vehicle

Yi Liu, Yanlai Zhang[(⊠)], and Jianghao Wu

School of Transportation Science and Engineering,
Beihang University, Beijing 100191, China
zhangyanlai@buaa.edu.cn

Abstract. Flapping Wing Micro Air Vehicles (FWMAVs), which are inspired by nature's flyers and mimic their flight, have numerous advantages compared with conventional fixed wing and rotating wing aircrafts at small scale and low Reynolds numbers, such as ability of hovering and anti-disturbance, high lift aerodynamic performance, good maneuverability. Due to the dimensional constrains and demands of compact design and low power consumption, suitable angles of attack and size of wings have to be tested to reach high aerodynamic efficiency. This paper presents results of experimental investigation of angle of attack (changed by a root deflection angle) and wingspan on the aerodynamics and power of a FWMAV. Based on the measured results, a quasi-steady model for lift and power estimation is suggested. The difference between the estimated and measured lifts and power is less than 10%, which reveals that the quasi-steady model is reasonable for a preliminary design. The results indicate that the wing with a root deflection angle around 15° shows the highest aerodynamic efficiency, and larger wings are preferred to reach a higher ratio of lift to power, implying that larger wing could give a higher lift under certain power input. Finally, we have obtained a best performance wing, which can generate about 40 g of lift at a power input of about 8 W when flapping at 22 Hz.

Keywords: Flapping Wing Micro Air Vehicles · Experimental measurement · Aerodynamics · Power consumption

1 Introduction

Micro air vehicles (MAVs) are a subclass of unmanned aerial vehicles (UAVs), limited to small size and light weight, with remotely piloted or autonomous operation. The bionic flapping-wing micro-aircraft has superior aerodynamic performance, manoeuvrability, and anti-disturbance ability compared with conventional fixed wing and rotating wing aircrafts at small scale and low Reynolds numbers [1]. Its small size, easy portability and strong concealment make it widely used in military investigations and civil counter-terrorism disaster relief. Therefore, research on flapping wing micro-aircraft is an important and meaningful event.

For these reasons, engineers have focused more on developing flapping-wing micro air vehicles (FWMAVs) [2–6] in recent years. A common FWMAV consists of a wing actuation mechanism, a pair of wings, a micro control system, a motor and a battery.

© Springer Nature Singapore Pte Ltd. 2019
X. Zhang (Ed.): APISAT 2018, LNEE 459, pp. 1330–1342, 2019.
https://doi.org/10.1007/978-981-13-3305-7_106

For flight, the FWMAV should be designed to generate a lift greater than its body weight. In order to optimize the FWMAV design, many studies have attempted to understand the effect of different parameters like wing area, angle of attack and flapping amplitude on the aerodynamic force generation and power consumption of flapping wing using CFD and quasi-steady methods [8–11], and results show that they all have a significant effect on aerodynamic performance. However, for the actual design of a FWMAV, due to the fact that the wings will inevitably have deformation during flapping and inaccurate assembly results in gaps between different mechanisms, it is difficult to accurately analyze the lift and power consumption of the system by numerical simulations, so it is necessary to use experimental measurement to further verify the simulation results.

The objective of this paper is to analyze the aerodynamic performance of a FWMAV by combining experimental and numerical method. We have learned from the previous works [5, 7] on the design of the mechanism and successfully used 3D printing technology to design and manufacture a flapping mechanism that can work at a high flapping frequency and operate for a long time. Subsequently, the aerodynamic lift and power consumption of the wing at different flapping angles of attack are investigated by experimental measurements. By changing the root deflection angle of wing to change angles of attack [15], the time course of wing deformation and wing flapping angles and angles of attack were extracted from images captured by a high speed camera. We found that different angles of attack and flapping frequencies will affect the wing flapping amplitude significantly. In addition, we used the quasi-steady aerodynamic model [9, 11] to estimate the lift and power consumption of those wings and also used a load cell to measure the corresponding lift and power and a vacuum chamber to measure inertial power. Then, the results of estimation and experimental measurement are compared, showing that the quasi-steady model is proper for initial design when considering real kinematics and deformations. Finally, using the quasi-steady model, we evaluated the lift and power consumption of different wings with different wingspans and analyzed the results applying to the design of FWMAV.

2 Flapping-Wing System and Experimental Method

2.1 Flapping Mechanism and Wing

The flapping mechanism, as shown in Fig. 1, is learned from Ref. [7]. It adopts a single-crank double rocker movement form, and the components are made of nylon material by 3D printing. With moderate density, high strength and stiffness, nylon material is suitable for processing flapping mechanism. The parts are connected by light aluminum rivets and the design flapping amplitude is 130°. During flapping, the maximum flapping amplitude can reach about 170° due to assembly gap and the structural elastic deformation. This design can also change the flapping amplitude easily by altering the length of the crank.

The design of the wing adopts a method of torsional deformation along the length of the wing. This design method is widely used in FWMAVs like Nano hummingbird and some studies [12–14] have shown that this wing design method exhibits better

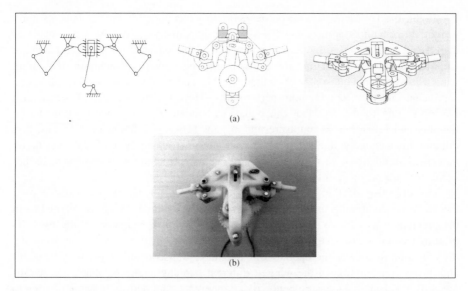

Fig. 1. (a) 3D design drawing of flapping mechanism (b) fabrication of the flapping mechanism

aerodynamic performance than the flat rigid wing and also has some advantages, such as easy processing and less vibration. As shown in Fig. 2, wings are manufactured of carbon fiber rods and PET. The carbon fiber rod with a diameter of 1 mm was selected as the leading edge of the wing, and the carbon rod with a diameter of 0.5 mm was used on surface to maintain the shape of the wing when flapping. The wing film was a PET film with a thickness of 0.02 mm. The membrane at wing root is bonded to the root carbon rod attached to the mechanism. When adjusting the angle of attack, the deflection angle of the membrane at wing root can be changed and the corresponding wing root bonding line is adhered to the fixed root carbon fiber (as shown in Fig. 2(c)). When the deflection angle is small, the wing membrane is tight and there will be a large angle of attack. On the contrary, when the deflection angle is large, the wing membrane will become slacker and the angle of attack will become smaller.

2.2   Measurement of Wing Kinematics, Lift and Power

In experiments, we need to obtain the flapping amplitude and angle of attack variation in a flapping period. A high-speed camera (MotionXtra HG-LE) is used to capture the wing while it is flapping at about 20 Hz – 30 Hz. Figure 3(a) shows the experimental platform used in the present study. The camera is horizontally placed and a mirror is tilted by 45° to horizontal to get a top view of the flapping-wing system. The filming frequency is 2500 fps which can capture the wing motion clearly. Figure 3(b) shows the captured images. Overlapping the start and end images in a stroke, we can measure the flapping amplitude. The angles of attack at different wing spanwise position can be measured by calculating the cosine of angle between the wing chord and its projection.

The study of the aerodynamic characteristics of the wing requires a relatively accurate measurement of lift. The load cell selected for lift force measurement is a

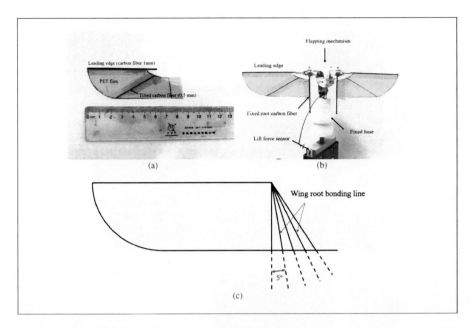

Fig. 2. (a) Experimental wing and (b) installation method (c) change the deflection angle to adjust the flapping angle

SMT1 sensor of American Interface Corporation. The maximum standard load is 10 N (1000 g) and the accuracy is 0.0001 N, the instantaneous safety overload can reach 10 times standard load. In the experiment, a fixed base with the flapping mechanism is vertically fixed on the load cell.

During force measurement, the power required by the FWMAV to flap the wings was recorded synchronously, similar to Ref. [15]. During flapping motion, the power input (P_{in}) was spent to produce an aerodynamic force (power of aerodynamic, P_{AE}), and a back and forth motion of the flapping mechanism (power of flapping mechanism, P_M), and to overcome the inertia of the wings during the flapping motion (wing inertia power, P_I). And some proportion of the power input is lost as the Joule heating loss in motor (power loss, P_{Loss}). Thus, the power input (P_{in}) can be broken down as follows:

$$P_{in} = P_{Loss} + P_M + P_{AE} + P_I \qquad (1.1)$$

In order to obtain the mechanism power consumption (P_M) of the flapping mechanism, we uninstalled the wings and let the mechanism flap only, then recorded the voltage and current between the two ends of the motor. The wing inertia power (P_I) was measured in a vacuum chamber, by subtracting the corresponding mechanism power (P_M) from the total input, as shown in Fig. 4.

Finally, the flapping mechanism tested in the vacuum chamber was placed in a normal pressure environment and fixed on the lift sensor to perform the same experiment. The lift was measured synchronously. From this step-by-step experiment, we can measure the power consumption of each part of the flapping MAVs.

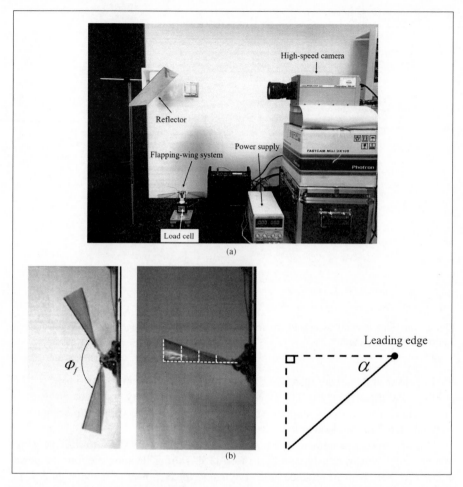

Fig. 3. (a) High-speed camera platform (b) flapping amplitude and flapping angle of attack obtained from high-speed camera images

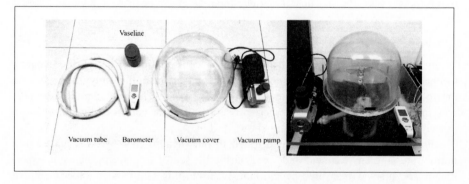

Fig. 4. Vacuum test platform

3   Results and Discussion

3.1   Effect of Angle of Attack on Aerodynamic Characteristics

From the aforementioned experimental steps of power consumption measurement, the mechanical power consumption of the mechanism was measured at several flapping frequencies (as shown in Fig. 5). And then we used an exponential function (Eq. 1.2) to fit these points. In the next experimental study, the fitting equation can be used to estimate the power consumption of the flapping mechanism.

$$P_M = 1.371 \times 10^{-4} f^{2.25} \tag{1.2}$$

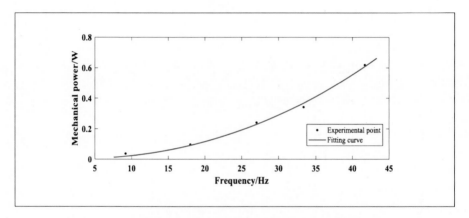

Fig. 5. Mechanical power consumption at different flapping frequency

As shown in Fig. 2(c), the wing angles of attack can be adjusted through changing the root deflection angle. In the present paper, we chose a wing with a wingspan of 8 cm (aspect ratio AR = 3) and measurements of five angles of attack were conducted by varying the root deflection angles (0°, 5°, 10°, 15°, and 20°). The larger the deflection angle is, the smaller the angle of attack is. The high-speed camera captured the angle of attack of the wing tip and the flapping angle of the wing (Fig. 6(a)). It can be seen that different root deflection angles indeed result in different angles of attack, and the time history of wing flapping angle in a flapping period agrees well with a sinusoidal function as designed. There exists a peak/valley at the end/start of a stroke, due to the fact that the inertia of the wing drives the wing to pitch continuously, resulting in a smaller angle of attack. In the next moment, the angle of attack of the wing with a larger root deflection angle will increase, this is because in the start of flapping, the flapping speed is slow, and the lift of the wing is small, as a result, the weight of the wing itself drives the wing to deflect downwards at the speed. After the lift increases, the wing will maintain a certain angle of attack.

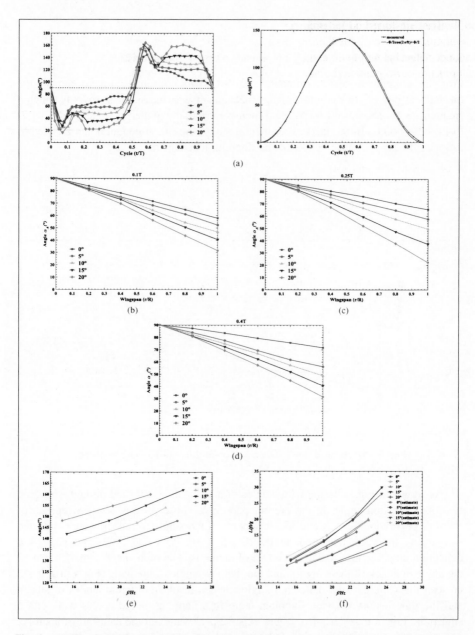

Fig. 6. (a) The angle of attack of the wing tip and the flapping angle of the wing in a flapping cycle. (b) (c) (d) average geometric angle of attack during translational stage (t/T = 0.1, 0.25 and 0.4) along the wingspan in one flapping cycle. (e) the flapping amplitude varies with flapping frequency and angle of attack. (f) the measured lift and estimate lift using quasi-steady theory.

From Fig. 6(b) (c) (d), we draw the average geometric angle of attack during translational stage ($t/T = 0.1$, 0.25 and 0.4) along the wingspan in one flapping cycle, they respectively correspond to the end of the wing flipping, the maximum flapping speed moment and the start of the wing flipping. We can see the angle of attack of different sections along the wingspan varies linearly. Therefore, the assumption of linear torsion along wing spanwise could be acceptable, which is used in the following quasi-steady aerodynamic model.

In experiments, we found that the flapping amplitude varies with the angle of attack and the flapping frequency. The design flapping amplitude of the flapping mechanism is $130°$, Fig. 6(e) shows that the flapping amplitude increase as the flapping frequency increases. It is found that the inertia force of the wing increases with the flapping frequency of, this will cause the deformation of the mechanism and the wing to increase due to the slight elastic deformation of the mechanism material and the wing itself. Therefore, the increase of the flapping frequency will result in a larger flapping amplitude. In addition, it can be seen from Fig. 6(e) that at the same flapping frequency, the flapping amplitude also increases with the wing root deflection angle. This is because if the wing's angle of attack is smaller, when the wing flips, the wing will have a greater flipping inertia force, which will increase the elastic deformation of the mechanism and the wing as well. So different flapping angles of attack will also cause slight differences in flapping amplitude. Therefore, for the real FWMAV design, these influencing factors need to be taken into account.

The input parameters of the quasi-steady model can be modified by the actual motion parameter. Figure 6(f) shows the measured lift and estimate lift using quasi-steady theory, we can see that when the root deflection angle is changed from $0°$ to $15°$, the lift force increases at the same flapping frequency, and the experimental results are in good agreement with the estimation results. The largest relative error is 7%. When the root deflection angle is $20°$, the experimental results are very different from the estimation results, and the experimental results are much smaller. From the flapping images, this is because when the root bonding angle is $20°$, the wing of the experimental model is too loose, and the wing lift cannot be conducted at the root, causing the flapping lift to be smaller than the theoretical estimation.

In the study of the angle of attack on aerodynamic characteristics, we are more concerned with the aerodynamic efficiency of different angles of attack, that is, the relationship between the aerodynamic power consumption (P_{AE}) and the lift force. Therefore, we measured the inertial power (P_I) consumption of the flapping motion, so that the aerodynamic power consumption under different flapping angles of attack can be obtained, and then the aerodynamic efficiency at different angles of attack can be analyzed. Figure 7(a) shows the measured inertial power of the wingspan. Furthermore, we have estimated the inertial power consumption based on the parameters and quality of the wing, and found that the trend of the test results is very close to the trend of the estimation results, and the actual measurement results are slightly larger than the estimation results by about 10%. Because the air pressure in the test vacuum environment is 5 kPa and it is difficult to reduce the vacuum air pressure furthermore, the wing still has a certain aerodynamic power consumption in the flapping motion, so the inertial power of the test will be larger than the actual inertial power. After measuring the inertial power consumption (P_I), the aerodynamic power consumption (P_{AE}) of the wing flapping is further obtained (as shown in Fig. 7(b)), and we plot the lift versus the aerodynamic power consumption of the wing in Fig. 7(c).

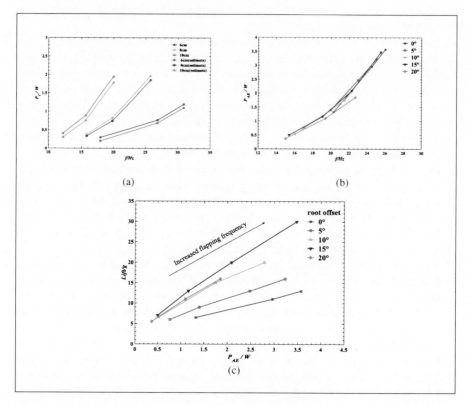

Fig. 7. (a) measured and estimated inertial power consumption. (b) aerodynamic efficiency at different flapping angles of attack. (c) lift versus the aerodynamic power consumption of the wing

Figure 7(b) shows that wings with different flapping angles of attack at the same flapping frequency consume almost the same aerodynamic power. As mentioned before, the angle of attack will also affect the flipping amplitude, so the combined effect of the two results in almost the same aerodynamic power consumption for different flapping angles of attack. Figure 7(c) shows that the optimal aerodynamic efficiency angle of attack is the wing whose wing root deflection angle is 15°. When the wing root deflection angle is larger, the wing lift force will decrease, so the aerodynamic efficiency will be reduced. When the wing root deflection angle is 15°, the angle of attack of the wing tip is about 35°, and the second moment of the wing has an angle of attack of about 55° which is similar to the result in [15]. Therefore, in the design of the angle of attack for the FWMAVs, the deflection of the wing root around 15° is a better choice, which has the highest aerodynamic efficiency.

Table 1 shows the comparison between the measured aerodynamic power at different angles of attack and the quasi-steady theory estimated power consumption. The result shows that the maximum relative difference is within 10%, which can be used to estimate the flapping aerodynamic power consumption effectively.

Table 1. Comparison of estimated and measured aerodynamic power consumption (P_{AE})

	Frequency (Hz)	Measured P_{AE}(W)	Estimate P_{AE}(W)	Difference (%)
8 cm 0°	20.3	1.32	1.3	1.5
	24.5	2.96	2.7	8.8
	26	3.57	3.36	5.8
8 cm 5°	17	0.76	0.73	4
	20	1.39	1.3	6.5
	25	3.22	3.1	3.7
8 cm 10°	16	0.52	0.56	7.7
	21.5	1.74	1.67	4
	24	2.78	2.67	4
8 cm 15°	15.4	0.5	0.5	0.1
	19.1	1.16	1.08	7
	25.5	3.47	3.4	2
8 cm 20°	15	0.37	0.4	-8
	19.4	1.1	1	9
	22.7	1.84	1.77	3.8

3.2   Effect of Wingspan on Aerodynamic Characteristics

Another research focus is the effect of wing area on aerodynamic characteristics. The choice of wing area has a significant effect on flapping wing power efficiency. A larger area wing will provide higher aerodynamic lift, but it will also require more aerodynamic and inertial power. Therefore, it is worth studying what wing area can provide higher power efficiency. In the previous section, we verified that the quasi-steady theory model can well estimate the aerodynamic lift, aerodynamic power consumption and inertial power consumption. So, in this section, in order to avoid a lot of experimental work, we first use the quasi-steady model to analyze the effect of wing area on aerodynamic characteristics, then we experimentally validate the wing with the highest power efficiency.

From the conclusion of the previous section, the deflection of the wing root around 15° has a higher aerodynamic efficiency, so we fixed the wing root deflection angle at 15° in this section, the angle of attack at the second moment of the wing is about 55°. The aspect ratio (AR) of the wing is fixed at 3, which has been proved to have a higher aerodynamic efficiency [16]. As a result, we chose the wingspan as the parameter of the study instead.

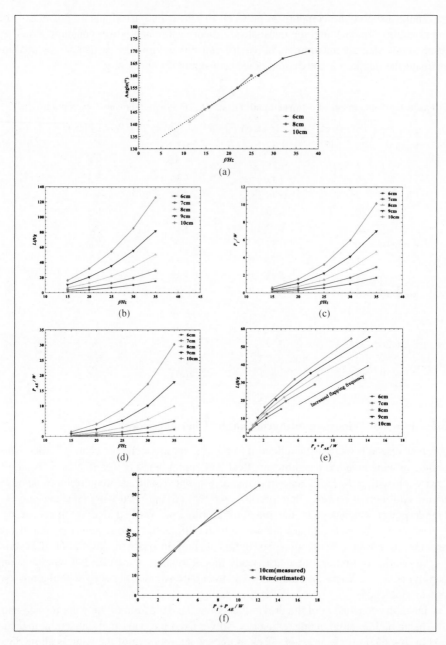

Fig. 8. (a) Flapping amplitude varies with frequency. (b) (c) (d) Lift, inertial power consumption, and aerodynamic power consumption of the wing. (e) Lift versus total power consumption of different wingspans. (f) Measured and estimated results of the 10 cm wing.

Since the estimation model requires the accurate flapping amplitude of FWMAV, the previous experiment found that the flapping amplitude increases with the flapping frequency. So we experimentally measured the flapping amplitude of different wingspans at the same root deflection angle, as shown in Fig. 8(a). It can be found that when the flapping frequency is within 35 Hz, the flapping amplitude changes substantially linearly with the flapping frequency. When the frequency is higher than 35 Hz, due to the constraints of the mechanism, it will basically reach the limit of the flapping amplitude about 170°. Since the FWMAV operation flapping frequency is basically within 35 Hz, a linear function can be used for fitting and can be written as follows:

$$\Phi_f = \frac{10}{7}f + 130 \tag{1.3}$$

We studied the wingspans from 6 cm to 10 cm and estimated the flapping lift, aerodynamic power consumption, and inertial power consumption of the wing at different frequencies, as shown in Fig. 8(b) (c) (d), it can be seen that the lift is basically proportional to the square of the flapping frequency, and the power consumption is proportional to the third power of the flapping frequency. The wings with larger wingspan can produce higher lift but at the same time consume more inertial power and aerodynamic power. From the estimation results, we can then obtain the Fig. 8(e), the relationship between the total power consumption $(P_{AE} + P_I)$ and the lift. It shows that the wing with larger wingspan can produce higher lift with the same total power consumption. The 10 cm wing has the highest power efficiency.

In order to verify the accuracy of the estimation results, we performed an experimental measurement on the 10 cm wing. The root deflection angle is 15° and the angle of attack at the second moment is still about 55°. The comparison between the measurement results and the estimation results is shown in Fig. 8(f). We can see that the experimental results are in good agreement with the estimation results. When generating a 40 g lift, it takes about 8 W of total power and the flapping frequency is 22 Hz.

4 Conclusion

This paper presented the results of experimental study of angle of attack and wingspan for a FWMAV. Experiments were carried out to measure the lift and aerodynamic power consumption of different flapping angles of attack. The inertial power consumption of wings with different wingspans were also measured in the vacuum test platform. The paper also uses the quasi-steady model to estimate the lift and aerodynamic power consumption and it is found that the experimental results are in good agreement with the estimation results. Thus, in the design of the FWMAV, the quasi-steady method can be well used for rapid engineering estimation.

The results of the paper show that when the wing root deflection is about 15°, the wing presents the highest the aerodynamic efficiency, which is a better design for the FWMAV. For wings with different wingspans, larger wing requires less power when producing the same lift. The larger the wing is, the higher the power efficiency is.

Therefore, through experiments and estimations, we have obtained a better wing design for a FWMAV, which has a 15° wing root deflection angle and 10 cm wingspan, and it can generate about 40 g of lift at a power input of about 8 W when flapping at 22 Hz.

Acknowledgements. This research was primarily supported by the National Natural Science Foundation of China (NSFC, Grant No. 11672022).

References

1. Muijres FT, Elzinga MJ, Melis JM, Dickinson MH (2014) Flies evade looming targets by executing rapid visually directed banked turns. Science 344:172–177
2. Keennon M T, Klingebiel K, Won H, Andriukov A (2012) Development of the nano hummingbird: a tailless flapping wing micro air vehicle. In: Proceedings of 50th AIAA Aerospace Sciences Meeting (Nashville, Tennessee)
3. Ma KY, Chirarattananon P, Fuller SB, Wood RJ (2013) Controlled flight of a biologically inspired insect-scale robot. Science 340:603
4. Phan HV, Nguyen QV, Truong QT, Truong VT, Park HC, Byun DY, Goo NS, Kim MJ (2012) Stable vertical takeoff of an insect-mimicking flapping-wing system without guide implementing inherent pitching stability. J Bionic Eng
5. Phan HV, Kang TS, Park HC (2017) Design and stable flight of a 21 g insect-like tailless flapping wing micro air vehicle with angular rates feedback control. Bioinspir Biomim 12:036006
6. Wood RJ (2008) The first takeoff of a biologically inspired at scale robotic insect. IEEE Trans Robot 24:1–7
7. Karásek M, Hua A, Nan Y, Lalami M, Preumont A (2014) Pitch and roll control mechanism for a hovering flapping wing MAV Int. J Micro Air Veh 6:253–264
8. Ansari SA, Knowles K, Zbikowski R (2008) Insect-like flapping wings in the hover part 2: effect of wing geometry. J Aircr 45:1976–1990
9. Truong QT, Nguyen QV, Truong VT, Park HC, Byun DY, Goo NS (2011) A modified blade element theory for estimation of forces generated by a beetle-mimicking flapping wing system. Bioinspir Biomim 6:036008
10. Phan HV, Truong QT, Park HC (2017) An experimental comparative study of the efficiency of twisted and flat flapping wings during hovering flight. Bioinspir Biomim 12:036009
11. Lee YJ, Lua KB, Lim TT et al (2016) A quasi-steady aerodynamic model for flapping flight with improved adaptability. Bioinspi Biomim 11(3):036005
12. Phan HV, Truong QT, Au TK, Park HC (2016) Optimal flapping wing for maximum vertical aerodynamic force in hover: twisted or flat? Bioinspir Biomim 11(4):046007
13. Zheng L, Hedrick TL, Mittal R (2013) Time-varying wingtwist improves aerodynamic efficiency of forward flight in butterflies. PLoS ONE 8:e53060
14. Du G, Sun M (2010) Effects of wing deformation on aerodynamic forces in hovering hoverflies. J Exp Biol 213: 2273–83)
15. Nguyen TA, Phan HV, Au TKL, Park HC (2016) Experimental study on thrust and power of flapping-wing system based on rack-pinion mechanism. Bioinspir Biomim 11:046001
16. Du G, Sun M (2010) Effects of wing deformation on aerodynamic forces in hovering hoverflies. J Exp Biol 213: 2273–2283

A New Concept of Compound Helicopter and Flight Tests

Yasutada Tanabe[1(✉)], Masahiko Sugiura[1], Noboru Kobiki[1],
and Hideaki Sugawara[2]

[1] Japan Aerospace Exploration Agengy (JAXA),
Osawa 6-13-1, Mitaka, Tokyo 181-0015, Japan
tan@chofu.jaxa.jp
[2] Ryoyu Systems Co., Ltd., Tokyo, Japan

Abstract. A new concept of compound helicopter to achieve high speed of two times of a conventional helicopter is described. This configuration consists of a single main rotor and a main fixed wing. Antitorque is performed through two electric driven propellers installed at the wing tips. Main propeller to achieve high speed is aft-mounted at the tail of the fuselage. Design of a flyable model of the compound helicopter proposed by JAXA is described. Following the flight test of the 1st concept demonstrator, it is found that the compound helicopter with single main rotor combined with a set of wing-tip propellers worked as anti-torque device is stable and can be controlled satisfactorily through the main rotor controls. However, it was difficult to simulate the high advance ratio flight conditions of the main rotor with the 1st concept demonstrator because the tail propeller was connected to the main rotor and driven by a same motor, thus decreasing the main rotor rotating speed also resulted in decrease of the tail propeller thrust. A new design which aims to achieve high speed flight and demonstrate the controllability of the aircraft even in high advance ratio is illustrated. Flight test results of the 2nd generation scaled-down model is reported. Advance ratio as high as more than 0.8 is achieved during test flight without any controls on the fixed wings.

Keywords: Compound helicopter · Electrically-driven anti-torque · Conceptual model · Flight test

1 Introduction

Conventional helicopters configured with a single main rotor and a tail rotor are widely used as a reliable aircraft with vertical-take-off and landing (VTOL) capabilities. The tail rotor produces side thrust to counter-act the main rotor torque. Propulsion and lift are produced by the main rotor at the same time during forward flight by tilting the main rotor tip-path-plane (TPP) through main rotor swashplate controls. However, the possible maximum flight speed of this type of helicopter is limited because of the shock-wave build-up on the rotor advancing side and stall occurrence on the retreating side. Increase of aerodynamic drag from the fuselage attitude change with the flight speed is often considered as another cause of maximum speed limitation of the conventional helicopters.

© Springer Nature Singapore Pte Ltd. 2019
X. Zhang (Ed.): APISAT 2018, LNEE 459, pp. 1343–1352, 2019.
https://doi.org/10.1007/978-981-13-3305-7_107

Several configurations to overcome the speed limits of the conventional helicopters are proposed so far [1, 2]. Most of them can be categorized into compound helicopters, tilt rotors, or tilt wings. Tilt-rotors represented by the famous Osprey V-22 use the main rotors to produce required lift for hovering and tilt the rotors forward to produce thrust as propellers during forward high-speed flight. It is often argued for its poor hovering capability and possible instabilities during the rotor tilting process. The tilt-wings are less successful historically because of their poor maneuverability during vertical take-off and vertical landing. However, many current electric VTOL designs adopts this type of techniques by tilting the main wing together with the tail (or canard) wing, both with a plural number of propellers installed, thanks to the autonomous flight control advances in recent years for the multiple rotors. The compound helicopters, which retain the main rotors, add propellers to generate required thrust for high-speed flight. Two types of compound helicopters are under development now. One utilizes co-axial rotors to generate required lift also during high-speed forward flight, represented by Sikorsky X2. Another one adds a wing to supplement the lift to a single main rotor during the forward flight, represented by Airbus Helicopters X3. Both demonstrated maximum flight speeds more than 250 kts, nearly comparable to the tilt-rotors. The co-axial rotor type compounds can be compact and hover-efficient, and with lift-offset or ABC (advancing blade concept) techniques, the flight efficiency during forward flight can be improved. The concerns are the mechanically complicated rotor head and vibration and noise from the interactions of the rotors during flight. The single main rotor type compounds require a fixed wing to unload the main rotor during high-speed flight where the rotor advance ratio may exceed 0.6. Airbus Helicopters utilizes a set of propellers installed on the wing-tips to generate anti-torque and thrust at the same time. Technical issues with this type of compound helicopters are wing download and extra-required power to generate anti-torque especially during hovering.

The authors have proposed a new concept of compound helicopter in 2014 [3]. Basic ideas are utilizing a set of electrically driven propellers on the wing-tips just for anti-torque of the main rotor. Thrust required for fast flight is provided by an aft-mounted propeller driven by turbo-shaft engines directly. Since then, conceptual studies utilizing small-scale flyable models [4] and numerical simulations of the complex flow at high advance ratios [5] and interactions between the main rotor and the fixed wing [6, 7] have been performed. Optimal design of the main rotor for high-mu (advance ratio) flight is also underway [8].

In this report, the conceptual studies with a series of small-scale flyable models are highlighted. Based on the experiences learned from the 1st proof-of-concept model, a 2nd generation scaled-down model which is capable to simulate the high-advance ratio flight is designed and flight-tested.

2   A New Concept of Compound Helicopter

As result of a conceptual study of high-speed rotorcraft, a new configuration of compound helicopter is proposed for emergency medical service (EMS) usage [3]. As shown in Fig. 1, a real-size of 4 ton class, carrying two crews with a doctor, a nurse, a patient and an accompanying person, together with 50 kg of medical devices are the designated payloads. Design target is to increase the maximum flight to about 500 km/hr, nearly twice of the conventional helicopter. Taking the EMS market in Japan as an example, as of 2014, there are 36 hub hospitals throughout Japan, located mainly in the central-cities of local prefectures. From the view point of emergency medical operations, it is desired to reach a site of accidents with mass bleeding within 15 min. However, as shown in Fig. 2, only about 60% of land can be covered currently from existing hub hospitals using current conventional helicopters. For a nearly 100% coverage, the number of hub hospitals has to be doubled. The cost for a hub hospital is in the order of tens of million US dollars. The running costs to cover the flight crews and emergency medical operations are also very expensive. It is much cost-efficient to double the flight speed of the helicopters if possible.

Fig. 1. A conceptual design of a high-speed compound helicopter

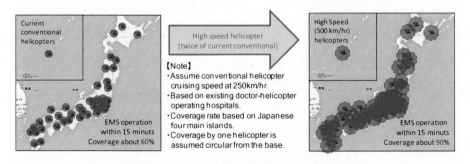

Fig. 2. EMS helicopter requirement in Japan

There are several important technical issues to realize a rotorcraft of such high speed. First of all, it is important to prove such a configuration of helicopter can fly at so much high speed, or non-dimensionally, at an advance ratio more than 0.8, without additional controls on the fixed wings.

3   Flight-Test of a Proof-of-Concept Model

Modified from an existing electric RC helicopter, a wing with two wing-tip propellers is added, and the tail rotor is converted to a tail pusher propeller as shown in Fig. 3. Flight tests with this model were carried out in October 2015. This model is stabilized with three-axis SAS. The operator is used to conventional RC helicopters. Only with brief instructions of expected new flight characteristics, e.g. use thruster slider instead of nose-down increase the collective pitch angle for forward speed acceleration, the operator is capable to fly this new model compound helicopter. From the operator's comments, during hover and low speed flight, the new model is similar to conventional helicopters. There is no side tilt at hover. At high speed, the model is more stabilized and response is similar to a fixed wing aircraft. Attitude-hold and other maneuvers can be satisfactorily achieved through the main rotor controls and the anti-torque controls. The achieved maximum flight speed was about 90 kts corresponding to an advance ratio of 0.5. This is not high enough as expected and even lower than the maximum speed of the base RC helicopter. Details about the flight test of this 1st generation proof-of-concept model can be found in Reference [4].

Fig. 3. Photographs of the proof-of-concept model

4   Design Changes for the 2nd Generation Scaled-Down Model

The 1st proof-of-concept model was modified from and existing conventional RC helicopter, the geometries are not optimal for high speed flight. The aft-mounted pusher propeller is mechanically linked with the main rotor, so it becomes difficult to simulate the slowed rotor high mu flight conditions while the tail propeller lost rotational speed at the same time and enough propulsion cannot be provided. Furthermore, the anti-torque and the tail pusher propellers utilized the existing helicopter's tail rotor blade, the blade shape and pitch-angle settings are not optimal for high speed flight. It is decided to design a new model to achieve high-mu flight conditions (Fig. 4).

Compared to the 1st proof-of-concept model, the 2nd generation model has an independently electric driven pusher propeller. The drive motor is larger than that for the main rotor. The main wing is re-sized and 20% smaller than that on the 1st model. The pitch angle settings for the anti-torque propellers are also adjusted so it will not become a wind-mill brake during fast flight. Finally, the fuselage is re-built to be more streamlined and to resemble the conceptual drawing.

Fig. 4. The 2nd generation scaled-down model design

The pusher propeller is optimally designed for high speed. The geometry is shown in Fig. 5, where r is in radial direction and y is in chordwise direction (pointing to leading edge), c means chord-length and θ_t is the blade twist angle, α is the angle of attack (AoA). With corrected twist distributions, the AoA becomes near flat which is expected to produce optimal thrust at high speed.

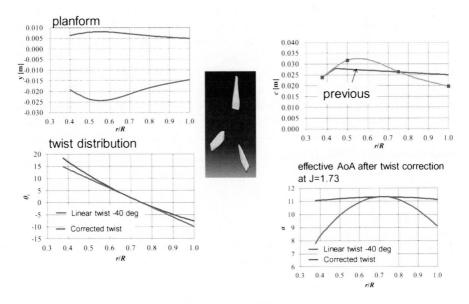

Fig. 5. New propeller blade geometry

The calculated thrust and required power of the aft-mounted pusher propeller is shown in Fig. 6 where $J = V/nD$ is the propeller advance ratio and $\theta 0$ is the propeller pitch angle. The generated thrust is enough to balance to estimated drag of the whole new model at $\mu = 0.7$ while the main rotor tip speed is kept at 118 m/s.

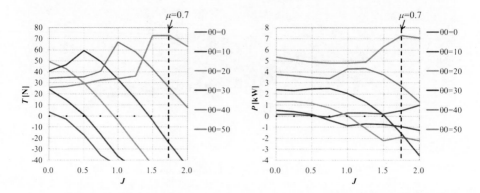

Fig. 6. Estimation of the new propeller performance

A three-view drawing of the 2nd generation scaled-down compound helicopter model is shown in Fig. 7. Main design parameters are listed in Table 1. The main rotor diameter is 1530 mm with 4 blades. The rotor hub is rigid. The maximum rotational speed of the main rotor is 2000 rpm. However, during this test, nominal rotational speed is about 1200 rpm to produce a tip speed about 100 m/s. During high speed flight, it is reduced to 75% to a tip speed of 75 m/s to simulate a high mu (about 0.8) flight condition. The tail pusher propeller is driven by a higher output motor than the main rotor to provide the major part of required thrust during high-speed flight. The side propellers have a lower maximum rotational speed. When the pitch angles of the side propellers are not high enough, it may fall into a wind-mill condition and has a remarkably high aerodynamic drag which may become the major limitation of the top speed of the whole aircraft. Detailed flight data analysis and comparison with wind-tunnel testing data are underway. Design changes may be required to the side propellers to achieve a higher flight speed.

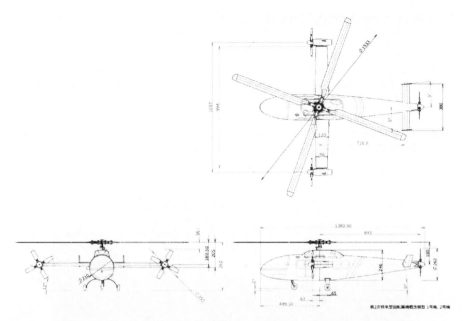

Fig. 7. Three-view drawing of the 2nd generation scaled-down compound helicopter model

Table 1. Main design parameters of the 2nd generation scaled-down compound helicopter model

• Main Rotor	• Tail pusher propeller
– Diameter: 1530 mm	– Diameter: 262 mm
– Number of blade: 4	– Number of blade: 3
– Pitch angle range: −8 ~ +14°	– Pitch angle range: −27 ~ +50°
– Rotational speed: 1500–2000 rpm	– Maximum rotational speed: 14000 rpm
– Drive: 5175 W brushless motor	– Drive: 7000 W brushless motor
• Side propellers (2 on wing-tips)	• Main wing
– Diameter: 250 mm	– Airfoil: NACA0020
– Number of blade: 4	– Wing area (incl. fuselage part): 0.107 m²
– Pitch angle range: −27 ~ +50°	– Wing span: 994 mm
	– Chord length at root: 120 mm
– Rotational speed: 6000–8000 rpm	– Chord length at tip: 96 mm
– Drive: 850 W brushless motor	– Incidence angle: 5° ± 3°
	• Weight
	– Gross weight (incl Bat. 6 cell, 5000mAhx4): 10.7 kg

5 Flight Test Result with the 2nd Generation Scaled-Down Model

Flight test of the 2nd generation scaled-down compound helicopter model was performed in October 2017 as shown in Fig. 8. The flight trajectory is shown in Fig. 9. The time histories of the flight directions and speeds are shown in Fig. 10. During level flight, a top speed of 120 kts is achieved and the maximum main rotor advance ratio exceeded 0.8, which is the same non-dimensional number when the real-sized helicopter reaches a top speed of 500 km/hr.

Fig. 8. Flight test of the 2nd generation scaled-down model

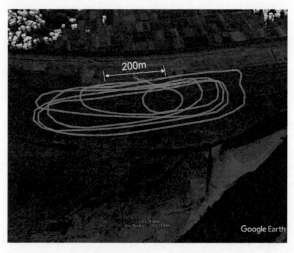

Fig. 9. Flight trajectory of the 2nd generation model at Ota flight site

The main rotor rotational speed was reduced to about 75% of that during hovering which also reduced the rotor noise significantly. Main rotor advance ratio increased due

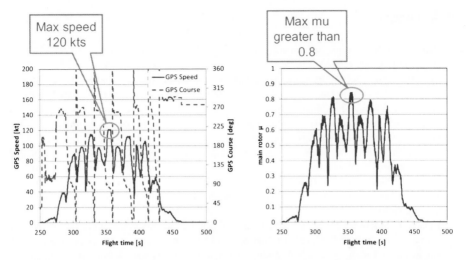

Fig. 10. Time histories of flight directions and speeds and main rotor advance ratios

to lower rotor speed. It is demonstrated that the main rotor controls are sufficient for this kind of compound helicopter up to this high advance ratio to perform level attitude hold and other flight maneuvers.

6 Summary

A new concept of compound helicopter proposed by JAXA is illustrated. Based on the flight test results of the 1st proof-of-concept model, a 2nd generation scaled-down model is designed and flight-tested.

The aft-mounted pusher propeller is optimally re-designed and independently driven. The main wing is re-sized with 20% wing area reduced. The wing-tip anti-torque propellers are re-adjusted to avoid wind-mill braking during high speed. A low drag fuselage is built to resemble the conceptual drawing.

During flight test of the 2nd generation scaled-down model, a top speed of 120 kts and an advance ratio of more than 0.8 are achieved. Main rotor controls are proved to be sufficient for this kind of compound helicopter up to this high advance ratio to perform attitude hold and other flight maneuvers.

Wind-tunnel testing of the 2nd generation scaled-down model has been carried out. Detailed comparisons and analyses between the wind tunnel testing and flight test data will be performed.

Further modification to improve the fidelity of the scaled-down model will be carried out. Retractable landing gears and wing-flaps will be installed to further reduce the aerodynamic drag of the aircraft. Optimal high-mu rotor under design using high fidelity CFD tools will be tested on this rotorcraft in the future.

References

1. Harris FD (2008) Rotor performance at high advance ratio: theory versus test, NASA CR-2008-215370, October 2008
2. Ormiston RA (2016) Revitalizing advanced rotorcraft research—and the compound helicopter: 35th AHS Alexander A Nikolsky Honorary Lecture. J Am Helicopter Soc 61(1)
3. Tanabe Y, Aoyama T, Kobiki N, Sugiura M, Miyashita R, Sunada S, Kawachi K, Nagao M (2014) A conceptual study of high speed rotorcraft, 40th European Rotorcraft Forum, Southampton, UK, 02–05 September 2014
4. Sugiura M, Tanabe Y, Kobiki N, Aoyama T (2016) Flight test of a model size compound helicopter. In: 5th Asian/Australian rotorcraft forum, Singapore, 17–18 November 2016
5. Tanabe Y, Sugawara H (2015) Aerodynamic validation of rFlow3D code with UH-60A data including high advance ratios. In: 41st European rotorcraft forum, Munich, Germany, 1–4 September 2015
6. Tanabe Y, Sugawara H (2015) Numerical simulation of aerodynamic interaction between a rotor and a wing. In: 5th Asian/Australian rotorcraft forum, Singapore, 17–18 November 2016
7. Sugawara H, Tanabe Y (2017) A study of rotor/wing aerodynamic interaction at high speed flight on a compound helicopter. In: 6th Asian/Australian rotorcraft forum & Heli Japan 2017, Kanazawa, Ishikawa, Japan, 7–9 November 2017
8. Sugiura M, Tanabe Y, Sugawara H (2017) Optimal aerodynamic design of main rotor blade for a high-speed compound helicopter. In: 6th Asian/Australian rotorcraft forum & Heli Japan 2017, Kanazawa, Ishikawa, Japan, 7–9 November 2017

Research on Morphing Scheme and Forward-Swept Wing Parameters Based on a Forward-Swept Wing Morphing Aircraft

Xuefei Li[1,2(✉)], Zhansen Qian[1,2], Chunpeng Li[1,2],
Xianhong Xiang[1,2], and Pengbo Xu[1,2]

[1] AVIC Aerodynamics Research Institute, Shenyang 110034, China
1034774512@qq.com
[2] National Laboratory for Computational Fluid Dynamics,
School of Aeronautic Science and Engineering, Shenyang 110034, China

Abstract. The morphing aircraft can change the shape of the vehicle by local or whole to improve the aerodynamic efficiency of the aircraft in a wide speed range, so as to achieve multi-task functions or multiple control purposes. Based on the self-developed numerical simulation platform ARI_CFD, a three-dimensional numerical simulation method is used to study the aerodynamic influence of the variable forward sweeping schemes and the main parameters of the forward-swept wing on the aircraft. The results show that, the two morphing schemes including "The forward-swept wing embedding in the main wing" and "the main wing rotation" both have advantages and disadvantages in the design aspects of the forward-swept wing and its morphing mechanism, in addition, the aerodynamic evaluation shows that the aerodynamic characteristics of the two morphing schemes are not very different and can be selected according to the specific condition as appropriate. With the increase of the exposed area of the forward-swept wing, the maximum lift-drag ratio of the whole aircraft increases first and then decreases with a non-monotonic changing law. Backward shifting of the forward-swept wing position is beneficial to improving the lift and lift-drag ratio of the whole aircraft, and is beneficial to increase the variable forward-swept wing size. By changing relative thickness of the forward-swept wing airfoil from 0.04 to 0.06, the lift-drag ratio of the forward-swept wing itself increases from 33.55 to 37.72, making transonic lift-drag ratio of the whole aircraft increases from 11.29 to 11.60. The use of a high-lift airfoil with bigger camber on the forward-swept wing can effectively increase the lift-drag ratio of the layout, and the aerodynamic efficiency gain of the whole aircraft can reach up to 13.2%. The optimization of the wing tip planform can enhance the aerodynamic efficiency near the wing tip, but the effect is very limited. The results show that a well-designed morphing aircraft with forward-swept wing can significantly improve transonic aerodynamic performance. It is verified that forward-swept wing morphing aircraft has much development potential as a multi-task vehicle within a wide speed range. This study can provide references for the design of the morphing aircraft.

Keywords: Morphing aircraft · Variable forward-swept wing ·
Aerodynamic characteristics · Morphing scheme · Wing parameter

© Springer Nature Singapore Pte Ltd. 2019
X. Zhang (Ed.): APISAT 2018, LNEE 459, pp. 1353–1364, 2019.
https://doi.org/10.1007/978-981-13-3305-7_108

1   Introduction

Morphing aircraft can change its shape locally or as a whole, so that the aircraft can adapt to various flight environments and task demands in high efficiency, high performance and real-time, improve the aircraft manufacturing/maintenance economy, and solve the contradiction between the endurance performance and the maneuverability of the aircraft. It is one of the important and cutting-edge research directions to expand the speed and mission scope of aircrafts presently at home and abroad. The research and development of "morphing aircraft" generally covers all the various methods used in aviation history to change the shape of the aircraft. Most of these deformation methods currently increase the cost of structural weight and complexity [1–4]. The design of morphing aircraft with swept-wing is not a new design concept, and it reached its technology development climax in the 1960s and 1970s, and various types of variable backward-swept aircrafts from various countries were put into active service [5]. However, variable forward-swept wing technology has relatively high requirements for integrated flight control technology, and the increase in the structural weight caused by the aeroelastic divergence problem makes its development lag behind. The United States X-29 forward-swept wing technology verification aircraft and Russian S-37 forward-swept wing research aircraft have been successfully conducted flight tests, fully demonstrating the excellent aerodynamic performance and great research potential of the forward-swept wing layout [6–8]. However, there is less systematic and detailed research work that has been done on the forward-swept wing morphing aircraft at home and abroad, except the United States and Russia. According to the author's limited knowledge and incomplete statistics, the scholars mostly focus on the concept design of the forward-swept wing layout or the typical state flight performance of the aircraft, and few detailed studies have been made on the influence of the morphing schemes and the main parameters of the forward-swept wing.

Based on the self-developed aerodynamic numerical simulation platform ARI_CFD, a three-dimensional compressible RANS equation numerical simulation method is used to study the influence of the morphing schemes and the main parameters of the forward-swept wing, but the morphing mechanism and its interference with the main wing are not considered. Two morphing schemes including "forward-swept wing embedded the main wing" and "the main wing rotation" are compared and analyzed. A detailed understanding of aerodynamic influence of main forward-swept wing parameters on the whole aircraft is gained, including its area, forward-swept angle at 1/4 chord line, relative thickness of airfoil profile, relative potion to the main wing, airfoil profile and planform of wing tip. This paper explores forward-swept wing morphing aircraft of high aerodynamic performance within a wide speed range, and provide references for morphing scheme and design of forward-swept wing.

2   Numerical Method

All calculations are conducted on self-developed aerodynamic numerical simulation platform ARI_CFD, which has been verified by a large number of examples it can provide accurate CFD results from low subsonic to hypersonic flow [9]. Far field boundary is about 20 times length of the fuselage. Half-module unstructured grids with the same density are adopted, and the total number of mesh elements are about 18 millions, and the height of the first layer of the wall surface ensures that y+ is approximately 1 referencing to Fig. 1. The main control equations are three-dimensional compressible RANS equations, the turbulence model is K-W SST, multi-grid technology is used to accelerate convergence process, the space discrete is two order precision AUSM format, and the time advance adopts the implicit two time step method (Table 1).

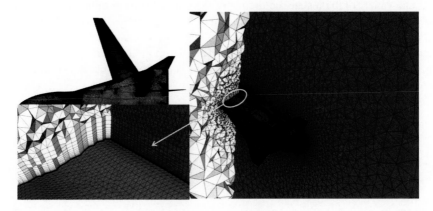

Fig. 1. Mesh distribution

Table 1. Simulation conditions

M	H	AOA
0.8	10 km	−2°, 0°, 2°, 3°, 4°, 8°, 12°
0.95	10 km	−2°, 0°, 2°, 3°, 4°, 8°, 12°
1.2	10 km	−2°, 0°, 2°, 3°, 4°, 8°, 12°

With reference to the actual flight conditions, the calculation conditions are set as follows:

Where M is the Mach number, H is the altitude, and AOA is the angle of attack. Through comparative analysis of multiple groups of simulations, it is found that the maximum lift-drag ratio of different cases corresponds to 3° angle of attack, so 3° angle of attack cases are used as the typical analysis state.

3 Discussion of Results

3.1 Morphing Schemes

Two kinds of morphing schemes are determined. Scheme 1: the forward-swept wing embedded in the main wing scheme, that is, the forward-swept wing is separated from the main wing and is placed on one side or inside of the main wing, and so that the main wing planform can be maintained when rotating. The advantage is that the main wing and the forward-swept wing airfoil can be restrained. The disadvantage is that the morphing mechanism is difficult to design. Scheme 2: The main wing rotation scheme, that is, the forward-swept wing and the main wing are the same component. In the folded state, the trailing edge of the forward-swept wing becomes partial leading edge of the main wing, and the leading edge is inserted into the fuselage to form part of the main wing. In the unfolded state, part of the main wing disappears. The advantage is that the morphing mechanism is simple to design. The disadvantage is that the wing airfoil needs to meet a variety of flight state requirements, therefore, it is more difficult to design, and further more the movable lifting surface accounts for larger proportion which leading the aerodynamic and gravity center to change greatly, and also brings about higher requirements of the morphing mechanism. The two morphing schemes are shown in Fig. 2, and the forward-swept wing in the folded state shows in red dotted lines.

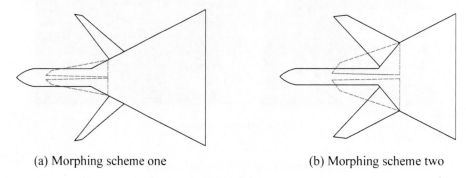

(a) Morphing scheme one   (b) Morphing scheme two

Fig. 2. Morphing schemes sketch map

3.2   Main Parameters of the Variable Forward-Swept Wing

Without consideration of the morphing details, it is relatively simple to design for morphing scheme 1, and thus it is used to study the influence of main forward-swept wing parameters. The forward-swept wing adopts a transonic airfoil with a high lift-drag ratio and relative thickness of 4%–6%, and follows the backward-swept wing design method twisting the forward-swept wing tip upward by 4°. The forward-swept wing is connected with the main wing at the leading edge of the main wing, and the root of the forward swept wing is twisted to ensure that the front and rear edges of the forward-swept wing coincide with the leading edge of the main wing, and the forward-swept wing has a cathedral angle to ensure that the root of the forward-swept wing is located in the main wing. The forward-swept wing needs to be folded, accordingly, it should match the main wing shape as much as possible which causes its aspect ratio and taper ratio are relatively fixed. Therefore, the main parameters of the forward-swept wing are chosen to be studied as follows: wing area-Ar, forward-swept angle at 1/4 chord line-$\psi 2$, relative thickness of airfoil profile-C/b and relative position to the main wing-Xa. Based on the above parameters, 8 forward-swept wing layouts from FWA1 to FWA8 are determined, and FWB1is an aircraft configuration of morphing scheme two.

The specific parameters are shown in Table 2. In this paper, Ae is the exposed area of the forward-swept wing, CL is the lift coefficient, CD is the drag coefficient, and K is the lift-drag ratio.

Table 2. Parameters of the forward-swept wing morphing aircrafts

Layout	$\psi1$	$\psi2$	Ar	Ae	C/b	Xa
Unit	Degree	Degree	Square meter	Square meter		meter
FWA1	22	25.5	115	60	0.06	19.3
FWA2	22	25.5	140	80	0.06	19.2
FWA3	21	25.5	111	60	0.06	16.7
FWA4	28	32.6	125	60	0.06	19.6
FWA5	18	22.7	92.3	45.6	0.06	17.3
FWA6	22	26.5	95.7	43.7	0.04	20.5
FWA7	22	26.5	95.7	43.7	0.06	20.5
FWA8	18	22.4	96.8	46.9	0.06	19.7
FWB1	18	22.5	87.4	42.7	0.06	20.7

All the cases are divided into four groups according to the main change of parameters in which other parameters are not much different. Ae group includes FWA1, FWA2 and FWA7 layouts, $\psi2$ group includes FWA1 and FWA4, FWA7 and FWA8 layouts, Xa group includes FWA1 and FWA3, FWA5 and FWA8, and C/b group includes FWA6 and FWA7 layouts. The names of components are defined as Fig. 3. It's worth noting that the inlet entrance and nozzle outlet are closed by a flat wall respectively, and the closed plane of the inlet is named "Inlet", and the closed plane of the nozzle is named "Nozzle" below (Table 3).

Fig. 3. The names of aircraft components displaying in half model

In the Ae group, as the exposed area of the forward-swept wing increases, the maximum lift-drag ratio of the whole aircraft shows a non-monotonic change law that increases first and then decreases. From the aspect of component force, the larger the exposed area is, the smaller the lift-drag ratio is for the forward-swept wing itself. The lift-drag ratio of the outer wing decreases with the increase of forward-swept wing area due to its influence. The presence of the forward-swept wing will change the pressure distribution of the airframe and the main wing, and its influence laws on the airframe

Table 3. Aerodynamic characteristics (M = 0.8, AOA = 3°)

Layout	Aircraft	Airframe	Wing	EWing	FSWing	Aircraft	Airframe	Wing	EWing	FSWing
	CL	CL	CL	CL	CL	K	K	K	K	K
FWA1	0.3662	0.0831	0.0604	0.0020	0.2207	12.62	4.61	19.52	5.88	29.25
FWA2	0.4485	0.0836	0.0621	0.0028	0.2999	11.88	3.88	14.77	4.89	26.32
FWA3	0.3431	0.0784	0.0435	0.0035	0.2176	12.33	4.24	17.97	9.00	33.49
FWA4	0.3306	0.0783	0.0576	0.0025	0.1922	12.21	4.38	18.22	7.08	33.71
FWA5	0.2945	0.0753	0.0437	0.0031	0.1723	11.46	4.24	16.99	8.15	34.73
FWA6	0.2920	0.0743	0.0598	0.0028	0.1551	11.29	4.24	17.91	7.62	33.55
FWA7	0.2929	0.0746	0.0599	0.0028	0.1557	11.60	4.27	18.12	7.55	37.72
FWA8	0.3183	0.0776	0.0582	0.0029	0.1796	12.15	4.43	18.18	7.96	35.17

and the main wing are basically the same, but from the perspective of the airframe lift exceeding that of the main wing, the forward-swept wing has a greater impact on the airframe. As the area of the forward-swept wing increases, the lift-drag ratio increases first and then decreases, but the lift increases always. In summary, it is determined that the forward-swept wing has an exposed area in the range of 45–60 m^2. Taking into account the folding requirements of the variable forward-swept wing, its area should be as small as possible (Fig. 4).

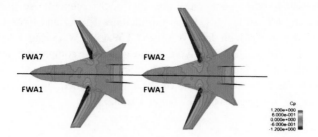

Fig. 4. Surface pressure distribution of group Ae (M = 0.8, AOA = 3°)

In the $\psi2$ group, compared with the FWA1 and FWA4 layouts, with the increase of forward-sweep angle $\psi2$,the low-pressure area at the leading edge of the forward-swept wing decreases. Since the forward-swept wing moves forward due to the increase of $\psi2$, its adverse interference to the outer wing is reduced and so that the lift-drag ratio and lift performance of the outer wing are all improved. Because of the relative increase of $\psi2$, the favorable interference of the forward-swept wing to the main wing and fuselage weakens, and both the lift and lift-drag ratio of the airframe and the main wing decrease. Compared with FWA7 and FWA8, the influence of $\psi2$ variation on the forward-swept wing and the airframe are the same as that of the FWA1 and FWA4 group. However, it exerts opposite influence to the main wing and the outer wing. Considering that other parameters are not completely consistent, the influence of $\psi2$ on the main wing and the outer wing needs to be further studied. But generally speaking, the lift and drag characteristics of the main wing and the outer wing of this group are not significantly different, and it does not affect the analysis of the whole aircraft (Fig. 5).

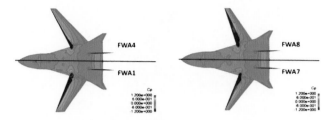

Fig. 5. Surface pressure distribution of group $\psi2$ (M = 0.8, AOA = 3°)

According to the Xa group, the lift force and lift-drag ratio of the airframe and the main wing increase as the forward-swept wing moves backward, but the lift force and lift-drag ratio of the outer wing are reduced due to the enhanced downwash effect. The change of Xa has little effect on the forward-swept wing itself. In general, the backward placement of the forward-swept wing is beneficial to improve aerodynamic performance of the whole aircraft and is also advantageous to increase the size of the variable forward-swept wing. In addition, from the perspective of the pitching moment, none of the current cases in group Xa produces a large negative pitching moment. Therefore, the Xa should be moved backward as far as possible under the premise that it imposes little impact on the external wing (Fig. 6).

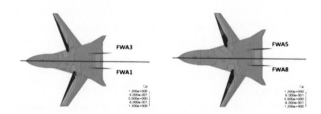

Fig. 6. Surface pressure distribution of group Xa (M = 0.8, AOA = 3°)

The C/b change in the C/b group has little effect on lift force, but has great influence on aerodynamic efficiency. Using the airfoil with C/b = 0.06 replacing that with C/b = 0.04 increase the forward-swept wing lift-drag ratio from 33.55 to 37.72, making the whole aircraft lift-drag ratio increase from 11.29 to 11.60. Therefore, the airfoil design of the forward-swept wing should be taken seriously (Fig. 7).

In summary, the influence rule and range of the forward-swept wing parameters are preliminarily found out, and Ae ranges between 45 and 50 m^2, ψ^2 is about 22°, Xa is about 20 m, C/b adopts 0.06, and FWA8 is evaluated as the preferred layout of morphing scheme 1.

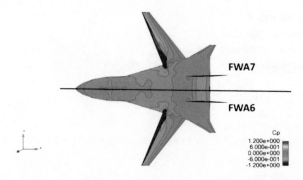

Fig. 7. Surface pressure distribution of group C/b (M = 0.8, AOA = 3°)

3.3 Design Optimization of Forward-Swept Wing Based on Airfoil Camber and Wing Tip Modification

According to the above study, the currently forward-swept wing layout has two problems: First, the forward-swept wing has a weak lift capability as a high lift-drag ratio component, and thus its area must be large enough to meet lift-drag ratio requirement of the whole aircraft, which not only increases the bending moment at the wing root, but also is not conducive to the morphing mechanism design. Second, the low pressure region of the wing tip only exists near its leading edge, and thus the wing tip has poor lift characteristics and aerodynamic efficiency. Therefore, the research on the improvement of the forward-swept wing aircraft from the airfoil and wing tip modification are carried out to further reduce the exposed area of the forward-swept wing and the resulting wing root bending moment.

3.3.1 Aerofoil Camber

In order to increase the lift force of the forward-swept wing, the airfoil improvement focuses on the effects of airfoil camber. Based on FA41 scheme, the airfoil camber at the leading and trailing edge of wing tip is modified forming FA46 scheme, which adopts the origin FA41 airfoil to build the inner section of the forward swept wing until 56.78% wing span and uses FA46 airfoil to form the outer wing section to simulate the adaptive variable camber at the leading and trailing edge of the outer wing. Based on the scheme FA41, the wing tip section is replaced by the NACA64A204 airfoil, and the upward twist angle is increased by 1°, and the swept wing root section keeps t the same with the original airfoil profile, which forms the scheme FA462 (Fig. 8).

Fig. 8. Different camber of airfoil profiles

After increasing the airfoil camber, the lift-drag ratio of the whole aircraft is increased to varying degrees due to the increase of the lift force of the forward swept wing itself. The FA46 wing tip profile is not quite different from that of FA462, and the range of camber change is limited to the outer wing section that contributes less lift force. It has less influence on the whole aircraft, the whole lift coefficient increases by only 0.01, and the lift-drag ratio only increases by 0.17. The wing surface pressure distribution of FA46 is similar to the FA41 scheme. The increase in the outer wing camber does not significantly improve the lift-drag ratio characteristics of the outer wing section. For FA462, the influence of the airfoil camber can be clearly seen from the pressure distribution. The negative pressure area of the upper wing surface increases and the negative pressure extremum decreases in the camber change area, although it leads to the decrease of the lift-drag ratio of the forward-sweep wing itself, but its lift force increases by about 0.05 compared with that of the FA41, making the whole lift-drag ratio increase by 1.36 to reach 11.66. In summary, adopting the high lift airfoil with camber on the forward-swept wing is one of the effective ways to improve the lift-drag ratio of the layout (Figs. 9 and 10).

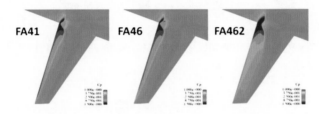

Fig. 9. Surface pressure distribution (M = 0.8, AOA = 3°)

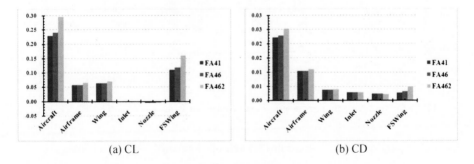

(a) CL   (b) CD

Fig. 10. Aerodynamic force coefficients of different airfoils (M = 0.8, AOA = 3°)

3.3.2 Wingtip Planform Modification

The Airfoil modification study shows that the wing tip aerodynamic efficiency of the current forward-swept wing is low, so the improvement of the wing tip planform is carried out. FA45 scheme adds a large sweep angle wingtip to FA41. Keeping the forward-swept wing exposed area the same with that of FA41, FA47 is got by obliquely cut the wing tip trailing edge of FA41. FA472 sweeps backward the obliquely cut wing tip of FA47.

It can be seen from Fig. 11 that the large swept wing tip increased by FA45 does not significantly change the flow characteristics of the wing tip, and the lift-drag characteristic is almost the same as FA41. The method of increasing the leading edge sweep angle for the wingtip does not have a flow control effect. Compared with FA41, FA47, and FA472, the method of increasing wing span by cutting the trailing edge of the wing can improve the lift-drag characteristic of the wing, and the sweepback of the modified wing tip can further improve its aerodynamic performance, but to a limited extent. The rolling moment coefficient is usually used to evaluate the wing root bending moment. When the forward-swept wing exposed area is the same, the forward-swept wing rolling moment coefficients of FA41/FA47/FA472 are 1.2298, 1.2475, 1.2662, respectively. Taking into account the lift-drag characteristics and the wing root bending moment, it is recommended to adopt the FA47 wingtip modification method (Fig. 12).

Fig. 11. Surface pressure distribution of different wingtip planforms (M = 0.8, AOA = 3°)

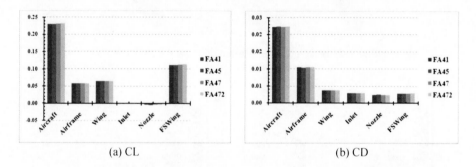

(a) CL   (b) CD

Fig. 12. Aerodynamic force coefficients of different wingtips (M = 0.8, AOA = 3°)

3.4   Comparison and Evaluation of Variable Forward-Sweep Schemes

According to the influence rule of forward-swept wing parameters, two kinds of variable forward sweeping schemes are designed, one is FWA8 scheme and the other is FWB1 scheme. As shown in Fig. 13, the transonic performance of the two morphing schemes have their own advantages and disadvantages. The maximum lift-drag ratio for both of the two morphing methods are greater than 12 at M0.8. The FWA8 scheme retains the main wing after the forward-swept wing deployed, has a larger lift area, and has a higher slope of the lift force with angle of attack curve. As the angle of attack increases, the lift coefficient at the same angle of attack is greater than that of the FWB1 scheme. The change law of the lift coefficient of the two schemes with the angle of attack is basically the same. Since FWA8 scheme has a bigger effective sweep angle, the maximum lift-drag ratio at M0.8 is slightly lower than that of the FWB1 scheme, while the maximum lift-drag ratio at M1.2 is slightly higher, but the difference is not significant. At M0.8, the maximum lift-drag ratio of FWA8 scheme is 12.11, and FWB1 scheme is 12.56, and the maximum lift-drag ratio of both the two schemes at M0.9 are 7.77. At M1.2, the maximum lift-drag ratio of the FWA8 scheme is 4.74, and the FWB1 scheme is 4.59.

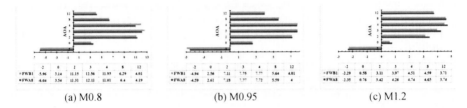

(a) M0.8   (b) M0.95   (c) M1.2

Fig. 13. Transonic lift-drag ratio vs. angle of attack of FWA8 and FWB1

4   Conclusion

Based on a double-sweptback wing aircraft, the three-dimensional numerical simulation method is used to study morphing schemes of variable forward-swept wing aircraft and main parameters of the forward swept wing. The research shows that:

1. Well designed morphing aircraft with the variable forward-swept wing can significantly improve the transonic aerodynamic efficiency, achieving a maximum lift-drag ratio at M0.8 greater than 12.
2. The second major finding is to get an important insights into aerodynamic influence on the morphing aircraft for the forward-swept wing's main parameters including its area, forward-swept angle at 1/4 chord line, relative thickness of airfoil profile, relative position to the main wing, airfoil profile and planform of wing tip.
3. This paper completes two kinds of morphing scheme design, qualitatively analyzes their advantages and disadvantages in the design of the morphing mechanism, and quantitatively analyzes the aerodynamic characteristics difference of the two kinds of morphing schemes at transonic speed.

As a young aviation researcher, my knowledge and experience are very limited. This study does not consider the implementation of the morphing mechanism and its interference with the main wing. In the future, the dynamic aerodynamic characteristics in the morphing process will be studied in depth.

References

1. Li JF, Ai JQ, Dong HF (2009) Research on the development of aircraft morphing technologies. Res Dev Aircr Morphing Technol 02. (in Chinese). 李军府, 艾俊强, 董海锋. 飞机变形技术发展探究[J]. 航空科学技术, 2009, (02)
2. Rodriguez AR (2007) Morphing aircraft technology survey, Reno, Nevada. In: 45th AIAA aerospace sciences meeting and exhibit, 8–11 January 2007
3. Lu YP, He Z, Lv Y (2008) Morphing aircraft technology. Aeronaut Manufact Technol (22). (in Chinese). 陆宇平, 何真, 吕毅, 变体飞行器技术[J]. 航空制造技术, 2008, (22)
4. Chen Q, Bai P, Li F (2012) Morphing aircraft wing variable-sweep: two practical methods and their aerodynamic characteristics. Acta Aerodynamica Sinica (05). (in Chinese). 陈钱, 白鹏, 李锋, 可变形飞行器机翼两种变后掠方式及其气动特性机理[J]. 空气动力学学报, 2012, (05)
5. Li C, Wu Z, Qi Y (2005) Design and wind tunnel experimental investigation of middling sweepback wing plane parameter. Acta Aerodynamica Sinica 26(6):627–654. (in Chinese). 李晨, 武哲, 祁彦杰. 中等后掠机翼平面参数设计与实验研究[J]. 航空学报, 2005, 26(6):627–654
6. Liu W, Wang X, Mi K (2009) A new aerodynamic configuration of UAV with variable forward-swept wing. Acta Aerodynamica Sinica (05). (in Chinese). 刘文法, 王旭, 米康., 一种新的变前掠翼无人机气动布局[J]. 航空学报, 2009, (05)
7. Shi Y, Liu Q, Bai P (2012) A canard morphing aircraft with area variable forward-swept wing. In: The eighteenth annual academic conference of Beijing mechanics association. 石永彬, 刘强, 白鹏, 一种变面积鸭式前掠翼飞行器气动布局研究[C]. 北京力学会第18届学术年会, 2012
8. Xu Y (2017) Research on the development and key technology of smart morphing aircraft. Tactical Missile Technol (02). 许云涛, 智能变形飞行器发展及关键技术研究[J]. 战术导弹技术, 2017, (02)
9. Computational investigation of a translating-throat variable single expansion ramp nozzle. In: 4th Symposium on Fluid-Structure-Sound Interactions and Control (Tokyo) (2017)

Empirical Correlations for Geometry Build-Up of Fixed Wing Unmanned Air Vehicles

Falk Götten[1,2(✉)], D. F. Finger[1,2], C. Braun[1], M. Havermann[1],
C. Bil[2], and F. Gómez[2]

[1] FH Aachen University of Applied Sciences,
Hohenstaufenallee 6, 52064 Aachen, Germany
goetten@fh-aachen.de
[2] RMIT University, GPO Box 2476, Melbourne, VIC 3001, Australia

Abstract. The results of a statistical investigation of 42 fixed-wing, small to medium sized (20 kg–1000 kg) reconnaissance unmanned air vehicles (UAVs) are presented. Regression analyses are used to identify correlations of the most relevant geometry dimensions with the UAV's maximum take-off mass. The findings allow an empirical based geometry-build up for a complete unmanned aircraft by referring to its take-off mass only. This provides a bridge between very early design stages (initial sizing) and the later determination of shapes and dimensions. The correlations might be integrated into a UAV sizing environment and allow designers to implement more sophisticated drag and weight estimation methods in this process. Additional information on correlation factors for a rough drag estimation methodology indicate how this technique can significantly enhance the accuracy of early design iterations.

Keywords: Unmanned Air Vehicle · Geometry · Correlations · Statistics · Drag

Abbreviations

AR	= aspect ratio
b_{ref}	= reference wing span
C_{D0}	= zero-lift drag coefficient
C_{fe}	= equivalent skin friction coefficient
c_{ref}	= reference chord length
cross	= cross sectional
D	= diameter
EO/IR	= electro-optical and infrared
FR	= fineness ratio
fus	= fuselage
HT	= horizontal tail
L,l	= length
LG	= landing gear
MTOM	= maximum take-off mass
Ref	= reference

© Springer Nature Singapore Pte Ltd. 2019
X. Zhang (Ed.): APISAT 2018, LNEE 459, pp. 1365–1381, 2019.
https://doi.org/10.1007/978-981-13-3305-7_109

S   = area
TB   = tail boom
UAV   = unmanned air vehicle
V   = tail volume coefficient
VT   = vertical tail
W   = width
wet   = wetted

1   Introduction

The design and development of Unmanned Air Vehicles (UAVs) has come a long way since their first appearance as remote control aircraft around 1918 [1]. With the significant micronization of electronic components and the advances in computer technology, the UAV market has seen a significant growth in the past years. UAVs are nowadays used for both military and civil missions including but not limited to science, reconnaissance, agriculture or mapping [11]. Traditionally, engineers used their knowledge of conventional aircraft design processes and adapted these to unmanned aircraft. This transition is questionable to some point, as the requirements and mission scenarios of unmanned and manned aircraft differ significantly. Today's state-of-the art UAVs have matured to an independent aircraft category with very specific design properties [10].

A conceptual aircraft design process requires the analysis of several configurations and is an iterative process with multiple refinement stages [8]. In the first stage (initial sizing), wing loading and thrust- or power-to-weight ratio are commonly determined by using the so called "matching diagram" [7]. In the second stage, the maximum take-off mass (MTOM) as a summation of empty mass, payload mass and fuel mass is estimated by empirical regression analyses of similar aircraft. The geometrical shape of the aircraft is determined in later stages using simple parameters like wing area, wingspan, aspect ratio, fuselage length and so forth. Those parameters are subject to change during the iteration process. With empirical regressions found for aircraft of a similar class, such parameters are initially estimated based on a correlation with the maximum take-off mass. This initial guess is crucial for the following stages as it can affect the number of iterations. Due to long experience with manned aircraft, empirical correlations are available for most of the common aircraft classes and are presented, for instance, in Roskam [14].

The available data for UAVs is restricted to basic correlations necessary for an initial sizing process. Several authors (Verstraete et al. [16], Finger [2] and Gundlach [8]) aim to provide such information. Verstraete et al. [16] performed regression analyses of UAVs ranging from 0.1 kg to 40,000 kg MTOM and developed correlations for payload fraction, endurance, empty mass, wing loading and power loading. Finger [2] investigated empty mass correlations for UAVs between 2 kg and 1000 kg. Gundlach [8] gives several correlations for basic sizing with a limited number of UAVs. With the currently available data for unmanned aircraft, an initial sizing process can be established equally well for unmanned aircraft as for traditional manned aircraft.

However, no data is available which links the UAV's maximum take-off mass to more sophisticated geometry parameters like fuselage length or landing gear size, to name only a few. The presented paper aims to close this gap for small to medium sized unmanned aircraft by providing detailed statistical data of 42 fixed-wing reconnaissance UAVs between 20 kg and 1000 kg maximum take-off mass. The correlations can be used to perform a complete correlation based geometry build-up that provides a starting point for further design iterations. Such correlations are of fundamental importance as they close the gap between initial sizing, where only the basic parameters of the aircraft are defined (wing-loading, power to weight ratio, MTOM), and later design stages in which the complete outer shape is determined.

This paper is structured in the following way: Sect. 2 describes the data acquisition methodology while Sect. 3 presents the evaluation results, divided into several subparagraphs for individual components. Section 4 gives a distinct conclusion.

2   Methodology

The basis for the analysis is a detailed review of 42 reconnaissance UAVs ranging from 20 kg up to 1000 kg maximum take-off mass. This specific range was chosen as it represents the "small to medium sized" unmanned aircraft category in which the general layout, propulsion systems and mission scenarios are similar. Very small UAVs (<20 kg) are often equipped with electric propulsion systems and feature more special configurations like flying wings or blended wing bodies [1]. On the contrary, very large UAVs can often be treated with correlations for commercial aircraft (see Verstraete [16]) and are therefore excluded in the presented analysis.

Correlations with geometry parameters can only be derived if a sufficient number of UAVs feature the same components and if the aircraft configuration as well as the use case are similar. Only UAVs which were at some point or are currently produced in significant numbers and have seen actual mission deployment are included in the database. No experimental or technology demonstration aircraft were analyzed. Such UAVs may never see mission use due to technical difficulties or changing requirements. Additionally, such aircraft could also be designed exclusively for testing purposes of specific components and it often cannot be determined whether instead of it they were sized to a specific design mission. Excluding these aircraft increases the data accuracy for correlations of actual mission proven aircraft but also limits the total number of UAVs.

An overview about all UAV configurations included in the study is given in Fig. 1, while detailed information can be found in appendix Table 2. It was found that about 88% of all analyzed UAVs are of tail boom or standard configuration. This paper therefore focuses on these configurations and neglects more exotic ones like flying wings or blended wing bodies. "Jane's All the World's Aircraft–Unmanned" [15] is used as the main reference for our research concerning the available UAVs.

This reference however, only provides basic geometry or mass information like wingspan, total length or maximum take-off mass. To gain detailed information on individual component sizes, high-quality images or three-view drawings are used, mainly provided by the UAV manufactures themselves. With basic geometry information, the dimensions of nearly all components can be extrapolated from such images. Care was taken to include angular- and distortion corrections to improve the data. Tests with UAVs where images and detailed CAD models were available indicate that the accuracy for individual component dimensions is in the order of 6%.

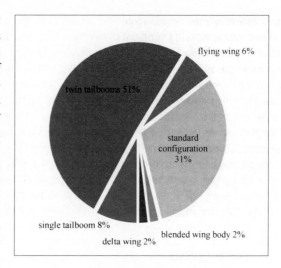

Fig. 1. UAV configurations in the database

A spreadsheet based geometry break-down is developed which divides UAVs into standard components and allows for an automated calculation of geometrical properties like cross sectional area, wetted area, aspect ratio and so forth. For this purpose, the geometry is simplified and represented by shapes for which analytical equations are available. Depending on the component, multiple geometrical representations are offered and the most realistic one is chosen by the user. This simplification leads to slight deviations, especially for wetted area calculations. Taking the data acquisition accuracy into account, the total deviations were found to have a maximum order of about 10%, which is adequate for the desired purpose. Each UAV requires a manual input of at least 150 parameters on average. The complete workflow process is visualized in Fig. 2 while Fig. 3 shows an example geometry simplification for a landing gear fairing.

The results are consolidated in diagrams (see Sect. 3) including all applicable UAVs. Those diagrams show the respective geometrical parameter in dependency of the UAV's maximum take-off mass. Regression analysis by means of a least squares fit is used to derive empirical equations representing the best average of the available data.

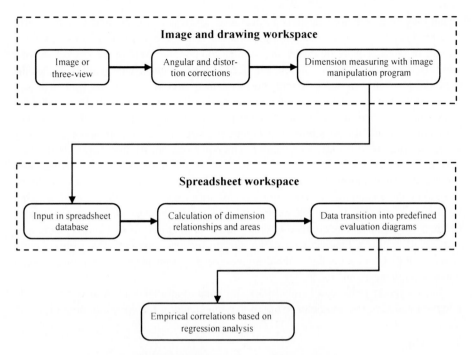

Fig. 2. Data acquisition workflow process

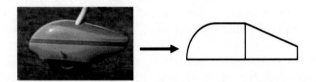

Fig. 3. Example of landing gear fairing geometry simplification, image from [4] (CC0 universal public domain)

3   Results

The following section describes the analysis results of the study in forms of diagrams and their corresponding regression analyses. The presentation follows the typical order in which UAV components are sized and allows for a structured evaluation. Due to the enormous amount of data, the correlations are reduced to the most important ones for each individual component. Wetted or cross sectional areas are only provided for components if an accurate calculation is not possible by the already presented data.

Data scatter is more or less significant depending on the individual correlation. This is to be expected, as the design space for UAVs is much bigger than for manned aircraft, which is a result of higher risk tolerance, reduced costs and easier certification [10]. With an increased number of development companies comes increased variation in component shape and design. Conventional manned aircraft companies tend to reuse technology from previous designs to decrease development costs and simplify certification [12]. Such reuse of technology naturally leads to better correlations as several aircraft of one company might feature similar components. Within the market study it was observed that this does not hold true for UAV companies. This again increases the expected geometry variations compared to manned aircraft.

Even though some of the derived equations might show a relatively low coefficient of correlation, they are still of high relevance for the design process of UAVs. The correlations enable the possibility to estimate the size of UAV components given only the maximum take-off mass as an input. The equations are trimmed towards the best average of the available data. This average leads to the most promising starting point for the design iterations in further development stages. The correlations might be integrated into an automated sizing process (for instance shown by Finger et al. [3]) and be utilized to perform a complete geometry build–up in very early design stages. Such a geometry build-up can be coupled to drag estimation methods or structural calculations and greatly improve the accuracy of the design feedback system. This consequently enhances the overall accuracy of the design process as it provides a direct link between initial sizing and geometry design.

If not explicitly stated, all equations require maximum take-off mass as an input in kilogram and give the geometry value in the unit stated in the corresponding diagram.

3.1 Wing

The diagrams below (Figs. 4 and 5) show correlations of the UAV's wing area and aspect ratio versus maximum take-off mass. Both regression analyses indicate a linear relationship and are presented in Eqs. 3.1 and 3.2. This is especially evident for the wing area, as it yields a rather constant wing loading for UAV's between 20 kg and 1000 kg.

All chosen UAVs feature reconnaissance mission scenarios which leads to similar requirements; thus designers might aim at similar wing loadings. Data scatter for aspect ratio is significant, however, the trend indicates that heavier UAVs tend towards higher values. The study also showed that larger UAVs are very much trimmed towards extreme endurance missions. These mission scenarios require aircraft with large aspect ratios increasing the achievable lift-to-drag ratio. Smaller UAVs might also be designed for very high endurance, but most of them have a more diversified use case [1]. As such, the UAV is more adaptable with a reduced aspect ratio. Additionally, with a small wing area, high aspect ratio wings lead to short chord lengths, which affects torsional stability and could intensify aero-elastic effects. An attempt was made to introduce the UAV's endurance as an additional physical parameter into the correlation but actually increased data scatter for smaller UAVs due to their diversified use case.

$$S_{wing} = 1.0339 \cdot 10^{-2} \cdot m_{MTOM} + 1.1585 \qquad (3.1)$$

$$AR_{wing} = 8.1658 \cdot 10^{-3} \cdot m_{MTOM} + 8.7720 \qquad (3.2)$$

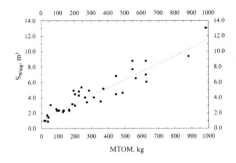

Fig. 4. Wing area against MTOM

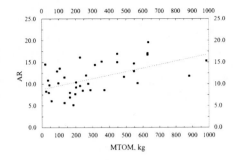

Fig. 5. Aspect ratio against MTOM

3.2   Stabilizer Surfaces

Within the very first stages of an aircraft design process, stabilizer surfaces are commonly estimated by employing empirically found tail volume coefficients, presented in Eqs. 3.3 and 3.4 [7]. These coefficients both incorporate the reference (wing) area and the reference chord or span as well as tail lever arms measured from the wing's quarter chord to the tail's quarter chord.

Typical coefficients can be identified for specific aircraft classes. The present study found the average horizontal tail volume coefficient to be 0.65, which is a value also common for homebuilt and general aviation aircraft. Marshall et al. [10] found a slightly lower coefficient of 0.5 for small fixed wing UAVs. Trends of the present study indicate that the coefficient slightly increases with increasing aircraft mass. This corresponds to larger horizontal tail surfaces and increased stability.

The average vertical tail volume coefficient was found to be 0.042 comparable to the findings of Marshall et al. [10]. The vertical tail volume coefficient shows a tendency to decrease with increasing aircraft size, which leads to decreased lateral stability.

$$S_{HT} = \frac{V_{HT} \cdot S_{ref} \cdot C_{ref}}{l_{HT}} \tag{3.3}$$

$$S_{VT} = \frac{V_{VT} \cdot S_{ref} \cdot b_{ref}}{l_{VT}} \tag{3.4}$$

3.3   Fuselage and Tail Booms

Correlations of fuselage length and fuselage fineness ratio are shown in Figs. 6 and 7 together with the results of the regression analysis in Eqs. 3.5 and 3.6. Fuselage length varies from 1 m for small UAVs up to 5.5 m for bigger ones. The length follows a power law trend, where the highest gradients are found between 20 kg and 150 kg MTOM.

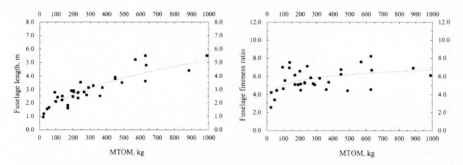

Fig. 6. Fuselage length against MTOM

Fig. 7. Fuselage fineness ratio against MTOM

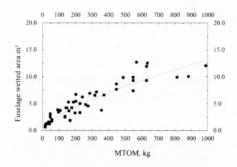

Fig. 8. Fuselage wetted area against MTOM

These observations are similar for the fuselage fineness ratio, which provides a link between diameter and length. However, the regression analyses reveals that a logarithmic trend provides a better curve fit. The fineness ratio varies especially between tail boom and standard configuration aircraft. Fuselages of tail boom aircraft are naturally shorter as the necessary tail lever arm is provided by the booms. These UAVs also feature smaller fineness ratios. The maximum fineness ratio was found to be 8.3 for a standard configuration UAV, while the majority of fineness ratios is in the order of 4 to 6.

A correlation of the fuselage wetted area is shown in Fig. 8 together with the regression analysis in Eq. 3.7. Fuselage wetted area is given here separately as an accurate estimation is not possible by knowledge of length and fineness ratio only. Instead, fuselage wetted area is calculated by subdividing the fuselage into four individual segments for which linear relationships are assumed. This yields to a very accurate calculation. Fuselage wetted area shows a very slight power law trend.

$$L_{fus} = 0.2825 \cdot m_{MTOM}^{0.4206} \tag{3.5}$$

$$FR_{fus} = 0.7342 \cdot ln(m_{MTOM}) + 1.6589 \tag{3.6}$$

$$S_{wet-fus} = 0.1219 \cdot m_{MTOM}^{0.6773} \tag{3.7}$$

The present study indicates that tail boom length is directly correlated to the UAV's fuselage length via a constant relationship. The results of the analyses are shown in Figs. 9, 10, and 11, revealing that the average tail boom length is about 0.75 × fuselage length. For the sake of completeness, a regression analyses of tail boom length against maximum take-off mass is shown Eq. 3.8. As tail boom cross sections are mostly sized towards the acting loads, their diameter might be of higher interest than the fineness ratio and is directly correlated with the maximum take-off mass (see Eq. 3.9).

$$L_{TB} = 0.5935 \cdot ln(m_{MTOM}) - 1.2687 \tag{3.8}$$

$$D_{TB} = 0.01604 \cdot ln(m_{MTOM}) - 0.00512 \tag{3.9}$$

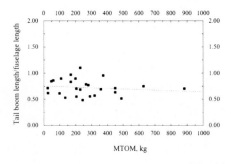

Fig. 9. Relative tail boom length against MTOM

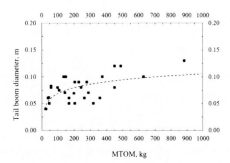

Fig. 10. Tail boom length against MTOM

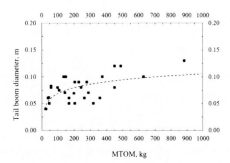

Fig. 11. Tail boom diameter against MTOM

3.4 Payload

The payload of small to medium sized reconnaissance UAVs is normally an electro-optical and infrared (EO/IR) gyro-stabilized gimbal pod attached to the lower side of the fuselage. Besides its overall impact on weight and communication requirements, Götten et al. [6] showed that payload drag can be significant for UAVs of the presented class and is largely driven by the respective cross sectional area.

Payload shapes range from spheres or half-spheres to combinations of spheres and cylinders. A geometrical representation based only on their diameter is therefore only partially sufficient. Taking the drag impact into account the cross sectional area is correlated against the maximum take-off mass, too.

Figures 12 and 13 show that EO/IR payload size grows with the UAV's take-off mass up to about 450 kg. Cubic functions give the best data fit and are presented in Eqs. 3.10 and 3.11. For higher take-off masses both payload diameter and cross sectional area stay nearly constant. This behavior was closely investigated and found to have a distinct origin. Verstraete [16] found a constant empty mass and payload fraction for a wide variety of UAVs, indicating that the primary origin for the findings in this study is not the UAV's structure or mission capabilities, but rather the payload itself.

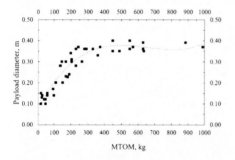

Fig. 12. Payload diameter against MTOM

Fig. 13. Payload cross sectional area against MTOM

$$D_{PL} = 1.2816 \cdot 10^{-9} \cdot (m_{MTOM})^3 - 2.5110 \cdot 10^{-6} \cdot (m_{MTOM})^2 + 1.5465 \cdot 10^{-3} \\ \cdot m_{MTOM} + 7.1638 \cdot 10^{-2} \tag{3.10}$$

$$S_{cross-PL} = 1.9604 \cdot 10^{-10} \cdot (m_{MTOM})^3 - 4.0708 \cdot 10^{-7} \cdot (m_{MTOM})^2 + 2.8406 \cdot 10^{-4} \\ \cdot m_{MTOM} - 4.1643 \cdot 10^{-3}$$

$$\tag{3.11}$$

A study of 43 EO/IR gimbals revealed that the packaging density of such gimbals increases with increasing gimbal mass, as shown in Figs. 14 and 15. This density increase is significant and can be as high as 100% comparing payloads between 1 kg and 100 kg. Heavier payloads therefore require proportionally less volume than lighter ones. With a constant payload fraction, heavier UAVs also carry heavier payloads, but payload volume and cross sectional area may stagnate due to the increase in payload density. This leads to the fact that the aerodynamic impact of EO/IR payloads decreases with increasing aircraft size. A critical region is identified for UAVs with maximum take-off masses between 150 kg and 400 kg (see Fig. 15). In this range, the payload's

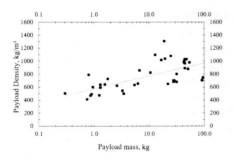

Fig. 14. Payload density against payload mass

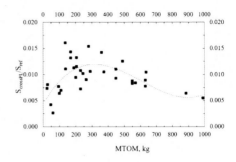

Fig. 15. Payload relative cross sectional area against MTOM

cross sectional area is maximized in relation to the overall UAV size (represented by the reference area). The payload's aerodynamic impact is especially high in this region.

3.5 Landing Gear

The layout of the landing gear is strongly dependent on the individual aircraft configuration and might be influenced by factors like propeller ground clearance or center of gravity shift. Therefore, no attempt is made to find correlations for positioning of individual gears or their detailed layout.

However, what is especially interesting for the designer are parameters like tire size or the landing gear's total cross sectional area. With this information, basic estimations of the landing gear's weight and drag are possible and might be used in the design feedback system. A power law correlation is found linking tire diameter to maximum

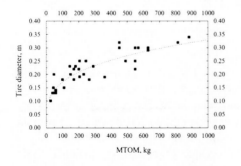

Fig. 16. Tire diameter against MTOM

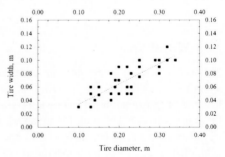

Fig. 17. Tire width against tire diameter

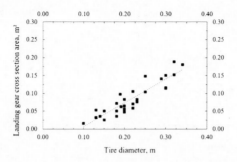

Fig. 18. Landing gear cross sectional area against tire diameter

take-off mass (Fig. 16 and Eq. 3.12), while a linear relationship between tire diameter and tire width can be identified (Fig. 17 and Eq. 3.13). Total landing gear cross sectional area again scales linearly with the tire diameter (Fig. 18 and Eq. 3.14).

The last two correlations prove the common methodology of determining total landing gear drag as a function of only one tire's cross sectional area as valid for the investigated UAV class. This methodology dates back to NACA Report 485 [9] and was also found to be applicable on UAVs [6], though the coefficients found in the prescribed NACA report should be modified.

$$D_{tire} = 0.04653 \cdot m_{MTOM}^{0.28344} \qquad (3.12)$$

$$W_{tire} = 0.30489 \cdot D_{tire} + 0.00368 \qquad (3.13)$$

$$S_{cross-LG} = 0.66490 \cdot D_{Tire} + 0.05981 \qquad (3.14)$$

3.6   Wetted Area

The wetted area of an aircraft can be used to provide and initial estimate of the total aircraft's zero-lift drag by applying the equivalent skin friction method presented in Raymer [13]. Equation 3.15 shows the method and the necessary input factors. Zero-lift drag is estimated by multiplying an equivalent skin friction coefficient (C_{fe}) with the aircraft's ratio of wetted area to reference (wing) area. The equivalent skin friction coefficient is a constant for a specific aircraft class and estimated by regression analysis. Factors for UAVs are not given in Raymer [13].

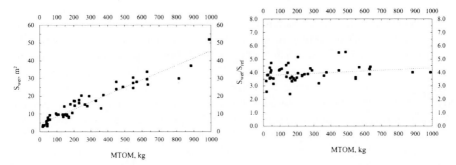

Fig. 19. Wetted area against MTOM

Fig. 20. Relative wetted area against MTOM

Figure 19 shows correlations of wetted area versus maximum take-off mass as found in the present study. Wetted area shows a strongly linear dependency on take-off mass described in Eq. 3.16. The ratio of wetted area to reference area (Fig. 20) is mostly constant with the average around 3.8.

Götten et al. [6] provide zero-lift drag values for four reconnaissance UAVs of the small to medium size class which were also analyzed in the presented study. Their findings are summarized in Table 1 and used to provide an initial estimate of the factor C_{fe} for UAVs of the presented class. C_{fe} varies between 0.00724 and 0.01163 with an average value of 0.00916. The difference between the highest and lowest values is influenced by differing landing gear configurations affecting zero-lift drag [6]. Slightly different configurations might significant affect zero-lift drag as outlined in [5]. It is noteworthy that the values for C_{fe} are significantly higher than for every aircraft category found in Raymer [13]. It is evident that zero-lift drag of UAVs is higher compared to other aircraft categories.

$$C_{D0} = C_{fe} \cdot \frac{S_{wet}}{S_{ref}} \qquad (3.15)$$

$$S_{wet} = 0.04125 \cdot m_{MTOM} + 4.46529 \qquad (3.16)$$

Table 1. UAV data as given in Götten et al. [6] with calculated C_{fe} values

UAV	Ref. area, m^2	S_{wet}, m^2	S_{wet}/S_{ref}	C_{D0}	C_{fe}
1	1.32	4.76	3.61	0.03152	0.00873
2	2.37	8.57	3.62	0.04211	0.01163
3	4.74	17.26	3.64	0.03291	0.00904
4	8.70	36.16	4.16	0.03012	0.00724

4 Conclusion

The presented correlations allow an estimation of the size of the most important geometrical properties of small to medium size reconnaissance UAVs by referring only to their maximum take-off mass. The findings are valid for both tail boom and standard configuration aircraft with masses between 20 kg and 1000 kg. Regression analyses determined equations representing the correlations to a degree which is adequate for early geometry estimations. These equations can be easily integrated into an aircraft design and sizing environment. They provide a bridge between the very first design stage, in which power loading, wing loading and take-off mass are sized and the following stages during which the shapes and dimensions of components are designed.

The correlations might be used to perform both empirical drag and weight estimations already within the initial sizing process and provide a valuable increase in accuracy in the flight performance estimations. This enhances the accuracy of the design feedback system and thus the overall precision of the sizing process.

Appendix

Table 2. UAVs used in the presented study sorted by MTOM, (dash indicates lack of information)

No.	Name	Manufacturer	MTOM, kg	Payload mass, kg	Endurance, h	Length overall, m	Span overall, m
1	Aerosonde Mk 4.7	Textron	25	–	14	1.70	3.60
2	Manta	Raytheon	28	5	6	1.90	2.66
3	Luna	EMT Penzberg	40	5	5	2.36	4.17
4	Sparrow	EMIT	45	12	6	2.14	2.44
5	Atlantic	SCR	45	7	6	2.80	3.80
6	Strix	Aerodreams	48	18	15	2.90	3.60
7	S4 Ehecatel	Hydra Technologies	60	–	–	2.90	4.20
8	T-20	Arcturus	84	–	–	2.90	5.33
9	Jump-20	Arcturus	95	–	–	2.90	5.60
10	GRIF-1	558 ARP	100	30	8	3.50	4.80
11	Hermes 90	Elbit	110	10	15	4.20	5.50
12	Outlaw SeaHunter	Griffon Aerospace	136	40	4	3.01	4.87
13	Pchela-1T	Yakovlev	138	–	–	2.78	3.25
14	Skylynx II	BAE Systems	150	31	15	4.23	5.60
15		Textron	170	45	7	3.40	4.30

(continued)

Table 2. (*continued*)

No.	Name	Manufacturer	MTOM, kg	Payload mass, kg	Endurance, h	Length overall, m	Span overall, m
	RQ-7B Shadow						
16	Shadow 200	AAI	170	27	6	3.40	4.30
17	Sentry HP	Leonardo DRS	190	31	8	3.35	3.90
18	ZALA 421-20	Zala Aero	200	40	7	5.00	6.00
19	Pioneer RQ2-2A	IAI	205	34	6	4.40	5.10
20	Tiger Shark	NavMar	205	34	8	4.55	6.70
21	Aerostrar	Aeronautics Systems	230	50	14	4.50	8.70
22	Pegaz	MIT	230	40	12	5.40	6.34
23	Flamingo	SATUMA	245	35	8	5.18	6.61
24	Shadow 600	AAI	265	41	14	4.80	6.83
25	Ranger	RUAG	280	45	4	4.61	5.71
26	RQ-101 Night Intruder	Korea Aerospace Industries	290	45	6	4.70	6.40
27	Xian ASN-209	Aisheng	320	50	14	4.28	7.50
28	F-720	UMS	360	70	12	4.80	7.20
29	Nishant	DRDO	375	45	5	4.63	6.57
30	Searcher MK II	IAI	450	120	15	5.85	8.55
31	Seeker 400	Denel	450	100	16	5.77	10.00
32	Falco	Leonardo	490	70	9	5.25	7.20
33	Hermes 450	Elbit	550	150	20	6.10	10.50
34	Yabhon-R	Adcom	550	100	27	5.00	6.50
35	Karayel	Vestel	550	70	20	6.50	10.50
36	Atlante	Airbus	570	100	15	5.47	8.00
37	Bayraktar Tactical	Baykar	630	55	24	6.50	12.00
38	Falcao	Avibras	630	150	16	5.90	10.80
39	GNAT 750	General Atomics	635	63	–	5.33	10.76
40	Rustom I	DRDO	815	75	12	5.12	7.90

(*continued*)

Table 2. (*continued*)

No.	Name	Manufacturer	MTOM, kg	Payload mass, kg	Endurance, h	Length overall, m	Span overall, m
41	MQ-5B Hunter	Northrop Grumman	885	113	12	7.01	10.44
42	Wing Loong	CAIG	990	200	–	9.05	14.00

References

1. Austin R (2010) Unmanned aircraft systems: UAVS design, development and deployment, 2nd edn. Wiley, Chichester, Hoboken
2. Finger DF (2016) Comparative performance and benefit assessment of VTOL- and CTOL-UAVs. In: Deutscher Luft-und Raumfahrtkongress
3. Finger DF, Braun C, Bil C (2018) An initial sizing methodology for hybrid-electric light aircraft. In: 18th AIAA Aviation Technology Integration and Operations Conference; Atlanta. American Institute of Aeronautics and Astronautics, Reston
4. Hodan G. Ultralight Airplane. https://www.publicdomainpictures.net/en/view-image.php?image=179889&picture=ultralight-airplane. Accessed 6 Jun 2018
5. Götten F, Finger DF, Havermann M, Braun C, Gómez F, Bil C (2018) On the flight performance impact of landing gear drag reduction methods for unmanned air vehicles. In: German Aerospace Congress 2018
6. Götten F, Havermann M, Braun C, Gómez F, Bil C (2018) On the applicability of empirical drag estimation methods for unmanned air vehicle design. In: 18th AIAA Aviation Technology Integration and Operations Conference; Atlanta. American Institute of Aeronautics and Astronautics, Reston
7. Gudmundsson S (2014) General aviation aircraft design: applied methods and procedures. Butterworth-Heinemann, Oxford
8. Gundlach J (2014) Designing Unmanned Aircraft Systems: A Comprehensive Approach, 2nd edn. AIAA Education Series, Washington, DC
9. Herrenstein W, Biermann D (1934). The Drag of Airplane Wheels, Wheel Fairings and Landing Gears I: NACA Report 485
10. Marshall DM, Barnhart RK, Shappee E, Most M (2016) Introduction to Unmanned Aircraft Systems, Second Edition, 2nd edn. CRC Press, Boca Raton
11. Newcome LR (2004) Unmanned Aviation: A Brief History of Unmanned Aerial Vehicles. American Institute of Aeronautics and Astronautics, Reston
12. Nicolai LM, Carichner G (2010) Aircraft design. AIAA American Inst. of Aeronautics and Astronautics, Reston
13. Raymer DP (2012). Aircraft design: a conceptual approach. 5th ed. AIAA Education Series, Reston
14. Roskam J (1987) Airplane Design. Roskam Aviation and Engineering Corporation, Ottawa, Kansas

15. Streetly M (2015) IHS Jane's All the world's aircraft - Unmanned: 2015-2016. IHS, Coulsdon
16. Verstraete D, Palmer JL, Hornung M (2018) Preliminary sizing correlations for fixed-wing unmanned aerial vehicle characteristics. J Aircr. 55:715–726. https://doi.org/10.2514/1. C034199

A Classification and Summary
of Degradation Process Model

Long Li[1(✉)], Tianxiang Yu[2], Bifeng Song[2], Yijian Chen[2],
and Bolin Shang[2]

[1] Northwestern Polytechnical University,
Xi'an 710072, Shaanxi, People's Republic of China
leedragon@mail.nwpu.edu.cn
[2] School of Aeronautics, Northwestern Polytechnical University,
Xi'an 710072, People's Republic of China

Abstract. A reasonable and precise degradation model plays a vital role during the process of structure degrade research. Enlightened with the practical engineering examples, we systematically review the consideration and classification about the contemporary research of degradation model, which in terms of data type and its ample extent. First of all, this paper expound the model set process, their suitable occasion and research status, then, some relate questions and application comparison of these models were made to explain their characters during the practical engineering. Finally, some feasible research directions and challenges were emphasized in the conclusion.

Keywords: Probabilistic statistic model physics-statistic model ·
Physical of failure model · Degradation model

1 Introduction

High reliability and long life request have become the basic requirements in sophisticate manufacturing industry fields, aerospace, nuclear instruments, aviation, for instance. Moreover, along with the progress of new technology and product technique, the reliability and safety index of products are being became an indispensable index of practical engineering or scientific testing, due to these reasons, the researchers motivate to find a new way to tackle the reliability accurate estimate problem or probabilistic failure evaluate, namely: degradation.

Conventionally degradation research own some characters as follows:

- The existed degradation model aims sole component failure of system. The objects were isolated from system to make degradation analysis, and set a specific model to suit the failure or degradation process, however, this method neglects the connection or interior relationship of the consistent unit, on the other hand, it creates a misleading that the independent unit enable represent the entire system degradation, actually, its impartial.

© Springer Nature Singapore Pte Ltd. 2019
X. Zhang (Ed.): APISAT 2018, LNEE 459, pp. 1382–1397, 2019.
https://doi.org/10.1007/978-981-13-3305-7_110

- The current research concentrate on stress, strength or other reference index. Such as use the mounting press or loading to test the structure degrade ratio, but a review relate on degradation and its classification to failure that according to the mechanism or pattern is not abundant at present.
- In order to simplify the mechanism and patterns identify of degradation or failure process, the existed research was not make model pre-screening and method classification in normal case, this paper summarize the model choose methods and data screening ways.
- Degradation factors have been taken into consideration during the practical technique and estimation process of present research, nevertheless, the majority work regard the single failure model as object, which can't represent the whole degradation situation of system, therefore, it is necessary to make a relative work prediction and recommend to the new research direction.

Degradation model provided a new strategy and manage method on the structure reliability estimate and products' life prediction. Furthermore, based on this model, reliability research can be widen to some new area, such as medical instrument product in micro-nano fabrication, some huge and super large-scale equipment like the sea drill platform. On account of complicate construction of these instruments accompany with intricate reliability problems, there are increasingly researchers put forward to use degradation data to evaluate the performance of production reliability or its lifetime predict.

Compared with the existed reliability research or performance testing trial lay too much emphasis on how to get failure data without concern failure mechanism, degradation model can overcome these shortcomings, this model can trace the failure reasons in original failure process.

With the increasing attention on high reliability and long life research, investigators found degradation phenomena accompany the whole life periodic of product. On the basis of these facts, some researchers suggested that the whole degradation process can be described by set a degradation model, further, make an evaluation or prediction to product's degradation through this model.

To identify and distinguish the degradation model set process, researchers often rely on empirical assumption or observation data to make a structure performance forecast during the whole degradation process. In this article, we provide a concise background of degradation model and its application scope. Firstly, the three major kinds of degradation process were set through constructive method which classified by the type of information, data amount etc. Also, we make a summary of these models category and analyse their advantages and disadvantages. Lastly, some prospects on the hot issues of degradation research were given and projected some relevant ideas in this field.

2 Resent Research and Current Models

How to build degradation model is a tough work due to its complex degrade process and stochastic factors influence. To make a performance evaluate or residual life predict are relate with how to set a degradation model accurate and effective, directly, because the different degradation models may lead a different results. Based on amount of research experience and existed articles, no matter the common degradation process or the accelerated process (ADT), according to the degradation data acquisition methods, degradation property, different degradation ration, and so on, degradation model divided into three major branches: statistic based model, physics-statistic model and physics failure model [1].

2.1 Probabilistic Statistic Model

Parameters and non-parameters factors are two main aspects of probabilistic statistic model. Degradation data can be collected by repeated recording or measure ways during the degradation process. Probabilistic statistic model regards degrade unit and degrade parameters follows a concrete function distribution or abide by some hypothesis distribution, however, the non-parameters degradation trajectory is unknown or given in partial, what a fact has been realized that the nature of degrade process is a stochastic process, therefore, a variety simulate methods can be chose to get degrade trajectory, evaluate parameters, and so on. The existed articles showed that current research on common degrade process concentrated on its trajectory simulate, time-to-relate parameter estimate, and stochastic process analysis, respectively. All the key points of these methods are focus on how to find a reasonable degrade trajectory to fit the actual degrade process. The main branches of probabilistic statistic model include linear or non-linear model, time-statistic fitting model, stress-statistic fitting model and stochastic process model, etc.

Linear degradation model suit for the simple degrade situation (such as mechanism wear process), this model based on degrade trajectory, which can be presented by deterministic function consisted with a stochastic parameters and fixed effect, it is stand for degrade trajectory that describe the variety of degrade value with time, a degrade path is consist of three factors: degrade value, time and stochastic effect. Its math equation can be written as follows;

$$X(t) = \varphi(t, \boldsymbol{\beta}, \boldsymbol{\alpha}) = \varphi\left(t_{ij}, \beta_1 \ldots \beta_n, \alpha_1 \ldots \alpha_n\right) \tag{1}$$

Where, $\boldsymbol{\alpha}=(\alpha_1 \ldots \alpha_n)$ is a stochastic vectors of n dimensions, which can depict the character of each parameter, $\boldsymbol{\beta}=(\beta_1 \ldots \beta_n)$ represents the fixed effect, it owns the same parameter to the whole units, φ stands for the specific function relation, this equation showed that every stochastic process $\{X(t_{ij}), t \geq 0\}$ predominately determined by $\boldsymbol{\beta}$ and $\boldsymbol{\alpha}$ correspondingly, in general case, degrade trajectory must take the noise into consideration, so, it is given by:

$$y_{ij} = X(t_{ij}) + \varepsilon_{ij} = \varphi(t_{ij}, \beta_1 \ldots \beta_n, \alpha_1 \ldots \alpha_n) + \varepsilon_{ij}, (i, j = 1 \ldots, n) \tag{2}$$

Where, $y_{ij}, X(t_{ij})$ are the measurement value and actual degrade value of the **i**th sample that record on time t_{ij} of degrade trajectory, respectively. ε_{ij} refers to measurement error, which is an independent value and follows the normal distribution $\varepsilon_{ij} \sim N(0, \sigma_\varepsilon^2)$. A typical character of statistical model set process can be presented through the Fig. 1.

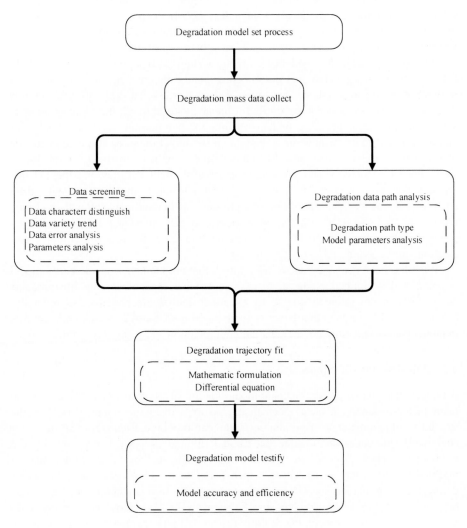

Fig. 1. The set process of statistic based model

Lu [2] makes a research on linear degrade process analysis by use a model with random regression coefficient and standard-deviation function, his research also makes an estimate of model parameters and use simulation to testify the relativity of estimated confidence interval. Si [3] used the non-linear degradation process to make estimation of remaining useful life, this research treats constant threshold and nonlinear drift coefficient play a central role about the PHM of system, especially for safety-critical system. Paper [4] makes an investigation on auto-regression model to predict degrade process, the application of auto-regression model can improve the predict precise, especially, the simulation method of this approach take the least-squared algorithm and Yule-Walker equation into consideration, these two algorithms combine the specific samples, in addition, demonstrate the effectiveness of the auto-regression model for the system diagnosis and degradation process prediction.

Generally speaking, the random effect of degradation models will increase the difficulty on parameters estimate, two stage degraded model provide a research method on this topic, to some extent, it can't assure the accuracy evaluation, however, convenient to the practical engineering use., especially in some multi-stage situation, Ni [5] use the two-stage degraded model to study the system degradation process, their study focus on different degenerate distribution correspond different stage that may appeared in practice multi-stage system. Due to the Wiener process can deal with the continue degradation process, paper [6] proposed a linear model through the Skew-Wiener in statistic lifetime inference, this model overcame the non-normal, asymmetric behavior among different units, meanwhile, it can attain the broad range of lase data. Yan first investigate a semi-stochastic process to handle the complicate of structural reliability, which owns the stochastic and temporal property of strength and stress in paper [7], this article makes a progress of time-variation reliability to satisfy the practical engineering condition, it consider the random of strength degradation and conventional stochastic process of stress to deal with the parameters that may affect the accuracy of independent increments, through this improved method the redundant and complex calculation can be avoid and get a precise result.

What has been discussed above tells a truth that probabilistic statistic based model is being widely applied in the aerospace, electronics, and other long life products degradation model research due to its excellent data tackle patterns and reliability evaluation ability, on the other hand, statistic based model also plays a significant role in model parameters estimate, trail design, life prediction and model discriminate fields.

2.2   Physics-Statistic Model

A typical character of physics-statistic model fits the situation that can attribute to two main factors, namely: a clear mechanism and experiment basement, therefore, the structure and parameters of degradation model can rely on existed empirical model or combined with degradation value and model optimal property test to set. Failure mechanism analysis provides an efficient way to investigate the component degradation mechanism and its failure causative, which compose the two aspects of components during their degradation stage. The performance value of degradation parameters can present the physics failure mechanism of systems units, also, provide a healthy management and function inspect, the recently research of physics-statistic model focus on how to set up model degradation path and assess the measurement accuracy of parameters during the statistic process.

The physic-statistic model set process can be described as follows (Fig. 2);

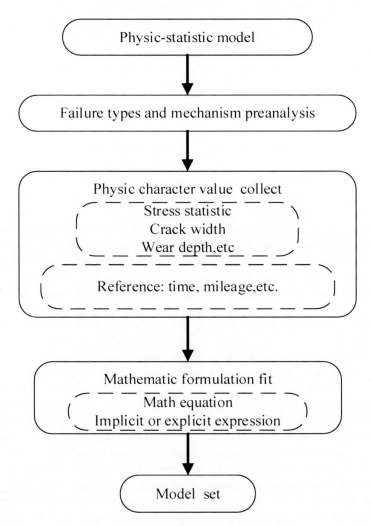

Fig. 2. Physic statistic model set process

Chen [8] made a lifetime prediction of epoxy resin adhesive product, they got the lifetime prediction method by acceleration factor and activation energy calculation method which determined by temperature and humidity, these physics statistic data can help researchers to find out the failure mechanism. Paper [9] focus on the physics-based analytical model of resistance in the linear transport and its application field, they presented a new approach to analysis the relationship between degradation rate and drain-current curves of electric devices, the statistic data between voltage and different ambient temperature, finally, this model was used to investigate the device degradation.

In the work [10], using physics statistic based model to make a reliability prediction, this model consider the randomness of the same product and proved the possibility of degradation units, which were over time used to obtain the parameters by statistic data that collect from an accelerated degradation test.

Xu [11] aimed at products reliability evaluation and bivariate degradation model that based on physics statistic model, this research can describe the situation that degradation of performance characteristic dependent or own multiple performance degradation process, their work proposed a new method to handle the bivariate degradation.

Paper [12] uses statistic-based degradation analysis and physics-based model to investigated a degradation path research method, this breakthrough way can be applied to the actual degradation path fit process of some electrical products degradation in a high efficient way, this research, indeed, suit the situation that products in different scales.

It is impossible to eliminate the measurement error or directly obtain its relevant physical degrade values in practical principle, which generated by the statistic method. Hence, how to get the accuracy statistic information during the physical process is a tough work, which always face the different situation, moreover, it needs to deal with the variable values or extrude the degradation value from the test, on the other hand, how to make a preprocess of statistic data should be consider, such as the test method, test equipment, noisy, abnormal condition, and so on. Paper [13] made a deep analysis on multi-sensor degradation data by analyze the degradation path which was caused by different scale of noise or interpretation, this paper focused on how to use a unified degradation path to test physics-based model or collect statistic data, meanwhile, it provide an useful way to simulate the actual degradation path in a high efficient way.

Due to the path change tendency can reflect the whole term development of products degradation characters, which has been caused attention among researchers. Mei made an elaborate analysis on physic and statistic of kinetics systems in paper [14], they state the difference and similarity of dielectric broken mechanism about special devices, this paper stands on the statistic and physical view, which different stage and distribution were involved to compare with the traditional planar analogues.

It is undeniable that the failure mechanism of degradation process, particularly, in some mixture process or models is the key point of degradation researching and criterion evaluate. Nevertheless, distinguish every mechanism is unrealistic, in addition, it was confined by many strictly conditions in some traditional data collection methods, such as use the Akaike information to deal with the complicate data relationship, traditionally, Akaike information not only based on threshold definition and its criterion was used to evaluate the mixture regression quality, but also suggest the parameters follow normal distribution, in the majority physics statistic model we need to stimulate the true degradation trace, however, Akaike information must fit some special distribution and require to get more parameters as possible, the abundant parameters can improve the fit precise extent undoubtedly, conversely, with the parameters amount increased, the extent of fit complication also appeared, which will be strictly confined the application of similar method on data statistic, otherwise, if this method applied in a multi-system, fit precise extent will show a plummeting that caused by parameters mount increase.

Recently, researchers want to find a extend way to overcome the overestimate method and decreased the minimum freedom parameters amount, indeed, this method can simplify the trajectory simulate during the physic statistic model establish process, for instance, in the chemistry of biology research field. Paper [15] made a research on the variable environment condition and mass data coexist situation during degradation process, this paper made a trial on how to deal with the frequent changing data, such as environment information or interaction of degradation mechanism. Paper [16] applied hierarchical Bayesian model to analyze geometry degradation, Bayesian statistic model use plenty prior information to describe the mechanism and degradation track, this statistic data can be applied to maintenance or predict the structure update, but this paper was not give a specific degradation index to evaluate the degradation process. Paper [17] makes a general classify of geometry degradation, some degradation data based on physic statistic such as profiles, gage, twist, alignment, displacement, and so on, these physic degradation performance provided an evaluation index to measure the degradation data, moreover, mechanistic and stochastic model were be classified by specific collective data, this paper aims to how the degradation rate from the acceptable level convert to deterioration level before interval time. Paper [18] used Weibull approach to make an analysis of how maintenance was affected by distribution time in the specified states, the statistic data that was simulated by a specific distribution can tell the degradation fluctuate trend and future performance, but in some certain extent, this model depend on its distribution time parameters, this will be limit its application in degradation evaluation, furthermore, even the statistic data of some parameters can reflect the deteriorate rate, which was also restricted by the physic condition, such as parameters distribution change.

In all, physics-statistic model often rely on abundant parameters data or relevant function cumulative distribution, but, these requirements hardly fulfilled in practical, the optimize research way is to find the relationship between degradation model and statistic data.

2.3   Physics of Failure Model

The principle of physics of failure model suit for the situation that failure property and process are clear, the degradation characters and process can provide a differential equation of degradation rate, on the other hand, the degradation rate reflects the property or function of products, these differential equations are the basement during the degradation model establish, a proper degradation model is vital to the next stage of degradation process analysis, such as how to determine the relation between degradation rate and time, how to use degradation trajectory or path to depict performance trend and uncertainty influence, to some extent, degradation model formulation may change with failure uncertainty, random effect, temporal uncertainty or serial correlation, otherwise, the description of equation may cause an influence during this model set. The physical of failure model set process can present as follows (Fig. 3).

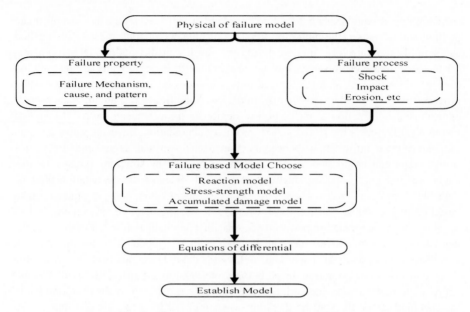

Fig. 3. The process of physical failure model set

Paper [19] applied a probabilistic physics of failure model to pitting and corrosion-fatigue in the PHM field of structure subject, the proposed model fit the situation that products work in the critical environment stresses and applied to prognostic the degradation information on reliability, moreover, it can make a development to the current inspection method and replacement ways. Physics of failure model also needs the failure data information and the special deal method, the tiny damage will have a consecutive change during the degradation process, regarding this situation, the dynamic and multi-state degradation was proposed.

In paper [20], Jin deployed a research about physics failure-based degradation model and prediction of lifetime under a dynamic covariate environment, MLE method and failure data used to estimate the parameters of this model, this application overcame the limitation condition of dynamic covariate environment.

This model focus on seek the original failure factors or nature of system, unfortunately, in the sophisticated system, property and process of degradation failure is complicated, such as chemical failure, physical failure or a couple of mix factors. On the other side, researchers are focus on degradation process identification, Luo [21] made a physical of failure analysis and diagnose method research on the chip degradation failure, this research can used to explain the failure in physical level that caused by hot carrier, but, this paper did not expound the causative of how the device interactive occur during the degradation.

As the degradation process comprise too many failure factors, hence, the drawback of physical of failure model concentrated on the data accuracy and probabilistic method limit, in some occasions, the probabilistic method was not suit the practical situation, reversely, non-probabilistic method or the random process have some merit on the data collective and distribution function seek process, Chookah [19], for example, his research based on the probabilistic condition of physical failure property, this model use the Bayesian method to represents and evaluate parameters, however, the practical pitting and erosion situation are more complicate than the experiment condition, the proposed model can't provide all-around information about the system degradation under the probabilistic circumstance.

Undeniably, failure property confirm process include the failure mechanism analysis and its type classify, physic of failure model set process has some crucial point in the degrade research but it was still unfolded.

3   Related Work and Recommend

Aforementioned work of these three models have their own merits and flaws, the different model suit the different engineering fields, some preliminary work should be done before the degradation model set, consult the existed data and research theory, an outline framework of degradation model choose process was made as Fig. 4 shows:

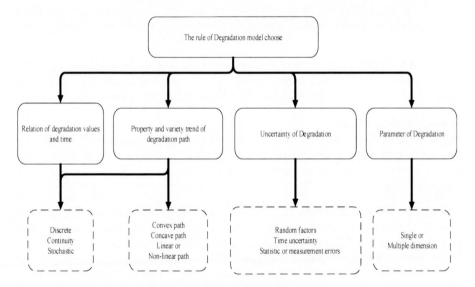

Fig. 4.   Degradation model choose rules

After the model choose, a suitable model plan can be used during the degradation research, however, some advices and recommend should be considered during this process.

Probabilistic Statistic Model:

1. **Data collect method.** Plenty of data is the foundation of statistic based model, except the traditional data collect methods, which mainly comprised by statistic ways, repeat experiment measurement, truncation and so on, nevertheless, these methods more likely encounter hamper in some occasions, an obvious case is the small sample or extreme sample trail and sophisticate instrument degradation research., which got a mass information or degradation data in short period is impractical, rather, under the non-probabilistic condition, as such, it is necessary to development a new data collect way. Recently, some new theories, convex model was used to deal with the non-probabilistic data, for example.

2 **Data screening rules.** After pooling the mass data, the data screening is advice to follow the principle as follows:
   1. It is necessary to eliminate the unreasonable information which have an obvious deviation when compared with the average level.
   2. If the truncation method is took to deal the experiment data, some rules should be formulated, which can avoid the useful or key point information was eliminated by cognitive fault.
   3. Find an appropriate math calculate method and distribution are important. Identify the data characters among the linear or non-linear, probabilistic or non-probabilistic, convex or concave type.

Not only the probabilistic statistic model, other two models should adopt the similar method during the data screening, either. All the mentioned aspects should pay attention in order to avoid the calculate error during the degradation model set process.

Physics-Statistic Model:

Physics-statistic model needs a clear failure mechanism and experiment data, but as to the analysis of complicated mechanical system failure, frankly, mechanism is not clearness during the degradation research, some suggestions are provided, correspondingly:

1. **Failure mechanism and failure pattern.** These two definitions are belong to the failure property of system, which are two aspects of degradation process, sometimes, failure mechanism is unknown, but the failure pattern can be observed.
   (a) Before the model set, what we suggest is to make a deep and comprehensive observation of different degradation performance patterns, further, in order to assure the relation of reliability and safety of system between failure mechanism and whole components through the physical of failure analysis, a scrutiny investigate on the variety of failure pattern may help understand the connation of each degradation stage.
   (b) Focus the performance pattern of degradation and mind the change of main character value, simultaneously, record the degradation time. Once the inductive reasons (working load, exterior stress, time, etc.) confirmed, the exclusive strategy of failure analysis can be assured.

2. **Physics statistic data.**

The mass data comes from the repeat physics trial or a long time observation during the degradation process, however, the failure mechanism and pattern of components were not constant, therefore, the physic statistic data was not suggest to use twice in different failure pattern trial. On the other hand, an efficient feedback come from the products' failure information, which should involve the general mechanism of physics failure.

Physics of Failure Model

A clear failure property and process are basis of Physics of failure model need, some guidelines may useful to this model set process:

1. Identify the potential status during the degradation process. Some potential status such as (thermal, chemical, magnetic, etc.) may influence the degradation process or failure performance. On the other side, distinguish the inducement and failure time are important.
2. Set the differential equation. Firstly, character parameters and character information value should have a representative, which can reflect the crucial point change of degradation. Especially, the amount of parameters in the complex system may increase in an exponential from, the traditional deal methods, for instance, regression analysis or linear fit method were not suitable any more, how to attain the differential equation need a new idea.

3.1   Model Evaluate Criterion and Application Result Comparison

All aforementioned models have their application fields and practical using effect, choose a proper degradation model and data process method will maximize improve the reliability analysis efficiency and save the degradation analysis time, hence, the degradation model evaluate criterion and application results can abide some rules as follows:

1. Model test criterion should comply with the information minimize rules, that is equivalent to saying, the information rules chose to evaluate the models should make the model change trajectory fit the practical degrade trace closely.
2. The fit index chose to evaluate the fit extent should reflect the degradation data change trend and make the model have a precise calculate results.
3. The degradation model should benefit to extrude the data's change cycle, besides, it is necessary to make model optimal property test before the model choose.

The practical application results comparison about these three models have appeared in some papers, for instance, paper [22, 23] use the probabilistic statistical model to predict the parameters change and products' performance degradation, compared with the specific distribution, the probabilistic statistical model can use the degradation data more fully and precisely. On the other hand, probabilistic statistic model usually give an accurate calculate results than other models, however, in the majority practical occasions, probabilistic statistic model is an empirical method, some cognitive error caused by the engineer is inevitable. When the failure mechanism clear and experiment data is abundance, physics-statistic model should be chose to set the degradation model. For example, Hummer [24] set a LMEM model by virtue of physics-statistics method, which have a better predict accurate on minimum acceptable retro-reflectivity values. Under the sophisticate degradation analysis condition, physics of failure model can provide a failure property explanation to the products and help mathematical equation search. A case in point is paper [25] make a reliability assessment by use physics of failure model, which has an obvious excellent to deal with the large degradation data.

3.2   Recommendation

Along with the new sophisticate system appeared, the degradation model set process of structure analysis fields are facing some new challenges as follow:

1 **Degradation process description of complicate structure.** The degradation depict of complicate structure is still an unsolved problem, the majority existed analysis method make a suggestion that there is single degradation pattern in system, however, this assumption was unrealistic in practical, due to loading, exterior condition, pressure, and so on, these factors be likely influence the degradation ration in a single or multiple ways, hence, the comprehensive of system degradation needs to recognize in an all-round aspects.
2 **Parameters uncertainty hypothesis.** Most failure models, in order to simplify the arithmetic process, the researchers may make a hypothesis on parameters, they ignore the parameters uncertainty and fluctuate during the degradation, indeed, the parameters uncertainty is an important aspect of degradation process and will affect the model set precise. Some non-probabilistic methods have put forward to deal this problem.
3 **Competing failure.** The majority products or system fail can attribute to different patterns, especially, in the complicate system, known as competing risk or failure are general exist in the equipment running, due to the different manufacturing process, production, material, working loading or other factors, the failure time is different. The conventional research analysis hold that the degradation ratio, failure time and failure threshold were determined by the weakness component, seemingly, it is true that the failure time or structure reliability was determined by the components which has a low quality or working ability in a crucial circumstance, however, in the recently research some conversely theory has been raised, a kind of opinions suggest that in the practical application some components' degradation ratio were more higher than others were not caused by the quality problem or product drawbacks, on the contrary, some sophisticate parts are more likely to cause the degradation or failed first due to they have a higher frequent using and the work conditions were not require the normal level, but this parts own a high processing and product quality.

The relevant research fields on competing failure include the degradation threshold, failure pattern, and so on. Recently, research showed that the same product may possess a multiple degradation pattern, which caused the finally failure to components, competing failure research has become a hot focus in the degradation fields.

4 **The issue of degradation failure pattern.** The failure patterns were divide into hard failure and soft failure according the exist theory, the relation between products failure and degradation failure may independent or not, when they are independent, the components of system can be regarded as series model, on the contrary, it is parallel model, distinguish the degradation failure pattern is a necessary work during the degradation model set process.

5 **The degradation value.** How to certain the degradation value is the evaluate criteria products failure, meanwhile, degradation value is the foundation and premise of degradation model set or reliability analysis, the current degradation value can be classified into three branches as follows:

(a) **Physical degradation value:** such as shock frequency, spectrum, character physic value, and so on.

(b) **Product structure value:** for instance, rigid, crack length, wear extent.

(c) **Mathematic value:** variety statistic value, matric vectors, for example.

What has been discussed above mentioned suggest that some other degradation index should raise attention, such as working time, chemical degradation value, or other factors which may cause the product degradation, next, the degradation research should take the multiple system status and coupling relation into consideration.

4 Conclusion

Degradation model has been become a vital tool in the structure degradation analysis and reliability improvement fields, even though the degradation theory and model have been proposed and applied for a couple of decades, the modern technique development push for a faster trend on the product and system progress, higher reliability of products are facing the latest new requirement in this field. Accurate degradation identify and suitable model set planning are significant to a successful degradation model set process.

This paper made a generally review on degradation model set method and its probability development trends, classifying the three degradation model set method and make some relative recommend to the degradation model set process.

Some feasible research directions can continue to do in the future. First of all, structure failure or degradation should not only limited in the low quality parts or components, on the other hand, its opposite side may be worth to study. Secondly, competing failure is a hot point in the degradation model set process, some unsolved problem need a deepen research. Lastly, there is a big difference between degradation and failure, the results and model set process provide by this paper is not specific, it is need a further research.

Acknowledgement. This work is supported by The Natural Science Foundation of China under grant No. 51675428.

References

1. Elsayed EA, Chen CK (1998) Recent research and current issues in accelerated testing. IEEE Int Conf Syst 5:4704–4709
2. Lu JC, Park J, Yang Q (2012) Statistics inference of a time to failure distribution derived from linear degradation data. Technometrics 39(4):391–400
3. Si XS, Wang W, Hu CH, Zhou DH, Pecht MG (2012) Remaining useful life estimation based on a nonlinear diffusion degradation process. IEEE Trans Reliab 61(1):50–67
4. Wang J, Zhang T (2008) Degradation prediction method by use of autoregressive algorithm. IEEE Int Conf Ind Technol 1–6
5. Ni X, Zhao J, Zhang X, Lv X (2014) System degradation process modelling for two-stage degraded model. Prognostics Syst Health Manag Conf 186–189
6. Peng CY, Tseng ST (2013) Statistical lifetime inference with skew-wiener linear degradation models. IEEE Trans Reliab 62(2):338–350
7. Yan M, Sun B, Li Z et al (2016) An improved time-variant reliability method for structural components based on gamma degradation process. Prognostics Syst Health Manag Conf (PHM-Chengdu) 1–6
8. Chen CY, Wu RF, Huang CY, Chung KJ (2015) Lifetime prediction of the epoxy resin adhesive under the optical performance degradation process. Microsyst Packag Assembly Circuits Technol Conf 213–216
9. Poli S, Denison M et al (2011) Physics-based analytical model for HCS degradation in STI-LDMOS Transistors. IEEE Trans Electron Devices 58(9):3072–3080
10. Xu D, Wei Q, Chen Y, Kang R (2015) Reliability prediction using physics–statistics-based degradation model. IEEE Trans Compon Packag Manuf Technol 5(11):1573–1581
11. Xu D, Sui S (2016) Bivariate degradation modeling and reliability evaluation of accelerometer based on physics-Statistics model and copula function. First Int Conf Reliab Syst Eng 1–7
12. Agbo I, Taouil M, Hamdioui S, Weckx P, Cosemans S et al (2016) Test Symposium (ETS). In: 2016 21st IEEE European
13. Hua D, Al-Khalifa KN, Hamouda AS, Elsayed EA (2013) Multi-sensor degradation data analysis. IEEE Conf Prognostics Syst Healthy Manag 31–36
14. Mei S, Raghavan N, Bosman M, Pey KL (2017) Statistical basis and physical evidence for clustering model in FINFET degradation. IEEE Int Reliab Phys Symp 3C-1.1–3C-1.6
15. Zhang Z, Xian J, Zhang C, Fu D (2017) Degradation of creatinine using boron-doped diamond electrode: statistical modelling and degradation mechanism. Chemosphere 182:441–449
16. Andrade AR, Teixeira PF (2015) Statistical modelling of railway track geometry degradation using Hierarchical Bayesian models. Reliab Eng Syst Saf 142:169–183
17. Cárdenas-Gallo I, Sarmiento CA, Morales GA, Bolivar MA, Akhavan-Tabatabaei R (2017) An ensemble classifier to predict track geometry degradation. Reliab Eng Syst Saf 161:53–60
18. Audley M, Andrews JD (2013) The effects of tamping on railway track geometry degradation. Proc Inst Mech Eng part J Rail Rapid Transit 227(4):376–391
19. Chookah M, Nuhi M, Modarresn M (2011) A probabilistic physics-of-failure model for prognostic health management of structures subject to pitting and corrosion-fatigue. Reliab Eng Syst Saf 96:1601–1610
20. Jin G, Matthews D, Fan Y, Liu Q (2013) Physics of failure-based degradation modeling and lifetime prediction of the momentum wheel in a dynamic covariate environment. Eng Fail Anal 28(3):222–240

21. Luo Y, Wang Y (2016) Chip physical failure analysis and diagnose method. Int Conf Mech Aerosp Eng 300–303
22. Li LP, Ma J, Zhao N, Zhao Z, Liu JZ (2009). Probabilistic model-based degradation diagnosing of thermal system and simulation test. Int Conf Mach Learn Cybernetics 1483–1486
23. Iyer G, Gorur RS, Krivda A, Mahonen P (2009) Statistical modeling for predicting degradation of medium voltage outdoor equipment. Annu Rep Conf Electr Insul Dielectr Phenom
24. Hummer JE, Rasdorf W, Zhang G (2011) Linear mixed-effects model for paint pavement-making retro-reflectivity data. J Transp Eng 137(10):705–716
25. Peng BH, Zhou JL (2010) Reliability assessment for product with wiener process degradation based on marker data. In: IEEE International Conference on Industrial Engineering and Engineering Management, p. 2394–2397

Hybrid Unstructured Mesh Deformation Based on Massive Parallel Processors

Hongyang Liu[1], Jiangtao Huang[1], Qing Zhong[2($\boxtimes$)], and Jing Yu[1]

[1] Computational Aerodynamics Institute,
China Aerodynamics Research and Development Center, Mianyang, China
`lhy_flow_nudt@163.com`
[2] Luoyang Electronic Equipment Test Center, Luoyang, China
`qingks@aliyun.com`

Abstract. According to radial basis functions, the greedy method and the subspace method are used to develop a deformation solver for hybrid unstructured mesh. The solver is constructed with massive parallel processors to improve the deformation efficiency of complex boundary. ONERA M6 wing of million mesh magnitudes and X48B flying wing of ten million mesh magnitudes are selected as test cases. The parallel acceleration, robustness and efficiency of the solver are validated and compared with different CPU cores and basis functions. The results indicate that the mesh quality can be guaranteed after deformation, and the deformation efficiency can be increased more than 80 times with massive parallel system.

Keywords: Massive parallel processors · Hybrid unstructured mesh ·
Mesh deformation · Radial basis functions

1 Introduction

Mesh deformation is one of the key methods in the numerical simulation of the free surface, multi-body separation, forced vibration and fluid-structure interaction. The principle is to calculate the displacements of mesh nodes based on the deformation of structural boundary without adding or removing nodes and keeping the original mesh topology unchanged [1]. Unstructured mesh is widely used in computational fluid and solid dynamics as it is adaptable to the complex configuration and easy to realize the mesh partition and parallel computation. The deformation of hybrid unstructured mesh used in viscous flow simulation is more difficult because of the complicated topology.

The existing unstructured mesh deformation methods are mainly divided into three types: physical model method, mathematical interpolation method and their mixed method [2, 3]. The physical model method compares elements such as nodes, lines, and cells in a mesh to elements in a physical model. Mesh deformation can be realized according to physical laws. The physical model method mainly includes spring method and elasticity analogy. The capability of spring method for large mesh deformation is weak. It has disadvantages of complicated data structure, large storage capacity and low deformation efficiency because it needs to clarify the connection between mesh nodes [4, 5]. Elasticity analogy has to solve large-scale equations, which leads the efficiency too low to satisfy 3D massive mesh deformation [6, 7]. The mathematical interpolation method distributes the boundary motion uniformly to the spatial nodes through interpolation, including transfinite interpolation, Delaunay graph method and radial basis function method. Transfinite interpolation is only applicable to structure mesh deformation [8]. Delaunay graph method does not need to solve large-scale equations iteratively. It is easy to implement and applicable to different mesh types. This method is dependent on the background grid strongly. It is only effective for convex domain [9, 10]. The basic idea of radial basis functions method (RBF) is to interpolate the displacements of boundary nodes and distribute the displacement effect smoothly to the entire computational mesh using the RBF sequence constructed. The dimension of equations is related to the boundary nodes number. Its amount of calculation is far less than spring method and elasticity analogy. RBF interpolation has a simple data structure and is easy to realize parallel calculation. It can be extended directly from 2D to 3D case [11–13]. These methods also can be mixed to obtain higher efficiency and stronger deformation capability [14].

Massive parallel environment has become one of the indispensable means to improve the efficiency of mesh deformation with the rapid development of computer performance [15]. According to the robustness and efficiency of RBF, hybrid unstructured mesh deformation on parallel system is realized through MPI communication mode with greedy method proposed by Rendall and Allen [16] and subspace method proposed by Wang [17]. Taking M6 airfoil and X48B flying wing as examples, the universality, robustness and parallel efficiency of greedy method and subspace method in combination with different types of basis functions are studied and compared.

2   Data Reduction Algorithm Based on RBF

The basic form of radial basis function interpolation is

$$S(\mathbf{r}) = \sum_{i=1}^{N_b} \omega_i \varphi(\|\mathbf{r} - \mathbf{r}_{bi}\|) \tag{1}$$

$S(\mathbf{r})$ is the interpolation function. ω_i is the weight coefficient of interpolation basis I. $\mathbf{r}_{bi}$ is the position of interpolation basis, that is the coordinates of a surface node. $\|\mathbf{r} - \mathbf{r}_{bi}\|$ denotes the distance between the position vector $\mathbf{r}$ and the interpolation basis $\mathbf{r}_{bi}$. N_b is the total number of interpolation basis. $\varphi(\|\mathbf{r} - \mathbf{r}_{bi}\|)$ is the general form of RBF, which has three types of commonly used functions: global, local, and compactly supported as shown in Table 1.

Table 1. Basis functions of different types

Global and local basis functions		Compactly supported basis functions	
Name	$\varphi(\eta)[= 10^{-5} \sim 10^{-3}]$	Name	$\varphi(\xi) = \varphi(\eta/d)$
TPS	$\eta^2 \log(\eta)$	CPC^0	$(1 - \xi)^2$
MQB	$\sqrt{a^2 + \eta^2}, a = 10^{-5} \sim 10^{-3}$	CPC^2	$(1 - \xi)^4(4\xi - 1)$
IMQB	$\sqrt{1/a^2 + \eta^2}, a = 10^{-5} \sim 10^{-3}$	CPC^4	$(1 - \xi)^6(35/3\xi + 6\xi + 1)$
QB	$1 + \eta^2$	CPC^6	$(1 - \xi)^8(32\xi^3 + 25\xi^2 + 8\xi + 1)$
IQB	$1/(1 + \eta^2)$	CTPS C^0	$(1 - \xi)^5$
Gauss	$e^{-\eta^2}$	CTPS C^1	$1 + 80/3\xi^2 - 40\xi^3 + 15\xi^4 - 8/3\xi^5 + 20\xi^2 \log(\xi)$

The mesh deformation with RBF can be mainly divided into two steps: (1) the RBF weight coefficient sequence is constructed according to the displacements of surface nodes; (2) the obtained RBF sequence is used to calculate the displacements of spatial nodes.

Equation (1) can be rewritten in Cartesian coordinates as followings:

$$\Delta x_j = \sum_{i=N_1}^{N_b} \omega_i^x \varphi(\|\mathbf{r}_j - \mathbf{r}_{bi}\|)$$

$$\Delta y_j = \sum_{i=N_1}^{N_b} \omega_i^y \varphi(\|\mathbf{r}_j - \mathbf{r}_{bi}\|) \qquad (2)$$

$$\Delta z_j = \sum_{i=N_1}^{N_b} \omega_i^z \varphi(\|\mathbf{r}_j - \mathbf{r}_{bi}\|)$$

The displacements of surface nodes along three coordinate directions are known. Function $\varphi()$ can be constructed based on the coordinates of the nodes. The RBF weight coefficient sequence $\mathbf{W}_X = \left\{ w_{N_1}^x, \cdots, w_{N_b}^x \right\}^{\mathrm{T}}$, $\mathbf{W}_Y = \left\{ w_{N_1}^y, \cdots, w_{N_b}^y \right\}^{\mathrm{T}}$ 和 $\mathbf{W}_Z = \left\{ w_{N_1}^z, \cdots, w_{N_b}^z \right\}^{\mathrm{T}}$ can be obtained by solving the matrix equations

$$\Delta \mathbf{X}_S = \mathbf{\Phi} \mathbf{W}_X$$
$$\Delta \mathbf{Y}_S = \mathbf{\Phi} \mathbf{W}_Y \qquad (3)$$
$$\Delta \mathbf{Z}_S = \mathbf{\Phi} \mathbf{W}_Z$$

Then construct function $\varphi()$ based on the spatial nodes and surface nodes, and the displacements of the entire computational mesh can be determined.

If all of the surface nodes are regarded as the interpolation basis, it can lead to large memory consumption and complicated calculation because of the large mesh size of 3D complex configuration. It is necessary to introduce data reduction algorithm to improve the deformation efficiency. The reduction object is the matrix Φ, and the dimensional reduction is realized on the premise of meeting the deformation precision by selecting parts of surface nodes.

The greedy method is proposed by Rendall and Allen. Firstly, the interpolation is made with p random surface nodes. Then the node with the maximum displacement error on surface is added to the set of interpolation support points. Repeat the process of interpolation until the required deformation precision is satisfied or the specified number of points is reached.

The subspace method is introduced by Wang Gang on the basis of greedy method. Firstly, the greedy method is used to select N_0 surface nodes for less dimensional interpolation. The corresponding weight coefficient sequence $\mathbf{W}^{(0)}$ of RBF and the displacement errors of all the surface nodes $\Delta \mathbf{S}^{(0)}$ can be obtained. Secondly, $\Delta \mathbf{S}^{(0)}$ is regarded as the interpolation function of the next subspace. N_1 surface nodes are selected for interpolation by the greedy method. The corresponding weight coefficient sequence $\mathbf{W}^{(1)}$ of RBF and the displacement errors of all the surface nodes $\Delta \mathbf{S}^{(1)}$ can be obtained again. Repeat the above process until the deformation precision is satisfied. Finally, the weight coefficients obtained in each step are superimposed to calculate the displacements of spatial nodes and realize the mesh deformation.

3 Parallel Computation

3.1 Mesh Partition

Mesh deformation is computed through the parallel strategy of space decomposition. The whole computational mesh is split into several subspaces, the number of which equals to that of CPU cores used. Code of the research group is applied, and its theory is the same as the METIS software [18]. Firstly, the surface is extracted from the entire mesh. Secondly, the mesh is decomposed and the subspaces are allocated to different processes for the following computation.

3.2 Computational Process

The parallel computation is implemented through the distributed storage based on the communication mode of the standard message passing interface (MPI) [19]. Taking the subspace method as an example, the computational process of mesh deformation is described as followings:

(1) Enter the mesh deformation parameters, the displacement of the boundary and the initial surface mesh. Each process reads the assigned spatial mesh separately. The mesh deformation parameters include the type of radial basis functions, the dimension of the subspace, the deformation precision and so on. After reading the information of surface mesh and displacement, the program will partition them according to the number of CPU cores and assign them to different processes.

(2) Select p surface mesh nodes randomly as the initial ones to create a subspace.

(3) The RBF interpolation is constructed to solve the weight coefficient sequence.

(4) The displacement errors of the assigned surface nodes are calculated by each process. As the maximum error and its node number can be determined in each process, the maximum error and its number among all the surface nodes can be obtained through inter-process communication. This is the process of selecting points by greedy method.

(5) Judge whether the maximum error meets the mesh deformation precision. If not, judge whether the dimension of the current subspace reaches the prescribed one. If not, add the node with maximum error obtained in step (4) to this subspace, and repeat step (3) and (4). If the prescribed dimension has been reached, regard the displacement errors of all the surface nodes in the current subspace as the inter-polation function of the next subspace.

(6) Step (2), (3) and (4) is repeated until the deformation precision is satisfied. The weight coefficients are superimposed and the supporting points selected are counted in each subspace. Each process interpolates separately to calculate the displacement change of spatial points assigned, and finally updates the new spatial mesh through inter-process communication.

If the subspace dimension is specified as 1000 and the program ends after 1000 surface mesh nodes are selected, the mesh deformation is computed based on the greedy method. If the subspace dimension is less than 1000 and multiple subspaces are created, it is based on the subspace method (Fig. 1).

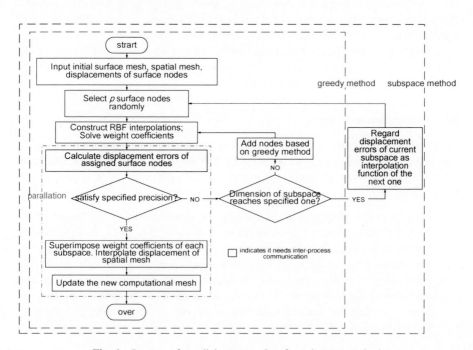

Fig. 1. Process of parallel computation for subspace method

4   Validation and Results

4.1   M6 Wing

Figure 2 is the hybrid unstructured mesh of M6 wing. It consists of 5,793,001 elements, 2,323,727 mesh nodes and 72,999 surface nodes. The displacement of the surface is described as:

$$\Delta z = 0.1y^2 \sin(4\pi y) \tag{4}$$

The mesh deformation is calculated by the greedy method and the subspace method respectively. The deformation precision is prescribed as $1.0e^{-4}$ m.

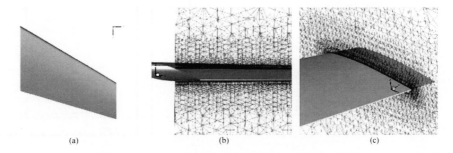

(a)   (b)   (c)

Fig. 2. Hybrid unstructured grid of M6

Figure 3 shows the order of surface nodes and spatial nodes assigned in each process during parallel computation. Table 2 lists the deformation time of the greedy method and the subspace method with different CPU numbers when choosing CPC^2 as basis function. Computation with the greedy method is converged after selecting 338 surface nodes. The subspace method converges after iterating 4 subspaces of 200-dimension. As the time of mesh deformation is mainly used to select basis nodes and update spatial nodes, the time taken for these two parts in calculation is given in detail before the middle number in Table 2. Figure 4 compares the time and parallel acceleration ratio of the greedy method and the subspace method, where the acceleration ratio is calculated based on the deformation time of one CPU core. The results show that the deformation efficiency of the greedy method is higher than that of the subspace method. Although the residual error of the greedy method oscillates slightly, the overall trend is downward as the point selected in each iteration is of the largest deformation error. The computation with the greedy method converges quickly as shown in Fig. 5. When the computation of subspace 1 ends, the residual error has dropped to point A. When the computation of subspace 2 begins, the residual jumps up to point B. The residual at the end of subspace 2 is close to that of point A, but it makes some useless efforts as 200 more surface points are selected for interpolation during the current subspace as shown in Fig. 6. For the same case, the greedy method needs 338 points to realize the residual convergence, while the subspace method needs 750 points, resulting in larger memory consumption, calculation amount and a longer time.

Table 2. Comparison of computational time for different methods with basis function CPC^2

CPU	Greedy method + CPC^2	Subspace method + CPC^2
1	40 + 54.57 = 94.57 s	43 + 120.24 = 163.24 s
8	7 + 7.70 = 14.70 s	7 + 15.40 = 22.40 s
16	5 + 3.50 = 8.50 s	3 + 8.32 = 11.32 s
64	2 + 1.74 = 3.74 s	1 + 2.25 = 3.25 s

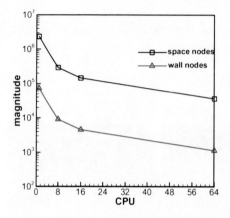

Fig. 3. Order of mesh nodes for each process

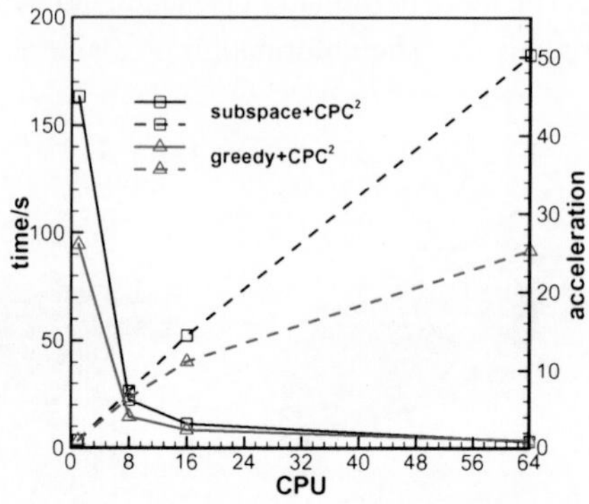

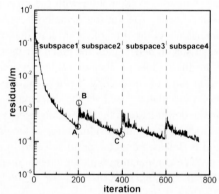

Fig. 4. Comparison of time and accelerator for different methods with basis function CPC^2

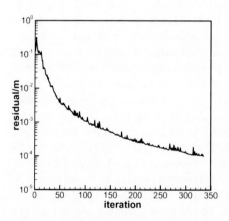

Fig. 5. Residual convergence of "greedy method + CPC^2"

Fig. 6. Residual convergence of "subspace method + CPC^2"

Table 3 lists the deformation time of the greedy method and the subspace method with different CPU numbers when choosing MQB as basis function. Computation with the greedy method stops after selecting 1000 surface nodes, and the residual drops to 2.0e-4 m. The subspace method can effectively complete the mesh deformation by iterating 21 subspaces of 200-dimension. Figure 7 shows the computational time and parallel acceleration ratio of the subspace method with function CPC^2 and MQB. Figure 8 gives out the partial residual convergence, where the residual with MQB at the end of subspace 1 is larger than $1.0e^{-3}$ m, while that with CPC^2 is less than $1.0e^{-3}$ m in Fig. 6. It shows that CPC^2 is more adaptable to large deformation. CPC^2 has higher deformation efficiency and stronger deformation capacity than function MQB. Figures 9 and 10 show the sectional view and mesh quality after deformation with different basis functions respectively. There is no negative volume in the mesh, and the distribution obtained by MQB is more uniform than that by CPC^2.

Table 3. Comparison of computational time for different methods with basis function MQB

CPU	Greedy method + MQB	Subspace method + MQB
1	>1000 s, 2e-4 m	300 + 356.71 = 656.71 s
8	>660 s, 2e-4 m	45 + 45.26 = 90.26 s
16	>600 s, 2e-4 m	26 + 21.86 = 47.86 s
64	>600 s, 2e-4 m	9 + 5.50 = 14.50 s

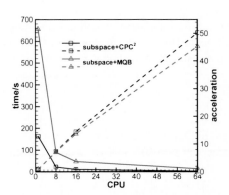

Fig. 7. Comparison of time and accelerator for subspace method with different basis function

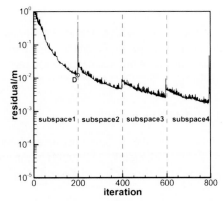

Fig. 8. Residual convergence of "subspace method + MQB"

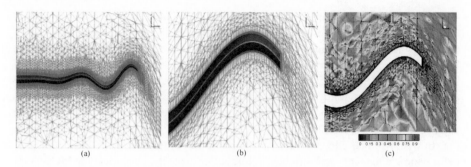

Fig. 9. Sectional view and mesh quality after deformation with basis function CPC^2

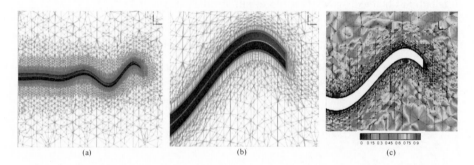

Fig. 10. Sectional view and mesh quality after deformation with basis function MQB

Above all, massive parallel environment improves the deformation efficiency significantly and the mesh quality is preserved well. The greedy method with CPC^2 is more efficient, while the deformation mesh by the subspace method with MQB is more uniform.

4.2   X48B Flying Wing

Figure 11 is the hybrid unstructured mesh of X48B flying wing. It consists of 52,129,903 elements, 24,350,801 mesh nodes and 766,508 surface nodes. The displacement of the surface is described as:

$$\text{Model A: } \Delta z = 0.1y^2 \sin(2\pi y) \tag{5}$$

$$\text{Model B: } \Delta z = 0.2 \cdot y^2 \tag{6}$$

The mesh deformation is calculated by the greedy method and the subspace method respectively. The deformation precision is prescribed as $1.0e^{-4}$ m.

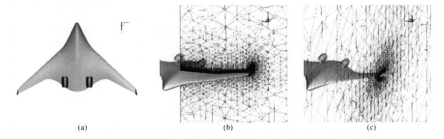

(a)   (b)   (c)

Fig. 11. Hybrid unstructured mesh of X48B

As can be seen from Sect. 4.1, when using CPC^2 as the interpolation basis function, the greedy method is preferred for mesh deformation, and when using MQB as the basis function, the subspace method should be chosen. According to the combination of "greedy method + CPC^2" and "subspace method + MQB", the deformation of X48B flying wing is implemented in two displacement modes. Figure 12 shows the order of surface nodes (from 1,000 to 1,000,000) and spatial nodes (from 100,000 to 10,000,000) assigned in each process during parallel computation.

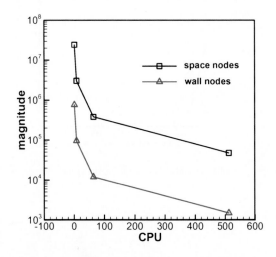

Fig. 12. Order of mesh nodes for each process during parallel computation

4.2.1   Model A

Table 4 lists the deformation time of model A with the greedy method and the subspace method under different CPU numbers. Computation with the greedy method is converged after selecting 493 surface nodes. The subspace method converges after iterating 24 subspaces of 300-dimension. Table 4 gives out the time cost by choosing surface nodes (Part 1) and updating spatial nodes (Part 2) during deformation. The parallel efficiency of Part 1 is lower than that of Part 2 with the increase of CPU cores, which is related to the total number of selected surface points and the order of magnitude of the surface mesh allocated by each process. Figure 13 compares the time and parallel acceleration ratio of the two methods. It can be seen that the parallel efficiency of greedy method is higher. Figure 14 shows the sectional view and grid quality with model A.

Table 4. Time cost by module A

CPU	Greedy method + CPC^2	Subspace method + MQB
1	609 + 1029.32 = 1638.32 s	9231 + 8505.43 = 17736.43 s
8	93 + 127.39 = 220.39 s	782 + 1033.27 = 1815.27 s
64	26 + 14.98 = 40.98 s	182 + 135.43 = 317.43 s
512	20 + 0.47 = 20.47 s	155 + 13.27 = 168.27 s

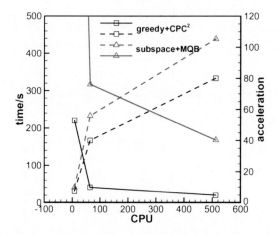

Fig. 13. Comparison of time and accelerator for different methods with module A

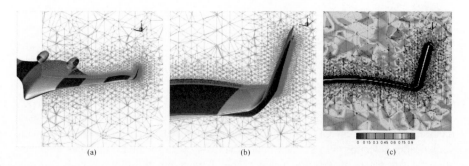

Fig. 14. Sectional view and mesh quality after deformation with module A

4.2.2   Model B

Table 5 lists the deformation time of model B with the greedy method and the subspace method under different CPU numbers. Computation with the greedy method is converged after selecting 124 surface nodes. The subspace method converges after iterating 40 subspaces of 100-dimension. Figure 15 compares the time and parallel acceleration ratio of the two methods. Figure 16 shows the sectional view and mesh quality with model B.

Table 5. Time cost by module B

CPU	Greedy method + CPC^2	Subspace method + MQB
1	44 + 208.74 = 252.74 s	1200 + 5236.62 = 6436.62 s
8	6 + 26.96 = 14.70 s	178 + 600.11 = 778.11 s
64	0 + 4.15 = 4.15 s	40 + 76.03 = 116.03 s
512	0 + 0.60 = 0.60 s	35 + 8.77 = 43.77 s

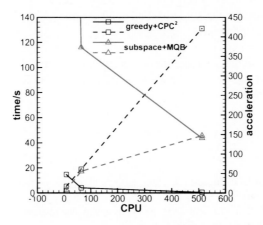

Fig. 15. Comparison of time and accelerator for different methods with module B

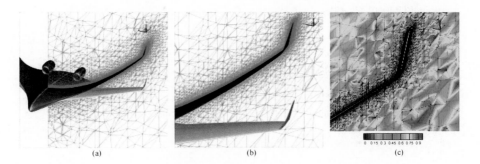

Fig. 16. Sectional view and mesh quality after deformation with module B

5 Conclusions

The hybrid unstructured mesh deformation in massive parallel environment is studied. The greedy method and subspace method are implemented in parallel calculation based on MPI communication mode. The deformation efficiency of these two methods combined with different types of basis functions is compared. The following conclusions are drawn on the premise of ensuring the deformation mesh quality:

(1) The mesh deformation efficiency of complex configuration is improved significantly by realizing parallel computation in parts of selecting surface nodes with the greedy method and interpolating the spatial grid. It also solves the problems of memory consumption and computation.

(2) The modes of mesh deformation used are all described with regular mathematical functions. The deformation efficiency of "greedy method + CPC" is superior to that of "subspace method + CPC^2" is superior to that of "subspace method + MQB" is superior to that of "greedy method + MQB", and "greedy method + MQB" cannot converge. The grid after deformation with "subspace method + MQB" is more uniform than that with "greedy method/subspace method + CPC^2". It can provide some experience for the study of fluid-solid coupling, aerodynamic optimization and free surface problems.

(3) The later research can start from two aspects to further improve the efficiency and capacity of mesh deformation: the low parallel efficiency of selecting points with greedy method and the "useless work" in the process of residual convergence with subspace method.

References

1. Schuster DH, Liu DD, Huttsell LJ (2003) Computational aeroelasticity: success, progress, challenge. J Airc 40(5):843–856
2. Zhou X, Li SX, Sun SL et al (2011) Advancers in the research on unstructured mesh deformation. Adv Mech 41(5):547–561 (in Chinese)
3. Zhang WW, Gao CQ, Ye ZY (2014) Research progress on mesh deformation method in computational aeroelasticity. Acta Aeronautica et Astronautica Sinica 35(2):303–319 (in Chinese)
4. Blom FJ (2000) Considerations on the spring analogy. Int J Numer Meth Fluids 32(6): 647–668
5. Banita JT (1990) Unsteady Euler airfoil solutions using unstructured dynamic meshes. AIAA J 28(8):1381–1388
6. Tezduyar TE (1992) Stabilized finite element formulations for incompressible flow computations. Adv Appl Mech 28(1):1–44
7. Bar-Yoseph PZ, Mereu S, Chippada S et al (2001) Automatic monitoring of element shape quality in 2D and 3D computational mesh dynamics. Comput Mech 27(5):378–395
8. Gaitonde AL, Fiddes SP (1993) A moving mesh system for the calculation of unsteady flows: AIAA-1993–0641. AIAA, Reston

9. Zhang LP, Duan XP, Chang XH et al (2009) A hybrid dynamic grid generation technique for morphing bodies based on Delaunay graph and local remeshing. Acta Aerodynamica Sinica 27(1):32–40 (in Chinese)
10. Liu XQ, Qin N, Xia H (2006) Fast dynamic grid deformation based on Delaunay graph mapping. J Comput Phys 211(2):405–423
11. Rendall TCS, Allen CB (2008) Unified fluid-structure interpolation and mesh motion using radial basis functions. Int J Numer Method Eng 74(10):1519–1559
12. Boer A, Schoot MS, Faculty HB (2007) Mesh deformation based on radial basis function interpolation. Comput Struct 85(11):784–795
13. Lin YZ, Chen B, Xu X (2012) Radial basis function interpolation in moving mesh technique. Chin J Comput Phys 29(2):191–197 (in Chinese)
14. Zhang B, Han JL (2011) Spring-TFI hybrid dynamic mesh method with rotation correction. Acta Areonautica et Astronautica Sinica 32(10):1815–1823 (in Chinese)
15. Chen GL, Sun GZ, Xun Y et al (2009) Study situation and development trend of parallel computation. Sci China Press 54(8):1043–1049 (in Chinese)
16. Rendall TCS, Allen CB (2009) Efficient mesh motion using radial basis functions with data reduction algorithms. J Comput Phys 228(5):6231–6249
17. Wang G, Lei BQ, Ye ZY (2011) An efficient deformation technique for hybrid unstructured grid using radial basis functions. J Northwest Polytechnical Univ 29(5):783–788 (in Chinese)
18. Liu W, Zhang LL, Wang YX et al (2013) Foundations of Computation Aerodynamics Parallel Programming. Beijing: National Defense Industry Press (in Chinese)
19. Snir M, Otto S, Huss-Lederman S et al (1996) MPI: the complete reference. The MIT Press, London

Inverse Airfoil Design Algorithm Based on Multi-output Least-Squares Support Vector Regression Machines

Xinqi Zhu$^{(\boxtimes)}$ and Zhenghong Gao

School of Aeronautics, Northwestern Polytechnical University,
Xi'an 710072, China
zhuxq3594@gmail.com, zgao@nwpu.edu.cn

Abstract. Inverse airfoil design algorithm can obtain the airfoil geometry according to the target pressure coefficient (Cp) distribution. Recently, the rapid development of machine learning method provides new idea to solve engineering problem. Multi-output least-squares support vector regression machines (MLS-SVR) is a multi-output regression machine learning method which can make prediction for several outputs simultaneously through learning a mapping from a multivariate input feature space to a multivariate output space. In this paper, MLS-SVR is used to learn the mapping from the Cp distribution to the geometry, which can be seen as a multi-output regression problem. Through iteratively adding the predicted airfoil geometry and its pressure coefficient distribution into the sample database, the precision of MLS-SVR to predict the right airfoil geometry corresponding to the target Cp distribution is improved. A low speed, transonic and supersonic airfoil inverse design problem are used to validate the efficiency of the proposed algorithm, and the experimental results show that the proposed algorithm can save 34.1% and 58.6% CFD evaluations for low speed and transonic cases respectively to obtain satisfactory airfoil.

Keywords: Airfoil inverse design · Support vector machines ·
Multi-output regression · Machine learning

1 Introduction

Airfoil design method can be generally categorized into direct design and inverse design methods. Airfoil direct design methods aim at obtaining the airfoil geometry that has the optimum performance index (i.e. lift-to-drag ratio) using optimization algorithm. Despite the simplicity of direct design methods, they generally requires large number of computational fluid dynamics (CFD) evaluation, which is time consuming [1–3]. Airfoil inverse design methods aim at obtaining the airfoil geometry according to a target pressure coefficient (Cp) distribution [1–9]. Generally, the target Cp distribution can be obtained through optimization, or adjusting an existing Cp distribution according to the designer's experience. Airfoil inverse design method can be used to further improve the performance of the existing airfoil, and it is popular in airfoil design.

Existing airfoil inverse design methods can be categorized into the following three categories: (1) residual-correction method which corrects the airfoil geometry according

X. Zhang (Ed.): APISAT 2018, LNEE 459, pp. 1412–1426, 2019.
https://doi.org/10.1007/978-981-13-3305-7_112

to the Cp distribution obtained through solving the transonic small perturbation integral equation [4]. However, since this method is based on the small perturbation velocity potential equation, it is restricted when applied to engineering problem; (2) airfoil inverse optimization design which uses optimization algorithm to search for the airfoil whose Cp distribution best suits the target one [5, 6, 7]. The efficiency of this method is limited by the optimization algorithm; (3) gappy proper orthogonal decomposition (POD) method which transforms solving the inverse design problem to solving a special data reconstruction problem [2, 8, 9]. However, Gappy POD is essentially a linear method, and it cannot simulate the nonlinear relationship between the airfoil geometry and the Cp distribution. When the sample airfoils differ greatly from the target one, or there exists strong shock wave in the flow field, the error of this method can be large.

Airfoil inverse design requires to learn a mapping from the Cp distribution to the geometry, in other words, to determine the mapping between two distributions. To this end, we have to choose a certain number of points on the airfoil surface, and use the pressure coefficient and the geometry position of these points to discretely describe the Cp distribution and the airfoil geometry. In order to describe the Cp distribution and the geometry precisely, we generally have to choose hundreds of surface points, making the problem a hundreds of inputs to hundreds of outputs regression problem. Moreover, the Cp of the surface point is not independent from each other, instead, there exists strong correlation among the Cp of the surface points that next to or near each other. Therefore, the mapping from the Cp distribution to the airfoil geometry is a distributed multi-output regression problem, and there exists correlation among the multiple outputs. How to deal with this problem determines the efficiency and precision of the inverse design method.

Recent years, the rapid development of the machine learning method has provided new idea to solve the engineering problems [10]. Multi-task learning refers to the machine learning method that can learn multiple tasks simultaneously [11]. Research shows that when there are multiple learning tasks, dealing with them simultaneously can obtain better results than separately [11−12]. Multi-output least-squares support vector regression machines (MLS-SVR) [13] is a multi-task learning method based on kernel method. Since MLS-SVR can consider the potential correlation among the multiple outputs (learning task), and it is a kernel based method, MLS-SVR is able to handle nonlinear multi-output regression problem.

In this paper, a new airfoil inverse design algorithm based on MLS-SVR is proposed. At first, a certain number of surface point is selected to discretely describe the Cp distribution and the geometry. Computational fluid dynamic (CFD) is used to numerically calculate the Cp distribution of sample airfoils, and these Cp distribution and the geometry make up the initial sample database. MLS-SVR is constructed using the Cp distribution as inputs and geometry as outputs based on the database. This MLS-SVR is then used to predict the airfoil corresponding to the target Cp distribution, and the Cp distribution of this predicted airfoil is evaluated using CFD. The obtained Cp distribution and geometry is are added to the sample database to further refine the precision of MLS-SVR predicting the airfoil of the target Cp distribution. By iteratively repeat these steps, satisfactory airfoil can be obtained. A low speed and a transonic airfoil inverse design cases are used to validate the performance of the proposed algorithm, and the results show the proposed algorithm can solve these problems more efficient.

2 MLS-SVR

In this section, the basic concepts of MLS-SVR are introduced. Firstly, the single-output least-squares support vector regression machines (SLS-SVR) [14] is introduced. Single-output regression problem refers to the problem which needs to predict the scalar output $y \in \mathbb{R}$ given a d dimension vector $x \in \mathbb{R}^d$. SLS-SVR utilizes a nonlinear mapping $\varphi(\cdot):\mathbb{R}^d \to \mathbb{R}^n$ to transform x to a higher (maybe infinite) dimensional feature space, and then simulate the real function using the following linear model

$$\hat{y} = w^T \varphi(x) + b \tag{1}$$

where w is the weight vector of each dimension of the variable vector $\varphi(x)$ in the feature spae, b is a constant that needs to determine. When $w = 0$, Eq. (1) uses constant b to simulate the real function, which is the simplest form. As the number of non-zero item of w increases, the linear model of Eq. (1) becomes more and more complex. Hence, we can say that w represents the complexity of the model. Because the feature space can be very high dimensional, even infinite, we can always find such a nonlinear mapping that in the feature space defined by this mapping, we can use the linear model (1) to approximate the real function well.

The prediction of SLS-SVR is based on a number of samples. Given a set of l samples $\{x^{(k)}, y^{(k)}\}_{k=1}^{l}$, where $x^{(k)} \in \mathbb{R}^d$ is the input vector, $y^{(k)} \in \mathbb{R}$ is the output scalar. SLS-SVR obtains w and b through solving the following optimization problem

$$\min_{w,b} J(w,e) = \tfrac{1}{2}w^T w + \tfrac{\gamma}{2}e^T e$$
$$\text{s.t.} \, y = Z^T w + bl_l + e \tag{2}$$

where $y = (y^{(1)}, \cdots, y^{(l)})^T$ is the vector of output of all the samples, $Z = (\varphi(x^{(1)}), \cdots, \varphi(x^{(l)}))^T$, $\mathbf{1}_l$ is a vector of 1 s and length l, $e \in \mathbb{R}^l$ is the vector of training error of all the samples. $\gamma > 0$ is the regularization parameter. In Eq. (2), the larger $w^T w$ is, the more complex the model will become, resulting in smaller training error $e^T e$ and the larger risk of over-fitting. On the other hand, smaller $w^T w$ is, the smaller the over-fitting risk will become, but the training error $e^T e$ can be larger. Therefore, in Eq. (2), γ adjusts the proportion of model complexity ($w^T w$) and the precision at the sample points ($e^T e$) in the objective function in order to decrease the risk of over-fitting without losing precision much, and SLS-SVR solves the optimization problem in Eq. (2) to obtain the model parameter.

Multi-output regression problem requires to predict the output vector $y \in \mathbb{R}^m$ given the input vector $x \in \mathbb{R}^d$. For multi-output regression problem, there always exists potential correlations among the multiple outputs. We can still use the model in Eq. (1) to predict the outputs

$$\hat{y}(i) = w_i^T \varphi(x) + b_i, i = 1, 2, \cdots, m \tag{3}$$

However, solving Eq. (3) directly equals to construct SLS-SVR separately for each output, which will result in low precision since the potential correlations are neglected.

As reference [11] points out, the most difficult part for solving multi-output regression problem is how to define the correlations among the outputs. In order to take into consideration the correlations when construct the model, MLS-SVR assumes that w_i can be written as $w_i = w_0 + v_i$, $i = 1, 2, \cdots, m$, where w_0 carries the information of the commonality of all the outputs and v_i carries the information of the specialty of each output. Therefore, v_i are "small" when the different outputs are similar to each other, otherwise the mean vector w_0 is "small". To solve w_0, v_i and $b = (b_1, b_2, \ldots, b_l)^T$, one can solve the following optimization problem

$$\min J = \frac{1}{2} w_0^T w_0 + \frac{1}{2} \frac{\lambda}{m} \sum_{i=1}^{m} v_i^T v_i + \frac{\gamma}{2} \sum_{i=1}^{m} e^{(i)T} e^{(i)}$$

$$\text{s.t.} \, Y = Z^T W + H + E \tag{4}$$

wherein $W = (w_1, w_2, \cdots, w_l)^T \in \mathbb{R}^{n \times m}$ is the weight matrix, $e^{(k)} \in \mathbb{R}^m$ is the training error vector of the kth output, $E = (e^{(1)}, e^{(2)}, \cdots, e^{(l)})^T \in \mathbb{R}^{l \times m}$ is the training error matrix, $H = (b, b, \cdots, b)^T \in \mathbb{R}^{l \times m}$, and $Y = (y^{(1)}, y^{(2)}, \cdots, y^{(l)})^T \in \mathbb{R}^{l \times m}$ is the output matrix of all the samples.

From Eq. (4), we can see that MLS-SVR balances the following three kinds of information through two positive real regularization parameters λ and γ: (1) the commonality information of all the outputs ($w_0^T w_0$); (2) the sum of all the specialty information of each outputs ($\sum v_i^T v_i$); (3) the total training error of all the samples ($\sum e^{(i)T} e^{(i)}$). This is a classical quadratic programming problem with equality constrain, and it can be solved through Lagrange method. The Lagrange function of Eq. (4) is

$$L = J - tr(A^T (Z^T W + H + E - Y)) \tag{5}$$

where $A = (\alpha_1, \alpha_2, \cdots, \alpha_m) \in \mathbb{R}^{l \times m}$ is the Lagrange coefficient matrix. According to the optimality condition we can obtain the following set of linear equations

$$\begin{cases} \frac{\partial L}{\partial w_0} = 0 \Rightarrow w_0 = \sum_{i=1}^{m} Z\alpha_i \\ \frac{\partial L}{\partial V} = 0 \Rightarrow V = \frac{m}{\lambda} ZA \\ \frac{\partial L}{\partial b} = 0 \Rightarrow A^T I_l = 0_l \\ \frac{\partial L}{\partial E} = 0 \Rightarrow A = \gamma E \\ \frac{\partial L}{\partial A} = 0 \Rightarrow Z^T W + repmat(b^T, l, 1) + E - Y = 0_{l \times m} \end{cases} \tag{6}$$

In this paper, the training algorithm in reference [15] is used to solve Eq. (6) to obtain $\boldsymbol{\alpha}_i$ and $\boldsymbol{b}$. Thereafter, given a new input vector $\boldsymbol{x}$, the corresponding outputs can be predicted through

$$
\begin{aligned}
\hat{\boldsymbol{y}}(i) = \boldsymbol{w}_i^T \varphi(\boldsymbol{x}) + \boldsymbol{b}(i) &= (\boldsymbol{w}_0 + \boldsymbol{v}_i)^T \varphi(\boldsymbol{x}) + \boldsymbol{b}(i) \\
&= \left(\sum_{j=1}^{m} \boldsymbol{\alpha}_j^T + \frac{m}{\lambda} \boldsymbol{\alpha}_i^T \right) \boldsymbol{k} + \boldsymbol{b}(i), \ i = 1, \cdots, m
\end{aligned}
\tag{7}
$$

where $\boldsymbol{k}$ is the kernel vector of $\boldsymbol{x}$ with respect to all the samples.

MLS-SVR is a kernel based method, and the relation between two input vectors is expressed through kernel function. Kernel function defines a inner product between two variable vectors, and therefore there is no need to know explicitly the form of the nonlinear mapping φ. Since MLS-SVR handles the multi-output regression problem in a higher dimensional feature space, it is capable to solve nonlinear problems [12, 13]. The popular radial basis function kernel function is used in this paper, and the hyper-parameters of the kernel function and λ, γ is determined through the model selection algorithm in reference [15].

3   Precision Test of MLS-SVR to Predict Airfoil Geometry

In this section, the Cp distribution is used as the input and the geometry as the output. The precision of MLS-SVR and three other machine learning methods SLS-SVR, Kriging [16] and Radial Basis Function network (RBF net) [17] to solve the airfoil geometry prediction problem are tested. SLS-SVR and Kriging are single-output regression method, while RBF net is multi-output regression method. The precision comparison of these four methods is used to illustrate the advantage of using MLS-SVR to predict airfoil.

3.1   Determine the Number of Surface Points

Since we can only deal with finite data in reality, we have to use some discrete data to represent the continuous Cp distribution and airfoil geometry. In this paper, n_s surface points of the airfoil is selected, and the pressure coefficient and geometry position of these points are used to represent the Cp distribution and geometry. Then, the mapping from Cp distribution to geometry becomes a multi-output regression problem of n_s dimensional input to n_s dimensional output. In order to describe the Cp distribution and geometry more precisely, the selected points should be dense on the surface where the curvature changes greatly, and where the Cp changes greatly (near the shock wave). Hence, the selected points should be dense at the leading edge of the airfoil. However, since the position of the shock wave is different for different airfoil or at different flight condition, we should select a relatively large number of surface points. Figure 1 shows the discrete Cp distribution of airfoil RAE2822 at Mach number 0.729 using different n_s surface point. It can be seen that the larger the n_s, the more precise the discrete Cp distribution is. However, on the other hand, larger n_s indicates larger computational burden when constructing the model. Therefore $n_s = 316$ is used in this paper.

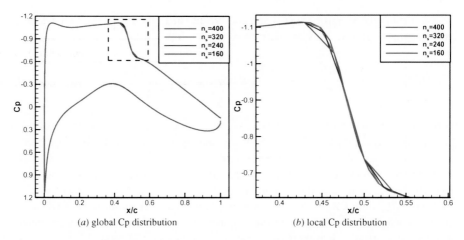

(a) global Cp distribution   (b) local Cp distribution

Fig. 1. Comparison of RAE2822 pressure coefficient distribution under different sampling number

3.2   CST Airfoil Parameterization Method

In this paper, class/function transformation (CST) parameterization method [18] is used to describe the airfoil geometry numerically. CST method describes the upper and lower surface of an airfoil through

$$y(x) = C(x)S(x) + x \cdot \Delta \tag{8}$$

where (x, y) is the non-dimensional coordinate of the points on the airfoil surface, $C(x)$ is the class function, $S(x)$ is the shape function and Δ represents the height of the trailing edge. Class function defines the overall feature of the airfoil, and the shape function determines the local character. The mathematical form of class function and shape function can be found in reference [18]. In application, other than using Eq. (8) directly, we can add the CST function to the base airfoil to obtain new airfoil geometry:

$$y(x) = y_{base}(x) + C(x)S(x) \tag{9}$$

where $y_{base}(x)$ is the coordinate of the base airfoil. In this paper, the CST method of Eq. (9) is used to parameterize the airfoil geometry.

3.3   Test Results

The NACA0012 airfoil is taken as the base airfoil, and CST with 16 variables is used to parameterize the airfoil. Latin hypercube sample method [19] is used sample 50 airfoils, and the open source CFD code CFL3D [20] is used to obtain the Cp distribution. The first 30 airfoils are used as the training samples, and the rest 20 airfoils are used as the test samples.

The mean relative error (MRE_y) and root mean squared error ($RMSE_y$) are used as the performance index. They are defined as

$$MRE_y = \frac{1}{n_{tst}} \frac{1}{n_s} \sum_{j=1}^{n_{tst}} \sum_{i=1}^{n_s} \left| \frac{y_{i,j}^{pre} - y_{i,j}^{tst}}{y_{i,j}^{tst}} \right|, \quad RMSE_y = \frac{1}{n_{tst}} \sum_{j=1}^{n_{tst}} \sqrt{\sum_{i=1}^{n_s} \left| y_{i,j}^{pre} - y_{i,j}^{tst} \right|^2} \quad (10)$$

where n_{tst} is the number of test samples, $y_{i,j}^{pre}$ and $y_{i,j}^{tst}$ are predicted and real position of the ith surface points of the jth test airfoil respectively. The smaller the two performance indexes are, the more precise the model is.

The two problems in Table 3 are tested by these four methods. Tables 1 and 2 give the performance indexes of the two problems. We can see that MLS-SVR obtained the smallest MRE_y and $RMSE_y$ for the two problems. The performance of RBF net is slightly worse than MLS-SVR, but both of these two multi-output regression method obtained better results than the two single-output regression method.

Table 1. Comparison of MRE_y for all the models

Flight condition	MRE_y (%)			
	MLSSVR	SLS-SVR	Kriging	RBFnet
Low speed	0.31	2.50	3.18	0.36
Transonic	0.85	2.94	3.55	0.98

Table 2. Comparison of $RMSE_y$ for all the models

Flight condition	$RMSE_y$			
	MLSSVR	SLS-SVR	Kriging	RBFnet
Low speed	0.0014	0.0136	0.0213	0.0015
Transonic	0.0038	0.0142	0.0237	0.0040

4 Airfoil Inverse Design Algorithm Based on MLS-SVR

From the test results of the previous section, we can conclude that the multi-output regression method is more suit able to be used to predict the airfoil distribution using the Cp distribution as the input. Therefore, we propose a new airfoil inverse design algorithm based on the multi-output regression method MLS-SVR. At first, a number of sample airfoils are selected using Latin hypercube sample method, and their Cp distributions are calculated using CFL3D. The obtained discrete airfoil geometry and Cp distribution make up the initial sample database. The database is used to train MLS-SVR, and this model is used to predict the airfoil corresponding to the target Cp distribution. The obtained airfoil is evaluated using CFL3D to obtain the Cp distribution. Then the obtained airfoil and Cp distribution are added to the sample database. The previous steps are iterated until the stopping condition is satisfied. Figure 2 gives the flow chart of the proposed algorithm. The predicted airfoil is an approximation to the target airfoil, and through adding the predicted airfoil and its Cp distribution, the precision of MLS-SVR to predict the target airfoil is improved. By iteratively improving the prediction precision, the proposed algorithm approximates the target airfoil corresponding to the target Cp distribution step by step.

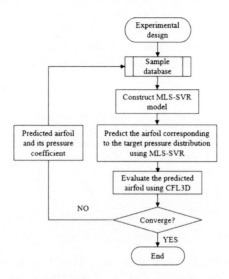

Fig. 2. The process of the airfoil inverse design algorithm based on MLS-SVR

The index to judge whether the algorithm has converged or not is e_{cp}:

$$e_{cp} = \sum_{i=1}^{n_s} \left| cp_i^{tar} - cp_i^{des} \right|^2 \tag{11}$$

where n_s is the number of the selected surface points, Cp_i^{tar} and Cp_i^{des} are the target and design pressure coefficient of the ith surface point. The smaller e_{cp} is, the better the obtained airfoil approximates the target one.

5 Experiment

In this section, three inverse design problem are tested using the proposed algorithm. The first problem is at low speed flight condition, which can be seen as a linear problem. The second problem is transonic problem, and it is a nonlinear problem since there is shock wave in the Cp distribution. The third problem is a supersonic problem. Table 3 gives the parameters of the two problems. In both cases, the parameterization method is CST with 16 variables, and the initial sample database consists 50 sample airfoils obtained through Latin Hypercube method, which is the same as the initial sample airfoil number in reference [1]. In order to validate the performance of the proposed algorithm, two other algorithms are used for comparison. The first algorithm is successive sample-approximating gappy proper orthogonal decomposition (SSAG-POD) which is proposed in reference [1] and [9]. SSAGPOD is a POD based algorithm, and it utilizes a sample-approximating strategy to reduce nonlinear errors. The second algorithm is the same as the proposed algorithm except that the utilized multi-output regression method is RBF net. The proposed algorithm is denoted as AID-MLSSVR and the third algorithm is denoted as AID-RBFnet, where AID stands for airfoil inverse design. The convergence condition is taken from reference [1], which is e_{cp} no larger than 0.0093, 0.009 and 0.002 for the low speed, transonic and supersonic case respectively.

Table 3. Parameters for the two inverse design examples

	Base foil	Target foil	Ma	Re(1e6)	State
Low speed	NACA0012	NACA4412	0.2	4.5	Cl = 0.5
Transonic	NACA0012	RAE2822	0.72	10	AoA = 2.3°
Supersonic	NACA0004	NACA64A204	1.5	30	AoA = 2°

5.1 Low Speed Case

The flight condition is Mach number Ma = 0.2, Reynolds number Re = 4.5e6 and lift coefficient Cl = 0.5. The base airfoil is NACA0012, and the target airfoil is NACA4412. The large differences between the geometry and Cp distribution of the two airfoil makes the inverse design problem difficult to solve. Figure 3 gives the geometric and pressure coefficient distribution comparison between the base and target airfoil for the low speed case. Figure 4 gives the geometric and pressure coefficient distribution comparison between the initial sample airfoils and target airfoil for the low speed case. The solid lines in Fig. 4 are the geometry and Cp distribution of the sample airfoils respectively.

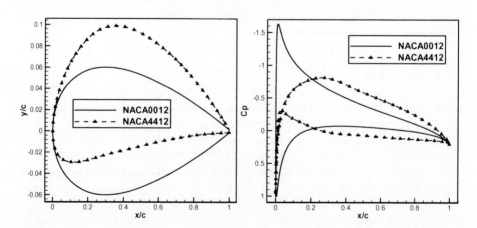

Fig. 3. Geometric and pressure coefficient distribution comparison between the base and target airfoil for the low speed case

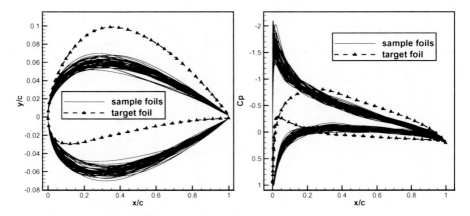

Fig. 4. Geometric and pressure coefficient distribution comparison between the initial sample airfoils and target airfoil for the low speed case

Table 4 gives the comparison of the design results of the three algorithms for the low speed inverse design case. We can see that the computation cost of AID-MLSSVR is the smallest, which needs only 54 total CFD evaluation. In other words, based on the 50 initial sample airfoils, AID-MLSSVR needs only 4 iterations before converge. On the other hand, the total CFD evaluation number for AID-RBFnet and SSAGPOD is 66 and 82 respectively. Still, we can see that the computational cost of AID-RBFnet is smaller than SSAGPOD. AID-MLSSVR and AID-RBFnet reduces the computational cost by 34.1% and 19.5% respectively compared with SSAGPOD. Figure 5 shows the comparison between the designed airfoil of AID-MLSSVR and the target airfoil for the low speed case. We can see that the obtained airfoil by AID-MLSSVR approximates the target airfoil well.

Table 4. Comparison of the design result for the low speed airfoil

Algorithm	e_{cp}	Total CFD evaluation number
SSAGPOD	0.0093	82
AID-MLSSVR	0.0088	54
AID-RBFnet	0.0080	66

5.2 Transonic Case

The transonic airfoil RAE2822 is used as the target airfoil, and the base airfoil is still NACA0012. The flight condition is Mach number Ma = 0.72, Reynolds number Re = 10e6 and angle of attack AoA = 2.3°. The target Cp distribution is the Cp distribution of RAE2822 at the corresponding flight condition. Since there exists a strong shock wave at this flight condition, this inverse design problem can be seen as a nonlinear problem. Figure 6 shows the geometric and pressure coefficient distribution comparison between the base and target airfoil for the transonic case. The two airfoils

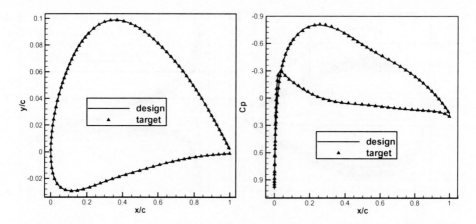

Fig. 5. Comparison between the designed airfoil of AID-MLSSVR and the target airfoil for the low speed case

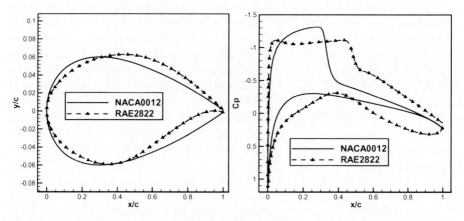

Fig. 6. Geometric and pressure coefficient distribution comparison between the base and target airfoil for the transonic case

have the same thickness, but the curvature differs from each other. There exists shock waves at the Cp distribution of these two airfoils, and the distribution have different characteristics. Figure 7 shows the geometric and pressure coefficient distribution comparison between the initial sample airfoils and target airfoil for the transonic case.

Table 5 gives the comparison of the design results of the three algorithms for the transonic inverse design case. We can see that both AID-MLSSVR and AID-RBFnet obtain better results than SSAGPOD and their computational cost is smaller. Although there exists strong nonlinear relations in this design case, AID-MLSSVR needs only 5 iterations before converge, which indicates the proposed algorithm is suitable to solve nonlinear inverse design problems. AID-MLSSVR and AID-RBFnet reduces the computational cost by 58.6% and 28.6% respectively compared with SSAGPOD.

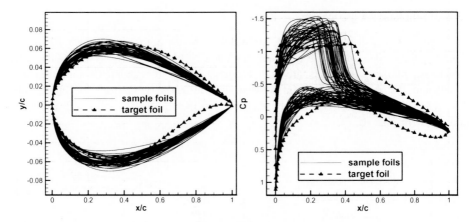

Fig. 7. Geometric and pressure coefficient distribution comparison between the initial sample airfoils and target airfoil for the transonic case

Figure 8 shows the comparison between the designed airfoil of AID-MLSSVR and the target airfoil for the transonic case. We can see that the obtained airfoil by AID-MLSSVR approximates the target airfoil well.

Table 5. Comparison of the design result for the transonic airfoil

Algorithm	e_{cp}	Total CFD evaluation number
SSAGPOD	0.0090	133
AID-MLSSVR	0.0057	55
AID-RBFnet	0.0041	95

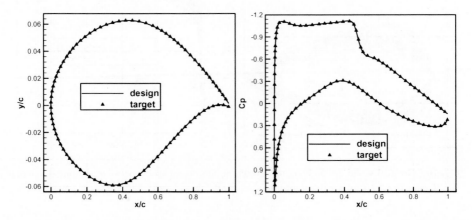

Fig. 8. Comparison between the designed airfoil of AID-MLSSVR and the target airfoil for the transonic case

5.3   Supersonic Case

The target Cp distribution is NACA64A204 at Mach number Ma = 1.5 Reynolds number Re = 30e6 and angle of attack AoA = 2°. The base foil is NACA0004. The thickness of these two foils are 4%. Figure 9 shows the geometric and pressure coefficient distribution comparison between the base and target airfoil. 50 sample foils are selected using Latin Hyper-cube method, and their geometry and Cp distribution are show in Fig. 10.

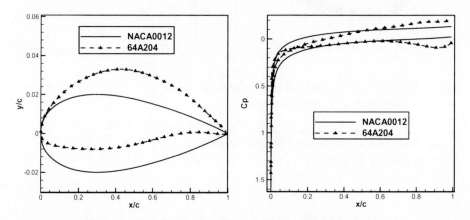

Fig. 9. Geometric and pressure coefficient distribution comparison between the base and target airfoil for the transonic case

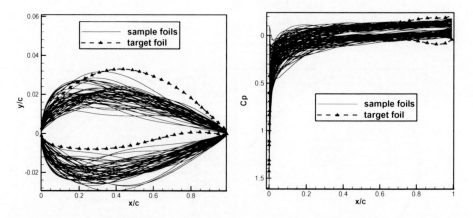

Fig. 10. Geometric and pressure coefficient distribution comparison between the initial sample airfoils and target airfoil for the transonic case

Table 6 gives the comparison of the design result for the supersonic airfoil case. Since in the reference SSAGPOD does not calculate the supersonic case, here we only compare the results of AID-MLSSVR and AID-RBRnet. We can see that to reach the same e_{cp} value, AID-MLSSVR needs only 5 iterations, while AID-RBFnet needs 27 iterations, which indicates the higher efficiency of AID-MLSSVR. Figure 11 shows the comparison between the designed airfoil of AID-MLSSVR and the target airfoil for the supersonic case. We can see that the obtained airfoil by AID-MLSSVR approximates the target airfoil well.

Table 6. Comparison of the design result for the transonic airfoil

Algorithm	e_{cp}	Total CFD evaluation number
AID-MLSSVR	0.0019	55
AID-RBFnet	0.0019	77

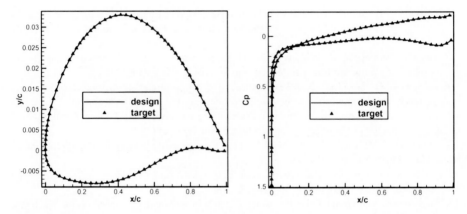

Fig. 11. Comparison between the designed airfoil of AID-MLSSVR and the target airfoil for the supersonic case.

6   Conclusion

In this paper, a new airfoil inverse design algorithm based on multi-output least-squares support vector regression machines is proposed. MLS-SVR is a multi-output regression method based on kernel method, making it capable to handle nonlinear problems, and it can predict multiple outputs simultaneously while taking into consideration the potential correlations among the outputs. The proposed algorithm utilizes sample data to construct MLS-SVR model, and this model is used to predict the airfoil geometry corresponding to the target Cp distribution. Iteratively, the predicted airfoil and its Cp distribution are added to the sample data to improve the precision of predicting the target airfoil. Experimental results show that the proposed algorithm is able to solve linear or nonlinear airfoil inverse problem, and its computational cost is the smallest.

References

1. Li S (2017) Research on POD-based inverse airfoil design methods. Northwestern Politechnical University, Xi'an (in Chinese)
2. Bai J, Qiu Y, Hua J (2013) Improved airfoil inverse design method based on Gappy POD. Acta Aeronautica et Astronautica Sinica 34(4):762–771 (in Chinese)
3. Zhan H, Zhu J, Bai J et al (2006) Improving design of airfoil with genetic algorithm (GA) Combined with inverse design method. J Northwest Polytechnical Univ 24(5):541–543
4. Takanashi S (1985) Iterative three-dimensional transonic wing design using integral equations. J Aircr 22(8):655–660
5. Leifsson L, Koziel S (2014) Inverse airfoil design using variable-resolution models and shape-preserving response prediction. Aerosp Sci Technol 39:513–522
6. Tandis E, Assareh E (2016) Inverse design of airfoils via an intelligent hybrid optimization technique. Eng Comput 1–14
7. Leifsson L, Koziel S, Ogurtsov S (2011) Inverse design of transonic airfoils using variable-resolution modeling and pressure distribution alignment. Procedia Comput Sci 4(4):1234–1243
8. Tan BT, Damodaran M, Willcox KE (2004) Aerodynamic data reconstruction and inverse design using proper orthogonal decomposition. AIAA J 42(8):1505–1516
9. Li S, Gao Z, Gao C et al (2017) A successive gappy proper orthogonal decomposition approach and its application to inverse airfoil design. AIAA Aerosp Sci Meet
10. Robert C (2012) Machine Learning, a Probabilistic Perspective. MIT Press, Cambridge
11. Caruana R (1997) Multitask learning. Mach Learn 28(1):41–75
12. Evgeniou T, Micchelli CA, Pontil M (2014) Learning multiple tasks with kernel methods. J Mach Learn Res 6(4):615–637
13. Xu S, An X, Qiao X et al (2013) Multi-output least-squares support vector regression machines. Pattern Recogn Lett 34(9):1078–1084
14. Suykens JA, Van Gestel T, De Brabanter J (2002) Least Squares Support Vector Machines, World Scientific
15. Zhu X, Gao Z (2018) An efficient gradient-based model selection algorithm for multi-output least-squares support vector regression machines. Pattern Recogn Lett 111:16–22
16. Sacks J, Welch WJ, Mitchell TJ et al (1989) Design and analysis of computer experiments. Stat Sci 4(4):409–423
17. Moody J, Darken CJ (1989) Fast learning in networks of locally-tuned processing units. Neural Comput 1(2):281–294
18. Kulfan BM (2008) Universal parametric geometry representation method. J Aircr 45(1):142–158
19. Mckay MD, Becjnag RJ, Conover WJ (1979) A comparison of three methods for selecting values of input variables in the analysis of output from a computer code. Technometrics 21(2):239–245
20. https://github.com/nasa/CFL3D

Radar Cross Section Gradient Calculation Based on Adjoint Equation of Method of Moment

Lin Zhou[1(✉)], Jiangtao Huang[2(✉)], and Zhenghong Gao[1(✉)]

[1] School of Aeronautics, Northwestern Polytechnical University,
Xi'an 710072, China
zhoulin199304@163.com, zgao@nwpu.edu.cn
[2] Computational Aerodynamics Institute,
China Aerodynamics Research and Development Center,
Mianyang 621000, China
hjtcyf@163.com

Abstract. A radar cross section gradient calculation method based on adjoint equation of method of moment (MoM) is proposed. The adjoint equation of method of moment as well as the variation of radar cross section is derived. Backscattering RCS of cylinder and missile model are firstly calculated and compared with measured data or FEKO MoM result to verify the reliability of the program adopted. Then, the cylinder and missile are parameterized and disturbed with domain element method. The gradient of design variables are computed with both adjoint method and finite-difference method. Numerical results of both methods are in good agreement proving that adjoint method has high reliability and precision. RCS gradient calculation based on adjoint equation has high efficiency, accuracy and can be applied in the calculation of complex geometry. It forms the basis of the gradient-based aerodynamic-stealth optimization platform.

Keywords: Radar cross section · Gradient · Adjoint method ·
Method of moment

1 Introduction

Radar cross section (RCS) represents the strength of echo signal of a target, and it is one of the most important criteria of stealth performance. Stealth aircraft generally adopts RCS as one major design object. The calculation of RCS in aircraft design mostly adopts high-frequency approximate methods such as geometric optics methods, physical optics, geometric theory of diffraction and physical theory of diffraction, etc. [1–4]. High-frequency approximate methods calculate surface current based on local incident field. It is efficient, economical in CPU and memory and therefore be wildly applied in engineering. However, local assumption adopted in high-frequency approximate methods may neglect important interactions between parts. And their accuracy decrease when

© Springer Nature Singapore Pte Ltd. 2019
X. Zhang (Ed.): APISAT 2018, LNEE 459, pp. 1427–1445, 2019.
https://doi.org/10.1007/978-981-13-3305-7_113

large and small length scales co-exists [5, 6]. Consequently, with the increasing demands on RCS evaluation accuracy, high-frequency approximation methods are not able to meet the requirements of low-observable aircraft design. Method of moment (MoM) proposed by Harrington [7] in 1968 is based on integral equation formulation of Maxwell's equations. MoM expresses the surface currents in terms of a set of basis functions, and calculates the unknown weight coefficients by solving linear systems of equations. It represents the solution for an exact formulation, and therefore is able to accurately calculate the RCS of complex objects. Its primary difficulty in engineering application before is its large CPU and memory consumptions. However, with the rapid development of high-performance computers this difficulty is reducing gradually, and therefore the importance of MoM is becoming more evident in engineering applications.

Aircraft stealth performance is closely related to its configuration, and its requirements on configuration are sometimes contradict with the requirements of aerodynamics [8, 9]. Therefore, stealth performance and aerodynamic performance should be balanced and considered simultaneously during configuration design. In previous studies, aerodynamic and stealth coupled design methods are generally based on intelligent algorithms, such as particle swarm optimization (PSO), genetic algorithm, and neural network algorithm [8, 10, 11], etc. Intelligent algorithms are able to converge to the global optimum theoretically, and their coding are relatively easy compared with gradient-based algorithms. However, the efficiency of intelligent algorithms is relatively low compared with gradient-based algorithms, and the computational cost increases drastically with design variables. Therefore, optimization with MoM and intelligent algorithms are extremely demanding on CPU and memory, and thus are not very suitable for engineering applications. Gradient-based optimization methods requires less RCS evaluations, and hence more efficient [12] and applicable in engineering. Its main difficulties lie on the efficient gradient calculation of design variables. Adjoint equation method is a very efficient way for gradient calculation, and it has been wildly adopted in gradient calculations in aerodynamics [13], structure [14], etc. Adjoint equation of MoM is first derived by Natalia K in 2002 [15, 16], and is adopted in input impedance optimization of Yagi-Uda array. Adjoint method based on MoM is able to obtain gradient information of all design variables by solving two linear equation systems. Its computational cost is significantly smaller than finite difference methods, and has huge potential in gradient-based aerodynamics-stealth coupled optimization.

In this paper, adjoint equation of MoM and the variation of RCS are firstly derived. Backscattering RCS of cylinder and missile model are calculated and compared with measured data or FEKO MoM results to verify the reliability of the MoM program adopted. Cylinder and missile models are parameterized and disturbed by domain element method. Gradients of design variables are computed by both adjoint method and finite difference method. Gradients calculated by both methods are in good agreement, and the reliability and precision of adjoint method is proved. Numerical results show that adjoint-based RCS gradient calculation has high efficiency and accuracy and can be applied in the calculation of complex geometry.

2   Solution of Method of Moment

In this paper, method of moment based on electric field integral equation and RWG basis function is adopted. RWG basis function is proposed by Rao in 1982 [17]. It is wildly adopted in MoM calculations due to its high efficiency and accuracy. RWG basis function adopts triangular panels which are able to apply to arbitrary surface, making it a very suitable method in engineering applications.

Let S denote the surface of an open or closed perfectly conducing scatter. An electric field E^i, defined to be the field due to the impressed source in the absence of the scatter, is incident on and incudes surface currents J on S. And the scattered electric field E^S can be computed from the surface current by

$$E^S = -j\omega A - \nabla \Phi \tag{1.1}$$

with the magnetic vector potential defined as

$$A(r) = \frac{\mu}{4\pi} \iint_S J(r') \frac{e^{-jkR}}{R} dS' \tag{1.2}$$

and the scalar potential as

$$\Phi(r) = \frac{-1}{4\pi j\omega\varepsilon} \iint_S \nabla'_S \cdot J(r') \frac{e^{-jkR}}{R} dS' \tag{1.3}$$

A harmonic time dependence $\exp(j\omega t)$ is assumed and suppressed, and $k = \omega\sqrt{\mu\varepsilon} = 2\pi/\lambda$, where λ is the wavelength. The permeability and permittivity of the surrounding medium are μ and ε, respectively, and $R = |r - r'|$ is the distance between an arbitrarily located observation point r and a source point r' on S. Both r and r' are defined with respect to a global coordinate. The surface charge density σ is related to the surface divergence of J through the equation of continuity,

$$\nabla_s \cdot J = -j\omega\sigma \tag{1.4}$$

The boundary condition of perfect conductor is

$$\hat{n} \times (E^i + E^S) = 0 \tag{1.5}$$

Combine Eqs. (1.5) and (1.1), obtaining

$$-E^i_{tan} = [-j\omega A(r) - \nabla\Phi(r)]_{tan} \quad r \in S \tag{1.6}$$

RWG basis function is defined in the two triangles attached to one edge. Figure 1 shows two such triangles T_n^+ and T_n^- corresponding to the nth edge of a triangulated surface. The vector basis function associated with the nth edge is

$$f_n(r) = \begin{cases} \frac{l_n}{2A_n^+}\rho_n^+ & r \in T_n^+ \\ \frac{l_n}{2A_n^-}\rho_n^- & r \in T_n^- \\ 0 & others \end{cases} \tag{1.7}$$

Where l_n is the length of the nth edge and $A_n^{\pm}$ is the area of triangle $T_n^{\pm}$; $\rho_n^+ = r - r_{n+}$ and $\rho_n^- = r_{n-} - r$.

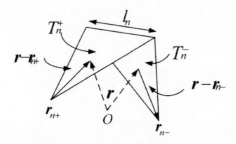

Fig. 1. RWG basis function [18]

The current on S may be approximated in terms of f_n as

$$J \approx \sum_{n=1}^{Nedge} I_n f_n(r) \tag{1.8}$$

Where $Nedge$ is the number of edges, and I_n is the weight coefficient of the nth basis function. I_n can be interpreted as the normal component of current density flowing past the nth edge. Test electric field integral equation with f_m, yielding

$$\langle E^i, f_m \rangle = j\omega \langle A, f_m \rangle + \langle \nabla \phi, f_m \rangle \tag{1.9}$$

The symmetric product is defined as

$$\langle f, g \rangle \equiv \int_S f \cdot g \, dS \tag{1.10}$$

Equation (1.9) can be written in matrix form as

$$ZI = V \tag{1.11}$$

Where Z is an *Nedge* $\times$ *Nedge* matrix; I and V are column vectors of length *Nedge*. Elements of Z_{mn} can be given by

$$Z_{mn} = l_m \left[j\omega \left(A_{mn}^+ \cdot \frac{\rho_n^{c+}}{2} + A_{mn}^- \cdot \frac{\rho_n^{c-}}{2} \right) + \Phi_{mn}^- - \Phi_{mn}^+ \right] \tag{1.12}$$

$$V_m = l_m \left(E_m^+ \cdot \frac{\rho_n^{c+}}{2} + E_m^- \cdot \frac{\rho_n^{c-}}{2} \right) \tag{1.13}$$

Where

$$A_{mn}^\pm = \frac{\mu}{4\pi} \int_S f_n(r') \frac{e^{-jkR_m^\pm}}{R_m^\pm} dS' \tag{1.14}$$

$$\Phi_{mn}^\pm = -\frac{1}{4\pi j\omega\varepsilon} \int_S \nabla_s' \cdot f_n(r') \frac{e^{-jkR_m^\pm}}{R_m^\pm} dS' \tag{1.15}$$

And $R_m^\pm = \left| r_m^{c\pm} - r' \right|$ and $E_m^\pm = E^i(r_m^{c\pm})$. The unknown weight coefficients I can be obtained by solving Eq. (1.11). And RCS can be calculated by

$$\sigma = \left| 2\sqrt{\pi} \frac{jk_0}{4\pi} \eta_0 \int_S J(r') e^{jk_s \cdot r'} ds' \cdot p_s \right|^2 \tag{1.16}$$

Where p_s is the polarization direction of scattered waves, and k_s is the scattering direction. For backscattering we have $k_s = -k_i$.

3   Adjoint Equation and Object Variation

In this section, the adjoint equation and the variation of RCS are derived. Denote design variables vector as x; objective function as $f = f(x, I)$. The objective function is decided by design variables (geometry) and induced surface current. The gradients of objective function about design variables is

$$df/dx \quad subject\ to \quad ZI = V \tag{1.17}$$

Apply the chain rule yield

$$\frac{df}{dx} = \frac{\partial f}{\partial x} + \frac{\partial f}{\partial I} \frac{\partial I}{\partial x} \tag{1.18}$$

When gradients df/dx are computed by finite difference method, each design variable x_i should be disturbed, then recalculate MoM matrix, and solve the linear system to re-evaluate objective function f. The computation cost of finite difference method is directly proportion to the number of design variables, which is highly inefficient due to the high expense of linear equation solving. Adjoint-based gradient calculation is derived in the followings.

Derive both sides of $ZI = V$ with respect to x, we have

$$\frac{\partial V}{\partial x} - \frac{\partial Z\bar{I}}{\partial x} - Z\frac{\partial I}{\partial x} = 0 \qquad (1.19)$$

Where $\bar{I}$ is the induced current before design variables are disturbed. Multiply Eq. (1.19) with adjoint variable φ and add to Eq. (1.18) yields

$$\begin{aligned}
\frac{df}{dx} &= \frac{\partial f}{\partial x} + \frac{\partial f}{\partial I}\frac{\partial I}{\partial x} + \varphi(\frac{\partial V}{\partial x} - \frac{\partial Z\bar{I}}{\partial x} - Z\frac{\partial I}{\partial x}) \\
&= \frac{\partial f}{\partial x} + \varphi\frac{\partial V}{\partial x} - \varphi\frac{\partial Z\bar{I}}{\partial x} + (\frac{\partial f}{\partial I} - \varphi Z)\frac{\partial I}{\partial x}
\end{aligned} \qquad (1.20)$$

The forth term in Eq. (1.20) includes $\partial I/dx$, and the calculation of this term needs to solver linear system. In order to avoid linear system solution, let the forth term to be zero, and the adjoint equation of method of moment is obtained.

$$\frac{\partial f}{\partial I} - \varphi Z = 0$$
$$Z^T\varphi^T = (\frac{\partial f}{\partial I})^T \qquad (1.21)$$

Gradient calculation based on adjoint equation is

$$\frac{df}{dx} = \frac{\partial f}{\partial x} + \varphi(\frac{\partial V}{\partial x} - \frac{\partial Z\bar{I}}{\partial x}) \qquad (1.22)$$

where $\partial f/\partial x$, $\partial V/\partial x$ and $\partial Z\bar{I}/\partial x$ are calculated by the original induced current (before design variable disturbance) and are calculated by finite difference methods by

$$\begin{aligned}
\frac{\partial f}{\partial x} &\approx \frac{f(x + \Delta x, \bar{I}) - f(x, \bar{I})}{\Delta x} \\
\frac{\partial V}{\partial x} &\approx \frac{V(x + \Delta x, \bar{I}) - V(x, \bar{I})}{\Delta x} \\
\frac{\partial Z\bar{I}}{\partial x} &\approx \frac{[Z(x + \Delta x) - Z(x)]\bar{I}}{\Delta x}
\end{aligned} \qquad (1.23)$$

Solve the adjoint equation Eq. (1.21) to obtain adjoint variable φ and calculate Eq. (1.22) to get gradients. Note that matrix Z based on RWG basis function is symmetric, therefore the matrix in adjoint equation is the same as in method of moment equation, and no extra memory is required to store adjoint matrix.

Let RCS be the objective function.

$$f = \sigma = \left| 2\sqrt{\pi}\frac{jk_0}{4\pi}\eta_0 \int_S \boldsymbol{J}(\boldsymbol{r'})e^{jk_s \cdot \boldsymbol{r'}}ds' \cdot \boldsymbol{p}_s \right|^2 \tag{1.24}$$

Note that σ is the square of a complex scalar, and it can be calculated by

$$f = g \cdot g^* \tag{1.25}$$

where * denotes the conjugate, and the expression of g is

$$g = 2\sqrt{\pi}\frac{jk_0}{4\pi}\eta_0 \int_S \boldsymbol{J}(\boldsymbol{r'})e^{jk_s \cdot \boldsymbol{r'}}ds' \cdot \boldsymbol{p}_s \tag{1.26}$$

Then $\partial f / \partial I_n$ can be calculated by chain rule

$$\frac{\partial f}{\partial I_n} = g^* \frac{\partial g}{\partial I_n} + g\frac{\partial g^*}{\partial I_n} \tag{1.27}$$

Note that I_n is a complex number, and the derivation has two parts, i.e.

$$\begin{aligned}\frac{\partial f}{\partial \mathrm{Re}(I_n)} &= g^* \frac{\partial g}{\partial \mathrm{Re}(I_n)} + g\frac{\partial g^*}{\partial \mathrm{Re}(I_n)} \\ \frac{\partial f}{\partial \mathrm{Im}(I_n)} &= g^* \frac{\partial g}{\partial \mathrm{Im}(I_n)} + g\frac{\partial g^*}{\partial \mathrm{Im}(I_n)}\end{aligned} \tag{1.28}$$

The expression of g in based on RWG basis functions is

$$g = 2\sqrt{\pi}\frac{jk_0}{4\pi}\eta_0 \sum_{n=1}^{Npanel} \boldsymbol{J}(\boldsymbol{r}_n^c)e^{jk_s \cdot \boldsymbol{r}_n^c}\Delta S_n \cdot \boldsymbol{p}_s \tag{1.29}$$

In which, *Npanel* is the number of triangle panels, and *Nedge* = 1.5 *Npanel* for closed surface; $\boldsymbol{r}_n^c$ is the position vector of the center of the nth panel; ΔS_n is the area of the nth panel; $\boldsymbol{J}(\boldsymbol{r}_n^c)$ is current at $\boldsymbol{r}_n^c$, and it can be expressed by RWG basis functions defined on the three edges of the nth triangle f_n^e by

$$\boldsymbol{J}(\boldsymbol{r}_n^c) = \sum_{e=1}^{3} I_n^e f_n^e(\boldsymbol{r}_n^c) \tag{1.30}$$

Substitute Eq. (1.30) into Eq. (1.29) yields

$$g = 2\sqrt{\pi} \frac{-jk_0}{4\pi} \eta_0 \sum_{n=1}^{Npanel} \sum_{e=1}^{3} I_n^e f_n^e (\boldsymbol{r}_n^c) e^{jk_s \cdot \boldsymbol{r}_n^c} \Delta S_n \cdot \boldsymbol{p}_s \tag{1.31}$$

The loop in Eq. (1.31) is about panel, while in Eq. (1.27) the derivation is about edge current I_n. Therefore, it is better to transform Eq. (1.31) into a loop about edge. Note that current on the nth edge I_n only affects two panels, i.e. T_n^+ and T_n^-. Therefore, there are two terms for each edge while edge is looped, and Eq. (1.31) can be transformed into

$$g = \boldsymbol{p}_s \cdot 2\sqrt{\pi} \frac{-jk_0}{4\pi} \eta_0 \sum_{n=1}^{Nedge} [I_n(\boldsymbol{f}_n(\boldsymbol{\rho}_n^+) e^{jk_s \cdot \boldsymbol{r}_n^{c+}} \Delta S_n^+ + \boldsymbol{f}_n(\boldsymbol{\rho}_n^-) e^{jk_s \cdot \boldsymbol{r}_n^{c-}} \Delta S_n^-)] \tag{1.32}$$

Then $\partial g / \partial I_n$ can be directly written as

$$\frac{\partial g}{\partial I_n} = \boldsymbol{p}_s \cdot 2\sqrt{\pi} \frac{-jk_0}{4\pi} \eta_0 [(\boldsymbol{f}_n(\boldsymbol{\rho}_n^+) e^{jk_s \cdot \boldsymbol{r}_n^{c+}} \Delta S_n^+ + \boldsymbol{f}_n(\boldsymbol{\rho}_n^-) e^{jk_s \cdot \boldsymbol{r}_n^{c-}} \Delta S_n^-)] \tag{1.33}$$

And substitute Eq. (1.7) into Eq. (1.33) yields

$$\frac{\partial g}{\partial I_n} = \boldsymbol{p}_s \cdot 2\sqrt{\pi} \frac{-jk_0}{4\pi} \eta_0 [I_n(\frac{\boldsymbol{\rho}_n^{c+}}{2} e^{jk_s \cdot \boldsymbol{r}_n^{c+}} + \frac{\boldsymbol{\rho}_n^{c-}}{2} e^{jk_s \cdot \boldsymbol{r}_n^{c-}})] \tag{1.34}$$

Note for backscattering ($\boldsymbol{k}_s = -\boldsymbol{k}_i$) Eq. (1.34) can be simplified into

$$\frac{\partial g}{\partial I_n} = 2\sqrt{\pi} \frac{-jk_0}{4\pi} \eta_0 V_n^* \tag{1.35}$$

Substitute Eq. (1.34) or Eq. (1.35) into Eq. (1.27) and $\partial f / \partial I_n$ is obtained.

4   Numerical Results

In this paper, linear system is solved by parallel conjugate gradient method [19–21]. Radar cross section of cylinder and missile model are calculated and compared with experiment results or MoM calculations of FEKO. Geometries are parameterized by domain element method [22], and design variables are defined as the control points of domain element method. Gradients of design variables calculated by adjoint method and finite difference method are compared.

4.1 Numerical Methods

Parallel conjugate gradient method is adopted to solve complex linear system $\overline{\overline{A}} \cdot \bar{x} = \bar{b}$. The diagram of conjugate gradient method is shown in Fig. 2. In the diagram, for two given vectors $\bar{x} = [x_1, x_2, \cdots, x_n]^T$, $\bar{y} = [x_1, x_2, \cdots, x_n]^T$, the definition of inner product is $\langle \bar{x}, \bar{y} \rangle = \sum_{i=1}^{n} (x_i y_i^*)$ and the definition of norm is $\|\bar{x}\| = \sqrt{\langle \bar{x}, \bar{x} \rangle} = \sqrt{\sum_{i=1}^{n} |x_i|^2}$. And "*", "+" represent conjugate and conjugate transpose respectively.

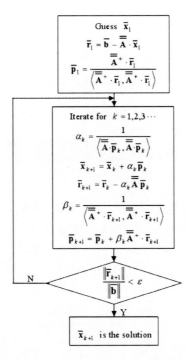

Fig. 2. Conjugate gradient method [20]

Domain element method is adopted to parameterize the geometry. Domain element method is based on radial basis functions (RBF) to construct a R^3 to R^3 mapping from domain element points to mesh points. And mesh points are moved along with domain element points through this mapping. Domain element method does not require local coordinates or connectivity information, and can therefore be applied to structured or unstructured mesh. The solution of an interpolation problem using RBFs can be written as:

$$s(r) = \sum_{i=1}^{N} \beta_i \phi(\|r - r_i\|) + p(r)$$
$$p(r) = \alpha_0 + \alpha_x x + \alpha_y y + \alpha_z z$$

(1.36)

where $s(r)$ is the approximated function; r_i is the location of the centre for RBFs (the domain-element nodes); $p(r)$ is a linear polynomial to recover translation and rotation exactly. β_i can be calculated by solving

$$\begin{cases} X_{DE} = Ca_x \\ Y_{DE} = Ca_y \\ Z_{DE} = Ca_z \end{cases} \tag{1.37}$$

where

$$X_{DE} = \begin{pmatrix} 0 \\ 0 \\ 0 \\ 0 \\ x_{DE_1} \\ \vdots \\ x_{DE_N} \end{pmatrix} \quad C = \begin{pmatrix} 0 & 0 & 0 & 0 & 1 & \cdots & 1 \\ 0 & 0 & 0 & 0 & x_{DE_1} & \cdots & x_{DE_N} \\ 0 & 0 & 0 & 0 & y_{DE_1} & \cdots & y_{DE_N} \\ 0 & 0 & 0 & 0 & z_{DE_1} & \cdots & z_{DE_N} \\ 1 & x_{DE_1} & y_{DE_1} & z_{DE_1} & \varphi_{DE_1 DE_1} & \cdots & \varphi_{DE_1 DE_N} \\ \vdots & \vdots & \vdots & \vdots & \vdots & \ddots & \vdots \\ 1 & x_{DE_N} & y_{DE_N} & z_{DE_N} & \varphi_{DE_N DE_1} & \cdots & \varphi_{DE_N DE_N} \end{pmatrix} \quad a_x = \begin{pmatrix} \alpha_0^x \\ \alpha_x^x \\ \alpha_x^x \\ \alpha_z^x \\ \beta_{DE_1}^x \\ \vdots \\ \beta_{DE_N}^x \end{pmatrix} \tag{1.38}$$

Analogous definitions are hold for Y_{DE}, Z_{DE} and for a_y and a_z, with $\varphi_{DE_1 DE_2} = \phi\|r_{DE_1} - r_{DE_2}\|$. The position of mesh points, given by vectors X_a, Y_a and Z_a are calculated by

$$\begin{cases} X_a = Aa_x = AC^{-1}X_{DE} = HX_{DE} \\ Y_a = Aa_y = AC^{-1}Y_{DE} = HY_{DE} \\ Z_a = Aa_z = AC^{-1}Z_{DE} = HZ_{DE} \end{cases} \tag{1.39}$$

where

$$A = \begin{pmatrix} 1 & x_{a_1} & y_{a_1} & z_{a_1} & \varphi_{a_1 DE_1} & \cdots & \varphi_{a_1 DE_N} \\ \vdots & \vdots & \vdots & \vdots & \vdots & \ddots & \vdots \\ 1 & x_{a_M} & y_{a_M} & z_{a_M} & \varphi_{a_M DE_1} & \cdots & \varphi_{a_M DE_N} \end{pmatrix} \tag{1.40}$$

The RBF adopted in this paper is the compact function of Wendland [23]

$$\phi(\|r\|) = \begin{cases} (1 - \|r\|)^4 (4\|r\| + 1) & \|r\| < 1.0 \\ 0 & \|r\| \geq 1.0 \end{cases} \tag{1.41}$$

Where $\|r\|$ is the Euclidean norm, and is actually calculated by $\|r\|/SR$, where SR is the support radius. In this paper, domain element method is only employed to disturb design variables, and therefore the linear polynomial $p(r)$ is neglected.

4.2   Cylinder

PEC cylinder is one of the benchmark cases in electromagnetics and is wildly adopted in numerical verification. Incident frequency is 3 GHz, and the height of the cylinder is $H = 2.76\lambda$; diameter is $D = 0.432\lambda$. Define the positive direction of z axis as 0°and the negative direction of z axis as 180°. The unstructured mesh has 730 panels (1095 edges); twenty-four design variables are employed in this test case. The mesh and lattice is shown in Fig. 3.

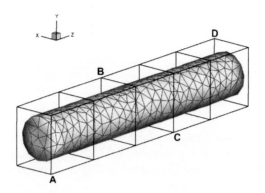

Fig. 3. Surface mesh of cylinder

Radar cross section results are shown in Figs. 4 and 5. Comparisons with measured data and FEKO MoM results prove the reliabilities of RCS evaluations.

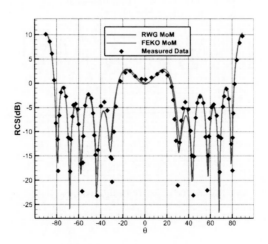

Fig. 4. Horizontal polarization

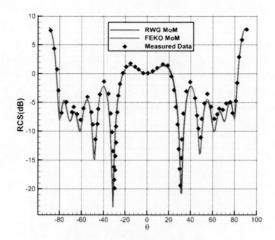

Fig. 5. Vertical polarization

Disturb the *x* coordinate, and the disturbance on design variables is $\Delta x = 1E - 4$ ($\Delta x/D = 0.036\%$); the support radius SR = 0.05. Adjoint gradient and difference gradient of variable A, B, C and D (defined in Fig. 3) at horizontal polarization are shown in Figs. 6, 7, 8 and 9. Adjoint gradients are in good agreement with finite difference results.

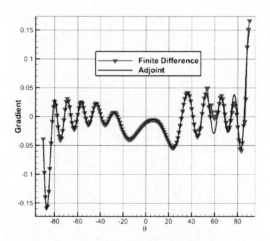

Fig. 6. Gradient of design variable A

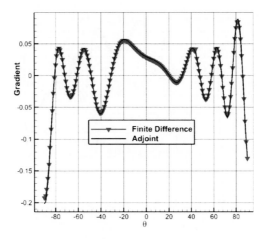

Fig. 7. Gradient of design variable B

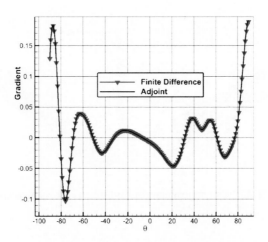

Fig. 8. Gradient of design variable C

Contour of induced current magnitude and adjoint variable magnitude ($\theta = 90°$ vertical polarization) are shown in Figs. 10 and 11. It can be seen that of surface current has very similar distribution with adjoint variable.

Time consumption in each stage is tabulated in Table 1. Though solution is fast, parallel is not very efficient for small mesh.

For adjoint gradient calculation, MoM at every incident angle is firstly calculated to get both RCS and information required in adjoint calculation, then adjoint equation at every incident angle is calculated to obtain adjoint variables, and finally adjoint gradients are computed. To obtain RCS and gradient of each design variable (24 in this case) at every incident angle (180 in this case) needs to solve 2×180 times linear systems (180 times for MoM and 180 times for adjoint) and to fill 25 impedance matrix

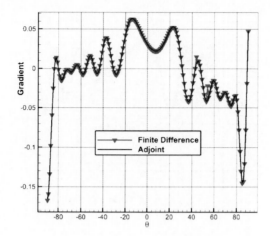

Fig. 9. Gradient of design variable D

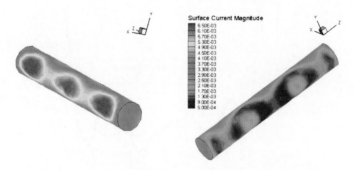

Fig. 10. Current magnitude ($\theta = 90°$ vertical polarization)

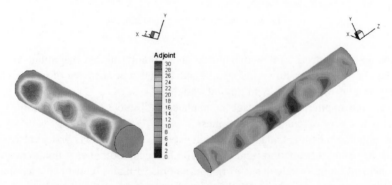

Fig. 11. Adjoint variable magnitude ($\theta = 90°$ vertical polarization)

Table 1. Time consumption

Number of CPU	Matrix filling (s)	Linear system solving (s)
1	0.59	0.767
5	0.17	0.389
10	0.13	0.289

(24 times for gradient calculation and 1 time for MoM). For finite difference calculation, 25 times MoM calculations are required in order to obtain gradient at one incident angle, and the total computational expense is 25×180 linear system solving and 25 matrix filling. The advantage of adjoint gradient calculation over finite difference is evident.

4.3   Missile Model

Missile model is complex geometry frequently met in engineer applications. It has both large and small structure, which is very challenging for high-frequency approximation methods. The length of missile model adopted in this paper is 3.99 m, diameter of base area is 0.143 m. The unstructured mesh has 17,438 panels (26,157 edges) as shown in Fig. 12 Lattice and surface mesh. The frequency of incident wave is 1 GHz. Define the negative direction of Z axis as $0°$, positive direction as $180°$. Thirty-two design variables are adopted, and the lattice and surface mesh are plotted in Fig. 12.

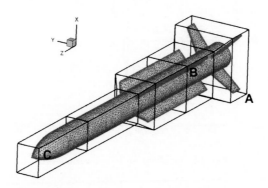

Fig. 12. Lattice and surface mesh

The horizontal polarization RCS is plotted in Fig. 13. Numerical results of in-house code are in very good agreement with FEKO MoM results, and its ability on complex geometry RCS calculation is proved. Disturb the x coordinate, and the disturbance on design variables is $\Delta x = 1E - 2$, and the support radius $SR = 0.5$ in domain element method. Adjoint gradient and finite difference gradient of design variable A, B and C (indicated in Fig. 12) are compared in Figs. 14, 15 and 16. Gradients near $90°$ (normal to z axis in y-z plane) are much higher than near $0°$ and $180°$ for all three design variables. Adjoint gradient are consistent with finite difference results, and the reliability of adjoint method in complex geometries are therefore proved.

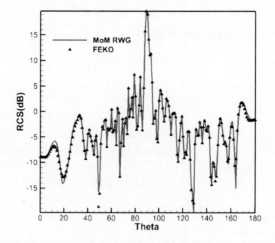

Fig. 13. Comparison of finite difference and adjoint (horizontal polarization)

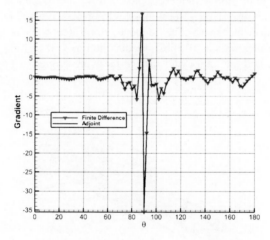

Fig. 14. Gradient of design variable A

The surface current magnitude at 0° and 90° are plotted in Figs. 17 and 18. The strength of scatter can be clearly seen in current magnitude contour, which provides helpful insight for further optimization.

Time consumption for one matrix filling and one linear system solving are tabulated in Table 2. Parallel computation proves to be very efficient for large linear system.

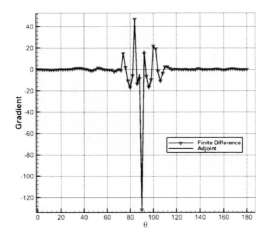

Fig. 15. Gradient of design variable B

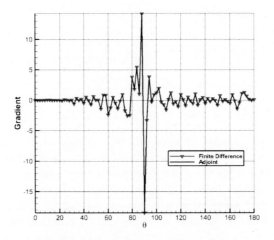

Fig. 16. Gradient of design variable C

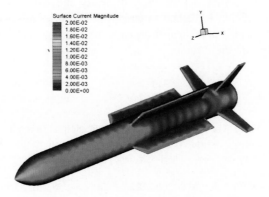

Fig. 17. Surface current magnitude ($\theta = 0°$)

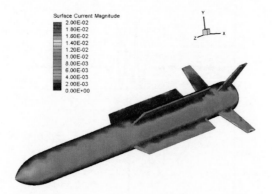

Fig. 18. Surface current magnitude ($\theta = 90°$)

Table 2. Time consumption

Number of CPU	Matrix filling (s)	Linear system solving (s)
1	357	4185
10	68	683
20	35	269

5 Conclusions

In this paper, adjoint equation of MoM and the variation of RCS are firstly derived. Backscattering RCS of cylinder and missile model are calculated and compared with measured data or FEKO MoM results. Comparison proves the reliability and precision of the program adopted. Domain element method is adopted to parameterize the geometry, and control points of domain element method are adopted as design variables. Gradients of design variables are computed by both adjoint method and finite difference method. Numerical results of both methods are consistent with each other, and the reliability and precision of adjoint method is proved. Results in this paper prove the accuracy of mathematics derivation, and clearly show the advantage of adjoint method over finite difference method in terms of efficiency. Adjoint-based RCS gradient calculation presented in this paper lays a foundation for stealth and aerodynamics coupled optimization in the future.

References

1. Aronstein DC (1996) The development and application of aircraft radar cross section prediction methodology. World Aviation Congress & Exposition, pp 1361–1374
2. Knott EF (1985) A progression of high-frequency RCS prediction techniques. Proc IEEE 73 (2):252–264
3. Knott EF, Shaeffer JF, Tuley MT (2004) Radar Cross Section, IET Digital Library

4. Ruan YZ (1998) Radar Cross Section and Stealth Technology. National Defense Industry Press, Beijing (in Chinese)
5. Gao ZH, Wang ML (2008) An efficient algorithm for calculating aircraft RCS based on the geometrical characteristics. Chin J Aeronaut 21(4):296–303
6. Li J, Liu ZH, Wang Y (2012) Research on the applicability of the approach for calculating the RCS of aircraft target, vol 1, pp 38–42 (in Chinese)
7. Harrington RF, Harrington JL (1968) Field Computation by Moment Methods. Macmillan
8. He KF, Qian WQ, Liu G (2006) Aircraft multi-objective design of aerodynamic and stealthy performance. Acta Aerodynamica Sinica 24(2):169–174
9. Zheng CY, Huang JT, Zhou Z, Liu G, Gao ZH, Xu Y (2017) Multidisciplinary optimization design of high dimensional target space for flying wing airfoil. Acta Aerodynamica Sinica 35 (4):587–597 (in Chinese)
10. Zhu JL (2011) Stealth research of armed helicopter based on RCS numerical calculation and optimization. Nanjing University of Aeronautics and Astronautics (in Chinese)
11. Wang R, Yan M, Bai P (2017) Optimization design of aerodynamics and stealth for a flying-wing UAV platform, vol 38, No 1, pp 77–84 (in Chinese)
12. Zingg DW, Nemec M, Pulliam TH (2008) A comparative evaluation of genetic and gradient-based algorithms applied to aerodynamic optimization. Européenne De Mécanique Numérique, Finis, vol. 17, No. 1–2, pp 103–126
13. Jameson A (2004) Efficient aerodynamic shape optimization. In: 10th AIAA/ISSMO Multidisciplinary Analysis and Optimization Conference, pp 2004–4369
14. Kenway GKW, Martins JRRA (2014) Multi-point high-fidelity aero structural optimization of a transport aircraft configuration. J Aircr 51(1):144–160
15. Georgieva NK, Glavic S, Bakr MH, Bandler JW (2002) Feasible adjoint sensitivity technique for EM design optimization. IEEE Trans Microw Theory Tech 50(12):2751–2758
16. Nikolova NK, Safian R, Soliman EA, Bakr MH, Bandler JW (2004) Accelerated gradient based optimization using adjoint sensitivities. IEEE Trans Antennas Propag 52(8):2147–2157
17. Rao SM, Wilton DR, Glisson AW (1982) Electromagnetic scattering by surfaces of arbitrary shape. IEEE Trans Antennas Propag 30(3):409–418
18. Zhang Y (2006) Parallel Computation in Electromagnetics. Xidian University Press, Xi'an (in Chinese)
19. Jin JM (1998) Finite Element Method in Electromagnetic Field, Xidian University Press (in Chinese)
20. Ma J (2009) The application of the method of moment and its parallel computation in EM-scattering from the rough surface and the target. Xidian University (in Chinese)
21. Guo XL, Wanf AQ, Han XB (2009) Investigation on electromagnetic scattering from complex target with parallel method of moments based on MPI of PC clusters. Aero Weaponry 5:20–25 (in Chinese)
22. Morris AM, Allen CB, Rendall TCS (2009) Domain-element method for aerodynamic shape optimization applied to modern transport wing. AIAA J 47(7):1647–1659
23. Wendland H (2005) Scattered Data Approximation. Cambridge University Press, Cambridge University Press

Large Eddy Simulation of Supersonic Open-Cavity Flows

Feng Feng[✉]

China Academy of Aerospace Aerodynamics,
Yungang West Road, No. 17, Beijing 100074, China
ffeng911@126.com

Abstract. A three-dimensional supersonic flow at Mach 1.4 over a rectangular open cavity of length-to-depth ratio $L/D = 6$, width-to-depth ratio $W/D = 2$ is studied with large eddy simulation. The numerical results are validated against the experimental data provided in the literature of Dudley and Ukeiley [1] through comparing time-averaged flow patterns and turbulent velocity in the cavity flow. The agreement is reasonable. Furthermore, we analyze the essential dynamics of the cavity flow, e.g. the resonant vortex-acoustic interactions, the dominant flow oscillation. The compressibility effects on the base flow and the pressure and velocity fluctuations are studied subsequently. Additionally, a two-dimensional open-cavity flow at the identical Mach number, Reynolds number and length-to-depth ratio is simulated with direct numerical simulation for reference. The two-dimensional result shows the vorticities are accumulated gradually in the initial stage and then switch the flow pattern eventually. Conversely, the flow in the three-dimensional open cavity could achieve a "stable" turbulence instantly and then maintain the flow pattern.

Keywords: Cavity flow · Large eddy simulation · Acoustic feedback · Supersonic

1 Introduction

Supersonic flow over open or shallow cavities is a subject of high current interest. It is mainly motivated by the aircraft applications included weapons bays, landing gear wells. For instance, the intense noise levels in open cavities on aircraft can reach levels exceeding 160 dB, which can destruct the structure of the aircraft and disturb the separation and accurate delivery of weapons. In addition, the intense radiated noise would also degrade the stealth performance of the military aircraft.

The cavity flow has been studied since the 1950s with the work of Krishanmurty [2], who also identified essential features of the cavity resonant feedback loop. Later, Rossiter [3] developed a semi-empirical formula to provide a complete description of the resonant process. Then, in order to meet the development of aircraft employing internal weapons, especially for the recent interest in the flight at supersonic conditions, a lot of experimental efforts are devoted to reduce the pressure fluctuations inside the cavity through actuation near the leading cavity edge. However, more essential dynamics for the cavity flow, e.g. the interactions among the acoustic wave, the shear

© Springer Nature Singapore Pte Ltd. 2019
X. Zhang (Ed.): APISAT 2018, LNEE 459, pp. 1446–1456, 2019.
https://doi.org/10.1007/978-981-13-3305-7_114

layer and the cavity, are still not well understood, which slow down the development of efficient control strategies.

On the other hand, computational efforts can play an important role in providing physical insight on the cavity flow physics. Especially, large eddy simulation shows great promise to reveal the underlying mechanism for the supersonic cavity flows, since it can profoundly depict the dominance of large scale vortices in the instability process. In earlier studies, Sinha et al. [4] and Smith et al. [5] respectively employed three-dimensional large eddy simulation with Smagorinsky model to compute the supersonic flow over a rectangular cavity of the same length-to-depth ratio $L/D = 4.5$, width-to-depth ratio $W/D = 4.5$, Reynolds number $Re_L = 4500000$, however, different Mach numbers $Ma = 2$ and $Ma = 1.5$. In particular, The results of Smith et al. [5] shown the peak amplitudes in the pressure spectra is in reasonable agreement with the corresponding experiments. Rizzetta and Visbal [6] also performed large eddy simulation with Smagorinsky dynamic subgrid-scale model of supersonic flow over full-span cavity of length-to-depth ratio $L/D = 5$, Reynolds number $Re_L = 200000$, $Ma = 1.19$ using a high-order numerical method and a massive grid size. The simulated pressure spectra at various positions along the cavity bottom, agrees well with the experiment results in both magnitude and trends.

During the past decades, researchers paid the most attention to suppress the oscillations through passive and active flow control strategies. However, since the baseline of cavity flow changes with different operating parameters, there is still not any general technology for noise control. Recently, researchers devote more efforts to the baseline uncontrolled flow physics again to improve the design of noise control strategies. For example, Brès and Colonius [7] analyzed the spanwise instabilities of the low-Reynolds-number, subsonic cavity flows using direct numerical simulation and global instability analysis. They found that the three-dimensional mode has a spanwise wavelength of approximately one cavity depth and oscillates with a frequency about one order of magnitude lower than two-dimensional Rossiter instabilities. More recently, Sun et al. [8] performed direct numerical simulations and large eddy simulations for finite width open-cavity flows at $Re_D = 10000$, to numerically examine the effects of cavity width, sidewall boundary conditions, free stream Mach numbers, and Reynolds numbers on open rectangular cavity flows. Nevertheless, we notice that a few three-dimensional large eddy simulations were employed to examine the characteristic of the supersonic flow over the shallow, finite-span, weapon bay-liked cavity.

In the present work, we investigate a three-dimensional supersonic flow at Mach 1.4 over a rectangular open cavity of $L/D = 6$, $W/D = 2$ with large eddy simulation to highlight the flow nature. The numerical results are validated by the companion experiments at $Re_D = 100000$. The resonant vorticity-acoustic interactions are studied in detail to explore the essential dynamics of the cavity flow. The different flow patterns in the two-dimensional and three-dimensional open cavity are presented to clarify the flow mechanism. The governing equations, the large-eddy simulation model, and the numerical method are described in Sect. 2. The results of the simulations are presented in Sect. 3. Conclusions are given in Sect. 4.

2   Numerical Model

2.1   Governing Equations and Subgrid Scale Model

Three-dimensional compressible conservative Navier-Stokes equations are filtered spatially in the LES approach, which read

$$\frac{\partial \bar{\rho}}{\partial t} + \frac{\partial \bar{\rho}\tilde{u}_i}{\partial x_i} = 0 \tag{1}$$

$$\frac{\partial \bar{\rho}\tilde{u}_i}{\partial t} + \frac{\partial \bar{\rho}\tilde{u}_i\tilde{u}_j}{\partial x_j} + \frac{\partial \bar{p}}{\partial x_i} - \frac{\partial}{\partial x_j}(\tilde{\tau}_{ij} - \tau_{ij}^{SGS} + D_{ij}^{SGS}) = 0 \tag{2}$$

$$\frac{\partial \bar{e}}{\partial t} + \frac{\partial [(\bar{e}+\bar{p})\tilde{u}_i]}{\partial x_i} - \frac{\partial}{\partial x_i}\left(\tilde{u}_j\tilde{\tau}_{ij} - \tilde{u}_j\tau_{ij}^{SGS} + \sigma_i^{SGS} - \tilde{q}_i - Q_i^{SGS} - H_i^{SGS}\right) = 0 \tag{3}$$

where $\bar{e}$ and $\hat{q}_i$ are the Favre-filtered total energy and heat flux, $\hat{\sigma}_{ij}$ and $\tilde{S}_{ij}$ are the Favre-filtered viscous stress tensor and strain rate tensor. The standard constant-coefficient Smagorinsky subgrid-scale model is implemented to close up the filtered equations. The subgrid-scale viscous stress tensor is modeled as

$$\tau_{ij}^{SGS} = \bar{\rho}\left(\widetilde{u_iu_j} - \tilde{u}_i\tilde{u}_j\right) = -2C_S\bar{\rho}\Delta^2\tilde{S}_M\left(\tilde{S}_{ij} - \frac{1}{3}\tilde{S}_{kk}\delta_{ij}\right) + \frac{2}{3}C_I\bar{\rho}\Delta^2\tilde{S}_M^2\delta_{ij} \tag{4}$$

where $\tilde{S}_M = \left(2\tilde{S}_{ij}\tilde{S}_{ij}\right)^{1/2}$. The subgrid-scale heat flux is modeled as

$$Q_i^{SGS} = c_p\left(\overline{\rho u_iT} - \bar{\rho}\tilde{u}_i\tilde{T}\right) = \frac{-C_S\bar{\rho}\Delta^2\tilde{S}_M}{Pr_t}\frac{\partial \tilde{T}}{\partial x_i} \tag{5}$$

The constants C, C_I and Pr_t are Smagorinsky model coefficients, we set $C = 0.012$, $C_I = 0.0066$, $Pr_t = 0.9$. The filtered state equation is used in the form of

$$\bar{p} = \bar{\rho}\tilde{T}/\gamma \tag{6}$$

where the ratio of specific heat $\gamma = 1.4$. Note that the length scales are nondimensionalized by the cavity depth D^*. The reference density and velocity are free-stream density ρ_∞^* and free-stream speed of sound a_∞^*.

2.2   Numerical Discretization

The nondimensional Eqs. (1–3) are discretized in generalized coordinates on orthogonal meshes with nonuniform grid spacing using a three dimensional, finite differences solver. The code has been previously used and validated for the simulation of jet flows [9].

To accurate depict the hydrodynamic and acoustic field in the strong turbulent cavity flow, the low-dispersion and low-dissipation DRP (dispersion-relation-preserving) is used to discrete the spatial derivatives [10]. Time integration is performed by the low storage optimization a six-step fourth-order accuracy Runge-Kutta scheme [11]. The selective damping and adaptive spatial filtering methods are used to assure the stability of the computation of the intense turbulent fluctuations and the shocked flow [10, 12].

On all solid boundaries, the no-slip condition $u_i = 0$, where $i = 1, 2, 3$ is imposed with $\partial p / \partial n = 0$, where n is the direction normal to the rigid surface [13]. The wall temperature is calculated by using the adiabatic condition. For the sharp corners formed by the intersection of planar cavity surfaces, the variables are determined by using the interior scheme, To avoid the ambiguity regarding the normal direction, e.g. the sharp corners, the finite-difference stencil for the convective terms is progressively reduced down to the second order. The slip boundary condition is applied along the far-field sidewalls of the computational domain. At the upstream and upper boundaries of the computational domain, the supersonic free-stream boundary conditions are prescribed. At the outlet, the characteristic non-reflection outflow boundary condition is used. In addition, a sponge zone is attached to the end of the computational domain to dissipate the vortices present in the flow field before they hit the outflow boundary.

2.3 Simulation Setup

As shown in Fig. 1, the structured mesh with 9 million points is used to resolve the flow. The length-to-depth ratio of the domain is $L/D = 2$. Its width-to-depth ratio is $W/D = 2$. The freestream Mach number is $Ma = 1.4$. The incoming laminar boundary layer thickness is $\delta = 0.167$. The Reynolds number based on the depth of the cavity is $Re_D = 10000$.

The physical domain extends from $x_1 = -3$ to 13, $x_2 = -1$ to 9, $x_3 = -2.5$ to 2.5. The nonuniform Cartesian mesh uses $199 \times 103 \times 91$ points inside the cavity and $339 \times 141 \times 183$ outside. The minimum grid size in each direction are $\Delta x_1 = 0.002$,

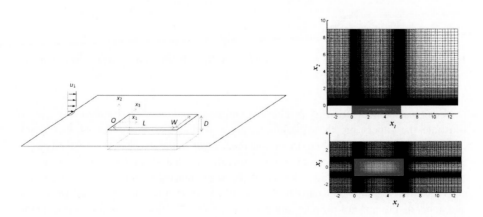

Fig. 1. Computational grid around the cavity model.

$\Delta x_2 = 0.001$, $\Delta x_3 = 0.002$. The simulation runs 5 hundred thousand iterations with a time step $\Delta t = 0.0005$, and statistical process being performed after 1 hundred thousand iterations. To speed the computation, 84 processors are employed to run the case in the CAAA workstation with the MPI parallel codes. With a CPU time of 0.5 μs per grid point per iteration, the total computation requires around 20 days.

3   Results

3.1   Instantaneous Flow Field

Figure 2 shows the instantaneous Q criterion iso-surface to depict the features of the unsteady cavity flow field. Observing from top view of the cavity flow, the shear layer rolls up into spanwise vortices after the flow passes the leading edge. The vortices lose coherence along the spanwise direction and then break down when they convect downstream. Many small turbulent vortical structures appear in the rear part of cavity because of the mixing of the free stream and the inner flows. The cavity flows become fully turbulence after the shear layer impact on the trailing edge. Due to the restriction of the sidewalls disappearing, the flow spreads out the cavity and extends in the spanwise direction when passing through the cavity.

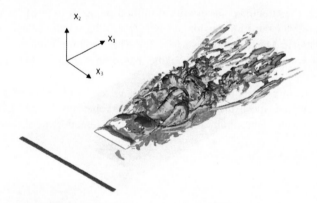

Fig. 2. The instantaneous Q criterion iso-surface (Q = 3) are shown with color contours representing streamwise velocity.

Four instantaneous density gradient magnitude contours are presented in Fig. 3 to illustrate the supersonic flow evolution. We can find that shock waves and strong compression waves are formed above the shear layer in the rear part of the cavity. The waves are induced by the large vortex structures in the shear layer obstructing the supersonic flow. Most of the vorticity of the shear layer pass through the cavity and convect downstream. Other parts of the shear layer impact on the aft wall of the cavity and generate the upward-propagating acoustic waves. The acoustic waves propagate to the forward direction inside the cavity and interfere to the initial shear layer. The shear layer roll up new spanwise vortices sustainably under the regular acoustic excitation. Then, the acoustic resonant feedback loop is formed.

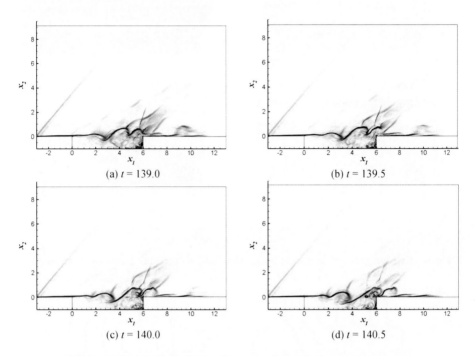

Fig. 3. Evolution of density gradient magnitude (shadowgraph) contours.

3.2   Validation

The comparison of the non-dimensional mean streamwise and turbulent velocity contours between the present LES results and the experimental and DES data published by Dudley and Ukeiley [1] are shown in Fig. 4. The mean streamwise velocity contours with vectors are qualitatively compared in Fig. 4(a) and (b). It is evident that the mean streamwise velocity in the shear layer and the flow penetration depth inside the cavity of the present LES result are in good agreement with the experimental data. The turbulent velocities from our simulation also agree well with the DES result computed from Dudley and Ukeiley [1]. It is evident that the turbulent intensity of the cavity flow gradually arrives to the maximum level along with the shear layer in the rear part of cavity.

Mean streamwise velocity profiles and mean streamwise turbulent velocities are quantitatively compared in Fig. 5(c) and (d) respectively at seven selected streamwise locations in the cavity. The simulations and experimental observations agree very well. The increased error at the cavity floor near the trailing edge may be due to the difficulties encountered in the experimental measurement close to the wall and the low resolution data measured.

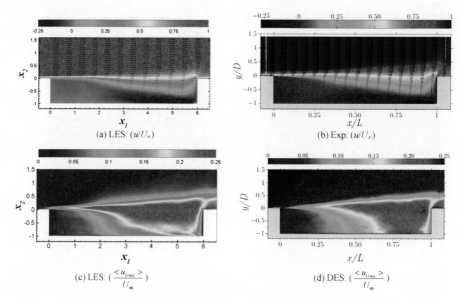

Fig. 4. Experimental and numerical mean streamwise and turbulent velocity comparison.

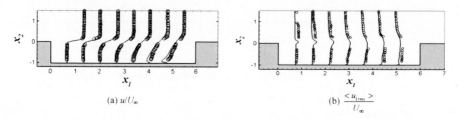

Fig. 5. Comparison of mean streamwise velocity and turbulent velocity profiles at different streamwise locations along cavity: LES (solid line), Experiment (circle).

3.3   Fluctuations in Flow Fields

The fluctuating pressure and pressure spectrums at the center points of the shear layer ($x_1 = 3$, $x_2 = 0$, $x_3 = 0$) and the cavity ($x_1 = 3$, $x_2 = 0.5$, $x_3 = 0$) are shown in Fig. 6. Multiple peaks are presented in the spectra. Two dominant frequencies are marked as f_0 and f_1. Through examining the flow field carefully, we find that the tone f_1 associating with the feedback acoustic wave and the f_0 seems like a function of the vorticity in the rear part of the cavity. The similar tones are also observed in the work of Dudley and Ukeiley [1] and Sheta et al. [15]. In addition, as the dashed line shown in the Fig. 6(b) and (d), these tones can be well predicted using the Rossiter equation [14] with constants K = 0.66 and α = 0.25.

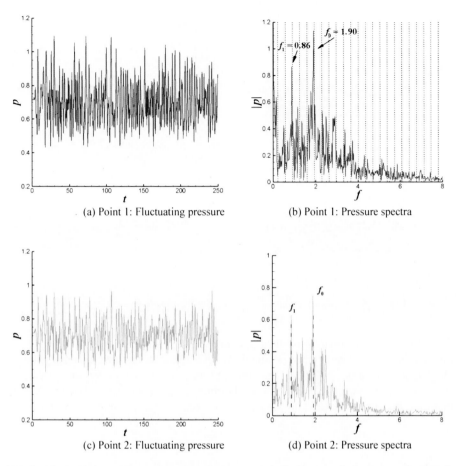

(a) Point 1: Fluctuating pressure

(b) Point 1: Pressure spectra

(c) Point 2: Fluctuating pressure

(d) Point 2: Pressure spectra

Fig. 6. Fluctuating pressure and the pressure spectra obtained at the Point 1 ($x_1 = 3$, $x_2 = 0$, $x_3 = 0$) and Point 2 ($x_1 = 3$, $x_2 = 0.5$, $x_3 = 0$) in the cavity.

3.4 Two-Dimensional Simulation

A two-dimensional simulation with the same setup of the previous three-dimensional supersonic cavity flow, except for without the spanwise dimension, is included in our previous work. As shown in Fig. 7, the vorticity is obviously accumulated in the forward part of the cavity in the initial stage. The vorticity accumulation could switch the flow pattern instantaneously and affect the fluctuating pressure in the cavity eventually. Conversely, This phenomenon is not evident for the three-dimensional cavity flow. The flow in the three-dimensional open cavity could achieves a "stable" turbulence instantly and maintain the flow pattern all the while.

As shown in Fig. 8, the two dominant frequencies corresponding to the acoustic wave and the flow mode are also observed in the two-dimensional simulation. Furthermore, the two tones f_0 and f_1 of the spectrums in the two-dimensional case are in fairly good agreement with three-dimensional simulation. Comparing to the pressure

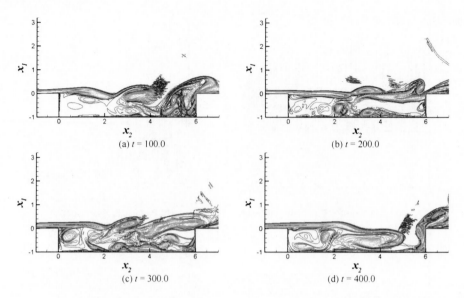

Fig. 7. Snapshots of instantaneous vorticity magnitude $|\omega|$ at four instants.

field of three-dimensional cavity flow, the level of the fluctuating pressure is higher for the two-dimensional cavity flow. Particularly, in contrast to the three-dimensional case, the amplitude of tone f_1 is higher than that of tone f_0 in the two-dimensional case. We speculate that the spanwise variation in the three-dimensional cavity flow could suppress the fluctuating pressure in the cavity and weaken the resonant feedback effects.

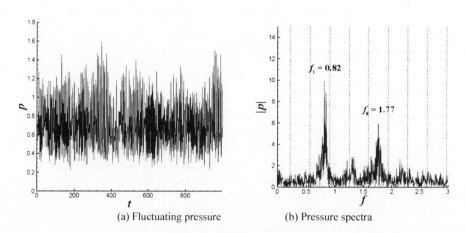

Fig. 8. Fluctuating pressure and pressure spectra at the point $(x_1 = 3, x_2 = 0)$ in the two-dimensional cavity.

4   Conclusion

Large eddy simulation method is developed to study a three-dimensional supersonic flow at Mach 1.4 over a rectangular open cavity of length-to-depth ratio $L/D = 6$, width-to-depth ratio $W/D = 2$, Reynolds number $Re_D = 10000$. The numerical result is validated against the experimental and DES data. The results show that the shear layer impact on the aft wall of the cavity and generate the upward-propagating acoustic waves. The acoustic waves propagate to the forward and interfere to the initial shear layer to generate new spanwise vortices. Then, the cavity resonant feedback loop is formed. Two dominant frequencies corresponding to the acoustic wave and the flow pattern are observed in the simulation. The two-dimensional open-cavity flow at the identical Mach number, Reynolds number and length-to-depth ratio is simulated. The two-dimensional result shows that two identical dominant frequencies are found as in the three-dimensional case. However, the amplitude of the fluctuating pressure is considerable high and the dominant frequency f_1 is predominated in the two-dimensional cavity flow.

Acknowledgments. This work was supported by the Foundation of the Equipment Development Department of China under Grant number 6140206040103 and the National Natural Science Foundation of China under Grant number 11302215.

References

1. Dudley JG, Ukeiley L (2011) Detached eddy simulation of a supersonic cavity flow with and without passive flow control. AIAA Paper 2011-3844
2. Krishnamurty K (1955) Acoustic Radiation from Two-Dimensional Rectangular Cutouts in Aerodynamic Surfaces. Technical Report CR-3487, NACA
3. Rossiter JE (1964) Wind-tunnel experiments on the flow over rectangular cavities at subsonic and transonic speeds. Technical Memoranda No. 3438, Aeronautical Research Council
4. Sinha N, Arunajatesan S, Ukeiley LS (2000) High fidelity simulation of weapons bay aeroacoustics and active flow control. AIAA Paper 2000-1968
5. Smith BR, Jordan JR, Bender EE, Rizk SN, Shaw LL (2000) Computational simulation of active control of cavity acoustics. AIAA Paper 2000-1927
6. Rizzetta DP, Visbal MR (2002) Large-eddy simulation of supersonic cavity flow fields including flow control. AIAA Paper 2002-2853
7. Bres GA, Colonius T (2008) Three-dimensional instabilities in compressible flow over open cavities. J Fluid Mech 599
8. Sun YY, Zhang Y, Taira K, Cattafesta LN, George B, Ukeiley LS (2016) Width and sidewall effects on high speed cavity flows. AIAA Paper 2016-1343
9. Feng F, Wang Q (2015) Large-scale structure evolution and sound generation in a hot jet. AIAA 2015-2534
10. Tam CKW, Webb JC (1993) Dispersion-relation-preserving finite difference schemes for computational acoustics. J Comp Phys 107(2)
11. Berland J, Bogey C, Bailly C (2004) Optimized explicit schemes: matching and boundary schemes and 4th-order Runge-Kutta algorithm. AIAA Paper 2004-2814

12. Bogey C, de Cacqueray N, Bailly C (2009) A shock-capturing methodology based on adaptative spatial filtering for high-order non-linear computations. J Comp Phys 228(5)
13. Gloerfelt X (2004) Large-eddy simulation of a high Reynolds number flow over a cavity including radiated noise. AIAA 2004-2863
14. Hughes G, Mamo T, Dala L (2009) Use of active surface waviness for control of cavity acoustics in subsonic flows. AIAA 2009-3202
15. Sheta EF, Harris RE, Luke EA, Ukeiley LS (2015) Hybrid RANS/LES acoustics prediction in supersonic weapons cavity. AIAA 2015-0009

Ranking Method for the Importance of Aircraft Damage Spare Parts

Qian Zhao[✉], Yang Pei, Peng Hou, and Chen Tian

School of Aeronautics, Northwestern Polytechnical University,
Xi'an 710072, China
zhaoqian1192@mail.nwpu.edu.cn

Abstract. Whether war wounded aircraft can be repaired quickly and effectively determines the operational capability of aircraft in the war environment. In order to ensure the combat readiness and continuous strength of the aircraft, it is necessary to scientifically determine the spare parts requirement of the war wounded aircraft. In this paper, a ranking method for the importance of combat aircraft damage spare parts is established. Firstly, the model of aircraft combat spare parts is established to determine the type of combat spare parts to be ranked. Secondly, the kill probability and survival probability of aircraft components in one mission are analyzed and determined. Thirdly, the independent existence state of aircraft and the occurrence probability and system performance of each independent state are determined. Finally, Griffith degree of importance of aircraft residual lethal parts was calculated and the order of war damage spare parts was conducted.

Keywords: War wound · Spare parts · Degree of importance · Survival · Kill

1 Introduction

Combat damage, referred to as war wounds, refers to a state in which the aircraft is damaged during operation or its function is seriously damaged due to the attack of the local weapon system. The quickest and most effective way to repair and troubleshoot a war-torn aircraft is to change parts [1]. The aircraft can be dispatched quickly and delivered with high strength by supplying sufficient spare parts quickly [2]. Determining the parts to be backed up and the reasonable amount of spare parts according to the rules of the war damage of the aircraft components can not only ensure that the aircraft has enough repair spare parts in the event of war damage, but also can ensure the smooth progress of the repair work, and avoid the waste caused by the excessive reserve of spare parts [3].

In the late 80 s, the United States Air Force adopted the method of combining computer simulation and solid missile and studied the war wound situation of different types of aircraft, the labour time and manpower of the war wound repair and the spare parts needed to solve the types and quantities of spare parts required during wartime [4]. The results of this research can be used to analyse various operational situations and combat conditions flexibly, and determine the requirements for the supply of spare parts in wartime, which is very beneficial to ensure the timely supply of spare parts.

© Springer Nature Singapore Pte Ltd. 2019
X. Zhang (Ed.): APISAT 2018, LNEE 459, pp. 1457–1468, 2019.
https://doi.org/10.1007/978-981-13-3305-7_115

After 1994, the Chinese Air Force formulated the "Research Plan of Aircraft War Damage Repair" [5]. The content includes basic theoretical research, etc. It is required to establish the prediction model of the combat damage of our military main fighter, study the distribution law of war injury and the evaluation standard of war damage and war injury, estimate the proportion of war injury and war damage under the condition of modern war, and predict the consumption of spare parts.

In recent years, many scholars have proposed performance-related importance to analyse the impact of components on system performance improvement [6]. Griffith introduces the concept of system performance in detail [7]. Wu and Chan define a new important utility function, and introduce the optimization method of the performance function of the polymorphic system [8]. Zio and other scholars use the importance degree method to decide the priority of railways, so as to improve the performance of railway network [9]. Pei investigates how importance measures in the field of reliability can be accounted for in the ranking of vulnerable components [10]. Hu Yifan of Northwestern Polytechnical University combined the threat characteristics of the anti-terrorism system, the probability of component warfare, and the type of spare parts in the "Aircraft Warfare Spare Parts Demand Model", and obtained the probability of a single-shot component war damage. The demand model of aircraft combat damage spare parts is established by using probability theory. According to the demand model, the number of combat damage spare parts is given.

In order to ensure the combat readiness and continuous strength of the aircraft, it is necessary to scientifically determine the spare parts requirement of the war wound aircraft. At present, fewer factors are considered in the given various methods for determining the demand for war wound spare parts, which leads to the use of very cumbersome algorithms to determine the demand for war wound spare parts. Therefore, this paper establishes a ranking method for the importance degree of combat aircraft, which can not only consider the middle state of the aircraft but also consider the performance of the aircraft, and the ranking results are more objective and accurate. After ranking the importance of aircraft combat damage spare parts, it helps to determine the spare parts demand of war wound aircraft more simply and quickly.

2   Ranking Model of the Importance of Aircraft Damage Spare Parts

The specific steps of the model for ranking the importance of aircraft damage spare parts proposed in this paper include: (1) Establish a model of the war wound spare parts of the aircraft, determine the type of war wound spare parts to be ranked; (2) Analyse and determine the killing probability and survival probability of the components of the aircraft in one mission; (3) Determine the independent existence state of the aircraft and the occurrence probability and system performance of each independent state; (4) Determine the Griffith importance of the redundant, critical components of the aircraft; (5) Rank the importance of war wound spare parts.

The theoretical model algorithms associated with this step are described below.

2.1 Aircraft Damage Spare Parts

Aircraft components are divided into critical components and non-critical components; critical components are further divided into redundant components and non-redundant components; among them, redundant critical components and non-critical components belong to combat spare parts to be ranked.

Assume that the number of critical components of redundancy is x, which are sequentially recorded as: $C_1, C_2, ..., C_x$; where x is an even number, and component C_1, and component C_2 are mutually redundant components, and component C_3 and component C_4 are mutually redundant components, and component C_{x-1} and component C_x are mutually redundant components.

Determine the minimum cut sets of the critical component of the redundancy: $\{\{C_1, C_2\}, \{C_3, C_4\}, ..., \{C_{x-1}, C_x\}\}$. Draw the aircraft kill tree according to the minimum cut set of the critical component of the redundancy.

Assume that the number of non-critical components is y, which are recorded as: $C_{x+1}, C_{x+2}, ..., C_{x+y}$.

2.2 Kill Probability and Survival Probability of Each Component

Determine the direction in which the aircraft is attacked according to the threat environment in which the aircraft performs its mission. Then define the initial parameters of component killing based on the direction. The initial parameters include: N-the number of hits the aircraft suffered in one mission and P_{k_i} -the probability that the i-th component of the aircraft C_i was killed ($i \in (1, 2, ..., x, ..., x+y)$.

The survival probability of each component in the aircraft after N hits is calculated using the binomial method,

$$P^N_{s_i} = \left(1 - P_{k_i}\right)^N \tag{1}$$

The probability of killing a component after the aircraft has been subjected to N hits is:

$$P^N_{k_i} = 1 - \left(1 - P_{k_i}\right)^N \tag{2}$$

2.3 The Independent Existence State of the Aircraft and the Probability of Occurrence of Each Independent State

All killing states of aircraft are determined by using permutation and combination method. Since there are a total of x critical components, the total independent state of the aircraft is 2^x. Only when the aircraft survives and the components are killed, will the war aircraft damage spare parts be considered, and the probability of each independent state of the aircraft after N hits is obtained. Among them, for x redundant critical components, there are independent existence states of which the aircraft survives and the components are killed are as follows:

(1) Only one redundant critical component is killed, assuming that only the C_v component is killed, and the other x–1 components are intact. At this time, the

probability value of the independent state of the aircraft after N hits is the product of the killing probability of the component C_v after N hits ($p_{k_v}^N$) and the survival probability of the other $x{-}1$ intact components after N hits.

(2) Only two redundant critical components are killed, and the killed parts are not complementary parts, assuming that only part C_v and component C_u are killed, and the other $x{-}2$ components are intact. At this time, the probability value of the independent state of the aircraft after N hits is: the product of the killing probability of the component C_v after N hits ($p_{k_v}^N$), the killing probability of the component C_u after N hits ($p_{k_u}^N$), and the survival probability of the other $x{-}2$ intact components after N hits.

The rest can be done in the same manner.

2.4   The Griffith Importance of the Redundant Critical Components of the Aircraft

The availability value of the i-th component (C_i) of the aircraft is 1 before the aircraft is in combat. The state space of the redundant critical component is $\{0, 1\}$, where 0 is the kill state and 1 is the survival state. The state space of the aircraft is $\{0, 1, 2, \ldots, M\}$. The Griffith importance of the i-th component (C_i) is determined by Eq. 3,

$$WI_1^G(i) = \sum_{j=1}^{M} a_j[\mathrm{Pr}_N(\Phi(1_i, X) = j) - \mathrm{Pr}_N(\Phi(0_i, X) = j)] \qquad (3)$$

Where $\mathrm{Pr}_N(\Phi(1_i, X) = j)$ represents the probability that the aircraft is the j-th independent state of existence when the i-th component (C_i) is intact, $\mathrm{Pr}_N(\Phi(0_i, X) = j)$ represents the probability that the aircraft is the j-th independent state of existence when the i-th component (C_i) is killed, X represents the component vector, $\Phi(1_i, X)$ represents the aircraft structure function, j represents the j-th independent existence state of the aircraft, a_j represents the system performance $(0 \leq a_j \leq 1)$, M represents the total number of independent states of the aircraft.

The calculation method of $\mathrm{Pr}_N(\Phi(1_i, X) = j)$ is as follows. Assume that in the j-th independent state, there are $x1$ components in intact state, and $x{-}x1$ components are in a killing state. For the components of the intact state, the survival probability after N hits is 1, and the kill probability after N hits is 0. For the components of the killing state, the survival probability after N hits is calculated by Eq. 1, and the kill probability after N hits is calculated by Eq. 2.

The calculation method of $\mathrm{Pr}_N(\Phi(0_i, X) = j)$ is as follows. Assume that in the j-th independent state, there are $x1$ components in intact state, and $x{-}x1$ components are in a killing state. For the components of the killing state, the survival probability after N hits is 0, and the kill probability after N hits is 1. For the components of the intact state, the survival probability after N hits is calculated by Eq. 1, and the kill probability after N hits is calculated by Eq. 2.

2.5   Order of Importance for the Aircraft Damage Spare Parts

Redundant critical components are arranged from high to low in order of importance, that is, ranking according to Griffith importance values from large to small. Non-critical components are arranged from high to low in order of importance, that is, ranking according to the kill survivability of each component after N hits. The non-critical components are arranged behind all redundant critical components, thereby obtaining the final sorting results of the components.

3   Computing Example and Result Analysis

Taking a certain aircraft as an example, a simplified aircraft model overview diagram is shown in Fig. 1, where B, C, D, F, G, H, I, J, K are the symbols of the aircraft components. B represents the radar, C represents the flight control computer No. 1, D represents the flight control computer No. 2, F represents the left inlet, G represents the right inlet, H represents the left fuel tank, I represents the right fuel tank, and J represents the left Engine, K represents the right engine.

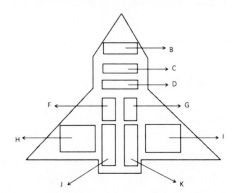

Fig. 1. Simplified aircraft model overview diagram

The aircraft components are divided into critical parts and non-critical components. B, C, D, H, I, J, and K are critical components. F and G are non-critical components. The aircraft will be destroyed once B is killed because it's non-critical component, so there is no need to consider it in order. C and D are mutually redundant components, H and I are mutually redundant components, and J and K are mutually redundant components. Identify the war damage spare parts of aircraft, that is, redundant critical components and non-critical components. The war damage spare parts are C, D, H, I, J, K, F and G.

The minimum kill set of all components causing aircraft damage is determined according to redundant critical components, which is $\{\{C, D\}, \{H, I\}, \{J, K\}\}$ in the example.

The threat environment for an aircraft to perform a task is shown in Fig. 2, where a target aircraft is fired by an anti-aircraft artillery at height of 100 m and velocity of 300 m/s. AAA is the position of 35 mm anti-aircraft artillery, l is the horizontal distance between the aircraft and the artillery, a is the offset distance of the aircraft relative to the artillery, β is the complementary angle between the direction of the aircraft being attacked and the aircraft's forward direction.

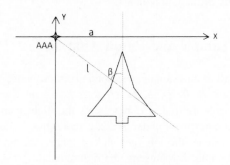

Fig. 2. The threat environment for an aircraft

The P_{k_i} value of each component is shown in Table 1.

Table 1. P_{k_i} values of each component

Component	C	D	F	G	H	I	J	K
P_{k_i}	0.0000695	0.0000695	0.00008	0.0000735	0.0001475	0.00013	0.0001063	0.0000875

The number of hits that the given aircraft has suffered in this mission is 20, that is, N = 20. The kill probability and survival probability of each component after N hits are calculated and shown in Table 2.

Table 2. Values of $p_{k_i}^N$ and $p_{s_i}^N$ of each component

Component ID	Value	
	Kill	Survival
C	0.00138908	0.99861092
D	0.00138908	0.99861092
F	0.00159878	0.99840122
G	0.00146897	0.99853103
H	0.00294587	0.99705413
I	0.00259679	0.99740321
J	0.00212286	0.99787714
K	0.00174855	0.99825145

In this example, the aircraft has eight combat spare parts, of which there are six critical components. For the six critical components, the total independent existence state of aircraft is $2^6 = 64$.

The state of a component killed, two components killed, three components killed and more than three components killed is analyzed separately. The state of aircraft killing is merged. Finally, the state of aircraft independent of combat damage spare parts is 18, that is, $j = 1, 2, \ldots 18$. Since the P_{k_i} values of the components C and D are the same, only the part C needs to be considered in the analysis, and the calculation method of the part D is the same as C.

The probability values of the state in which each of the aircraft independently exists after N hits are given in Table 3, and the a_j value of each state of the system is determined as shown in Table 3.

Table 3. Probability values of each state

State ID	Probability values of each state		a_j
I	Only C is killed: $C \cap \overline{D} \cap \overline{H} \cap \overline{I} \cap \overline{J} \cap \overline{K}$	$P_{K_I}^N = p_{k_1}^N p_{s_2}^N p_{s_5}^N p_{s_6}^N p_{s_7}^N p_{s_8}^N$	0.8
II	Only H is killed: $\overline{C} \cap \overline{D} \cap H \cap \overline{I} \cap \overline{J} \cap \overline{K}$	$P_{K_{II}}^N = p_{k_5}^N p_{s_1}^N p_{s_2}^N p_{s_6}^N p_{s_7}^N p_{s_8}^N$	0.75
III	Only I is killed: $\overline{C} \cap \overline{D} \cap \overline{H} \cap I \cap \overline{J} \cap \overline{K}$	$P_{K_{III}}^N = p_{k_6}^N p_{s_1}^N p_{s_2}^N p_{s_5}^N p_{s_7}^N p_{s_8}^N$	0.75
IV	Only J is killed: $\overline{C} \cap \overline{D} \cap \overline{H} \cap \overline{I} \cap J \cap \overline{K}$	$P_{K_{IV}}^N = p_{k_7}^N p_{s_1}^N p_{s_2}^N p_{s_5}^N p_{s_6}^N p_{s_8}^N$	0.6
V	Only K is killed: $\overline{C} \cap \overline{D} \cap \overline{H} \cap \overline{I} \cap \overline{J} \cap K$	$P_{K_V}^N = p_{k_8}^N p_{s_1}^N p_{s_2}^N p_{s_5}^N p_{s_6}^N p_{s_7}^N$	0.6
VI	Only C and H are killed: $C \cap \overline{D} \cap H \cap \overline{I} \cap \overline{J} \cap \overline{K}$	$P_{K_{VI}}^N = p_{k_1}^N p_{k_5}^N p_{s_2}^N p_{s_6}^N p_{s_7}^N p_{s_8}^N$	0.3
VII	Only C and I are killed: $C \cap \overline{D} \cap \overline{H} \cap I \cap \overline{J} \cap \overline{K}$	$P_{K_{VII}}^N = p_{k_1}^N p_{k_6}^N p_{s_2}^N p_{s_5}^N p_{s_7}^N p_{s_8}^N$	0.3
VIII	Only C and J are killed: $C \cap \overline{D} \cap \overline{H} \cap \overline{I} \cap J \cap \overline{K}$	$P_{K_{VIII}}^N = p_{k_1}^N p_{k_7}^N p_{s_2}^N p_{s_5}^N p_{s_6}^N p_{s_8}^N$	0.3
IX	Only C and K are killed: $C \cap \overline{D} \cap \overline{H} \cap \overline{I} \cap J \cap K$	$P_{K_{IX}}^N = p_{k_1}^N p_{k_8}^N p_{s_2}^N p_{s_5}^N p_{s_6}^N p_{s_7}^N$	0.3
X	Only H and J are killed: $\overline{C} \cap \overline{D} \cap H \cap \overline{I} \cap J \cap \overline{K}$	$P_{K_X}^N = p_{k_5}^N p_{k_7}^N p_{s_1}^N p_{s_2}^N p_{s_6}^N p_{s_8}^N$	0.25
XI	Only H and K are killed: $\overline{C} \cap \overline{D} \cap H \cap \overline{I} \cap \overline{J} \cap K$	$P_{K_{XI}}^N = p_{k_5}^N p_{k_8}^N p_{s_1}^N p_{s_2}^N p_{s_6}^N p_{s_7}^N$	0.25
XII	Only I and J are killed: $\overline{C} \cap \overline{D} \cap \overline{H} \cap I \cap J \cap \overline{K}$	$P_{K_{XII}}^N = p_{k_6}^N p_{k_7}^N p_{s_1}^N p_{s_2}^N p_{s_5}^N p_{s_8}^N$	0.25
XIII	Only I and K are killed: $\overline{C} \cap \overline{D} \cap \overline{H} \cap I \cap \overline{J} \cap K$	$P_{K_{XIII}}^N = p_{k_6}^N p_{k_8}^N p_{s_1}^N p_{s_2}^N p_{s_5}^N p_{s_7}^N$	0.25
XIV	Only C, H and J are killed: $C \cap \overline{D} \cap H \cap \overline{I} \cap J \cap \overline{K}$	$P_{K_{XIV}}^N = p_{k_1}^N p_{k_5}^N p_{k_7}^N p_{s_2}^N p_{s_6}^N p_{s_8}^N$	0.18
XV	Only C, H and K are killed: $C \cap \overline{D} \cap H \cap \overline{I} \cap \overline{J} \cap K$	$P_{K_{XV}}^N = p_{k_1}^N p_{k_5}^N p_{k_8}^N p_{s_2}^N p_{s_6}^N p_{s_7}^N$	0.16
XVI	Only C, I and J are killed: $C \cap \overline{D} \cap \overline{H} \cap I \cap J \cap \overline{K}$	$P_{K_{XVI}}^N = p_{k_1}^N p_{k_6}^N p_{k_7}^N p_{s_2}^N p_{s_5}^N p_{s_8}^N$	0.12
XVII	Only C, I and K are killed: $C \cap \overline{D} \cap \overline{H} \cap I \cap \overline{J} \cap K$	$P_{K_{XVII}}^N = p_{k_1}^N p_{k_6}^N p_{k_8}^N p_{s_2}^N p_{s_5}^N p_{s_7}^N$	0.1
XVIII	No component is killed: $\overline{C} \cap \overline{D} \cap \overline{H} \cap \overline{I} \cap \overline{J} \cap \overline{K}$	$P_{K_{XVIII}}^N = p_{s_1}^N p_{s_2}^N p_{s_5}^N p_{s_6}^N p_{s_7}^N p_{s_8}^N$	1

For component C, $\mathrm{Pr}_N(\Phi(1_1, X) = j)$ is the probability that each state exists when C is intact. In this state, $P_{k_1} = 0$, and the P_{k_i} values of the other components are the same as in Table 2. The values of $\mathrm{Pr}_N(\Phi(1_1, X) = j)$ are calculated by substituting theses values into Table 3 and shown in Table 4. Similarly, $\mathrm{Pr}_N(\Phi(0_1, X) = j)$ is the probability that each state exists when C is killed. In this state, $P_{k_1} = 1$, and the P_{k_i} values of the other components are the same as in Table 2. The values of $\Pr_N(\Phi(0_1, X) = j)$ are calculated by substituting theses values into Table 3 and shown in Table 4.

Table 4. The calculation results of each state value of C

State ID	P_{i1}	a_j	$\Pr_N(\Phi(1_1, X) = j)$	$\Pr_N(\Phi(0_1, X) = j)$
I	1	0.8	0	0.98924265
II		0.75	0.00292279	0
III		0.75	0.00258457	0
IV		0.6	0.00210449	0
V		0.6	0. 00173277	0
VI		0.3	0	0.00292279
VII		0.3	0	0.00257554
VIII		0.3	0	0.00210449
IX		0.3	0	0.00173277
X		0.25	0.00000622	0
XI		0.25	0.00000512	0
XII		0.25	0.00000548	0
XIII		0.25	0.00000451	0
XIV		0.18	0	0.00000622
XV		0.16	0	0.00000512
XVI		0.12	0	0.00000548
XVII		0.1	0	0.00000451
XVIII		1	0.98924265	0

In this way, the state values of the remaining components are determined, and the results are shown in Tables 5, 6, 7, 8 and 9.

Table 5. The calculation results of each state value of D

State ID	P_{i1}	a_j	$\Pr_N(\Phi(1_1, X) = j)$	$\Pr_N(\Phi(0_1, X) = j)$
I	1	0.8	0	0.98924265
II		0.75	0.00292279	0
III		0.75	0.00258457	0
IV		0.6	0.00210449	0
V		0.6	0. 00173277	0
VI		0.3	0	0.00292279
VII		0.3	0	0.00257554
VIII		0.3	0	0.00210449
IX		0.3	0	0.00173277
X		0.25	0.00000622	0
XI		0.25	0.00000512	0
XII		0.25	0.00000548	0
XIII		0.25	0.00000451	0
XIV		0.18	0	0.00000622
XV		0.16	0	0.00000512
XVI		0.12	0	0.00000548
XVII		0.1	0	0.00000451
XVIII		1	0.98924265	0

Table 6. The calculation results of each state value of H

State ID	P_{il}	a_j	$\Pr_N(\Phi(1_1, X) = j)$	$\Pr_N(\Phi(0_1, X) = j)$
I	1	0.8	0.00137820	0
II		0.75	0	0.99078724
III		0.75	0. 00257956	0
IV		0.6	0. 00210777	0
V		0.6	0. 00173548	0
VI		0.3	0	0.00137820
VII		0.3	0.00000359	0
VIII		0.3	0.00000293	0
IX		0.3	0.00000241	0
X		0.25	0	0.00210778
XI		0.25	0	0.00173548
XII		0.25	0.00000549	0
XIII		0.25	0.00000193	0
XIV		0.18	0	0.00000293
XV		0.16	0	0.00000241
XVI		0.12	0.000000008	0
XVII		0.1	0.000000006	0
XVIII		1	0.99078724	0

Table 7. The calculation results of each state value of I

State ID	P_{il}	a_j	$\Pr_N(\Phi(1_1, X) = j)$	$\Pr_N(\Phi(0_1, X) = j)$
I	1	0.8	0. 00137771	0
II		0.75	0.00292633	0
III		0.75	0	0.99044048
IV		0.6	0.00210704	0
V		0.6	0.00173487	0
VI		0.3	0.00000407	0
VII		0.3	0	0.00137771
VIII		0.3	0.00000293	0
IX		0.3	0.00000241	0
X		0.25	0.00000623	0
XI		0.25	0.00000513	0
XII		0.25	0	0.00210704
XIII		0.25	0	0.00173487
XIV		0.18	0.000000009	0
XV		0.16	0.000000007	0
XVI		0.12	0	0.00000293
XVII		0.1	0	0.00000241
XVIII		1	0.99044048	0

Table 8. The calculation results of each state value of J

State ID	P_{i1}	a_j	$\mathrm{Pr}_N(\Phi(1_1, X) = j)$	$\mathrm{Pr}_N(\Phi(0_1, X) = j)$
I	1	0.8	0.00137706	0
II		0.75	0.00292494	0
III		0.75	0.00257744	0
IV		0.6	0	0.98970082
V		0.6	0.00173404	0
VI		0.3	0.00000407	0
VII		0.3	0.00000359	0
VIII		0.3	0	0.00137706
IX		0.3	0.00000241	0
X		0.25	0	0.00292494
XI		0.25	0.00000512	0
XII		0.25	0	0.00257744
XIII		0.25	0.00000451	0
XIV		0.18	0	0.00000407
XV		0.16	0.000000007	0
XVI		0.12	0	0.00000359
XVII		0.1	0.000000006	0
XVIII		1	0.98997008	0

Table 9. The calculation results of each state value of K

State ID	P_{i1}	a_j	$\mathrm{Pr}_N(\Phi(1_1, X) = j)$	$\mathrm{Pr}_N(\Phi(0_1, X) = j)$
I	1	0.8	0.00137654	0
II		0.75	0.00292384	0
III		0.75	0.00257647	0
IV		0.6	0.00210525	0
V		0.6	0	0.98959888
VI		0.3	0.00000407	0
VII		0.3	0.00000358	0
VIII		0.3	0.00000293	0
IX		0.3	0	0.00137654
X		0.25	0.00000622	0
XI		0.25	0	0.00292384
XII		0.25	0.00000548	0
XIII		0.25	0	0.00257647
XIV		0.18	0.000000009	0
XV		0.16	0	0.00000408
XVI		0.12	0.000000008	0
XVII		0.1	0	0.00000358
XVIII		1	0.98959888	0

The Griffith importance and order of each critical component is calculated by substituting the above data into Eq. 3 and shown in Table 10.

Table 10. The Griffith importance and order of each critical component

Component ID	$WI_1^G(i)$	Order number
C	0.201483	5
D	0.201483	5
H	0.251669	4
I	0.251843	3
J	0.400634	1
K	0.400546	2

Redundant critical components are arranged in order of importance from high to low by the value of $WI_1^G(i)$. Non-critical components are arranged in order of importance from high to low by the value of $p_{k_i}^N$. The non-critical components are arranged behind all the critical components, resulting in the final sorting results of the components.

In this example, the importance of combat spare parts from high to low are left engine J, right engine K, right fuel tank I, left fuel tank H, flight control computer No. 1 C or No. 2 D, left inlet F, right inlet G.

4 Conclusion

In this paper, a ranking method for the importance degree of aircraft damage spare parts is established by introducing the concept of Griffith importance to the field of aircraft spare parts, in which both the intermediate state and the performance of aircraft are considered. The ranking results given are more objective and accurate. After ranking the importance of aircraft combat damage spare parts, it helps to determine the spare parts demand of war wound aircraft more simply and quickly.

After computing and analysing, the importance of combat spare parts of the aircraft from high to low are left engine, right engine, right fuel tank, left fuel tank, flight control computer No. 1 or No. 2, left inlet, right inlet. It helps to determine the spare parts requirements of the war wound aircraft more simply and quickly after sorting the importance of aircraft war damage spare parts.

References

1. Zhao PZ, Ji BL, Wei HK (2014) A brief review on modeling and simulation in aircraft battle damage repair studies. Electron Opt Control 21(02):55–59
2. Yao WW (2016) Study on spare part requirement prediction model of aircraft battle damage assessment and repair based on layer analysis method. New Technol New Process 10:28–30

3. Hu YF, Song BF, Wang X (2009) Ascertaining of spares requirements for aircraft battle damage. Acta Aeronautica ET Astronautica Sinica 30(03):450–455
4. Hu YF (2007) Research on the demand model of aircraft battle damage spare parts. Northwestern Polytechnical University
5. Zhang JH (1995) The air force vigorously carried out research on aircraft war damage repair. Ordnance Eng College Newspaper 7(3)
6. Kuo W, Zhu XY (2012) Some recent advances on importance measures in reliability. IEEE Trans Reliab 61(02):344–360
7. Griffith WS (1980) Multi-state reliability models. J Appl Probab 17(03):735–744
8. Wu S, Chan L (2003) Performance utility-analysis of multi-state systems. IEEE Trans Reliab 52(01):14–21
9. Zio E, Marella M, Podofillini L (2007) Importance measures-based prioritization for improving the performance of multi-state systems: application to the railway industry. Reliab Eng Syst Saf 92(10):1303–1314
10. Pei Y, Cheng T (2014) Importance measure method for ranking the aircraft component vulnerability. J Aircr 51(1):273–279

Heading Load Dynamic Simulation of Landing Gear Test

Zihao Zhang[1($\boxtimes$)], Xiaohui Wei[2,3], and Qi Ye[2,3]

[1] Nanjing University of Aeronautics and Astronautics, Nanjing, China
863193203@qq.com
[2] Key Laboratory of Fundamental Science for National Defense Advanced
Design Technology of Flight Vehicle, Nanjing University of Aeronautics
and Astronautics, Nanjing 210016, Jiangsu, China
[3] State Key Laboratory of Mechanics and Control of Mechanical Structures,
Nanjing University of Aeronautics and Astronautics,
Nanjing 210016, Jiangsu, China

Abstract. In this paper, a force-measuring platform which is used to test the
heading load of landing gear is analyzed, as well as the influences of its inertial
forces on the accuracy of the test results. Two sets of drop test simulation models,
considering or not considering the dynamometer platform, are built based on the
dynamic simulation platform named LMS Virtual.lab motion. The installation gap
between the heading force sensor and the dynamometer platform is simulated
through the form of spherical contact. The gap size was changed at two subsidence
speeds of 1 m/s and 1.5 m/s, and the relationship between the gap size and the
heading inertia force of the force-measuring platform was analyzed. According to
the design requirements of the drop test in Book 14 of the Aircraft Design Manual,
a drop test bench was built. The landing gear load of a high-speed UAV landing
gear was measured and compared with the simulation results.

Keywords: Force-measuring platform · LMS Virtual.lab motion ·
Inertial force · Heading load · Simulation

1 Introduction

The design of aircraft landing gear is very important in the whole aircraft design, in
which the weight of the landing gear accounts for 15% to 20% of the total weight of the
aircraft structure [1]. Therefore, reasonable landing gear design can reduce the weight
of the whole aircraft when it meets the normal use of the aircraft. At present, the
method to determine whether the design of the landing gear meets the requirements of
use is to conduct the landing test before the installation. Data such as the cushioning
effect and structural strength of the landing gear were collected by the drop test.
Therefore, the accurate collection of test data can provide the best advice for the design
of landing gear. Since the internal output force of the landing gear damper cannot be
directly measured, most of the current earthquake drop tests in China use the indirect
measurement method, that is, the ground load of the landing gear is collected through
the three-point supporting force measuring platform [2].

© Springer Nature Singapore Pte Ltd. 2019
X. Zhang (Ed.): APISAT 2018, LNEE 459, pp. 1469–1476, 2019.
https://doi.org/10.1007/978-981-13-3305-7_116

However, this method of measuring load is imperfect [3, 4]. For example, the measured load also includes the platform inertial force [5]. With the gradual improvement of aircraft speed, the requirements for aircraft landing gear are getting higher and higher. High-speed aircraft landing its turn on springback load also corresponding increase a lot, so can increase the stiffness of the landing gear to ensure no landing gear structure deformation, to make the gear to work properly, and meet the requirements of the use of its high-speed take-off and landing. The premise that the landing gear achieves its cushioning function is that its buffer can work normally, which through the buffer of relative movement between the outer cylinder, piston rod and buffer and wasted the landing impact load [6]. When the landing gear structure is deformed, the landing gear buffer is stuck and the buffer is lost, which may directly lead to the landing failure of the aircraft, thereby destroying the overall structure of the aircraft and causing serious safety accidents. However, increasing the strength and rigidity of the landing gear raises the problem of increased weight of the landing gear. It is therefore important to accurately measure the load on the landing gear.

In order to explore the influence of the inertial force of the force platform on the heading load of the landing gear, this paper based on the multi-body dynamics modeling method, using the LMS Virtual. Lab Motion simulation software to establish the kinematics model of the falling seismic force platform and simulate the drop test.

2   Characteristic Analysis of the Force Measurement Platform of the Drop Test

In the drop test, the ground load measurement method is a general measurement method for load measurement in the current drop test. The loads to be measured in the general drop test include vertical load, heading load and lateral load. At present, the three-point bracing platform is mainly used in the drop test in China. During the test time wheel impact platform, the vertical load of the landing gear is measured by three force sensors at the bottom, and the heading load of the landing gear is measured by the horizontal force sensor.

The landing gear is fixed to the hanging basket during the landing test, and the hanging basket falls on the measuring platform with a large impact load along the guide rail. At the same time, due to the high-speed rotation of the wheel and the contact with the platform, the rotation is stopped within a short time under the action of the friction force. The load distribution on the force measuring platform is shown in Fig. 1.

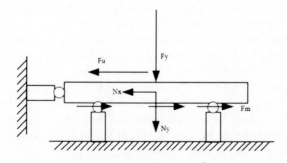

Fig. 1. Load distribution on the force measuring platform

Vertical load on the force measuring platform:

$$F_{cy} = F_y + N_y \qquad (2.1)$$

heading load on the force measuring platform:

$$F_{cx} = F_\mu + N_x + F_m \qquad (2.2)$$

While:

$$\left\{ \mu = \begin{cases} F_\mu = F_y * \mu \\ 5.62 S_{gi} & -0.13 < S_{gi} < 0.13 \\ 0.77 - 0.32 S_{gi} & 0.13 < S_{gi} < 1.0 \end{cases} \right. \qquad (2.3)$$

$$\begin{cases} S_{gi} = \frac{v - R_{ei}\omega}{v} \\ R_{ei} = R_{0i} - \frac{1}{3}\delta_i \end{cases} \qquad (2.4)$$

$$\begin{cases} N_x = M * a_x \\ N_y = M * a_y \end{cases} \qquad (2.5)$$

Where F_y is the actual vertical load on the force measuring platform, F_{cy} and F_{cx} are respectively vertical and heading load measured by the sensor, N_y and N_x are respectively the vertical and heading inertia of the platform, F_μ is the friction between aircraft tires and platform, F_m is the friction between the platform and the vertical support sensor, M is the weight of the platform, μ is the friction coefficient between platform and tire, ω is angular speed of the aircraft wheel.

3   Establish a Virtual Drop Test Model

LMS Virtual.lab has been widely used in aerospace, automotive and mechanical engineering. It not only provides a complete tool for the multidisciplinary design analysis team, but also has automatic capture and management functions for the design process, and fully implements parameter-driven [7].

In this paper, the contact command in motion is used to simulate the gap size between the heading force sensor and the force measuring platform. Establish a spherical contact at the junction of the force measuring platform and the heading force sensor, as shown in Fig. 2. The stretching surface is a cylindrical surface for simulating the hole surface of the hole at the joint, and the diameter of the hole is simulated by the diameter D of the stretching surface. Create a massless point inside, establish a contact ball at the point, and simulate the bolt diameter with the diameter d of the ball. The value of d is subtracted from D to simulate the gap size of the hole axisgap fit.

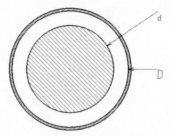

Fig. 2. Spherical contact

The simulation analysis of the drop test is based on the reduced mass method [8]. The simulation parameters of the drop test are determined by the reduced mass method as shown in Table 1.

Table 1. Drop test simulation parameters

Operating conditions	Sinking speed m/s	Throw weight Kg
D1	1	113.7
D2	1.5	136.7

4 Comparative Analysis of Simulation and Experimental Results

According to the requirements of the 14th volume of the aircraft design manual, the drop test bench was built and tested according to the simulation design conditions. Between the heading load sensor and the force measuring platform is a hole shaftgap fit, connect with M20 polished rod bolts. The ground load of 1 m/s and 1.5 m/s sinking speed was measured by the test of a high-speed unmanned aerial vehicle landing gear and compared with the simulation results. In order to obtain a typical lifting rebound load, different surface finishes must be used [9]. At 1 m/s sinking speed, the pavement is made of patterned steel plate, and the smooth steel plate is used for simulation analysis at 1.5 m/s sinking speed.

The simulation results of the heading load at 1 m/s sinking speed in the model without the force-measuring platform are compared with the experimental results as shown in Fig. 3.

At 1 m/s sinking speed: the maximum heading load of the test is 1.41 Kn, the simulation result is 1.28 Kn, and the error between simulation and test results is 9.22%. Since the heading stiffness of the UAV landing gear is too large, the fluctuation of the heading load is assumed to be caused by the inertia force of the platform, without considering the deformation of the propeller heading. Construct a drop test model containing a

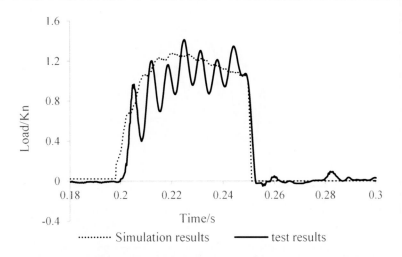

Fig. 3. Heading load without inertial force at 1 m/s sinking speed

force-measuring platform for simulation analysis, The heading inertial force of the platform measured at 1 m/s sinking speed is shown in Fig. 4, The comparison between the heading load obtained from the simulation and the test data is shown in Fig. 5.

At 1 m/s sinking speed: considering the inertial force of the platform in the simulation analysis, the simulation result is 1.37 Kn, and the error between simulation and test results is 3.43%. The maximum value of the inertial force of the platform occurs when the bolt and the hole face collide for the first time, and the maximum value is 0.33 Kn. When the heading load reaches the maximum value, the inertial force is

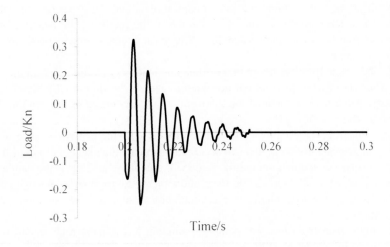

Fig. 4. Heading inertial force of the force measuring platform

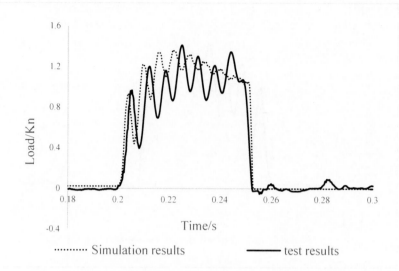

Fig. 5. Heading load with inertial force at 1 m/s sinking speed

0.09 Kn, and the inertial force accounts for 6.57% of the heading load. The simulation model containing the force platform verifies that the heading load fluctuation is caused by the inertia force of the platform.

At 1.5 m/s sinking speed: Without considering the inertial force of the platform in the simulation analysis, the maximum heading load of the test is 0.97 Kn, the simulation result is 0.84 Kn, and the error between simulation and test results is 13.40%. Considering the inertial force of the platform in the simulation analysis, the simulation result is 0.91 Kn, and the error between simulation and test results is 7.68%, the inertial force is 0.07 Kn, and the inertial force accounts for 7.69% of the heading load.

At the same time, comparing the simulation data and the experimental data, it is found that when considering the inertial force of the platform, the heading load has hysteresis compared with the vertical load (Fig. 6) and the lag time increases as the gap amount increases.

It can be seen from the figure that the heading load is synchronized with the vertical load when the inertial force is not considered, and the heading load is delayed when considering the inertial force of the platform, so the hysteresis of the heading load is caused by the platform gap.

By comparing the established simulation model with the experimental results, it is shown that the established model can accurately simulate the load condition of the drop test, so it is speculated that the inertial force will gradually increase as the gap becomes larger. By adjusting the gap size in the simulation model to observe the change of the inertial force of the load platform, the simulation results at 1 m/s sinking speed are shown in Fig. 7.

It can be seen from Fig. 7 that the gap amount has a linear relationship with the maximum value of the heading acceleration.

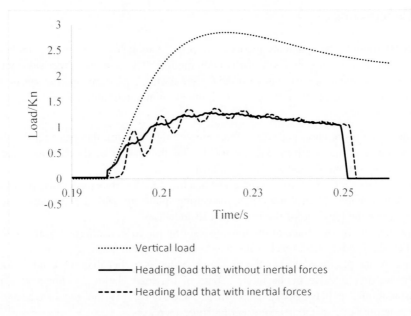

Fig. 6. Analysis of load hysteresis under sinking speed of 1 m/s

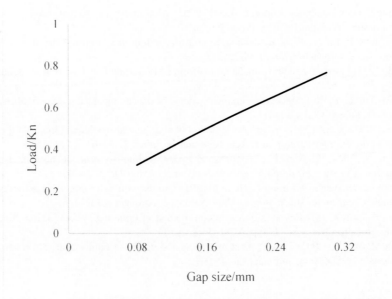

Fig. 7. Maximum inertia force under different gaps

5 Conclusions

Based on multi-body dynamics theory, this paper establishes a dynamic simulation model of the drop test in LMS Virtual.lab motion. Through the spherical contact command to simulate the gap between the force measuring platform and the sensor, the inertial force of the platform heading is accurately simulated, and the following conclusions are obtained:

1. Due to the installation gap of the force measuring platform, the measured heading load includes the platform inertial force. The fluctuation of the heading load is caused by the inertia force of the platform.
2. Due to the existence of inertial force, the heading load exhibits hysteresis, and as the matching gap between the force measuring platform and the heading sensor increases, the lag time of the heading load increases.
3. The heading inertial force of the force measuring platform is different from the time when the heading load reaches the maximum value, and the time when the inertia force of the platform reaches the maximum value is earlier than the heading load.
4. The maximum value of the inertial force of the load platform is linear with the installation gap between the platform and the sensor. As the gap increases, the maximum value of the heading inertial force increases.

References

1. Aircraft Design Handbook Editorial Board (2002) Aircraft design manual: takeoff landing system design. Aviation Industry Press, Beijing
2. Du J, Meng F, Lu X (2018) Evaluation criteria for landing gear drop test based on energy method. Acta Aeronautica Sinica 39:221375
3. Lin H (2005) Innovative hybrid simulation platform LMS Virtual.Lab. CAD/CAM Manuf Inf 10:46–47
4. Yali M, Yumei Y, Bo J (2002) Development scheme of the earthquake test system. Shenyang Higher Vocational Coll 4:39–41
5. Haiwen S, Daqian Z (2001) Effect of simulation of starting and rebounding loads on the test results in the earthquake drop test. Acta Aeronautica Sinica 22:39–41
6. Wenhai S, Yang Shuxun F, Yonghong ZD (2002) Improvement of horizontal load measurement method for dropping test machine wheel. Aircr Des 9:24–27
7. Xiaohui W, Xiaochen S, Lirong L, Hong N (2014) Study on the dynamic critical friction of the landing gear buffer stuck. Beijing Univ Aeronaut Astronaut 6:732–736
8. Jianbo Y, Jia R (2017) Research on correction method of buffering system loading force in landing gear drop test. Chin J Appl Mech 34:329–334
9. Daqian Z, Wei L (2002) Development and present situation of falling gear test system for landing gear. J Shenyang Aeronaut Ind 4:10–12

Research on Edge Computing Architecture for Intelligent NC Machining Monitoring CPS

Shaochun Sui, Xiaohua Li, and Wenyi Li[✉]

Chengdu Aircraft Industrial (Group) Co., Ltd., Chengdu 610091, Sichuan, China
18180866088@qq.com

Abstract. On the basis of summarizing the research results of CPS and CPS architecture for NC machining process, the application of the edge computing architecture and its characteristics, the relationship between cloud computing and fog computing, and the application value are discussed. Based on the application scenarios of the construction technology and the edge computing technology, an edge computing architecture based on the intelligent monitoring and control CPS built in the NC machining process is proposed, and its function positioning, function model and key technologies are discussed.

Keywords: NC-machining processes · Intelligent monitoring ·
Cyber-Physical Systems · Edge computing ·
Information and Communication Technology · Heterogeneous computing

1 NC Machining Processes Monitoring Technology

As society's diversified demands for products have increased, product categories have increased, and replacements have become faster. This has led to the increasing use of numerically-controlled machine tools in production. At the same time, NC systems and servos for numerically-controlled machine tools have become increasingly popular [1]. The drive system and the host structure put forward higher requirements. In this context, various types of NC machine tool monitoring technologies have flourished. With the development of digital control technology, drive technology, and sensor technology, there are several typical trends in the development of NC machining technologies, such as high speed, high precision, high flexibility, integration, and intelligence. Based on the development of OT and ICT technology, the monitoring technology for NC machining process has also shifted from traditional data acquisition and KPI statistical analysis to real-time status monitoring, fault early warning diagnosis, and intelligent monitoring [2].

2 A Summary of Intelligent Monitoring CPS for NC Machining Process

2.1 CPS Architecture

Usually, the basic Cyber-Physical System consists of a sensing unit, a decision-control unit and an execution unit. Sensing unit is Responsible for monitoring signals and

© Springer Nature Singapore Pte Ltd. 2019
X. Zhang (Ed.): APISAT 2018, LNEE 459, pp. 1477–1485, 2019.
https://doi.org/10.1007/978-981-13-3305-7_117

conditions of the equipment. decision-control unit is responsible for generating control logic according to user's definition and execution unit is responsible for receiving the control instruction. Combined with the feedback loop control principle, the physical components makes the basic monitoring and control functions of Cyber-Physical System [3]. As shown in Fig. 1 below.

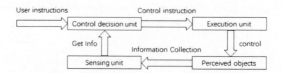

Fig. 1. Basic physical composition of Cyber-Physical System

2.2 The Challenges of NC Machining Process Monitoring CPS to Traditional OICT Technology

The "Industry 4.0", considered as the "fourth industrial revolution", was first proposed by Germany in 2013. Its concept is the integration of Information and Communication Technology (ICT) and OT (Operation Technology) through the cyberspace physical system. (CPS) Achieve independent exchange of information, collaborative work and autonomous control among manufacturing units throughout the product lifecycle, and gradually realize intelligent transformation of manufacturing. He Jifeng pointed out that CPS is mainly characterized by tight integration and coordination between the computing process and the physical process. It emphasizes that physical devices are interconnected through various network means, so that physical devices have calculations, communications, precise control, and remote coordination. And five major functions such as autonomy. It can be seen that, compared with the traditional system, CPS has more obvious requirements for high-speed storage performance, computational real-time performance, and low latency of communication.

2.3 The Reference Architecture of NC Machining Intelligent Monitoring CPS

2.3.1 Definition of Intelligent Monitoring CPS of NC Machining Process
Intelligent monitoring CPS of NC machining Cyber-Physical System is used for the actual production monitoring of CNC machine base on the real-time monitoring of NC machining process and adaptive control capacity, connected with the high performance network, which has better calculating performance, accuracy of controlling and highly autonomy.

2.3.2 The Hierarchy of Intelligent Monitoring CPS of NC Machining Process
Reference to the abstract architecture of CPS, Considering the characteristic of intelligent monitoring of NC machining process, A SOA based intelligent monitoring CPS Architecture of NC machining process is proposed [4], as shown on Fig. 2. In this

architecture, there are four layers, the sensor and controlling layer, the network layer, the resource service layer and decision application layer.

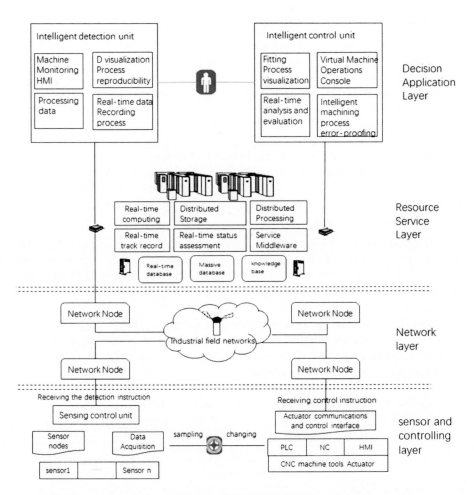

Fig. 2. System architecture of intelligent monitoring CPS of NC machining process

3 Edge Computing Overview

3.1 Edge Computing

In 2004, Professor HH Pang, published the first paper on "edge computing" at the Institute of infocomm Research of Singapore, the European Telecommunications Standards Institute (ETSI) established the Mobile Edge Computing (MEC) working group. Renamed Multi-site Edge Computing, in 2016, Huawei and 6 other organizations jointly initiated the establishment of the Edge Computing Consortium. In October 2016, the IEEE/ACM Symposium on Edge Computing was formally established by

IEEE and ACM. The application value of edge computing, research direction to carry out research and discussion [5].

3.2   Edge Computing Reference Architecture

Edge computing is a distributed open platform which combines network, computing, storage and application core capabilities near the source of objects or data. It provides edge intelligent services to meet the key needs of industry digitization in agile connection, real-time service, data optimization, security and privacy protection. It can serve as a bridge connecting physical and digital world, enabling intelligent assets, intelligent gateways, intelligent systems and intelligent services [6].

To promote edge computing technology and industry development, ECC and AII formally proposed the Edge Computing Reference Architecture 2.0 in 2017. As shown in Fig. 3, the architecture includes four levels of intelligent services, business fabrics, connected computing fabrics, and edge computing nodes. Define modeled open interfaces at each level to achieve global openness of the architecture. The main functions of each level include [7]:

- Intelligent services: Based on the model-driven unified service framework, unified software development interfaces and deployment operations automation are achieved through a unified development framework and deployment framework.
- Service Fabric: Defines end-to-end business processes through the service fabric to support service agility;
- Connection Computing: The Complexity and Computing Fabric (CCF) masks the complexity of the edge-based distributed architecture through functional abstraction and modularization, and supports automation and visualization of OT and ICT infrastructure deployment operations.
- Edge Computing Node: Edge Computing Node (ECN) implements heterogeneous connectivity, supports real-time processing, and low latency response.

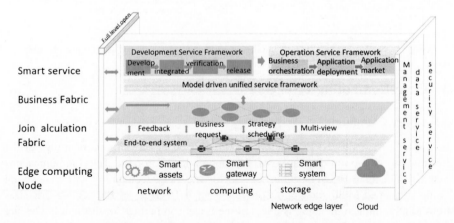

Fig. 3. Edge computing reference architecture [7]

4 An Edge Computing Achitecture for Intelligent Monitoring CPS of NC Machining

4.1 The Value Analysis of Edge Computing Architecture in the Construction of CPS

In the process of digital and intelligent transformation of manufacturing companies, hundreds of millions of smart devices and machines are being networked, and the storage, analysis, and utilization of the massive data they generate will become an important challenge for CPS construction. The construction of CPS with secure, reliable, and low-cost ICT technology forms a data-driven distributed intelligent control and drives the convergence of OT and ICT is an inevitable choice. The edge computing formally utilizes edge-end embedded computing capabilities to implement edge intelligence and self-managing in distributed information processing to support distributed sensing, decision-making, and control at the edge. Therefore, it can be considered that the edge calculation will be the core of CPS construction and will also become the core of "Industry 4.0" core.

4.2 The Position of Edge Computing Layer in Enterprise Information System

From the perspective of the overall architecture of the enterprise information system, in the future digital and intelligent transformation of a enterprise, the edge computing layer will become the key fusion connection layer between the digital world (Cyber) and the physical world (Physical) to realize the integration of ICT and OT [5], as shown in Fig. 4.

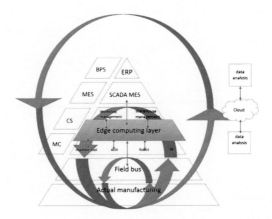

Fig. 4. The positioning of edge computing layer in enterprise information system

4.3   Edge Computing Architecture for Intelligent Monitoring CPS in NC Machining Process

Based on the theoretical research and the verification of the system prototype for the intelligent monitoring and control of the NC machining process, the edge computing reference architecture is integrated, and an edge computing architecture for intelligent monitoring of CPS in the CNC machining process is proposed. As shown in Fig. 5, the architecture consists of four layers: a sensing layer (physical device), an edge layer (edge computing), a service layer (connection and calculation fabric), and an application layer (service fabric).

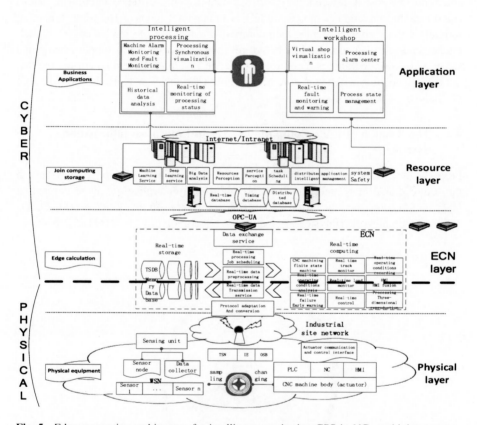

Fig. 5. Edge computing architecture for intelligent monitoring CPS in NC machining process

4.4   Functional Model

4.4.1   Physical Layer

This layer is the industrial control equipment and network layer, belongs to the physical layer category of CPS, this layer needs to solve three problems. Protocol adaptation and conversion: The protocol gateway supported by multiple communication protocols, based on OPC-UA, realizes the integration of sensing and communication with the edge layer; Sensor network and sensor fusion: Sensor networking and communication management for workshop/factory floor independent of processing equipment; Industrial field heterogeneous network convergence: For the current status of heterogeneous communication networks coexisting, an industrial field converged communications network based on TSN and SDN technologies is built to connect CNC machine tools, controllers, HMI, AGV, Robotics and other devices to achieve high speed and low latency, high reliability industrial process communication.

4.4.2 Edge Computing Node Layer

According to the application scenario and technical requirements of NC machining process intelligent monitoring, integrate network, computing and storage capabilities. Based on TSN and SDN, the high performance network capability of the edge end is built to realize the low delay transmission of the real time working condition, the track data and the process state control instruction in the NC machining process. Based on the storage, processing and analysis capability of the industrial large data of TSDB and MMDB, the implementation of the NC machining process and the real time track record are realized. Based on the CPU and GPU computing architecture, the integration of heterogeneous computing power is carried out to realize the real time analysis, evaluation, fault warning and diagnosis of the NC machining process, and realize the 3D reappearance of NC machining process and the application of the 3D visualization in the process.

4.4.3 Resource Layer

According to the logical function architecture of edge computing, the resource layer is composed of computing, network, and storage components, and the unified management of edge computing nodes is realized by virtualized service. It supports intelligent control and autonomous management of edge computing nodes, provides industrial big data storage and analysis, machine learning, deep learning and other resources, and provides a unified mechanism for ECN configuration, status monitoring, and security management.

4.4.4 Application Layer

Based on the resource layer, it builds monitoring applications for specific monitoring scenarios, such as smart machine intelligent monitoring applications for single CNC machine tools, intelligent monitoring applications for production lines, and intelligent monitoring applications for manufacturing plants/factories.

4.5 Key Technology

4.5.1 Real Time and Continuous Industrial Field Data Storage and Processing Technology

Due to the high-speed and high-precision requirements of the NC machining process, the amount of data such as real-time operating condition data, sensor data, and numerical control machining tool trajectory generated during the NC machining process is very large. Using heterogeneous database (memory database, real-time database, timing database, etc.) fusion application architecture to support the storage, processing, and analysis of real-time, high-throughput object-aware data is one of the key technologies of the architecture.

4.5.2 Industrial Big Data High Performance Analysis Technology for Industrial Control Processes

Intelligent monitoring application scenarios for NC machining processes include analysis and prediction of operating conditions, intelligent identification of abnormal conditions, intelligent fault early warning, and diagnostics. The industrial big data intelligent analysis will be the key to the realization of these application scenarios. In addition to the traditional data analysis model and expert knowledge base technology, industrial big data analysis based on machine learning and deep learning technology will be the key to the breakthrough of smart analysis applications in the future.

4.5.3 High-Performance, Low-Cost Heterogeneous Computing Architecture for Multiple Computing Scenarios

With the deepening of the application of numerical control process monitoring and control, it can flexibly support various types of computing scenarios. Research on heterogeneous computing architectures that meet the performance and cost requirements has broad application prospects. The main goal of the heterogeneous computing architecture is to provide the most efficient, balanced, and low-cost computing solution on the premise of resource-constrained and cost-constrained. At the same time, cross-platform applications are realized through an open and unified programming interface. The key technologies of heterogeneous computing include: memory processing optimization, task scheduling optimization, and integrated tool chain.

5 Conclusion

At present, the digital and intelligent transformation with the goal of implementing intelligent manufacturing is ongoing. CPS and edge computing technology, as important technologies supporting intelligent manufacturing applications, will certainly play an increasingly important role. Just like the challenges faced by CPS technology development itself, TSN, SDN, TSDB, heterogeneous computing convergence, etc, will become the focus areas of intelligent manufacturing research.

References

1. Wei H (2011) Numerical control technology and application. Tsinghua University Press, Beijing
2. Wenli H (2000) Research on the development strategy of CNC machine tools technology. National Machinery Industry Bureau
3. Raj Kumar R, Lee I, Sha L, et al (2010) Cyber-physical systems: the next computing revolution. In: Proceedings of the 47th ACM/IEEE design automation conference. IEEE Press, Anaheim

4. Li X, Li W (2016) The research on intelligent monitoring technology of NC machining process. In: 9th international conference on digital enterprise technology (DET2016)-intelligent manufacturing in the knowledge economy era
5. Yang L (2018) Some hot issues in the development of edge computing and reflections. In: Industrial internet summit, March
6. AII, ECC (2017) Edge Computing Reference Architecture 2.0, Edge Computing Consortium (ECC) and Industrial Internet Industry Alliance (AII) Joint release, Beijing, November
7. Edge Computing Consortium (ECC) (2018) Edge computing architecture. Automation Panorama, January

High Subsonic NLF Airfoil Design
at Low Reynolds Number

Jing Li[1(✉)] and Zhenghong Gao[2]

[1] School of Mechanics and Engineering Sciences, Zhengzhou University,
Zhengzhou, China
lijingself@zzu.edu.cn
[2] School of Aeronautics, Northwestern Polytechnical University, Xi'an, China
zgao@zzu.edu.cn

Abstract. We developed an optimization design approach which can be used to the high subsonic Natural Laminar Flow (NLF) airfoil design at low Reynolds number. The aerodynamic characteristics are evaluated by the γ-$Re_{\theta t}$ transition model, and a modified CST method based on the disturbing CST basis function is introduced in the airfoil parameterization. After that, we build up the optimization system which employs the MCPSO (Multi-Groups Cooperative Particle Swarm Algorithm) as the search algorithm, and then apply the optimization methodology to the design of a propeller tip airfoil. More specifically, the laminar separation bubble and transition location are controlled in the optimization process. Our simulation results indicate that the optimized distribution shows better ability to control separation position and reattachment position, and the pressure drag can be greatly reduced when the laminar separation bubble is weakened. We demonstrate that it is a reasonable design idea that the friction drag increment is proposed as a constraint condition. These design experiences can provide valuable reference data for the design of the high subsonic NLF airfoil at low Reynolds number.

Keywords: High subsonic · NLF airfoil · CST · Low Reynolds number · PSO

1 Introduction

The solar energy UAV, which has been a hotspot in recent years, generally flied at an altitude of 10 km–40 km with low atmospheric density. However, a low atmospheric density could lead to the low Reynolds numbers for the high-altitude cruise, and also can give rise to a rapid drop of the propeller's efficiency. To data, the designers have been focusing on the improvement of the aerodynamic efficiency of the propeller at the low Reynolds numbers [1]. On the other hand, the local velocity of the blades is proportional to the radius of propeller blades. In this context, the velocity nearby the blade tip would be high subsonic or transonic, when the velocity of the root region is low. More recently, the distribution of a blade's Mach number and Reynolds numbers at different altitude has been discussed in the reference [2], in which the Mach number of the blade tip and root region has demonstrated to be 0.1 and 0.05 at an altitude of 0 km, and to be 0.85 and 0.3 at an altitude of 25 km. Based on the above mentioned

© Springer Nature Singapore Pte Ltd. 2019
X. Zhang (Ed.): APISAT 2018, LNEE 459, pp. 1486–1498, 2019.
https://doi.org/10.1007/978-981-13-3305-7_118

two features, the aerodynamic design for the propeller blade airfoil of a solar energy UAV can be properly divided into low speed NLF airfoil design at low Reynolds numbers and high subsonic NLF design at low Reynolds numbers.

For low Reynolds numbers, the effect of air viscosity is very remarkable. The aerodynamic characteristics and the structure of the flow field are much different from those of the middle and high Reynolds numbers. These obvious differences mainly exist in the laminar separation and the laminar separation bubble transition. Accurate predictions of transition and separation are the fundamental basis for the study of the low Reynolds number problem. This has received many attentions from many researchers who have been working on the structure and aerodynamic characteristics of the flow field under the low Reynolds number. Gratifying results have been achieved by many researchers. For example, Morgado [3] simulated the E387 airfoil aerodynamic performance at low Reynolds numbers by using the Xfoil code, the refurbished $k - k_L - \omega$ transition model and the Shear Stress Transport $k - \omega$ turbulence model. It is showed that the Xfoil code can provide the sufficient accuracy for the preliminary design of the propeller, and the refurbished $k - k_L - \omega$ transition model can also provide promising results in the prediction of the lift coefficient and drag coefficient; Chen [4] have integrated the Michel transition criterion and γ-$Re_{\theta t}$ transition model to conduct the numerical analysis for the E387 airfoil at low Reynolds numbers. It is demonstrated that the prediction of the transition location and the laminar separation from Michel transition criterion is more accurate than those of γ-$Re_{\theta t}$ transition model; Genc [5] focused on the numerical simulation on laminar separation bubble over a NACA2415 aerofoil at low Reynolds numbers. It is demonstrated that the transition model has stronger ability to capture the laminar separation bubbles than that of the turbulence model.

Nowadays, most designers concentrate on the low speed NLF airfoil design at low Reynolds numbers. Chen [6] carried out the research on the single point optimization and multipoint optimization based on SD7032 airfoil at low Reynolds number. Liu [7] simulated the influence of the crystalline silicon solar cells by using the broken line airfoils-modeled with FX 63-137 airfoil, and the results show that the broken line airfoils' aerodynamic characteristics are better than the baseline airfoil at low Reynolds numbers. Wang [8] employed the $k - k_L - \omega$ transition model to carry out the multi-objective optimization design of SD7073 airfoil, and put forward an momentous design concept, which is changing the boundary layer by controlling the transition position. Zhao [9] developed a robust design optimization method on the basis of an adjusted polynomial chao's expansion, which is used to carry out the robust design on a NLF airfoil. The results show that the optimized airfoil exhibits an outstanding robust-performance over a range of flight conditions. Kamari [10] optimized the suction mode and the suction position of the boundary layer for SD7003 airfoil. The results show that the flow separation could be reduced by the application of active flow control methods, and the suction was more effective than the blowing.

It can be seen from the above-mentioned studies that, the low speed NLF airfoil aerodynamic characteristics at low Reynolds numbers are mainly influenced by the laminar separation and transition of the boundary layer. Based on the reasonable pressure distribution, the fundament idea of these designs is proposed to weaken or eliminate the laminar separation bubbles. However, for the region nearby the propeller tip, the local Mach number is kept at a high subsonic speed and maybe at a transonic speed. In this case, there may be a mutual interference between the shock waves and laminar boundary layer on the upper surface of the wing. Also, it can influence the position of laminar separation bubble and transition position, or lead to an unsteady aerodynamic force. Therefore, at the low Reynolds numbers, the high subsonic NLF airfoil design is more complex and difficult than the low speed NLF airfoil. To date, the research on the high subsonic at low Reynolds number problem of the propeller tip area has been less studied. In this regard, the purpose of this study is to perform an aero-dynamic design of the high subsonic NLF airfoil at low Reynolds numbers, in order to provide a new design concept and engineering value for the design of the propeller blade tip of the solar UAV.

2 Aerodynamic Optimization Design Methodology

2.1 Numerical Simulation Method for Aerodynamic Characteristics

The numerical simulations of low Reynolds number flow field around the airfoils using e^N, γ-$Re_{\theta t}$ transition model and LES, has been described in detail [2–5], in which the ability of these methods to predict the separation transition and the accuracy of sim-ulating the aerodynamic characteristics is compared. On the basis of these studies, the γ-$Re_{\theta t}$ transition model is used to evaluate the aerodynamic characteristics of high subsonic NLF airfoils at low Reynolds numbers.

Langtry and Menter have established the γ-$Re_{\theta t}$ transition model based on the local variable of SST $k - \omega$ turbulence model [11]. The proposed model combined transition experience relationship with intermittent function. It can control the distribution of intermittent function in boundary layer by the relevance function of transition momentum thickness Reynolds number, and also can control turbulence by the inter-mittent function. Furthermore, the calculation of momentum thickness can be avoided by relating the momentum thickness Reynolds number to the strain rate, which con-tributes to a higher accuracy in predicting the separation transition [12].

2.2 Multi-Groups Cooperative Particle Swarm Algorithm (MCPSO)

The evolution information of individual only arise from single group for traditional particle swarm optimization (GPSO) algorithm. Finally, the evolutionary ability of the algorithm can be weakened and thus leads to an algorithm premature in local optimum. Based on the biological systems symbiosis, a Multi-groups cooperative particle swarm optimization (MCPSO) algorithm is proposed in this paper. The algorithm considered both the influence of the group itself and the information of other symbiotic groups. The populations are divided into several subgroups, and the symbiotic relationship between subgroups is described by a master-slave structure. Subgroups and master group are responsible for the global and local search, respectively.

The basic principle of the algorithm is described as follows: The populations are divided into several subgroups, and they are denoted as the 1st subgroup, 2nd sub-group, 3rd subgroup and 4th subgroup. The 1st subgroup is chosen to be the master group, mainly responsible for local search, and the others are specified to be the subordinate groups, which are responsible for the global search. The standard PSO or improved PSO algorithm is used by the above-mentioned slave groups [13–15]. The best individual information is sent to the 1st subgroup until all the groups have been updated. The speed and position of the 1st subgroup is updated according to the Eq. (1):

$$
\begin{cases}
\overrightarrow{v}_i(t+1) = \omega \overrightarrow{v}_i(t) + c_1 r_1 \left(\overrightarrow{p}^1_i(t) - \overrightarrow{x}_i(t) \right) + c_2 r_2 \left(\overrightarrow{p}^1_g(t) - \overrightarrow{x}_i(t) \right) \\
\qquad + \phi c_3 r_3 (\overrightarrow{p}^{2,\cdots}_g(t) - \overrightarrow{x}_i(t)) \\
\overrightarrow{x}_i(t+1) = \overrightarrow{x}_i(t) + \overrightarrow{v}_i(t+1)
\end{cases}
\tag{1}
$$

where $\overrightarrow{p}^1_g$ is the best particle of the 1st subgroup; $\overrightarrow{p}^{2,\cdots}_g$ is the best particle of other subordinate groups; $\phi \in (0,1)$ is the migration factor.

The performance of the proposed optimization method is tested by using the Ackley function (Eq. (2)) and the Rosenbrock function (Eq. (3)).

$$
f(x) = -20 \cdot \exp\left(-0.2 \sqrt{\frac{1}{n} \sum_{i=1}^{n} x_i^2} \right) - \exp\left(\frac{1}{n} \sum_{i=1}^{n} \cos(2\pi x_i) \right) + 20 + e
\tag{2}
$$

$$
f(x) = \sum_{i=1}^{n-1} \left(100(x_i^2 - x_{i+1})^2 + (x_i - 1)^2 \right)
\tag{3}
$$

The population size of the two algorithms (GPSO and MCPSO) counts 120 particles. Each tests runs 30 times independently, and the maximum iterations of the convergence condition of the algorithm is set to be 1000. The damping strategy is employed in the boundary treatment method [16]. The optimization results of GPSO and MCPSO are shown in Fig. 1 and Fig. 2, respectively. Both of them show that

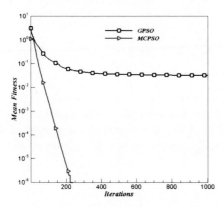

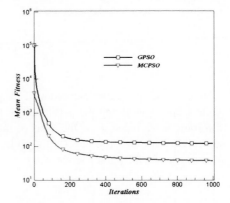

Fig. 1. Convergent curves of the Ackley function

Fig. 2. Convergent curves of the Rosenbrock function

MCPSO has less probability of getting into the local optimal than that of GPSO in the 30 optimization processes. Moreover, the convergence speed is much faster than that of GPSO. Therefore, the optimal solution of aerodynamic design can be positioned in the space quickly and accurately.

2.3   Airfoil Parameterization Method

The disturbing basis function, based on the class/shape function transformation (CST) [17], is employed for the airfoil parameterization. The parameterization method established here can preserve the advantages of the original CST parameterization method. Furthermore, it avoids efficiently the errors which are caused by fitting the initial parameters. The CST function is regarded to be a disturbing function for the airfoil parameterization. Its basic processes are listed as follows:

The original CST method is shown in the Eq. (4):

$$\zeta(\psi) = \psi^{0.5}(1 - \psi) \sum_{i=0}^{n} A_i K_i \psi^i (1 - \psi)^{n-i} + \psi \zeta_T \tag{4}$$

Where, ψ is the x-coordinates of the airfoil; $\zeta(\psi)$ is the y-coordinates of the airfoil; ζ_T is the y-coordinate of the trailing edge point; A_i is the design variables; $K_i = C_n^i = \frac{n!}{i!(n-i)!}$.

The perturbation ΔA_i is added to the initial design variable A_i, and then Eq. (4) can be written as

$$\begin{aligned}\zeta(\psi)' &= \psi^{0.5}(1 - \psi) \sum_{i=0}^{n} (A_i + \Delta A_i) K_i \psi^i (1 - \psi)^{n-i} + \psi \zeta_T \\ &= \zeta(\psi) + \Delta\zeta(\psi)\end{aligned} \tag{5}$$

Where $\zeta(\psi)'$ are the y-coordinates of the new airfoil; $\zeta(\psi)$ is Eq. (4), which represents the y-coordinates of the original airfoil; $\Delta\zeta(\psi)$ are the perturbations of y-coordinates, which can be expressed as

$$\Delta\zeta(\psi) = \psi^{0.5}(1 - \psi) \sum_{i=0}^{n} \Delta A_i K_i \psi^i (1 - \psi)^{n-i} \tag{6}$$

In the other word, the y-coordinates of the new airfoil can be decomposed into the original ones ($\zeta(\psi)$) and the perturbations ($\Delta\zeta(\psi)$). Equation (6) can be regard as the disturbing basis function on the basis of the CST. Thus, this parameterization method not only avoids the trouble of fitting initial design variables, but also preserves the advantages of CST parameterization.

3   Aerodynamic Design Problem

3.1   Optimization Definition

The length of the laminar separation bubble and the transition location should be taken into consideration during the design of high subsonic NLF airfoil at low Reynolds numbers. On the one hand, a backward transition position could reduce the friction drag significantly. However, a laminar flow region with extra length would introduce increase of the adverse pressure gradient at the pressure recovery section obviously, thus leading to a larger laminar separation bubble. Accordingly, the pressure drag is increased significantly because of the larger laminar separation bubble. On the other hand, the classical laminar bubble separation theory [18] suggests that the transition always takes place at the front of the laminar separation bubble at low Reynolds number. Therefore, a forward transition position and reattachment position can weaken the laminar separation bubble, and thus reduce the pressure drag. In turn, the friction drag will be increased by the forward transition. We demonstrate that these factors must be balanced in the aerodynamic design.

A propeller tip airfoil, is selected as the original airfoil. There is a large laminar separation bubble on the airfoil's upper surface at the design point $Ma = 0.72, C_L = 0.5, \mathrm{Re} = 1.1 \times 10^5$. Three fourths of the total drag is the pressure drag, which may be reduced by weakening the laminar separation bubble. The flight turbulence intensity is $Tu = 0.2\%$, and turbulent viscosity ratio is 10. The objective is to reduce the pressure drag coefficient. There are some constraints that the increase of friction drag coefficient is less than ten percent and the maximum thickness of airfoil is greater than 8.92%. The optimization design problem is expressed as:

$$\text{Objective}: Min \quad C_{Dp}$$

$$\text{Subjective to}: \begin{cases} thickness_{\max} \geq 0.0892 \\ \dfrac{C_{Dv}^{optimized} - C_{Dv}^{original}}{C_{Dv}^{original}} \leq 0.1 \end{cases}$$

Where C_{Dp} is the pressure drag coefficient, $thickness_{\max}$ is the maximum thickness of the airfoil, $C_{Dv}^{optimized}$ and $C_{Dv}^{original}$ is the friction drag coefficient of the optimized and original airfoil, respectively.

We adopted ten design variables in all for each airfoil, and five design variables for the upper and lower surface. The population of the algorithm is set to be 30 particles per generation, and the maximum iteration step is set to be 60. The cell numbers of the grid is 617 * 109. Figure 3 and Fig. 4 show the grid distribution of the leading edge grid and the trailing edge, respectively.

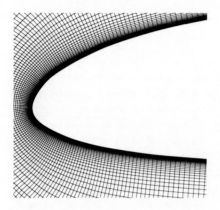

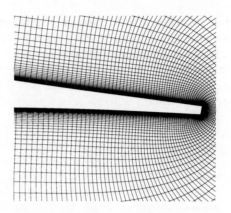

Fig. 3. Grid distribution of the leading edge

Fig. 4. Grid distribution of the trailing edge

3.2   Optimization Results and Discussion

The aerodynamic force coefficients for the original and optimized airfoils are shown in Table 1. By comparison, it can be seen that the drag coefficient decreases by 48 counts at the design condition, the lift-drag ratio increases by 28.7%, the friction drag coefficient increases by 7.4%, and the pitching moment coefficient decreases by 26.6%. The geometric characteristics for the original and optimized airfoils are shown in Table 2. The location of the max thickness moves backward at the optimized airfoil. The Max camber of the optimized airfoil is smaller than that of original, with a backward location of the max camber. The leading edge radius becomes smaller compares the optimized airfoil and the original one.

Table 1. Aerodynamic force coefficients for the original and optimized airfoils ($Ma = 0.72, C_L = 0.5, \mathrm{Re} = 1.1 \times 10^5$)

	C_L	C_D	$K = C_L/C_D$	C_{Dp}	C_{Dv}	C_m
Original airfoil	0.5	0.0222	22.5	0.0168	0.0054	−0.0785
Optimized airfoil	0.5	0.0174	28.7	0.0116	0.0058	−0.0576

Table 2. Geometric characteristics for the original and optimized airfoils

	Max thickness	Location of Max thickness	Max camber	Location of Max camber	Leading edge radius
Original airfoil	0.08914	0.274	0.01936	0.491	0.00775
Optimized airfoil	0.08915	0.306	0.01682	0.697	0.00351

Figure 5 compares the original airfoil with that optimized. The optimized airfoil has a flat upper surface, with a smaller camber before the 60% chord, and larger camber after the 60% chord. Figure 6 compares the Drag divergence curves of original with that of optimized airfoil. The optimized airfoil has smaller drag coefficient over the range of Mach number from 0.68 to 0.78. Figure 7 and Fig. 8 show the comparison results from the pressure drag and the friction coefficient, respectively. Although the friction drag of optimized airfoil shows a slight increase, the pressure drag exhibits a substantial decrease.

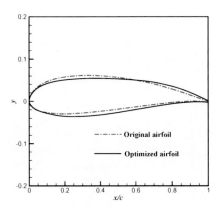

Fig. 5. The original and optimized airfoil

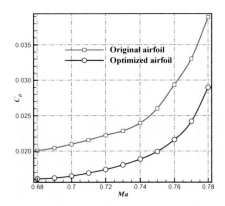

Fig. 6. Drag divergence curves of original and optimized airfoil

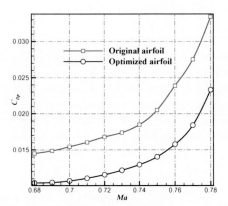

Fig. 7. Pressure drag divergence curves of original and optimized airfoil

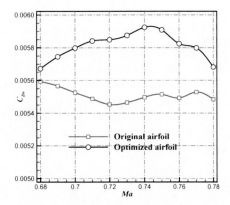

Fig. 8. Friction drag divergence curves of original and optimized airfoil

Figure 9 compares the moment coefficients curves varying with the Mach number, and it is revealed that the pitching moment characteristic of the optimized airfoil is better than the comparative airfoil. Figure 10 shows the comparison of the original pressure distributions with that optimized. The separation position (SP) and reattachment position (RP) are also marked in Fig. 10. The optimized airfoil with larger suction peak surfers a stronger inverse pressure gradient about from 5% chord to 20% chord, and then a weaker inverse pressure gradient about from 50% chord to 75% chord. The SP and RP is moved forward at the position of 15% and 60% chord, respectively. The stronger inverse pressure gradient on the leading edge results in a forward transition location.

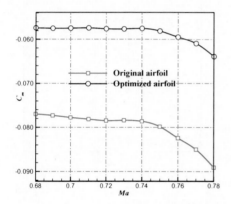

Fig. 9. The moment coefficient curves with the Mach number

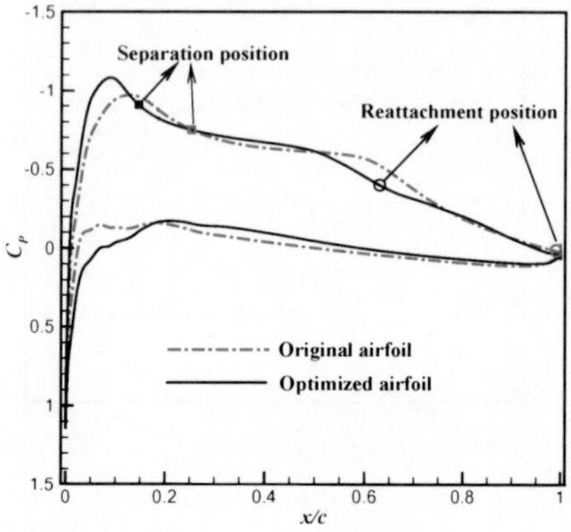

Fig. 10. Pressure distributions of original and optimized airfoil $(Ma = 0.72, \text{Re} = 1.1 \times 10^5, C_L = 0.5)$

Figure 11 shows the friction distributions of original and optimized airfoil. Figures 12 and 14 show the pressure distribution contours of the original and optimized airfoil. The separation positions (SP) and reattachment positions (RP) as function of the Mach numbers are plotted in Fig. 13. The laminar separation bubble of the optimized airfoil is smaller than the original one, as shown in Figs. 11, 12, 13 and 14, and thus leads to a larger reduction of the pressure drag but a slighter increase of the friction drag.

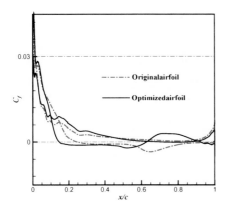

Fig. 11. Friction distributions of original and optimized airfoil ($Ma = 0.72, \text{Re} = 1.1 \times 10^5, C_L = 0.5$)

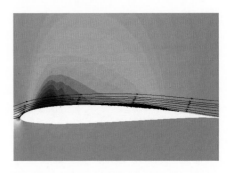

Fig. 12. Pressure contours and streamlines of the original airfoil ($Ma = 0.72, \text{Re} = 1.1 \times 10^5, C_L = 0.5$)

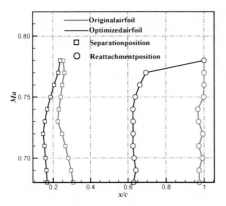

Fig. 13. The SP and RP varying with Mach number ($Ma = 0.72, C_L = 0.5$)

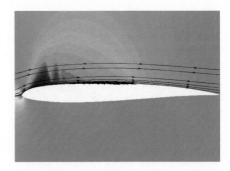

Fig. 14. Pressure contours and streamlines of the optimized airfoil ($Ma = 0.72, \text{Re} = 1.1 \times 10^5, C_L = 0.5$)

The pressure distribution of original and optimized airfoil varying with the Mach number is shown in Fig. 15 and Fig. 16, respectively. The stronger inverse pressure gradient on the leading edge of the optimized airfoil results in a forward transition location. There is an obvious flat pressure distribution in the location of the separation bubble and an inverse pressure gradient in the RP. These characteristics indicate that the laminar separation of optimized airfoil is consistent with the classical laminar separation bubble theory. The SP of original airfoil move forward when the Mach number becomes larger, but the RP is difficult to be detected on the upper surface. These characteristics are suggested to be consistent with the trailing edge separation bubble theory.

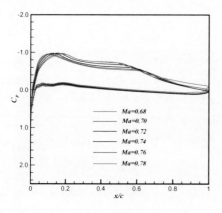

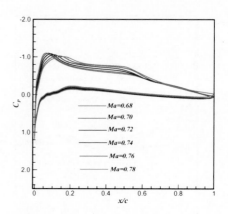

Fig. 15. Pressure distribution of original airfoil

Fig. 16. Pressure distribution of optimized airfoil

The lift coefficient curves as the function of the AoA is shown in Fig. 17. Because of the smaller camber, the lift coefficient decreases at the same AoA. the Lift-Drag coefficient curves of the original and optimized airfoil is plotted in Fig. 18, in which there is a low drag region around the design lift coefficient. Figures 19 and 20 show the change of the lift coefficient as the pressure drag coefficient and friction drag coefficient varies. The low drag region taken place due to the sharp drop of the pressure drag.

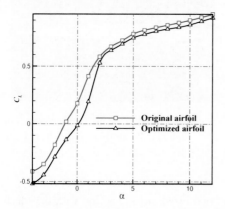

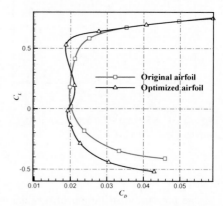

Fig. 17. Lift coefficient curves varying with AOA ($Ma = 0.72, \mathrm{Re} = 1.1 \times 10^5$)

Fig. 18. Lift-Drag coefficient curves of the original and optimized airfoil ($Ma = 0.72$, $\mathrm{Re} = 1.1 \times 10^5$)

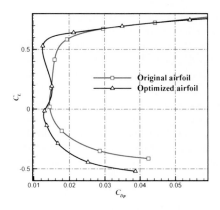 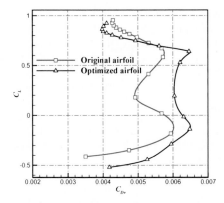

Fig. 19. Lift coefficient varying with pressure drag coefficient ($Ma = 0.72, \mathrm{Re} = 1.1 \times 10^5$)

Fig. 20. Lift coefficient varying with friction drag coefficient ($Ma = 0.72, \mathrm{Re} = 1.1 \times 10^5$)

4 Conclusions

1. Multi-Groups Cooperative Particle Swarm Algorithm (MCPSO) is established, and it is demonstrated to be less probability of getting into the local optimal than GPSO. The optimal solution of aerodynamic design can be captured in the space quickly and accurately. Based on the CST, the disturbing basis function method avoids the trouble of fitting initial design variables, and preserves the advantages of CST parameterization.
2. We select the friction drag increment as a constraint condition, aiming to obtain the minimized pressure drag. It is found that the separation position and reattachment position of the airfoil move forwards to the leading edge by virtue of the optimized pressure distribution. The laminar separation bubble of the optimized airfoil is smaller than the original one, and thus leads to a larger reduction of the pressure drag but a slighter increase of the friction drag. The optimized distribution show better ability to control the laminar separation bubble and transition.

Acknowledgements. This research is sponsored by the National Science Foundation of China (NSFC) under grant No. 11602226 and the University Science Project of Henan Province under grant No. 16A130001.

References

1. Gao G, Li Z, Song B (2010) Key technologies of solar powered unmanned air vehicle. Flight Dyn 28(1):1–4
2. Li F, Bai P et al (2017) Aircraft aerodynamic at low Reynolds number. China Aerospace Press, Beijing, pp 146–147

3. Morgado J, Vizinho R, Silvestre MAR et al (2016) XFOIL vs CFD performance predictions for high lift low Reynolds number airfoils. Aerosp Sci Technol 52:207–214
4. Chen LL, Guo Z (2016) Numerical analysis for low Reynolds number airfoil based on γ-$Re_{\theta t}$ transition model. Acta Aeronautica et Astronautica Sinica 37(4):1114–1126
5. Karasu I, Açikel HH, Genc S (2013) Numerical study on low Reynolds number flows over an aerofoil. J Appl Mech Eng 2(5):309–321
6. Chen XK, Guo Z, Yi F et al (2014) Aerodynamic shape optimization and design of airfoils with Reynolds number. ACTA Aerodynamic Sinica 32(3):300–307
7. Liu XC, Zhu XP, Zhou Z et al (2017) Research on low Reynolds number airfoils based on application of solar-powered aircraft. Acta Aeronautica et Astronautica Sinica 38(4):120459
8. Wang K, Zhu X et al (2015) Studying optimization design of low Reynolds number airfoil using transition model. J Northwest Polytechnical Univ 33(4):580–587
9. Zhao H, Gao Z, Gao Y et al (2017) Effective robust design of high lift NLF airfoil under multi-parameter uncertainty. Aerosp Sci Technol 68:530–542
10. Kamari D, Tadjfar M, Madadi A (2018) Optimization of SD7003 airfoil performance using TBL and CBL at low Reynolds numbers. Aerosp Sci Technol 79:199–211
11. Langtry RB, Menter FR (2009) Correlation-based transition modeling for unstructured parallelized computational fluid dynamics codes. AIAA J 47(12):2849–3116
12. Li J, Gao Z, Huang J et al (2013) Robust design of NLF airfoil. Chin J Aeronaut 26(2):309–318
13. Kennedy J (1997) The particle swarm: social adaptation of knowledge. In: IEEE international conference on evolutionary computation, pp 303–308
14. Li J, Gao Z et al (2011) Airfoil optimization based on distributed particle swarm algorithm. Acta Aerodynamica Sinica 29(4):464–469
15. Li J, Gao Z et al (2012) Aerodynamic optimization system based on CST technique. Acta Aerodynamica Sinica 30(4):443–449
16. Li J (2014) Research on the high performance aircraft aerodynamic design. PhD thesis of Northwestern Polytechnical University, Xi'an
17. Kulfan BM (2008) Universal parametric geometry representation method. J Aircr 45(1):142–158
18. Horton HP (1968) Laminar separation bubbles in two and three dimensional incompressible flow. University of London, Queen Mary

The Investigation of the Maximum Possible Drag Reduction of the Winglet Under the Limitation of Wing Root Bending Moment

Yi Liu[1(✉)], Shaoxiu Ouyang[2], and Xiaoxia Zhao[2]

[1] Shanfei Research Institute, Wuyi Road 1, Hanzhong 723000, Shaanxi, China
evanliuyi@hotmail.com
[2] AVIC Aircraft Co., Ltd., Hanzhong Branch, Hanzhong, Shaanxi, China

Abstract. When reducing the drag of the aircraft using winglet, the size of the winglet is limited mainly under the wing root bending moment increment (ΔBM) due to the winglet. The existence of the maximum possible drag reduction (ΔC_D) at certain ΔBM is not touched by the majority of public resources. 24 winglet plans are evaluated in Computation Fluid Dynamic (CFD) to study the impact of 5 major geometric parameters (span, cant angle, incident angle, twist angle, sweptback angle) on the ΔC_D and ΔCm. It is found that for a fixed span, a linear trend line exists between ΔC_D and ΔCm for various winglet plans that the $\Delta C_D/\Delta$BM is the highest. This trend line is defined as the maximum ΔC_D achievable at certain ΔBM at a fixed span. With increasing span, the obtainable ΔC_D is higher but the $\Delta C_D/\Delta$BM decreases. The concept of the $\Delta C_D/\Delta$BM trend line as the drag reduction limit will greatly simplify the optimization process of winglet design.

Keywords: Winglet · Wing root bending moment · Drag reduction · CFD

1 Introduction

The winglet is a common design feature for modern transport aircraft due to its drag reduction ability since the invention of winglet by Whitcomb at 1976. The winglets are efficient in reducing the induced drag, and for equal ΔBM the ΔC_D can be 50% to 150% higher compared with the simple wing extension (Whicomb 1976; Heyson et al. 1977; Flechner and Jacobs 1978). Winglets are widely studied and applied to various types of aircraft: from early high subsonic transport aircrafts such as KC-135A, DC-10, L-1011, Boeing 747 to regional aircrafts such as Gulfstream III, Bombardier 170/190, ARJ21, not to mention aircrafts of other categories such as EMBraer 202, ERJ145 Early Warning Aircraft, delta wings, etc. Modern airliners such as Boeing 737-800, 787, and Airbus A380 all incorporate winglets of various shapes (Taylor 1986; Boeing 1980; Smith et al. 1996; Mattos et al. 2003). Different wing tip devices are also developed other than Whitcomb winglet, such as vortex diffuser, blended wingtip, shark fin, stepped wingtip devices, tip-fence, etc. These devices have the common feature in reducing the induced drag, and depressing the strength of wingtip vortex. The other impacts of winglet such as wing root bending moment, lateral and directional

X. Zhang (Ed.): APISAT 2018, LNEE 459, pp. 1499–1507, 2019.
https://doi.org/10.1007/978-981-13-3305-7_119

stability, buffeting characteristics are also studied by many researchers (Zhang et al. 2011; Qian et al. 2012; Gong et al. 2011; Jiang and Li 2011; Weng and Xia 2011; Li et al. 2013; Si et al. 2011).

The Multidisciplinary Design Optimization (MDO) technology and surrogate model methods are applied to optimize the parameters of winglet in recent years. The maximum ΔC_D at design lift coefficient (C_L) are usually set as the objective, while the ΔBM is the usual restriction factor. Bai et al. (2014) predict 2520 data based on 100 CFD results utilizing an improved RBF neural network model, and achieve drag reduction of 5.58% at the cost of 3% increase of wing root BM. Li (2015), Jiang (2010), Weng et al. (2013), Jiang et al. (2012) perform optimization design of the winglet parameters such as height, cant angle, sweptback angle etc. at given limiting conditions with Kriging surrogate model, response surface model, genetic algorithms, etc., and reasonable benefits on drag characteristics are achieved.

The above optimization design methods focus on the search of the winglet geometry that gives the maximum drag reduction, however the overall ΔC_D-ΔBM characteristics of the winglet plans are not studied, hence ignore the existence of the limit of drag reduction by winglets. This research studied the ΔC_D-ΔBM characteristics of 24 winglet plans with combination of 5 parameters, and suggested a novel theory on winglet performance by analyzing the overall performance of various winglet plans.

2   Geometry and Numerical Method

The research is based on a transport aircraft with a straight wing, with aspect ratio of 12. Winglet has many geometric parameters whose combinations are difficult to be completely studied. The following simplifications are made to control the number of plans: (1) the sectional airfoils of winglet plans are the same (12% thickness, moderately cambered), which use tip and root sections as control sections. (2) based on the affecting parameters on the performance of finite wing, 5 parameters (span, cant angle, incident angle, twist angle and sweptback angle) are chosen to perform the sensitivity study, which only change one parameter once based on a default winglet. 24 winglet plans and 4 wingtip extension plans (for comparison purpose) are studied, as shown in Fig. 1 and Table 1.

The winglet parameters are defined as following: The cant angle is defined as the angle between wing plane and winglet plane, and when they are parallel the angle is 0°. The span is given as the ratio over the wingtip chord length. The winglet incident angle is the rotated angle around the vector towards its tip, and when it rotates to increase its inwards lift the angle is defined as positive. The twist angle is the incremental incident angle of the winglet tip section over the root section, with larger positive value for higher incident angle of tip section. The sweptback angle is measured at the winglet's leading edge. When constructing the 3-dimensional winglet the wing extends and intercepts with the winglet, with sharp corners rounded to avoid extra pressure drag.

(a) The configuration of the aircraft

(b) Typical winglet geometry

Fig. 1. The geometry of the aircraft and winglet plans

Table 1. The geometric parameters of winglet and wingtip extension plans

Series	Parameter	Value
1	Cant angle (CA)	0°/15°/30°/45°/60°/75°
2	Winglet span (WS)	0.53 Ct/0.76 Ct/0.88 Ct/1.00 Ct/1.12 Ct
3	Incident angle (IA)	−12°/−9°/−6°/−3°/0°/3°
4	Twist angle (TA)	−3°/0°/3°/6°/9°/12°
5	Sweptback angle (SA)	15°/20°/25°/30°/35°
6	Wingtip extension (WE)	0.35 Ct/0.53 Ct/0.71 Ct/0.88 Ct

Note: The default parameter for winglet is CA = 45°, WS = 1.00 Ct,
IA = −9°, TA = 0°.
Ct is the chord length of wingtip section.

The aerodynamic characteristics of the winglet and wing extension plans are obtained by solving the Reynolds Averaged Navior-Stokes equation (RANS). The computing meshes are generated in commercial software ICEM CFD with fidelity of same level for all plans. Hybrid unstructured grids are generated in the computing domain while 15 layers of prism are created normal to the aircraft surface. Higher mesh density is applied at areas of complex flow structure, and the mesh number is around 13 million. The SST k-ω two equation turbulent model is adopted to close the RANS equation. The wall function is used to bridge the physical parameters between the near-wall region and outer flow, with the y+ value around 30. The convection term is discritized by second order upwind scheme, and the kinetic energy k and dissipation rate ω are discritized by upwind scheme. The iterations are done in FLUENT, with equal settings for all winglet plans to ensure the consistency of results.

The credibility of the CFD strategy is validated by wind tunnel test at first, and the mesh and the test model of the aircraft are shown in Fig. 2. The scale of the numerical mesh is of the same scale compared to the test model, both without the winglet. The comparison of drag characteristics is shown in Fig. 3. The data from CFD agree

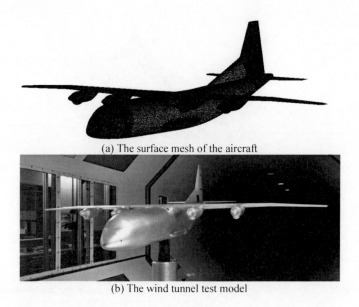

(a) The surface mesh of the aircraft

(b) The wind tunnel test model

Fig. 2. The picture of computational mesh and wind tunnel test model

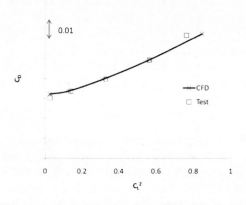

Fig. 3. The comparison of the data from CFD and wind tunnel test

well with the test at the cruise C_L range, which indicate that the CFD method is accurate enough for the research.

3   Results and Analysis

3.1   The Linear Relationship of ΔC_D Versus ΔBM

For wingtip devices, the maximum ΔC_D at design C_L at the cost of limited ΔBM is the ultimate goal for their optimum design. Obviously for certain type of wingtip device with ΔBM limited (say about 3%), the ΔC_D must has the boundary, which can be demonstrated with the simple case: wingtip extension. The wingtip extension essentially has only one effective factor: the span. The aircraft wing is extended for 4 lengths (0.35 Ct/0.53 Ct/0.71 Ct/0.88 Ct), with same airfoil section, leading edge and trailing edge sweptback angle. The ΔC_D of the 4 plans at typical cruise C_L (0.6, 0.8, 1.0) are plotted in Fig. 4. We can easily draw the conclusion that for each C_L the ΔC_D varies linearly with ΔBM, with the R^2 parameter higher than 0.966. This trend line can be defined as the maximum ΔC_D obtainable for certain ΔBM. We can apply the concept to find that for ΔBM of 3%, and $C_L = 0.8$, the wingtip extension can give a ΔC_D of −0.0010.

The scenario could also happen to the winglet that there is a ΔC_D-ΔBM curve that gives the highest ΔC_D at given ΔBM. The parameters for winglet and their combinations are much more complicated, which can yield inferior results easily due to reasons such as local flow separation. By plotting the ΔC_D-ΔBM points of the 24 plans together as shown in Fig. 5, the linear relationship between ΔC_D-ΔBM is found by omitting the bad points, which do not give enough ΔC_D at certain ΔBM. If we check the ΔC_D-ΔBM trend line of winglet, we can conclude that for ΔBM of 3%, and $C_L = 0.8$, the winglet is likely to yield the ΔC_D of −0.0030, three times the value of wingtip extension, while the winglet to fulfill the objective is not unique.

The author would give more comments about the ΔC_D-ΔBM trend line. Firstly, the samples for the conclusion are far from complete, and winglet of more sophistication might give higher ΔC_Ds. Secondly the noise of the data can not be completely avoided from the uncertainty of CFD. Thirdly whether the research based on the large aspect ratio and straight wing being applicable to other aircraft types is not checked. However it is tempting to advance and create such a trend line that settles the boundary of the ΔC_D for an aircraft at given ΔBM, which can also gives more freedom for the definition of winglet geometry, because as long as the parameters are in proper range, they can be equally effective in reducing the drag.

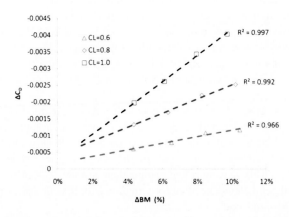

Fig. 4. The ΔC_D-ΔBM linear correlation of different wing tip extension plans

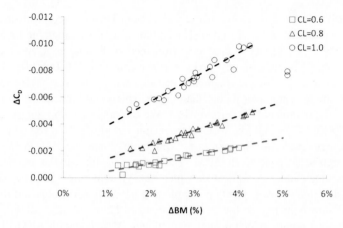

Fig. 5. The ΔC_D-ΔBM linear correlation of different winglet plans

3.2   The Detailed Analysis About Winglet Parameters

The 5 chosen winglet parameters are analyzed in detail about their ΔC_D-ΔBM characteristics. For various cant angles, the increase of its value (winglet more vertically installed) generates less ΔC_D and less ΔBM, with ΔC_D-ΔBM points on the trend line, which can be seen in Fig. 6. It can be concluded that the cant angle does not affect the efficiency parameter $\Delta C_D/\Delta$BM. This feature means that with a higher budget of ΔBM, we can use a lower cant angle to generate more ΔC_D at the same winglet span, or we can tailor the lateral stability which is sensitive to winglet cant angle.

When the winglet incident angle varying from $-12°$ to $3°$, the ΔBM increase proportionally with IA, while only the points of IA between $-9°$ to $-3°$ are on the ΔC_D trend line (Fig. 7). This is straightforward to explain that when the IA is too large or too small, local separation may occur and produce extra drag. However the range of $6°$ being the effective winglet IA gives much freedom in choosing the IA.

The twist angle is the extra incident angle of the winglet tip section compared with the root section, and has similar effect as IA (Fig. 8). When the winglet TA increases, both the ΔC_D and ΔBM increases, and TA range of $0°$ to $6°$ is on the ΔC_D-ΔBM trend line.

The sweptback angle of the winglet does not has a clear trend on the ΔC_D-ΔBM characteristics of the winglet, and the SAs from $25°$ to $35°$ are roughly on the trend line (Fig. 9). For different SA value the winglet geometry is constructed to maintain the span along the wing, therefore the span of the winglet along its leading edge line increases with SA. This may introduce some inconsistency for parameter sensitivity study, but we still get the evidence that for winglets with SAs in a proper range they are on the ΔC_D-ΔBM trend line.

Lastly the ΔC_D-ΔBM points of winglets with different spans are shown in Fig. 10. With smaller spans, the ΔC_D is higher than the trend line, which means a higher efficiency in reducing the drag with a given ΔBM. This is possibly due to the fact that the wingtip vortex is stronger when the location is closer to the wingtip, and the tip vortex of shorter winglet is more efficient in dissipating the wingtip vortex.

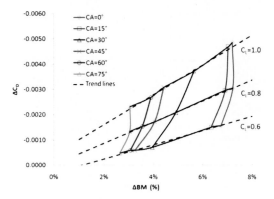

Fig. 6. The ΔC_D-ΔBM curve of winglets with different cant angles

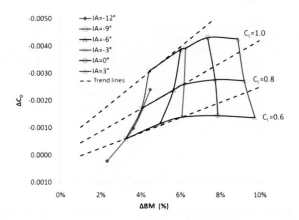

Fig. 7. The ΔC_D-ΔBM curve of winglets with different incident angles

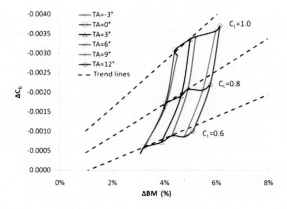

Fig. 8. The ΔC_D-ΔBM curve of winglets with different twist angles

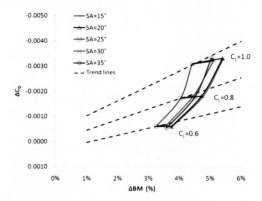

Fig. 9. The ΔC_D-ΔBM curve of winglets with different leading edge sweptback angles

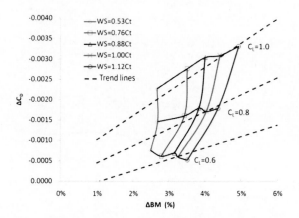

Fig. 10. The ΔC_D-ΔBM curve of winglets with different span

4   Conclusion

The research suggests the existence of the ΔC_D-ΔBM trend line which is the maximum possible ΔC_D for a given ΔBM, and the winglet parameters with relatively wide range would have the same efficiency in reducing C_D at a given ΔBM. The theory is useful for the optimization of winglet that enables the winglet parameters to be adjusted with certain freedom to meet the requirement of lateral stability, stall, buffet etc. while has the same ΔC_D with defined ΔBM. Additionally we can get the slightly higher ΔC_D with a winglet with shorter span, lower cant angle, properly higher incident angle and twist angle at the same cost of ΔBM.

As mentioned before, the samples for the theory is far from complete, for winglet with more sophisticated geometry, and different wing or wing airfoils, the theory needs to be validated and refined further.

References

Bai JQ, Wang D, He XL et al (2014) Application of an improved RBF neural network on aircraft winglet optimization design. Acta Aeronautica et Astronautica Sinica 35(7):1865–1873 (in Chinese)

Boeing Commercial Airplane Company (1980) Selected advanced aerodynamics and active controls technology concepts development on a derivative B-747 aircraft summary report. NASA CR-3295

Flechner SG, Jacobs PF (1978) Experimental results of winglets on first, second and third generation jet transports. NASA TM-72674

Gong ZB, Yang SP, Zhang H et al (2011) Research on aerodynamic properties of advanced wingtip devices for civil transport aircraft. Flight Dyn 29(5):27–30 (in Chinese)

Heyson HH, Riebe GD, Fulton CL (1977) Theoretical parametric study of the relative advantages of winglets and wing-tip extensions. NASA TP-1020

Jiang B, Li J (2011) Aerodynamic analysis of civil aircraft equipped with winglet on numerical simulation. Aeronaut Comput Tech 41(1):38–43 (in Chinese)

Jiang W, Jin HB, Sun WP (2010) Aerodynamic optimization for winglets based on multi-level response surface model. Acta Aeronautica et Astronautica Sinica 31(9):1746–1751 (in Chinese)

Jiang W, Jin HB, Shu KS (2012) Wing-and-winglet integrated design of an amphibious aircraft. Aircr Des 32(5):36–39 (in Chinese)

Li YF, Bai JQ, Guo BZ et al (2015) Studying design of wingtip devices with FFD (Free Form Deformation) technology. J Northwest Polytechnical Univ 33(4):533–539 (in Chinese)

Li ZK, Wu M, Liu C (2013) Aerodynamic analysis of advanced wingtip devices on numerical simulation. Chin J Appl Mech 30(4):498–503 (in Chinese)

Mattos BS, Macedo AP, Silva Filho DH (2003) Considerations about winglet design. In: AIAA, pp 2003–3502

Qian GP, Liu PQ, Yang SP et al (2012) A comprehensive study on wingtip devices in large civil aircraft. Acta Aeronautica et Astronautica Sinica 33(4):634–639 (in Chinese)

Si L, Wang HP, Gong CC (2011) Investigation of effects of winglets on wing's aerodynamic and structural behavior. Acta Aerodynamica Sinica 29(2):177–181 (in Chinese)

Smith LA, Campbell RL (1996) Effects of winglets on the drag of a low-aspect-ratio configuration. NASA TP-3563

Taylor AB (1986) Selected winglet and mixed-flow long-duct nacelle development for DC-10 derivative aircraft summary report. NASA CR-3296

Weng CT, Xia L (2011) Study of civil airplane wing tip design. Flight Dyn 29(2):13–16 (in Chinese)

Weng CT, Xia L, Li D (2013) The optimization design of winglet for civil aircraft. Acta Aerodynamica Sinica 31(1):56–63 (in Chinese)

Whitcomb RT (1976) A design approach and selected wind-tunnel results at high subsonic speeds for wing-tip mounted winglets. NASA TN D-8260

Zhang JJ, Yang SP, Si JT (2011) Study on aircraft lateral-directional character with different winglets. Flight Dyn 29(4):41–44 (in Chinese)

Drag Reduction Effect of a Variable Camber Wing of a Transport Aircraft Based on Trailing Edge Flap Deflection of Small Angles

Yi Liu[1]([✉]), Shaoxiu Ouyang[2], and Xiaoxia Zhao[2]

[1] Shanfei Research Institute, Wuyi Road 1, Hanzhong 723000, Shaanxi, China
evanliuyi@hotmail.com
[2] AVIC Aircraft Co., Ltd., Hanzhong Branch, Hanzhong, Shaanxi, China

Abstract. The improvement of the lift to drag ratio (L/D), and the decrease of the required power for flight, are important measures to enhance the range and endurance of propeller aircrafts. The research investigates the effect of variable camber wing on the flight performance of a transport aircraft based on wind tunnel tests. The wing camber is altered by deflect the trailing edge flap in small angles. The tests reveal that the maximum L/D reaches its peak when the flap deflects 6°, and is 4.0% higher than the configuration with flap retracted. With flap deflecting 6°, the lift coefficient for the maximum L/D is 16.1% higher. The flight performance analysis shows that the range and the endurance increase by 4.0% and 13.7% respectively at the selected flap angle. By flying at larger lift coefficient and larger L/D, the required power for endurance flight decreases significantly, which is the reason for the improvement of the range and endurance performance.

Keywords: Variable camber wing · Drag reduction · Flight performance · Wind tunnel test

1 Introduction

Modern transport aircrafts generally perform their aerodynamic optimum design according to one or several design cruise points, and fly with fixed configuration during cruise stage. Loss of performance appears when the flight is not at design points. The morphing aircraft concept is advanced to overcome the drawback by changing the geometry of the aircraft during flight. For specific flight condition the aircraft morphs to the corresponding optimum geometry, thus the performance is improved. NASA, Boeing, the US military and other organizations have investigated several morphing aircraft concept since 1970s, including Mission Adaptive Wing, Active Flexible Wing, Active Aero-elastic Wing, Smart Wing et al. with certain achievements (Zhang et al. 2013). The Boeing research suggests that by applying variable camber wing technology, the fuel saving for international flight is as high as 4.2% (Boeing 1980). The flight tests of F-111 aircraft adopting adaptive wing which change its camber with continuous and smooth skin validate the improvement of flight performance at non-design status

© Springer Nature Singapore Pte Ltd. 2019
X. Zhang (Ed.): APISAT 2018, LNEE 459, pp. 1508–1514, 2019.
https://doi.org/10.1007/978-981-13-3305-7_120

(Powers 1992; Powers 1997). The wind tunnel test of a fighter type wing-body con-figuration shows that by increasing the camber of the wing at leading and trailing edge, the drag is reduced at larger lift coefficients, and the lift coefficient for buffet onset is significantly delayed (Ferris 1977).

The wing camber can also be modified by deflecting the traditional trailing edge flap, aileron, leading edge flap, et al., and the drag reduction effect is obtainable. The flight performance analysis based on the wind tunnel test data of L-1011 aircraft demonstrates that 1% to 3% increment of L/D ratio is achieved at cruise condition, and the benefit of L/D reaches 10% at non-standard, high lift flight conditions, by deflecting its flap and aileron in small angles (Bolonkin et al. 1999). The numerical study of 2-dimensional GAW-1 airfoil reveals that by deflecting the trailing edge Fowler flap and aileron in small angles, the L/D ratio is enhanced at relative large lift coefficients (Lu et al. 2016). The wind tunnel test study of the F-8 fighter aircraft shows that the conformal leading and trailing edge flaps have better performance in drag reduction and maximum lift enhancement compared with simple hinged flaps (Boltz et al. 1977).

To change the wing camber with continuous and smooth skin is not applied to aircrafts in mass production until now due to the weight penalty, structure complexity, and other factors. On the other hand, the drag reduction by the deflection of existing flap and aileron systems is more practical in wider circumstances. Furthermore, for propeller aircrafts executing reconnaissance or surveillance missions, the endurance is of higher importance, and the parameters for flight performance optimization comprise not only L/D, but also $C_L^{1.5}/C_D$, which is not widely touched by researchers. This research investigates the aerodynamic effect of trailing edge flap deflected in small angles by wind tunnel test, and analyzes its impact on propeller aircraft flight range and endurance.

2   Geometry and Wind Tunnel Test

The propeller transport aircraft is equipped with two-slot trailing edge Fowler flap system. The outer shape of the flap is clean, with the supporting structure and actuating system contained within the contour of the airfoil section. When deflected by small angles, the chord and the camber of the wing are increased simultaneously, and the gap between the flap and the main wing remains small. On the planform view, the flap occupied the area from 10% to 70% of the wing semi-span. The airfoil section, planform, and test model of the flap can be seen in Fig. 1.

The deflection of the flap will increase the lift and drag of the wing at the same time, and the quantity of the $\Delta C_L, \Delta C_D$ is crucial for the overall L/D. For a 3 dimensional wing, the ΔC_D due to flap deflection comprises three terms: the form drag, the induced drag and the interference drag. The form drag is determined by the type and geometric parameters of the flap section, and for Fowler flap the form drag is relatively small. The interference drag is a fraction of the form drag. The induced drag is relevant to the planform arrangement of the flap, and can be reduced by increasing the ratio of the wing span with the flap over the complete wing span, and decreasing the gap between left and right flap (Committee of Aircraft Design Manual 2002). The type and planform arrangement of the flap on the aircraft has relatively small ΔC_D at small ΔC_L, therefore the potential for drag reduction with small flap deflection angles can be expected. The wind tunnel test of the aircraft model is carried out with flap deflected for 2°, 4°, 6°, 8°, 10°, as well as retracted configuration (the "baseline").

The 0.07 scale model of the transport aircraft is tested in FL-12 wind tunnel of China Aerodynamics Research and Development Center (CARDC). The FL-12 tunnel is a single return low speed tunnel with a closed test section. The test section size is 8 m * 4 m * 3 m (length * width * height). The wind speed range can vary from 50 to 100 m/s, with turbulent degree less than 0.12%. The wind speed for the test is 70 m/s, with Reynolds number based on mean aerodynamic chord around 1 million.

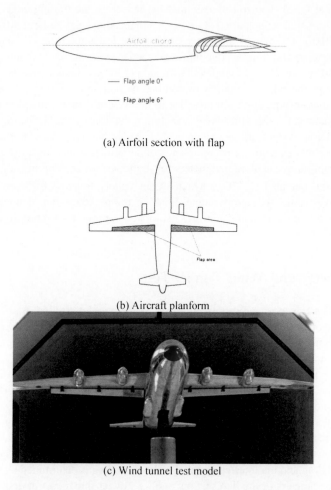

(a) Airfoil section with flap

(b) Aircraft planform

(c) Wind tunnel test model

Fig. 1. The airfoil section, planform, and test model of the flap

3   Aerodynamic Characteristics

The lift and drag characteristics of the transport aircraft with flap angle (δ_f) of 0° to 10° are shown in Figs. 2, 3, 4 and Table 1. The C_L and the maximum lift coefficient C_{Lmax} increase proportionally to flap angle, with stall angle almost unchanged. The drag polar indicates that at smaller C_Ls, the C_D increases as the flap angle enlarges, but above certain C_L the configurations with flap deflected have less C_D at the same C_L compared with the flap retracted configuration. It is clearer in the L/D curve that when the flap is deflected in small angles, the L/D_{max} increases in different extent, and for flap angle of 6° the L/D_{max} is highest with 4% increment compared with the baseline. Also noticeable is that the C_L for L/D_{max} increases proportionally with the flap, in higher rate compared to the increase of C_{Lmax}. This is somewhat detrimental for flight since the margin between the C_L for L/D_{max} and C_{Lmax} is smaller, and the flight speed is lower.

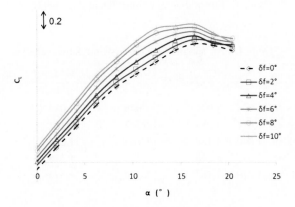

Fig. 2. Lift curve of different flap angles

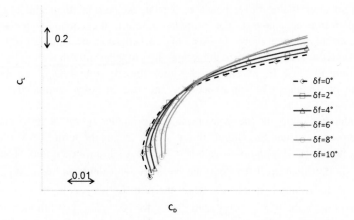

Fig. 3. Drag polar of different flap angles

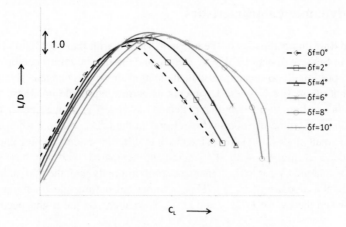

Fig. 4. Lift to drag ratio of different flap angles

Table 1. Lift and drag parameters (in relative value)

δ_f	$C_{L\max}$	$L/D_{\max}$	$(C_L^{1.5}/C_D)_{\max}$	C_L for $L/D_{\max}$	C_L for $(C_L^{1.5}/C_D)_{\max}$
0°	1	1	1	1	1
2°	1.016	1.006	1.041	1.045	1.045
4°	1.034	1.021	1.087	1.103	1.131
6°	1.059	1.040	1.138	1.166	1.227
8°	1.085	1.027	1.148	1.238	1.318
10°	1.105	1.020	1.161	1.274	1.375

4 Flight Performance Analysis

For transport aircraft the cruise phase is the largest part in flight and is chosen to perform performance analysis. For propeller aircraft, the equations for range and endurance calculation are shown in Eqs. (1) and (2), given by Pamadi (2004).

It can be seen that to maximize the range, the propeller aircraft should fly at the C_L corresponding to the $L/D_{\max}$. The range varies linearly with $L/D_{\max}$, so the aircraft flying with flap angle of 6° can expect 4% increase in range compared to baseline configuration.

For endurance flight, the fuel consumption of propeller aircraft is determined by the required power to sustain the aircraft in air, which means the term $D * V$ (drag times velocity) should be minimum to achieve the longest airborne time. The drag is reversely proportional to L/D, and the velocity is calculated as $(2W/\rho SC_L)^{1/2}$, therefore the endurance is maximum when the term $C_L^{1.5}/C_D$ is highest. As shown in Table 1 and Fig. 5, the $(C_L^{1.5}/C_D)_{\max}$ and the C_L for $(C_L^{1.5}/C_D)_{\max}$ both increase almost linearly with flap angle. The flap angle of 6° is a good choice for endurance flight in that the endurance can be increased by 13.8% compared to the baseline while the C_L is not too close to C_{Lmax}. Further larger flap angles could generate slightly higher $(C_L^{1.5}/C_D)_{max}$, but can not justify the drawbacks that the aircraft being required to flight at lower speeds and closer to stall point. The endurance of the propeller aircraft is increased in higher magnitude compared to the range, which is attributable to the superior performance of the configuration with flap deflected for small angles at higher C_Ls. The conclusion does not hold true for turbo fan engine aircraft whose specific fuel consumption is proportional to the thrust.

$$\text{Range} = \frac{\eta_P}{c} \cdot \frac{L}{D} \cdot \ln\frac{W_0}{W_1} \tag{1}$$

$$\text{Endurance} = \frac{2\eta_p}{C} \cdot \frac{C_L^{1.5}}{C_D} \cdot \sqrt{\frac{\rho S}{2}} \cdot \left(\sqrt{\frac{1}{W_1}} - \sqrt{\frac{1}{W_0}} \right) \tag{2}$$

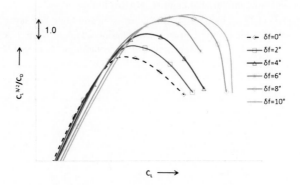

Fig. 5. The endurance parameter $C_L^{1.5}/C_D$

5   Conclusion

Aerodynamic characteristics and flight performance of the propeller transport aircraft with flap deflected for small angles are analyzed. The flap angle of 6° is the optimum which increases the range and endurance of the aircraft for 4% and 13.8% respectively.

Further work might be done to investigate the combined deflection of the flap and aileron, which will increase the camber of the wing till the wingtip. Less energy loss is expected due to the continuous spanwise lift distribution of the plan, hence higher lift to drag ratio.

References

Boeing (1980) Assessment of variable camber for application to transport aircraft. NASA-CR-158930

Bolonkin A, Gilyard GB (1999) Estimated benefits of variable-geometry wing camber control for transport aircraft. NASA-TM-1999-206586

Boltz FW, Pena DF (1977) Aerodynamic characteristics of an F-8 aircraft configuration with a variable camber wing at Mach numbers from 0.70 to 1.15. NASA-TM-78432

Committee of Aircraft Design Manual (2002) Aircraft design manual volume 6 aerodynamic design. Aviation Industry Press, Beijing (in Chinese)

Ferris JC (1977) Wind-tunnel investigation of a variable camber and twist wing. NASA-TN-D-8475

Lu WS, Tian Y, Liu PQ et al (2016) Aerodynamic performance of GAW-1 airfoil leading-edge and trailing-edge variable camber. Acta Aeronautica et Astronautica Sinica 37(2):437–450 (in Chinese)

Pamadi BN (2004) Performance, stability, dynamics, and control of airplanes, 2nd edn. AIAA, Reston

Powers SG, Webb LD, Friend EL et al (1992) Flight test results from a supercritical mission adaptive wing with smooth variable camber. NASA-TM-4415

Powers SG, Webb LD (1997) Flight wing surface pressure and boundary-layer data report from the F-111 smooth variable-camber supercritical mission adaptive wing. NASA-TM-4789

Zhang YB, Liu MD, Xiong JJ (2013) Morphing aircraft technology. Aeronaut Sci Technol 8:64–68 (in Chinese)

The Research on the Drag Reduction of Transport Aircraft Using Ventral Fins

Shaoxiu Ouyang[1(✉)], Yi Liu[2], Xiaoxia Zhao[2], and Xiao Zhang[2]

[1] Shanfei Research Institute, Wuyi Road 1, Hanzhong 723000, Shaanxi, China
85923917@qq.com
[2] AVIC Aircraft Co., Ltd., Hanzhong Branch, Hanzhong, Shaanxi, China

Abstract. The aerodynamic characteristics of the upswept tail of a transport aircraft are studied and the optimum arrangements of ventral fins to reduce the drag are evaluated. Two symmetric vortices emerge under the fuselage after-body, which are the major causal factor of the additional pressure drag. A pair of fins installed under the fuselage, extruding the core of the vortices effectively damp the vortex. Parametric study shows that the length, height, location and yaw angle of the fins are the sensitive factors of drag reduction. Drag reduction of 0.0021 is achieved in wind tunnel test for typical cruise angle of attack (AoA). The pitching moment has nose down tendency and the longitudinal stability is reduced. The reason is that the pressure recovery of the bottom surface of the tail is improved by adding the fins, which is dependent on the AoA.

Keywords: Upswept tail · Drag reduction · Ventral fin · Vortex flow · Pressure recovery

1 Introduction

The tails of conventional airliner or cargo aircraft all have upswept angles of certain degrees. For airliners the upswept tail minimizes the undercarriage height required by ground clearance, while for cargo or military aircraft there is an additional requirement of locating cargo gates under the tail, which allows airdropping during flight or dropping down as ramp on ground in hostile environments. There are 2 types of concepts about the tails with cargo door: the earlier transport aircrafts with turbo-propeller engines such as C-130, C-160 and AN-12 adopt large upswept angles to meet the requirement of cross section for cargo with minimum cutout length, and the associated drag penalty is high. The aircrafts with turbofan engines such as C-141A, C-5A and Il-76 have smaller upswept angles, smoother tail shape and larger tail fineness ratio, which reduce the drag with the cost of longer cutout length (Committe of Aircraft Design Manual 2002).

The upswept tail generates additional drag and complex flow structure in surrounding space. The drag increment can be estimated by cross flow theory, which assumes that the pressure drag can be calculated by the flow vector perpendicular to the fuselage which is determined by AoA and the tail upswept angle. For circular cross section the drag coefficient based on cross flow dynamic pressure is about 1, while the

© Springer Nature Singapore Pte Ltd. 2019
X. Zhang (Ed.): APISAT 2018, LNEE 459, pp. 1515–1525, 2019.
https://doi.org/10.1007/978-981-13-3305-7_121

drag coefficient for rectangular section with rounded corner reaches 1.5 to 2 (Raymer 2006; Torenbeek 1982). Researches show that the section shape, the variation of the cross section shape along the fuselage axis, upswept angle, fineness ratio and contraction ratio are the major parameters that affecting the pressure drag of the afterbody. The frictional drag is nearly constant when the AoA increases as long as the AoA is less than the upswept angle, but the pressure drag decreases at the same AoA range (Kong et al. 2003a; Mao et al. 2011; Zhang et al. 2010). The characteristics of flowfield around fuselage tail have been investigate by wind tunnel tests (using oil flow, smoke, laser, et al.) and Computational Fluid Dynamics (CFD) by many researchers, and the major conclusions are: cross flow emerges under the fuselage tail, and a pair of symmetric vortex are observed in the space near the body and move downstream. The vortex system shifts from lower vortexes, none vortex to upper vortexes when the AoA changes from negative to positive. Higher upswept angle enhances the afterbody cross flow and the separation of flow (Kong et al. 2002; Huang et al. 2003; Kong et al. 2003b; Wang 2009; Zhang et al. 2010).

The reduction of the drag of fuselage afterbody has drawn wide attention around the world, as it usually accounts for 15% to 20% of the total drag of the fuselage, or 5% to 7% of total aircraft drag. Two methodologies are commonly adopted to reduce the afterbody drag. The first is the optimization of the tail parameter and shape, the second is the installation of flow control devices on the afterbody to change the flow structure. The first method suggests lower upswept angle, optimized cross section, larger fineness ratio, smaller contraction ratio, et al. to avoid or suppress vortex flow and reduce the drag, and is applicable to preliminary stage of fuselage design (Zhang et al. 2010; Mao et al. 2011; Shan et al. 2013). The second method installs flow control devices on the baseline afterbody to reduce the drag, hence is applicable to wider situations. Wu et al. installed a pair of chine on the tail at the starting point of vortex flow, which depressed the separation vortex and reduced the drag for 0.0009 at cruise condition (2012). Yu et al. (2011), Du et al. (2012) investigated the effects of vortex generators (VGs) at different location and incident angle on the control of afterbody separation, and reduced the drag of the aircraft by around 1%. Wortman (1999) validated the effects VGs on Boeing 747 and C-5A in wind tunnel tests and the drag reduction rates based on fuselage drag was 3% and 6% respectively. Xia et al. (1991) studied the effect of different flow control devices on aircraft afterbody, and the small plates reduces the fuselage drag by 10.8% at $0°$ of AOA, but the drag reduction dropped to 5.7% when the horizontal tail was installed.

The above mentioned researches reveal detailed information about the flow structure of the upswept tail and achieve significant progress in drag reduction of the tail. However, improvement in research is still needed in that most researches are based on single fuselage, with the ignorance of the complex interference with the other components of aircraft. The suggested flow control devices are generally small in size and the drag reduction rate is less than 10% of the drag of fuselage. This report studies the flow characteristics of the afterbody of a complete aircraft in production, and designs a pair of long fins to significantly damp the vortex flow, achieving much higher drag reduction rate, hence is highly attractive to engineering application.

2   General Information About the Research

The transport aircraft for research has flat bottom under its tail for cargo door, and the upswept angle is 11°. The typical cross section of the tail consists an upper arc, a straight line in the bottom, and the connecting arcs of smaller radius, as shown in Fig. 1. Based on the geometric properties, strong vortex flows are expected under the tail, and the pressure drag could be high. The suggested fin plans in the research are installed on the flat portion of the tail bottom and are convenient to be manufactured.

The effects of the fins on aircraft aerodynamic characteristics are studied by numerical simulations and wind tunnel tests. The numerical simulation is based on finite volume method, and solves Reynolds Averaged Navior-Stokes (RANS) equation to get the aerodynamic force and flowfield. Mixed unstructured mesh is adopted to discretize the flow domain, and mesh density is set higher in critical areas where vortex or separation flows exist. Prisms are created normal to wall boundaries and the y+ value is between 30 and 100 at the calculated flow conditions. Wall function method is used to blend the variables between boundary layer and outer flow, which reduces the mesh number with reasonable accuracy. The total mesh number for complete aircraft simulation is around 15 million, and the sketch of the surface mesh is shown in Fig. 2. The cases are iterated in commercial software FLUENT using pressure based coupled solver. The advection term is discretized by second order upwind scheme. The SST k-ω two equation turbulent model is adopted to model the viscosity effect, with turbulent energy k and dissipation rate ω discretized by upwind method.

The optimal fins selected by CFD are validated by wind tunnel test of a 0.07 scale model in the FL-12 tunnel of China Aerodynamics Research and Development Centre (CARDC). It is a single return low speed tunnel with a closed test section. The cross section is 4 m in width and 3 m in height. The wind speed can vary from 50 to 100 m/s, with turbulent degree less than 0.12%. In order to minimize the interference from supporting stings the test model is reversely installed, as shown in Fig. 2.

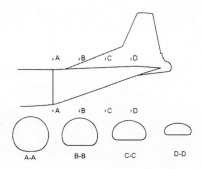

Fig. 1. The sketch of the upswept tail of the transport aircraft

Fig. 2. Sketch of the computing mesh and the test model

3   The Flowfield Characteristics of the Upswept Tail

The tail of large upswept angle and flat bottom promotes strong cross flow around it as shown Fig. 3. The cross flow velocity increases as the air moves downstream, and eventually forms a pair of vortex flow. The core of the vortex is close to the bottom surface of the tail, which makes it possible to install effective flow control devices on the fuselage. The existence of vortex flow structure creates low pressure areas on nearby fuselage, hence increases the pressure drag.

   The vorticity which identifies the core of vortex flow demonstrates that the intensity of vortex flow is dependent on the AoA, as shown in Fig. 4. As the AoA drops, the width of the vortex core increases, extending longer and lower downstream. The enhanced vortex flow leads to larger energy dissipation and higher pressure drag. The drag of fuselage part at different AoAs shows the same trend that higher drag is observed at lower AoAs, as shown in Fig. 5. It can be seen that when AoA is less than 8° the drag increases rapidly. As the vicious drag is proportional to the wetted area which is almost constant in the case, it can be concluded that the drag increment is from the contribution of the pressure drag. For typical cruise AoA of 4° the drag is higher than the minimum drag of the fuselage by 0.0023, which suggests that the potential of drag reduction is rather high.

Fig. 3. Projected velocity vector in cross section of aircraft tail (cruise angle, half model)

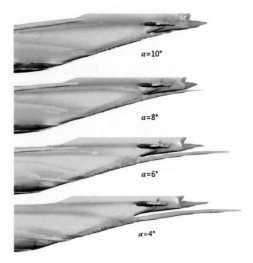

Fig. 4. Vorticity of equal value around fuselage tail

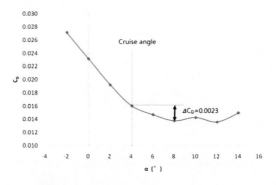

Fig. 5. Drag characteristics of the fuselage (wind tunnel test)

4 Parametric Study of the Fins

The geometric parameters including length, height, installation position and yaw angle are studied to find the sensitive parameters, and the arrangement of the fins has to fulfill the constraints of structure at the same time. The selected parameters are varied systematically and evaluated in CFD to find the sensitivity on drag reduction, and dozens of fin plans are studied to find the optimum. The fins have 3 different sizes as shown in Fig. 6, and the lengths are 2.5 m, 4.5 m and 6.0 m. The 2.5 m fins are installed after the cargo gate and are constrained in length by structure limit. The 4.5 m and 6.0 m fins are installed partly on the cargo gate and partly on the fuselage, and need to be separated to 2 sections for cargo gate operation. Selected fin plans are shown in Fig. 9. All the fin plans are rounded flat plate with thickness of 50 mm for real aircraft.

The effect of the spanwise position on drag reduction is studied based on 4.5 m fins, and the results can be seen in Fig. 7. The best value for spanwise location Z is 1.5 m, where the fins intersect with the vortex core at the upstream of vortex and almost coincide with the vortex core. If the fins move to a Z value either smaller or larger the damping effect of them are reduced hence the drag reduction is less. Different

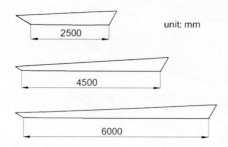

Fig. 6. Geometry of the drag reduction fins

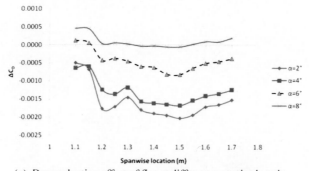

(a) Drag reduction effect of fins at different spanwise location

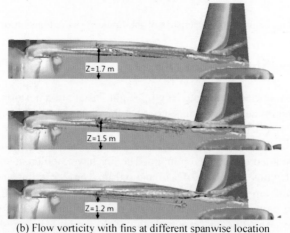

(b) Flow vorticity with fins at different spanwise location

Fig. 7. Drag reduction and vorticity with fins at different spanwise location

heights of the fins significantly affect the effect of drag reduction, as shown in Fig. 8. When the height is higher than 0.5 m the drag reduction decreases in linear relation with it. It can be seen that when the height increases the secondary vortex is created by the fin itself, which generates extra energy lost and higher drag. The height of the fins should be less than 0.5 m to achieve reasonable drag reduction.

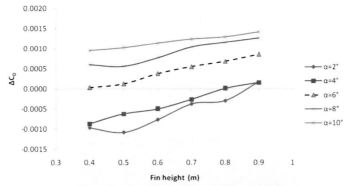

(a) Drag reduction effect of fins of different heights

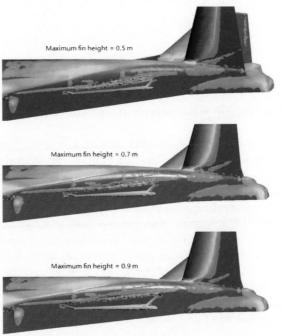

(b) Flow vorticity with fins of different heights

Fig. 8. Drag reduction and vorticity of fins of different heights

Typical fin plans and the corresponding drag reduction at cruise AoA are shown in Fig. 9. The fin plans with no yaw angles have poor performance in drag reduction as they are not perfectly aligned with the direction of the vortex core. The yawed fins achieve certain degree of drag reduction, but the effect is limited. The longest 6.0 m fins show the best effect in drag reduction, and the drag reduction reaches 0.0028 when properly arranged in CFD.

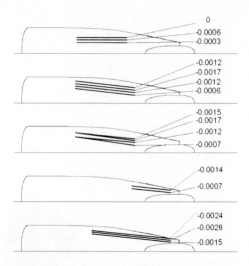

Fig. 9. Drag reduction of different fins ($\alpha = 4°$)

5   The Aerodynamic Characteristics of the Optimal Fins

The optimum paired fins selected by CFD are tested in wind tunnels, and the aerodynamic characteristics of them are shown in Figs. 10, 11, 12 and 13. The picture of the vorticity shows that the fins effectively depress the strong vortex flow of the baseline aircraft. The core of the vortex flow can no longer be seen with fins installed, indicating that the energy dissipation is reduced significantly. The drag reduction from wind tunnel test is 0.0021 at cruise AoA and has same trend with CFD, which shows that the CFD is fairly valid in simulate this type of flow. The discrepancy between the test and CFD is possibly caused by the modeling capacity of the numerical simulations, because the fidelity of the unstructured mesh is not high enough, introducing numerical viscosity, and the interference between the vortex and the fin might be too complex to be fully predicted by RANS models. On the other hand, the bolts that fix the fin model on the aircraft generate extra drag, which also make the test data inferior to CFD.

The fins also affect the pitching moment of the aircraft, showing reduced longitudinal stability and nose down tendency. The major reason is that the fins recover the pressure of the tail, generating positive lift and nose-down pitching moment. This effect decreases as the AoA is larger, which creates negative lift slope on the tail and reduces the longitudinal stability. The comparison of the pressure distribution before and after the installation of the fin from CFD shows the trend of the increased pressure on the bottom of the tail.

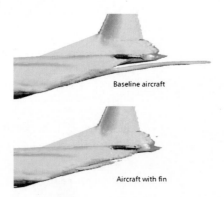

Baseline aircraft

Aircraft with fin

Fig. 10. The effect of fins on flow vorticity ($\alpha = 4°$)

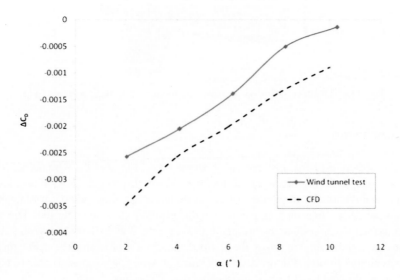

Fig. 11. The quantity of drag reduction from CFD and wind tunnel test

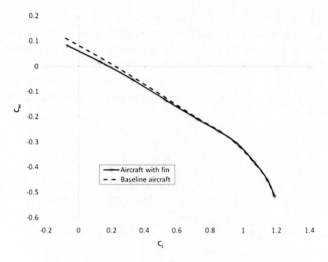

Fig. 12. The effect of fins on pitching moment (test)

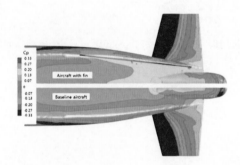

Fig. 13. The effect of fins on pressure distribution of fuselage bottom (CFD)

6 Conclusion

The flow structure of the upswept tail is studied by CFD, and the symmetric vortex flow is found to be the major causal factor of the high pressure drag. Fin plans are designed parametrically to suppress or eliminate the vortex flow, and the parameters of the optimal paired fins such as length, height, position et al. are selected. The wind tunnel test demonstrates that the optimal fins reduce the drag by 0.0021 at cruise AOA, which can significantly improve the performance of the aircraft. The CFD method achieves similar results of drag reduction compared with tests, and is capable to obtain the detailed flowfield and to find the mechanism of the fins.

References

Committe of Aircraft Design Manual (2002) Aircraft design manual volume 6 aerodynamic design. Aviation Industry Press, Beijing (in Chinese)

Du XQ, Jiang ZY, Tong SX et al (2012) Experimental study on control of separated flow over aft-body and drag reduction mechanism by using vortex generator. Eng Mech 29(8):360–365

Huang T, Deng XY, Wang YK et al (2003) An experimental study of flow patterns over the upswept after-body of a civil aircraft. Acta Aerodynamica Sinica 21(1):29–37 (in Chinese)

Kong FM, Hua J, Deng XY et al (2003a) Analysis of the flows and the drag about upswept afterbodies. Acta Aerodynamica Sinica 21(1):67–74 (in Chinese)

Kong FM, Hua J, Feng YN et al (2002) Investigation of the flow mechanism on the afterbody with larger upswept angle. Acta Aerodynamica Sinica 20(3):326–330 (in Chinese)

Kong FM, Hua J, Feng YN et al (2003b) Effects of geometry parameters and flow parameters on drag coefficient of upswept afterbodies. J Beijing Univ Aeronaut Astronaut 29(1):39–42 (in Chinese)

Mao XM, Zhang BQ, Wang YY (2011) Research on parameters influence of fuselage afterbody with large upswept angle of transport aircraft. Aeronaut Comput Tech 41(1):78–81 (in Chinese)

Raymer DP (2006) Aircraft design: a conceptual approach, 4th edn. AIAA Education Series, Reston

Shan D, Wang ZP (2013) Research on aerodynamic characteristics of rear of transport aircrafts. Aeronaut Comput Tech 43(4):79–82 (in Chinese)

Torenbeek E (1982) Synthesis of subsonic airplane design. Delft University Press, Delft

Wang J (2009) Aerotransport profile design based on UG secondly development and analysis of the aftbody's flow distribution. Nanjing University of Aeronautics and Astronautics (in Chinese)

Wortman A (1999) Reduction of fuselage form drag by vortex flows. J Aircr 36(3):501–506

Wu N, Duan ZY, Liao ZR et al (2012) Research on chine of aerotransport after-body for drag reduction and stability enhancement. Acta Aerodynamica Sinica 30(2):223–227 (in Chinese)

Xia XJ, Ma SL (1991) The effect of afterbody devices on drag reduction. Acta Aeronoutica et Astronautica Sinica 12(8):435–438 (in Chinese)

Yu YZ, Liu JF, Jiang ZY et al (2011) The investigation of flow control and drag reduction mechanism for transport airplane aft-body. Acta Aerodynamica Sinica 29(5):640–644 (in Chinese)

Zhang BQ, Wang YY, Duan ZY et al (2010) Design method for large upswept afterbody of transport aircraft. Acta Aeronoutica et Astronautica Sinica 31(11):1933–1939 (in Chinese)

Zheng LQ, Yang Y (2010) Influence of landing gear cabin on drag of a transport aircraft body. Aeronaut Comput Tech 40(4):56–58 (in Chinese)

A Study on Aerodynamic Drag Reduction for High Speed Helicopter Airframe

Noboru Kobiki[1]([✉]), Yasutada Tanabe[1], Masahiko Sugiura[1],
and Hideaki Sugawara[2]

[1] JAXA, 6-13-1 Osawa, Mitaka, Tokyo 181-0015, Japan
kobiki.noboru@jaxa.jp
[2] Ryoyu Systems Co., Ltd., Tokyo, Japan

Abstract. JAXA is proposing a concept of a high speed EMS compound helicopter with the maximum cruise speed 500 km/h which is twice as fast as conventional helicopters. In order to realize this high speed within an envelope of the existing propulsion system, it is essential to develop a very low drag airframe. Furthermore, the compound helicopter has several rotor/propellers and lifting surfaces in the vicinity, which may cause the aerodynamic interference problems. This paper describes the review to the mechanism of the high drag of helicopter airframe followed by the resolution proposals for the airframe with the low drag and the less severe aerodynamic interference.

Keywords: Compound helicopter · Airframe drag · Aerodynamic interference

1 Introduction

A concept of a high speed compound helicopter assuming the use of the emergency medical service (EMS) is proposed by JAXA as shown Fig. 1.

Fig. 1. A concept of a high speed EMS compound helicopter

© Springer Nature Singapore Pte Ltd. 2019
X. Zhang (Ed.): APISAT 2018, LNEE 459, pp. 1526–1534, 2019.
https://doi.org/10.1007/978-981-13-3305-7_122

The target maximum cruise speed is 500 km/h which is twice as fast as conventional helicopters. In order to realize this type of the rotorcraft, the major technical achievements as shown in Fig. 2 are essential [1, 2]. Especially, developing a very low drag airframe is mandatory for the high speed compound helicopter within an envelope of the existing propulsion system. This high speed compound helicopter generates the lift supporting 4 ton gross weight by the main rotor + the wing and the thrust overwhelming the huge drag by the main rotor + both side propellers + tail pusher propeller on a high speed cruise condition [1, 2].

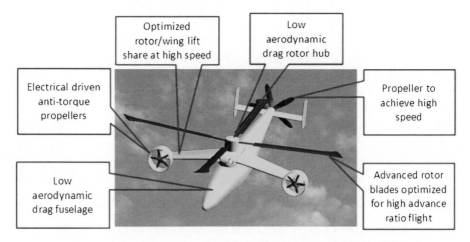

Fig. 2. Major technical topics for a high speed compound helicopter [1, 2]

Furthermore, the compound helicopter has several rotor/propellers and lifting surfaces in the vicinity as shown in Fig. 3, which may cause the significant aerodynamic interference problems. This paper describes the review to the mechanism of the

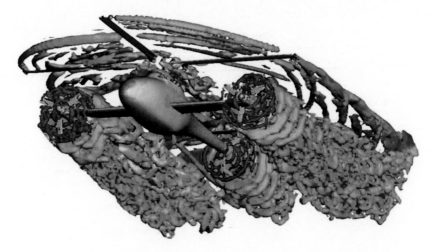

Fig. 3. Analytical result of a flow field of the high speed compound helicopter [2]

high drag helicopter airframe followed by the resolution proposals for the airframe with the low drag and the less severe aerodynamic interference.

2   Aerodynamic Problems Particular to Rotorcraft Airframe

In order to simplify the basic idea to build the countermeasures for the high drag config-uration of the conventional helicopters as well as the compound helicopters, it is helpful to review the drag and the aerodynamic interference of the airframe of the rotorcraft.

2.1   Drag of Airframe

The airframe drag consists of the friction drag and pressure drag, because the target maximum speed of the compound helicopter is 500 km/h which is in the subsonic range. Figure 4 shows several geometries and their drags with breaking down into friction and pressure drags. The friction drag is dominant to the streamlined shape as shown in Fig. 4(c), the pressure drag caused by the separation is dominant to the flat

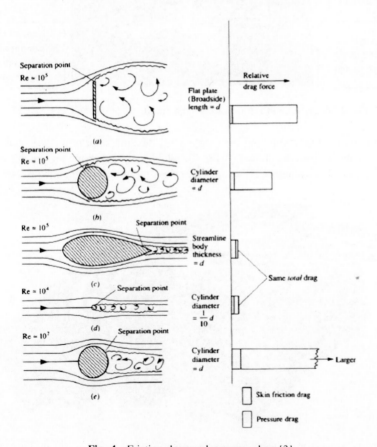

Fig. 4. Friction drag and pressure drag [3]

plate or the cylinder on the other hand. The comparison between Fig. 4(c) and (d) indicates that the cylinder has the same drag as the streamlined shape with ten times larger reference length. The rotorcraft has many exposed parts and components which are not streamlined, then induces the separation followed by the large pressure drag in the wide range of flight condition unlike the fixed wing aircraft.

It is well known that the total drag of a generic helicopter is almost equally divided into three components, namely a fuselage, a rotor hub and a landing gear system. The existing drag trend of each component is very useful for the aerodynamic design of the advanced types of the rotorcraft such as the high speed compound helicopters. As JAXA proposing concept of a high speed EMS compound helicopter assumes the retractable landing gears, there is no description for the landing gear drag considerations in this paper.

Figure 5 shows the trend between gross weight and airframe drag of the existing helicopters. "Utility" helicopters are on the trend of having the larger airframe drag than "Clean" helicopters. "Utility" helicopters mainly contain transport helicopters equipped with clam shell doors or ramp doors which make it easier to load/unload the personnel and cargo from the rear side of the airframe. This characteristic of the airframe shape, as shown in Fig. 5 as the silhouette of "Utility" helicopters, is easy to induce the severe separation at the rear part of the airframe and generate the large pressure drag. On the other hand, the core members of "Clean" helicopters are commercial/civil helicopters on which the stylishness is emphasized for sales activities. The airframe shape, as shown in Fig. 5 as the silhouette of "Clean" helicopters, is streamlined to reduce the drag and increase the fuel efficiency and the cruise speed.

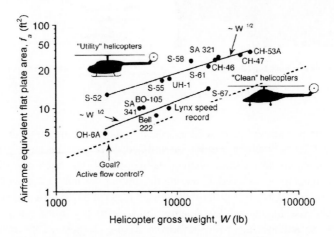

Fig. 5. Trend between gross weight and airframe drag [4]

The trend between gross weight and hub drag of the existing helicopters is shown in Fig. 6. An example of the structure around rotor hub is shown in Fig. 7(a) and the sketch of generic rotor is depicted in Fig. 7(b).

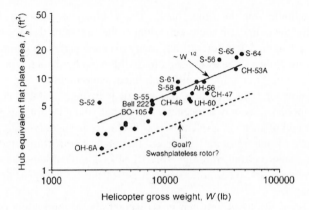

Fig. 6. Trend between gross weight and hub drag [4]

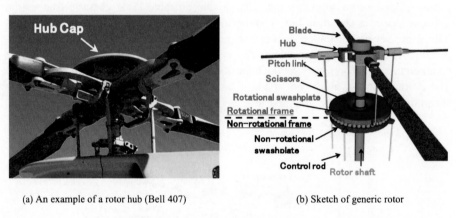

(a) An example of a rotor hub (Bell 407)   (b) Sketch of generic rotor

Fig. 7. Structure around rotor hub

There are many exposed parts and components on the rotating frame, namely a hub, pitch links and scissors which are not streamlined, thus generate large pressure drag around the rotor hub. From the advent of helicopters, it has been one of the major technical problems to be resolved. Covering the entire rotor hub to dramatically reduce the drag was proposed, but this proposal had been rejected based on a trade off studying the production cost, maintainability (mainly the accessibility to the pitch control systems). But these attitudes are being improved by the progress of the manufacturing technology and the strong requirement for necessity toward the high speed capability as shown in Fig. 8 which is an example of a helicopter equipped with a hub fairing (EC135).

Fig. 8. An example of a helicopter equipped with a hub fairing (EC135)

2.2   Aerodynamic Interference

One of the well known aerodynamic interferences is a tail shake as shown in Fig. 9. The airflow passing around the rotor hub is made turbulent with some dominant frequencies, usually 1/rev excited by a scissors or b/rev by pitch links, where b is the number of the blades, propagates and impinges on the empennage of the rotorcraft. If one of the structural natural frequencies of the empennage is sufficiently close to the turbulent frequencies, the severe vibration occurs as the tail shake [5, 6].

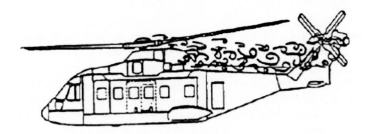

Fig. 9. Tail shake mechanism [6]

Installing a hub cap on top of the rotor hub as shown in Fig. 7(a) to deflect the turbulent flow far downward the empennage is the common and well proved countermeasure for the tail shake problem.

More general aerodynamic interference overview including the tail shake is depicted in Fig. 10 [7]. UTTAS (Utility Tactical Transport Aircraft System), US Army renewal program of UH-1 in 1970's, required the embarkation capability inside C-130 cargo bay [8], which inevitably made the distance between the rotor and the fuselage shorter than ever. This initiates the opportunity to notice the aerodynamic interference as one of the major technical challenges for rotorcraft.

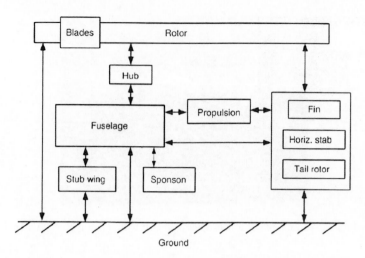

Fig. 10. Aerodynamic interferences patterns among the components of a helicopter [7]

3   Proposals for Airframe with Low Drag and Less Severe Aerodynamic Interference

The proposals for the airframe of JAXA high speed compound helicopter with lower drag and less severe aerodynamic interference are as follows.

(1) Assuming the retractable landing gear system to eliminate the large part of the entire airframe drag.

(2) Optimization of the contraction rate of the rear fuselage cross sectional area to avoid the flow separation by compromising to ensure the volume inside the fuselage for EMS mission.

(3) Covering the rotor hub, a rotor shaft and pitch links by an integrated hub fairing conformal to the pylon, as shown in Fig. 11, to avoid these non-streamlined components to be exposed. Optimization of the indices defining the geometries of the hub fairing.

(a) Hub fairing (diameter, side view aspect ratio)
(b) Pylon (width, side and plan view aspect ratio)
(c) Hub wake evader (incident angle, length)
(d) Hub wake/pylon flow diverter (curvature)

Hub Fairing

Hub wake/Pylon flow diverter

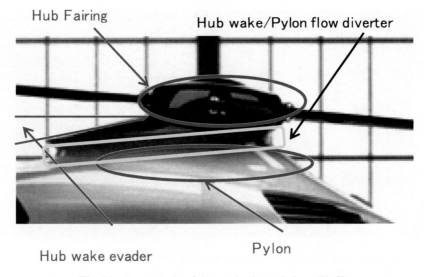

Hub wake evader

Pylon

Fig. 11. An example of drag reduction solutions (H160)

The main part of the optimization process is carried out by utilizing CFD code, rFlow3D developed by JAXA [2], and the candidates of the several configuration are evaluated to select the final configuration by wind tunnel tests.

4 Summary

The mechanism of a large drag and severe aerodynamic interference characterizing the rotorcraft airframe is reviewed and the solutions for the airframe of JAXA high speed compound helicopter are proposed as follows.

(1) Retractable landing gear system
(2) Optimization of the contraction rate of the rear fuselage cross sectional area
(3) An integrated hub fairing conformal to the pylon

5 Future Works

The solution proposals are designed to incorporate the compound helicopter configuration and evaluated in the wind tunnel testing as shown in Fig. 12 which is now going on.

Fig. 12. High speed compound helicopter in JAXA 2 m × 2 m Low Speed Wind Tunnel

References

1. Tanabe Y, Aoyama T, Kobiki N, Sugiura, M, Miyashita R, Sunada S, Kawachi K, Nagao M (2014) A conceptual study of high speed rotorcraft. In: 40th European Rotorcraft Forum, Southampton, UK, 2–5 September 2014
2. Sugawara H (2017) A study of rotor/wing aerodynamic interaction at high speed flight on a compound helicopter. In: 6th Asian/Australian Rotorcraft Forum & Heli Japan 2017, Kanazawa, Japan 7–9 November 2017
3. Talay TA (1975) Introduction to the Aerodynamics of Flight, NASA SP-367
4. Leishman JG (2007) The Helicopter Thinking Forward, Looking Back, College Park Press
5. Roesch P, Dequin AM (1983) Experimental Research on Helicopter Fuselage and Rotor Hub Wake Turbulence. 59th Annual Forum of AHS
6. Ishak IS, Mansor S, Lazim TM (2008) Experimental Research on Helicopter Tail Shake Phenomenon. Jurnal Mekanikal 26:107–118
7. Sheridan P, Smith R (1979) Interactional Aerodynamics–A New Challenge to Helicopter Technology. In: 35th Annual Forum of AHS, Washington DC, 21–23 May 1979
8. http://www.sikorskyarchives.com/S-70%20(YUH-60A%20UTTAS).PHP

Effects of Distributed Propulsion Crucial Variables on Aerodynamic and Propulsive Performance of Small UAV

Yiyuan Ma, Wei Zhang$^{(\boxtimes)}$, Yizhe Zhang, Ke Li, and Yiding Wang

School of Aeronautics, Northwestern Polytechnical University,
Xi'an 710072, China
weizhangxian@nwpu.edu.cn

Abstract. The distributed propulsion (DP) with the effect of boundary layer ingestion (BLI) can improve the flight performance of unmanned aerial vehicles (UAV), and the mounting parameters of DP can further affects the propulsive efficiency and the aerodynamic performance of airfoil. To explore the effect of DP mounting parameters, computational fluid dynamics (CFD), sensitivity analysis and Kriging surrogate models methods were used, results indicate that in a wide range of design space, the effect of BLI significantly improves the aerodynamic performance and propulsive efficiency at medium/small angle of attack (2° and 6°), but the undesirable configuration will exert adverse effects on the propulsive efficiency and pitching moment, lower mounting height or larger setting angle of DP will produce the aerodynamic effect of the reflex airfoil, which is suitable for flying-wing UAV.

Keywords: Boundary layer ingestion (BLI) · Distributed propulsion (DP) · Small unmanned aerial vehicles · Computational fluid dynamics (CFD) · Propulsive efficiency

1 Introduction

With the progress of flight control system in recent years, small unmanned aerial vehicles (UAV) have not only been limited to military applications, but also developed rapidly in the civilian field such as traffic surveillance, law enforcement, and resource exploration [1, 2]. However, The flight performance and endurance of small UAV still need to be improved due to the current bottleneck of battery technology and the limitations of design methods, and there is a urgent need to find a breakthrough [3].

Distributed propulsion (DP) refers to a propulsion system of multiple small engines mounted on the rear area of fuselage or wing section of UAV rather than traditional large-size engines. Researches have shown that DP has advantages of reduced noise, improved safety, reduced consumption, and improved endurance [4]. In particular, comparing to conventional internal combustion engines, electric engine is a more suitable choice when developing DP system for small UAV because it is quieter, more reliable, and has less vibration [5–8].

© Springer Nature Singapore Pte Ltd. 2019
X. Zhang (Ed.): APISAT 2018, LNEE 459, pp. 1535–1550, 2019.
https://doi.org/10.1007/978-981-13-3305-7_123

Researches of DP mainly focuses on the next generation of civil aircraft. For example, Wanfang Yan et al. [9] used computational fluid dynamics (CFD) method to study the effect of the different mass flow rate of DP on the aerodynamic performance of aircraft with the conditions of cruise and take-off. Leifsson et al. [10] used multi-disciplinary design optimization (MDO) method to optimize two aircraft design schemes of conventional and DP. The results showed that DP can reduce the weight of the wing structure and improve the flight performance. Blumenthal et al. [11] used the CFD method to analyze the aerodynamic and propulsive performance benefit from the boundary layer ingestion (BLI) effect. The results showed that BLI effect can reduce the power requirements for cruise by 8.7%. However, the civil aircraft use turbofan engines, and operate at high Reynolds and high Mach numbers ($Re > 2 \times 10^6$ and $Ma > 0.7$), which have significantly different operating conditions compared to small electric UAV operating at lower Reynolds and lower Mach numbers ($Re < 10^6$ and $Ma < 0.1$). Therefore, the research conclusions about the civil aircraft with DP can not be directly applied to the small electric UAV [12]. Valencia et al. [13] used semi-empirical correlations to study the influence of different ducted fan sizes on the aerodynamic and propulsive performance of the small UAV with DP. However, the decoupling analysis of aerodynamic and propulsion can not quantify the coupling effect on the performance of the UAV.

In this paper, a mathematical model of the propulsive efficiency of the small UAV with DP is established. Firstly, The sensitivity of the design parameters of DP to aerodynamic and propulsive performance is analyzed with the methods of CFD and Kriging surrogate models. Then the typical sensitive design variables are analyzed in detail. Finally, the physical reasons for the influence of the design variables on the performance parameters are explained.

2   Research Object

The emergence of small electric ducted fan and electronic speed controller makes it possible to design DP systems for small UAV. In order to improve the flight performance and tap the potential of UAV's aerodynamic, propulsion and so on, Valencia et al. [13] proposed a small flying-wing UAV with DP. It is a highly integrated configuration which is superior in aerodynamics, noise and safety.

The study case in this work is a small UAV using distributed electric ducted fans as propulsion system, as shown in Fig. 1. Due to the large width, DP system can be simplified to a two-dimensional system for analysis. For simplicity, the NACA 0012 airfoil is selected and the chord is set to 0.35 m. The cowling model of the propulsion system and the definition method of its parameters can be found in [13, 14].

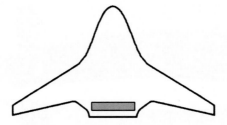

Fig. 1. Distributed propulsion UAV

The location of the ducted fan on the airfoil is shown in Fig. 2. The chord position coefficient L_f^* of the propulsion system is defined as the ratio of the L_f (the horizontal distance from the midpoint of the intake and exhaust boundary to the leading edge of the airfoil) to the chord length of the airfoil. The height position coefficient H_f^* is defined as the ratio of the H_f (the height from the mid-point of the intake and exhaust boundary to the airfoil upper surface) to the chord length. The setting angle τ is defined as the angle between the axis of the propulsion system and the horizontal line, and the τ is positive when the jet of propulsion system is downward.

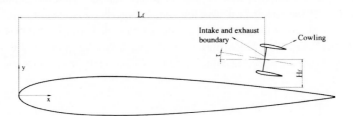

Fig. 2. Schematic of airfoil and ducted fan position

3   Methodology

3.1   Aerodynamic Model and Verification

The numerical calculation of aerodynamic was carried out by using CFD. The flow simulations used the finite-volume Navier-Stokes solver Fluent. The Reynolds-averaged Navier-Stokes (RANS) equations were solved, coupled with the Spalart-Allmaras turbulence model [15]. A second-order upwind scheme was used for both the discretized momentum and turbulence model equations. Structure grid of flow field was generated by using ICEM. The upper and side boundaries of the flow field are $15c$ away from the airfoil, and the distance from the downstream boundary to the airfoil is $20c$. Mesh near the airfoil surface and the trailing edge was refined as shown in Fig. 3, the total number of multiblock mesh is about 140,000 cells.

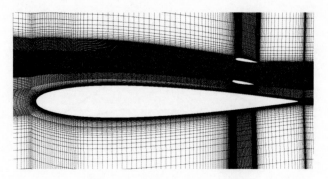

Fig. 3. Schematic of CFD mesh

The aerodynamic assessment was done for cruising at sea level, the static temperature and pressure of freestream is 288.15 K and 212500 Pa, Mach number 0.088, Reynolds number 7.19×10^5. The fan was simplified to an equivalent actuator disk which accelerates the slipstream of the fan by setting pressure increments, and the pressure increments generated reaction force which can be regarded as an external force applied to the entire system through the equivalent actuator disk.

Self-propelled assumption was adopted to ensure that the result has generality, that is, the net drag of the airfoil in the design state should be close to zero [16].

To validate the proposed methodology, numerical calculations were performed for the NACA 0012 airfoil at Mach number 0.19 and Reynolds number 7.0×10^5. As shown in Fig. 4, the curves corresponding to the CFD results is very close to the wind tunnel data [17].

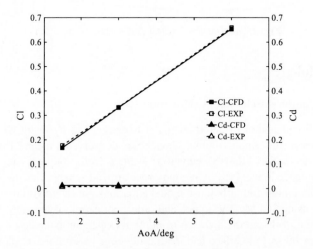

Fig. 4. C_l and C_d comparison of NACA 0012

3.2   Propulsive Efficiency

The overwing DP system has the potential to increase the efficiency of an integrated propulsion system by ingesting slow boundary layer flow, and different DP mounting positions have different effects on propulsive efficiency. Therefore, it is necessary to establish a mathematical model of the propulsive efficiency for analysis.

Propulsive efficiency η_p represents the effectiveness of the propulsion system as a thruster. The explanation for η_p increase through BLI is given by the consideration of energy losses in the flow field [18]. For self-propelled systems, the viscous drag is balanced by the engine thrust. The wake of the airfoil needs to be filled with net kinetic energy, leading to a reduction in propulsion efficiency [10]. The wake could be characterized and quantified by the non-uniformities in the velocity profiles behind the airfoil shown in Fig. 5.

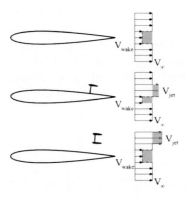

Fig. 5. Velocity profiles of different configurations

The viscous drag of the airfoil comes from the non-uniformities in the velocity profiles. Therefore, the more uniform the velocity profile is, the lower the viscous drag, the higher the propulsive efficiency. As shown in Fig. 5, the region sizes of the velocity profile and the wake of the different ducted fan mounting positions are all different. Therefore, it is necessary to establish a mathematical model of propulsive efficiency to quantitatively evaluate the propulsive efficiency with different configurations.

The total drag of a UAV includes profile drag (the friction and form drag) and lift induced drag. This means that in an ideal DP system, the jet will perfectly 'fill in' the wake and create a uniform velocity profile, making the upper limit of the propulsive efficiency is as below.

$$\eta_{p,\max} = \frac{D_{viscous}}{D_{viscous} + D_{induced}} \times 100\% \tag{1}$$

where $\eta_{p,\max}$ is the maximum of propulsive efficiency in theory, $D_{viscous}$ is the viscous drag, and $D_{induced}$ is the induced drag.

The wake displacement area reflects the magnitude of the viscous drag. Therefore, due to the jet of the overwing DP system, the propulsive efficiency can be expressed as follows.

$$\eta_p = \eta_{p,Non} + \Delta\eta_p \tag{2}$$

$$\Delta\eta_p = k * \frac{\left(S_{Non,wake} - S_{i,wake}\right)/S_{Non,wake} * D_{viscous}}{D_{viscous} + D_{induced}} \times 100\% \tag{3}$$

$$k = 1 - \eta_{p,Non} \tag{4}$$

where $\eta_{p,Non}$ is the propulsive efficiency without BLI (determined to be 70% for a small UAV with electric ducted fan by Valencia et al. [13]), $\Delta\eta_p$ is the increment of the propulsive efficiency due to BLI, k is a scale factor, $S_{Non,wake}$ is the wake displacement area without BLI, and $S_{i,wake}$ is the wake displacement area with BLI.

By ignoring the constant induced drag in the cruising condition, the two-dimensional self-propelled model utilized in this paper is simplified. The velocity profiles and the airfoil pressure distributions with different configurations are shown in Figs. 6 and 7. The propulsive efficiency of each configuration is obtained by substituting $S_{Non,wake}$ and $S_{i,wake}$ (obtained by integral the area of the wake displacement area) into formulas (2) to (4).

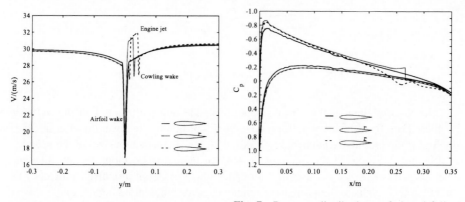

Fig. 6. Velocity profiles behind the airfoil at $\alpha = 2°$

Fig. 7. Pressure distributions of the airfoil at $\alpha = 2°$

As shown in Fig. 6, different configurations lead to different velocity profiles. The overwing DP system would increase the wake displacement area below the airfoil and reduces the wake displacement area above the airfoil. Therefore, the effect of BLI on propulsion efficiency is a combined result of changes in the upper and lower wake displacement area of the airfoil, and undesirable mounting configuration may lead to lower propulsive efficiency than normal.

3.3   Kriging Surrogate Models

The surrogate model method can reduce the computation time and improve the design efficiency. Kriging models is unbiased estimation models with minimal variance estimation. It has good approximation capability and error estimation ability for nonlinear functions. It is one of the most representative and potential surrogate models [19].

Kriging model is an interpolation function that linearly weights the response of known sample functions, it is expressed as below.

$$\hat{y}(x) = \sum_{i=1}^{n} \omega^{(i)} y^{(i)} \tag{5}$$

where $\hat{y}(x)$ is the Kriging model, $\omega^{(i)}$ is the weighting factor, and $y^{(i)}$ is the sample function response value.

Compared with other interpolation techniques, Kriging models introduce statistical assumptions into the $\omega^{(i)}$ calculation, and consider the unknown functions as specific implementation of a Gaussian static random process. Based on this assumption, Kriging models search for the optimal weighting factor ω to minimize the mean square error

$$\text{MSE}[\hat{y}(x)] = E\left[\left(\omega^{\text{T}} Y_s - Y(x)\right)^2\right] \tag{6}$$

while satisfying the unbiased condition.

$$E\left[\sum_{i=1}^{n} \omega^{(i)} Y\left(x^{(i)}\right)\right] = E[Y(x)] \tag{7}$$

where Y_s is the set of n function response values, and $Y(x)$ is the static random process.

Details of the definition method of the unknown functions and the searching method of the optimal weighting factor ω can be found in [20].

4   Results and Discussion

Latin hypercube design was used to generate 30 sets of experimental points in design space, the CFD calculations were carried out for these 30 different configurations of flow fields at incoming angles of attack 2° and 6°. Kriging models ware used to approximate the aerodynamic parameters and propulsive efficiency.

For the small UAV with overwing DP system in this paper, we are more interested in the variation of the aerodynamic parameters with BLI effect. Through the analysis of $\Delta L/D$ and ΔCm, the aerodynamic design laws can be summarized. The design laws of overwing DP system can be summarized through the comparison of the propulsive efficiency of different configurations.

4.1   Sensitivity Analysis of Crucial Variables

Sensitivity analysis refers to the fitting of the numerical calculation results through surrogate models, and comparing the partial and total variances of the L_f^*, H_f^*, and τ to get analysis of the importance of each variable to the target variables [21]. The sensitivity of surrogate models is defined as below.

$$S_i^{local} = V_i/V, S_i^{global} = (V_i + V_{iz})/V \tag{8}$$

where V_i is the partial variances of the variable i, V is the total variance of the target variable, V_{iz} is the partial variance that considers the interaction between variables, S_i^{local} is the local sensitivity and reflects the influence of variables on target variables, and S_i^{global} is the global sensitivity and reflects the influence of variables on target variables considering the interaction.

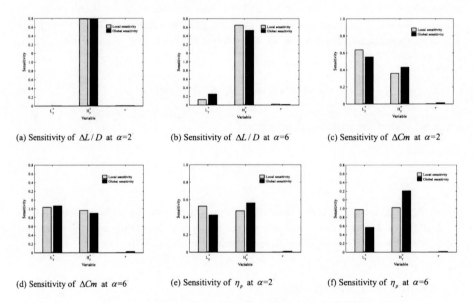

(a) Sensitivity of $\Delta L/D$ at $\alpha=2$ (b) Sensitivity of $\Delta L/D$ at $\alpha=6$ (c) Sensitivity of ΔCm at $\alpha=2$

(d) Sensitivity of ΔCm at $\alpha=6$ (e) Sensitivity of η_p at $\alpha=2$ (f) Sensitivity of η_p at $\alpha=6$

Fig. 8. Sensitivity indices of variables to target variables

As shown in Fig. 8, although the local and global sensitivity indices of L_f^*, H_f^* and τ are basically the same, they are weighted differently. The interaction between those variables is significant, thus they should not be regarded as independent variables. The increment of lift-drag ratio $\Delta L/D$ is mainly affected by the height position of the ducted fan, the increment of pitching moment coefficient ΔCm and propulsive efficiency η_p are affected by the chord position and height position of the ducted fan.

The study in this paper has many parameters and large amount of data. Therefore, only sensitive variables are selected for further research according to the results of the sensitivity analysis.

4.2   Aerodynamic Performance

Because L_f^*, H_f^* and τ are not independent variables, The performance parameters were selected for analysis.

4.2.1   Increment of the Lift-Drag Ratio

The increment of lift-drag ratio $\Delta L/D$ of the system with or without BLI is mainly affected by the height position of the ducted fan. Kriging surrogate models is used to fit the calculation results of the sample points to obtain the tendency of $\Delta L/D$ varied with H_f^*.

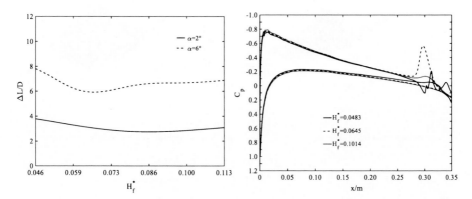

Fig. 9. $\Delta L/D$ varied with H_f^* **Fig. 10.** Pressure distributions with different H_f^* at $\alpha = 2°$

As shown in Fig. 9, the BLI effect of the overwing DP system leads to a significant improvement on the $\Delta L/D$. The variation of $\Delta L/D$ with H_f^* at different angles of attack is approximately a quadratic function, and the $\Delta L/D$ obtains the maximum at the lower boundary of H_f^*.

From Fig. 10, it can be seen that H_f^* has a great influence on the pressure distributions at the trailing edge of the airfoil. When H_f^* is small, the cowling of the ducted fan approaches the upper surface of the airfoil, creating a blockage to the airflow and causing the suction of the upper surface at the trailing edge drops significantly. As H_f^* increases, a throat is formed between the cowling and the upper surface, and the airflow passing there is accelerated. A new suction peak is formed at the trailing edge of the airfoil due to this throat. With a large adverse pressure gradient appears after the suction peak, the airflow is prone to separate, resulting in increased drag. When H_f^* is further increased, the throat is widened, the acceleration of the airflow is reduced, the peak of suction is reduced and the adverse pressure gradient decreases after the peak, resulting in the ΔC_d (increment of drag coefficient) decreases. Overall, with the increase of H_f^*, the ΔC_d increases and then decreases compared with that without BLI. The trend of the lift is similar to the drag but the increment of the lift coefficient is smaller. Therefore, the $\Delta L/D$ shows a trend of decreasing first and then increases slowly.

4.2.2 Increment of Pitching Moment Coefficient

The increment of pitching moment coefficient ΔCm is influenced by the L_f^* and H_f^* of the overwing DP system. That is, the tendency of the ΔCm varied with a variable will change with the change of other variables. Figure 11 shows the ΔCm varied with L_f^* and H_f^*.

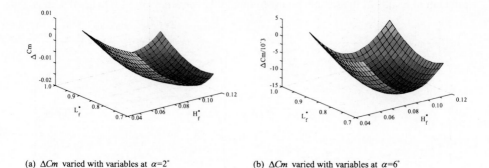

(a) ΔCm varied with variables at $\alpha=2°$ (b) ΔCm varied with variables at $\alpha=6°$

Fig. 11. ΔCm varied with variables

Pressure distributions of the airfoil are shown in Figs. 7 and 10, it can be seen that the pressure difference between the upper and lower surface in the front part of the airfoil is increased due to the BLI effect, and the area surrounded by the C_p curves increases. From the above chapter, the area surrounded by the C_p curves in the part of trailing edge increases first and then decreases gradually with the increase of H_f^*. Therefore, the ΔCm decreases with H_f^* first and then pick up, as shown in Fig. 11. When H_f^* is constant, The more rearward the ducted fan is mounted, the larger the increment of the area surrounded by the C_p curves in the front part of the airfoil is, and the ΔCm is gradually increasing. This tendency is similar to the results in [9].

4.3 Propulsive Efficiency

The sensitivity analysis of the propulsive efficiency η_p is influenced by the L_f^* and H_f^* of the overwing DP system, similar to the result of ΔCm.

As shown in Fig. 12, the propulsive efficiency η_p has the same tendency with L_f^* and H_f^* at $\alpha = 2°$ and $\alpha = 6°$. The BLI effect reduces the airflow velocity near the lower surface of the airfoil, and the larger the L_f^* or the smaller the H_f^*, the stronger the BLI effect, the lower the airflow velocity near the lower surface of the airfoil, meaning that BLI effect will broaden the wake of the airfoil. When L_f^* is too large or H_f^* is too small, the area of the weak displacement is significantly increased, resulting in a lower η_p than

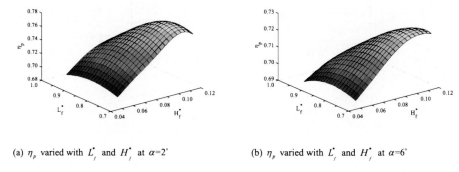

(a) η_p varied with L_f^* and H_f^* at $\alpha=2°$ (b) η_p varied with L_f^* and H_f^* at $\alpha=6°$

Fig. 12. η_p varied with variables

normal. With the decrease of L_f^* or the increase of H_f^*, the effect of BLI is weakened, which makes the η_p increase significantly, exceeding the $\eta_{p,Non}$. In particular, the η_p can reach the maximum at the specific pair of L_f^* and H_f^*.

4.4 Influence of Setting Angle

Sensitivity analysis showed that the performance parameters are less affected by the setting angle τ of propulsion system. However, in order to find the design rules of the overwing DP system and further improve the flight performance of the UAV, the performance parameters, ΔCm and η_p, which have a relatively large affected by the τ, are selected for analysis.

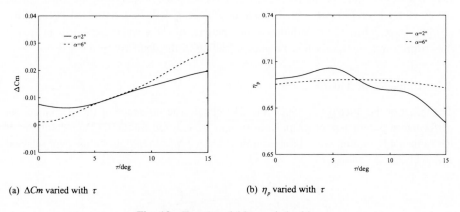

(a) ΔCm varied with τ (b) η_p varied with τ

Fig. 13. Target variables varied with τ

As shown in Fig. 13, the ΔCm increases monotonously with τ, which is similar to the results in [22]. The trend of the η_p with τ is similar to the results in [16], which approximately subjects to the quadratic curve relationship, and the maximum value of η_p is obtained near $\tau = 5°$.

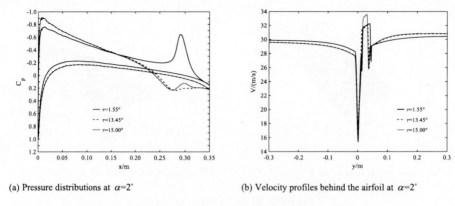

(a) Pressure distributions at $\alpha=2°$ (b) Velocity profiles behind the airfoil at $\alpha=2°$

Fig. 14. Pressure distributions and velocity profiles of the airfoil

As τ increases, the distance between the trailing edge of the ducted fan cowling and the airfoil decreases, forming a blockage to the airflow there, lowering the suction here, and reducing the area surrounded by the curves of C_p at the trailing edge of the airfoil, as shown in Fig. 14(a). Whereas the jet has a downward deflection angle, the airflow near the upper surface of the airfoil will accelerate downwardly at the trailing edge, increasing the airfoil's camber and circulation, play the role of flap, and increases the area surrounded by the C_p curves in the front part of the airfoil, which is similar to the results in [16]. The combined effect of the changes in the area surrounded by the front and back parts of the C_p curves leads to the ΔCm increases with τ. The angle of the jet increases with the increase of τ, thus the airflow near the lower surface is blocked and the velocity is reduced, which broadens the wake of the airfoil. However, when the ducted fan has a setting angle, the jet will completely fill the wake produced by the lower part of the cowling. Therefore, at a small angle of attack, η_p will be maximized because the cowling wake is filled. But with the further increase of τ, the wake displacement area of the airfoil lower surface further broadens, leading to the decrease of the η_p.

4.5 Influence of Thrust

The thrust of the ducted fan affects the BLI effect. The influence of thrust changes on performance parameters of propulsion system is analyzed by the CFD calculations of the same configuration with different thrust. Three typical configurations were selected and analyzed, as shown in Fig. 15.

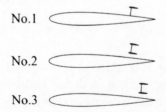

Fig. 15. Configurations for research

Taking the thrust based on the self-propelled assumption as the reference thrust T_0, new thrust T_i is obtained for the T_0 changes of -50%, $+50\%$, and $+100\%$ at $\alpha = 2°$ and $\alpha = 6°$, respectively, and the system with the influence of T_i is numerically calculated.

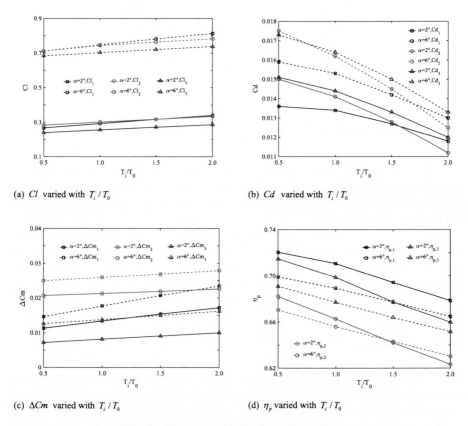

(a) Cl varied with T_i / T_0

(b) Cd varied with T_i / T_0

(c) ΔCm varied with T_i / T_0

(d) η_p varied with T_i / T_0

Fig. 16. Target variables varied with the thrust

As shown in Fig. 16, The variation of each performance parameter with the thrust is basically the same for the three configurations, and the pressure distributions of the airfoil with different thrust of the No. 2 configuration is analyzed.

As shown in Fig. 17, with the increase of thrust, the suction on the upper surface of the airfoil increases, the suction peak rises significantly, and the pressure on the lower surface increases slightly. In addition, the higher the thrust is, the smaller the adverse pressure gradient, the more difficult the airflow to separate and the smaller the drag generated. As shown in Fig. 16, the lift coefficient Cl varies linearly with thrust, and the drag coefficient Cd decreases exponentially with thrust, which is similar to the results in [23].

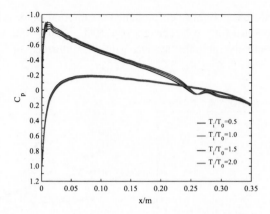

Fig. 17. Pressure distributions of the No. 2 configuration at $\alpha = 2°$

   The larger the thrust, the larger the area surrounded by the C_p curves in the front part of the airfoil, but the area surrounded by the C_p curves at the trailing edge changes very little. Therefore, same as Sect. 3.2, the increase of thrust will increase the pitching up moment of the airfoil.

   Same as Sect. 3.3, the pressure on the lower surface of the airfoil increases with the increase of thrust, which expands the wake displacement area, but the jet can only fill the wake displacement area above the trailing edge of the airfoil. Therefore, η_p decreases as the thrust increases, and the larger the angle of attack, the faster the decline, which is similar to the trend of the η_p in [13].

5 Conclusion

In this paper, the NACA 0012 airfoil was used as a simplified model for the wing or fuselage of a UAV with overwing DP system. The ducted fan was integrated across on the upper-surface trailing edge. The CFD, sensitivity analysis and Kriging surrogate models were used to explore the influence of different DP mounting positions and different thrust on aerodynamic and propulsive performance of the airfoil with BLI. Conclusions are drawn as follows:

(1) The sensitivity analysis of the crucial variables showed that the height position has the greatest influence on each performance parameter, followed by the chord position, and the setting angle has little influence on the performance parameters.
(2) With different angles of attack, the BLI has the same effect on the aerodynamic and propulsive performance, though varied in specific values. Specifically, the BLI effect has more significant influence on the aerodynamic performance at the moderate angle of attack, while the improvement of the propulsive performance is more notable at the small angle of attack.

(3) Lower altitude or larger setting angle of the ducted fan will produce the aero-dynamic effect of the reflex airfoil, so this mode of DP mounted is suitable for flying-wing UAV. The increase of the height or the reduction of the setting angle of the ducted fan will result in a greater increment of pitching down moment and additional drag, reducing the aerodynamic revenue but improving the propulsive efficiency.

References

1. Saeed AS, Younes AB, Cai C, Cai G (2018) A survey of hybrid unmanned aerial vehicles. Prog Aerosp Sci 98:91–105. https://doi.org/10.1016/j.paerosci.2018.03.007
2. Hening S, Baumgartner J, Walden C, Kirmayer R, Teodorescu M, Nguyen N, Ippolito C (2013) Distributed sampling using small unmanned aerial vehicles (UAVs) for scientific missions. In: AIAA Infotech@Aerospace Conference, Boston, MA. https://doi.org/10.2514/6.2013-4734
3. Schoemann J, Hornung M (2012) Modeling of hybrid electric propulsion systems for small unmanned aerial vehicles. In: 12th AIAA Aviation Technology, Integration, and Operations (ATIO) Conference and 14th AIAA/ISSMO Multidisciplinary Analysis and Optimization Conference, Indianapolis, Indiana. https://doi.org/10.2514/6.2012-5610
4. Schetz JA, Hosder S, Dippold V, Walker J (2010) Propulsion and aerodynamic performance evaluation of jet-wing distributed propulsion. Aerosp Sci Technol 14(1):1–10. https://doi.org/10.1016/j.ast.2009.06.010
5. Huang J, Yang F (2016) Development and challenges of electric aircraft with new energies. Acta Aeronautica Et Astronautica Sinica 37(1):57–68 (in Chinese)
6. Gohardani AS, Doulgeris G, Singh R (2011) Challenges of future aircraft propulsion: a review of distributed propulsion technology and its potential application for the all electric commercial aircraft. Prog Aerosp Sci 47(5):369–391. https://doi.org/10.1016/j.paerosci.2010.09.001
7. Kong X, Zhang Z, Lu J, Li J, Yu L (2018) Review of electric power system of distributed electric propulsion aircraft. Acta Aeronautica Et Astronautica Sinica 35(1):46–62 (in Chinese)
8. Ko A, Ohanian OJ, Gelhausen P (2007) Ducted fan UAV modeling and simulation in preliminary design. Paper presented at the AIAA Modeling & Simulation Technologies Conference, Hilton Head, South Carolina
9. Yan W, Wu J, Zhang Y (2015) Effects of distributed propulsion crucial variables on aerodynamic performance of blended wing body aircraft. J Beijing Univ Aeronaut Astronaut 41(6):1055–1065 (in Chinese)
10. Leifsson L, Ko A, Mason WH, Schetz JA, Grossman B, Haftka RT (2013) Multidisciplinary design optimization of blended-wing-body transport aircraft with distributed propulsion. Aerosp Sci Technol 25(1):16–28. https://doi.org/10.1016/j.ast.2011.12.004
11. Blumenthal BT, Elmiligui AA, Geiselhart KA, Campbell RL, Maughmer MD, Schmitz S (2018) Computational Investigation of a Boundary-Layer-Ingestion Propulsion System. J Aircr
12. Valencia E, Hidalgo V, Cisneros J (2016) Design point analysis of a distributed propulsion system with boundary layer ingestion implemented in UAV's for agriculture in the Andean region. In: 52nd AIAA/SAE/ASEE Joint Propulsion Conference, Salt Lake City, UT. https://doi.org/10.2514/6.2016-4799

13. Valencia E, Saa J, Alulema V, Hidalgo V (2018) Parametric study of aerodynamic integration issues in highly coupled blended wing body configurations implemented in UAVs. Paper presented at the 2018 AIAA Information Systems-AIAA Infotech. https://doi.org/10.2514/6.2018-0746

14. Xiang Y, Wu J, Zhang Y (2016) Effects of cowling design on aerodynamic performance of airfoil with BLI. J Beijing Univ Aeronaut Astronaut 42(5):945–952 (in Chinese)

15. Spalart PR, Allmaras SR (1992) A One-Equation Turbulence Model for Aerodynamic Flows. AIAA 92-0439

16. Duan J, Yuan W, Li Q (2015) Numerical investigation on trailing edge jet and boundary layer ingestion in distributed propulsion system. J Aerosp Power 30(3):571–579 (in Chinese)

17. Sheldahl RE, Klimas PC (1981) Aerodynamic characteristics of seven airfoil sections through 180 degrees angle of attack for use in aerodynamic analysis of vertical axis wind turbines. SAND80-2114

18. Drela M (2009) Power balance in aerodynamic flows. AIAA J 47(7):1761–1771

19. Queipo NV, Haftka RT, Shyy W, Goel T, Vaidyanathan R, Kevin Tucker P (2005) Surrogate-based analysis and optimization. Prog Aerosp Sci 41(1):1–28. https://doi.org/10.1016/j.paerosci.2005.02.001

20. Han Z (2016) Kriging surrogate model and its application to design optimization: a review of recent progress. Acta Aeronautica Et Astronautica Sinica 37(11):3197–3225 (in Chinese)

21. Xiong Y, Liu Y, Zhao X, Sun H, Wang G, Gao D (2016) Analysis of cavitation performance of an aviation fuel pump based on surrogate model. Acta Aeronautica Et Astronautica Sinica 37(10):2952–2960 (in Chinese)

22. Kerho M (2015) Aero-propulsive coupling of an embedded, distributed propulsion system. In: AIAA Applied Aerodynamics Conference, Dallas, TX. https://doi.org/10.2514/6.2015-3162

23. Perry AT, Ansell PJ, Kerho MF (2018) Aero-propulsive and propulsor cross-coupling effects on a distributed propulsion system. In: AIAA Aerospace Sciences Meeting, Kissimmee, Florida. https://doi.org/10.2514/6.2018-2051

Wing Selection and Dynamic Derivative Estimation of a Tailless UAV

Tianji Ma[1], Da Huang[1(✉)], and Lihui Zhang[2]

[1] College of Aerospace Engineering,
Nanjing University of Aeronautics and Astronautics,
Nanjing 210016, People's Republic of China
njdhcn@nuaa.edu.cn
[2] Sichuan Special Vehicle Science and Technology Co., Ltd., Mianyang 622752,
People's Republic of China

Abstract. Two alternative designs of a flying wing configuration unmanned aerial vehicle (UAV) with different aspect ratios were investigated by the numerical calculation. With the comparisons of lift-drag characteristics, lateral-directional stability and efficiency of control surfaces, the scheme of the smaller aspect ratio was adopted. Then the wind tunnel tests of the selected configuration were conducted to investigate the aerodynamic characteristics. In order to obtain more dynamic parameters desired in the flight simulation, a simple and useful method for the engineering estimation of dynamic derivatives was proposed. The longitudinal and lateral-directional dynamic derivatives of the UAV respectively calculated by the estimation method and CFD method were then compared. The comparison results showed that the estimation method can accurately reveal the dynamic aerodynamic characteristics of the aircraft.

Keywords: Tailless wing flying layout · Wind tunnel test ·
Rigid moving mesh technique · Estimation of dynamic derivatives

1 Introduction

With the development of UAV design technology, the typical layout of current flying wing drones can be roughly divided into two categories: tailless wing and double vertical-tail wing. The investigations of Yang [1] and Wang [2] showed that the tailless layout with composite control surfaces cannot provide satisfactory aerodynamic characteristics, due to the effects of cross-coupling of the ailerons and elevators on control efficiency. In order to solve this problem, a high-efficiency control law is commonly required. However, the design process of such a control system is difficult and time consuming. Therefore, the winglet is proposed to enhance the lateral stability. By placing the rudders on winglets, the control system can be decoupled and greatly simplified. However, the weak control power of the rudders on the winglets is another problem which bothers designers.

In order to solve above problems, the CFD method [3, 4] was used to study the aerodynamic characteristics of a flying wing UAV with two different wing spans in this investigation. Then, the wing configuration with better lift-drag features and lateral

© Springer Nature Singapore Pte Ltd. 2019
X. Zhang (Ed.): APISAT 2018, LNEE 459, pp. 1551–1565, 2019.
https://doi.org/10.1007/978-981-13-3305-7_124

stability was determined, and was verified by wind tunnel experiments. In addition, an engineering method for the estimation of dynamic aerodynamic characteristics of an aircraft was proposed. The estimation method can be used to calculate the dynamic derivatives of the UAV with the selected configuration. Then the results of numerical calculation and experiments were compared. The comparison results show that the estimation method works well.

2   Selection of Wing Span

2.1   Calculation Method

Two configurations which have the same fuselage and airfoil but different aspect ratios were designed to investigate the difference between lift-drag characteristics and lateral-directional stability.

After determining the numerical calculation model for CFD, the aerodynamic coefficients of the aircraft were calculated by Fluent software. The flow speed was set as 40 m/s. In the calculation process of CFD, the variables were selected as angle of attack (α), side slip angle (β) and control surface deflection (δ). In Fluent each of the states of variables was calculated by means of the one-equation (Spalart - Allmaras) model in turbulence model. Calculation meets the following transport equation

$$\frac{\partial}{\partial t} \iiint_V W dV + \iint_{\partial V} F \cdot n \, ds = \frac{1}{Re} \iint_{\partial V} F_V \cdot n \, ds \tag{1}$$

Where V is any control volume, W is a conservative variable, F is the non-viscous flux vector, F_V is the viscous flux, n is the normal vector outside the boundary unit of control volume, Re is the Reynolds number of calculation.

The model is shown in Figs. 1 and 2.

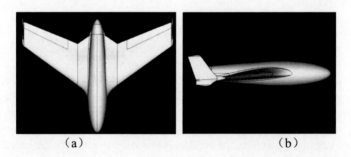

(a)   (b)

Fig. 1. Top and side views of low aspect ratio numerical calculation model

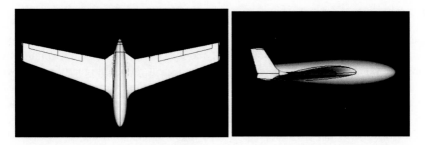

Fig. 2. Top and side views of high aspect ratio numerical calculation model

In this investigation, the origin O of the body axis system is fixed at the center of gravity (CG) of the aircraft. The $(OX_b Y_b)$ plane is coincident with the symmetry plane of the aircraft as illustrated in Fig. 3. The OX_b axis is parallel to the geometrical horizontal fuselage datum and points to the nose, the OY_b axis is directed upwards, and the OZ_b axis is directed to starboard.

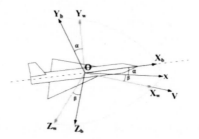

Fig. 3. Coordinate system definition

For a symmetrical aircraft, the efficiency of all the control surfaces can be simply measured by deflecting the surfaces on one side of the aircraft to improve the economy.

Figure 4 shows the surface mesh of the semi-body of the low-aspect-ratio aircraft with no-deflection of control surfaces. The total grid is 400 W. The far field is 10 times the size of the reference length. The boundary conditions are set as pressure far field.

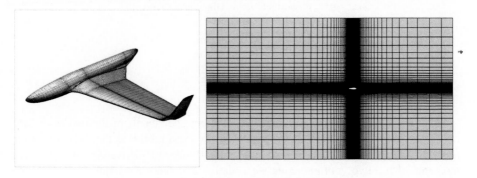

Fig. 4. Calculation grid for low aspect ratio configuration

2.2   Comparison of Calculation Results

Figure 5 shows the curves of lift coefficients of the two configurations with different aspect ratios. As illustrated, the shapes of the curves of the two configurations are almost the same. The lift coefficients of the low aspect ratio configuration are always higher than those of the high aspect ratio configuration at the same angle of attack. The critical stall angle of attack of the aspect ratio one is 12°, much higher than that of the high aspect ratio one which is 8°. Furthermore, the maximum lift coefficient is 1.266 for the low aspect ratio one and 0.997 for the high one. Overall, the lift characteristics of the low aspect ratio wing configuration are superior to the high aspect ratio wing configuration.

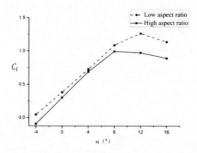

Fig. 5. Comparison of lift coefficient for two kinds of aspect ratio aircraft

Table 1 shows the lift-drag ratios of the two configurations at different angles of attack. Obvious difference of the two configurations can be observed, especially at the angle of attack about 4° which is the cruise angle of attack of the aircraft. At this angle of attack, the lift-drag ratio of the low aspect ratio configuration is about 10% higher than that of the high aspect ratio configuration. The advantages of low aspect ratio configurations are obvious.

Table 1. Lift-to-drag ratio for different configurations

α	Low aspect ratio	High aspect ratio
0°	11.888	11.282
4°	11.548	10.118
8°	9.515	6.713
12°	4.841	3.957

Figure 6 shows the static aerodynamic moment coefficients of the two configurations. The left column shows the moments of low aspect ratio, as the right shows the moments of high aspect ratio. It is obvious that all the static derivatives of these three moments of the two configurations are negative and meet the requirement of static stability. For the directional static stability derivative m_y^β, which is the focus of this investigation, the value of the low aspect ratio configuration is $-0.00075/°$ and is more stable than $-0.00045/°$ of the high aspect ratio. The yaw static stability of the low aspect ratio configuration is better than that of the high aspect ratio.

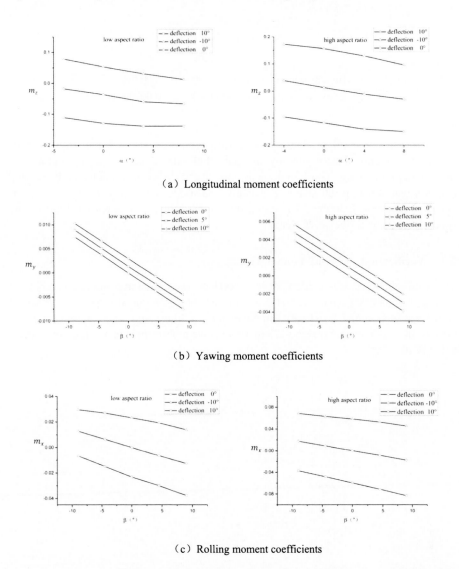

(a) Longitudinal moment coefficients

(b) Yawing moment coefficients

(c) Rolling moment coefficients

Fig. 6. Comparison of moment characteristics for the whole aircraft of two configurations

In above range of angle of attack and sideslip angle, the efficiency of each control surface, as shown in Fig. 6, has no insignificant difference. The extra aerodynamic moments produced by each control surface are symmetrical about zero deflection.

Control surface efficiency of the two configurations can be calculated by $\Delta m_i / \delta \, (i = x, y, z)$, and are shown in Table 2. The rudder efficiency of the aircraft with low aspect ratio is much better than that of the high aspect ratio one.

Table 2. Control surface efficiency of two configurations

	$m_z^{\delta z}$	$m_y^{\delta y}$	$m_x^{\delta x}$
High-aspect-ratio	0.00876	0.000193	0.00399
Low-aspect-ratio	0.00743	0.000321	0.00233

Above all, the configuration with low-aspect-ratio wings has better lift-drag characteristics, directional static stability and control efficiency than the configuration with high-aspect-ratio wing. The longitudinal, lateral and directional static stability of the two configurations are all stable. Especially, the rudder efficiency of the low-aspect-ratio configuration is higher. Therefore, the configuration with low-aspect-ratio wing was selected and further investigated.

2.3   Verification of Wind Tunnel Test

The wind tunnel test was carried out in the NH-2 wind tunnel at Nanjing University of Aeronautics and Astronautics. NH-2 is a closed return circuit low-speed wind tunnel with a test section of 2.5 m high, 3 m wide and 6 m long, with a maximum velocity of 90 m/s. In this investigation, all the experiments were performed at a tunnel free-stream velocity of 40 m/s.

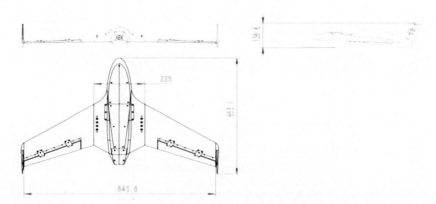

Fig. 7. Three views of experimental model

The experimental model is all-steel whose three views are shown in Fig. 7. The moment reference point of the test model is consistent with the calculation model, and is 240 mm away from the tail of the fuselage. The reference area of the model is

$0.14\,\mathrm{m}^2$, the longitudinal reference length is 166 mm, and the lateral reference length is 843.8 mm. The model was installed by tail support. In this experiment, a Φ20 internal six-component strain balance was used to measure the aerodynamic forces and moments of the model. The load range and correction accuracy of the balance are given in Table 3.

Table 3. Load range and accuracy of six-component balance static calibration

Balance unit	X(kg)	Y(kg)	Z(kg)	M_X(kg.m)	M_y(kg.m)	M_z(kg.m)
Design load	25	75	30	1.5	2.0	4.5
Correction accuracy %	0.12	0.1	0.34	0.47	0.38	0.16

Because the moment reference point of the model and the calibration center of the balance do not match while installing, the center conversion correction is used to improve the accuracy during data collection. All moments corresponding to the center were converted to the model aerodynamic center. According to the traditional correction method of wind tunnel test, the wall interference correction was performed on the test results. With the corrected results of the balance, the angle caused by the elastic deformation of the balance and that of the support rod were corrected. The model in the wind tunnel test is shown in Fig. 8 [5].

Fig. 8. Installation of experimental model

In order to verify the calculation results of CFD, the wind tunnel test project included:

(1) Elevator efficiency test. The downward deflection of elevator is defined as positive. The experiments in the range of angles of attack from −4 to 20° are conducted when the elevator is deflected to −10°, 0°, and 10° at $\beta = 0°$.

(2) Aileron efficiency test. The downward deflection of right aileron is defined as positive. The experiments in the range of side slip angle from −9 to 9° are conducted when the aileron is deflected to −10°, 0°, and 10° at $\alpha = 0°$.
(3) Rudder efficiency test. The rightward deflection of rudder is positive. The experiments in the range of side slip angle from −9 to 9° are conducted when the aileron is deflected to −10°, 0°, and 10° at $\alpha = 0°$.

Figures 9 and 10 show the comparison curves between the experimental and calculation results of the selected configuration.

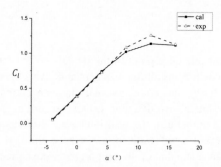

Fig. 9. Comparison of calculation and the experimental results

As illustrated in Fig. 9, the lifting coefficient of the selected model is well consistent with the experimental results in the linear segment. When the angle of attack is higher than 8°, the difference between the results of experiment and calculation is obvious, which was probably caused by the uncertainty due to the flow separation in the CFD results.

The comparison of the elevator efficiency is shown in Fig. 10(a). The experimental results are slightly higher than the calculation results. The difference is caused by the influence of the tail support on the flow filed around the elevator in the experiments. The efficiency of aileron is illustrated in Fig. 10. The static stability derivative of the rolling moment obtained from experiments is higher than the calculation results. The experimental curve of m_x has a little deviation at $\beta = 0°$. This negligible error is caused by a very small mounting roll angle of the experimental model.

Overall, the efficiency of control surfaces of the experimental results is in good agreement with those of calculation results. The main control derivatives of each control surface are shown in Table 4.

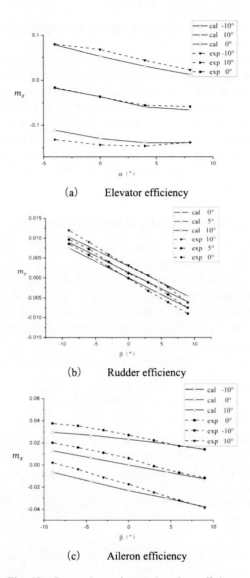

(a)   Elevator efficiency

(b)   Rudder efficiency

(c)   Aileron efficiency

Fig. 10. Comparison of control surface efficiency

Table 4. Comparison of the results of calculation and experiment for low-aspect-ratio configuration

	$m_z^{\delta z}$	$m_y^{\delta y}$	$m_x^{\delta x}$
Experiment	0.00889	0.000312	0.00225
Calculation	0.00743	0.000321	0.00233

3   Calculation of Pitching and Yawing Damping Derivatives

The dynamic derivatives of an aircraft are important dynamic aerodynamic parameters for flight simulation. The commonly used methods to obtain dynamic derivatives are dynamic derivative experiments in wind tunnel and numerical simulations by CFD. The dynamic derivative experiment is complicated and has a high requirement for the experimental model. Therefore, the dynamic derivative tests are usually time-consuming and costly. For the numerical simulation of CFD method, it also requires a long period to obtain all the dynamic derivatives due to the complicated calculation process. In this investigation, an engineering estimation method for the dynamic derivatives, which is high-efficiency, is proposed.

3.1   Engineering Estimation Method of Dynamic Derivatives

Taking the pitch-damping derivative as an example, the key step of the estimation process is the measurement of the hysteresis loop of m_z in a small-amplitude pitching oscillation. The pitching motion is assumed as a sinusoidal motion with a center point of α_0, amplitude of α_m and reduced frequency k. As the model is rotated around the CG, a velocity is generated which is perpendicular to the horizontal plane. Thus, the velocity vector can be divided into two parts. One of them is parallel to the flow direction and expressed as v_2. Another is vertical to the flow direction and defined as v_1. Due to the effects of v_1, an additional angle of attack $\Delta\alpha$ is added on α_0.

$$\Delta\alpha = -arctan\left(\frac{v_1}{v_\infty}\right) \tag{2}$$

Thus, the real-time angle of attack is $\alpha_0 + \Delta\alpha$. Then, the pitching moment coefficient at the real-time angle can be interpolated from the curve of the static pitching moment coefficient. Then the hysteresis loop of the pitching moment coefficient can be obtained.

After obtaining the hysteresis loop of the moment coefficient, the integral process of time-average [6] is used to obtain the pitch-damping derivative at α_0.

$$m_{\dot\alpha} + m_{q\dot\alpha} = \frac{\oint m_z d\alpha}{\pi \alpha_m^2 k} = \frac{\frac{2}{T}\int_0^T m_z \sin(\omega t)dt}{\pi \alpha_m k} \tag{3}$$

Where α_m is the amplitude, T is the period of oscillation. The yaw-damping derivative can be calculated similarly.

3.2   Numerical Method of Dynamic Derivatives

In the process of the numerical calculation of dynamic derivatives, the hysteresis loop in the pitching motion should also be obtained first. Assuming that the aircraft performs the same motion as discussed above, the most important parameter is the reduced frequency k. The calculation result is not sensitive to the quality of the boundary layer mesh during the calculation in the unsteady flow. Thus, an unstructured mesh was used to save computing time. The total number of the grid was about 320 W. Commonly, two typical methods, rigid dynamic mesh technology and grid reconstruction technology, are used to calculate the dynamic derivatives of the aircraft [6–12]. As there is no relative rotation between the aircraft components, the rigid mesh technology is the better method to calculate the dynamic derivative quickly. Therefore, the rigid dynamic mesh technique was chosen and used in the numerical calculation. The turbulent model used in calculation is the inverse SST model because of the higher accuracy of the simulation of pressure gradient [13]. Figure 11 shows the surface mesh and the grid of far field grid.

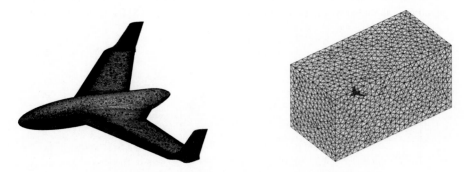

Fig. 11. Grids for the surface and far field

The pitching moments during the calculation process are shown in Fig. 12. Then the pitch-damping derivative can be calculated by Eq. 3.

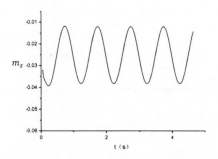

Fig. 12. Time history of numerical calculation

3.3 Comparison of the Dynamic Derivatives Calculated by Different Methods

The estimation and CFD methods are used for the calculation of dynamic derivatives, respectively. The pitching motion can be described as

$$\alpha = \alpha_0 - 2^\circ cos\omega t \tag{4}$$

The reduced frequency k can be expressed as

$$k = \frac{\omega C_{ref}}{2V_\infty} \tag{5}$$

C_{ref} is the reference length, V_∞ is the free-stream velocity.

Figures 13 and 14 show the hysteresis loops of the pitching moment separately obtained from the engineering estimation method and the numerical calculation method, at the angles of attack of 0°, 2°, 4°, 6°, and 8°. And the pitch-damping derivatives calculated by the two methods are shown in Fig. 15. Est is the abbreviation of estimation.

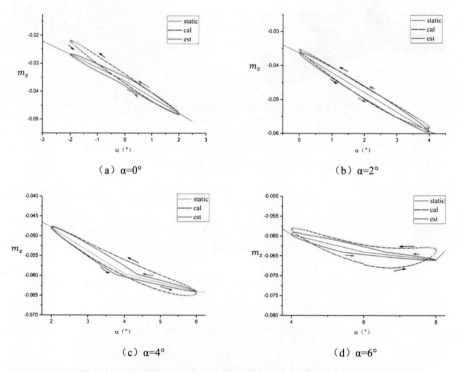

(a) α=0°   (b) α=2°

(c) α=4°   (d) α=6°

Fig. 13. Comparison of unsteady pitching moment coefficients

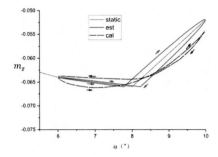

Fig. 14. Comparison of unsteady pitching moment coefficients at 8°

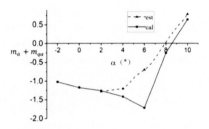

Fig. 15. $m_{\dot{\alpha}} + m_{q\alpha}$ versus α

As shown in Figs. 13 and 14, the directions of the hysteresis loops obtained from different methods are consistent at each angle of attack. In the motion with a center angle of attack of 8°, the hysteresis loop calculated by the engineering estimation method shows the shape of "8". The intersection is located at 7.8°. When the angle of attack is less than 7.8°, the hysteresis loops are counterclockwise. When the angle of attack is higher than 7.8°, the hysteresis loops are clockwise. The shape of the hysteresis loop given by the numerical simulation method is consistent with the estimating loop, but the intersection moves to 8.3°. In general, the dynamic hysteresis characteristics revealed by these two methods are substantially consistent.

As illustrated in Fig. 15, the pitch-damping derivatives calculated by the two methods are basically the same at small angles of attack. However, when the angle of attack is 6°, the estimation result of the numerical simulation method is two times as that of the numerical method. It is caused by the difference of the hysteretic loop areas with respect to those two methods as shown in Fig. 13(d).

In general, the damping characteristics revealed by the damping derivatives calculated by the engineering estimation method and the numerical simulation method are basically the same.

The yaw-damping derivatives can be obtained by the same approach. The results are shown in Fig. 16.

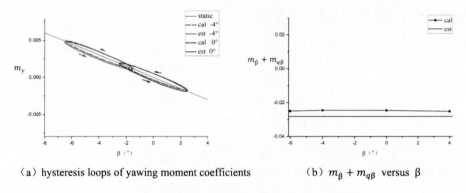

(a) hysteresis loops of yawing moment coefficients   (b) $m_{\dot{\beta}} + m_{q\beta}$ versus β

Fig. 16. Yaw-damping characteristics

For the directional dynamic derivatives, there is no change in sign of the slopes of yawing moment coefficients. The results of the two methods are consistent. The engineering estimation method proposed in this investigation is applicable.

4 Conclusion

The aerodynamic characteristics of different wingspans of a flying wing layout drone were studied in this investigation by numerical simulation method. According to the lift-drag characteristics and efficiency of rudder, the configuration is selected and verified by the wind tunnel test. In addition, a simple and practical engineering estimation method for the calculation of dynamic derivatives is proposed. It is concluded that

(1) For the flying wing layouts in this investigation, the low-aspect-ratio wing has the better lift-drag characteristics. For the rudder placed on the winglet, the low-aspect-ratio configuration has higher rudder efficiency and the better directional static stability than the high-aspect-ratio configuration.
(2) The estimation method of dynamic derivatives proposed in this paper is simple and practical. The results of the estimation method are approximately the same with the numerical simulation results and therefore credible.

References

1. Yang G (2007) The Concept Design of Composite Rudder for Large-aspect-ratio Tailless Layout. Master, Northwestern Polytechnical University, Xi'an, People's Republic of China (in Chinese)
2. Yuanyuan W, Binqian Z, Dong S (2005) Exploring combined capability of aerodynamic control surfaces of W-Shaped tailless configuration. J Northwest Polytech Univ 26(6):698–703 (in Chinese)

3. Jian G (2013) The Numerical Research on Transient Aerodynamic Performances and Configurations of Low Speed UAVs. Master, National University of Defense Technology, Changsha, People's Republic of China (in Chinese)
4. Jiechuan F (2002) Handbook of Wind Tunnel Test, 1st edn, pp 647–684. Aviation Industry Press, Beijing, People's Republic of China (in Chinese)
5. Huang D, Wu G (2006) Unsteady rolling moment characteristics for a fighter oscillating with yawing-rolling coupled motion. J Aircr 43(5):1570–1573
6. Liu X, Liu W, Chai Z, Xiaoliang Y (2016) Research progress of numerical method of dynamic stability derivatives of aircraft. Acta Aeronautica et Astronautica Sinica 37 (8):2348–2360. https://doi.org/10.7527/s1000-6893.2016.0098 (in Chinese)
7. Ren Y, Lei G (2009) Theory and Calculation Method of Aircraft Stability Parameter. Proceeding of the Chinese Conference of Theoretical and Applied Mechanics, Beijing, The Chinese Society of Theoretical and Applied Mechanics, pp 231–245 (in Chinese)
8. Liu W, Qu Z (1998) Calculation of damping-in-yaw derivatives by forced oscillation method. J Propul Technol 19(2):30–32 (in Chinese)
9. Yuan X, Zhang H, Xie Y (2005) Pitch damping derivative calculation method based on unsteady flow field numerical simulation. In: Proceeding of the Chinese Conference of Theoretical and Applied Mechanics, pp 301–309, Beijing (in Chinese)
10. Yuan X, Xie Y, Chen L (2013) Study on the Unsteady Numerical Simulation of the Static Stability Of The Aircraft. In: Proceeding of the Chinese Conference of Theoretical and Applied Mechanics, p 181, Beijing (in Chinese)
11. Yuan X, Zhang H, Xie Y (2005) The pitching static/dynamic derivatives computation based on CFD methods. Acta Aerodynamica Sinca 23(4):458–463 (In Chinese)
12. Mi B, Zhan H, Wang B (2005) Numerical simulation of dynamic derivatives based on rigid moving mesh technique. J Aerosp Power 29(11):2659–2664. https://doi.org/10.13224/j.cnki.jasp.2014.11.016 (in Chinese)
13. Mi B, Zhan H (2016) New Calculation Methods for Dynamic Derivatives of Advanced Flight Vehicles. J Shanghai Jiao Tong Univ 50(4):619–635. https://doi.org/10.16183/jcnki.jsjtu.2016.04.023 (in Chinese)

Design and Flight Test Validation
of a Rotor/Fixed-Wing UAV

Peixing Niu[1], Yu Zheng[2], Xu Zeng[3], and Xiaoguang Li[1(✉)]

[1] College of Aerospace Engineering, NUAA, Yudao Street 29, Nanjing, China
npxl23@yeah.net, lxg@nuaa.edu.cn
[2] College of Automation Science and Electrical Engineering,
BUAA, Beijing, China
yzheng@buaa.edu.cn
[3] College of Automation Engineering, NUAA, Yudao Street 29, Nanjing, China
492462168@qq.com

Abstract. Vertical take-off and landing aircraft (VTOL) has been a hot research topic in the aeronautics. This paper discusses the feasibility of a new concept of rotor/fixed-wing which can switch the flying mode during flying. In this design, wings on both sides of the UAV could rotate respectively around the pitch axis to switch the flying mode, so as to achieve the purpose of VTOL or the cruising flight. The paper made an analysis on the aerodynamic characteristics and the mechanical characteristics of this kind of UAV. Based on these study, a physical prototype were made and conducted a flying test, which the recorded date was basically in line with the design goals, the power to weight ratio of the rotor/fixed-wing is significantly greater than the VTOL aircraft which use traditional propeller. It is proved that the analysis on the mechanical characteristics and the aerodynamic characteristics is valid. It illustrates that this aircraft configuration had certain feasibility and paved the way for potential application prospect in the future.

Keywords: Fixed wing · Rotor · Rotor/fixed wing switch · Transform · VTOL

1 Introduction

To combination the fixed wing and rotor aircraft efficiently to make it cruising at high-speed as a fixed wing, and hover or take off and land vertically like a rotor is a hotspot in aeronautics. If it can be achieved, not only has high economic and military value, but also may open up a new situation for the aircraft market [8].

The principle of most vertical take-off and landing technologies is to change the thrust direction of the power device, so that convert the thrust of the power device into the lift which supporting the aircraft's weight directly. Whether the power device is a propeller or a jet engine, it conforms to the basic rotor momentum theory, that is, when a certain lift force is generated, the rotor diameter is inversely proportional to the required power [9]. The large rotor diameter requires less power of the engine. However, the blade pitch of the large rotor is lower than that of the small rotor, which will affect the speed of horizontal flight. Therefore, the key of this technology lies in the selection of rotor diameter during vertical take-off and landing.

© Springer Nature Singapore Pte Ltd. 2019
X. Zhang (Ed.): APISAT 2018, LNEE 459, pp. 1566–1575, 2019.
https://doi.org/10.1007/978-981-13-3305-7_125

The V-22 tilt-rotor aircraft is a successful example. In order to reduce the power required in hover, the diameter of the V-22 propeller is large, and it also has a maximum speed of 250 kn/h r [1].

X50-A, by contrast, takes a different approach. Its airfoil shape is symmetrical in front and back, and the airflow can generate the same lift whether the air flows through the front or the back of the airfoil [6]. In hover, the wings of X50-A rotating around the center axis in high-speed, acts as the rotor. In the fixed wing state, the rotor is locked in the location of 90°, and the UAV flying as a fixed wing [6]. But although this project was not went to deeper, the rotor/fixed-wing offers a positive thought for VTOL to reduce disc load, reduce the requirement of the engine power, and fly horizontally at a high speed.

2 Basic Characteristics

The rotor/fixed-wing in this paper is a tailless UAV. The wings could deflect respectively, and then switch to rotor, as shown in Fig. 1. Unlike X50-A, the conventional upper and lower symmetric airfoils are adopted for the rotor/fixed-wing UAV in this paper. There is an electrically driven propeller at the tip of each wing, so that the direction of propulsion changes as the wing deflects. When the UAV changes into rotor, the torque generated by the two propellers causes the rotate of whole UAV, and the wing acts as the rotor to generate lift. When the UAV flying by fixed-wing, both propellers are face forward, driving the plane to fly horizontally. By this stage the directional stability is provided by the difference of thrust between the two propellers. The trailing edge of the wing is provided with a control surface to control the aircraft's pitch and roll.

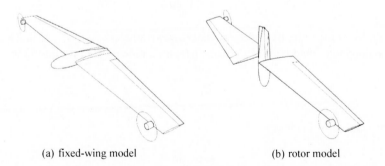

(a) fixed-wing model   (b) rotor model

Fig. 1. Two model of the rotor/fixed-wing UAV. The propellers driven by motors located in the tip of wings respectively. The wings can deflect so that the direction of propulsion changes to generate the torque which drives the wings rotation

2.1   Aerodynamic Characteristics

Due to the wings would deflect when changing the flying model, the top and bottom surface of the wing will exchange after the deflection, as shown in Fig. 2. Considering the aerodynamic balance, both sides of the wings adopt symmetrical airfoil.

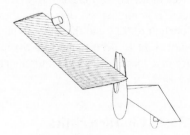

(a) the bottom surface of wings both face

(b) one of the bottom surface is face down, the other one is face up

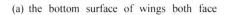

Fig. 2. The shadow represents the lower surface of the wing. After deflection, the surface of one wing will exchange the top and bottom

In order to reduce the induced drag under the fixed-wing model, trapezoidal wings were adopted instead of the flat wings commonly used by rotor blades [5]. According to the rotor momentum method [9], the wing length of the aircraft is R, the induced velocity at the rotor disc is

$$v_1 = \sqrt{\frac{T}{2\rho A}} \tag{1.1}$$

For most of the rotor, the inflow angle φ is generally less than $10°$, so that the small angle assumption could be taken to calculate, the inflow angle of wing tip is

$$\varphi_t = \frac{v_1}{v_t} \tag{1.2}$$

The velocity of blade tip v_t is

$$v_t = \omega R \tag{1.3}$$

where ω is the angular velocity of the wing, calculated in radians per second. According to the blade element method [9] of hover flight, the formula for calculating lift can be rewritten as:

$$L = \rho v_t^2 b_t Ra \left(\frac{\eta + 3}{24} \theta - \frac{2\eta + 1}{24} \varphi_t \right) \tag{1.4}$$

which L for unilateral wing lift, ρ for air density, b_t for blade tip chord length, η for the root tip, θ for the angle of airfoil installation, a is the lifting line slope for each radian, can approximate to each radian $5.73°$ for low speed airfoils. The corresponding formula of rotor resistance is:

$$Q = \frac{1}{20}\rho b_1 v_t^2 R^2 (c_l\varphi_t + c_d)(\eta + 4) \tag{1.5}$$

where c_l and c_d are the lift coefficient and drag coefficient of the airfoil respectively.

In the fixed-wing model the wing's lift in accordance with the general equation of the regular wing [5]. Put v_w is the flying speed of uniformly and in a straight line, c_{lW} is the lift coefficient, the lift L_W is

$$L_W = \frac{1}{2}\rho v_w^2 (\eta + 1) b_1 R c_{lW} \tag{1.6}$$

2.2   Flight Stability

Rotational Stability. Different from the traditional fixed-axis rotorcraft, the UAV in this paper is essentially a rigid body which rotating at a fixed point (ignoring small deformation) in the rotor state. When the rotor rotates stably around the inertia principal axis in the air, the external torque is zero. According to the rotational stability of Euler gyroscope [3], the rotating rigid body can maintain stable only by rotating around the principal axis of maximum or minimum inertia. That means when the rotor rotates stably around the z axis, the condition to be satisfied is $I_x < I_z$ and $I_y < I_z$.

In order to meet the conditions above, the balance weight lever is added to both wings. Balance weight is set on the outermost of the balance weight lever, which can increase I_z, as shown in Fig. 3. In addition, the length of the fuselage should be shortened as much as possible, and the horizontal and vertical tail fins should be eliminated, so that I_x and I_y can be minimized. This is also the reason why the rotor/fixed-wing in this paper adopt the tailless configuration. According to the vertical axis theorem, $I_z = I_x + I_y$, which satisfies the stability requirement of rigid body rotation at fixed point.

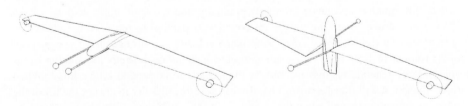

(a) the balance weight is in the front of plane,   (b) the inertia of the z axis is expanded
which move the center of gravity forward.   by balance weight in rotor model

Fig. 3. The configuration of the balance weight lever. There is balance weight on the outermost of the lever

Stability of Tailless Aircraft. By above knowable, in order to maintain the stability of the rotor model, the UAV need to shorten the length of the fuselage and cancel the tails, so the tailless configuration is adopted naturally. Without tail fins, the structure of the vehicle can be greatly simplified, and flight efficiency is improved. However, the absence of tail fins also has some influence on the aircraft's roll and pitch stability [7]. In order to increase stability, the following measures are adopted:

For Directional Stability.

1. Use cracked rudder [2] to correct the course using the moment of air resistance. This requires the addition of a pair of dedicated control mechanisms.
2. To control engines rotation in differential speed, and the thrust difference of both engines generates steering control moment [4]. This approach has high requirements for control signals and cannot be completed by manual remote control. Therefore, it is necessary to integrate a flight control system within the UAV. This is exactly the approach adopted by the UAV in this paper, because compared with the first method, there is no need to add complex mechanical structure, and the two engines can be used effective.

For Pitching Stability.

1. Move forward the gravity center of tailless aircraft can increase course stability. Therefore, the balance weight lever is set on the wings, which can deflect with the wings. When the UAV is in fixed-wing mode, the weight is also moved to the head of the aircraft to move the center of gravity forward.
2. The traditional flying wings usually adopt the "S" airfoil, which is upwarped at the trailing edge. During flight, the lifting torque is provided. However, according to the analysis in the chapter on aerodynamic characteristics, the UAV in this paper needs to adopt the upper and lower symmetric airfoil. Therefore, the same effect can be achieved when the rudder face of the trailing edge is turned up in horizontal flight and the lifting torque can be provided artificially to keep the UAV pitching stable.

2.3   Flight Process

As shown in Fig. 4, the UAV is rotor model when take-off, after climbing to a certain height, the rotor enters the transition stage (rotor translates to fixed-wing). As the wings deflect, the rotation resistance increases gradually and the spin angle velocity of the UAV slows down. After the wings are completely parallel to the fuselage, the UAV enters the wing flying phase. There is no plane velocity and lift because it was just out of the rotor state. So it is go into a subduction to accumulate the velocity, and the wings are going to generate enough lift and fix the attitude to horizontal cruise flight. When ready to land, it will translates from fixed-wing to rotor. In order to ensure the lift of the rotor is vertically upward after the switched to the rotor model, the head should be pulled up in the fixed wing state first, and the conversion should be vertical to the ground before starting. As the wing deflection, the rotation speed of the UAV increases gradually, and finally becomes rotor. The basic flow of a takeoff and landing is shown in Fig. 4.

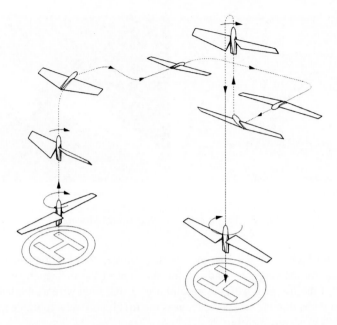

Fig. 4. A process of flight. The UAV is rotor model when take-off and landing

3 Prototype and Flight Test Validation

3.1 Prototype

Using the above analysis results, we designed and built a prototype to verify the correctness of the above analysis and the feasibility of rotor/fixed-wing. The basic parameters of the vehicle are as follows (Table 1):

Table 1. Parameters of prototype

Weight	0.7 kg	Propeller size	3.5×4.5 in.
Wingspan	1.2 m	Maximum level speed	25 m/s
Hovering angular velocity	4 r/s	Maximum take off weight	900 g
Climbing angular velocity	6 r/s	Battery capacity	1300 mAh
The ratio of energy consumption	7.37 g/w	Endurance	10 min

The fuselage of the prototype is framed by plywood which thickness is 1 mm, with electronics attached to the fuselage. The wings are cut from styrofoam. The wing beam structure is made by 8 mm carbon fiber tube, and there are bearings in the part where

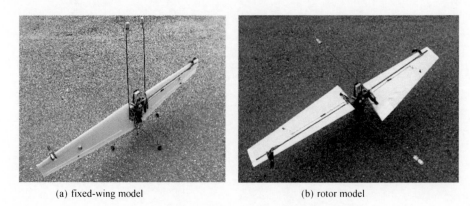

(a) fixed-wing model   (b) rotor model

Fig. 5. The prototype of the rotor/fixed-wing

the beam combined with the fuselage, which reduces the friction when the wing deflects. A worm steering gear and the linkage control the deflection of wings. The contact part of the fuselage with the ground uses 2 mm steel wire as the landing gear. The wheel at the end of the steel wire reduces the friction between fuselage and ground. The entire UAV is shown in Fig. 5.

3.2   Flight Test Validation

The test flight took place in a windless, open area. The UAV takes off vertically in rotor model. After reaching a height of 30 m, it became fixed-wing model and entered a subduction. After a subduction of about 10 m, it enters horizontal flight and cruising like a regular plane. The flight process picture is shown in Fig. 6. The flight process from take-off to the fixed wing can be viewed by scanning the two-dimensional code in Fig. 6(h).

Unfortunately, due to the immaturity of control methods, the UAV has not fulfilled controllable hovering and vertical landing in the rotor model. In the changing process, the aircraft failed to maintain a suitable attitude (the rotor disk parallel to the ground), finally the aircraft fell to the ground in a spiral shape and crashed under the action of gyro torque (Fig. 6). The stability and attitude control of the rotor are still under study (Fig. 7).

In addition to the test flight, the ratio of energy consumption of the UAV was measured. Compared with the traditional VTOL configuration that use propellers to lift

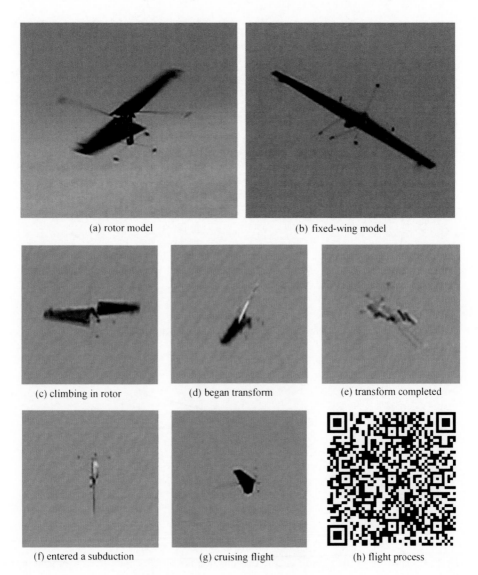

(a) rotor model

(b) fixed-wing model

(c) climbing in rotor

(d) began transform

(e) transform completed

(f) entered a subduction

(g) cruising flight

(h) flight process

Fig. 6. The flight process of the prototype

the aircraft, the power required of the rotor/fixed-wing is less. In the case of the prototype, the total weight was 700 g, only 95 W power was used to hovering, and the ratio of energy consumption reached 7.37. Figure 8 shows the comparison of the ratio of energy consumption between the rotor/fixed-wing and the traditional propeller [10].

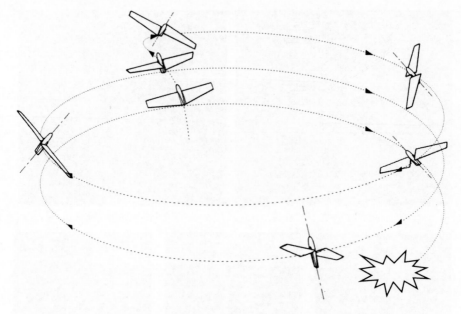

Fig. 7. The aircraft fell to the ground and crashed in a spiral shape

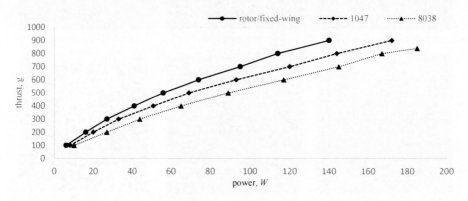

Fig. 8. Comparison of the ratio of energy consumption between the rotor/fixed-wing and the traditional propeller

4 Summary and Prospect

This article analysed some key features of the rotor/fixed-wing UAV. According to the test of prototype, it has more advantages than traditional VTOL configuration in the ratio of energy consumption, it can fly longer under the same energy.

The future study may focus on the aerodynamic efficiency and control method to make it more efficient during the flight, and be able to control the attitude in the rotor model.

References

1. Bolkcom C (2006) V-22 Osprey tilt-rotor aircraft. In: Congressional Research Service Reports. Congressional Research Service, Library of Congress
2. Buffington J, Buffington J (1997) Tailless aircraft control allocation. In: Guidance, Navigation, and Control Conference
3. Fung YC, Drucker DC (1980) Foundation of solid mechanics. Acta Mech Solida Sin 33 (1):238
4. Gillard W, Dorsett K, Gillard W et al (2006) Directional control for tailless aircraft using all moving wing tips. In: Atmospheric Flight Mechanics Conference
5. Katz J, Plotkin A (2004) Low-Speed Aerodynamics, 2nd Ed. J Fluids Eng 126(2)
6. Mitchell C, Vogel B (2013) The canard rotor wing (CRW) aircraft - a new way to fly. In: AIAA International Air and Space Symposium and Exposition: The Next 100 Years
7. Northrop JK, Sears WR (1946) Tailless aircraft: US, US2412646
8. Nickel K, Wohlfahrt M (1994) Tailless aircraft in theory and practice. Am Inst Aeronaut Astronaut E Arnold
9. Prouty RW (1986) Helicopter performance, stability, and control. PWS. Co
10. Zhongshan Langyu Model Co., Ltd. Propeller thrust manual [EB/OL]. http://www.rcsunnysky.com/content/157.html. 25-06-2018

Uncertainty-Based Design Optimization of NLF Airfoil Based on Polynomial Chaos Expansion

Huan Zhao[✉] 🄳 and Zhenghong Gao[✉]

School of Aeronautics, Northwestern Polytechnical University,
Xi'an, People's Republic of China
huanzhao_aero@163.com, zgao@nwpu.edu.cn

Abstract. The high probability of the occurrence of separation bubbles or shocks and early transition to turbulence on surfaces of airfoil makes it very difficult to design high lift and high-speed Natural-Laminar-Flow (NLF) airfoil for high altitude long endurance unmanned air vehicles. To resolve this issue, a framework of uncertainty-based design optimization (UBDO) is developed based on the polynomial chaos expansion method. The $\gamma - \overline{Re}_{\theta t}$ transition model combined with the shear stress transport $k - \omega$ turbulence model is used to predict the laminar-turbulent transition. The particle swarm optimization algorithm and surrogate model are integrated to search for the optimal NLF airfoil. Using proposed UBDO framework, the aforementioned problem has been regularized to achieve the optimal airfoil with a tradeoff of aerodynamic performances under fully-turbulent and free transition conditions. The tradeoff is to make sure its good performance when early transition to turbulence on surfaces of NLF airfoil happens. The results indicate UBDO of NLF airfoil considering Mach number and lift coefficient uncertainty under free transition condition shows a significant deterioration when complicated flight conditions lead to early transition to turbulence. Meanwhile, UBDO of NLF airfoil with a tradeoff of performances under fully-turbulent and free transition conditions holds robust and reliable aerodynamic performance under complicated flight conditions.

Keywords: Uncertainty-based design optimization (UBDO) ·
Natural-Laminar-Flow (NLF) airfoil · Fully-turbulent · Free-transition ·
High altitude long endurance unmanned air vehicles (HALE UAVs)

Nomenclature

Thickness	= maximum thickness of the airfoil
C_l	= lift coefficient
C_D	= drag coefficient
C_m	= pitching moment coefficient
C_f	= skin friction coefficient
C_P	= pressure coefficient
AoA	= angle of attack
Ma	= Mach number of the flow
Re	= Reynolds number of the flow
K	= lift to drag ratio

© Springer Nature Singapore Pte Ltd. 2019
X. Zhang (Ed.): APISAT 2018, LNEE 459, pp. 1576–1592, 2019.
https://doi.org/10.1007/978-981-13-3305-7_126

1   Introduction

The new high-altitude long-endurance (HALE) unmanned air vehicles (UAVs) have increasingly high demands for certain NLF airfoil. Such NLF airfoil shows some very special aerodynamic characteristics, for example, low Reynolds number together under transonic flight, high lift coefficient, low drag coefficients, necessary thickness. However, this kind of NLF airfoil tends to generate separation bubbles or shocks on its surfaces, especially the rapid movement of shocks with changing Mach number and angle of attack (AoA), greatly increasing its sensitivity of airfoil performance to the uncertain flight conditions. What's worse, the strong adverse pressure gradients associated with the shocks and the effect of compressibility on the size and location of the bubbles complicates the design of this kind of NLF airfoil. Till now, very few this kind of NLF airfoil data [1, 2] is available.

Even though numerous NLF designs have been demonstrated by many researchers [3–6], there are only a few implementations on current commercial aircrafts, i.e. the NLF body and wing on the HA-420 business jet [7] and the NLF wing on the Global Hawk UAV [1]. One of the reasons that NLF designs are rarely used is the strict restrictions on the manufactory precision and complicated flow condition. As the manufacturing level improves, geometry induced laminar instabilities will no longer be an obstacle to the application of NLF designs, while flight conditions uncertainty still cannot be eliminated. Traditional deterministic optimization method [8] lacks the ability to restrain the drag at off-design points for the highly sensitive NLF airfoil, even though the performances at on-design points have been optimized well. So that the resulting optimal design by traditional deterministic optimization method shows sensitive performance when the values of some parameters of flight conditions are uncertain or may deviate from the design values. Thus, the optimal design should take into account the variability or uncertainties of such parameters by minimizing an overall measure of the performance over all possible values of the uncertain parameters and the sensitivity of performance to uncertainties, for example, uncertainty-based design optimization (UBDO) [9]. UBDO should be developed by the stochastic approach [10] because the deterministic concept cannot include the uncertainties. The reference [11, 12] made some detailed comparisons and discussions of the differences of deterministic optimization and UBDO about this kind of NLF airfoil, which validates the practicability and effectiveness of UBDO for this kind of NLF airfoil.

The UBDO method is made up of three main parts [12]. The first stage consists of identifying and quantifying the uncertain parameter associated with the problem definition and the analysis modules. Probability distribution function (PDF) is often used to quantify uncertainties of these parameters. The second phase focuses on the uncertainty quantification (UQ) through the analysis approach to obtain probabilistic descriptions of the objective functions and constraints. Robust descriptions of the objectives depend on the numerical approximation of their statistic moments including expectation and variance, and reliability forms of constraints rely on the probability the constraint is violated or exceeds a reference value. Finally, the third stage defines the mathematic model of UBDO and search for the optimal shape with robust performance. UBDO for aerodynamic shape minimizes the mean and the standard deviation of objective subject to reliability constraints over the range of possible values of these uncertain parameters.

This paper focuses on building an effective UBDO framework to design the high-performance NLF airfoil for complicated flight conditions. It mainly resolves some issues as following.

Expensive computational cost and unsatisfactory accuracy are main difficulties for UQ in second stage of UBDO. Existing approaches to UQ [12] includes Monte Carlo simulation (MCS), Taylor-based moment propagation, sigma-point, stochastic expansion, etc. However, MCS method [13] causes very expensive cost for UQ and usually needs at least 10^6 samples for accurate estimate of mean and variance of C_D. The Taylor-based moment propagation method [14] or sigma-point method [15] unless restricts the analysis to very small values of the input variance its accuracy is often unsatisfactory in the frequent case of highly nonlinear function, so that it is difficult to be used for aerodynamic shape optimization widely. Polynomial chaos expansion (PCE) [16] and Stochastic collocation (SC) [17] methods are popular stochastic expansion approaches. Due to the orthogonality of polynomial terms, PCE method is a promising approach to perform this task for efficient and accurate UQ in aerodynamic shape optimization. However, applying the PCE method for UQ, the challenging problem is how to select as few collocation points as possible and proper polynomial order to obtain accurate estimate. For each of collocation points, a true value of aerodynamic performance is given to determine the unknown polynomial coefficients.

Proper mathematic model of UBDO is the most critical step in the third stage of UBDO. More recent UBDO work [18–22] only considers Mach number uncertainty and lacks reliability consideration for constraints, which is very hard to meet the design demand of the NLF airfoil for HALE UAVs. The NLF airfoil shows a strong sensitivity to many parameters including Mach number, AoA, Reynolds number, turbulence, etc. For example, the performance of NLF airfoil deteriorates when lift coefficient or AOA slightly exceeds the design value, due to the occurrence of separation bubbles on its surfaces. Another fact, the NLF airfoil with over-extended laminar region has short pressure recovery region, which easily leads to shocks or separation flow with changes of Mach number and AoA. Thus, the UBDO problem should be formulated simultaneously considering these parameters uncertainties. Uncertainties should also be taken into account for the estimation of the constraints involved in design optimization [9–11]. The constraints should be satisfied for all possible values of uncertain parameters, which is not practical for aerodynamic shape optimization problems. Alternatively, the probability of violating the constraints can be restricted within a range as small as possible, formulating the constraint in terms of reliability. The reliability constraint ensures that the optimal shape is within the feasible region of constraints with a range of tolerances. Thus, a mathematic model of UBDO considering Mach number, lift coefficient and transition location uncertainty is built subject to reliability constraints, which combines some cases of NLF airfoil design to demonstrate the most effective UBDO method for the NLF airfoil design of HALE UAVs.

The remainder of this paper is structured as follows. Section 2 validated a high-fidelity numerical transition method by some computational cases. Then, the sparse PC representations were built to meet the requirements. Finally, we built a framework of UBDO based an adjusted PCE method. This framework combines a global particle swarm optimization (PSO) algorithm and surrogate model to optimize NLF airfoil.

In Sects. 3 and 4, we combined the discussions and comparisons of two representative cases to find the most effective design method of NLF airfoil for HALE UAVs under complicated flight condition. These two cases include UBDO of NLF airfoil considering Mach number and lift coefficient uncertainty under free transition condition as well as UBDO of NLF airfoil considering Mach number and lift coefficient uncertainty under both fully-turbulent and free transition conditions. By the study of UBDO method of NLF airfoil and the comparisons of two cases of NLF airfoil design, some important conclusions were summarized in Sect. 5.

2 Methodology

2.1 Validation of Numerical Method

The transition prediction capability of the flow solver is validated by comparison to experimental data for the NLF-0416 airfoil and PSU 94-097 airfoil. The $\gamma-\overline{Re}_{\theta t}$ transition model [23] is combined with the SST turbulence model to predict laminar-turbulent transition. The optimized grid is used [24]. The NLF-0416 [25] is a Low-Reynolds-number airfoil and its experimental data comes from Langley Low-Turbulence Pressure Tunnel (LTPT) [25]. Figures 1 and 2 give the comparison of the lift-drag curves between the computational result and experiment data separately at 2×10^6 and 4×10^6 Reynolds number. The result shows good agreements especially at the large AOA.

PSU 94-097 [26] airfoil is a winglet airfoil and has been experimentally verified in the Pennsylvania State University Low-Speed, Low-Turbulence Wind Tunnel (LSLTT) [26]. Figures 3 and 4 give the comparison of computational and experimental pressure distributions for the PSU 94-097 winglet airfoil separately at $2°$ and $5°$. The results show good agreements. What's more, the exact predictions of separation bubbles at 61% chord for $2°$ and the suction peak at 5% chord for $5°$ will largely contribute to the NLF design.

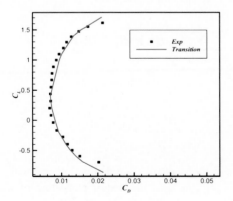
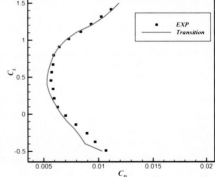

Fig. 1. The comparison of lift-drag curves of NLF-0416 between experiment and CFD (Re $= 2.0 \times 10^6$)

Fig. 2. The comparison of lift-drag curves of NLF-0416 between experiment and CFD (Re $= 4.0 \times 10^6$)

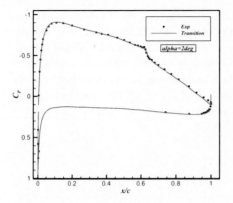

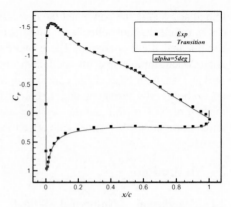

Fig. 3. Predicted and measured pressure distributions for the PSU 94-097 winglet airfoil $(\mathrm{Re} = 1.0 \times 10^6, AOA = 2^o)$

Fig. 4. Predicted and measured pressure distributions for the PSU 94-097 winglet airfoil $(\mathrm{Re} = 1.0 \times 10^6, AOA = 5^o)$

2.2   Polynomial Chaos Expansion

As this work differs from MCS method which applies sampling method for UQ, the PCE method can accurately approximate the stochastic behavior of a solution and apply such an approximation to estimate the statistical characteristics of the solution with random input variables. In this paper, the nonintrusive PCE method is used, which estimates the polynomial coefficients without modifying the governing equations. Original Hermite polynomial chaos [27] employs the orthogonal Hermite polynomials in terms of Gaussian random variables inputs. For more general cases, the generalized polynomial chaos expansion [28], the Askey-chaos, as a generalization of different kinds of random variables inputs, was proposed. The wiener-Askey polynomial chaos and homologous random inputs are listed in Table 1. Thus, a general random process $X(\xi(\theta))$, viewed as a function of θ, i.e. the random event, can be represented in the form

$$X(\xi(\theta)) = a_0 H_0 + \sum_{i_1=1}^{\infty} a_{i_1} H_1(\xi_{i_1}(\theta)) + \sum_{i_1=1}^{\infty} \sum_{i_2=1}^{i_1} a_{i_1,i_2} H_2(\xi_{i_1}(\theta), \xi_{i_2}(\theta)) + \cdots$$

$$= \sum_{j=0}^{\infty} b_j \psi_j(\xi) \tag{1}$$

$$H_n(\xi_{i_1}, \cdots, \xi_{i_1}) = e^{\frac{1}{2}\xi^T \xi}(-1)^n \frac{\partial^n}{\partial \xi_{i_1} \cdots \partial \xi_{i_n}} e^{-\frac{1}{2}\xi^T \xi} \tag{2}$$

where H_n are Hermite polynomials in terms of the standard Gaussian variables ξ with zero mean and unit variance. Here $(\xi_{i_1}, \cdots, \xi_{i_n})$ denotes the vector consisting of n uncorralated Gaussian variables ξ. If the Hermite polynomials order is p, the number $N_p + 1$ of polynomials chaos terms can be expressed as:

$$N_P + 1 = \frac{(p+n)!}{p!n!} \tag{3}$$

Equation (3) can be written as the simplified form

$$X(\xi) = \sum_{j=1}^{N_P+1} b_j \psi_j(\xi) \tag{4}$$

The polynomial chaos forms a complete orthogonal basis in the L_2 space of the Gaussian random variables:

$$\langle \psi_i, \psi_j \rangle = \langle \psi_i^2 \rangle \delta_{ij} \tag{5}$$

where δ_{ij} is the Kronecker delta and $\langle .,. \rangle$ denotes the ensemble average. The inner product in the Hilbert space of the Gaussian random variables can be expressed as

$$\langle f(\xi), g(\xi) \rangle = \int f(\xi) g(\xi) w(\xi) d\xi \tag{6}$$

$$w(\xi) = \frac{1}{\sqrt{(2\pi)^n}} e^{-\frac{1}{2}\xi^T \xi} \tag{7}$$

The expectation and variance of $X(\xi(\theta))$ can be expressed as

$$\mu(X(\xi)) = E(X(\xi)) = b_1 \tag{8}$$

$$\sigma^2(X(\xi)) = E(X(\xi) - E(X(\xi)))^2 = \sum_{j=2}^{N_P+1} b_j^2 \langle \psi_j^2 \rangle \tag{9}$$

Obtain the polynomial coefficients based on the Gaussian quadrature rule

$$b_j = \frac{\langle X(\xi), \psi_j \rangle}{\langle \psi_j^2 \rangle} = \frac{\int X(\xi) \psi_j(\xi) w(\xi) d\xi}{\langle \psi_j^2 \rangle}$$

$$= \frac{\sum_{i_1=1}^{k} \cdots \sum_{i_N=1}^{k} A_{i_1} \ldots A_{i_N} X(\zeta_{i_1}, \ldots \zeta_{i_N}) \psi_j(\zeta_{i_1}, \ldots \zeta_{i_N})}{\langle \psi_j^2 \rangle}, \quad j = 1, 2, \ldots N_P + 1 \tag{10}$$

For Eq. (11), A_i is the Gaussian-Hermite quadrature weight coefficient, and ζ_i is the Gaussian-Hermite quadrature point. The number of collocation points is $M = k^n$.
Obtain the polynomial coefficients by least square method

$$\mathbf{B} = \left(\mathbf{\Psi}^T \mathbf{\Psi} \right)^{-1} \mathbf{\Psi}^T \mathbf{Y} \tag{11}$$

For Eq. (12), M is the number of collocation points, and $N_P + 1$ is the number of Hermite polynomials chaos terms. The minimum number M of collocation points is $N_P + 1$.

The PCE by Gauss quadrature rule (Eq. 11) can determine where the points should be collocated in the stochastic space, which effectively avoid unphysical oscillations from the least square method. Ke et al. [29] applied Gauss quadrature rule to obtain the coefficients of PCE, and selected Gauss points for the random inputs. However, some collocation points from Gauss quadrature rule may exceed the valid range of input variables for specific problem. For example, high Mach number from Gauss points exceeding design range may cause strong shocks or large separations for transonic airfoil, which results in large computational errors. Moreover, the number of collocation points from Gauss quadrature rule increases exponentially with the number of uncertainty parameters. Obviously, for the UQ of aerodynamic performance under multi-parameter uncertainty, the PCE based on Gauss quadrature rule leads to large computational cost. Distinctly, the coefficients of PCE can be computed by the least square method [16] (Eq. 12), which can avoid the weakness of the PCE method based on Gauss quadrature rule. However, the number and locations of collocation points in the stochastic space and the polynomial order have an obvious effect on the UQ accuracy for the PCE method based on Gauss quadrature rule, e.g. too high polynomial order causes overfitting or Runge's phenomena [30]. How to select as few collocation points as possible and proper polynomial order for accurate UQ is still challenging. [11] provides an effective adjustment strategy for PCE method based on least square with very few collocation points for sufficient efficient and accurate UQ. In this paper, the adjustment strategy for PCE method is applied, and the sampling method of collocation points in the stochastic space is MCS [31].

Table 1. The Wiener–Askey polynomial chaos and the corresponding type of continuous random inputs

Polynomial chaos	Random inputs
Hermite-chaos	Gaussian
Laguerre-chaos	Gamma
Jacobi-chaos	Beta
Legendre-chaos	Uniform

2.3   Sparse PC Representation

Sparse PC reconstruction can be regularized as l_1-minimization problem [32], also referred to as basis pursuit denoising (BPDN), as shown in

$$\min_{\boldsymbol{\beta}} \|\boldsymbol{\beta}\|_1, \quad s.t. \ \|\boldsymbol{\psi}\boldsymbol{\beta} - \mathbf{Y}\|_2^2 \leq \varepsilon, \tag{12}$$

wherein, Eq. (12) can be solved by a large number of efficient algorithms. These algorithms include basis pursuit and greedy algorithms, e.g., OMP and least angle regressions (LARs) are widely used greedy algorithms.

OMP includes two important steps namely basis selection and coefficients update. Starting from initial active set $A^{(0)} = \phi$ and initial residual vector $\gamma^{(0)} = \mathbf{Y}$, at each iteration k, OMP identifies only the most correlated basis with the current residual from the remaining set $\Omega - A^{(k)}$ namely chosen bases removed from the universal set Ω, to be added to the active set. The residual vector $\gamma^{(k+1)}$ is updated by subtracting the contribution of chosen bases of active set from the output vector. The iteration procedure is continued until the residual tolerance ε is achieved. The following depicts a step-by-step implementation of the OMP algorithm (Table 2).

Table 2. Orthogonal matching pursuit algorithm

Algorithm1: OMP
Input: $\mathbf{Y} = \{y^{(j)}\}_{j=1}^{N}$, $\{\Xi^{(j)} = (\xi_1^{(j)}, \xi_2^{(j)}, \cdots, \xi_n^{(j)})\}_{j=1}^{N}$, $\varepsilon > 0$.
Output: $A^{(k)}$, $\boldsymbol{\beta}^{(k)}$
Initialization: $k = 0$, $A^{(0)} = \phi$, $\gamma^{(0)} = \mathbf{Y}$, $\Omega = \{1,2,\cdots M\}$,
$\{\boldsymbol{\psi}_1, \boldsymbol{\psi}_2, \cdots, \boldsymbol{\psi}_M\} = \{\psi_i(\Xi^{(j)})\}_{N \times M}$
While $k < M$ & $\|\gamma^{(k)}\| \geq \varepsilon$ **do**
$\quad i_+^{(k+1)} = \arg\max_i \|\boldsymbol{\Psi}_i^T \gamma^{(k)}\| / \|\boldsymbol{\psi}_i\|_2 \ (i \in \Omega/A^{(k)})$
$\quad A^{(k+1)} = A^{(k)} \cup \{i_+^{(k+1)}\}$
$\quad \boldsymbol{\beta}^{(k+1)} = \arg\min_{\boldsymbol{\beta}} \|\mathbf{Y} - \boldsymbol{\Psi}_{A^{(k+1)}} \boldsymbol{\beta}\|$
$\quad \gamma^{(k+1)} = \mathbf{Y} - \boldsymbol{\Psi}_{A^{(k+1)}} \boldsymbol{\beta}^{(k+1)}$
$\quad k \leftarrow k + 1$
end while

A complex scalar function with two independent random inputs is formulated as

$$f(x_1, x_2) = \ln(1 + x_1^2) \sin 5x_2 \tag{13}$$

where x_1, x_2 are assumed to be normally distributed with the mean $\mu = 2.0$ and standard deviation $\sigma = 0.4$. We set standard normally distributed variables $\xi_i \in N(0, 1^2)$, and inputs can be expressed by $x_i = 2.0 + 0.4\xi_i, (i = 1, 2)$. The reference values of mean (μ_f) and standard deviation (σ_f) are -0.115369 and 1.141393, respectively. To examine the performance of OMP, we estimate the stochastic performance of f by OMP algorithm and full PC representation, respectively. Figure 5 gives the true response of the two-dimensional function. The convergence procedures of relative error of mean and standard deviation with increasing number of collocation points for the two methods are shown in Figs. 6 and 7, respectively. Their results show that OMP algorithm achieves the faster reduction rate of error than full PC metamodel. When the number of collocation points reaches 20, OMP achieves $O(10^{-1})$ and $O(10^{-2})$ magnitude for relative error of mean and standard deviation, respectively. As a contrast, based on 20 collocation points, the relative error of mean and standard deviation by full PC are $O(10^0)$

and $O(10^0)$, respectively. Figure 8 gives the comparisons of estimated PDF of building PC metamodel by OMP and full PC. The results show that sparse PC representations achieve more consistency with that by MCS than full PC. Hence, these results indicate that the OMP method can enhance the sparsity of PCE and achieve more accurate estimate of the stochastic behavior of a solution. On the other hand, the OMP method needs less number of collocation points compared with full PC when required the same prediction accuracy, largely improving the efficiency of probabilistic UQ.

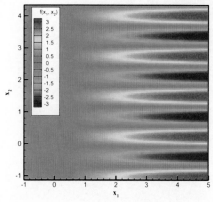

Fig. 5. True response of testing function f

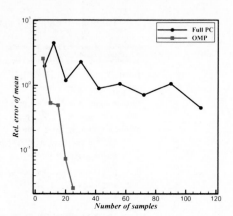

Fig. 6. The comparison of relative error in estimate mean

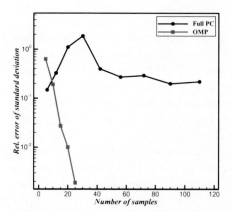

Fig. 7. The comparison of relative error in estimate standard deviation

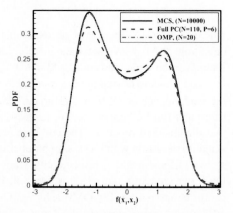

Fig. 8. The comparison of estimate PDF by OMP and full PC

2.4 Building of UBDO Framework

A UBDO framework based on an adjusted PCE method is proposed to design reliable and robust NLF airfoils for the HALE UAVs, as shown in Fig. 9. Compared with UBDO method based on MCS method, the proposed UBDO method based on an adjusted PCE method will largely reduce the computational cost. The UBDO method includes three important steps.

Step 1 is the basis of the method. The small changes of some parameters may lead to large fluctuations of drag coefficients of NLF airfoil. So, these parameters should be identified and quantified for complicated flight conditions, such as Mach number, AoA, lift coefficient, Reynolds number, transition location, etc. In this paper, these uncertain parameters, including Mach number, lift coefficient, transition location, are considered for the UBDO of NLF airfoils. Next, the effective mathematic model of UBDO is defined, as shown in Eq. (1). In Eq. (1), X represents the d dimensional geometric variables of NLF airfoil. The parameter C means the parameter vector of flight conditions including Mach number, lift coefficient. The parameter Z with a Gaussian distribution represents the noise around C. The objectives of UBDO include mean and standard deviation of drag coefficients. The mean is used as a performance measure of the drag coefficient that takes into account the uncertainties of flight conditions. The standard deviation is used as a measure of the sensitivity of the drag coefficient to uncertainties of flight conditions, providing optimal shapes that are robust to variability in the flight conditions. The C_m of Eq. (1) is a reliability constraint [9] defined by the probability that the pitching moment coefficient is lower than a reference value C_{m0}.

$$
\begin{aligned}
&Find &&\mathbf{X} \in R^d \\
&Minmize &&\sigma_{C_D}(\mathbf{X}, \mathbf{C} + \mathbf{Z}),\ \mu_{C_D}(\mathbf{X},\ \mathbf{C} + \mathbf{Z}) \\
&S.\,t. &&P(C_m(\mathbf{X}, \mathbf{C} + \mathbf{Z}) < C_{m0}) \leq P_0,\ \ t(\mathbf{X}) \geq t_0
\end{aligned}
\tag{14}
$$

Step 2 is the most important step in the process of UBDO. This step builds an efficient UQ method for the UBDO. In this paper, the sparse PC representation by orthogonal matching pursuit (OMP) is used for sufficient accurate UQ with low computational cost.

Step 3 applies global particle swarm optimization (PSO) algorithm [33, 34] to search for the optimal NLF airfoil. As this work differs from the deterministic optimization, the UQ of aerodynamic performance of NLF airfoil need to be estimated at each step of the evaluation of UBDO. The optimal NLF airfoil is a trade-off of mean and standard deviation of drag coefficients. The particle swarm optimization (PSO) algorithm and surrogate model are integrated to search for the optimal NLF airfoil. The CST and B-spline geometry parameterization methods are used to deform the NLF airfoil.

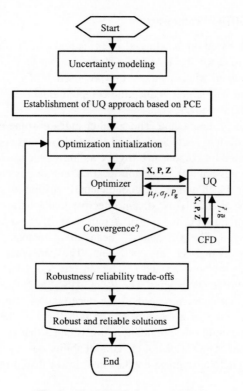

Fig. 9. The UBDO framework [12]

3   UBDO of NLF Airfoil

When the NLF airfoil is designed by UBDO under free transition condition, the aerodynamic performance of the NLF airfoil is insensitive to Mach numbers and lift coefficients [11]. However, when the complicated flight conditions lead to early transition to turbulence, the performance of the NLF airfoil from robust design shows a significant deterioration, which is harmful to engineering use.

To solve the aforementioned issue, an effective optimization method is provided by proposed methodologies to design high-performance NLF airfoils. The mathematic optimization formulations of OPT1 and OPT2 are Eqs. (13) and (14) respectively. An airfoil's aerodynamic performances under free transition and fully-turbulent conditions are balanced to make sure its good performance under both flow conditions. CDTRI stands for the drag coefficient of NLF airfoil under free transition condition and CDTUR is the drag coefficient of NLF airfoil under fully-turbulent condition. The w1 is a weight. Mach number and lift coefficient are supposed to be normal distribution. Transition and turbulence refer to free transition and fully-turbulent conditions respectively. The pitching moment coefficients of OPT1 and OPT2 constrained for reliability-based design optimization. OPT1 is the NLF airfoil by UBDO under free transition condition, which applies 32 design variables for B-spline geometry parameterization method and iterates 1200 steps for PSO algorithm to achieve convergence criterion. Opt2 is the NLF airfoil by UBDO under both free transition and fully-turbulent conditions, which applies 32 design variables for B-spline geometry parameterization method and iterates 1600 steps for PSO algorithm to achieve convergence criterion.

Flight condition :   *Free transition*

Find : $X \in R^d$

Min : $\left(\mu_{C_D}(Ma, C_L), \sigma^2_{C_D}(Ma, C_L)\right)$ (15)

S.t. $t \geq t_0, \ \mu_{C_m} - 2\sigma_{C_m} \geq C_{m0}, \ C_D = C_{DTRI}$

$Ma \sim N\left(0.60, \ 0.02^2\right), \ C_L \sim N\left(1.0, \ 0.1^2\right), \ \mathrm{Re} = 0.5 \times 10^6$

Flight condition :   *Free transition and Fully − turbulent*

Find : $X \in R^d$

Min : $\left(\mu_{C_D}(Ma, C_L), \sigma^2_{C_D}(Ma, C_L)\right)$ (16)

S.t. $t \geq t_0, \ \mu_{C_m} - 2\sigma_{C_m} \geq C_{m0}, \ C_D = w_1 C_{DTRI} + (1 - w_1)C_{DTUR}$

$Ma \sim N\left(0.55, \ 0.02^2\right), \ C_L \sim N\left(0.9, \ 0.1^2\right), \ \mathrm{Re} = 0.5 \times 10^6$

4   Results and Discussion

Figure 10 compares the initial airfoil and the optimal airfoil. The comparative airfoil is LRN1015 which has been applied in the Global Hawk UAV [1] and considered as the classic high speed and high lift NLF airfoil. The design point of LRN1015 airfoil is $\mathrm{Re} = 0.5 \times 10^6, Ma = 0.55, C_L = 1.0$. Figure 11 gives the skin friction distribution of the surface of initial and optimized airfoil under free transition condition which indicates further downstream transition points of Opt1 and Opt2 on the upper surface compared with that of LRN1015 on the upper surface. Figure 12 gives the pressure distribution of the surface of initial and optimized airfoil under free transition condition which shows that Opt1 and Opt2 have higher suction peak and slower pressure recovery on the upper surface than LRN1015, while the upper surface of LRN1015 forms a separation bubble. Figure 13 gives the pressure distribution of initial and optimized airfoil under fully-turbulent condition which indicates that Opt2 shows slower pressure recovery on the upper surface, while LRN1015 shows significant separation on the upper surface.

Figure 14 gives the lift-to-drag ratio curves of initial and optimized airfoils under free transition condition, which shows that Opt2 has a higher lift-to-drag ratio over a range of lift coefficients from 0.8–1.1 than Opt1 and LRN1015, while Opt1 shows slightly larger stall angle of attack than Opt2. Figure 15 gives the lift-to-drag ratio curves of initial and optimized airfoils under fully-turbulent condition. Figure 15 indicates that the lift-to-drag ratio of Opt1 shows an obvious drop near maximum lift compared with LRN1015; while Opt2 shows a significant increase of both maximum lift-drag ratio and stall angle of attack.

Figure 16 gives the drag-divergence curves of Opt1, Opt2 and LRN1015 under free transition condition, which indicates shock formation of LRN1015 on the upper surface when Mach number exceeds 0.56. Opt1 and Opt2 show lower drag coefficients over the range of Mach number from 0.50 to 0.60 than LRN1015 under free transition condition. The sophisticated comparison of the drag-divergence curves of Opt1 and Opt2 shows that the drag coefficient of Opt2 is on average 15 counts lower than that of Opt1 over a range of Mach number from 0.5 to 0.60.

Figure 17 gives the drag-divergence curves of Opt1, Opt2 and LRN1015 under fully-turbulent condition, which shows a significant deterioration of aerodynamic performance of Opt1 and a 50 counts lower drag coefficient of Opt2 compared with that of Opt1 over a range of Mach number from 0.5 to 0.60. Figures 18 and 19 give the pitching moment coefficients curves with variation of Mach number under free transition and fully-turbulent conditions separately. Obviously, it provides a very practical approach for meeting the pitching moment coefficient constraints in a probabilistic manner, ensuring that the optimal shape is within the feasible region of pitching moment coefficient constraints with a range of tolerances.

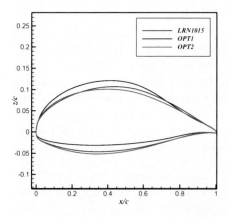

Fig. 10. Initial and optimized airfoils

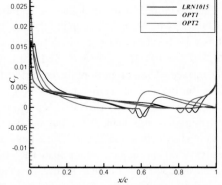

Fig. 11. The skin friction distributions of initial and optimized airfoils under free transition condition $(Ma = 0.55, \mathrm{Re} = 0.5 \times 10^6, C_L = 1.0)$

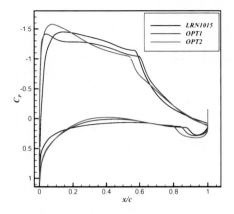

Fig. 12. The pressure distributions of initial and optimized airfoils under free transition condition $(Ma = 0.55, \mathrm{Re} = 0.5 \times 10^6, C_L = 1.0)$

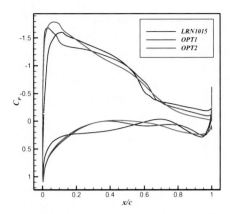

Fig. 13. The pressure distributions of initial and optimized airfoils under fully-turbulent condition $(Ma = 0.55, \mathrm{Re} = 0.5 \times 10^6, C_L = 1.0)$

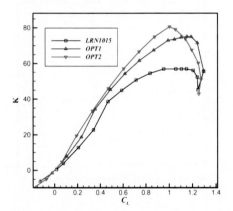

Fig. 14. Lift-to-drag ratio curves of initial and optimized airfoils under free transition condition $(\mathrm{Re} = 0.5 \times 10^6, Ma = 0.55)$

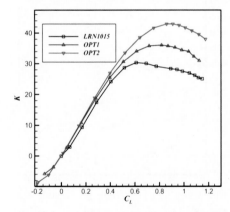

Fig. 15. Lift-to-drag ratio curves of initial and optimized airfoils under fully-turbulent condition $(\mathrm{Re} = 0.5 \times 10^6, Ma = 0.55)$

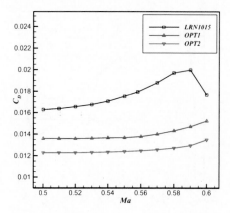

Fig. 16. Drag-divergence curves of initial and optimized airfoils under free transition condition (Re $= 0.5 \times 10^6, C_L = 1.0$)

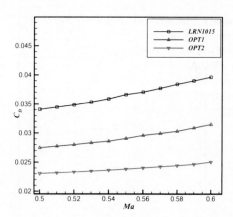

Fig. 17. Drag-divergence curves of initial and optimized airfoils under fully-turbulent condition (Re $= 0.5 \times 10^6, C_L = 1.0$)

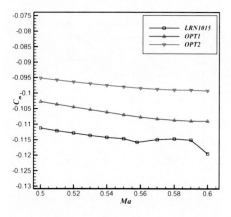

Fig. 18. The pitching moment curves of initial and optimized airfoils under free transition condition (Re $= 0.5 \times 10^6, C_L = 1.0$)

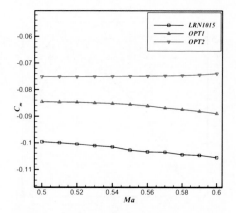

Fig. 19. The pitching moment curves of initial and optimized airfoils under fully-turbulent condition (Re $= 0.5 \times 10^6, C_L = 1.0$)

5 Conclusion

In this paper, a reliable numerical method is built for providing sufficient accuracy of NLF airfoil, including some validations of the accuracy by $\gamma-\overline{Re}_{\theta t}$ transition model.

This paper focuses on proposing an effective UBDO framework based on sparse PCE method, which can consider multi-parameter uncertainty. The optimal design minimizes a measure of the drag coefficient that is robust to uncertainties, subject to the pitching moment coefficient reliability constraint. Applying the proposed effective

UBDO framework, two high-performance NLF airfoils are designed and discussed to meet the high demand of NLF airfoil for the HALE UAV. The results of two design NLF airfoil draw some important conclusions, as follows.

A. UBDO provides an insensitive design considering free stream Mach number and lift coefficient uncertainty under free transition condition, but the lack of consideration of complicated flight condition which may lead to early transition to turbulence results in a significant deterioration of aerodynamic performance of the design airfoil (e.g. OPT1).

B. The complicated flight conditions which may lead to early transition to turbulence should be considered for the design of NLF airfoil. Effective UBDO of NLF airfoil should simultaneously consider Mach number, lift coefficient and transition location uncertainty.

References

1. Hicks RM, Cliff SE (1991) An evaluation of three two-dimensional computational fluid dynamics codes including low Reynolds numbers and transonic Mach numbers. NASA TM-102840
2. Zhao H, Gao Z, Wang C, Yuan G (2017) Robust design of high speed natural-laminar-flow airfoil for high lift. In: 55th AIAA Aerospace Sciences Meeting, p 1414
3. Driver J, Zingg D (2006) Optimized natural-laminar-flow airfoils. In: 44th AIAA Aerospace Sciences Meeting and Exhibit, p 247
4. Rashad R, Zingg DW (2013) Toward high-fidelity aerodynamic shape optimization for natural laminar flow. In: 21st AIAA Computational Fluid Dynamics Conference, p 2583
5. Cameron L, Early J, McRoberts R (2011) Metamodel assisted multi-objective global optimisation of natural laminar flow aerofoils. In: 29th AIAA Applied Aerodynamics Conference, p 3001
6. Khayatzadeh P, Nadarajah S (2012) Aerodynamic shape optimization of natural laminar flow (NLF) airfoils. In: 50th AIAA Aerospace Sciences Meeting including the New Horizons Forum and Aerospace Exposition, p 61
7. Fujino M, Yoshizaki Y, Kawamura Y (2003) Natural-laminar-flow airfoil development for a lightweight business jet. J Aircr 40(4):609–615
8. Nemec M, Zingg DW, Pulliam TH (2004) Multipoint and multi-objective aerodynamic shape optimization. AIAA J 42(6):1057–1065
9. Papadimitriou DI, Papadimitriou C (2016) Aerodynamic shape optimization for minimum robust drag and lift reliability constraint. Aerosp Sci Technol 55:24–33
10. Park G-J, Lee T-H, Lee KH, Hwang K-H (2006) Robust design: an overview. AIAA J 44 (1):181–191
11. Zhao H, Gao Z, Gao Y, Wang C (2017) Effective robust design of high lift NLF airfoil under multi-parameter uncertainty. Aerosp Sci Technol 68:530–542
12. Huan Z, Zhenghong G, Fang X, Yidian Z (2018) Review of robust aerodynamic design optimization for air vehicles. Arch Comput Methods Eng 1–48. https://doi.org/10.1007/s11831-018-9259-2
13. Keane AJ (2012) Cokriging for robust design optimization. AIAA J 50(11):2351–2364
14. Padulo M, Campobasso MS, Guenov MD (2011) Novel uncertainty propagation method for robust aerodynamic design. AIAA J 49(3):530–543

15. Padulo M, Maginot J, Guenov M, Holden C (2009) Airfoil design under uncertainty with robust geometric parameterization. In: 50th AIAA/ASME/ASCE/AHS/ASC Structures, Structural Dynamics, and Materials Conference 17th AIAA/ASME/AHS Adaptive Structures Conference 11th AIAA No, p 2270

16. Kim NH, Wang H, Queipo NV (2006) Efficient shape optimization under uncertainty using polynomial chaos expansions and local sensitivities. AIAA J 44(5):1112–1116

17. Eldred M (2009) Recent advances in non-intrusive polynomial chaos and stochastic collocation methods for uncertainty analysis and design. In: 50th AIAA/ASME/ASCE/AHS/ASC Structures, Structural Dynamics, and Materials Conference. Structures, Structural Dynamics, and Materials and Co-located Conferences. American Institute of Aeronautics and Astronautics. https://doi.org/10.2514/6.2009-2274

18. Li W, Huyse L, Padula S (2002) Robust airfoil optimization to achieve drag reduction over a range of mach numbers. Struct Multidisciplinary Optim 24(1):38–50

19. Croicu A-M, Hussaini MY, Jameson A, Klopfer G (2012) Robust airfoil optimization using maximum expected value and expected maximum value approaches. AIAA J 50(9):1905–1919

20. Duvigneau R (2007) Robust design of a transonic wing with uncertain mach number. In: EUROGEN 2007, Evolutionary Methods for Design. Optimization and Control, Jyvaskyla, Finland

21. Li J, Gao Z, Huang J, Zhao K (2013) Robust design of NLF airfoils. Chin J Aeronaut 26(2):309–318

22. Zhao K, Z-h G, J-t H (2014) Robust design of natural laminar flow supercritical airfoil by multi-objective evolution method. Appl Math Mech 35(2):191–202

23. Langtry RB, Menter FR (2009) Correlation-based transition modeling for unstructured parallelized computational fluid dynamics codes. AIAA J 47(12):2894–2906

24. Zhao H, Gao Z, Wang C, Gao Y (2018) Research on the computing grid of high speed laminar airfoil. Chin J Appl Mech 35(2):351–357

25. Somers DM (1981) Design and experimental results for a flapped natural-laminar-flow airfoil for general aviation applications

26. Maughmer MD, Swan TS, Willits SM (2002) Design and testing of a winglet airfoil for low-speed aircraft. J Aircr 39(4):654–661

27. Wiener N (1938) The homogeneous chaos. Am J Math 60(4):897–936

28. Xiu D, Karniadakis GE (2003) Modeling uncertainty in flow simulations via generalized polynomial chaos. J Comput Phys 187(1):137–167

29. Zhao K, Gao ZH, Huang JT, Jing L (2013) Airfoil flow uncertainty quantification and robust optimization based on polynomial chaos technique. Acta Mech Sin 46(1):10–19

30. Shimoyama K, Inoue A (2016) Uncertainty quantification by the nonintrusive polynomial chaos expansion with an adjustment strategy. AIAA J 54(10):3107–3116

31. Hosder S, Walters R, Balch M (2007) Efficient sampling for non-intrusive polynomial chaos applications with multiple uncertain input variables. In: 48th AIAA/ASME/ASCE/AHS/ASC Structures, Structural Dynamics, and Materials Conference, p 1939

32. Rauhut H, Ward R (2012) Sparse legendre expansions via ℓ1-minimization. J Approx Theory 164(5):517–533

33. Kennedy J, Eberhart R (1995) Particle swarm optimization. In: IEEE International Conference on Neural Networks, Proceedings, vol 1944, pp 1942–1948

34. Zhao H, Gao Z, Gao Y (2017) Design optimization of natural-laminar-flow airfoil for complicated flight conditions. 35th AIAA Applied Aerodynamics Conference. AIAA AVIATION Forum. American Institute of Aeronautics and Astronautics, p 3060

Research on Scheme of Maglev-Rotor UAV

Yantao Liu[⊠], Qiang Sun, Weigui Zhong, Yongfei Yang,
and Wang Xie

China Helicopter Research and Development Institute, Central Road 39#B,
Airport Economic Area, Tianjin, Jingdezhen 333001, Jiangxi, China
LYT18505050@avic.com

Abstract. The scheme of maglev-rotor unmanned aerial vehicle (UAV) is presented based on research of the existing multi-rotor UAV with vertical take-off and landing (VTOL) capability in this paper. Besides calculation and CFD analysis of aerodynamic characters of the maglev-rotor system and the transition between the VTOL mode and the forward flight mode of the maglev-rotor UAV, test flight are carried out to verify the calculation method and the feasibility of the scheme. The prototype of conventional electric motors is specially designed and built on this purpose. The attack angle of fuselage, the tilt angle of rear duct, the power of lift duct and the power of rear tilt duct are measured at different wind speeds. It is confident that the transition mode is available since the changes of the fundamental parameters, such as the attack angle of fuselage, the power of the lift duct fan and the rear tilt duct fan, the tilt angle of the rear duct fan, go on smoothly and continuously.

Keywords: Maglev-rotor · Lifting body with duct fan · Tilt duct fan ·
Lift combination adjusting

The multi-rotor Unmanned Aerial Vehicle (UAV) has vertical take-off and landing (VTOL) capability, and it does not need any launching device or runway, which is typical for the fixed-wing UAVs. Current types of multi-rotor UAV include conventional multi-rotor UAV, composite multi-rotor UAV and tilt-rotor UAV; however, those types are restrained from low flight speed and poor safety performance.

The conventional multi-rotor UAV is equipped with more than three (incl.) rotors in order to receive lift. The flight azimuth is controlled by adjusting these rotors for differing lift, and the steering control is practiced by torque differential of rotors. The flight of conventional multi-rotor UAV is controlled by tilting the fuselage to provide thrust for expanding the windward area, thus the drag increases and forward speed declines. The maximum speed of this type of UAV could reach 80 km/h.

Developing from the conventional multi-rotor UAV, the composite multi-rotor UAV adopts horizontal propeller to provide thrust and reduce the tilt angle of fuselage for diminishing drag; however, the speed improvement is not significant, and the maximum speed is no more than 100 km/h.

© Springer Nature Singapore Pte Ltd. 2019
X. Zhang (Ed.): APISAT 2018, LNEE 459, pp. 1593–1601, 2019.
https://doi.org/10.1007/978-981-13-3305-7_127

The tilt-rotor UAV is equipped with rotated propulsion system. By adjusting the angle between the axis of the propulsion system and the fuselage axis, the transition between helicopter mode and airplane mode is achieved. During vertical-take-off, the axis of propulsion systems stay vertical, and in cruising flight, the axis of propulsion systems turn to be horizontal. The diameter of propulsion rotor increases since the propulsion system of tilt-rotor UAV needs to be operated at both vertical taking-off and forward flight; therefore, the drag increases at forward flight mode and affects the forward speed - the maximum speed could reach 150 km/h (Fig. 1).

Fig. 1. The conventional multi-rotor UAV and the tilt-rotor UAV

The rotors of the multi-rotor UAV are exposed in the air, and it may causes collision with obstacles during low-altitude operation, and decreases flight safety. Moreover, it has potential threat to human beings safety. Additionally, the forced landing capability of the multi-rotor UAV is quite unsatisfying and it cannot perform autorotation due to its small rotor diameter, plus the wing area is too small to make glide landing.

1   Overview of the Scheme of Maglev-Rotor UAV

Considered as an improvement to low-speed and poor-safety performance of the existing VTOL multi-rotor UAV, the maglev-rotor UAV combines with a maglev rotor, a lifting body fuselage and a tilt rear duct fan. The front center of its fuselage is equipped with a lift duct fan which adopts two counter-rotating electromagnetic drive rotor. The lift duct could be shut down in high-speed forward flight, and the fuselage forms integral lifting body; a pair of symmetrical rotated duct fan is installed at rear fuselage, and the direction of thrust can be converted between vertical and horizontal, which is capable of providing lift during VTOL or thrust at high-speed forward. The configuration of maglev-rotor UAV achieves the transition of flight mode by tilting the rear duct fan, and makes the fuselage posture without tilt. Similar to fixed-wing UAVs, the lift duct fan of maglev rotor UAV turns off at high-speed flight mode to form the fuselage into a whole lifting body to increase forward speed dramatically. Generated from the duct fan, the rotor is enveloped to improve flight safety, and rotor flight efficiency is significantly elevated as well (Fig. 2).

Fig. 2. Flight mode of maglev-rotor UAV (The flight mode from left to right is VTOL, transition and forward flight)

Maglev-rotor comprises the orbit ring and the rotary ring embedded in the body, and the rotor in the lift duct fan. The orbit ring is fixed with the fuselage as the stator, the rotary ring is magnetic suspended as the rotor in the slots of the duct fan, lift is passed to the fuselage through the way of magnetic levitation, and the UAV flight is supported by the magnetic levitation force.

The maglev-rotor system can avoid mechanical frictional loss and power loss and improve transmission efficiency. It integrates the traditional engine, transmission and rotor system, simplifies the system and improves the maintainability. The non-clearance duct fan system at the outer end can further increase the rotor lift and improve the aerodynamic efficiency. The outer end support mode of blade can reduce bending moment and improve fatigue life. Magnetic levitation also has the effect of active vibration isolation, reduce vibration and noise, and improve work efficiency of reconnaissance, inspection, monitoring, surveying and mapping tasks. such as the permanent magnetic material has developed three generations of new rare earth permanent magnet (Nd - FeB) compared with the magnetism of magnetic steel used in the 1990s, and the performance has been improved more than 100 times, the rapid development of permanent magnetic materials provides a good solution to weight reducing problem of maglev-rotor system (Fig. 3).

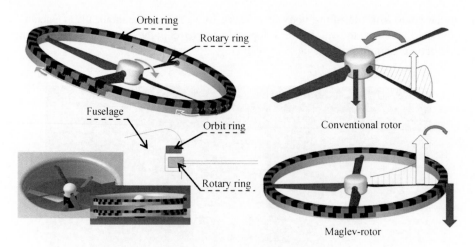

Fig. 3. Schematic diagram of maglev-rotor system

The power to weight ratio of maglev-rotor is expected to reach practical level 15 years later, at the same time, the maglev-rotor UAV has a lifting body design, which could support the forced landing, makes the security 2 times higher than the multi-rotor UAV and the tilt-rotor UAV; and the noise levels could be reduced by as much as a quarter of the noise of the smallest four seats helicopter on the market, i.e. 62 dB. For long-distance commuting and business flight, the target maglev-rotor UAV with a maximum cruise speed of 400 km/h, can fly 200 km in half an hour for best time efficiency. On the other hand, this type of UAV is very suitable for implementing various military missions because of its high speed quiet flight capability and low radar reflection area (Fig. 4).

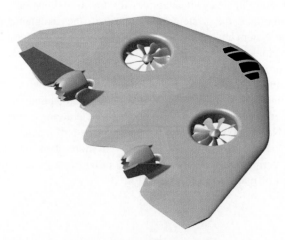

Fig. 4. Future maglev-rotor UAV

2   Scheme of the Maglev-Rotor UAV

Maglev-rotor UAV achieves the flight mode transfer by adjusting lift among the maglev-rotor, the lifting body with duct fan and the rear tilt duct fan, the adjustment of the transition becomes smoother, can improve the operation stability of vertical take-off and landing high-speed UAV and flight safety effectively. To test and verify the stability of lift adjusting and transition of maglev-rotor UAV, the author carried out CFD analysis and prototype test method for the validation work, maglev-rotor aerodynamic model and the lifting body fuselage aerodynamic model is set up, combined with the model calculation and analysis results, the aerodynamic characteristics and the flight mode transition is analysed, 30 kg level prototype is designed and manufactured of conventional motor driving system, the stability of the transition is verified by the test flight.

2.1   Analysis of Maglev-Rotor Duct Fan

Gaps exist between conventional rotor blade and the duct wall, but the outer end of the rotor blade of the maglev-rotor is installed directly on the rotary ring, suspended in the duct wall, the gaps between the blade and the duct wall are cancelled, the three dimensional flow effect of the blade tip is reduced, the efficiency of the hovering is proved, the analysis results show that:

1. With the increase of thrust, the hover efficiency increases gradually, and the high efficiency section maintained in a thrust interval, and then decreases with the increase of thrust.
2. The maximum hover efficiency of maglev-rotors can reach 0.82; it is 5.1% higher than that of conventional duct fan and 7.8% higher than that of conventional rotors under the same blade configuration.
3. The higher efficient working range of maglev-rotor is wider; interval width is relative to the conventional duct fan expanded by 28%, relative to the conventional rotor expanded nearly 1 times (Fig. 5).

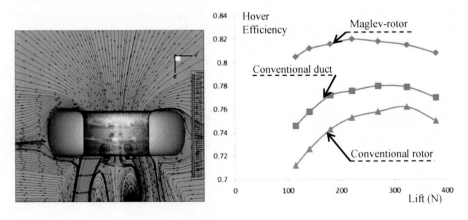

Fig. 5. CFD analysis of maglev-rotor system

2.2   Aerodynamic Analysis of Lifting Body Fuselage

The take-off weight of 30 kg level prototype is 33 kg, the fuselage configuration is confirmed according to the design demand, the typical Angle of attack of fuselage (1° and 3°) under the flight mode transfer speed range is analysed by CFD, to determine the flight mode transition way; drag features of the high speed forward flight will be calculated according to the balancing attack angle of fuselage, the results are validated by test. Calculation results show that the lift and drag of fuselage are increased with the increase of the speed, lift and drag increase faster when Angle is larger (3°) (Fig. 6).

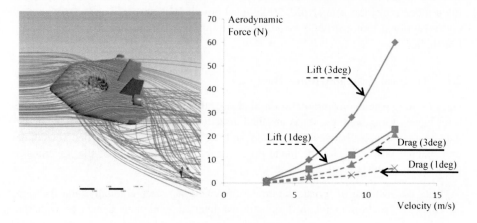

Fig. 6. CFD analysis of the Fuselage

2.3   General Parameters of Prototype

By integrating the aerodynamic characteristics of the maglev-rotor and the lifting body fuselage with duct fan, and applying the engineering calculation model of the reference [3] to the rear tilt duct fan aerodynamic calculation, the prototype's general parameters are defined by trimming calculation, which is equipped with small UAVs motor as power plant and the V type tail for balance and torque control in the light of demand study. The general parameters are shown in Table 1, and the general three views are shown in Fig. 7.

Table 1. General parameters of prototype

Parameter		Value
Length, mm		960
Width, mm		1200
Height, mm		360
Lift duct	Diameter, mm	300
	Twist angle, °/R	−16
Wing	Span, mm	1200
	Area, m²	0.4
Rear duct	Span, mm	120
	Twist angle, °/R	−30
Tail	Area, m²	0.03

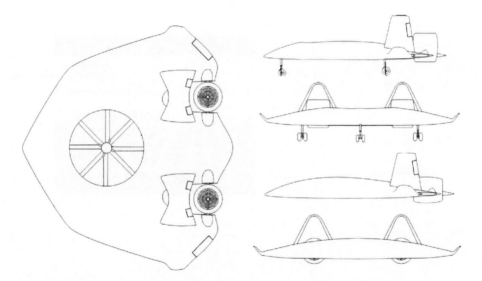

Fig. 7. Three views of the prototype

3   Trimming in the Transition

According to the smooth convert requirements of lift, power and stability, the parameters in the transition are design in Table 2, the transition is achieved through the rear duct fan tilting forward. Angle attack of the fuselage is reduced to reduce the drag and increase speed further. The parameters in the transition of the lift, attack angle, tilt angle and power changing at the transform nodes are shown in Fig. 8.

Table 2. Parameters in the Transition

Transform node	Attack angle (degree)	Velocity (m/s)	Lift duct (N)	Rear duct (N)	Tilt angle of rear duct (degree)
Take-off	0	0	215.6	107.8	0
Low speed	3	3	213.8	106.9	0.6
Transition	3	12	163.2	86.8	19.2
End of transition	3	22.5	0	130.7	89.6
Forward flight	2	28.1	0	107.3	89.6
Forward flight	1	38.2	0	145.4	89.5
Forward flight	0	64.5	0	179.4	89.9

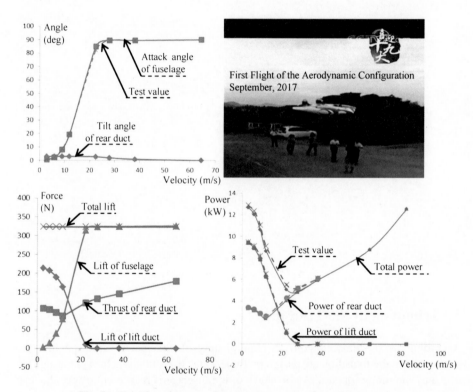

Fig. 8. The lift and power changing status at each transform node

As can be seen from Fig. 8, the power demand of the prototype decreases gradually as the speed increases. After the completion of the conversion, the total power drops to the lowest point, and then increases with the speed. Flight test values consistent with the calculated value trend, the power tested is slightly taller than the calculated value in the states of take-off, landing states and transitions. And it is due to that the prototype adopts the conventional duct fan whose efficiency is lower, and then power difference decreases with the completion of flight state transitions, because that the forward flight efficiency is more close to the calculated value. The change of attack angle of the fuselage is smooth and consistent with the calculated value trend, and the transition between the take-off, landing and forward flight state keep smoothing, the test values consistent with the calculated value trend.

4 Conclusion

Calculation and test results show that the maglev-rotor technology not only has high integration, efficiency transmission, high safety, high fatigue life, and the advantages of the smooth and quiet, also has the advantages of high aerodynamic efficiency; At the same time, the configuration of maglev-rotor UAV has VTOL and high-speed forward flight ability, present design value up to 232 km/h, and this configuration could transfer into lifting body fixed wing aircraft mode with lift duct fan shut down, speed can be higher; The maglev-rotor UAV has good stability, transition between each flight state act smooth.

5 Future Works

The next work is to optimize the structure of maglev-rotor and reduce the system weight. At the same time, the aerodynamic characteristics of the closed lift duct are further studied to improve the speed of maglev-rotor UAV configuration.

References

1. Lemont HE (2000) Technologies for future vertical flight concepts. In: American Helicopter Society 56th Annual Forum. American Helicopter Society, Virginia
2. Huanjin W (2002) Aerodynamic design of rotorcycle and research on the feasibility of its high-speed scheme, Nanjing University of Aeronautics & Astronautics, Nanjing (in Chinese)
3. Li JB, Gao Z (2003) Aerodynamic modelling and experiment for simulation of helicopter manoeuvre flight. Acta Aeronautica et Astronautica (in Chinese)
4. Prouty RW (1986) Helicopter performance, stability and control. PWS Press, Singapore
5. Zhou H-b, Duan J-n (2010) Levitation mechanism modelling for maglev transportation system. J Central South Univ Technol (English Edition), Changsha
6. Sinha PK, Pechev AN (1999) Model reference adaptive control of a maglev system with stable maximum descent criterion. Automatica
7. Sha Z, Wu G, Sun H, Zhu W, Zhang X (2014) Application of maglev bearing technology in wind turbines. Mach Tool Hydraul (in Chinese)
8. Weihong K, Renliang C (2008) Aerodynamic characteristics of rotor/duct/fan system in forward flight. J Nanjing Univ Aeronaut Astronaut Nanjing (in Chinese)
9. Ruzicka GC, Strawn RC (2004) Discrete blade CFD analysis of ducted tail fan flow. In: 42nd AIAA Aerospace Sciences Meeting and Exhibit. Reno, AIAA, Necada

Experimental Investigation on Ground Effect of Ducted Fan System for VTOL UAV

Yangping Deng[(✉)]

School of Aeronautics, Northwestern Polytechnical University,
Xi'an 710072, Shaanxi, China
flyhighdyp@nwpu.edu.cn

Abstract. Unlike traditional helicopters, ground effect has great influences on the performance of vertical take-off and landing (VTOL) ducted fan unmanned aerial vehicles (UAVs) during take-off and landing. A testing rig including a scaling ducted fan system and an imitation floor is designed and manufactured to investigate the influence of ground distance on the thrust and required power of the ducted fan system. The testing results show that the thrust is decreased and the driving power is increased because of the ground effect, and the loss of thrust and the increase of power become larger when the ground distance decreases. Finally, an acceptable distance is provided according to the testing results, and the ground effect should be specially considered during the design of ducted fan VTOV UAVs.

Keywords: Ground effect · Ducted fan · VTOL UAV

1 Introduction

A ducted fan system is a widely used propulsion system in unmanned aerial vehicles (UAVs), and it gives UAVs the capability to take off and land vertically, as well as to hover in one location, providing surveillance data continuously over a specific area. The duct diameter of these vehicles currently ranges from less than 12 in. (small enough to fit in backpack) to over 5.9 ft (such as Israel AIRMULE unmanned aerial system). Researches on ducted fan UAVs have shown that ducted fan systems have many advantages over isolated propellers, and ducted fan systems typically produce larger static thrusts than isolated propellers with the same diameter and power loading. The tip losses normally associated with unducted propellers are minimized with the small clearance between the tip of the fan blade and the inner duct wall. In addition, ducted fan systems offer a supplementary safety feature attributed to enclosing the rotating fan in the duct. This is very useful when the vehicle is flying in tight quarters as well as when it is launched and recovered by the soldiers. Figure 1 shows some examples of such vehicles that are currently under development and testing stages.

Numerous experimental and theoretical studies on ducted fan systems have been carried out in order to quantify the effects of blade pitch, propeller and duct shape such as chord, diameter ratio, duct profile, leading edge radius, on static operation (i.e., hovering flight) and axial flight, and the effect of angle of attack on lift, drag and pitching moment of ducted fans [1–7]. But few studies investigate the ground effect of

© Springer Nature Singapore Pte Ltd. 2019
X. Zhang (Ed.): APISAT 2018, LNEE 459, pp. 1602–1609, 2019.
https://doi.org/10.1007/978-981-13-3305-7_128

ducted fan systems when the vertical take-off and landing (VTOL) UAV is in takeoff or landing condition. Actually, the ground effect has big influences on takeoff and landing performance of ducted fan UAVs.

For the complex viscous flow characteristics, experimental investigation is still the major approach to study the aerodynamic characteristics of the ducted fan systems. In this paper, a testing rig is designed and manufactured to investigate the influences of ground distance on the thrust and required power of ducted fan systems, and the results can help to design ducted fan VTOL UAVs.

Fig. 1. Ducted fan UAVs under development

2 Experimental Equipment and Methods

2.1 Ducted Fan Model

Table 1 shows the parameters of the testing ducted fan system. The four-blades propeller is made with metal for high manufacturing accuracy and strength, which can be seen from Fig. 2. The pitch angle of the propeller can be regulated manually.

Table 1. Geometry parameters of test model

Item	Size/Quantity
Number of blade	4
Propeller diameter (D_P)	654 mm
Duct inner diameter (D_I)	660 mm
Duct outer diameter (D_O)	801 mm
Duct chord(mm)	398 mm
Tip clearance(mm)	3 mm

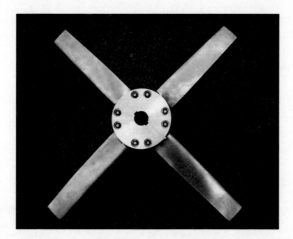

Fig. 2. Four blades metal propeller for test

2.2   Installation

Figure 3 shows the installation of the designed ducted fan system. The propeller of the ducted fan system is directly driven by an induction motor, and the rotary axis of the motor and propeller is coincidental and horizontal. The motor and the ducted fan system are rigidly fixed to a support flat, and the flat is mounted on a frame balance. A 2- by 4-meter wood floor is mounted vertically beside the outlet of testing duct, to simulate the ground when the ducted fan UAV hovering near the ground, and the distance between the outlet of the duct and the imitation floor can be adjusted.

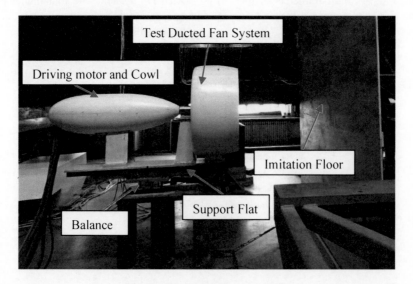

Fig. 3. The installation of test rig

2.3   Instrumentation and Data Acquisition

Forces on the testing ducted fan system are measured using the balance system. The propeller rotational speed is measured using an infrared trigger. The motor importing voltage and electric current is measured using voltmeters and ammeters, and the import power of the motor can be calculated by the measured voltage and current. According to the plotting motor driving characteristics, the export power of the motor can be obtained from the import power.

2.4   Data Treatment

All experimental data are converted to non-dimensional form, and the thrust and power coefficient are defined as:

$$C_T = \frac{T}{\rho n^2 D^4} \tag{1.1}$$

$$C_P = \frac{P}{\rho n^3 D^5} \tag{1.2}$$

where T and P are the ducted fan thrust and shaft power, C_T and C_P are the thrust and power coefficient, ρ is the air density, n is the rotational speed, and D is the duct inner diameter.

3   Experimental Results

The testing tip Mach number of the propeller is set as 0.65, and the pitch angle (ϕ) is adjusted with five different positions. Figure 4 shows the testing ducted fan system thrust coefficient as a function of propeller pitch angle at different distances (L_F) between the outlet of the duct and the imitation floor. It can be seen from this figure that compared to no floor condition, the trust loss of the ducted fan system increases when the distance decreases. Figure 5 shows the thrust loss, which is compared to no floor condition, as a function of disk load. When the duct is too close to the ground, the trust loss is very considerable, and the trust loss increases as the disk load increases.

According to the classical slipstream theory [8], the trust of ducted fan can be calculated as:

$$T = \dot{m}(V_2 - V_0) \tag{1.3}$$

where $\dot{m}$ is the mass flow through the duct, V_0 and V_2 are the flow velocities in number 0 and number 2 sections, which are shown in Fig. 6.

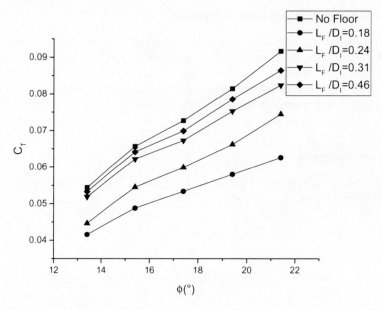

Fig. 4. Thrust coefficient as a function of pitch angle with different ground distances

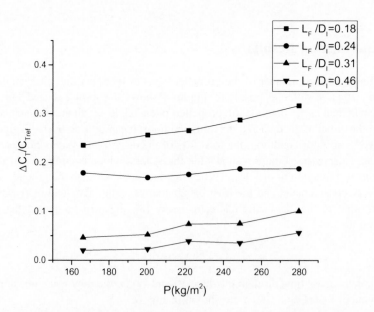

Fig. 5. Thrust coefficient loss as a function of disk load with different ground distances

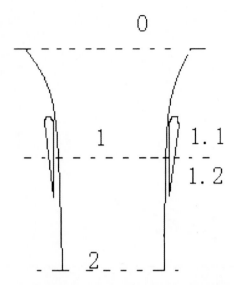

Fig. 6. Schematic of the ducted fan flow field

Obviously, the ground can obstructs the outflow of the ducted fan system when the duct outlet gets close to the ground, and this makes the propeller section (number 1 section in Fig. 6) induced velocity decrease. According to the blade element theory, the decrease of propeller induced velocity makes the angle of attack of propeller blade sections, the propeller thrust and shaft power increase. The outflow of the ducted fan system attacks the ground, disperses and reflects upwards. Most of the reflected airflow will reflow to the ducted fan system again due to the suction of duct inlet, and this makes the effective angle of attack of duct lip decrease, which leads to the decrease of suction peak of the flow over the duct lip and the reduction of duct thrust. For the ducted fan system, duct thrust is the major part in total generated thrust, and the increase of the propeller thrust is smaller than the reduction of the duct thrust, this makes the total generated thrust from the ducted fan system reduce when it gets close to the ground.

Figure 7 shows the testing ducted fan system power coefficient as a function of propeller pitch angle with different distances between the outlet of the duct and the imitation floor. It can be seen from this figure that, when the ground distance is larger than 30% of the duct inner diameter, the power coefficient is close to no floor condition with the same propeller pitch angle. When the ground distance is less than 30% of the duct inner diameter, the power coefficient will increase when the distance decreases. As analysed in the previous section, the ground stoppage of the outflow leads to the increase of the angle of attack of the propeller blade sections, and this makes the shaft power increase.

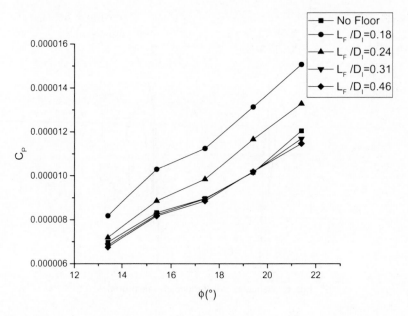

Fig. 7. Power coefficient as a function of pitch angle with different ground distances

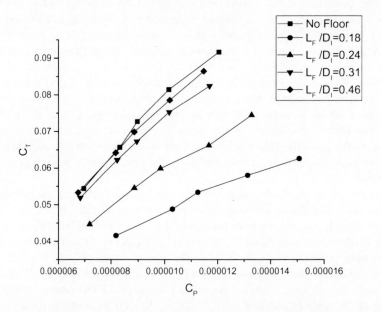

Fig. 8. Thrust coefficient as a function of power coefficient with different ground distances

Figure 8 shows the thrust coefficient as a function of power coefficient with different ground distances. When the duct is too close to the ground, the large loss of thrust and increase of power lead to enormous increase of required power with the same thrust or takeoff weight, which is adverse to the takeoff and landing performance of VTOL UAVs. According to the experimental results, a ground distance that is bigger than 30% of the duct inner diameter is required.

4 Conclusions and Future Works

An experimental investigation is completed to determine the ground effect of a ducted fan system. The results show that the thrust is decreased and the driving power is increased due to the ground effect. The loss of thrust and increase of power become larger when the ground distance decreases. Although an acceptable distance is provided according to the testing results, there are still some problems that need to be further studied, such as the influence of the duct shape on ground effect. In addition, more detailed flow characteristics also need to be studied based on experimental and computational investigations. For light and micro ducted fan UAVs, it will be easy to reduce the influence of adverse ground effect simply by increasing the length of landing gear; for heavy ducted fan UAVs, both the layout of ducted fans and the design of landing gear need to be considered during the preliminary design of this kind of UAVs.

References

1. Martin P, Tung C (2004) Performance and flow field measurements on a 10-inch ducted rotor VTOL UAV. In: Proceedings of the 60th Annual Forum of the American Helicopter Society, Alexandria, VA
2. Lind R, Nathman JK, Gilchrist I (2006) Ducted rotor performance calculations and comparisons with experimental data. In: Proceedings of the 44th AIAA Aerospace Sciences Meeting and Exhibit, p 1069
3. Graf W, Fleming J, Wing N (2008) Improving ducted fan UAV aerodynamics in forward flight. In: Proceedings of the 46th AIAA Aerospace Sciences Meeting and Exhibit, p 430
4. Chang IC, Rajagopalan RG (2003) CFD analysis for ducted fans with validation. In: Proceedings of the 21st AIAA Applied Aerodynamics Conference, p 4079
5. Ahn J, Lee KT (2004) Performance prediction and design of a ducted fan system. In: Proceedings of the 40th AIAA/ASME/SAE/ASEE Joint Propulsion Conference and Exhibit, p 4196
6. Ko A, Ohanian OJ, Gelhausen P (2007) Ducted fan UAV modeling and simulation in preliminary design. In: Proceedings of AIAA Modeling and Simulation Technologies Conference and Exhibit, p 6375
7. Guerrero I, Londenberg K, Gelhausen P, Myklebust A (2003) A powered lift aerodynamic analysis for the design of ducted fan UAVs. In: Proceedings of the 2nd AIAA Unmanned Unlimited Conference and Workshop & Exhibit, p 6567
8. Haipeng W, Huaming W (2006) A general summarize for the ducted fan computational methods. In: Proceedings of the 22nd China Helicopter Conference (in Chinese)

Optimization for Conceptual Design of Reconfigurable UAV Family

Haoyu Zhou[1(✉)], Ya Ding[2,3], Yalin Dai[2,3], and Xiaoqiang Qian[2,3]

[1] Nanjing University of Aeronautics and Astronautics,
Yudao Street 29, Nanjing, China
zhyzhy525@126.com
[2] Key Laboratory of Fundamental Science for National
Defense Advanced Design Technology of Flight Vehicle,
Nanjing University of Aeronautics and Astronautics,
Nanjing 210016, Jiangsu, China
[3] State Key Laboratory of Mechanics and Control of Mechanical Structures,
Nanjing University of Aeronautics and Astronautics,
Nanjing 210016, Jiangsu, China

Abstract. The aircraft family is a set of aircraft products that share common components but vary in configurations based on different mission performance and requirements. The reconfigurable UAV family using modular design can improve the efficiency of executing multiple missions and enable the acquisition cost benefits, which is one of key competitive edges in military applications. This paper aims to study an effective optimization method for conceptual design of the reconfigurable UAV family with interchangeable components that can be reconfigured between missions. First, a Flying-wing UAV family defined for combat and reconnaissance missions is used as an example for demonstration of the method, and an appropriate comprehensive analysis model is developed. Next, the optimization formulation for the conceptual design of UAV family is presented in detail, including design variables, design constraints, and objectives. A hybrid optimization strategy is then applied to solve the optimization problem of UAV family conceptual design. The results after optimization indicate that the mission performance and requirements for each configuration are satisfied, and costs are reasonably compromised for the UAV family.

Keywords: Flying-wing · Reconfigurable · UAV family · Conceptual design · Optimization

1 Introduction

1.1 Family and Reconfigurable Design

Product family with modularity and standardization allows designing a variety of products using the same modules of components called platforms [4]. Using platform-based design allows important family design savings and easy manufacturing. The principles of family design has also been applied to aircraft design. For instance, commercial aircraft manufacturers often adopt the aircraft family (or aircraft series)

© Springer Nature Singapore Pte Ltd. 2019
X. Zhang (Ed.): APISAT 2018, LNEE 459, pp. 1610–1619, 2019.
https://doi.org/10.1007/978-981-13-3305-7_129

strategy when developing aircraft products. However, the classical series aircraft design take a single aircraft as baseline to develop new members without considering the entire family design. As opposed, the real aircraft family design determined common characteristics at the very beginning of the entire family design period.

Although the aircraft family strategy has been widely adopted in the aircraft industry, there is still limited research on the method of optimizing the conceptual design of the aircraft family. Yong [9–11] discussed several problems of aircraft family design including configuration and layout, aerodynamic design and structural design by multi-disciplinary design optimization. Willcox and Wakayama [8] conducted a preliminary study on the multiple-aircraft family design of blended-wing-body (BWB). Zhang [12] proposed an optimization method for conceptual design of airliner family.

The concept of reconfigurable aircraft is derived from aircraft family based on modular design. Research on aircraft family establishes the concept of sharing platform, while reconfigurable design extends the module of interchangeable private components. Through offline reconfiguration achieved with different payload, multi-mission that could satisfy the requirement of different flight drones, which reduces the production, use and maintenance costs, increased flexibility (Figs. 1 and 2).

Fig. 1. Yakovlev PRORYV (Breakthrough)

Fig. 2. Reconfigurable family [6]

The Yakovlev design bureau has proposed the concept design of a mission-based modular reconfigurable UAV in which fuselage, wing, engine pods and mission modules are designed as interchangeable modules. The designers expect the UAV to be able to perform different missions such as reconnaissance, early warning and strike by matching the main body with different wings and power pods. The UAV family called PRORYV (Breakthrough) is shown in fig. Pate et al. [6, 7] studied the optimization method of modular reconfigurable UAV family and indicated that a "reconfigurable family of aircraft may offer improved mission performance compared to a fleet of fixed multi-mission aircraft, as the reconfigurable aircraft can be better customized for the unique performance needs of each mission". Chowdhury et al. perform the conceptual design of a new offline-reconfigurable Unmanned Aerial Vehicle (UAV) platform that can take up both a quadrotor UAV (QR-UAV) configuration and a fixed-wing UAV (FW-UAV) configuration [1].

In this paper, we consider the optimization of a reconfigurable UAV family with modular components that can be interchanged between mission sorties. The purpose of the optimization is to determine the key design parameters that satisfy the performance constraints and reach optimum in terms of cost. The next section provides a detailed design scheme for describing families of reconfigurable aircraft with modular components.

1.2   Mission-Based Modular Design Scheme

Unmanned combat aerial vehicle (UCAV) is a comprehensive aircraft platform capable of carrying advanced surveillance equipment and weapons. UCAVs can execute a variety of missions, requiring high maneuverability for air-to-air combat, long endurance for surveillance missions, and high speed for quick penetration into hostile terrain [5]. With the improvement of technology and design method, researches on aircraft with modular mission capability, and strong environmental adaptability gradually have received more attention.

We define a reconfigurable aircraft as a combination of shared platform and interchangeable parts. The shared platform serves as a common component in aircraft family design, as well it cannot change separately within one design scheme. The interchangeable parts are mission-based special components of different configurations, which shall be interchanged between mission sorts. In general, the fuselage is defined as shared platform while the wing and empennage are designed as interchangeable parts. To ensure commonality, interfaces (physical or system) between the platform and the interchangeable modules shall be considered, because each interface must be shared by all components.

The modular design of traditional aircraft is mainly divided into fuselage, wing, tail, engine, etc., of which the fuselage is defined as a shared platform, the rest as interchangeable modules. However, the interfaces between them must enforce necessary constraints to ensure that any free components can be installed on the shared platform.

This paper chooses surface combat air patrol missions and high-altitude reconnaissance missions as application scenarios, in which the design optimization scheme of reconfigurable module is discussed. Since it is almost impossible to design an aircraft that can accommodate both long flight endurance and high speed, we must make a trade-off on performance. In all variables, the sweepback angle and aspect ratio make the most difference upon performance of speed and lift-drag ratio. To achieve high speed, the inner wing (fuselage), as a shared platform, is designed with large sweep angle and small wingspan. To meet various mission requirements, the interchangeable outer wings are designed with large-sweep-low-aspect-ratio and small-sweep-high-aspect-ratio respectively. The two configurations are drawn in Fig. 3, left of which is for combat missions and the right for reconnaissance. In detail, the trimmed wingtip of UCAV is designed for better stealth capability and the transition section of URAV for reducing interference drag.

The two configurations share a same fuselage (inner wing) but differ in outer wings, therefore the fuselage must be designed to meet the most stringent strength demand. Note that the root chord length of the outer wing acts as physical interface, and the equality structural weight of the inner wing is taken as a reinforcement constraint.

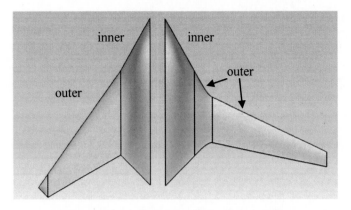

Fig. 3. UAV configurations for different missions

2 Conceptual Design and Analysis Framework

Conceptual design comprehensive analysis program is the basis of general parameter optimization, which comprehensively evaluates the aerodynamic, weight, performance and other characteristics of the preliminary design scheme. In the preliminary design stage, engineering algorithms are mainly adopted in the analysis models of various disciplines. Yet in this paper, the quick numerical algorithm is used to replace the engineering algorithm in the aerodynamic module. The stability and trim modules are also analyzed with the numerical results.

The whole analysis program mainly includes several analytical models of geometry, power, aerodynamic, weight, performance, stability and stealth as shown in Fig. 4.

Geometric analysis model describes the shape of the main components of an aircraft, including the dimension parameters of the fuselage platform and interchangeable wing components. After these parameters are determined, a three-sided diagram can be drawn to calculate the mean aerodynamic chord length, exposed area and tank volume of the aircraft.

Engine Generation model is based on the main design parameters of the engine (maximum static thrust, bypass ratio, sea level static thrust and total pressure ratio, etc.), to estimate the thrust of the engine, fuel consumption, and characteristics of size and weight.

Weight estimation model is to calculate the component structure, basic empty weight, employ empty weight, zero fuel weight and maximum take-off weight, and according to the changes of the actual fuel consumption, the iterative solution is given. The weight estimation method of UAV is a reference from Denis Howe's monograph *Aircraft conceptual design synthesis*.

Aerodynamic analysis model determines the aerodynamic characteristics with the Jameson-Caughey NYU Transonic Swept-Wing [3] (FLO 22) code based on full potential equation, which is chosen for calculation of induced drag and wave drag. Moreover, it takes methods based on form factors and wet area into computations of friction and from drag.

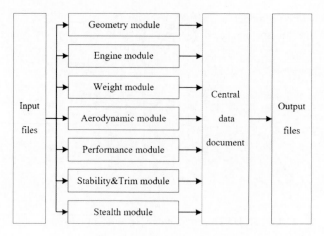

Fig. 4. Analysis framework

Stability and Trim analysis model is built to check the stability and trim capacity of aircraft, and feeds back the trim results to the aerodynamic database for correction. In this paper, a pair of flaps and elevons are defined for calculation on take-off & landing performance, and trim performance with elevons deflected only. The panel method program Panair [2] of NASA is used to calculate aerodynamic variations responding to deflection angles of control surface.

Performance analysis model is performed using an algorithm implemented in MATLAB to estimate fuel consumption and calculate take-off & landing performance, mission performance, and point performance based on the aerodynamic database, engine model and weight statement. The monograph *Airplane aerodynamics and performance* written by Roskam and Lan is taken as a reference in forming the performance analysis method.

3 Synthesis and Optimization

There are significant differences between the expression and solution methods for the optimization problem of UAV family and the single ones, because the optimization of the population parameters should consider two configurations. This section first defines the general parameter optimization problem of UAV, and then explains the solution method of the problem.

3.1 Definition of Optimization Problem

The three elements that define engineering design optimization problems are optimization objectives, design variables, and design constraints. For the general parameter optimization of UAV, the maximum takeoff weight is a proper choice to be the optimization objective, which to a certain extent represents the acquisition cost. The configuration feature of the wing and the sea level static thrust of the engine are taken as design variables. The performance indicators proposed in the mission requirements constitute design constraints.

The design variable and design constraint of UAV family optimization are defined as follows.

Parameters that have important influence on the optimization objective and design constraints are taken as design variables, while design variables of an UAV family may divide into common variables and private variables. This paper considers the fuselage a shared platform, hence the profile parameters of fuselage is also unchangeable and the only common variable is sea level static thrust of the engine buried within fuselage. The range is shown in Table 1. The Private variables of two configurations contain the projected area of entire vehicle, the aspect ratio, the sweptback angle of the front edge

Table 1. Common variable

Common variable	Value range
SLST of Engine	46 to 54 kN

and the twist angle of the wing tip, whose value range is shown in Table 2.

Design constraints derive from specified performance indicators of mission requirements, including mission radius, take-off and landing distance, the maximum rate of climb on sea level, the maximum level flight speed and stability margin etc. The low speed performance of reconnaissance UAVs is superior to that of combat UAVs,

Table 2. Private variable

Private variable	Combat		Reconnaissance		Units
	Low	High	Low	High	
Reference area/	60	72	64	76	m²
Sweep of outer wing/°	40	60	10	30	degrees
Aspect ratio	2.2	3.0	6.0	8.6	1
Wingtip twist angle/°	−3	3	−5	5	degrees

Table 3. Design constraint

Design constraint	Combat		Reconnaissance		Units
	Low	High	Low	High	
Mission radius	1999	2001	1999	2001	km
Takeoff distance	–	1100	–	1000	m
Sea Level climb rate	26	–	30	–	m/s
Landing distance	–	1000	–	900	m
Service ceiling	14	–	15	–	km
Static stability margin	0.08	0.12	0.1	0.15	1
Turning radius	–	1200	–	–	1
Maximum level speed	250	–	192	–	m/s
Tank capacity margin	0	–	0	–	1

whereas combat UAVs adopt higher demand than reconnaissance on manoeuvrability and maximum speed. The value range of design constraints are shown in Table 3.

3.2   Optimization Method

There are two optimization methods suitable for aircraft design, one of which is looped method and the other is seeded method. The looped method use a hierarchical optimization formulation with loop iteration of entire analysis process, which is computationally expensive. The seeded method consider a discrete subset of possible UAV variants, thus reducing the cost on computation by abandoning the strictly optimal solution.

In this paper, a single objective optimization problem is defined. The looped method is adopted for this research, but due to the high computational cost of running the entire analysis process directly for optimization, a surrogate model of the analysis function is created. Considering the design space and the response with high non-linearity, the optimization method chooses the hybrid optimization strategy of global optimization and gradient optimization. The global optimization chooses the multi-island genetic algorithm (MIGA). MIGA divides a large population into several sub-populations, which are called islands. Every certain algebra will select the individuals of each island in a certain proportion and transfer them to other islands, and executes selection, crossover and mutation according to the classical GA. The optimization technique parameters are listed in Table 4. There are two judging conditions for stopping an optimization procedure. Wither the optimal individuals in the population

Table 4. Optimization technique parameters

Optimization technique parameters	Value
Number of island	20
Sub-population size	10
Number of generations	15
Rate of crossover	0.8
Rate of mutation	0.01
Rate of migration	0.1

do not improve in successive generations, or the average fitness stops when successive generations do not improve at all, it will bring about stopping. In this issue, the MIGA has run for 2251 steps and the gradient optimization has achieved convergence within 200 steps.

3.3   Results and Comparison

Firstly, the general parameters of each model are optimized by using the traditional optimization method (which does not take into account the universality of components in the optimization of each model). Then, the general parameter optimization method

Table 5. Optimization results of independent configurations

Variables and objectives	Combat		Reconnaissance		Units
	Original	Optimized	Original	Optimized	
SLST of engine	49.05	48.60	49.05	47.03	kN
Reference area	68	65.92	70	73.86	m^2
Sweep of outer wing	55	46.65	20	24.9	degrees
Aspect ratio	2.5	2.79	7.2	7.69	1
Wingtip twist angle	0	−0.31	0	−3.0	degrees
Maximum takeoff weight	11441.55	9890.87	10569.24	9871.14	kg
Fuel weight	4880.36	3808.02	4898.62	4200.42	kg

Table 6. Optimization results of UAV family

Variables and objectives	Combat	Reconnaissance	Units
SLST of engine	48.42		kN
Reference area	65.18	74	m^2
Sweep of outer wing	46.05	25	degrees
Aspect ratio	2.8	7.61	1
Wingtip twist angle	−0.30	−3.0	degrees
Maximum takeoff weight	10056.03	9935.21	kg
Fuel weight	3942.74	4219.04	kg

for UAV family (considering the universality of components) was used to optimize the two configurations simultaneously. The aim is to compare the differences between the optimization results of two method.

For the two different configurations, minimizing the maximum takeoff weight was selected as the optimization target and optimized calculation was carried out respectively. The results are summarized in Table 5. It can be seen from the optimization results that the aspect ratio of the combat UAV is larger, and the swept back angle is smaller, in order to obtain a higher lift-drag ratio to reduce fuel consumption under the premise of meeting the requirement of maneuverability. In order to reduce the induced drag further, reference area and aspect ratio of Reconnaissance UAV increases. Moreover the increase of sweepback reduces wave drag, as well increasing the arm of force to reduce deflection angle of elevons, indirectly improves the aerodynamic efficiency.

When applying the general parameter optimization method for UAV family, the proper weight coefficient is selected to weight sum of two configurations, transforming the multi-objective optimization problem into a single-objective optimization problem. Considering that the design of this paper contains both combat and reconnaissance missions, the objective weight coefficient both is 0.5. Table 6 shows the optimization results of UAV family.

Contrasting the data in Tables 5 and 6, we can see the UAV family optimization results in the scheme is inferior to the single UAV optimization scheme (regardless of commonality) in economy. In consideration of commonality, the maximum take-off weight of the combat and reconnaissance types increases by 1.67% and 0.75% respectively. The single UAV optimization scheme can't fully meet the performance requirements and strict interchangeability, due to not taking the engine power and the fuselage structure weight of general requirement into account, and thus may increase the design and manufacturing cycle and cost. By considering the general design scheme of components commonality, it can significantly reduce the UAV family's design and manufacturing cycle and cost with a small loss in economy of each configuration. Within a tradeoff, the UAV family optimization scheme achieves penetration performance on combat configuration and long loiter time for surveillance on reconnaissance configuration by exchanging wing components offline between missions, merely causing a few increasing in take-off weight.

4 Conclusion

Based on the design concept of reconfigurable aircraft family, this paper studies a general parameter optimization method for UAV family. Taking the flying-wing UAV family (combat type and reconnaissance type) as an example, the general parameter optimization calculation is carried out. The optimized results indicate that,

1. In consideration of commonality, the optimized design parameters can not only meet the requirements of each mission performance, but also balance the design goals (maximum takeoff weight) of each configuration. Although each model has certain losses in economy, it earns significant improvements on the flexibility of mission by exchanging wing modules. It also reduces the design cycle and cost of the entire family of aircraft.
2. This paper only considers the thrust of engine as a common variable, while the planform characteristic of the fuselage can also be used as common variables such as sweepback angle, fuselage width and root chord length, etc., which can further increase the flexibility of the UAV family design.
3. This paper establishes a general parameter optimization method for UAV family and its computing environment can be used as a kind of auxiliary tool in conceptual design of UAV family, which helps designers to determine reasonable general parameters.

References

1. Chowdhury S, Maldonado V, Patel R (2014) Conceptual design of a multi-ability reconfigurable unmanned aerial vehicle (UAV) through a Synergy of 3D CAD and Modular Platform Planning. In: AIAA/ISSMO Multidisciplinary Analysis and Optimization Conference
2. Epton MA, Magnus AE (1990) Pan air - a computer program for predicting subsonic or supersonic linear potential flows about arbitrary configurations using a higher order panel method. Volume 1. theory document

3. Jameson A, Caughey DA, Newman PA, Davis RM (1976) A brief description of the Jameson-Caughey NYU transonic swept-wing computer program: FLO 22
4. Jose A, Tollenaere M (2005) Modular and platform methods for product family design: literature analysis. J Intell Manuf 16(3):371–390
5. Khalid M, Zhang F, Ball N (2005) A CFD Based Study of UCAV 1303 Model. In: AIAA Applied Aerodynamics Conference
6. Pate DJ, Patterson MD, German BJ (2011) Methods for Optimizing a Family of Reconfigurable Aircraft. In: AIAA 2011-6850
7. Pate DJ, Patterson MD, German BJ (2015) Optimizing families of reconfigurable aircraft for multiple missions. J Aircr 49(6):1988–2000
8. Willcox K, Wakayama S (2012) Simultaneous optimization of a multiple-aircraft family. J Aircr 40(4):616–622
9. Yong M, Yu X (2006) Aircraft family design using modular product platform methodology-an exploratory study. Aircr Des
10. Yong M, Yu X (2008) A structural optimization method for aircraft family. Acta Aeronautica Et Astronautica Sinica 29(3):664–669
11. Yong M, Yu X (2008) Wing aerodynamic optimization method for aircraft family design. J Nanjing Univ Aeronaut Astronaut
12. Yu X, Zhang S (2012) Optimization for conceptual design of airliner family. J Nanjing Univ Aeronaut Astronaut

Combustion and Propulsion

Laminar Transition over Airfoil: Numerical Simulation of Turbulence Models and Experiment Validation

Shuyue Wang[1], Gang Sun[1(✉)], Meng Wang[2], and Xinyu Wang[1]

[1] Department of Aeronautics and Astronautics, Fudan University,
220 Handan Road, Shanghai 200433, China
{sywang14,gang_sun,xinyuwang_15}@fudan.edu.cn
[2] AVIC Aerodynamics Research Institute,
1 Yangshan Road, Shenyang 110034, China
m.wang_sy@139.com

Abstract. The article reviews the necessity of simulation in laminarization as an important optimization technique in aircraft design. Transition models e.g. k-ω SST with γ-$Re_{\theta t}$ model and k-k_l-ω model are popular in industry and embedded in many optimization framework. The article aims to evaluate their accuracy in laminar transition predicting as well as in evaluating the effect of laminarization, i.e. the extended length of laminar brought by optimization in shape. The object for numerical simulation and experiment validation for laminar transition position predicting are one original airfoil and one laminarized airfoil under $Re \sim 2.62 \times 10^6$, $Ma \sim 0.785$. The mesh with boundary conditions, the criterion for judging laminar transition, and the method to monitor laminar/turbulence distribution over model surface in wind tunnel experiment are introduced. The result shows that due to many factors, the two models cannot very accurately predict the laminar transition position, but the effect of laminarization can be evaluated quite satisfactorily by the k-ω SST with γ-$Re_{\theta t}$ model. Therefore, the laminarization applied on three-dimensional aircraft part, e.g. wing, nacelle can be constructed using this model in the future optimization framework of the authors.

Keywords: Laminar · NLF · Transition model · RANS wind tunnel · Optimization

1 Introduction

The requirement in economy to produce faster transport aircraft has resulted in researches to improve aircraft aerodynamic performance [1]. In addition to improving aircraft operation and jet engine characteristics, a solution to address the challenge is to reduce aircraft drag during cruise [11]. One of the approaches is extending the portion of the aircraft surface over which there is laminar flow, which is able to decrease turbulent friction as well as flow separation region [2]. This makes possible a lower drag coefficient, which offers a large potential fuel saving and consequently economic savings and improvements in environmental protection [11].

© Springer Nature Singapore Pte Ltd. 2019
X. Zhang (Ed.): APISAT 2018, LNEE 459, pp. 1623–1632, 2019.
https://doi.org/10.1007/978-981-13-3305-7_130

Natural Laminar Flow (NLF) technique has been widely used to reduce skin friction drag by extending laminar length over aircraft surface [3]. NLF applies airfoils with good laminar length performance into construction of aircraft surface [11]. There are many related approaches proposed in the field of inverse design [15] and numerical optimization algorithm [8]. In every iteration of NLF optimization, CFD simulation of laminar transition position is quite necessary [4]. In essence, it is the difference of laminar positions of contiguous iterations during optimization that requires designer's interest, which is used to evaluate the effect of laminarization).

The mechanism on an aircraft in flight at moderate to high Reynolds number and low freestream turbulence levels (Tu $<$ 0:6%) which triggers the laminar transition mainly depends on the pressure gradient along the boundary layer, the surface profile, and the freestream Mach number. The most common transition mechanism under this situation is natural transition due to the growth of T-S waves [11]. Conventionally the e^N method (based on local linear stability theory and the parallel-flow assumption to calculate the growth of the disturbance amplitude from the boundary-layer neutral point to the transition location) has limitations in that it does not account for nonlinear effects, e.g., high freestream turbulence [16].

Navier–Stokes equations should be considered in the calculation of laminar transition position for accuracy. The relative ease of implementation and moderate computational cost of RANS (Reynolds-averaged Navier-Stokes equations) simulations has made them an attractive solution for aerodynamic design. The k-ω SST model combines the advantages of k-ω model and k-ε model, and has been applied successfully to a wide range of flows for many years [6]. Based on turbulence model, some transition models have been proposed. Walters and Leylek developed and implemented a new model (k-k_l-ω) for bypass and natural transition prediction, which uses an eddy-viscosity RANS framework [17]. Menter developed γ-$Re_{\theta t}$ transition model, which is increasingly accepted for transition prediction in industry. The model involves two additional transport equations for turbulence intermittency γ to trigger transition, and for transported transition momentum thickness Reynolds number $Re_{\theta t}$ [12]. These two models are popular in industry and is embedded in many optimization framework.

In this article, a numerical simulation is constructed in the case of laminar transition position predicting over airfoil (under Re $\sim$ 2.62 $\times$ 10^6, Ma $\sim$ 0.785) with two transition models, i.e. k-k_l-ω model vs. k-ω SST with γ- $Re_{\theta t}$ model. The airfoils chosen for the simulations are respectively a two-dimensional profile from an engine nacelle and another airfoil as the optimized product. The comparison between the two models was constructed by scholars but under lower Reynolds number situations [4]. Many numerical optimization methods require aerodynamic performance of airfoils that constitute aircraft's wing or nacelle, etc., because an airfoil not only forms the three-dimensional aircraft geometry, but also determines local aerodynamical performance of an aircraft [8]. The CFD simulation via turbulence models will be used to predict laminar transition position over leeward part of airfoils, the result of which is later compared with that in the wind tunnel experiment. Plus, the difference of laminar transition position between the original airfoil and optimized airfoil in the numerical simulation is also examined. The numerical simulation is constructed using ANSYS FLUENT® commercial software. The article attempts to validate the effectiveness of the two transition models by employing CFD simulation and wind tunnel experiment, and selects the better one for future optimization of three-dimensional cases, e.g. wings, engine nacelles, etc.

2   Mesh and Boundary Conditions

Among various kinds of meshes used for simulating the flow over an airfoil, a C-Mesh is used in this article [4]. The numerical simulation requires the mesh to represents features that are necessary for CFD accuracy. The airfoil should be located inside a large flow field, indicating that the pressure boundary condition is transferred to the airfoil from far field (by 25 chord lengths), which mimics the atmospheric environment during cruise. A denser mesh is used in the regions of interest, i.e. boundary-layer and near-wake regions. The near-wall region in the flow should be occupied by grids that are of high orthogonality and of y + approximately around 1.0. The latter requirement is especially proposed so as to obtain accurate laminar development [11, 12]. The curve matched to the airfoil geometry should be of high accuracy and the variation of grid size along the wall should be smooth, since drastic variation will result in artificial laminar transition. The mesh is created via ANSYS ICEM CFD® (Fig. 1).

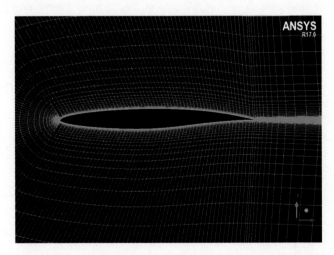

Fig. 1. The mesh around the geometry of the airfoil to be simulated.

The boundary conditions used during all simulations (two airfoils, original and optimized) are as follows. The airfoils are of 0.2 m long. On the airfoil, no slip conditions are applied for the velocity and Neumann boundary conditions for the pressure at far field (99614Pa). The freestream velocity is set at Mach number 0.785 with Reynold number approximately 3.4×10^6 (viscosity is 1.7894×10^{-5}). The air in the flow field is treated as ideal gas with temperature at 310 K at far field boundary. Turbulent intensity is 0.3% and turbulent viscosity ratio 10%. Total number of mesh grids is 49350.

3 Simulation Result and Comparison with Experiment

3.1 Numerical Result

The γ-$Re_{\theta t}$ model takes the following form. For intermittency γ:

$$\frac{\partial(\rho\gamma)}{\partial t} + \frac{\partial(\rho U_j \gamma)}{\partial x_j} = P_\gamma - E_\gamma + \frac{\partial}{\partial x_j}\left[\left(\mu + \frac{\mu_f}{\sigma_f}\right)\frac{\partial\gamma}{\partial x_j}\right]$$

For transition momentum thickness Reynolds number $Re_{\theta t}$:

$$\frac{\partial(\rho\overline{Re_{\theta t}})}{\partial t} + \frac{\partial(\rho U_j \overline{Re_{\theta t}})}{\partial x_j} = P_{\theta t} + \frac{\partial}{\partial x_j}\left[\sigma_{\theta t}(\mu + \mu_t)\frac{\partial\overline{Re_{\theta t}}}{\partial x_j}\right]$$

Each term to be solved above has its own formula. The common parameters in the formulae used in this article are: $\sigma_{\theta t} = 2.0, c_{\theta t} = 0.03, c_{e1} = 1.0, c_{e2} = 50, c_{a1} = 2.0,$ $c_{a2} = 0.06, \sigma_\gamma = 1.0$.

The k-kl-ω model takes the following form. For laminar kinetic energy,

$$\overline{U}\frac{\partial k_L}{\partial x} + \overline{V}\frac{\partial k_L}{\partial y} = -(\overline{u'v'})\frac{\partial\overline{u}}{\partial y} - \frac{\partial}{\partial y}\left(\overline{v'k_L} - v\frac{\partial k_L}{\partial y}\right) - \varepsilon + \left(\overline{u'\frac{\partial U'}{\partial t}}\right)$$

Similarly, each term to be solved above has its own formula. The common parameters in the formulae used in this article are: $\sigma_{k_L} = 0.0125, C_1 = 0.02974,$ $C_2 = 59.79, C_3 = 1.191, C_4 = 1.65 \times 10^{-13}$.

The CFD simulation is accomplished on two airfoils (original and optimized NLF airfoils) via k-kl-ω model and k-ω SST with γ-$Re_{\theta t}$ model. The operation is conducted at three angles of attack: $1°, 3°,$ and $4°$ (Figs. 2 and 3 for the original airfoil, and Figs. 4 and 5 for the optimized airfoil).

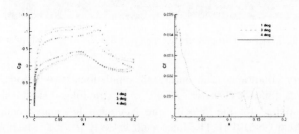

Fig. 2. Pressure distribution (overall) and skin friction distribution (leeward) via k-ω SST with γ-$Re_{\theta t}$ model

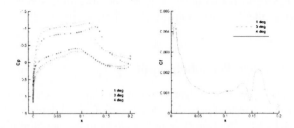

Fig. 3. Pressure distribution (overall) and skin friction distribution (leeward) via k-k$_l$-ω model

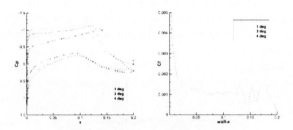

Fig. 4. Pressure distribution (overall) and skin friction distribution (leeward) via k-ω SST with γ-Re$_{\theta t}$ model

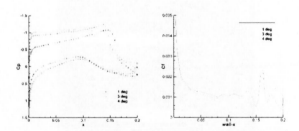

Fig. 5. Pressure distribution (overall) and skin friction distribution (leeward) via k-k$_l$-ω model

As for each airfoil (original/optimized), the two models predict pressure distribution very similarly, in terms of the position of large pressure gradient (i.e. weak shock occurrence) and variation corresponding to different angles of attack. There is slight difference between the maximum values of pressure peak between two models (the k-k$_l$-ω model predict results with comparatively higher pressure peak value). In the region in front of transition sensitive area (i.e. x = 0–0.1 m), the two models predict result quite similar. However, the difference occurs in the transition sensitive area (i.e. x > 0.1 m) in terms of the amplitude of skin friction value. Still the k-k$_l$-ω model predict comparatively higher skin friction peak value. In spite of this, the two models predict the tendency of skin friction of in transition area quite similarly. The detailed analysis of judging the laminar transition position is discussed in the following subsection.

3.2   Laminar Transition Point Judgement

In the CFD simulation, there are several output variables that can be used to identify transition: turbulence intermittency (γ), local turbulence intensity (TI), turbulence kinetic energy (TKE), skin friction (C_f), and changes in the pressure distribution [11]. Turbulence intermittency γ is the critical variable in k-ω SST with γ-$Re_{\theta t}$ model, triggering the switch between laminar and turbulent treatment in the solver [12]. It starts to increase at early stage of transition region in the boundary layer. The end of transition is regarded at the location where the maximum value of C_f occurs. In addition, C_f is shared by both two models in the article in the result. TKE and TI start to increase with the onset of turbulence intermittency in the boundary layer. The numerical comparison of above variables is based on simulation in Figs. 6 and 7. Therefore, this article regards the mid-point between the first lowest point of C_f and the first peak value of C_f along the airfoil as the laminar transition point.

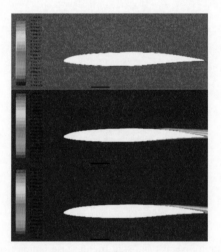

Fig. 6.   The contour of γ, TI, TKE

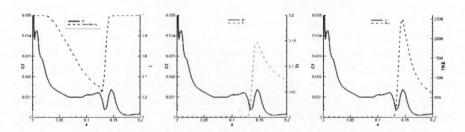

Fig. 7.   The xy-plot over airfoil (leeward) of C_f, γ, TI, and TKE

3.3   Comparison with Experiment: Laminar Transition Point

The experiment is conducted in the continuous transonic wind tunnel FL-61 at AVIC Aerodynamics Research Institute (Shenyang). Its sectional area is 0.6×0.6 m^2,where the available Mach number ranges from 0.15 to 1.6. The turbulence intensity is less than 0.2% when freestream Mach number is less than 0.8 under 1 atm pressure (Fig. 8).

Fig. 8. Picture of wind tunnel FL-61.

The detection of laminar/turbulence distribution is based on the fact that the convective heat exchange coefficients of laminar is different from that of turbulence. The difference of temperature between laminar flow and turbulent flow can be generated by the difference of temperature between the flow and the airfoil model. As for the flow considering compressibility, the temperature at laminar flow is less than that of turbulence under thermal equilibrium. The approach to generating the difference of temperature is as follows: firstly, utilize the circulation of flow in the wind tunnel to generate heat on the surface of airfoil models by friction, in order to increase the temperature on the surface. After the temperature rises to 45 °C (monitored by infrared cameras), the cooling system is started immediately to rapidly decrease the temperature of the flow. In this experiment, infrared cameras, FLIR SC7750L and FLIR A655sc, are utilized to monitor the surface temperature of airfoil models. Thus, the transition position can be detected [20]. Through the infrared photo the laminar flow region appears white and the turbulent flow region appears grey.

The comparison between the numerical result in form of C_f distribution over two airfoils (leeward) and the result from wind tunnel experiment in form of infrared photo is shown in Figs. 9, 10 and 11 at different angles of attack. The model of airfoil is three-dimensional with transverse length is much bigger than streamwise length so that the flow over the surface of model can be considered two-dimensional.

From Table 1 we can see that generally the transition positions predicted by both k-ω SST with γ-Re$_{\theta t}$ model and k-k$_l$-ω model occur at a distance from that in wind tunnel experiment. Comparatively, the k-k$_l$-ω model predicts laminar transition position later than that using k-ω SST with γ-Re$_{\theta t}$ model. The accuracy of laminar transition position prediction is not satisfactory. One of the main reasons lies in the inconsistency of boundary condition of the wind tunnel (e.g. fluctuating pressure, turbulence intensity, speed in real flow field, etc.) and that of the numerical simulation. The wind tunnel experiment itself has error brought by the advanced transition due to the roughness of

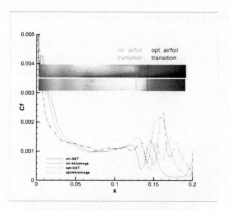

Fig. 9. C_f over two airfoils (leeward) compared with experimental result (1° angle of attack)

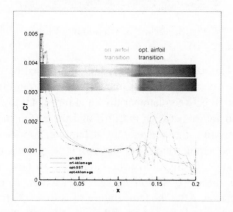

Fig. 10. C_f over two airfoils (leeward) compared with experimental result (3° angle of attack)

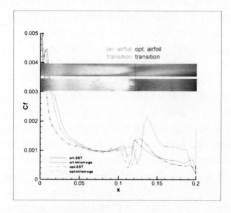

Fig. 11. C_f over two airfoils (leeward) compared with experimental result (4° angle of attack)

the model and the possible uncleanness in the airflow (cf. the zigzag transition line in the upper infrared photo in Fig. 10). (Therefore, the laminar transition for one certain airfoil under each angle of attack is judged as the maximum value of series of transition points on the surface of the model.) In engineering application, designers are more interested in the difference between original and optimized airfoils to evaluate the effect of optimization. Comparatively, the result predicted by k-ω SST with γ-Re$_{\theta t}$ model more agrees with the result of experiment. In a word, the accuracy of simulating laminar transition point itself is not satisfactorily reliable according to the numerical prediction by the two models. However, the function of evaluating the effect of laminarization as optimization is well accomplished by k-ω SST with γ-Re$_{\theta t}$ model.

Table 1. Comparison between numerical simulation and experimental result

		Transition position				Increment of laminar	
		Ori		Opt			
		Numerical	Exp	Numerical	Exp	Numerical	Exp
1°	SST	72.1%	71.2%	76.6%	75.8%	4.5%	4.6%
	kklomega	75.9%		78.6%		2.7%	
3°	SST	63.4%	62.0%	67.8%	66.5%	4.4%	4.5%
	kklomega	68.7%		72.8%		4.1%	
4°	SST	59.9%	60.1%	61.9%	62.3%	2.0%	2.2%
	kklomega	60.0%		62.5%		2.5%	

4   Conclusion

The development of transition models is discussed in the beginning, especially k-ω SST with γ-Re$_{\theta t}$ model and k-k$_l$-ω model, which are extensively embedded in many optimization frameworks. The article aims to evaluate their accuracy in laminar transition predicting and to verify their accuracy in evaluating the optimization effect of laminarization. The mesh with boundary conditions, the criterion for judging laminar transition, and the method to obtain laminar/turbulence distribution in wind tunnel experiment are introduced. The result shows that the effect of the laminarization can be evaluated quite accurately by k-ω SST with γ-Re$_{\theta t}$ model. Therefore, the laminarization on three-dimensional aircraft part, e.g. wing, nacelle can be constructed using this model in the future optimization framework of the authors.

References

1. Roveda L, Ghidoni S, Cotecchia S et al (2017) EURECA H2020 CleanSky 2: a multi-robot framework to enhance the fourth industrial revolution in the aerospace industry. In: ICRA 2017 - workshop IC3 – Industry of the future: collaborative, connected, cognitive. Novel approaches stemming from factory of the future & industry 4.0 initiatives
2. Joslin RD (1998) Aircraft Laminar Flow Control. NASA Langley Technical Report Server
3. Selig MS, Maughmer MD, Dan MS (2015) Natural-laminar-flow airfoil for general-aviation applications. J Aircr 32(32):710–715
4. Rahimi H, Medjroubi W, Stoevesandt B et al (2014) 2D numerical investigation of the laminar and turbulent flow over different airfoils using OpenFOAM
5. Dyson TE, Bogard DG, Bradshaw SD (2013) A CFD evaluation of multiple RANS turbulence models for prediction of boundary layer flows on a turbine vane. In: ASME turbo expo 2013: turbine technical conference and exposition, p V03CT14A014
6. Wilcox DC (2006) Turbulence Modeling for CFD. DCW Industries
7. Menter FR (1994) Two-equation eddy-viscosity turbulence models for engineering applications. AIAA J 32(8):1598–1605
8. Wang S, Sun G, Chen W et al (2018) Database self-expansion based on artificial neural network: an approach in aircraft design. Aerosp Sci Technol 72:77–83
9. Derksen RW, Rogalsky T (2010) Bezier-PARSEC: an optimized aerofoil parameterization for design. Adv Eng Softw 41(7):923–930
10. Fujino M (2010) Design and development of HondaJet. J Jpn Soc Precis Eng 42:755–764
11. Lin Y et al (2015) Implementation of Menter's transition model on an isolated natural laminar flow nacelle. AIAA J. 49(4):824–835
12. Langtry R, Menter F (2015) Transition modeling for general CFD applications in aeronautics. In: AIAA aerospace sciences meeting and exhibit
13. Cutrone L, Palma PD, Pascazio G et al (2007) An evaluation of bypass transition models for turbomachinery flows. Int J Heat Fluid Flow 28(1):161–177
14. Stock HW, Haase W (2000) Navier-Stokes airfoil computations with e transition prediction including transitional flow regions. AIAA J. 38:2059–2066
15. Ishikawa H, Tokugawa N, Ueda Y et al (2016) Development of inverse design system for supersonic natural laminar flow wing on high reynolds number condition. Nihon Kōkū Uchū Gakkai ronbunshū (J Jpn Soc Aeronaut Space Sci) 64(2):113–122
16. Drela M (1989) XFOIL: an analysis and design system for low reynolds number airfoils. In: Lecture notes in engineering, vol 54, pp 1–12
17. Leylek JH, Walters DK (2004) A new model for boundary layer transition using a single-point RANS approach. J Turbomachinery 126(1):193–202 2002-HT-32740
18. Pope SB (2001) Turbulent flows. Turbulent flows 12(11):806
19. Ashok, A (2008) Transonic airfoil aerodynamic characterisation by means of PIV. For obtaining the degree of Master of Science in Aerospace Engineering, Delft University of Technology
20. Joseph LA, Borgoltz A, Devenport W (2016) Infrared thermography for detection of laminar–turbulent transition in low-speed wind tunnel testing. Exp Fluids 57(5):1–13

Comparison of Electric Ducted Fans for Future Green Aircrafts

Roman Pankov[(⊠)] and Jiyong Tang

Nanjing University of Aeronautics and Astronautics,
29 Yudao Street, Nanjing 210016, China
pankovroman1995@outlook.com

Abstract. The market of light hybrid and all-electric aircraft is developing rapidly, many start-up companies have introduced their concepts and designs which are to enter to service in the next few years. At the same time, university and government research programs on hybrid and all-electric aircraft have focused more on regional, medium and long-haul aircraft. This article will emphasize the application of electric propulsion for eco-friendly light general aircraft. The goal of this paper is to introduce and develop aircraft and propulsor concepts, create methodology and a tool to calculate and make analysis of propulsor and electric motors that they contain. The Electric Ducted Fan (EDF) was chosen as a propulsor, four different types of EDF cores were introduced and developed in this article. For EDF parameters calculation analytical methods were chosen, additionally guidelines for creating a MATLAB code for making calculating easier will be presented. The design point was to achieve the same thrust for all types of EDF by changing design parameters to be able compare all other EDF's parameters. The results represent calculations for four core types of EDF and their analysis. Electric motors are developing rapidly, so research was held in regard to technological development. Based on the analysis guidelines of usage of different types EDF for role aircraft were presented.

Keywords: All-electric aircraft · Hybrid electric aircraft ·
Electric Ducted Fan · Compressor stage · Gas dynamic calculation

1 Introduction

The 21 century brings new requirements to aviation such the environmental and fuel price challenges. It goes without saying that environmental pollution and global warming are one of the most serious problems that mankind is facing. However, the total effect contributed by aviation to global warming is about 3% to 5% when taking into account such effects as emission of oxides and nitrogen from combustion to the atmosphere and contrail cirrus cloud formation. Future predictions of these contributions raise to the order of 15 to 30%, depending on the continuing growth of air traffic [1].

To regulate and reduce impact of aviation to the environment, some international programs were established and developed. One of the first in 21st century was Vision

X. Zhang (Ed.): APISAT 2018, LNEE 459, pp. 1633–1646, 2019.
https://doi.org/10.1007/978-981-13-3305-7_131

2020 of the Advisory Council for Aeronautical Research in Europe (ACARE) [2]. A more recent program by ACARE called Flightpath 2050 set an even more ambitious goal [3]. In the USA such research is held by NASA and described it the Strategic Implementation Plan Thrust 4: Transition to Alternative Propulsion and Energy [4]. This goal is going to be achieved by three fundamental steps and NASA believes that this step-by-step approach is the safest and fastest path for overall improvement in the future. At a global level, the International Air Transport Association (IATA) has also set ambitious targets to decrease carbon fuel consumption and emissions from aviation in its Carbon Neutral Growth initiative, which sets three goals before aviation industry: "(1) a 1.5% average annual improvement in fuel efficiency from 2009 to 2020; (2) carbon-neutral growth from 2020 and (3) a 50% absolute reduction in carbon emissions by 2050." [5]. In 2013, the 38th Session of the International Civil Aviation Organization (ICAO) Assembly, affirmed ambition for collective global aspirational goals for the international aviation sector to improve annual fuel efficiency by 2%, and to limit CO2 emissions at 2020 levels [6].

To meet new developing trends in aviation many universities and researchers keep working in this field of study. Bauhaus Luftfahrt is one of the leading research universities that focuses on eco-friendly aircraft, Pornet, Isikveren and others implemented comprehensive research about hybrid-electric aviation in articles [7–10]. In paper [7] they introduced concepts of hybrid and universally-electric aircraft and developed these ideas further in [8, 9] where pre-design strategies and methods for such aircraft was introduced. The article [10] was made focus on two and four engine hybrid-electric aircraft, which were compared to a A320 size projected conventional aircraft employing Geared TurboFan (GTF) propulsion targeting year 2035. It was concluded that the short-range/regional market would be the most suited for hybrid-electric concepts [10].

Another European university closely researching eco-friendly aircraft is Cranfield University in Great Britain. In article [11] the challenges of future aircraft propulsion were discussed and a review of turbo-electric distributed propulsion (TeDP) technology and its potential application for the all-electric commercial aircraft was given. Authors estimated that the majority of aircraft employing electric propulsion will be subsonic aircraft, further studies are needed to identify an optimized framework for an all-electric aircraft (AEA). Further research was done by Cranfield University in collaboration with Rolls-Royce Strategic Research Centre. This document is divided into two parts in which Part A [12], presents the conceptual design tool for evaluation of distributed hybrid electric propulsion on tube-and-wing A-350 sized aircraft configurations and Part B [13] investigates the potential for distributed propulsion on a Blended Wing Body (BWB) aircraft configuration and compares the performance with an advanced turbofan variant BWB of the same technology level.

More recent research was published by Borys Łukasik in source [14] where he stated that the feasibility of using the TeDP system will primarily depend on the technological advancement of electrical devices and that they also have a bigger potential for improvement than turbine engines. This article presents the results of analysis for 90 PAX class regional jets with turbofan engines and turboelectric distributed propulsion which proves that distributed propulsion is able to provide a great reduction in fuel consumption, while mission analysis shows the penalty of extra mass of electric appliances. As can be seen from an overview of international programs and universities research, research in field of eco-friendly aircraft was primarily developed for medium and long range aircraft. However, the focus of this research document will focus on light eco-friendly aircraft. In following paragraphs, a survey of experimental, in-production or near in-production light hybrid and universally-electric aircraft will be represented and discussed.

Currently, electro-mobility for manned aircraft mostly exists in the single/twin seater categories and are typically represented by conventional designs with reduced payload-range capability. Figure 1 exhibits experimental, in-production and near in-production light hybrid and universally electric aircraft in chronological order.

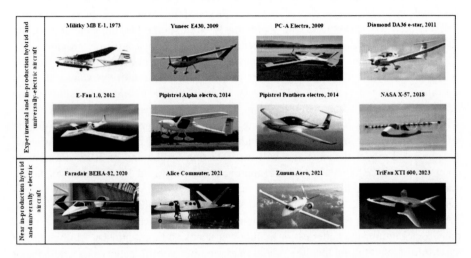

Fig. 1. List of experimental, in-production and near in-production light hybrid and universally-electric aircraft

The era of electric aircraft began with Brditschka's converted HB-3 renamed the MB-E1. The aircraft flew for just 9 min in 1973, was powered by a 10 kW Bosch motor and utilized Ni–Cd battery cells for energy [15]. The Yuneec International E430 is a Chinese two-seat electric aircraft; it was the first commercially available kit electric aircraft [16]. PC-A Electra was experiment of SolarWorld company to create long range electric glider, now the company claims that the Electra one Solar has a range of up to 1000 km and that the Electra Two Record boasts a range of more than 2000 km [17]. These aircrafts were mostly designed for experiment rather than for commercial use. Diamond and Pipistrel companies attempted to create commercial universally electric and hybrid aircraft made by installing Bosch electric motor on board of baseline aircraft. Airplanes Pipistrel Alfa and Pipistrel Panthera were certificated according to FAR-23 certification standards and available for purchase in all-electric and piston-powered versions for Alfa and in all-electric, hybrid-electric and piston-powered versions for Panthera [18].

Aircrafts Airbus E-Fan and NASA X-57 were designed and built for research purpose only. The first E-Fan aircraft was built in 2012 with two 30 KW engines just to prove possibility of all-electric flight; now Airbus, Rolls-Royce, and Siemens have teamed up for future partnership launches of the E-Fan X hybrid-electric flight demonstrator with 2 MW electric motor on board [19]. This project is aims to develop a near-term flight demonstrator which will be a significant step forward in implementation of hybrid-electric propulsion for commercial aircraft. NASA's X-57 was designed as a testbed for the Leading Edge Asynchronous Propeller Technology (LEAPTech) project, which intended to demonstrate technology to reduce fuel use, emissions, and noise [20]. This aircraft is based on the Italian-built Tecnam P2006T with 14 electric motors turning propellers which are integrated into a uniquely-designed wing. It's a key part of NASA's 10-year-long New Aviation Horizons initiative which was introduced with a previously cited source [4]; the general aviation-sized X-57 will take the first step in opening Strategic Thrust 4 of this program.

The next four aircrafts represented in Fig. 1 are near in-production eco-friendly aircrafts which use different approaches to minimize noise and emissions. The first aircraft represented is BEHA-82, a 6-seat aircraft with a hybrid propulsion system designed by UK based company Faradair. The aircraft utilizes a parallel hybrid engine with a 300 hp internal combustion engine (ICE) running on either JetA1 or Biofuel in combination with twin electric motors with power of 300 hp in a pusher configuration. A sub-scaled model demonstrator will begin testing by the end of 2018 and aims to enter service in 2020 [21]. Another approach is being used by Eviation, an Israeli startup team; the Alice aircraft is a full-electric full-composite aircraft which is designed to take 9 passengers up to 1,000 km at more than 240 KTAS and to follow the FAA's FAR 23 part regulations for commuter and On Demand operations. The company uses Lithium-ion batteries for high power needs and a proprietary Aluminum-Air system for range, hitting the 400 Wh/kg mark and more. Still, battery takes up 65% of the aircraft's weight, and it takes a state-of-the-art all composite body to reduce this penalty [22]. Like Faradair, the Eviation sub-scaled demonstrator will start flight tests soon; Eviation plans to make first flight in late 2018 and enter service in 2021.

The next aircraft is being designed by the American Seattle based startup company ZUNUM Aero, which was established by Boeing HorizonX and JetBlue Airways; the company is designing megawatt class hybrid electric aircraft that will be certificated in early 2020. The company then intends to expand the aircraft's range from 700 nm in the early 2020s to 1000 nm by 2030 and transition to full electric aircraft with the future developing of batteries. The company claims that aircraft is designed with batteries making up 12 to 20% of the total weight [23]. The company then expects to fly an existing twin-turboprop testbed in 2018 with commercial off-the-shelf motors. The last project is the TriFan 600 from the XTI Aircraft Company, the world's first hybrid-electric Vertical Takeoff & Landing (VTOL) business aircraft. It is expected to enter service in 2023. The XTI TriFan is a hybrid aircraft, utilizing a small Honeywell HTS900 turboshaft coupled to 3 generators that can generate 734 kW of electricity to replenish a bank of LiPo batteries that in-turn energize three ducted fan units [24]. The port & starboard fans are mounted on a gimbal system to provide lifting thrust that rotates to forward thrust as the wing enters translational lift. A third ducted fan is mounted in the fuselage and comes online for VTOL operation [24].

As can be seen from this review, the market of light hybrid and all-electric aircraft is developing rapidly, so this paper will focus exclusively on this type of aircraft. The goal of this paper is to introduce aircraft and engine concepts, create methodology and a tool to calculate and analyze Electric Ducted Fans (EDF) and the electric motors that they contain. In the next section, the conception of engines and tools for their conceptual design and preliminary analysis will be introduced.

2   Aircraft and Engine Concept

This paper is developing aircraft and engine concepts that were introduced in [25]. It could be concluded that the reference aircraft is a light general aircraft 6 PAX capacity with hybrid-electric power unit. This aircraft's powertrain contains one Diesel-Generator (DG) in the fuselage behind the passenger cockpit and two EDF on the fuselage. A turbocharger will be provided for DG to reduce the influence of flight level on the power output. A diesel engine was selected for the generator because it can use non-toxic, low emissions biofuel, which has already been tested and is being used for other applications like hybrid-electric cars and highly eco-friendly trucks. Additionally, kerosene and biofuel have approximately the same power density; hence weight penalty will be avoided. Finally, biodiesel fuel much is less fire hazardous and has lower toxic fumes when compared to jet fuels, which will make aircraft safer for passengers. Batteries and biofuel tanks will be installed in the wing between front and rear spar.

The propulsor concept also was introduced in [25]. Once air enters into the engine inlet, it is compressed by the fan, driven in rotation by synchronous permanent magnet motor which was estimated to be perspective in terms of mass and efficiency by Vratny et al. in [26]. After compression, the air is expanded in the nozzle of the engine, creating its thrust. In this EDF design it is assumed that pairs of permanent magnets will be located at the ends of the fan blades. This decision makes an electric engine with a larger diameter as was noticed before in [11]; the larger electric engine creates more torque hence better characteristics and performance. The stator of the electric motor and it electrical system will be mounted in the nacelle to reduce wire length and weight. In the area where stator of the electric motor embedded, a front mount flange is placed to fasten the nacelle to the fuselage. The second mount flange is placed in the area of the jet nozzle in the area of the power racks, which is the second bearing of the fan's rotor. The benefits and drawbacks of utilizing EDF will be discussed in the next paragraph.

The chosen scheme significantly simplifies the design of the engine and its maintenance compared with modern turbofan engines. The elimination of a combustion chamber will make this engine much more fire safe and will make engine installation much easier due to the lack of hot parts. Also, the EDF will have much better engine acceleration when compared to jet engines that require a few seconds to react to throttle change. Additionally, in contrast to conventional jet engines, the EDF will not depend on ambient temperature and pressure for starting the engine, therefore this engine will be the better choice for use at high altitude or with high temperature airports. Finally, the EDF is expected to have a lower weight than conventional jet engines because of the elimination of heavy fundamental parts for jet engines such as the combustion chamber and turbine. The EDF's weight estimation is beyond of the scope of this paper; however, mass estimation of the electric motor for the EDF will be presented. Nevertheless, the EDF requires a much higher current and voltage than a conventional aircraft's electric system which, in turn, requires much bigger and heavier wires. Another drawback is that this design requires heavy shielding to protect the structure and passengers from the electric current.

Some changes were made in this aircraft and propulsor concept. As could be seen from Fig. 2 in [25], the aircraft had a low-mounted swept wing and a T-tail. In the refined concept this aircraft has a low-mounted tapered wing to obtain Natural Laminar Flow (NLF) and a V-tail to reduce the wetted area and therefore increase aerodynamic performance. To the EDF concept chevrons were added to the nozzle to reduce the noise of the aircraft. The refined aircraft and engine concept is represented in Fig. 2.

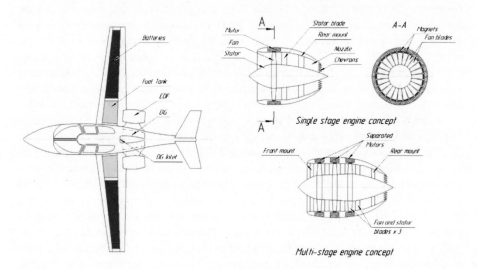

Fig. 2. Aircraft and EDF refined concepts.

Also in the previous concept, a high-pressure supersonic fan was used in the EDF. In this paper different types of core for the EDF will be investigated: single stage subsonic/transonic/supersonic and multiple stage subsonic. A subsonic fan was taken in single and multiple stage versions to investigate the difference in weight and drag due to diameter difference, a single stage fan is expected to have a larger diameter than a multiple stage fan. In the multiple stage variation electric motors will drive each stage separately to achieve optimal performance for each fan and motor. Another difference between single and multistage engines is that the latter has both front and rear mounts to support the bigger electric motor's weight.

3   Methods

For the conceptual design stage of this aircraft a tool for the preliminary calculation of different types of EDF concepts was developed. The main goal of this tool is to match engine performance to aircraft requirements in order to reduce design time. Another goal is to estimate electric the motor weight, propulsor drag in regard to different design cases and then to compare them in terms of aerodynamic and weight efficiency. To be able to compare them properly in this simulation, the same thrust of different types of engines must be determined by the changing of design parameters such as compression ratio, compressor efficiency and mass flow rate in regard to the type of compressor. This tool is written in MATLAB and based on formulas for the prediction of subsonic, transonic and supersonic compressor performance calculation for different efficiency, flight altitude and temperature from [27, 28]. Calculations were held only in the case of take-off, because it is the engine's hardest work regime and requires maximum thrust. However, this tool can be used to calculate engine performance for other regimes by changing parameters such as ambient temperature, pressure and aircraft velocity.

The first stage is determining constants and input parameters for the code. The input values are: mass flow rate, stage compression ratio and efficiency, electric motor power density in regard to the year of design. Constants in the code are shown in the Table 1.

Table 1. Constants for calculation

Parameter name	Nomenclature	Value
Inlet air pressure, Pa	P^*_{amb}	101325
Inlet air temperature, K	T^*_{amb}	288.15
Aircraft velocity	V_{aircr}	0
Inlet pressure recovery coefficient	G_{in}	0.97
Heat capacity ratio	K	1.4
EDF rotor efficiency	η_r	0.98
Jet nozzle speed coefficient	φ_c	0.98
Inner/outer diameter ratio	$\bar{d}$	0.5
Specific heat of the air, J/kg	C_p	1005

The second stage is calculation through engine cross sections which are: engine inlet, fan inlet, fan outlet and nozzle and finally thrust of one EDF. The first step is calculating total parameters at the fan inlet stage:

$$T^*_{in} = T^*_{amb} \tag{1}$$

$$P^*_{in} = P^*_{amb} \cdot G_{in} \tag{2}$$

Next, the total parameters at the fan outlet stage must be calculated:

$$P^*_{out} = P^*_{in} \cdot \pi^*_c \tag{3}$$

$$T^*_{out} = T^*_{in} \cdot \left(1 + \frac{\pi_{in}^{*\frac{k}{k-1}} - 1}{\eta^*_c} \right) \tag{4}$$

Where π^*_c is the compressor's pressure ratio and η^*_c is the compressor's efficiency, different types of axial compressors data was taken from [29].

The nozzle outlet velocity is calculated from [28] by the following formula:

$$C_n = \varphi_c \cdot \sqrt{2C_p \cdot T^*_{out} \cdot \left(1 - \frac{1}{\pi_n^{\frac{k-1}{k}}} \right)} \tag{5}$$

The total pressure ratio between nozzle inlet and outlet is calculated using formula 6:

$$\pi_n = \frac{P^*_{out}}{P_{amb}} \tag{6}$$

Formula 7 calculates velocity difference at the inlet and outlet of the engine:

$$V_p = C_n - V_{aircr} \tag{7}$$

Where V_{aircr} is aircraft velocity and equal to 0 for this case.

Thrust one EDF can provide:

$$Th = V_p \cdot \dot{m} \tag{8}$$

The following stage is calculating geometrical parameters of fan stage based on previous calculation and then uninstalled and installed thrust. Area of engine could be found using the following formula:

$$A_{in} = \frac{G_B \cdot \sqrt{T^*_{in}}}{K_m \cdot P^*_{in} \cdot q(\lambda_{in})} \tag{9}$$

In order to calculate the area of the engine inlet, the relative flux density $q(\lambda_{in})$ is needed based on the formula below:

$$q(\lambda_{in}) = \left(\frac{k+1}{2} \right)^{\frac{1}{k-1}} \cdot \lambda_{in} \cdot \left(1 - \frac{k-1}{k+1} \cdot \lambda_{in}^2 \right)^{\frac{1}{k-1}} \tag{10}$$

The relative flux density is a function of reduced speed λ_{in}:

$$\lambda_{in} = \frac{C_a}{\sqrt{\frac{2k}{k+1} \cdot R \cdot T_{in}^*}} \tag{11}$$

Where C_a is the component of absolute speed.

The flux coefficient depends on the adiabatic exponent and the gas constant:

$$K_m = \sqrt{\frac{k}{R} \cdot \left(\frac{2}{k-1}\right)^{\frac{k+1}{k-1}}} \tag{12}$$

Formulas 9–12 were taken from source [28]. Once area of engine is calculated, the outside diameter of the fan can be obtained with the following formula:

$$d_{out} = \sqrt{\frac{4 A_{in}}{\pi \cdot \left(1 - \overline{\overline{d^2}}\right)}} \tag{13}$$

Where inner/outer diameter ratio is constant and could be defined as:

$$\bar{d} = \frac{d_{in}}{d_{out}} \tag{14}$$

Where d_{in} is the inside diameter if the fan and d_{out} is the outside diameter of the fan. Hence outside diameter of the fan is:

$$d_{in} = \bar{d} - d_{out} \tag{15}$$

Thus the middle diameter of the fan can be calculated as:

$$d_m = 0.5 \cdot (d_{in} + d_{out}) \tag{16}$$

The height of the fan blade:

$$h_b = 0.5 \cdot (d_{out} - d_{in}) \tag{17}$$

The last stage is calculation of the electric motor's power requirements and motor weight based on power density of state-of-art motors. Specific work done by the compressor is:

$$L_c = \frac{C_p \left(T_{out}^* - T_{in}^*\right)}{\eta_r} \tag{18}$$

Where η_r is EDF rotor efficiency which takes into account losses in bearings.

Ideal power required by compressor could be found from [28] by formula:

$$N = \dot{m} \cdot L_c \tag{19}$$

In order to calculate the mass of the electric motors, power density of state-of-art motors is defined as the motor power divided by its mass:

$$\rho_m = \frac{N_m}{m_m} \tag{20}$$

The estimated values of power density for the time being, in 10 years and in 20 years were taken from [30] shown in the Table 2 below.

Table 2. Electric motor power density and efficiency

Time	The time being	In 10 years	In 20 years
Power density (kw/kg), ρ_m	5	13.2	20
Motor efficiency, η_m	0.965	0.975	0.99

The equivalent electric power of the real electric motor is:

$$N_m = \frac{N \cdot \eta_m}{\rho_m} \tag{21}$$

Where η_m is the motor efficiency. In the next section the results and their discussion will be represented.

4   Results and Discussion

The results of numerical simulations based on mathematical model that was presented in previous section are shown in Table 3 below with different compression ratios and compressor efficiencies. First three parameters are input design parameters that vary depending on the type of compressor. The compression ratio of subsonic three-stage EDF was described as 1.2 cubed because in multistage compressors the compression ratios of all stages multiply on each other. As can be seen from the fourth parameter, all EDFs were designed to achieve approximately the same thrust; indeed, significantly small deviations of this parameter just in 2 N were achieved.

The next two parameters in Table 3 show the dimensions of different EDFs. As can be seen from these results a higher compression ratio produces smaller dimensions: the fan diameter of the supersonic EDF is almost 30% smaller than the subsonic EDF, however the subsonic multistage EDF appears to be a good trade-off with the high efficiency of the subsonic core and the small dimensions that are associated with the supersonic EDF. Engine drag has an influence on the total thrust required by the aircraft, hence this parameter must be taken into account. As can be seen in Eq. 22,

Table 3. Performance and dimensions of different EDF cores

Parameters	Subsonic	Subsonic Multi-stage	Transonic	Supersonic
Compression ratio	1.2	$1.2^3 = 1.73$	1.6	1.8
Compressor efficiency	0.92	0.92	0.85	0.85
Mass flow rate (kg/s)	20.46	10.75	11.68	10.26
Thrust (N)	3121.3	3121.9	3121.4	3123.1
Fan diameter (m)	0.452	0.328	0.3416	0.32
Fan area (m^2)	0.12	0.0633	0.0688	0.0604
Electrical power (kW)	351.4	585.38	583.58	652.25

drag is dependent on the EDF's area, therefore the supersonic EDF has the lowest drag while subsonic EDF has the highest. As previously stated, subsonic multistage and transonic core offers a good compromise between the two.

The drag force of the engine is defined as:

$$D = \frac{1}{2} \cdot C_D \cdot A_{in} \cdot V_n^2 \qquad (22)$$

It can be concluded that the subsonic core EDF will have approximately 50% bigger drag than all other types. The supersonic core EDF will have lowest drag due to it having the highest compression ratio and smallest dimensions. However, the subsonic multistage core could be even smaller were it designed with more stages which would increase compression ratio.

The next results that were obtained are masses of electric motor in regard to technology development and are shown in Fig. 3. As can be seen from this figure, the subsonic EDF has the lowest weight of all EDF types. At the same time the subsonic multistage EDF has a significantly higher weight due to a much higher requirement of electrical power when compared to the single stage subsonic EDF. This can be seen in Table 3, transonic and supersonic EDFs also require approximately the same electrical power because they need to overcome shock waves that occur on the tips of the compressor blades due to their high rotational speed. As can be seen in the table below, the electric motor, which is the largest contributor to the EDF's weight, will become lighter as the technology progresses. It is predicted that for all EDF types motor weight will decrease approximately by 66% in 10 years and by 74% when compare to current status of technology.

The last parameter that influences engine selection for eco-friendly aircraft is noise. Noise reduction is also an important concern for all international programs that have set goals to reduce aviation's impact on the environment. This parameter was not precisely calculated in this paper; however, it could be stated that subsonic core EDFs will create lower noise pollution than the other two types of design. Transonic and supersonic core EDFs will have higher noise because of the formation of shock waves on the tips of the blades.

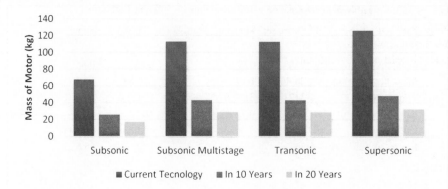

Fig. 3. Mass of electric motor in regard to time of design

5 Conclusion and Outlook

In this article, concepts of light hybrid electric aircraft and it's potential propulsors were presented and developed from a previous article by the same author [25]. The EDF propulsor was presented and its advantages and disadvantages were discussed. Four different types of cores were considered and methods and a tool for it's calculation were presented. These methods help to estimate engine performance, size and weight of the motor for each type of EDF. The design point was to achieve the same thrust for all types of propulsors by changing the design parameters to be able to compare all other EDF's parameters. As can be seen from results, the single stage subsonic EDF has a significantly lower motor weight when compared to all other variants, higher efficiency by 7% and lower noise when compared to transonic and supersonic EDFs. At the same time due to its larger diameter, the subsonic core EDF will have approximately 50% higher drag than all other EDF types for given thrust. The supersonic propulsor has the smallest size and the lowest drag, but at the same time, is less efficient and has the heaviest motor. The subsonic multistage and transonic EDF presents a compromise between the previous two versions; however, the subsonic multistage EDF has 8% less drag, as well as the advantages of high efficiency and low noise level. Another important conclusion is that the electric motor, which is the largest contributor to the EDF's weight, will become lighter as the technology progresses for all types of EDF. This changes will make EDF's application to aviation easier and more effective when compare to current situation.

Having different benefits and drawbacks, different EDFs could have different applications in aviation. It can be concluded from previous remarks, that that subsonic multistage EDF is optimal for use in general hybrid electric aircraft. Additionally, this conception has the most flexible parameters because its performance can be regulated by adding or reducing the number of compressor stages in it or increasing the mass flow rate. The optimization of parameters for the subsonic multistage EDF design could be a point of future studies. At the same time, the single stage EDF will be the most effective for light general aircraft with small take-off weight because it has the lowest motor weight, low noise and high efficiency. It can also be concluded that the use of the subsonic single stage EDF will be further complicated with an increase of aircraft take-off weight and required thrust. Therefore, if these aircrafts have two engines, they will require significantly larger engine with high drag characteristics. This problem could be solved by applying NLF technology to nacelle. Such experiments were done by NASA in the past and are currently being applied to Boeing 777X aircrafts. Another way of solving this problem would be the application of distributed propulsion technology, hence increasing number of engines. These approaches could also be a point of future research for the subsonic single stage EDF. Because transonic and supersonic core EDFs are smaller in size than the subsonic core EDF, they could be applied to military aircraft and UAV where high speed and performance are required and noise restriction is not a priority. Additionally, small size will benefit placing and lofting of the engine on the aircraft.

Acknowledgments. This work was funded by the project 11502112, 11432007 and 11672132 and supported by National Natural Science Foundation of China. Also this work was supported by a project funded by the Priority Academic Program Development of Jiangsu Higher Education Institutions of China. Editing assistance was provided by Andrew Guerra. Finally, the authors would like to thank Deng Feng for his fruitful discussion and valuable advice.

References

1. Jupp JA (2016) The design of future passenger aircraft – the environmental and fuel price challenges. Aeronaut J 120:37–60. https://doi.org/10.1017/aer.2015.4
2. Advisory Council for Aeronautical Research in Europe (ACARE) (2001) European aeronautics: a vision for 2020
3. European Commission (2011) Flightpath 2050 Europe's vision for aviation-report of the high level group on aviation research, Luxembourg
4. NASA new aviation horizons strategic implementation plan 2017 update. http://www.aeronautics.nasa.gov/strategic-plan.htm
5. International Air Transport Association (IATA) (2009) A global approach to reducing aviation emissions
6. 38th Session of the International Civil Aviation Organization (ICAO) (2013) Environmental report
7. Pornet C (2014) Hybrid and universally-electric aircraft concepts. Access Science, McGraw-Hill Yearbook of Sciences and Technology, McGraw-Hill Education. http://doi.org/10.1036/1097-8542.YB150553
8. Isikveren AT, Kaiser S, Pornet C, Vratny PC (2014) Pre-design strategies and sizing techniques for dual-energy aircraft. Aircr Eng Aerosp Technol 86(6):525–542
9. Pornet C, Gologan C, Vratny PC, Seitz A, Schmitz O, Isikveren AT, Hornung M (2015) Methodology for sizing and performance assessment of hybrid energy aircraft. J Aircr 52 (1):341–352. https://doi.org/10.2514/1.C032716
10. Pornet C, Isikveren AT (2015) Conceptual design of hybrid-electric transport aircraft. Prog Aerosp Sci 79:114–135. https://doi.org/10.1016/j.paerosci.2015.09.002
11. Gohardani AS, Doulgeris G, Singh R (2011) Challenges of future aircraft propulsion: a review of distributed propulsion technology and its potential application for the all-electric commercial aircraft. Prog Aerosp Sci 47:369–391. https://doi.org/10.1016/j.paerosci.2010.09.001

12. Kirner R et al (2015) An assessment of distributed propulsion: advanced propulsion system architectures for conventional aircraft configurations. Aerosp Sci Technol 46:42–50. https://doi.org/10.1016/j.ast.2015.06.022

13. Kirner R et al (2016) An assessment of distributed propulsion: Part B - advanced propulsion system architectures for blended wing body aircraft configurations. Aerosp Sci Technol 50:212–219. https://doi.org/10.1016/j.ast.2015.12.020

14. Łukasik B (2017) Turboelectric distributed propulsion system as a future replacement for turbofan engines. In: Proceedings of ASME Turbo Expo 2017: Turbomachinery Technical Conference and Exposition GT 2017, 26–30 June 2017, Charlotte

15. Moulton R (1973) An electric aeroplane, flight international

16. Yuneec official website. www.yuneec.com

17. PC-Aero GmbH official website. http://www.pc-aero.de/

18. Pipistrel official website. http://www.pipistrel.si/plane/alpha-electro/overview, http://www.pipistrel.si/plane/panthera/overview

19. Airbus Group official website. http://www.airbus.com/newsroom/press-releases/en/2017/11/airbus–rolls-royce–and-siemens-team-up-for-electric-future-par.html

20. NASA official website. https://www.nasa.gov/image-feature/nasas-x-57-electric-research-plane

21. Faradair official website. http://faradair.com/fwp/introducing-beha

22. Eviation official website. https://www.eviation.co/alice

23. Zunum Aero official website. http://zunum.aero/aircraft

24. XTI official website. http://www.xtiaircraft.com/trifan-600

25. Pankov R (2016) Схема легкого многоцелевого самолета с гибридной силовой установкой (Scheme of Light Multipurpose Plane with Hybrid Power Unit). Open information and computer integrated technologies, vol. 71, pp 62–72

26. Vratny PC, Forsbach P, Seitz A, Hornung M (2014) Investigation of universally electric propulsion systems for transport aircraft. In: 29th congress of the international council of the aeronautical sciences, St. Petersburg

27. Mattingly JD, Heiser WH, Pratt DT (2002) Aircraft engine design, 2nd edn. American Institute of Aeronautics and Astronautics, Reston

28. Nechaev UN, Fedorov RM (1977) Теория авиационных газотурбинных двигателей (Theory of aviation gas turbine engines). Mashinostroenie, Moscow

29. Boyce MP Axial-flow compressors. https://www.netl.doe.gov/File%20Library/Research/Coal/energy%20systems/turbines/handbook/2-0.pdf

30. Madavan N et al (2016) A NASA perspective on electric propulsion technologies for commercial aviation. In: Workshop on technology roadmap for large electric machines, University of Illinois Urbana-Champaign, 5–6 April 2016

Test Research on Operational Deflection Shape and Operational Modal Analysis of Aeroengine Blade

Chao Hang[(✉)], Qun Yan, Jian Xu, and Xiang Gao

Aircraft Strength Research Institute, Dianzi Er Road 86, Xi'an 710065, China
hangchaonwpu@163.com

Abstract. The blade is a key part of aeroengines. It is necessary to investigate the modal parameters of blades for guaranteeing structural integrity. Operational deflection shape (ODS) and operational modal analysis (OMA) are all modal test methods for aeroengine blades. First of all, the theory of ODS and OMA is introduced separately. Then, the ODS and OMA test systems are established with the high frequency vibration table and scanning laser Doppler vibrometer, by which the ODS and OMA of aeroengine blade are researched within the range of 10 kHz frequency respectively. The natural frequencies and mode shapes of the blade are educed through ODS data, while the natural frequencies, mode shapes and damping ratios are acquired through OMA data. Finally, the two test results obtained by ODS and OMA are contrasted, which indicated the two results are consistent. It is proved that the two test systems of ODS and OMA in this study are designed reasonably, and the test results are analyzed validly.

Keywords: Operational deflection shape · Operational modal analysis · Natural frequency · Mode shapes · Base excitation

1 Introduction

The new generation of aeroengine has important performance indexes such as high thrust-weight ratio, high reliability and long life. The safety and reliability of the engine is mainly dependent on the structural integrity of the engine components. The structural integrity is basically determined by the resistance of the system structure to the vibration stress generated by the forced vibration or the unstable aerodynamic disturbance [1]. As the core parts of aeroengine, the blades are easy to fail, which are affected by many loads including centrifugal force, aerodynamic force, vibration and heat and so on. The failure of blades caused by resonance is an important concern for designers, which has an important effect on the performance of the whole engine [2–4].

In order to ensure the reliability of blades, modal analysis of blades is needed to obtain its natural frequencies and vibration modes, so as to avoid or weaken the possible resonance of blades in the actual work. At present, modal analysis usually has two methods, which are finite element analysis used at the design stage and test analysis used at the verification stage respectively. In the design stage, the three-dimensional

© Springer Nature Singapore Pte Ltd. 2019
X. Zhang (Ed.): APISAT 2018, LNEE 459, pp. 1647–1665, 2019.
https://doi.org/10.1007/978-981-13-3305-7_132

numerical model of the blade is established. The natural frequencies and vibration modes of blades are calculated by finite element method. The blade design parameters are adjusted according to the calculation results so that the blade meets the design requirements [5, 6]. In the verification stage, the modal analysis test of blades is carried out, and the first several modal parameters of blades are measured to verify whether the modal parameters of blades meet the design requirements [7].

The commonly used experimental modal analysis (EMA) adopts phase resonance method based on pure modal test and phase separation method based on frequency response function matrix parameters identification. According to the ways of excitation, it can be divided into two types: exciter excitation method and force hammer excitation method. These test methods are widely used in structural modal analysis [8–10]. However, there are some shortcomings in the above test methods for the lightweight cantilever structure of blades type. The exciter excitation method needs to connect structures with a long pole which adds additional mass and stiffness for structures. It impacts modal test precision of lightweight cantilever structure. The energy of the hammer excitation method is concentrated in a short time, which may cause overload and nonlinear problems. The test results are affected by the human factors. For the modal test of blades in special environments, such as blades installed on the shaking table, the implementation of the above test methods is rather troublesome.

In 1987, Dossing [11] proposes the concept of operational deflection shape (ODS) to evaluate the dynamic performance of mechanical structure. Richardson [12] describes the differences and connections between the operational deflection shape and modal shape in the modal analysis theory and test. He points out that the operational deflection shape includes the resonant response and forced response of structures. It is the linear superposition of modal shape of structures. The modal shape only includes the resonant response of structures. When the structure is excited by a sinusoidal excitation, the operational deflection shape can be considered as modal shape at the frequency if the following 3 conditions are satisfied: (1) the excitation force is not located on the nodal lines of vibration mode; (2) the excitation frequency is close to the resonance frequency; (3) the operational deflection shape at the resonant frequency is only a mode dominated. Kromulski [13] analyzes the operational deflection shape of the corn separator, and verifies the results of ODS by system analysis. The results of the two methods were in good agreement. The correctness of ODS was proved. The ODS method is suitable for structural modal analysis under operational conditions.

In 1995, James [14] proves the feasibility of operational modal analysis (OMA) method based on the theory of natural excitation (NExT) for the first time. He applies the method to modal parameters identification of the wind turbine under operational conditions successfully. The theoretical derivation is based on the single white noise excitation hypothesis and real modal analysis. Brincker [15] generalizes the theory of OMA method to the general white noise excitation hypothesis and complex modal analysis. Peeters [16] proposes a modal parameters identification method for least square complex frequency domain method (LSCF, commercial name PolyMAX). It adopts the discrete time frequency domain model and the fast recursive algorithm, which has good anti-interference ability and clear steady state diagram. It is one of the best recognized methods of modal parameter identification at present. The operational PolyMAX method is based on the theory of least squares complex frequency domain method. The identification of operational modal parameters under unknown excitation is carried out by cross power spectrum. The ODS method is also suitable for structural modal analysis under operational conditions.

Aiming at blades modal analysis in vibration test environment, blades are often installed on shaking table. However, the blade is not in operational state. So the modal parameters of blade can not be directly analyzed by ODS or OMA method. In this paper, the white noise excitation in operational state of the blade is simulated by means of base excitation of the shaking table. At the same time, the vibration response of the blade is measured by a scanning laser vibration meter. This non-contact measurement method does not give additional mass to the blade, which is helpful to improve the precision of modal analysis. Then, the modal parameters of blade are analyzed by ODS and OMA methods respectively. Finally, the two test results are discussed and compared to verify the correctness of the two test methods. The two test methods in this paper provide a new way for modal analysis of blades in vibration test environment.

2 Modal Analysis Theory

Modality is the natural vibration characteristics of structures. Each mode has specific natural frequencies, damping ratios and mode shapes. Modal analysis is a method to study the dynamic characteristics of mechanical structures. It is the application of system identification method in engineering vibration field. The vibration mode is an inherent and integral characteristic of the elastic structure. Modal analysis decomposes the complex vibration of the structure into a number of simple and independent vibrations, which is characterized by a series of modal parameters. According to the principle of linear superposition, the complex vibration of structures is the result of superposition of innumerable modes. The ultimate goal of modal analysis is to identify the modal parameters of the system. It provides a basis for vibration characteristic analysis, vibration fault diagnosis and structural dynamic optimization design of structural system.

Traditional experimental modal analysis method usually adopts the phase separation method based on frequency response function matrix parameters identification. The basic principle of the phase separation method is to measure the exciting force and vibration response when the structure is excited under static state. The fast Fourier transform (FFT) of the excitation and response signals is used to get the transfer functions between the response points and the excitation points. The transfer functions of test are used to fit mathematical model of the transfer function of modal theory, so as to identify the modal parameters of structures. The modal parameters can be fitted in time domain or frequency domain. In order to solve the problem of blade modal analysis in vibration test environment, the theory of ODS and OMA are introduced below.

2.1 Operational Deflection Shape (ODS) Theory

Operational deflection shape refers to the deformation of a structure at a specified frequency. The more generalized operational deflection shape refers to the forced motion of two or more points in structure under operational state. The structural operational deflection shape can be obtained only by measuring the response signals of multiple points on the structure. Operational deflection shape includes structural dynamic performance parameters. ODS analysis is often used to identify structural modal parameters under unknown excitation loads.

The basic principle of structural modal parameters by ODS analysis is that the response signals of multiple points on the structure can be measured in the operational state and the operational deflection shape frequency response functions (ODS FRF) are calculated. Because the ODS at the resonance frequency is dominated by the only one mode, it can be considered that the ODS at this frequency is the mode shape. By selecting the resonant peak on the ODS FRF, the corresponding amplitude and phase of all response points at the resonant peak frequency are combined, and the corresponding modes at the resonance frequency are obtained.

The response of each point in the structure is related to the frequency response functions and excitation. The relationship is as follows.

$$F_y(j\omega) = \sum_{i=1}^{n} H_{y,fi}(j\omega)F_{fi}(j\omega) \tag{1}$$

Where $F_y(j\omega)$ is the Fourier spectrum of response signal. $H_{y,fi}(j\omega)$ is the frequency response function between the response and excitation. $F_{fi}(j\omega)$ is the Fourier spectrum of excitation signal.

If the each order mode shape of the structure have little influence on the y point, the total frequency response function $H_y(j\omega)$ of the y point is also small. So the response of the y point is mainly influenced by the environmental excitation, which can be approximated as the following approximate expression.

$$F_y(j\omega) \approx CF_f(j\omega) \tag{2}$$

Where C is a constant. Therefore, the y point response signal can be instead of the excitation signal to calculate the ODS FRF.

The methods for calculating ODS FRF are as follows. If the external force acting on the structure is unknown, the response signal which is affected slightly by the each order mode shape is selected as the reference signal. The response signals of other measuring points are taken as the flow signals. The ratio of each flow response signal to the reference signal is calculated respectively, and the ODS FRF is obtained. In order to reduce the error of unrelated noise and retain the correct phase information, ODS FRF is usually calculated by self power spectrum and cross power spectrum. Its mathematical expression is as follows.

$$ODS\,FRF(\omega) = \frac{F_x(\omega)}{F_y(\omega)} = \frac{F_x(\omega) * F_y^*(\omega)}{F_y(\omega) * F_y^*(\omega)} = \frac{G_{xy}(\omega)}{G_{yy}(\omega)} \tag{3}$$

In which the x represents the flow signal, and the y represents the reference signal. The $G_{xy}(\omega)$ represents the cross power spectrum of the flow signal and the reference signal, and the $G_{yy}(\omega)$ represents the self power spectrum of the reference signal.

From the above analysis, we can see that ODS FRF contains structural dynamic performance parameters, which is similar to structural frequency response functions. Therefore, the natural frequencies of the structure can be determined by selecting formants on ODS FRF. The corresponding ODS at this frequency is the vibration mode. The mode phase is the phase of response point relative to reference point. Compared with the traditional experimental modal analysis method, the ODS method does not need to measure the external force and does not need to fit the frequency response functions. The test and analysis process is simple, and the real-time performance is good. But the ODS method can only obtain the frequencies and vibration modes. The damping ratio can not be obtained.

2.2   Operational Modal Analysis (OMA) Theory

Compared to the traditional experimental modal analysis, operational modal analysis is to replace the frequency response functions by the cross power spectrum between the response signals under the excitation of the structure. The modal parameters of structures are obtained by fitting the cross power spectrum with the similar method in the experimental modal analysis. The theoretical basis of OMA method is that the cross correlation functions between the response points and the structure's impulse response functions have the same form under the white noise excitation, both the real mode and the complex mode [14, 15].

Taking the real modal analysis excited by white noise on the single point as an example, the impulse response function between the input point k and the output point i is expressed as

$$x_{ik}(t) = \sum_{r=1}^{n} \frac{\varphi_{ir}\varphi_{kr}}{m_r \omega_{dr}} \exp(-\xi_r \omega_{nr} t) \sin(\omega_{dr} t) \tag{4}$$

Where the φ_{ir} and φ_{ik} represent the i and k components of the r mode shape separately. The m_r represents the r undamped modal mass. The ξ_r represents the r modal damping ratio. The ω_{nr} represents the r undamped modal frequency. The ω_{dr} represents the r damped modal frequency. The cross correlation function of the response point i and the response point j is

$$R_{jk}(T) = \sum_{r=1}^{n} \frac{\varphi_{ir}A_{jr}}{m_r \omega_{dr}} \exp(-\xi_r \omega_{nr} T) \sin(\omega_{dr} T + \theta_r) \tag{5}$$

Where the A_{jr} is a constant associated with the j components of the r mode shape. The θ_r is a phase angle associated with the r modal frequency and damping ratio. From the formula (4) and (5), it can be seen that the cross correlation functions between the response points and impulse response functions have the same form, which are composed of a set of attenuated sinusoidal functions. For the complex mode analysis of general white noise excitation, the proof of the same form of impulse response functions and cross correlation functions is shown in Ref. [15].

Cross power spectrum is the Fourier transform of cross correlation function, and frequency response function is the Fourier transform of impulse response function. Therefore, the cross power spectrum between the response points in the structure has the same form as the structural frequency response function. Then, the mutual power spectrum of the response points can be identified by using the modal parameter identification method based on frequency domain, such as LSCF. The modal parameters of the structure are obtained which are including natural frequencies, mode shapes and damping ratios. Compared with the traditional experimental modal analysis method, the OMA method does not need to measure the external excitation force, and directly fit the cross power spectrum of the response points. It is suitable for structural modal analysis under the condition of unknown excitation.

3 ODS and OMA Test of Blade

Taking an engine blade as the research object, the dynamic characteristics of the blade are studied by means of the ODS method and the OMA method respectively. The modal parameters of blade in the frequency range of 10000 Hz are analyzed.

3.1 Test System

It is different from the vibration exciter and force hammer used in traditional experimental modal analysis. It is also different from the natural environment excitation in the operational modal analysis. In this paper, a high frequency vibrator is used as an incentive device. The blade is mounted on a vibrating table by a fixture. The blade is stimulated by a vibration table. The excitation mode is constant amplitude fast sine sweep frequency (also known as "chirp") excitation. Because the amplitude of chirp signal is constant in frequency domain, it has the same characteristics as white noise signal. Therefore, the chirp signal satisfies the requirements of ODS and OMA. The frequency range of sweep vibration frequency is 0–10 kHz and the excitation time is 1.28 s.

In order to measure the vibration response of the blade surface without adding the mass to the blade, a scanning Doppler laser vibrometer is used to measure the vibration response of blades. The laser vibration meter is based on the principle of Doppler. The vibration velocity of the structure surface can be measured directly. Scanning laser Doppler vibrometer has the advantages of multi-point, non-contact and high-precision measurement. It is very suitable for the modal analysis of the structure. In the experiment, the sampling frequency of laser vibrometer is 25.6 kHz.

The blade test system of ODS is similar to that of OMA. The test system principle of ODS and OMA is shown in Fig. 1, and the site photo is shown in Fig. 2. The difference between the two methods is that they use different modal parameters identification methods. The blade is fixed on the vibration table through a fixture. The signal generator sends out the chirp voltage signal. Then the shaking table produces a fast sinusoidal sweep vibration. At the same time, a laser vibrometer is used to measure the velocity signal of a point on the surface of the blade. The data acquisition instrument is triggered by the chirp signal, and the speed signal of the measuring point is recorded by the data acquisition instrument. The above method is repeated for each measurement point. Finally, different modal analysis methods are used to get the modal parameters of the blade.

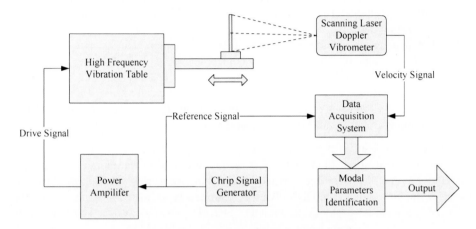

Fig. 1. Test system principle of ODS and OMA

Fig. 2. Photo of blade test site

3.2   Test Process

The test procedure of ODS is basically the same as that of OMA.

(A) Install the blade through the fixture on the shaking table, and connect the test equipment as shown in Fig. 1.
(B) As shown in Fig. 3, 55 measuring points are arranged on the surface of the blade with a scanning laser vibration meter. The points layout of the ODS and OMA test are exactly the same.
(C) The shaking table produces a chirp vibration excitation, and the laser vibrator is used to measure the vibration velocity of a point on the blade. The data acquisition instrument collects the chirp signal and the speed signal on the blade synchronously.
(D) The laser vibrometer changes a measuring point. Repeats step (C) until all the measuring points complete the measurement.

Modal parameters of the blade can be obtained by modal analysis of the above test results. The analysis method of ODS is as follows: Taking the chirp signal as the reference signal, the vibration velocity signal of the measuring point on the blade is used as the flow signal. The ODS FRF for each point is calculated in terms of formula (3). The peak value is selected on the ODS FRF. The ODS analysis are carried out to obtain the natural frequencies and mode shapes of the blade. The analysis method of OMA is as follows: the vibration velocity response of a certain measuring point is taken as reference signal. The cross power spectrum of the response signal and the reference signal of each measuring point is calculated. The cross frequency spectrum is fitted by least squares complex frequency domain method. The natural frequencies, mode shapes and damping ratios of the blades are obtained.

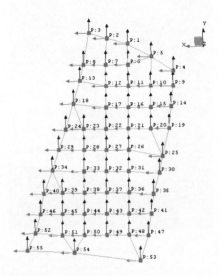

Fig. 3. Layout of measuring points on the blade surface

It needs to be explained that the true shape of the blade is a three-dimensional twisted structure. However, the complex configuration of blades was neglected in the experiment. The geometry of blades is simplified to a flat plate. This simplification will bring a little error, but it is basically consistent with the structural characteristics of the blade. Moreover, the vibration direction of the blade is mainly along the normal direction of the blade plane. So it can reflect the vibration pattern of the blade.

4   Analysis and Discussion of Test Results

4.1   Test Results of ODS

In the blade ODS test, the chirp signal is chosen as the reference signal. The ODS FRF for each measuring point is calculated. Because of the large number of measuring points, in order to display the curves clearly and without losing generality, the amplitude curves of the sum of ODS FRF of all measuring points are given in Fig. 4. When the natural frequencies are searched from the ODS FRF curve, the frequency points with sharp peak value are chosen as the natural frequencies of the blade. The corresponding mode shapes at the peak frequencies are extracted.

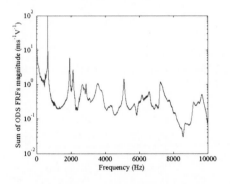

Fig. 4. Sum of ODS FRF amplitude of all measuring points

The blade is similar to the cantilever beam structure. Theoretically, the blade root is a fixed boundary condition, and the amplitude of points at the blade root should be zero. However, the blade is connected to the shaking table by means of fixtures in actual test. Shaking table and fixture will inevitably affect the blade vibration mode. According to the above dynamic characteristics of blades, the vibration modes with larger amplitude at the blade root are abandoned, which are considered as the coupled vibration modes of the blade and shaking table. It can not be used as the mode of the blade itself. The final blade ODS analysis obtains modal parameters as shown in Table 1, and the vibration modes is shown in Figs. 5, 6, 7, 8, 9 and 10.

Table 1. Modal parameters of the blade obtained by ODS analysis

Modal order	Frequency (Hz)	Mode shape description
1	657.8	First order bending
2	1930.6	First order twist (The nodal line is the connection between measuring point 1 and 34)
3	2120.0	First order twist (The nodal line is the connection between measuring point 1 and 35)
4	2874.9	Second order bending
5	5105.9	Second order twist
6	7237.9	Bending-twist coupling

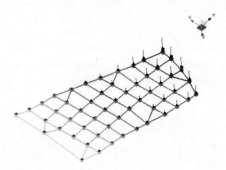

Fig. 5. First order vibration mode of blade obtained by ODS analysis

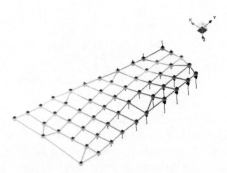

Fig. 6. Second order vibration mode of blade obtained by ODS analysis

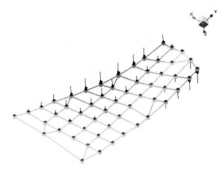

Fig. 7. Third order vibration mode of blade obtained by ODS analysis

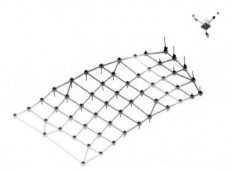

Fig. 8. Fourth order vibration mode of blade obtained by ODS analysis

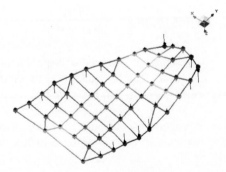

Fig. 9. Fifth order vibration mode of blade obtained by ODS analysis

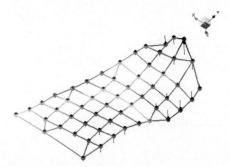

Fig. 10. Sixth order vibration mode of blade obtained by ODS analysis

As shown in Table 1 and Figs. 5, 6, 7, 8, 9 and 10, ODS method identifies the front six frequencies and vibration modes of the blade in the 10 kHz frequency range. The first order natural frequency of the blade is 657.8 Hz. The front six modes are in turn first bending, two kinds of first twist, second bending, second twist and bending-twist coupling. The vibration modes of the blade are mainly composed of bending, torsion, bending-twist coupling, which is consistent with the dynamic mechanical properties of blades. It should be noted that the mass distribution of the blade is uneven. So there are two different first order twist modes, the difference of which is the location of the nodal line.

In order to further evaluate the vibration modes quantitatively, modal assurance criteria (MAC) is used to verify the correlation of different order modes. The MAC can not only describe the correlation of two vectors of the same mode shape, but also compare the correlation of different mode shapes. The MAC function of the two modal vectors is defined as follows.

$$MAC(\varphi_r, \varphi_s) = \frac{\left|\varphi_r^T \varphi_s\right|^2}{\left|\varphi_r^T \varphi_r\right|\left|\varphi_s^T \varphi_s\right|} \tag{6}$$

In the formula, φ_r and φ_s represent two modal shape vectors respectively. If the MAC value is close to 1, it indicates that the similarity between the two modal vectors is very high. If the MAC value is close to 0, it is shown that the two modal vectors are orthogonal.

Under the assumption of proportional damping, the mode shapes of different modes have orthogonality in the condition of mass or stiffness matrix weighted.

$$\begin{cases} \varphi_r^T [M] \varphi_s = 0 \\ \varphi_r^T [K] \varphi_s = 0 \end{cases}, r \neq s \tag{7}$$

Assuming that the structural mass is uniform, it is distributed at all measuring points. The modal shapes of different orders are also orthogonal. That is, the MAC value of the vibration modes of different orders is 0.

Based on the above MAC theory, the MAC value between the front six modes of the blade measured by ODS method is calculated, and the result is shown in Fig. 11. According to the graph, the MAC values of the two vectors describing the same mode shape are all 1. The MAC values of two vectors of different mode shapes are all less than or equal to 0.2. Among them, the MAC values of first order and second order, second order and third order, third order and fourth order are slightly higher, which are 0.15, 0.18, 0.20, respectively. The MAC values between other orders are less than 0.1. The actual blade is not completely proportional damping, and the blade mass distribution is uneven. Therefore, the MAC values of different order modes are less than or equal to 0.2 within acceptable limits. The reliability of the modal results obtained by the ODS method is verified.

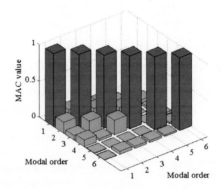

Fig. 11. MAC values of different order vibration modes obtained by ODS analysis

4.2   Test Results of OMA

In the OMA test of the blade, the speed response of each point on the blade is measured one by one under the excitation of the chirp signal. Taking the response of point No. 3 as the reference, the cross power spectrums between the reference signal and the response signal of each measuring point are calculated. Because of the large number of measuring points, in order to display the curves clearly and without losing generality, the amplitude curve of the sum of the cross power spectrums between all the measuring points and the reference points is given in Fig. 12. The least squares complex frequency domain method is used to fit the cross power spectrum. The natural frequencies, mode shapes and damping ratios of the front 6 orders of blades can be obtained, as shown in Table 2. The mode shapes of the blade are shown in Figs. 13, 14, 15, 16, 17 and 18.

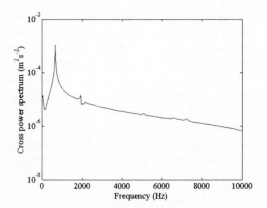

Fig. 12. Cross power spectrum sum amplitude curve between measuring points and reference point

Table 2. Modal parameters of the blade obtained by OMA

Modal order	Frequency (Hz)	Damping ratio	Mode shape description
1	657.7	0.13%	First order bending
2	1930.0	0.49%	First order twist (The nodal line is the connection between measuring point 1 and 34)
3	2130.2	0.94%	First order twist (The nodal line is the connection between measuring point 1 and 35)
4	2874.8	0.50%	Second order bending
5	5105.0	0.47%	Second order twist
6	7222.5	0.44%	Bending-twist coupling

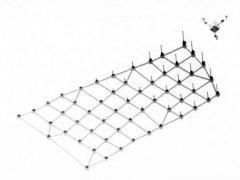

Fig. 13. First vibration mode obtained by OMA

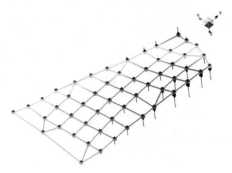

Fig. 14. Second vibration mode obtained by OMA

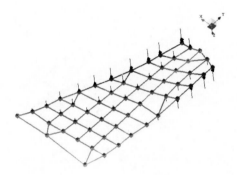

Fig. 15. Third vibration mode obtained by OMA

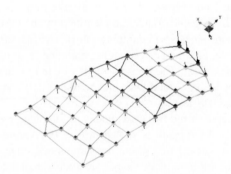

Fig. 16. Fourth vibration mode obtained by OMA

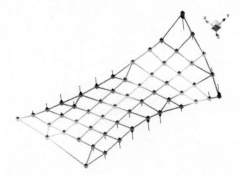

Fig. 17. Fifth vibration mode obtained by OMA

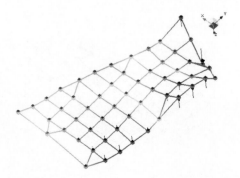

Fig. 18. Sixth vibration mode obtained by OMA

As shown in Table 1 and Figs. 5, 6, 7, 8, 9, 10, the front six modal parameters of the blade in the 10 kHz frequency range are identified by the OMA method, including natural frequencies, mode shapes and damping ratios. The first order natural frequency of the blade is 657.7 Hz, and the corresponding damping ratio is 0.13%, which is the smallest value in the six modes. The front six modes are in turn first bending, two kinds of first twist, second bending, second twist and bending-twist coupling. The results are similar to those of the ODS test.

The MAC value between the front six modes of the blade measured by the OMA method is calculated, and the result is shown in Fig. 19. According to the graph, the MAC values of the two vectors describing the same mode shape are all 1. The MAC values of two vectors of different mode shapes are all less than 0.2. The MAC values of first order and second order, second order and third order, third order and fourth order, first order and fourth order, second order and fourth order are slightly higher, which are 0.17, 0.14, 0.18, 0.14, 0.13, respectively. The MAC values between other orders are less than 0.1. The MAC values of different order modes are less than 0.2 in the acceptable range. The reliability of the modal results obtained by the OMA method is verified.

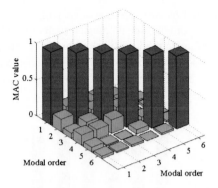

Fig. 19. MAC value of different order vibration modes obtained by OMA

4.3 Comparison and Analysis of ODS and OMA Test Results

The front six modes of the blade in the 10 kHz frequency range are measured by two methods of ODS and OMA. The results of two tests are compared.

Table 3 gives the natural frequencies of the blades measured by ODS and OMA methods. The relative error of the natural frequencies measured by the two methods is calculated. It can be seen from Table 3 that the natural frequencies of the blades identified by the two methods are very close. The relative error of all order natural frequencies is less than 0.5%, of which the third order frequency has the largest relative error, only 0.479%. It fully demonstrates that ODS and OMA have the same ability to identify the natural frequencies of structures.

Table 3. Comparison of the blade natural frequencies obtained by ODS and OMA

Modal order	ODS frequency (Hz)	OMA frequency (Hz)	Relative error
1	657.8	657.7	0.015%
2	1930.6	1930.0	0.031%
3	2120.0	2130.2	0.479%
4	2874.9	2874.8	0.003%
5	5105.9	5105.0	0.018%
6	7237.9	7222.5	0.213%

Through the description of blade vibration modes in Tables 1 and 2, and comparing the vibration modes in Figs. 5, 6, 7, 8, 9, 10 and 13, 14, 15, 16, 17, 18, we can see that the vibration modes measured by the two methods of ODS and OMA are basically the same. In order to further quantitatively describe the correlation of the vibration modes measured by the two methods, the MAC values of the corresponding order modes can be calculated in view of the same distribution of the measuring points in the two methods of ODS and OMA. Figure 20 is the MAC of corresponding mode shapes measured by ODS and OMA methods. The MAC values of the first and second order mode shapes are all greater than 0.99, which indicates that the similarity of the low order mode shapes measured by two methods is very high. The MAC values of mode shapes from third to sixth orders are all greater than 0.8, which indicates that the similarity of the high order mode shapes measured by two methods is relatively high. In summary, the vibration modes obtained by the two test methods are consistent.

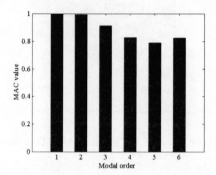

Fig. 20. MAC values of corresponding mode shapes measured by ODS and OMA

Through the comparison of the above test results, it can be seen that the relative error of the front six natural frequencies of the blade measured by the two methods is less than 0.5%. The MAC values of the low order mode shapes are greater than 0.99, and the MAC values of the high order mode shapes are greater than 0.8. The consistency between the two test results is good, which proves the reasonableness of the two test methods and the reliability of the test results.

5   Conclusion

In this paper, the ODS and OMA test systems are built using the high frequency vibration table and scanning laser vibrometer. The ODS and OMA of an engine blade in the 10 kHz frequency range are carried out by the above two test systems. The front six order modal parameters of the blade are measured. By comparing the results of the two test methods, the rationality of the two test methods and the reliability of the test results are proved.

Based on the results of this study, the main conclusions are as follows:

(1) The ODS test system based on the high frequency vibration table and the scanning laser vibrometer can measure the natural frequencies and mode shapes of the blade. The ODS method selects the peak value on ODS FRF to obtain the modal parameters, which avoid the complex frequency response function fitting process. The test and analysis process is simple. However, the ODS method can only get frequency and vibration mode. The damping ratios can not be obtained. It has subjectivity when selecting the peak value.

(2) The OMA test system based on the high frequency vibration table and the scanning laser vibrometer can measure the natural frequencies, mode shapes and damping ratios of the blade. The OMA method obtains the modal parameters by fitting the cross power spectrum between the response points. The results are reliable.

(3) The relative errors of the front six order natural frequencies obtained by the two methods of ODS and OMA are less than 0.5%. The low order mode MAC values are greater than 0.99, and the high order mode MAC values are all greater than 0.8. The two test results are in good agreement. The reliability of modal test results of the engine blade is proved.

(4) The blade has six order modes in the 10 kHz frequency range. The vibration modes are in turn the first order bending, the first order torsion (the nodal line is the connection between measuring point 1 and 34), the first order torsion (the nodal line is the connection between measuring point 1 and 35), the two order bending, the two order torsion, and the bending and torsion coupling.

References

1. Chen G (2006) Analysis of aeroengine structure design. Beijing University of Aeronautics and Astronautics Press, Beijing
2. Troshchenko VT, Prokopenko AV (2000) Fatigue strength of gas turbine compressor blades. Eng Fail Anal 7:209–220
3. Walls DP, Delaneuville RE, Cunningham SE (1997) Damage tolerance based life prediction in gas turbine engine blades under vibratory high cycle fatigue. J Eng Gas Turb Power 119:143–146
4. Naeem M, Singh R, Probert D (1999) Implication of engine deterioration for a high-pressure turbine blades low cycle fatigue (LCF) life-consumption. Int J Fatigue 21:831–847
5. Sun Q, Zhang Z, Chai Q et al (2004) Temperature effect on vibration frequency of aeroengine compressor blade. Chin J Appl Mech 2:137–139
6. Xie YH, Zhang D (2005) Numerical modal for vibration characteristic of steam turbine blade with damped shroud. Proc CSEE 25:86–90
7. Lin MW, Li KH, Yu BF (2013) Research on modal analysis method for rotor blades of axial flow compressors. Agri Equip Veh Eng 51:56–62
8. Fu ZF, Hua HX (2000) Modal analysis theory and application. Shanghai Jiaotong University Press, Shanghai
9. LMS International (2005) The LMS theory and background book - analysis and design, Leuven
10. Peeters B, Guillaume P (2004) Automotive and aerospace applications of the PolyMAX modal parameter estimation method. In: Proceedings of IMAC 26–29 January 2004
11. Dossing O (1988) Structural stroboscopy-measurement of operational deflection shapes. Sound Vib 22:18–26
12. Richarson MH (1997) Is it a mode shape, or an operating deflection shape? Sound Vib 31:54–61
13. Kromulski J, Hojan E (1996) An application of two experimental modal analysis methods for the determination of operational deflection shapes. J Sound Vib 196:429–438
14. James GH, Carne TG, Lauffer JP (1995) The natural excitation technique (NExT) for modal parameter extraction from operating structures. Int J Anal Exp Modal Anal 10:260–277
15. Brincker R (2017) On the application of correlation function matrices in OMA. Mech Syst Signal Process 87:17–22
16. Peeters B, Auweraer HV, Guillaume P et al (2004) The PolyMAX frequency domain method: a new standard for modal parameter estimation. Shock Vib 11:395–409

Application of Ray Tracing Method in Analyzing the Electromagnetic Scattering of Different Nozzles

Xiang Gao[1(✉)], Hong Zhou[1], Qingzhen Yang[2], and Wenjian Deng[1]

[1] AVIC the First Aircraft Design and Research Institute,
Xi'an 710089, People's Republic of China
amosgax@qq.com
[2] Northwestern Polytechnical University,
Xi'an 710072, People's Republic of China
qzyang@nwpu.edu.cn

Abstract. Combining iterative physical optic method (IPO) with equivalent edge currents method (EEC), a code for calculating the radar scattering characteristics of the nozzle was developed. The reliability of it has been validated by calculating a model introduced from a reference. Ray Tracing method was proposed to improve the efficiency of geometric blanking judgment. By comparing with traditional method, RTM can at least make computational efficiency tenfold in geometric blanking calculation. The aerodynamic and electromagnetic scattering characteristics of axially symmetrical nozzles S-shaped nozzle and double S-shaped nozzle were studied. The results show that the code developed is reliable. The performance of the nozzle was evaluated from the thrust, flow coefficient and total pressure recovery coefficient, and then the flow fields of the nozzles were analyzed. The S-shaped nozzle can effectively shorten the high temperature tail flame area, but it will bring about 2% of thrust loss. The electromagnetic scattering characteristics of three nozzles show that the S-bend structure can effectively reduce the RCS of the nozzle. Compared with the axisymmetric nozzle, double S-shaped nozzle can reduce RCS by at least 74.4%.

Keywords: Ray-tracing method · S-shaped nozzle ·
Iterative physical optics method · RCS

1 Introduction

The cavities are common configuration in the aircraft, and they can be a strong source of the scattering echo when the incident electromagnetic wave reflects many times between the inner surfaces. The nozzle, as a kind of cavities, its RCS accounts for a large proportion of the RCS of the aircraft. Therefore, carrying out the investigation of the nozzle's radar scattering characteristics is of great importance to improve the performance of the aircraft.

Iterative physical optics method (IPO) [1, 2] is a high-frequency asymptotic method to analyze the radar scattering characteristics of electrically large cavities. It can simulate the multiple reflections of the electromagnetic wave and can be programmed easily, with which the simulation result is precise. Equivalent Edge Currents Method (EEC) [3–5] is a general method to calculate the edge diffraction field. Combining IPO with EEC, the RCS of both cavities and edges of the nozzle can be calculated accurately. Forward and back-forward IPO method and open MP, MPI parallel computing technology were added to accelerate convergence and reduce computational time. Besides, the Ray Tracing method (RTM) was proposed and applied to improve the efficiency of geometric blanking judgment.

2   Numerical Approach

2.1   Iterative Physical Optics Method

Iterative physical optics method (IPO) proposed by F. Obelleiro-Basteiro et al. is an iterative method based on the physical optic approach which can simulate the multiple reflections of the electromagnetic wave in the Open-ended cavity [1]. With this method, the real electric current on the surface of the conductor is approached by the superposition of the optical current and correction current, and then the electromagnetic scattering characteristics of the target can be obtained [2].

With this method, the real electric current on the surface of the conductor is approached by the superposition of the optical current and correction current, and then the electromagnetic scattering characteristics of the target can be obtained.

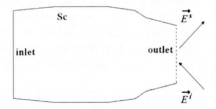

Fig. 1. Sketch map of an Open-ended cavity

The distance between the target and the radar is so far that the incident electromagnetic wave can be treated as uniform plane wave. The electromagnetic wave incidents into an Open-ended cavity which is presented in Fig. 1, and the initial electric current $\vec{J_0}(\vec{r_c})$ can be obtained by physical optic approach.

$$\vec{J_0}(\vec{r_c}) = \begin{cases} 2\vec{n} \times \vec{H_c^i}(\vec{r_c}) & \textit{lit region} \\ 0 & \textit{shadow region} \end{cases} \tag{1.1}$$

Where $\vec{n}$ is the unit surface normal vector pointing into the cavity. $\vec{r_c}$ is the position vector of a certain point on the inner surface Sc. $\vec{H_c^i}(\vec{r_c})$ denotes the incident magnetic field. The real electric current of the inner surface of the cavity can be obtained by the iterative calculation of magnetic field integral equation (MFIE).

$$\vec{J_n}(\vec{r_c}) = \vec{J_0}(\vec{r_c}) + 2\vec{n} \times P.V. \int_{Sc} \vec{J_{n-1}}(\vec{r_c}) \times \nabla G(\vec{r_c} - \vec{r_{c'}})dS \qquad (1.2)$$

Where n denotes iterative times. $P.V. \int$ denotes the principal value of the integral and ∇G is the gradient of the free space Green's function. Based on the irradiation relationship between face elements, the multiple reflections of the electromagnetic wave can be simulated by the iterative calculation of above equation. The forward-backward methodology is combined with relaxation factor $w(0 < w \leq 1)$ to accelerate the convergence and reduce the chance of divergence of the iterative solution. According to the experience, for the general cavity w is $0.8 \sim 0.9$, and for the complicated cavity w is $0.6 \sim 0.7$ [6]. The w is 0.8 during the calculation of current investigation.

2.2   Ray-Tracing Method (RTM)

The irradiation relationship between face elements has to be judged before the induction current is calculated with the IPO. The ray-tracing method is applied in the present investigation to judge the irradiation relationship. The process is as following.

(1) Generate the mesh in the cavity field. The boundary face elements of the mesh are used for IPO and the volume elements of the mesh are used for ray-tracing to judge the irradiation relationship between face elements. IPO has nothing to do with the cavity interior volume mesh, so the coarse volume elements can be generated to increase the efficiency [7]. The mesh of the cavity is shown in Fig. 2.

(2) For face elements A and B, if they can irradiate each other, the following conditions must be satisfied.

$$\vec{r_{21}} \cdot \vec{n_1} > 0 \quad and \quad \vec{r_{21}} \cdot \vec{n_2} < 0 \qquad (1.3)$$

Where $\vec{r_{21}}$ denotes the vector from the center point of B to A. $\vec{n_1}$ and $\vec{n_2}$ are unit normal vector of A and B respectively.

If above conditions cannot be satisfied, the irradiation relationship between A and B does not exist, otherwise the following step is performed.

(3) A radial is assumed to radiate from the element center of A to B. Then, search the volume cells in the direction of the radial one by one and find the boundary face elements of the volume cells intersected with the radial. If any one of the boundary elements is opaque, the irradiation relationship between A and B does not exist, as radial 2 shown in Fig. 3. Otherwise, the irradiation relationship exists, as radial 1 shown in Fig. 3.

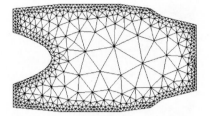

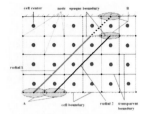

Fig. 2. Sketch of the volume mesh of the cavity

Fig. 3. Illustration of ray-tracing

2.3   Validation of the Code

To validate the accuracy of the program, the RCS for a triangle trihedral corner reflector [8] was simulated. Figure 4 shows the geometry and the mesh of the model, each edge of the triangle is 5λ, where λ is the wavelength. Figure 5 establishes the result of the horizontal polarization of the simulation and the experiment. As shown in Fig. 5, the simulation results agree with that of experiment within the range of the calculation angels. Therefore, the code is reliable and accurate.

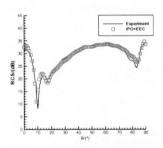

Fig. 4. Triangle trihedral corner reflector

Fig. 5. Horizontal polarization

2.4   The Benefits of RTM

For contrast, the model wall mesh was encrypted with a total of 14110 grids, equivalent to 81 grids per square wavelength. The dimensions of the model are in reference [9]. Because the IPO approach has nothing to do with the internal body grid of the cavity, a coarse internal interconnected grid could be generated to accelerate the ray tracing. The coarse grid in the Table 1 is generated as coarse as possible, as shown in Fig. 6(a); the fine grid is shown in Fig. 6(b). It should be noted that the electromagnetic boundary grids of the two body grids are the same. Therefore, whether the coarse grid or the fine grid is used, the results will be the same calculation precision.

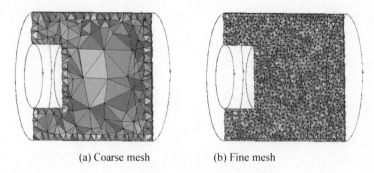

(a) Coarse mesh   (b) Fine mesh

Fig. 6. The section of different mesh

The acceleration ratio using open MP parallel method is 3.87, which is close to the linear acceleration ratio. The geometric blanking calculation using the traditional method, even at linear acceleration ratio, takes at least 10 times the time consuming by RTM using fine grid, which is 25 times by using coarse grid. In general, the geometric blanketing consumes at least the half time of the whole calculation process, so using the RTM method can greatly reduce the geometric blank time, significantly reducing the time-consuming calculation.

Table 1. The time cost of RTM and traditional method

Grid type	RTM			Traditional method	
	Coarse grid	Coarse grid	Fine gird	Fine gird	Fine gird
Thread count	1	4	4	1	Computing time/4
Computing time/s	727	188	479	19382	4832
Time ratio	3.87	1	2.5479	102.8085	25.7021

3 Geometry Model

The electromagnetic scattering characteristics and aerodynamic performance of different nozzles are studied. These nozzles have the same outlet shape, inner duct, external duct and so on. The nozzles models are shown in Fig. 7, the Design parameters of S-shaped nozzle are shown in Table 2. The law of the change of center line and area can be found in the reference [10].

axis symmetry nozzle   double S-shaped nozzle   single S-shaped nozzle

Fig. 7. The sketch of different nozzles

Table 2. Design parameters of S-shaped nozzle

Parameters	Double S-shaped nozzle	S-shaped nozzle
Length/Diameter	2.5	2.5
Law of centerline	Front centerline: Front acute, after slow After centerline: Uniform change	Uniform change
Law of area	Uniform change	Uniform change
Outlet shape	Rectangle, AR = 4	Rectangle, AR = 4
Middle section shape	Super-elliptical shape, a/b = 3:1	/
Offset position	2:3	/
Offset distance/Diameter	−0.3, 0.45	0.15

4   Discussion

CFD software is used for aerodynamic analysis of nozzles, the Thrust coefficient, Total pressure recovery factor and the temperature distribution of nozzles are obtained. Figure 8 is a non-dimensional temperature distribution cloud diagram of the three exhaust systems on the symmetry plane. In the dimensionless temperature T/T_t, T is the local static temperature, T_t is the highest value of the total temperature of the nozzle. In the figure, there is a high-temperature zone near the exit of the nozzle due to the alternating appearance of expansion waves and compression waves. Compared with the three nozzles, it cloud be clearly obtained by the dimensionless temperature distribution. The high temperature region of the S-shaped exhaust system is significantly smaller than the axisymmetric exhaust system. Taking the dimensionless temperature line 0.50 as an example, the length of the high temperature zone is reduced by at least 59%, while the single S-shaped nozzle and the double S-shaped nozzle has the similar high temperature range.

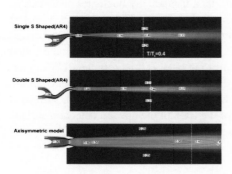

Fig. 8. The temperature distribution of different nozzles

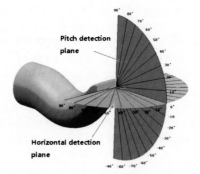

Fig. 9. Schematic diagram of the detection plane

Table 3 gives a comparison of the aerodynamic performance of the three nozzles. Taking the axisymmetric nozzle as the reference, the mass flow rate of the S-shaped nozzle is basically the same as that of the axisymmetric nozzle, and the thrust is lower than that of the axially symmetrical nozzle. Among them, the thrust of the double S-shaped nozzle is the smallest, which is 98% of the axial symmetrical nozzle. The total pressure recovery coefficient of the double S bend nozzle is slightly lower than the other two types of nozzles. This is due to the increase in flow losses in the nozzle pipe brought by the use of the S-bend structure.

Table 3. The aerodynamic characteristics of nozzle

	Axisymmetric nozzle	Double S shaped nozzle	S shaped nozzle
m_i/m_{axi}	1.0	0.997	0.996
F_i/F_{axi}	1.0	0.98	0.981
δ	0.98	0.975	0.977

The electromagnetic scattering characteristics under X-band are numerically studied for three nozzles. The calculated mesh densities are $9 \sim 12$ triangular facets per square wave, the calculated frequency f is 10 GHz, the wavelength λ is 30 mm, the total number of mesh facets is about 160,000–180,000, the residuals convergence criterion is calculated as 0.0001, the maximum The number of iterations is 10, and the calculated angle is $-40° \sim 40°$ on the pitch plane. The interval is $0° \sim 40°$ for the horizontal detection plane, and the angle interval is. The material of the nozzle is a pure metal effective conductor. The detection angle and the detection surface definition are shown in Fig. 9.

Figures 10 and 11 are the RCS angular distribution curves of the three nozzles in the pitch detection plane at the X-band. The curve in the figures clearly shows that single S-shaped nozzle and double S-shaped nozzle can effectively reduce the RCS of the exhaust system in the range of $-20°$ to $20°$. In the figure, the RCS of the axisymmetric nozzle is the largest at the detection angle of $0°$, there is a local increase in the range of $5°$ to $15°$ (or $-5°$ to $-15°$), and then it gradually fluctuates decreases as the detection angle deviates from $0°$. The main reason for this is that the axisymmetric nozzle at the $0°$ detection angle has a strong specular reflection area from the ended wall. When in the detection range of $5° \sim 15°$, the angular reflection between the ended wall and the wall of the nozzle will generate a strong radar echo. From the RCS angular distribution curve of the S-shaped nozzle, it can be seen that due to the presence of the S-shaped structure, the RCS distribution of the nozzle on the pitch detection surface is no longer symmetrical, and the overall level of the RCS of the upward detection surface is smaller than the top-view detection surface.

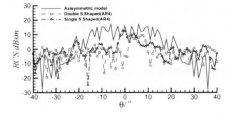

Fig. 10. The RCS distribution of different nozzle under vertical polarization

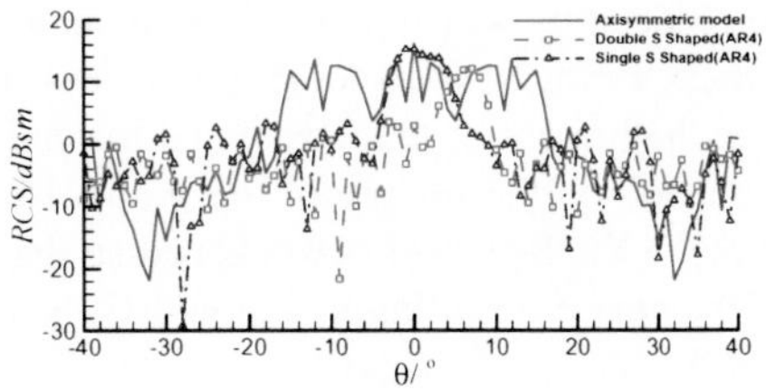

Fig. 11. The RCS distribution of different nozzle under horizontal polarization

In Fig. 12, there is a high induced current density area at the axis-symmetrical nozzle end, which is due to the corner reflection area formed by the terminal wall and the mixing chamber. The figure clearly shows that the double S-shaped nozzle has the smallest induced current density, followed by the single S-shaped nozzle, and the axially symmetrical nozzle has a large induced current density. At the end of the axisymmetric nozzle, there is a large area of induced current due to specular reflection. The S-shaped nozzle and the double S-shaped nozzle do not allow the radar wave to completely reach the terminal due to the effect of S-shaped, so that no large area of induced current is formed. This also indirectly shows that the S-shaped nozzle structure can effectively reduce the RCS of the exhaust system.

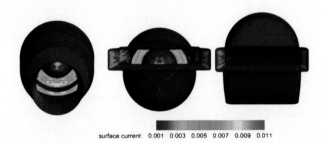

surface current   0.001   0.003   0.005   0.007   0.009   0.011

Fig. 12. Distribution of induced current density in cavities at azimuth angle of 9° under vertical polarization

Comparing the average RCS values of the nozzles of the three structures in the range of –40° to 40°, the average RCS of the double S-shaped nozzle and the single S-shaped nozzle is significantly lower than the average RCS of the axisymmetric nozzle. Compared with the axisymmetric nozzle, double S-shaped nozzle can reduce RCS by at least 74.4% (in square meters), and single S-shaped nozzle can also reduce at least 41.9%. Comparing the RCS mean values in other angle ranges, it was found that the mean RCS of single S-shaped nozzle and double S-shaped nozzle is smaller in both polarization modes. As the angular range decreases, the average RCS of the three nozzles gradually increases. In the detection range of –10° to 10°, the suppression effects of the two S-shaped nozzles are minimized, but the two types of nozzles can still be reduced by at least 70%. This is mainly due to the shielding effect of the S-shaped nozzle. In the upward looking detection plane, the radar wave cannot directly illuminate the interior of the nozzle and the end wall, thereby weakening the effect of mirror reflection on the surface of the probe. However, in the plan view plane, the shielding effect of the S-shaped is weakened so that the radar wave can directly reach the inside of the nozzle, resulting in a large RCS value in a small detection angle range (Tables 4 and 5).

Table 4. The dimensionless RCS average of different detection angle ranges in the vertical polarization mode of the pitch detection plane

Detection angle range	Axisymmetric nozzle	Double S shaped nozzle	S shaped nozzle
$-40° \sim 40°$	0.3709	0.0467	0.0769
$-30° \sim 30°$	0.4895	0.0535	0.0950
$-20° \sim 20°$	0.7017	0.0685	0.1257
$-10° \sim 10°$	1.0000	0.1097	0.2040

Table 5. The dimensionless RCS average of different detection angle ranges in the horizontal polarization mode of the pitch detection plane

Detection angle range	Axisymmetric nozzle	Double S shaped nozzle	S shaped nozzle
$-40° \sim 40°$	0.2219	0.0567	0.1289
$-30° \sim 30°$	0.2887	0.0693	0.1666
$-20° \sim 20°$	0.4250	0.0944	0.2329
$-10° \sim 10°$	0.5594	0.1658	0.4240

Figure 13 and 14 are the angular distribution curves of RCS under the two polarization modes of the three kinds of nozzles in the horizontal detection plane. Compared with the RCS angular distribution curve of the pitch detection plane, the double S-shaped nozzle has the smallest RCS value at the detection angle of 0°. Single S-shaped nozzle does not achieve a complete shield design. The RCS value at this angle is only slightly smaller than the axis symmetrical nozzle. In the range of $-15°$ to 15°, the single S-shaped nozzle has a good reduction effect, but it is still inferior to the double S-shaped nozzle. When the detection angle is greater than 15°, the RCS of the single S-shaped nozzle gradually increases, even exceeding the axisymmetric nozzle, in the vertical polarization mode of the horizontal detection plane. This is mainly due to the fact that the shape of the outlet of the S-shaped nozzle is rectangular, and it is easy to form angular reflection at a large angle of the horizontal detection surface, which will make the radar echo stronger.

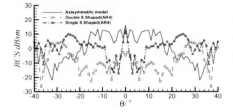

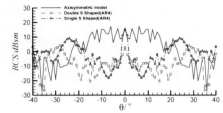

Fig. 13. The RCS distribution of different nozzle under vertical polarization

Fig. 14. The RCS distribution of different nozzle under horizontal polarization

From Tables 6 and 7, it can be obtained that the mean value of RCS of single S-shaped n nozzle and double S-shaped nozzle in both polarization modes is smaller than that of the axisymmetric nozzle. In the vertical polarization mode, the RCS mean value of the double S-shaped nozzle is equivalent to 4.86% of the axial symmetrical nozzle; in the horizontal polarization mode, the average RCS of the double S-shaped nozzle is equivalent to 9.6% of the axial symmetrical nozzle. In summary, compared to axisymmetric nozzles and single S-shaped nozzles, double S-shaped nozzles can greatly reduce the RCS of the exhaust system.

Table 6. The dimensionless RCS average of different detection angle ranges under vertical polarization in horizontal detection plane

Detection angle range	Axisymmetric nozzle	Double S shaped nozzle	S shaped nozzle
$0° \sim 40°$	0.3886	0.0189	0.1043

Table 7. The dimensionless RCS average of different detection angle ranges under horizontal polarization in horizontal detection plane

Detection angle range	Axisymmetric nozzle	Double S shaped nozzle	S shaped nozzle
$0° \sim 40°$	0.2409	0.0232	0.1114

5   Conclusions

(1) A code was developed to calculate the electromagnetic scattering of electrically large cavity based on the iterative physical optics method and equivalent edge currents method. It verified the accuracy and reliability of the program by comparing with the existing experimental data.

(2) Ray Tracing method was proposed to improve the efficiency of geometric blanking judgment. By comparing with traditional method, RTM can at least make computational efficiency tenfold in geometric blanking calculation.

(3) The mean RCS of double S-shaped nozzle and single-S-shaped nozzle is significantly lower than that of axisymmetric nozzle. Compared with axisymmetric nozzle, the mean RCS of double S-shaped nozzle can be reduced by 74.4%, the mean RCS single S-shaped nozzle reduced at least 41.9%. Combining the loss of aerodynamic performance and the gain of electromagnetic scattering performance of different nozzles, there is a balance point that can balance the both.

References

1. Obelleiro-Basteiro F, Rodriguez JL, Burkholder RJ (1995) An iterative physical optics approach for analyzing the elctromagnetic scattering by large open-ended cavities. IEEE Trans Antennas Propag 43(4):356–361
2. Burkholder RJ, Lundin T (2005) Forward-backward iterative physical optics algorithm for computing the RCS of open-ended cavities. IEEE Trans Antennas Propag 53:793–799
3. Ryan C, Peters L (2003) Evaluation of edge-diffracted fields including equivalent currents for the caustic regions. IEEE Trans Antennas Propag 17:292–299
4. Michaeli A (1984) Equivalent edge currents for arbitrary aspects of observation. IEEE Trans Antennas Propag 32:252–258
5. Michaeli A (1986) Elimination of infinities in equivalent edge currents, Part II: physical optics components. IEEE Trans Antennas Propag 34:1034–1037
6. Yan Y, Nie X (2001) An Improved IPO method applied to the analysis of EM scattering from a large open-ended cavity. J Microwaves
7. Chen L, Yang Q, Chen L, Cui J (2012) Numerical simulation of RCS for 2-D convergent nozzle with different trailing edges. J Aerosp Power 27:513–520
8. Polycarpou AC (1995) Radar cross section of trihedral corner reflectors using PO and MEC. Annales Des Télécommunications 50:510–516
9. Yan Y, Ge D (2001) An improved IPO method applied to the analysis of EM scattering from a large open-ended cavity. J Microwaves 35–39
10. Lee C, Boedicker C (1985) Subsonic diffuser design and performance for advanced fighter aircraft. In: Aircraft design systems and operations meeting, pp 69–73

Numerical Study of Reverse-Rotating Wave in the Hollow Rotating Detonation Engines

Xiang-Yang Liu, Yan-Liang Chen, Song-Bai Yao,
and Jian-Ping Wang[(⊠)]

Center for Combustion and Propulsion, CAPT and SKLTCS,
Department of Mechanics and Engineering Science, College of Engineering,
Peking University, Beijing 100871, China
{1801111625, wangjp}@pku.edu.cn

Abstract. This paper adopts the method of injection via an array of holes in three-dimensional numerical simulations of the rotating detonation engines (RDE) with hollow combustor using the premixed stoichiometric hydrogen-air mixture. The calculation is based on the Euler equations coupled with a one-step Arrhenius chemistry model. The array hole injection method is more practical than previous conventional simulations where ideal full area injection method is used. The wave structure of the flow field is composed of obverse-rotating waves (ORWs) propagating clockwise and reverse-rotating waves (RRWs) propagating counterclockwise. ORW is detonation wave (DW) near the outer solid wall while degenerate to shock wave (SW) near the nominally inner wall. This phenomenon is never found in the previse numerical studies.

Keywords: Hollow combustor · Rotating detonation · Reverse-rotating waves

1 Introduction

The rotating detonation engine (CRDE) is a concept engine using detonation as its power source. Experimental researches have been done widely around the world by Wolański et al. [1], Fotia et al. [2], Rankin et al. [3] and Bykovskii et al. [4]. Numerical simulations of RDE with annular chamber were performed by Shao et al. [5], Schwer et al. [6] and Frolov et al. [7].

To resolve the problem of overheating of the inner cylinder in the co-axial annular combustor model, a new model called hollow model was proposed by Tang et al. [8]. Figure 1 shows the schematic of a hollow combustor model of RDE [9]. Various number of detonation fronts was observed under different fuel injection area ratios. To get closer to experimental conditions, Yao's simulation [10] used array holes injection method which was close to actual injection structure. Multiple detonation fronts were observed in his simulation which was consistent with the multi-head experimental results. In the present paper, we perform three-dimensional numerical simulations of a hollow combustor RDE with array hole injectors which is comparable to actual injection method.

X. Zhang (Ed.): APISAT 2018, LNEE 459, pp. 1677–1684, 2019.
https://doi.org/10.1007/978-981-13-3305-7_134

2 Physical and Numerical Models

2.1 Governing Equations and Numerical Method

The three-dimensional Euler equations with source term were used as governing equations:

$$\frac{\partial \tilde{U}}{\partial t} + \frac{\partial \tilde{E}}{\partial \xi} + \frac{\partial \tilde{F}}{\partial \eta} + \frac{\partial \tilde{G}}{\partial \zeta} = \tilde{S},$$ (1)

where the variables are expressed as follows:

$$\tilde{U} = \frac{1}{J}\begin{bmatrix} \rho \\ \rho u \\ \rho v \\ \rho w \\ \rho e \\ \rho Z_1 \end{bmatrix} \quad \tilde{E} = \frac{1}{J}\begin{bmatrix} \rho \bar{U} \\ \rho \bar{U}u + p\xi_x \\ \rho \bar{U}v + p\xi_y \\ \rho \bar{U}w + p\xi_z \\ \bar{U}(p+\rho e) \\ \rho Z_1 \bar{U} \end{bmatrix} \quad \tilde{F} = \frac{1}{J}\begin{bmatrix} \rho \bar{V} \\ \rho \bar{V}u + p\eta_x \\ \rho \bar{V}v + p\eta_y \\ \rho \bar{V}w + p\eta_z \\ \bar{V}(p+\rho e) \\ \rho Z_1 \bar{V} \end{bmatrix} \quad \tilde{G} = \frac{1}{J}\begin{bmatrix} \rho \bar{W} \\ \rho \bar{W}u + p\zeta_x \\ \rho \bar{W}v + p\zeta_y \\ \rho \bar{W}w + p\zeta_z \\ \bar{W}(p+\rho e) \\ \rho Z_1 \bar{W} \end{bmatrix},$$ (2)

$$\tilde{S} = \frac{1}{J}[0,0,0,0,0,\dot{\omega}_1]^T,$$ (3)

$$\bar{U} = u\xi_x + v\xi_y + w\xi_z \quad \bar{V} = u\eta_x + v\eta_y + w\eta_z \quad \bar{W} = u\zeta_x + v\zeta_y + w\zeta_z.$$ (4)

The pressure p and total energy e are calculated through the equation of state

$$p = \rho RT,$$ (5)

$$e = \frac{p}{\rho(\gamma-1)} + \frac{1}{2}(u^2 + v^2 + w^2) + Z_1 q.$$ (6)

The specific heat ratio γ, gas constant R of the mixture and mass production rate $\dot{\omega}_1$ are calculated by the relation

$$R = \sum Z_i R_i \, i = 1, 2,$$ (7)

$$\gamma = \frac{\sum Z_i R_i \gamma_i / (\gamma_i - 1)}{\sum Z_i R_i / (\gamma_i - 1)} i = 1, 2.$$ (8)

The mass production rate of reactant is defined with the one-step kinetics model as

$$\dot{\omega}_1 = \frac{dZ_1}{dt} = -A\rho Z_1 \exp\left(-\frac{T_a}{T}\right).$$ (9)

Fifth-order weighted essentially non-oscillatory (WENO) scheme is used to split the flux vectors. The time integration process is performed by the third-order total-variation-diminishing (TVD) Runge–Kutta method. The one-step chemistry model parameters are the same as those Ma et al. [11] used.

2.2 Physical Model

The outer radius of the hollow chamber is R_{outer} = 6.0 cm. The radius of the inner region is R_{inner} = 3.0 cm (R_{inner} is just a dummy argument because there is no core cylinder in the hollow chamber). The outer region $R_{inner} < R < R_{outer}$ on the head wall is the fuel injection region where the combustible premix gas fed into the chamber by array holes. The structure of the head wall illustrated in Fig. 1. The combustor length is 8.0 cm.

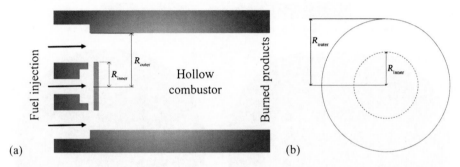

Fig. 1. (a) Schematic of RDE with hollow combustor (b) Schematic of injection surface

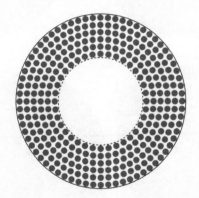

Fig. 2. The array hole injection schematic

The schematic of array hole injection method is shown in Fig. 2. The premixed stoichiometric hydrogen-air mixture enters to the chamber through an array of holes in the head wall. The inner and outer radius of the numerical simulation combustor are 3.0 cm and 6.0 cm, respectively.

3   Results and Discussion

3.1   The Reverse-Rotating Waves in the Flow Field

Figure 3 shows the pressure contours of the case. These figures illustrate that there are five detonation waves (DWs) propagate clockwise along outer wall while the DWs can't reach the inner wall. For instance, the magnified structure of DWs in Fig. 3b (presented in Fig. 3c) shows that a shock wave (SW) portion enclosed in yellow box is connected to peripheral DW enclosed in the red circle. From the contours, it also can be seen that the strength of DWs decreases as radius goes down. This means that the DWs degenerate to shock wave at certain radius position. While for the standard case

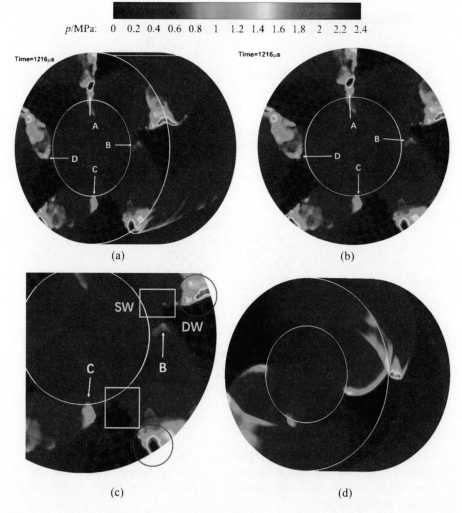

Fig. 3. A/B/C/D represent the four counter-rotating waves in the flow field. (a) Pressure contours of the whole flow field (b) Pressure contours on the head wall.

adopting full area injection method, only curved detonation fronts exist in the flow field just as shown in Fig. 3d. We call this kind of clockwise propagating wave structure represented in Fig. 3 which is composed of DW and SW as obverse-rotating wave (ORW). Four wave structures propagate along the nominally inner wall in the counter-clockwise direction (Label A-D). We call them reverse-rotating waves (RRWs) corresponding to ORWs. The RRW-A collides with a ORW at the time shown in Fig. 3 caused a high pressure and temperature area. RRWs have curved wave front on the inlet plane. But it should be noticed that the RRW has very complicated three-dimensional structure. The red dashed lines in Fig. 4 represent the wave front of RRW. It can be found that the wave phase is different from place to place. So the collision can't happen at the same time for different part of RRW.

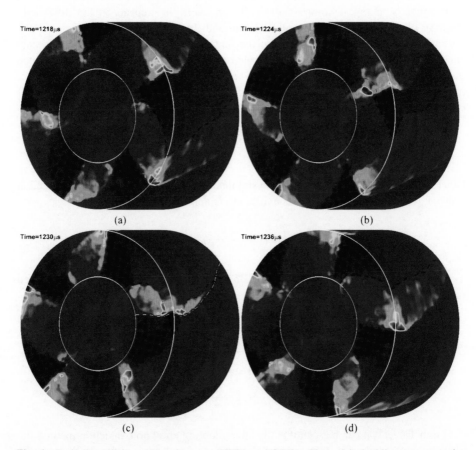

Fig. 4. Periodic collision process between RRWs and ORWs. The red dashed line represents the RRW-C in Fig. 3. Dash dot line in Fig. 4c represents RRW which is composed of DW and attached oblique shock wave.

RRWs and ORWs periodically collision with each other and achieve delicate dynamic equilibrium at the outer region of combustor. Flow field with array hole injectors is much more complicated because of the RRWs. But it has good symmetry. The collision pattern is almost identical at every time. Figure 4 shows the detailed collision process of RRW-C in Fig. 3. We use it to uncover the pattern of wave collision. RRW is about to collide with ORW at around 1218 μs. It's curved in the inlet plane and stretches from inlet plane to outlet plane just as shown by the red dash line on the outer wall. The phase lead part of RRW-3 firstly collides with the shock part of ORW forming high temperature and pressure zone as enclosed by the dash red circle in Fig. 4b. Figure 4c shows the intermediate state of the collision. It can be found that part of the RRW has finished the collision while other part is about to collide. The RRW in Fig. 4d has finished the collision process and prepares to collide with the next ORW. Four counters in Fig. 4 compose the collision cycle that repeats in the flow field. Because of the violent collision of RRW around the inner wall of combustor, DWs decouple near the inner wall and form the fantastic structure of ORWs. RRWs gain energy in the collision process which is released by fuel. So they can be self-sustained in the chamber. The energy is essentially released by the fuel injected in the outer region of combustor ($R_{inner} < R < R_{outer}$). So the RRWs and ORWs are mainly concentrating at the outer region. Only weak compression waves exist in the inner region ($R < R_{inner}$) because no fuel is accumulated in the inner region.

3.2   Propulsive Performance

Other size of combustor chamber is also simulated in our study which is summarized in Table 1. The thrust at the chamber exit plane is defined as

$$F = \oint_{exit} \left(\rho w^2 + p - p_\infty \right) dA.$$

The ambient pressure p_∞ is set to be 0.05 MPa. Fuel-based specific impulse is used for assessment which is defined as

$$I_{sp} = \frac{F}{g \dot{m}_f},$$

where $\dot{m}_f$ is the fuel (hydrogen in present study) mass flow rate

$$\dot{m}_f = \int \rho_f w dA.$$

From Table 1, it can be seen that the after adopting array hole injection method the total thrust of combustor decreases for the reduce of fuel mass fuel rate. But specific impulses for various size chamber don't reduce too much even higher for some cases. So the existence of RRWs doesn't impairing propulsion performance.

Table 1. Thrust performance for different chamber size

Injection method	Inner radius	Outer radius	$\dot{m}_f(kg/s)$	$F(kN)$	$I_{sp}(s)$
Full area	2.0	6.0	0.134	10.10	7691
Array hole	2.0	6.0	0.087	6.79	7964
Full area	3.0	6.0	0.106	7.38	7093
Array hole	3.0	6.0	0.073	5.13	7166
Full area	4.0	6.0	0.074	5.05	6964
Array hole	4.0	6.0	0.054	3.32	6274

4   Conclusions

By considering the more actual injection method of the hollow combustor of RDE, array holes injection method, we carried out the three-dimensional numerical simulations. The conclusion is as follows:

1. The flow field adopting full area injection method presents a typical three-dimensional feature. For the wave structures propagating clockwise (ORW), they're DWs in the periphery of the combustor but degenerate to SWs in the inner part of the outer region. The curved wave fronts adopting the full area injection method disappear and are replaced by the ORW structures which perpendicular to the outer wall of combustor.
2. Besides the detonation waves, four counter-rotating waves propagate along the inner wall which distributing symmetrically to the axis of chamber and are curved in the radial direction. The strength of RRWs gradually increase as the radius goes down and reaching peak values around the inner wall. Periodic collisions happen between ORWs and RRWs at the outer region.
3. The existence of RRWs in the combustor doesn't reduce the thrust performance of RDEs while weak the stability of engines.

References

1. Kindracki J, Wolański P, Gut Z (2011) Experimental research on the rotating detonation in gaseous fuels–oxygen mixtures. Shock Waves 21(2):75–84
2. Fotia ML, Hoke J, Schauer F (2017) Experimental performance scaling of rotating detonation engines operated on gaseous fuels. J Propul Power 33(5):1187–1196
3. Rankin BA, Richardson DR, Caswell AW, Naples A, Hoke J, Schauer F (2015) Imaging of OH* chemiluminescence in an optically accessible nonpremixed rotating detonation engine. In: 53rd AIAA aerospace sciences meeting
4. Bykovskii FA, Zhdan SA, Vedernikov EF (2006) Continuous spin detonations. J Propul Power 22(6):1204–1216
5. Shao Y-T, Liu M, Wang J-P (2010) Numerical investigation of rotating detonation engine propulsive performance. Combust Sci Technol 182(11–12):1586–1597

6. Schwer DA, Corrigan AT, Kailasanath K (2014) Towards efficient, unsteady, three-dimensional rotating detonation engine simulations. In: 52nd AIAA aerospace sciences meeting
7. Frolov SM, Dubrovskii AV, Ivanov VS (2012) Three-dimensional numerical simulation of the operation of the rotating-detonation chamber. Russ J Phys Chem B 6(2):276–288
8. Tang X-M, Wang J-P, Shao Y-T (2015) Three-dimensional numerical investigations of the rotating detonation engine with a hollow combustor. Combust Flame 162(4):997–1008
9. Yao S, Tang X, Luan M, Wang J (2017) Numerical study of hollow rotating detonation engine with different fuel injection area ratios. Proc Combust Inst 36(2):2649–2655
10. Yao S, Han X, Liu Y, Wang J (2016) Numerical study of rotating detonation engine with an array of injection holes. Shock Waves 27(3):467–476
11. Ma F, Choi J-Y, Yang V (2006) Propulsive performance of airbreathing pulse detonation engines. J Propul Power 22(6):1188–1203

The Transient Performance
of FLADE Variable Cycle Engine
During Mode Transition

Hong Zhou[1,3(✉)], Xiang Gao[1], Zhanxue Wang[2], and Wei Zhang[1]

[1] AVIC the First Aircraft Institute, Xi'an 710089, People's Republic of China
[2] Northwestern Polytechnical University,
Xi'an 710072, People's Republic of China
wangzx@nwpu.edu.cn
[3] AVIC the First Aircraft Design and Research Institute,
Xi'an 710089, People's Republic of China
675629936@qq.com

Abstract. Variable cycle engine with FLADE, abbreviation for Fan on Blade, is one of the research focus for future military and civilian aircraft power plants, showing outstanding performance advantages. A FLADE calculation method was established by calculating bypass flow and inner flow independently, to developing a steady-state performance simulation model for double bypass variable cycle engine with FLADE. The dynamic equations that can reflect the rotor inertia effect and component volume effect were added, to developing a transient performance simulation model. The transient characteristics of mode transition were analysed during opening/closing FLADE duct, putting emphasis upon the influence of geometric parameters adjustment and its different combinations. The results indicate that the mode transient characteristics of FLADE variable cycle engine are only influenced by FLADE vane angle and FLADE nozzle area, which should be increased or decreased synchronously. Both of the engine bypass flow and core flow are affected by opening or closing FLADE duct slightly.

Keywords: Variable cycle engine · FLADE · Mode transition ·
Transient characteristics · Geometry adjustment

Nomenclature

θ = inlet guide vanes angle
A = area
π = pressure ratio
B = bypass ratio
$\bar{n}_L$ = low-pressure rotor speed
S = surge margin
Fn = net thrust
Fs = specific net thrust

© Springer Nature Singapore Pte Ltd. 2019
X. Zhang (Ed.): APISAT 2018, LNEE 459, pp. 1685–1695, 2019.
https://doi.org/10.1007/978-981-13-3305-7_135

sfc	= specific fuel consumption
H	= height
Ma	= Mach number
t	= time

1   Introduction

Variable cycle engine (VCE), which could achieve excellent working cycle and low installed drag due to its ability to regulate the flow rate, has become one of the best options for civilian or military aircraft propulsion equipment [1, 2]. Double bypass VCE with FLADE is one of important structural forms, with some promising performance benefits below [3].

- The power of the FLADE is powerful because of larger tip line speed;
- The core engine thermal load can be dissipated by FLADE duct flow surrounding with relative low temperature;
- The greater adjustment ranges of engine inlet flow rate, with less impact on the core engine.

The objective of FLADE VCE mode transition is to adapt specific thrust and fuel consumption for the requirements of different flight phase, by adjusting engine inlet flow. Developing a regulation law which can achieve both of smooth mode transition and excellent performance of FLADE VCE is an important research issue [4–7].

2   FLADE VCE Construction

FLADE is an acronym for "fan on blade", which is characterized by an outer fan driven by a radially inner fan and discharging the air into an outer fan duct [8, 9]. The FLADE VCE instrumentation plane definition is shown in Fig. 1. The numbers shaded indicate the adjustable geometry, including fan inlet guide vane (IGV) angle θ_2, FLADE IGV angle θ_{112}, high pressure compressor (HPC) IGV angle θ_{25}, high pressure turbine (HPT) area A_{41}, low pressure turbine (LPT) area A_{47}, rear variable area bypass injector (RVABI) outer bypass area A_{16}, nozzle throat area A_8, nozzle exit area A_9, FLADE nozzle area A_{118}.

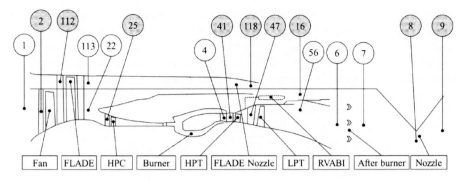

Fig. 1. FLADE VCE instrumentation plane definition

3   Computational Model

3.1   FLADE Part Numerical Simulation Method

Although the work of FLADE and fan interacts, in terms of physical parameters, only the rotor speed is same for both. FLADE is modeled as an independent component and is coaxial with the fan. FLADE inner flow does zero work is assumed, to eliminate the effect on fan inlet flow. That is, FLADE is an annular fan that is coaxial with the fan. Beyond that the calculation of FLADE and conventional fan are similar. A comparison between the FLADE calculation model and the conventional fan calculation model is shown in Fig. 2. X represents the section parameter, $f(X)$ represents the function of fan air compression, and $X21 = X1$ means that the airflow parameters are unchanged during flow through the inner duct.

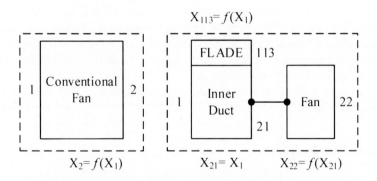

Fig. 2. Schematic diagram of FLADE part calculation model

3.2   FLADE VCE Numerical Simulation Method

Engine performance calculation nonlinear equations reflect the basic physical constraints of flow continuity, power conservation and static pressure balance in the engine operation. In terms of engine performance simulation, FLADE VCE is equivalent to adding a bypass fan to a turbofan. The iterative variables and residual variables of FLADE VCE in double bypass mode is shown in Table 1. The difference between the number of iteration variables and the number of residual variables is the number of control laws, and the control law can be selected from the iteration variables arbitrarily. In the table, the number of iteration variables is 11, and the number of residual variables is 10, which means the number of control variable is 1. When the FLADE duct is closed, the nonlinear equilibrium equations of FLADE VCE are the same as those of dual-rotor turbofan engine, with 9 iterations variables and 8 residual variables.

Table 1. Iteration and residual variables of FLADE VCE (in double bypass mode)

Part class	Part number	Iteration variable	Residual variables
Inlet	1	β	——
Compressor	3	β	Mass flow
Splitter	2	Bypass ratio	——
Burner	1	Exit temperature	——
Turbine	2	β	Mass flow
Mixer	1	——	Static pressure
Nozzle	2	——	Mass flow
Rotation shaft	2	Speed	Power
Total	14	11	10

The residual is the difference between the same variable calculated by two difference methods. For example, compressor mass flow residual is the difference between mass flow of upstream component outlet (after bleeding) and mass flow read from compressor characteristic map (before bleeding), and so on. β is a dimensionless mathematical variable, which can stand for the ratio of pressure ratio in reading compressor map, or mass flow ratio in reading turbine map. The iteration variable of burner can choose from the exit temperature, fuel consumption, or excess air coefficient. For each compressor, there will be an iterative variable of β, corresponding to a residual variable of mass flow, and so on.

The transition-state calculation method considering rotating shaft inertia effect and vessel volume effect was modeled [10].

4   Numerical Results

Once the FLADE operating point reaches the surge line, or the FLADE vane is closed fully ($\theta_{112} = 90°$), or the FLADE pressure ratio drops to 1.0, the FLADE duct will be treated as closed completely. There is no air flow through the FLADE duct. The following principles should be meet during mode transition:

• The change of fan, HPC operating parameters should be mild;
• The surge margin of FLADE, fan and HPC should be sufficient.

Table 2. Initial geometry of transition from double to single bypass

$\bar{n}_L/\%$	$\theta_{112}/°$	$\theta_{25}/°$	$A_{47}/\%$	$\Delta A_{16}/\%$	A_8/m^2	A_{118}/m^2
90	0	0	100	0	0.24	0.07

4.1   Transition from Double to Single Bypass Mode

Flight conditions is altitude $H = 11$ km, the flight Mach number $Ma_0 = 0.9$, the mode transition time $t = 2$ s assumed. The initial geometry of transition from double to single bypass is shown in Table 2. $\bar{n}_L$ represents the low pressure rotor speed.

Table 3. Initial geometry of transition from single to double bypass

$\bar{n}_L/\%$	$\theta_{112}/°$	$\theta_{25}/°$	$A_{47}/\%$	$\Delta A_{16}/\%$	A_8/m^2	A_{118}/m^2
90	90	0	100	0	0.24	0.058

With a special low-pressure rotor speed, the FLADE duct mass flow gradually decreases as the FLADE VCE transitions from double to single bypass mode. Both FLADE vane angle and FLADE nozzle area can affect FALDE duct flow, so FLADE vane angle, FLADE nozzle area and LPT guide area were selected to form the following three different regulation rules:

- Law 1: FLADE IGV angle θ_{112} and FLADE nozzle area A_{118} alternative variation, see Fig. 3(a);
- Law 2: FLADE IGV angle θ_{112} and FLADE nozzle area A_{118} synchronization variation, see Fig. 3(b);
- Law 3: FLADE IGV angle θ_{112} and FLADE nozzle area A_{118} synchronization variation, with LPT IGV area A_{47} gradually enlarges, see Fig. 3(b), (c).

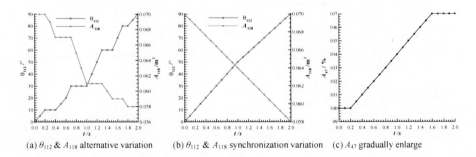

(a) θ_{112} & A_{118} alternative variation   (b) θ_{112} & A_{118} synchronization variation   (c) A_{47} gradually enlarge

Fig. 3. Geometry regulation law of transition from double to single bypass

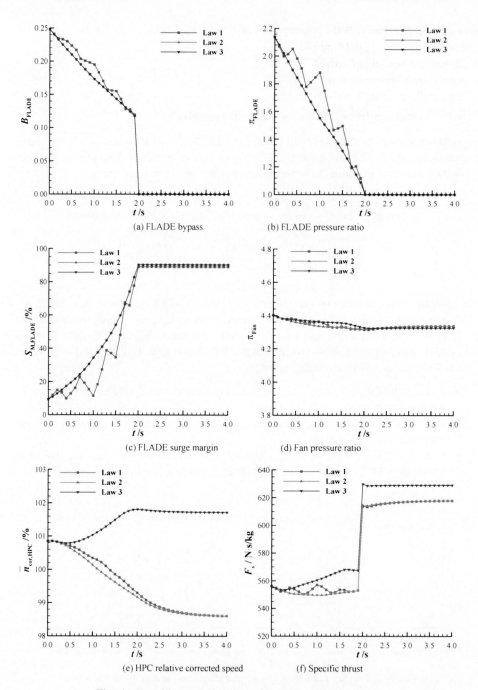

(a) FLADE bypass

(b) FLADE pressure ratio

(c) FLADE surge margin

(d) Fan pressure ratio

(e) HPC relative corrected speed

(f) Specific thrust

Fig. 4. Transition mapping data of double to single bypass

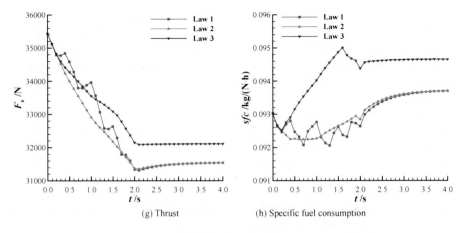

(g) Thrust   (h) Specific fuel consumption

Fig. 4. (*continued*)

The operating parameters of the FLADE VCE transition from double to single bypass mode is shown in Fig. 4, with the regulation laws above.

If θ_{112} and A_{118} are alternative variation (Law 1 in Fig. 4), when θ_{112} or A_{118} is closing, the FLADE bypass ratio B_{FLADE} is reduced, and HPC relative corrected speed $\bar{n}_{cor,HPC}$ is slightly reduced. Closing θ_{112} decreases the FLADE bypass ratio π_{FLADE} and increases the FLADE surge margin $S_{M,\ FLADE}$; while closing A_{118} increases π_{FLADE} and decreases $S_{M,\ FLADE}$. It is advisable to close θ_{112} first, and then close A_{118}, to ensure that FLADE has sufficient surge margin. When θ_{112} is closed to 90°, π_{FLADE} drops to 1.0, which means that the FLADE duct is completely closed, so B_{FLADE} drops to zero suddenly, the FLADE surge margin does not make sense. The adjustment of θ_{112} and A_{118} has less influence on the fan, mainly because of the small π_{FLADE}, changes in FLADE duct flow have less effect on the engine core and bypass flow Especially when θ_{112} is close to 90°, π_{FLADE} is close to 1.0, so even if B_{FLADE} suddenly drops to zero, it will not cause severe fluctuations in the operating parameters of the engine core and bypass flow. Engine thrust F_n increases with increasing π_{FLADE}, and vice versa. When B_{FLADE} drops to zero, the engine specific thrust F_s suddenly increases due to the sudden decrease of the engine inlet flow, which achieves the purpose of switching the FLADE VCE from the double bypass mode to the single bypass mode.

If θ_{112} and A_{118} are synchronously adjusted (Law 2 in Fig. 4), the fluctuation of engine parameters caused by alternate adjustment of θ_{112} and A_{118} can be eliminated. In short, the smooth transition of the FLADE VCE from double bypass mode to single bypass mode can be achieved by closing θ_{112} and A_{118}.

If θ_{112} and A_{118} are synchronously adjusted and A_{47} is gradually enlarged (Law 3 in Fig. 4), $\overline{n}_{\text{cor,HPC}}$ can be increased, so that the FLADE VCE has greater F_n and F_s in single bypass mode, and the specific fuel consumption sfc also increases. The coincidence of Law 2 curve and Law 3 curve in Fig. 4(a), (b) and (c) indicates that the adjustment of A_{47} has almost no effect on FLADE. Therefore, the adjustment of A_{47} is not necessary during FLADE VCE mode transition.

Compared to rotor shaft acceleration/deceleration or turning on/off afterburner, the variation of shaft speed, the pressure ratio of components except for FLADE, etc., is relatively small, when the FLADE VCE transforms from double bypass mode to single bypass mode. Under the condition of choosing a reasonable mode transition starting point, over temperature and over rotation will not occur during FLADE VCE mode transition. In addition, when the starting point $\overline{n}_{\text{L}}$ is different, the characteristics of the mode transition are basically the same, so the mode transition characteristics at different $\overline{n}_{\text{L}}$ are not analyzed in detail in this paper.

4.2   Transition from Single to Double Bypass Mode

The initial geometry parameter of transition from single to double bypass is shown in Table 3. The start point of single bypass mode below in Table 3 is the same as the end point of double bypass mode above.

When the FLADE VCE transforms from single to double bypass mode, the reverse regulation law in Fig. 3 is adopted, and LPT IGV area A_{47} is unchanged, that is, alternately opening θ_{112} and $A_{118,}$ or simultaneously opening θ_{112} and A_{118}. The operating parameters of transition from single to double bypass mode is shown in Fig. 5.

If θ_{112} and A_{118} are opened alternately (Law 1 in Fig. 5), when θ_{112} is turned on from $90°$, airflow is allowed to pass through the FLADE duct, causing the B_{FLADE} to increase rapidly from zero. Since π_{FLADE} still approaches 1.0, the increase of mass flow causing F_n to increase gradually, but F_s to decrease rapidly. Before the transition, the FLADE duct is closed, therefore $S_{\text{M, FLADE}}$ in Fig. 5(c) does not make sense at $t = 0$. In the early stages of transition, FLADE usually has sufficient surge margin because θ_{112} is still close to $90°$. In addition, π_{FLADE} will increase with the opening of θ_{112}, and decrease with the opening of A_{118}. In order to ensure the operation of FLADE at the beginning of the transition, it is preferable to first open θ_{112} and then open A_{118}. In the later stages of transition, it is not advisable to open θ_{112} too quickly in order to prevent $S_{\text{M, FLADE}}$ from falling too fast.

If θ_{112} and A_{118} are opened simultaneously (Law 2 in Fig. 5), the transition will be smoother. During the mode transition, the influence of fan inner and outer flow is still very small, $\overline{n}_{\text{cor,HPC}}$ is increased slightly, and sfc is increased due to the increase of the engine total bypass ratio.

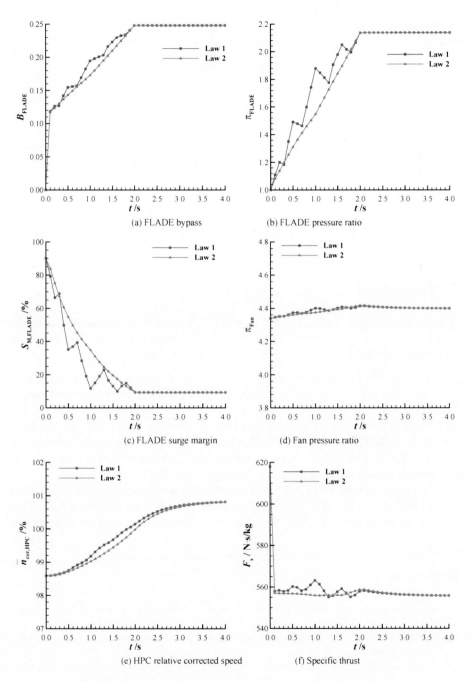

Fig. 5. Transition mapping data of single to double bypass mode

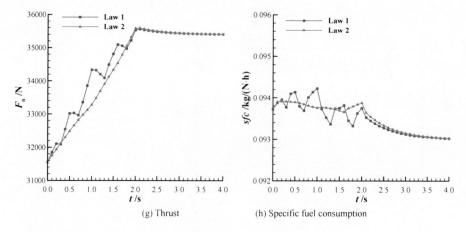

(g) Thrust

(h) Specific fuel consumption

Fig. 5. (*continued*)

5   Conclusions

A numerical simulation model of the FLADE VCE transient performance was established, to analyze its transient characteristic during mode transition. The following conclusions can be drawn.

(1)   The FLADE VCE mode transition process is only determined by the FLADE IGV angle and the FLADE nozzle exit area, with a special low-pressure rotor speed.

(2)   When opening the FLADE duct, the FLADE IGV should be opened, and the FLADE nozzle exit area should be increased simultaneously, and vice versa.

(3)   If the FLADE IGV angle and the FLADE nozzle exit area are adjusted alternately, the FLADE IGV should always be adjusted first, whether transition from single to double bypass mode or transition from double to single bypass mode.

(4)   Although a switch of FLADE duct would bring a sudden change of flow rate in FLADE duct, the other operating parameters would not fluctuate wildly. That means the switch of FLADE duct has less influences on fan duct or core engine.

References

1. Special Report (2005) Soaring ambitions – future of offensive air systems [EB/OL]. Jane's Defence Weekly, London
2. Jennings G (2010) USAF issues next-generation fighter request [EB/OL]. Jane's Defence Weekly, London
3. Rallabhandi SK, Mavris DN (2008) simultaneous airframe and propulsion cycle optimization for supersonic aircraft design, AIAA 2008-143, pp 1–29. AIAA, Reston
4. French MW, Allen GL (1981) NASA VCE test bed engine aerodynamic performance characteristics and test results, AIAA 1981-1594, pp 1–9. AIAA, Reston

5. Vdoviak JW, Knott PR, Ebacker JJ (1981) Aerodynamic/Acoustic performance of YJ101/ double bypass VCE with coannular plug nozzle. NASA CR-159869, pp 1–45, 347–356. NASA, Washington
6. Rock SM, De Hoff RL (1979) Variable Cycle Engine multivariable control synthesis interim report- control structure definition. AFAPL-TR-79-2043, pp 1–23. AFAPL
7. Przybylko SJ, Rock SM (1982) Evaluation of a multivariable control design on a variable cycle engine simulation. AIAA 1982-1077, pp 1–8. AIAA, Reston
8. Johnson JE (1995) Spillage drag and infrared reducing FLADE engine, US 5404713, 11 April 1995
9. Wadia AR (2010) FLADE fan with different inner and outer airfoil stagger angles at a shroud therebetween, US 7758303, 20 September 2010
10. Sellers JF, Daniele CJ (1975) DYNGEN - a program for calculating steady-state and transient performance of turbojet and turbofan engines. NASA TN D-7901. NASA, Washington, 24 April 1975

Numerical Simulation of Bird Strike on a S-Shaped Stealth Inlet

Kun-yang Li$^{(\boxtimes)}$, Xiang-hua Jiang, and Da-sheng Wei

School of Energy and Power Engineering, BUAA, Beijing, China
likunyang94@qq.com

Abstract. Bird strike is a great threat to the safety of aircraft. The bird's collision with the aero engine may lead to engine power loss, fires. Large blade debris in high-speed produced by the collision could even cause the disaster of aircraft crush. In this paper, the simulations of the bird striking a S-shaped inlet and blades were carried out to study the influences of the inlet to the blade's capability of crashworthiness. Besides, the influences of bird's incidence pitch angles to the bird strike inlet was studied. In the simulation, the SPH method and the linear Mie–Grüneisen equation of state were used in ABAQUS/Explicit to describe the bird's fluid-like behavior during the bird strike events. The SPH method and the material model were verified by simulating the bird strike on rigid plate. Compared the simulation of the bird strike on fan blades with and without a S-shaped inlet, which could not only weaken the bird's after-impact kinetic energy but also make the bird crush into pieces. Both of them can enable the engine fan blades withstand a heavier bird impact.

Keywords: Bird strike · S-shaped inlet · Incidence pitch angles ·
Crashworthiness of blade

1 Introduction

Modern aircraft planes are at risk of colliding with birds. Report [1] indicated that although most bird strikes collided with small birds did not cause structural failures or casualties, serious bird accidents caused by large birds collided with critical aircraft components, such as aero-engines, can always lead to economic damage and even heavy casualties. According to the report, 71% of the bird collisions occurred below 500 ft. Most of the bird strike cases occurred during the take-off or landing phases. The military fighter aeroplane may have a greater probability of bird strike when flying at a low altitude [2].

In order to improve the survivability, military fighter jets usually adopt special geometric shapes and stealth materials to reduce the radar reflection cross section (RCS) of the aircraft [3]. Usually the fan blade of the fighter is a strong scattering sources, the S-shaped inlet can be used to shield the fan blades to reduce the reflected electromagnetic intensity. When a bird strikes a military fighter, the bird body may collide with the inlet before it hits the fan blades, which would avert the bird's movement state or damage the inlet structure. The change of the bird's movement may prevent the blades from damage during bird strike. Therefore, the dynamic response of a bird strike on the S-shaped inlet is simulated in this paper by using finite element software ABAQUS/Explicit.

© Springer Nature Singapore Pte Ltd. 2019
X. Zhang (Ed.): APISAT 2018, LNEE 459, pp. 1696–1706, 2019.
https://doi.org/10.1007/978-981-13-3305-7_136

In the finite element simulation of the bird strike, traditionally there are three numerical methods to accurately simulate the fluid-like behavior of birds: Lagrangian method, Eulerian method, and smooth particle hydrodynamic method (SPH). Each of them has different advantages and disadvantages [4]: the Lagrangian method uses Lagrangian element for bird body. The disadvantage is that the bird element will produce serious mesh distortion during the simulation; In Eulerian method, the bird is modeled as a fluid-like body with Eulerian elements, which can avoid the mesh distortion, but require a much better fined mesh and computational cost; The SPH algorithm is a meshless method based on the Lagrangian technique, in which the finite elements are replaced by a set of discrete, interactive particles. This method has no mesh distortion problem and the computational cost is less than Eulerian method, while there is no tensile stress in the model, which could incur an easier destruction when subjecting to a strong impact. In this paper, the SPH method is employed for the simulation of bird strike.

This paper simulated the bird strike on a rigid plate to validate the SPH method and the constitutive model of the bird. It also simulated that the bird strikes blade directly and bird strikes inlet and blade successively to analyze the influences of the inlet. Moreover, the simulation of bird strike on a S-shaped inlet with different bird's incidence pitch angles were carried out to study the incidence angles' influences to the impact procedure.

2 Bird Strike Model

2.1 Bird Model

There are three substitute bird geometric models which are usually used in the simulation of bird strike [5] (Fig. 1): cylinder, cylinder with hemispherical ends, ellipsoid. The length-to-diameter aspect ratio is usually 2. In the experiment, because of the complexity of the real bird's structure compared with the substituted bird model of the simulation [6], there are some differences in the local pressure distribution, but the results of average pressure are similar. The result of the cylinder with hemispherical ends model is the closest to the experimental result, so the cylinder with hemispherical ends model is often used as an alternative model in the bird strike simulation. [7] The bird mass is 1.8 kg in the simulation. The bird model established by the SPH method is shown in Fig. 2.

Fig. 1. Three substitute bird impactor geometries

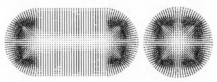

Fig. 2. Bird's SPH model in the simulation

The structure of a real bird body is complex, and the mechanical properties of the bird body material usually are anisotropic and non-homogeneous at a lower speed impact. However, as the impact speed increasing, the effect of this anisotropy and inhomogeneity decreases, while the phenomenon of rheology gradually emerged. When the speed is high enough, the bird body material can be regarded as a compressible fluid matter. Therefore, the classical constitutive equation of the homogeneous fluid can be used to describe the bird material:

$$\sigma_{ij} = -P\delta_{ij} + 2\rho\gamma\dot{e}_{ij} \tag{2.1}$$

In order to describe the fluid's hydrodynamic response accurately, the equation of state (EOS) of the bird materials is employed. The Murnaghan equation of state, the Mie–Grüneisen equation of state, and the polynomial equation of state are commonly used in recent paper. This paper uses the linear Mie–Grüneisen equation of state, also known as the Us-Up equation of state:

$$P_H = \frac{\rho_0 c_0^2 \eta}{(1 - s\eta)^2} \tag{2.2}$$

$$\eta = 1 - \frac{\rho}{\rho_0} \tag{2.3}$$

$$P = P_H(1 - \Gamma_0\eta/2) + \Gamma_0\rho_0 E_m \tag{2.4}$$

In which, c_0 is the sound velocity, c_0 and s define the linear relationship between the shock velocity u_s and the shock front particle velocity u_p:

$$u_s = c_0 + su_p \tag{2.5}$$

The material parameters of the bird [6]: $c_0 = 1480\,\text{m/s}$, $s = 1.92$, $\Gamma_0 = 0.1$, $\rho_0 = 942\,\text{kg/m}^3$, which are close to the water EOS parameters. At the same time, the bird material has also set a tensile cutoff pressure of 50 MPa [8].

The bird material model and the bird strike algorithm was verified by comparing the central pressure-time curve with the experimental results of a bird strike on a rigid plate. The widely used experiment data of this kind of experiment is the Wilbeck's [9] test data. According to the bird strike theory, the pressure at the contact center reaches a peak value P_H in a short period of time after the initial contact, and then decreases to a steady flow pressure P_S:

$$P_H = \rho_0 u_s u_p \tag{2.6}$$

$$P_S = \frac{1}{2}\rho_0 u_p^2 \tag{2.7}$$

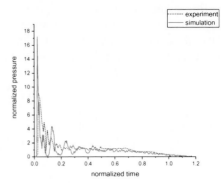

Fig. 3. Bird strike on rigid plate

Fig. 4. Central pressure of bird strike on a rigid plate

Figure 3 is the simulation of the bird strike on a rigid plate to verify the SPH method and the bird material model, in which the bird body perpendicularly impacts on the rigid plate at a velocity of 170 m/s. Figure 4 compares the center pressure of the SPH method's simulation with Wilbeck's experimental results. The theoretical values of P_H and P_S are 289.3 MPa and 13.1 MPa, respectively, and the normalized value are 22.1 and 1. As shown in this figure, the simulation results of the central steady pressure is similar to the experimental results; the peak pressure of simulation and experiment are all lower than the theoretical values, but the simulation's peak value is close to the theoretical value.

2.2 Inlet and Fan Blade Material Model

During the impact of bird strike on the inlet and blade, it is a transient process with high strain rate and large deformation, thus the material model needs to consider the influence of strain rate. This paper uses the widely-used Johnson-Cook constitutive and damage models, which parameters can be easily obtained by experiments.

The Johnson-Cook constitutive model considers the effects of strain hardening, strain rate hardening, and thermal softening on the relationship of equivalent plastic strain $\bar{\varepsilon}^p$ and stress $\bar{\sigma}$, and the formula is:

$$\bar{\sigma} = [A + B(\bar{\varepsilon}^p)^n]\left[1 + C\ln\frac{\dot{\bar{\varepsilon}}^p}{\dot{\bar{\varepsilon}}_0^p}\right]\left[1 - \left(\frac{T - T_{\text{room}}}{T_{\text{melt}} - T_{\text{room}}}\right)^m\right] \tag{2.8}$$

In which, A, B, C, n, m are the material constant, $\dot{\bar{\varepsilon}}^p$ is the equivalent plastic strain rate, $\dot{\bar{\varepsilon}}_0^p$ is the reference strain rate (usually 1), T_{melt} and T_{room} are the melting temperature and room temperature of the material, respectively.

The failure model includes two phase, the damage initiation and the damage evolution. In the simulation, the damage initiation is based on the damage parameter ω of the element integration point. When the damage parameter exceeds 1, the damage is considered to have been initiated, the damage parameter ω is defined as:

$$\omega = \sum \frac{\Delta \bar{\varepsilon}^p}{\bar{\varepsilon}_f^p} \tag{2.9}$$

$$\bar{\varepsilon}_f^p = [d_1 + d_2 \exp(-d_3 \chi)] \left[1 + d_4 \ln \frac{\dot{\bar{\varepsilon}}^p}{\dot{\bar{\varepsilon}}_0}\right] \left[1 + d_5 \frac{T - T_{\text{room}}}{T_{\text{melt}} - T_{\text{room}}}\right] \tag{2.10}$$

In which, χ is the ratio of hydrostatic pressure to equivalent stress: $\chi = -\frac{p}{\bar{\sigma}}$.

After the damage initiated, a material degradation model that considers the degradation of yield strength and the elastic strength, which is proportional to $1 - d$, was employed in the damage evolution phase. For the overall damage variable d, the linear damage form based on the fracture dissipation energy G_f is used in this paper:

$$\dot{d} = \frac{L \dot{\bar{\varepsilon}}^p}{\bar{u}_f^p} = \frac{\dot{\bar{u}}^p}{\bar{u}_f^p} \tag{2.11}$$

$$\bar{u}_f^p = \frac{2 G_f}{\sigma_{y0}} \tag{2.12}$$

σ_{y0} is the yield stress at the onset of damage. In some researches [10], the G_f of 50 mJ/mm^2 is employed.

The materials used for the fan and the inlet in the simulation are titanium alloy Ti-6Al-4V and aluminum alloy 2024-T3, respectively. The material parameters of Ti-6Al-4V and 2024-T3 are shown in Table 1.

Table 1. Material parameters of titanium alloy Ti-6Al-4V and aluminum alloy 2024-T3 [11]

	$\rho(\text{kg/m}^3)$	$E(\text{GPa})$	v	$T_{\text{melt}}(^\circ\text{C})$	$T_{\text{room}}(^\circ\text{C})$
Ti-6Al-4V	4440	110	0.34	1668	20
2024-T3	2850	73.1	0.33	650	20
	$A(\text{MPa})$	$B(\text{MPa})$	n	C	m
Ti-6Al-4V	1098	1092	0.93	0.14	1.1
2024-T3	369	684	0.73	0.0083	1.7
	d_1	d_2	d_3	d_4	d_5
Ti-6Al-4V	−0.09	0.27	0.48	0.014	3.87
2024-T3	0.112	0.123	1.5	0.007	0

2.3   Computational Model

When the bird impact blade, the blade rotates at a high speed (e.g., 6000 r/min). Therefore, the initial stress state of the blade should be considered in the bird strike simulation, and the stress distribution of the blade obtained by implicit static analysis is shown as Fig. 5. After the implicit analysis, the stress and strain of the blade are imported into the subsequent explicit dynamics analysis of the bird strike blade as the initial stress state of blade, which was shown in Fig. 6. The blade has an initial rotation speed of 6000 r/min, and the blade root has a constant rotation speed. The bird mass is 1.8 kg, has an initial moving speed of 100 m/s, and will impact at 75% height of the blade.

Fig. 5. Stress distribution of a rotating blade   **Fig. 6.** Initial state of the bird strike blade

In the simulation of the bird strike inlet, a kind of S-shaped ventral inlet was employed [3], as shown in Fig. 7, the entrance of inlet section is a rectangular on the left side of the figure, the exit section of inlet is a circular.

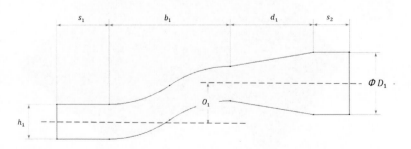

Fig. 7. Schematic of the S-shaped ventral inlet

3 Simulation Result

3.1 Bird Strikes Blades Directly

Figure 8 shows the impact procedure of a bird striking blades. The blade cuts the bird body and the slice of bird slides on the blade. Large deformation and plastic strain of the blade is produced in the blade during the impact. Failure elements can be found at the impact center of the blade leading edge. Figure 9 shows the equivalent plastic strain of the blade at the end of the simulation ($t = 3\,ms$). The deformation of the blade mainly occurs at the leading edge of impact zone, and there are a few failure elements around the center of the impact zone.

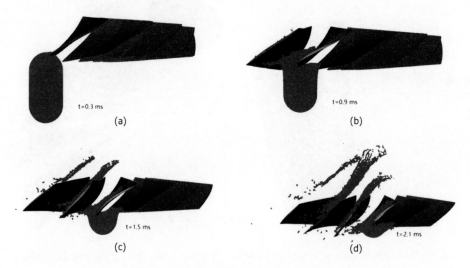

Fig. 8. Bird strike blade simulation results

Fig. 9. The equivalent plastic strain of blade after impact

3.2   Bird Strikes a S-Shaped Inlet

Figure 10 illustrates that the bird strikes a S-shaped inlet at the speed of 100 m/s, in which the interaction of airflow with bird body was not considered. As depicted in the figure, the bird will collide with the inlet twice during it passing through. When the bird enters horizontally, it will contact with the lower surface of the inlet firstly, and the velocity of the bird's SPH particles will decrease. Then it will continue to move along the tangential direction of the impact point until the bird collides with the upper surface of the expansion section of the inlet. After this two collisions, the bird body separate and the bird's total kinetic energy gets lower.

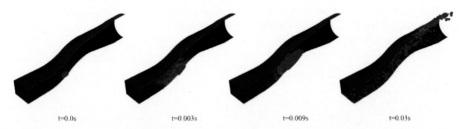

t=0.0s   t=0.003s   t=0.009s   t=0.03s

Fig. 10. Simulation results of bird strike S-shaped inlet

Figure 11 shows that the bird particles continue to hit the fan blades after striking the S-shaped inlet, and the bird's SPH particles are greatly separated after they hit the inlet. Figure 12 shows the equivalent plastic strain of the blade after the bird strike. The main contact area of the SPH particle to the blade is at the front tip of the blade, and the deformed area is smaller than Fig. 9 due to the separation of the particles and the lower kinetic energy after the impact with the inlet.

t=0.01s   t=0.015s

Fig. 11. Bird strikes the blade after hitting the inlet

Fig. 12. Equivalent plastic strain of the blade after bird strike inlet and blade

The incident pitch angles of the bird (Fig. 13) will influence the bird's movements when passing through the inlet, Fig. 14 illustrates that, (a) the bird velocity vs time with different pitch angles, (b) the bird's exiting-inlet velocity vs pitch angle, (c) the height of central impact point at the blade. The figure shows that the incident pitch angle of the bird and the shape of the inlet can affect the impact procedure. With the pitch angle changes from oblique downward to oblique upward, the final velocity first increases and then decreases. Oblique downward incidence will cause two collisions of bird with the inlet wall which will consume more bird's kinetic energy. The bird impact with $-30°$ pitch angle will consume about 51% bird's kinetic energy. While the bird impacts at a specific angle ($10°$) which was decided by the inlet geometric shape, the bird body would not collide with the inlet wall. According to the Fig. 14(a), when the relative vertical velocity of the bird respect to the inlet is getting greater, the kinetic energy of bird consumed by the collision becomes greater.

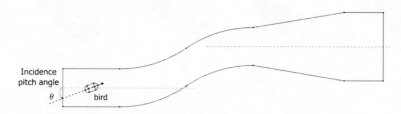

Fig. 13. Schematic of bird's incidence pitch angle

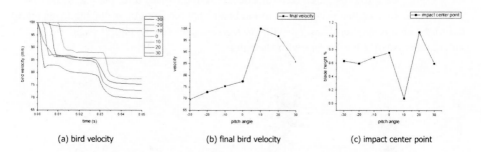

(a) bird velocity   (b) final bird velocity   (c) impact center point

Fig. 14. Bird velocity and impact center point vs incident pitch angle

3.3   Crashworthiness of the Blade Under Bird Strike

In order to evaluate the influences of bird-to-inlet strike to the subsequent bird-to-blade strike, the studies of blade's capability of crashworthiness were carried out under the two conditions in which bird strikes blade directly and bird strikes inlet and blade successively. The capability of crashworthiness is characterized by the blade's critical bird mass. And the critical bird mass is a maximum allowed bird mass, over which the blade cannot withstand and will cause failure.

Critical bird mass can be searched by an algorithm of dichotomy. First, define a range of bird mass, in which there would be no blade failure by bird strike with the minimum bird mass, and should have blade failure by bird strike with the maximum bird mass. Then, simulate the bird strike by a bird with intermediate mass and check the failure status of the blade by examining the damage dissipation energy of the blade, and then get a new maximum or minimum bird mass. Narrow down the bird mass range and repeat the above calculations, until the maximum and minimum mass difference is less than a given tolerance, then the critical bird mass at a given speed is found.

In this case, the critical bird mass of directly strike to blade is found to be 0.99 kg, and the critical bird mass of firstly hits the inlet is found to be 1.91 kg. Simulation results show that the inlet can reduce the velocity of the bird body, disperse the bird body and improve the blade's crashworthiness.

Figure 15 shows the initiation's damage parameter of the blade after critical bird mass impact in both cases. When the damage parameter of the elements reaches 1, the damage initiate and this element is considered to be failed at this time. Compared with Fig. 15(a), the failure element area of Fig. 15(b) is much smaller because of the separation of particles and the reduction of bird speed after the bird strike inlet. Thus, it can be considered that the inlet can block the bird body and improve the blade's crashworthiness.

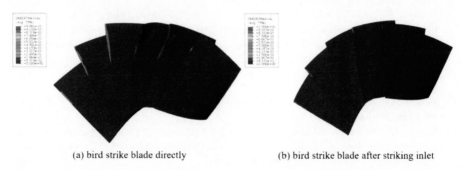

(a) bird strike blade directly   (b) bird strike blade after striking inlet

Fig. 15. The blade damage parameter of (a) the bird strike blade directly (b) the bird strike inlet and blade successively

4 Conclusion

The SPH algorithm can accurately predict the contact pressure of the bird strike, simulate the bird's movement status, and predict the deformation and the failure of the inlet and blade when the bird strikes inlet and blades. The simulation of the bird strike inlet shows that the S-shaped inlet will consume the bird's kinetic energy and make the bird crush into pieces during the collision. Thus, the inlet can improve the crashworthiness of the engine fan blades. The incidence pitch angle of impact bird could also affect the collision procedure through changing the relative velocity of the bird respect to the inlet. When the bird impacts at a specific angle (10°) determined by the geometry of the inlet, although the bird does not collide with the inlet, the bird will strike at the root of the blade, thereby improving the crashworthiness of the blade.

This study shows that the S-shaped inlet has the potential to allow the blade to withstand the heavy bird impact, which could be useful in the aircraft design.

References

1. Dolbeer RA, Weller JR, Anderson AL et al (2016) Wildlife Strikes to Civil Aircraft in the United States, 1990–2015. FAA National Wildlife Strike Database, Serial Report Number 22, November 2016
2. Heimbs S (2011) Computational methods for bird strike simulations: a review. Comput Struct 89(23–24):2093–2112
3. Li QP, Wang HP (2009) Research on electromagnetic scattering characteristics and radar cross section reduction of ventral inlet. Aeronaut Comput Tech 39(6):67–70
4. Ryabov AA, Romanov VI (2007) Fan blade bird strike analysis using Lagrangian, SPH and ALE approaches. In: 6th European LS-DYNA users' conference, pp 79–88
5. Budgey R (2000) The development of a substitute artificial bird by the international Bird strike Research Group for use in aircraft component testing. International Bird Strike Committee ISBC25/WP-IE3. International Bird Strike Committee, Amsterdam
6. Hedayati R, Madighi M, Mohammadi-Aghdam M (2014) On the difference of pressure readings from the numerical, experimental and theoretical results in different bird strike studies. Aerosp Sci Technol 32(1):260–266
7. Hedayati R, Ziaei-Rad S (2013) A new bird model and the effect of bird geometry in impacts from various orientations. Aerosp Sci Technol 28(1):9–20
8. Siddens A, Bayandor J (2013) Multidisciplinary impact damage prognosis methodology for hybrid structural propulsion systems. Comput Struct 122(2):178–191
9. Wilbeck JS (1978) Impact behavior of low strength projectiles. Wright-Patterson Air Force Base
10. Verleysen P, Peirs J (2017) Quasi-static and high strain rate fracture behaviour of Ti6Al4V. Int J Impact Eng 108:370–388
11. Kay G (2003) Failure Modeling of Titanium6Al4V and 2024-T3 Aluminum with the Johnson-Cook Material Model. Technical report DOT/FAA/AR-03/57, US Department of Transportation, Federal Aviation Administration

An Experimental Study
on Reducing Depositing on Turbine Vanes
with Transverse Trenches

Zhengang Liu, Fei Zhang, and Zhenxia Liu[(✉)]

School of Power and Energy, Northwestern Polytechnical University,
Xi'an, Shaanxi, China
zxliu@nwpu.edu.cn

Abstract. Particles depositing is a severe damage to turbine vanes. The depositing on upper surface for the plate models with film cooling configuration is experimentally studied for the attack angle $-5°$ with a kind of wax, which is melted and atomized to generate particles. Some models are trenched along the row of film cooling holes to study the effects of different trenches with different depths on the depositing. All the trenches could distribute the depositing on the near downstream of the row of film cooling holes more uniformly and the deeper trench could make the depositing on this area more uniform. When the blowing ratio is 0.98, all the trenches could decrease the depositing on the upper surface and the depositing mass decreases with the trench depth increasing. When the blowing ratio is 1.47, all the trenches increase the depositing on the upper surface although the increment is not vary large and the depositing mass increases with the trench depth increasing. However, even if the blowing ratio is 1.47, the trench may reduce the negative effect of the depositing on turbine vanes since it could delay the depositing to the area far from the row of film cooling holes.

Keywords: Depositing · Trench · Multiphase flow

1 Introduction

The particles such as sands and salts may be melted when they go through combustion chamber and may deposit on turbine vanes [1, 2]. This depositing is a severe damage to turbine vanes if it becomes too thick since it may increase the surface roughness and decrease the aerodynamic performance of turbine vanes [3] and the power of turbine [4]. The depositing also could decrease the cooling efficiency of turbine vanes [3]. In some cases, film cooling holes may be blocked by the depositing or some other dusts, turbine vanes could burn out [5]. Therefore, the depositing and how to prevent it are of interest in engineering.

Wenglarz and Fox experimentally studied deposition of particles with coal-water fuels and found that the particles depositing increases with increasing gas and surface temperature [6]. Crosby et al. experimentally studied effects of temperature and particle size on deposition in land based turbines with the TADF (Turbine Accelerated Deposition Facility) and they found that the depositing rate decreases with decreasing gas temperature and increasing mass flow of cooling air [7]. The surface roughness due

© Springer Nature Singapore Pte Ltd. 2019
X. Zhang (Ed.): APISAT 2018, LNEE 459, pp. 1707–1716, 2019.
https://doi.org/10.1007/978-981-13-3305-7_137

to deposition was also found to decrease with decreasing gas temperature and increasing cooling air [7]. Combined with the viscosity model suggested by Senior and Srinivasachar [8], Sreedharan and Tafti built a depositing model [9], which could be applied to numerically simulate depositing of particles. In this model, the depositing probability for a particle on the surface is related to its viscosity, which depends on its temperature. This depositing model seems to work well [9]. Yang and Zhu also used this model to numerically study the particles depositing inside turbine cascade [10].

The techniques to reduce the harm of depositing to turbine vanes have also been developed and investigated. One of these techniques is arranging transverse trenches on the surface of airfoil or turbine vane. Albert and Bogard experimentally studied the influence of transverse trench on particles deposition on turbine vane pressure surface and they found that the trench improves the cooling efficiency [3]. They also found that the particles depositing becomes thinner downstream the cooling holes due to the trench compared the case without trench [3]. However, they did not change the trench dimensions. Lawson and Thole studied the influence of trench on the particles depositing on endwall film cooling holes with experiments and found that the trench could reduce the negative effect of particles depositing on the cooling efficiency [11]. Vighneswara et al. also found the cooling efficiency improvement with surface trenches; however, they did not investigate the effect of the trenches on the particles depositing [12].

In this paper, the experiment is designed to study the effects of different transverse trenches with different depth on the depositing on the pressure surface of the model. As conducted in Ref. [3], a kind of wax is melted and atomized to generate the particles. All the models are made of plates with the same dimensions and have a film cooling configuration. There are totally 4 models, one of which has no trench on its upper surface, the others of which are arranged with transverse trench along the row of film cooling holes.

2 Experimental System and Parameters

In this section, the experimental system and procedures are firstly introduced and then the testing models are introduced. Finally, the main testing parameters are listed.

2.1 Experimental System

The tests are conducted in a wind tunnel, in which, however, the flow could be heated. Figure 1 shows the experimental system and procedures. The experimental mainly consists of three sub-systems: wind tunnel, wax spray system and cooling air system. Air is driven into the wind tunnel and heated by a heater to generate flow with higher temperature than ambient temperature. The flow then goes through turbulence grids and becomes turbulent. A hollow cylinder is installed downstream of the turbulence grids and on it a sprayer is installed. In the hollow cylinder, three pipes are arranged to connect the sprayer and the wax spray system, which will be introduced later in detailed. As mentioned later, the wax reaching the sprayer is melted and atomized in the sprayer by compressed air to generate wax particles, which are initially molten.

Therefore, the flow becomes a multiphase flow downstream of the hollow cylinder and the wax mass flow rate is controlled carefully such that the wax particles are dilute in the flow, which is a working condition for turbine vanes. The wax particles are driven by the flow, and some may reach and deposit on the test model, which is located in the test section of the wind tunnel. An infrared camera is used to record the temperature on the surface of the model during testing. It should be noted that an infrared window is installed to close the test section, as shown in Fig. 1(a). However, in this paper, the temperature is not focused on. The wax particles not depositing are transferred outside of the wind tunnel. The depositing mass on the model will be measured with an analytical balance.

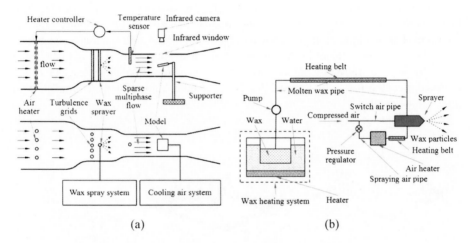

Fig. 1. (a) Experimental system and (b) wax spray system

The wax spray system is displayed in Fig. 1(b). The wax is firstly melted and to stabilize the temperature of wax, a heater firstly heats the water in a container and then the water heats the wax in the wax container surrounded by the water. The molten wax is pumped through the molten wax pipe to the sprayer and atomized by the compressed air reaching the sprayer to generate molten wax particles. The molten wax pipe, compressed air pipe and sprayer are wrapped in the heating belts and some adiabatic material, as shown in Fig. 1(b), to make the wax maintain molten in the wax spray system and facilitate restarting the system by preheating the wax left in the pipes last time.

The cooling air system provides cooling air for the test model, which has a film cooling configuration, as introduced in the following section. The cooling air is generated by evaporating liquid nitrogen and therefore is actually nitrogen. Nitrogen has similar physical properties to air and it is accordingly rational to substitute nitrogen for air.

The significant sources of uncertainty are summarized here. As mentioned above, the particles depositing mass is weighed with an analytical balance, i.e. METTLER TOLEDO XPE 206DR, whose measure range and repeatability are 220 g and 0.01 mg, respectively. The analytical balance is also used to measure the wax mass flow rate and the relative uncertainty of wax mass flow rate is 2.31% when wax mass flow rate is

10 g/min. A flowmeter is used for reading the flow rate of cooling air in the cooling air system and its relative uncertainty is 1.13%, indicating that the relative uncertainty of cooling air flow rate is about 1.13%.

2.2   The Testing Model and Testing Parameters

In this paper, two kinds of models are employed, as shown in Figs. 2(a) and (b), and both models are made based on the same plates and have film cooling configuration, as shown in Fig. 2(c). The models are 137 mm long, 150 mm wide and 13 mm thick and have the leading and trailing edges with the same diameter 13 mm. Totally 15 film cooling holes with the same diameter 1.5 mm are arranged with back inclination angle 45° along the spanwise direction on the models. The distance (i.e. L_{hole} in Fig. 2(a)) between the leading edge and the center line of the row of film cooling holes is 38.5 mm. The first kind of model, shown in Fig. 2(a), does not have a transverse trench and called smoothed model in this paper, while the second kind of model, shown in Fig. 2(b), have a transverse trench along the row of film cooling holes and called trenched model. It is should be noted that there are three kinds of trenched models with three trench depths.

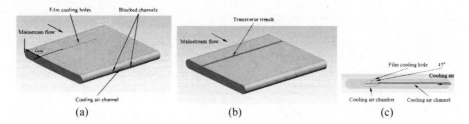

Fig. 2. (a) The smoothed model, (b) the trenched model and (c) the film cooling configuration

The film cooling configuration is shown in Fig. 2(c). The cooling air (actually nitrogen, as mentioned above) is transferred into the cooling air chamber through the cooling air channel and injected into the mainstream flow. Originally, three cooling air channels are designed and made; however, one channel is enough based on the testing and two channels are blocked in all tests, as shown in Fig. 2(a).

As introduced above, there are totally 4 models, one of which is smoothed model, the left of which are trenched models with different trench depths. The different models are named in Table 1 for the convenience of discussing the results. It should be noted that D is the diameter of the film cooling hole. For example, the $2.0D \times 0.50D$ model has the trench with width 2.0 D and depth $0.50D$.

Table 1. The configurations and names of models

No.	Trench width	Trench depth	Model name
1	—	—	Smoothed model
2	2.0D	0.50D	2.0D × 0.50D model
3	2.0D	0.75D	2.0D × 0.75D model
4	2.0D	1.00D	2.0D × 1.00D model

The main testing parameters are listed in Table 2. The attack angle α is $-5°$ and therefore the upper surface of the model is the pressure surface. The mainstream velocity u_∞ is 10 m/s with turbulence intensity I_∞ about 5% and the mainstream temperature T_∞ is 40 °C. The cooling air temperature T_c is -3 °C. The blowing ratio M is defined by $M = u_c\rho_c/(u_\infty\rho_\infty)$ with u_∞, ρ_∞, u_c and ρ_c being the mainstream velocity, mainstream density, cooling air velocity and cooling air density, respectively. The blowing ratio is 0.98 and 1.47, respectively. The melting temperature $T_{p,m}$ and density ρ_p of the wax used in the tests is 44 °C and 880 kg/m^3, and the wax mass rate $\dot{m}_p$ is 10 g/min. The molten wax is atomized to generate particles, whose size is mainly 10–20 μm. The Stokes number Stk is defined by $Stk = \rho_p d_p^2 u_\infty/(18\mu_\infty L)$ with d_p, μ_∞ and L being the particle diameter, mainstream viscosity, and reference length, respectively. The distance between the sprayer and the model is about 10 times length of the model and the corresponding Stk is about 0.002–0.007, which means that the wax particles could diffuse in the mainstream flow uniformly enough before they reach the model.

Table 2. The main experimental parameters

α (°)	u_∞ (m/s)	T_∞(°C)	I_∞(%)	M	$T_{p,m}$ (°C)	$\dot{m}_p$ (g/min)	T_c (°C)
-5	10	40	5.0	0.98, 1.47	44	10	-3

3   Results and Discussion

The tests are conducted for 40s and the effects of different trenches on the depositing on the upper surface (i.e. pressure surface) of the model are focused on and discussed in this paper. The effects of the trenches on the depositing distribution are firstly discussed. Figure 3 shows the depositing distribution on the upper surface for the four models when the blowing ratio is 0.98. It could be firstly found that the depositing mainly form on the part of the upper surface downstream of the row of film cooling holes and this allows to separately measure the depositing mass on the leading edge and the upper surface of the model. As shown in Fig. 3(a), for the smoothed model the depositing firstly forms on the near downstream of the film cooling hole and no visible depositing could be found on the near downstream of the space between the two film cooling holes. However, the depositing becomes thicker on the slightly far downstream of the space between the two film cooling holes. This pattern is very similar to that in Ref. [3]. On the middle and rear part of the upper surface, the depositing becomes thinner in the streamwise direction and more uniform in the spanwise direction. The depositing distribution pattern on the most parts for the 2.0D × 0.50D model is very similar to that for the smoothed model, as shown in Fig. 3(b). However, some different details should be noted. Firstly, for the $2.0D \times 0.50D$ model, the depositing forms on the near downstream of both the film cooling holes and the space between the two holes. Secondly, the depositing on the near downstream of the row of film cooling holes becomes more uniform compare to that for the smoothed model. This effect of trench on the depositing was also found experimentally by Albert and Bogard [3], however, they did not study the influence of trench depth on the depositing. If there is very thick depositing at one location, then the temperature at this location may be higher than that at other locations according to the temperature measure in the testing. Therefore the more uniform depositing is beneficial to reducing the negative influence of the depositing on turbine vanes. The depositing pattern for the $2.0D \times 0.75D$ and $2.0D \times 1.00D$ models as shown in Figs. 3(c) and (d) is similar to that for the $2.0D \times 0.50D$. One important feature that should be noted by comparing Figs. 3(b), (c) and (d) is that the depositing on the near downstream becomes more uniform with the trench depth increasing.

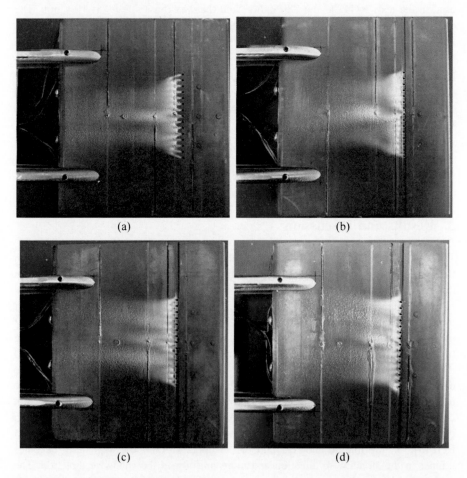

(a)

(b)

(c)

(d)

Fig. 3. The depositing distribution on the upper surface of (a) the smoothed model, (b) the $2.0D \times 0.50D$ model, (c) the $2.0D \times 0.75D$ model and (d) $2.0D \times 1.00D$ model when $M = 0.98$

The depositing distribution on the upper surface for the blowing ratio 1.47 is displayed in Fig. 4. It is visible that the depositing for each model when $M = 1.47$ is very similar to that for the same model when $M = 0.98$. Therefore, in current experimental conditions, increasing the blowing ratio has little influence on the depositing distribution pattern. It should be noted by comparing Figs. 4(b), (c) and (d) that the depositing on the near downstream also becomes more uniform with the trench depth increasing. This feature is the same as that for $M = 0.98$.

The effects of the trenches on the depositing distribution pattern have been discussed above and in the following parts of this section the depositing mass is investigated and discussed. The capture efficiency β is firstly defined by $\beta = m_{dep}/(\dot{m}_p \times t)$, where m_{dep} is the depositing mass, $\dot{m}_p$ is the wax mass flow rate (see Table 2) and t is the testing time. From the definition of capture efficiency, it can be seen that more capture efficiency means more depositing mass in the same testing conditions and time.

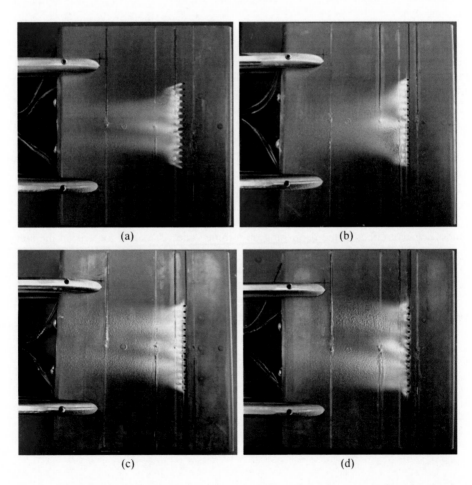

(a)   (b)

(c)   (d)

Fig. 4. The depositing distribution on the upper surface of (a) the smoothed model, (b) the $2.0D \times 0.50D$ model, (c) the $2.0D \times 0.75D$ model and (d) $2.0D \times 1.00D$ model when $M = 1.47$

As mentioned above, it is allowed to measure the depositing mass on the upper surface. The capture efficiency on the upper surface calculated based on the depositing mass for the models when $M = 0.98$ and $M = 1.47$ is given in Fig. 5, where the data marked by "Ref" means the capture efficiency of the smoothed model. When the blowing ratio is 0.98, the capture efficiency for the smoothed model is 0.82%. Under the same blowing ratio, the capture efficiency for the $2.0D \times 0.50D$, $2.0D \times 0.75D$ and $2.0D \times 1.00D$ models is 0.77%, 0.74 and 0.39%, respectively, and the relative increment with respect to the capture efficiency for the smoothed model is -6.1%, -9.8% and -52.4%, respectively. Basically, it can be found from Fig. 5(a) that the depositing mass decreases with the trench depth increasing and the trench with depth $1.0D$ has remarkable effect on reducing the depositing mass on the upper surface. As discussed above, the deepest trench could make the depositing on the near downstream of the row of film cooling holes most uniform and accordingly the deepest trench may be most beneficial to reducing the negative effect of the depositing on turbine vanes.

Figure 5(b) shows the capture efficiency when $M = 1.47$ and its feature is a little different from the case when $M = 0.98$ shown in Fig. 5(a). When $M = 1.47$, the capture efficiency on the upper surface for the smoothed model is also about 0.82%, indicating the blowing ratio has a little influence on the depositing. The capture efficiency for the $2.0D \times 0.50D$, $2.0D \times 0.75D$ and $2.0D \times 1.00D$ models is 0.87%, 0.90 and 0.93%, respectively, and the relative increment with respect to the capture efficiency for the smoothed model is 6.1%, 9.8% and 13.4%, respectively. Therefore, the capture efficiency increases with the trench depth increasing, although the increase amount is not remarkable. However, it cannot be deduced that the deeper has less effect on reducing the negative influence of the depositing on turbine vanes. It can be found by comparing Figs. 4(b) and (c) that more depositing forms on the near downstream of the row of film cooling holes for the $2.0D \times 0.50D$ model than the $2.0D \times 0.75D$ and less depositing forms on the middle and rear part of the upper surface for the former than the latter. The most negative effect of the depositing on turbine vanes is mainly located on the area near the film cooling holes and therefore it is better to delay the depositing to the area far from the film cooling film even if the total depositing mass increases a little. Accordingly, the trench with the depth $0.75D$ may be better for reducing the negative effect of the depositing. Similarly, it can also be deduced by comparing Figs. 4(c) and (d) that the trench with the depth $1.00D$ may be better than the trench $0.75D$.

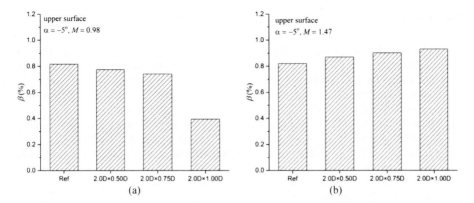

Fig. 5. The capture efficiency on the upper surface for the models when (a) $M = 0.98$ and (b) $M = 1.47$

4   Conclusions

The effects of different transverse trenches on the depositing distribution and mass on the upper surface (i.e. pressure surface) of the model are experimentally studied for the attack angle $-5°$ with a kind of wax. All the models have a film cooling configuration, but some models further are trenched along the row of film cooling holes. In the tests, the wax is melted and atomized to generate particles.

The effects of the trenches on the depositing distribution are firstly discussed. When $M = 0.98$, all the trenches could distribute the depositing on the near downstream of the row of film cooling holes more uniformly compared to that for the smoothed model. This is also valid when $M = 1.47$. The depositing on the near downstream of the row of film cooling holes becomes more uniform with the trench depth increasing both when $M = 0.98$ and $M = 1.47$. These influences are beneficial to reducing the negative effect of the depositing on the turbine vanes.

The effects of the trenches on the depositing mass are also discussed. When $M = 0.98$, all the trenches could decrease the depositing mass on the upper surface compared to that for the smoothed model and the depositing mass decreases with the trench depth increasing. When $M = 1.47$, all the trenches could increase the depositing mass on the upper surface and the depositing mass increase with the trench depth increasing. However, when $M = 1.47$, as similar to the case when $M = 0.98$, the deeper trench may be better for reducing the negative effect of the depositing on the turbine vanes since it may delay more depositing to the area far from the row of film cooling holes.

References

1. Webb J et al (2011) Coal ash deposition on nozzle guide vanes: part i – experimental characteristics of four coal ash types. In: Proceedings of ASME Turbo Expo 2011, Vancouver, British Columbia, Canada
2. Chambers JG (1982) The volcanic cloud encounter of a Rolls-Royce powered Boeing 747 of the British airways fleet 24 June 1982. Internal Rolls-Royce Report
3. Albert JE, Bogard DG (2013) Experimental simulation of contaminant deposition on a film-cooled turbine vane pressure side with a trench. ASME J Turbomach 135:051008
4. Wenglarz RA (1992) An approach for evaluation of gas turbine deposition. ASME J Eng Gas Turbines Power 114:230–234
5. Kim J et al (1993) Deposition of volcanic materials in the hot sections of two gas turbine engines. ASME J Eng Gas Turbines Power 115:641–651
6. Wenglarz RA, Fox RG (1990) Physical aspects of deposition from coal-water fuels under gas turbine conditions. ASME J Eng Gas Turbines Power 112:9–14
7. Crosby JM et al (2008) Effects of temperature and particle size on deposition in land based turbines. ASME J Eng Gas Turbines Power 130:51503
8. Senior SL, Srinivasachar S (1995) Viscosity of ash particles in combustion systems for prediction of particle sticking. Energy Fuels 9:277–283
9. Sreedharan SS, Tafti DK (2011) Composition dependent model for the prediction of syngas ash deposition in turbine gas hotpath. Int J Heat and Fluid Flow 32:201–211
10. Yang XJ, Zhu JX (2017) Numerical simulation of particle deposition process inside turbine cascade. Acta Aeronaut et Astronaut Sinica 38:120530
11. Lawson SA, Thole KA (2011) Simulations of multi-phase particle deposition on endwall film-cooling holes in transverse trenches. In: Proceedings of ASME Turbo Expo 2011, Vancouver, British Columbia, Canada
12. Vighneswara RK et al (2013) Enhanced film cooling effectiveness with surface trenches. In: Proceedings of ASME Turbo Expo 2013, San Antonio, Texas, USA

Numerical Study on the Influence of the Trailing Edge Overflow Holes on the Flow and Heat Transfer of the Inner Cooling Passage on the Trailing Edge of the Turbine Blade

Shun Zhao, Guanghua Zheng[(⊠)], and Chengcheng Hui

School of Power and Energy,
Northwestern Polytechnical University, Xi'an, China
zhengguanghua@nwpu.edu.cn

Abstract. In this paper, the influence of overflow outlet at the trailing edge of the turbine blade on the flow and heat transfer in the trailing edge cooling passage is studied by numerical simulation method. The influence of the number of overflow holes, overflow hole radius and overflow contact area of the overflow hole (change the number of holes and hole radius, and unchange the flow cross-sectional area) on the distribution of flow coefficient of the trailing edge overflow hole were compared. And the influence of the flow distribution from the trailing edge and the lower edge plate, the influence on the heat transfer of the inner cooling wall ware compared. The results show that the number, radius, and flow contact area of the trailing edge overflow hole have large influence on the distribution of the flow coefficient of the trailing edge overflow hole. The decrease of the number of holes, the increase of the hole radius, and the reduction of the flow contact area all make the flow coefficient of the edge overflow hole rises; Increase of the number of hole and hole radius both increase the outflow ratio at the trailing edge and increase the average heat transfer coefficient of the inner cooling passage. The overflow contact area of the overflow hole has little influence on the overflow outflow ratio and the average heat transfer coefficient.

Keywords: Overflow hole of trailing edge · Cooling passage ·
Flow coefficient · Flow distribution · Heat transfer coefficient ·
Numerical simulation

1 Preface

In order to improve the heat transfer of the turbine blade trailing edge, different types of spoiler are usually installed inside the inner cooling passage. In addition, the outflow from the trailing edge is also an important factor affecting the flow and heat transfer of the inner cooling passage. The common outflow types of the turbine blade trailing edge include squid outflow, overflow hole of trailing edge outflow, and upper and lower edge plate outflow.

© Springer Nature Singapore Pte Ltd. 2019
X. Zhang (Ed.): APISAT 2018, LNEE 459, pp. 1717–1729, 2019.
https://doi.org/10.1007/978-981-13-3305-7_138

The cooling air in the inner cooling passage flows out from the trailing edge overflow hole and cools the area behind the outlet. The cooling mechanism is the same as the film cooling. Huiren et al. [1] studied the effect of the geometric shape of the film hole, the Reynolds number of the secondary flow, and the blow ratio on the flow coefficient. The hole shapes used are sigma-shaped holes, conical holes and cylindrical holes; The experimental parameters range from the secondary flow Reynolds number Re = 10000 to 25000, the secondary flow blow ratio Ma = 0.3 to 2.0, and 26 cases in the above range were selected. The experimental results show that the blow ratio is an important parameter affecting the flow coefficients of various hole shapes, and the secondary flow Reynolds number has less influence on the flow coefficient; the conical hole has the largest flow coefficient. The second is the hole-shaped hole, and the smallest is the cylindrical hole. Lu [2] studied the heat transfer characteristics of the crescent-shaped hole, the slot-shaped hole and the funnel-shaped hole downstream. It was found that crescent-shaped holes and slot-shaped holes continuously ejected out-flow between adjacent holes, and the air-conditioning coverage was wide. The funnel-shaped hole has a higher cooling efficiency downstream of the orifice, and a lower efficiency along the blade height than the other two. Yu [9] studied the heat transfer characteristics of the convergent hole and compared it with a circular hole. Compared with the circular film hole, the cold air that converges the slit hole is more easily pressed by the mainstream to form a better film layer, and the thickness of the film layer is thicker as the blowing ratio and the Reynolds number increase. With stronger extension ability, it can better isolate the mainstream and improve the film cooling efficiency. Zhi et al. [10] used numerical simulation methods to study the effect of outflow of air film holes on the heat transfer characteristics of the flow passages in the blade. He focused on the influence of air hole spacing on the heat transfer character-istics. The results show that the increase of the film hole spacing ratio will increase the heat transfer coefficient of the film hole edge significantly, and the film hole outflow will increase the heat transfer coefficient of the film hole surface. Wenchao et al. [8] used the smooth cooling passage, the 90° overflow cooling passage and the 45° overflow cooling passage as the research object to study the flow around the overflow hole when the flow Mach number of the cold air passage changed. Mathematical simulations were carried out when the Mach number was 0.2 and 0.7 near the overflow orifice. It was found that when the transverse flow Mach number increases from 0.2 to 0.7 at the same inlet Reynolds number and crossflow ratio, the reflux area in the overflow hole decreases, and the flow resistance in the hole becomes smaller. When the cross-flow ratio changes, the cross-flow Mach number has a greater effect on the heat transfer of the cooling passage wall under low cross-flow ratio conditions. Meng et al. [3, 4] measured the flow coefficients of the film holes distributed along the main flow direction in the inner flow passage with a 60° rib wall. The results showed that: increasing the Reynolds number of the incoming flow and the total outflow of the passage the ratio increases the Reynolds number of the film hole, which increases the flow coefficient. When the hole Reynolds number increases to 15000. The flow coefficient remains around 0.7. At low Reynolds numbers, the distribution of the flow coefficient varies along the passage. Low Reynolds number. Tao et al. [6, 7] studied the distribution of rib angles, outlet orifice locations, Reynolds number of inlet passages, and total outlet ratios for the distribution of flow coefficients and the distribution of total pressure coefficients for ribbed and double exhaust flow passages. influences. Research shows that different rib angles and outlet hole positions do not change the distribution law of flow coefficient and total pressure coefficient along the flow direction. Schüler et al. [5] studied U-passages in the trailing edge region of the blade and studied the effect of side wall outflow on the flow characteristics of the passage. It was found that the outflow of the side wall has little effect on the pressure loss in the turning area, and the main effect is reflected in the turning area. Farther downstream.

It can be seen from the previous research that the trailing edge overflow hole has an effect on the flow and heat transfer in the downstream area, and also affects the flow and heat transfer in the inner cooling passage upstream of the overflow hole. There are currently few studies on the latter. This paper analyzes the number of overflow holes, the diameter of the overflow hole, and the overflow contact area (the number of holes and the hole diameter change at the same time, the flow area is constant) for the dual-outlet turbine blades at the trailing edge and the lower edge. We study the influence of the distribution of the flow coefficient of the edge overflow hole, the influence of the distribution of the cooling air out of the trailing edge and the lower edge of the plate, and the influence of the heat transfer on the wall surface of the inner cooling passage.

2   Numerical Method

2.1   Calculation Model

The calculation model is shown in Fig. 1. The chord length is 30.9 mm, the axial chord length is 24.6 mm, the blade height is 28 mm, and the blade pitch is 22.8 mm. The computational domain includes blade and inlet and outlet pre-extensions (b) and (c). The cooling air enters from the inlet (a) and flows out of the lower edge plate outlet orifice (d) and the trailing edge overflow orifice (f).

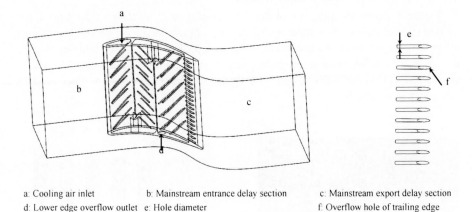

a: Cooling air inlet   b: Mainstream entrance delay section   c: Mainstream export delay section
d: Lower edge overflow outlet   e: Hole diameter   f: Overflow hole of trailing edge

Fig. 1. Calculation model diagram

The calculation model geometry parameters are shown in Table 1:

Table 1. Calculation model geometric parameters

Number of holes	20	15	10
Hole radius	0.2 mm	0.23 mm	0.26 mm
Flow contact area (Number of holes * Hole radius)	20*0.2 mm	15*0.23094 mm	10*0.28284 mm

2.2   Calculation Methods and Boundary Conditions

This paper uses CFD software ANSYS Fluent 18.0 for calculations. The turbulence model uses realizable k-ε and enhanced wall treatment. The solver uses a discrete implicit solver. The pressure and velocity are coupled using the SIMPLEC algorithm, and the convection terms are discretized using the second-order upwind style.

Mainstream and cooling air inlets are set to pressure inlet conditions. The mainstream inlet pressure is 104,000 Pa, the cooling air inlet pressure is 145600 Pa; The mainstream inlet temperature is T1 = 300 K, the cooling air inlet temperature is 400 K; The outlet is set as a pressure outlet condition, and the mainstream and the cooling air outlet pressure is 100000.

2.3   Grid and Independence Verification

The model mesh is modeled using ICEM software. The mesh is shown in Fig. 2. The unstructured mesh is used. The fluid and solid junction wall is the coupling wall surface. The boundary layer is set at the interface. The height of the first layer is 0.001 mm. The rate is 1.12, the number of boundary layer layers is 10, and the y^+ value of the wall surface is near 1. Encryption of fluid domains, ribs, and trailing edge overflow holes in the inner cooling passage.

Figure 3 shows the influence of the number of meshes on the outflow ratio at the trailing edge of the overflow hole. As the number of grids increases, the outflow ratio at the trailing edge tends to be consistent. After over 8.01 million, as the number of grids increases, the outflow ratio at the trailing edge hardly changes. At this time, the grid is considered to have no influence on the calculation results. The final number of grids selected was around 8 million.

Fig. 2. Computational grid

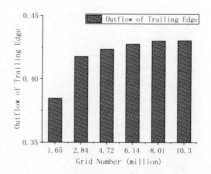

Fig. 3. Independence verification

3   Calculation Results and Analysis

3.1   Parameters Definition

Hole flow coefficient:

$$C_d = \frac{m_1}{m_2} = \frac{m_1}{A_2\sqrt{2\rho_2(P_t - P_s)}} \tag{1}$$

Where: m_1 is the actual mass flow through the orifice, the denominator is the ideal flow through the orifice; P_s is the static pressure at the trailing edge overflow orifice; P_t is the average total pressure upstream of the orifice inlet; ρ_2 is the density at the outlet of the orifice; A_2 is the cross-sectional area at the orifice outlet.

Total pressure coefficient:

$$C_{pt,i} = \frac{(P_{t,i} - P_{t,\infty})}{(\rho_\infty V_\infty^2/2)} \tag{2}$$

Where: $P_{t,\infty}$ is the total pressure at the inlet of the passage; $P_{t,i}$ is the total pressure at the position i; $\rho_\infty V_\infty^2/2$ is the dynamic pressure at the inlet of the passage.

3.2   Influence of the Trailing Edge Overflow Hole on Its Flow Coefficient

Figure 4 shows the influence of the trailing edge overflow hole on the distribution of its flow coefficient along the height of the blade Y/L is the relative height of the blade. (a) is the influence of the number of holes on the distribution of flow coefficient. The greater the number of overflow holes, the smaller the flow coefficient near the middle of the blade. This is because the change in the number of holes has a greater influence on the flow through the trailing edge. The more the number of holes, the less the actual flow rate of each hole, and the influence is closer to the middle of the blade; (b) is the influence of the hole radius on the distribution of the flow coefficient. The larger the hole radius, the larger the flow coefficient near the tip and root of the blade. (c) is keep the flow area constant while changing the number of holes and the hole radius. Equivalent to changes the flow contact area between the fluid and the hole. It can be seen that the smaller the flow contact area, the larger the hole flow coefficient. Because the smaller the flow contact area, the smaller the flow loss caused by the overflow hole, and the larger the ratio of the actual flow rate to the ideal flow rate.

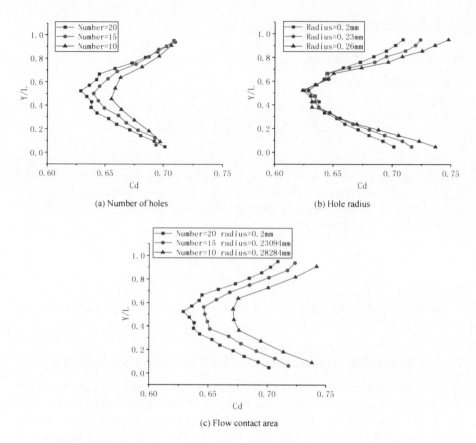

Fig. 4. Influence of the trailing edge overflow hole on its flow coefficient

3.3   Influence of the Trailing Edge Overflow Hole on Its Outflow Ratio

The trailing edge outflow ratio is the ratio of the mass flow of cooling air from the trailing edge to the total flow mass flow in the passage in the inner cooling passage. Figure 5 shows the comparison of the influence of the trailing edge overflow hole on the outflow ratio at the trailing edge. The values of (a), (b), and (c) correspond to the effect of the number of holes, the hole radius and the flow contact area on the outflow ratio at the trailing edge. The influence of the flow contact area on the outflow ratio at the trailing edge is significantly smaller than the number of holes and the hole radius. The number of holes increases, the hole radius increases, and the flow area of the trailing edge increases, thereby increasing the outflow. The flow contact area increases, and the tail-edge outflow ratio decreases slightly. Because the contact area increases, the number of holes increases, and the hole radius decreases. The influence of the two on the outflow ratio of the trailing edge offsets each other. The cross-sectional area is constant, the flow contact area is increased, and the flow resistance is increased. Therefore, when the flow contact area is changed, the tail-edge flow ratio changes little and slightly decreases.

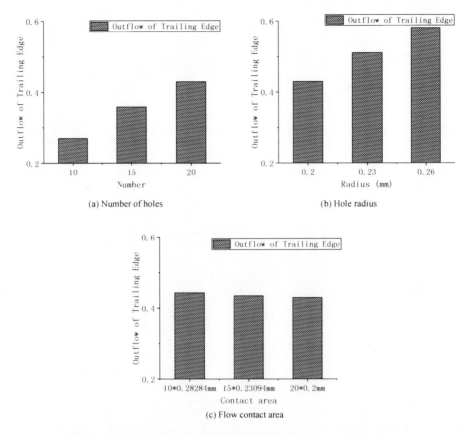

Fig. 5. Influence of the trailing edge overflow hole on its outflow ratio

3.4   Influence of the Trailing Edge Overflow Hole on the Pressure Loss in the Cooling Passage

The total pressure coefficient reflects the pressure loss inside the inner cooling passage. In Fig. 6, Y/L is the relative height of the blade. The flow direction is from the tip (Y/L = 1) of the blade to the root (Y/L = 0) of the blade. The total pressure curve is upside high and under low. As the number of holes increases and the hole radius increases, the lateral outflow of the cooling air flowing from the tip of the blade to the blade root increases, resulting in an increase in the pressure loss of the cooling air in the radial direction; The flow cross-sectional area remains unchanged, and the cooling air is changed when the number of holes and the hole radius change. The pressure loss along the radial direction changes very little, because the flow cross-sectional area is constant, the number of holes increases, the hole radius decreases, and the influence of the two on the pressure loss cancel each other out; The fluctuation of the total pressure coefficient is due to the influence of the ribs of the inner cooling passage wall, has nothing to do with the trailing edge overflow hole.

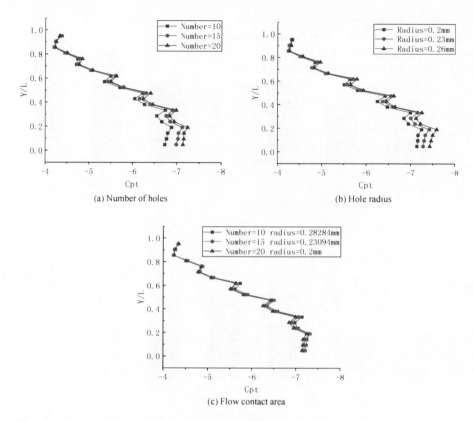

Fig. 6. Influence of the trailing edge overflow hole on the pressure loss in the cooling passage

3.5   Influence of the Trailing Edge Overflow Hole on the Heat Transfer Inside the Cooling Passage Wall

Figure 7(a) shows the heat transfer coefficient cloud picture of the inner wall of the cooling passage on the trailing edge of the pressure surface for different number of trailing edge overflow holes. With the increase of the number of holes, the flow of cooling air in the cooling passage increases, and the velocity of the air passing through the passage ribs increases. Therefore, the heat transfer is enhanced; Fig. 7(b) shows the heat transfer coefficient cloud picture of the inner cooling passage wall of the trailing edge of the pressure surface on the trailing edge of different trailing edge overflow hole radius. As the hole radius increases, the influence is consistent with the increase in the number of holes, so the heat transfer is enhanced. Figure 7(c) shows the same flow cross-sectional area and different flow contact areas correspond to the heat transfer coefficient cloud picture of the cooling passage wall on the trailing edge of the pressure surface. The number of holes increases and the hole radius decreases. The influence on the heat transfer effect is reversed. Therefore, the change of heat transfer effect of the flow contact area is not obvious.

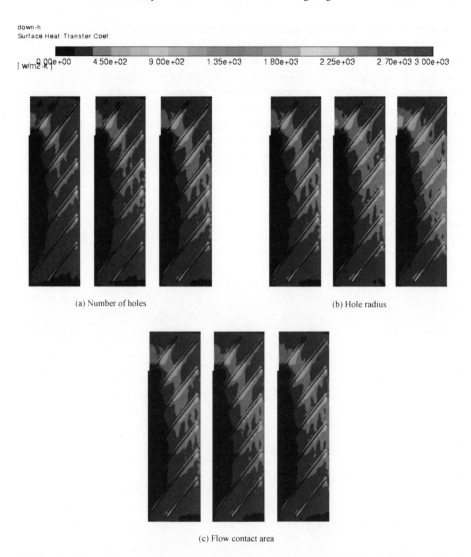

(a) Number of holes   (b) Hole radius

(c) Flow contact area

Fig. 7. Influence of the trailing edge overflow hole on the heat transfer inside the cooling passage wall

Figure 8 shows the distribution of the radial average heat transfer coefficient in the direction of the blade height of the inner cooling passage wall of the trailing edge on the side of the pressure surface of different trailing edge overflow holes. Y/L is the relative height of the blade. Cooling air flows from the 180-degree U-turn passage at the top of the midchord region to the trailing edge passage. The heat transfer coefficient increases rapidly at the inlet, reaches a maximum at a relative blade height of about 0.8, and then begins to decrease. Figure 8(a) and (b) show the distribution of radial average heat transfer coefficients of the inner cooling passage wall with different number of holes and different hole radius corresponding to the pressure side. The number of holes increases, the radius of the hole increases, and the flow rate of the cooling passage cooling air increases. As the speed increases, the heat transfer increases, and the cooling air changes its maximum near the relative blade height of 0.8, so the heat transfer coefficient changes the most. Figure 8(c) shows that the flow cross-sectional area is constant and only change flow contact area, the distribution of the radial average heat transfer coefficient of the cooling passage in the trailing edge of the pressure surface. The influence of the increase in the number of holes and the reduction in hole radius is opposite to the heat transfer. Therefore, changing the flow contact area has little effect on the radial average heat transfer coefficient distribution.

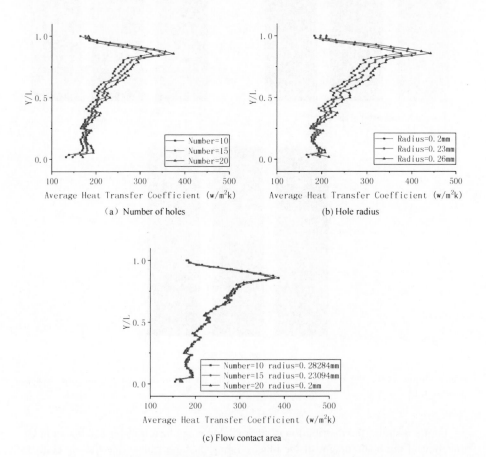

(a) Number of holes

(b) Hole radius

(c) Flow contact area

Fig. 8.   Radial average heat transfer coefficient

Figure 9 shows the distribution of spanwise average heat transfer coefficient of the inner cooling passage wall surface of the trailing edge on the side of the pressure surface of different trailing edge overflow holes. X/D is the relative position of the wall surface of the cooling passage on the trailing edge of the pressure surface. The closer to the midchord region, the lower the average heat transfer coefficient. The relative position of the spanwise is from 0 to 0.5, and the average heat transfer coefficient is gradually increased. Figure 9(a) and (b) show the spanwise average heat transfer coefficient of the wall face of the cooling passage on the side edge of the pressure surface corresponding to the number of holes and the hole radius. The spanwise average change is different from the radial average heat transfer coefficient. The overall radial average heat transfer coefficient is changing, but the change in the spanwise average heat transfer coefficient is mainly concentrated in the position where the relative position of the span is 0.5 to 1. With reference to Fig. 7, we can easily see that the heat transfer of the cooling air in the cooling passage is mainly concentrated in the position of the span is 0.5 to 1, the distribution of the spanwise average heat transfer coefficient is also concentrated in this area. Figure 9(c) shows that the spanwise average heat transfer coefficient distribution of the cooling passage in the side trailing edge of the pressure wall corresponding to the flow cross-sectional area is constant, contact area is changed. The influence on the spanwise average heat transfer coefficient is the same as radial average. Therefore, changing the flow contact area has little influence on the spanwise average heat transfer coefficient.

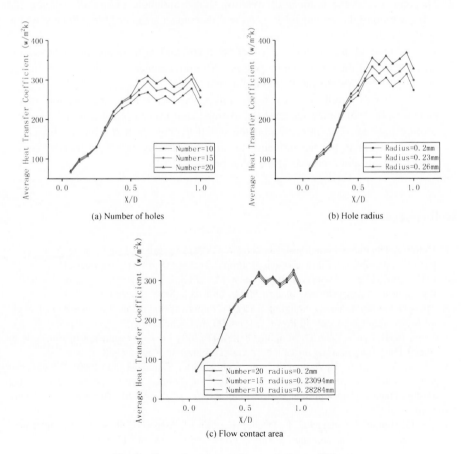

(a) Number of holes   (b) Hole radius

(c) Flow contact area

Fig. 9. Spanwise average heat transfer coefficient

4   Conclusion

In this paper, the flow and heat transfer characteristics of the trailing edge overflow hole on the trailing edge and the lower edge of the dual-flow turbine blade trailing edge in the cooling passage are studied by numerical simulation. The number of holes, the hole radius, and the flow contact area are compared and analyzed. Include the influence of the flow coefficient of the overflow hole, the outflow ratio at the trailing edge, the total pressure coefficient in the trailing edge passage, and the heat transfer characteristics of the inner wall of the trailing edge inner cooling passage. The main conclusions are as follows:

1. The number of holes and hole radius in the trailing edge have different influence on the flow coefficient of the trailing edge overflow hole. The influence of the number of holes is mainly concentrated in the middle of the blade, and the influence of the hole radius is mainly on the root and the top of the blade. Both increase in number of holes and hole increase the flow coefficient. The influence of the flow contact area on the flow coefficient is relatively even. The smaller the flow contact area, the larger the flow coefficient.
2. The increase of the number of overflow holes and hole radius all increase the outflow ratio of the trailing edge, and the flow contact area has little influence on the outflow ratio of the trailing edge.
3. The increase of the number of overflow holes and hole radius all increase the pressure loss of the inner cooling passage. The flow contact area has little influence on the pressure loss of the inner cooling passage.
4. The number of overflow holes in the trailing edge increases, and the hole radius increases, both of which increase the average heat transfer coefficient of the inner cooling passage. The influence on the radial average heat transfer coefficient is relatively average, but the influence on the spanwise average heat transfer coefficient is mainly concentrated at the span position 0.5 to 1.

References

1. Huiren Z, Duchun X, Songling L, Baolong W (1998) Experimental study on the influence of gas film pore shape on flow coefficient. Propul Technol 01:43–46 (in Chinese)
2. Lu Y (2007) Effect of hole configurations on film cooling from cylindrical inclined holes for the application to gas turbine blades. Ph.D., Louisiana State University
3. Meng N, Huiren Z, Yun Q, Songling L (2004) Numerical simulation of influence of rib angle on flow coefficient. J Aerosp Power 2:196–200 (in Chinese)
4. Meng N, Huiren Z, Yun Q, Songling L (2004) Study on flow coefficient of film hole in ribbed Inner flow passage (in Chinese). Gas Turbine Exp Res 1:24–28
5. Schüler M, Zehnder F, Weigand B, von Wolfersdorf J, Neumann SO (2009) The effect of turning vanes on pressure loss and heat transfer of a ribbed rectangular two-pass internal cooling channel. Paper presented at the Proceedings of the ASME Turbo Expo, Orlando, USA
6. Tao G, Huiren Z, Guangchao L, Duchun X (2009) Influence of rib angle and outlet hole position on flow characteristics. J Aerosp Power 3:537–541 (in Chinese)

7. Tao G, Huiren Z, Guangchao L, Duchun X (2007) Flow characteristics of ribbed and double exhaust orifice passages (in Chinese). Propul Technol 4:399–402
8. Wenchao S, Huiren Z, Jianhao N (2014) Influence of cross-flow Mach number on flow and heat transfer near overflow hole. Propul Technol 12:1645–1652 (in Chinese)
9. Yu Y (2010) Study on Film Cooling Characteristics of Convergent Slit Hole. Ph.D., Nanjing University of Aeronautics and Astronautics (in Chinese)
10. Zhi T, Hong W, Yi C (1997) Effect of film outflow on heat transfer coefficient of inner surface of blade (in Chinese). J Aerosp Power 04:78–80

MBSE Approach to Aero-Engine Turbine System Design and Requirements Management

Zhiying Chen[1], Yufeng Wang[1(✉)], Yuchen Zhang[2], and Teng Li[2]

[1] School of Energy and Power Engineering,
Beihang University, No.37 Xueyuan Road, Beijing, China
wangyufeng3@buaa.edu.cn
[2] AVIC Digital. Inc, Block E, Aviation Industry Building,
No. 5 Shuguang Xilijia College, Chaoyang District, Beijing, China

Abstract. In view of the problems of the traditional aero engine development, such as the design elements are difficult to reuse, the traceability of requirements is not good and lack of top-level logic verification, the model based systems engineering (MBSE) method is introduced. Apply the Harmony for Systems Engineering to the top-down design flow of aero-engine turbine system. The system requirement model is established through the creation and classification of turbine system requirements and the definition of system use cases. To establish the function model, the functional, expressive, interface, and other requirements are transformed into a clear description of the system function. The system architecture is analyzed and designed on the basis of the system function, then the function of each use case is decomposed and allocated into the subsystems and components of turbine system. Based on the system model, the requirements of the system are refined, traced and verified. The results show that the MBSE method can perfect the requirement definition, complete the mapping of the requirements to the system elements, realize the function logic verification, and support requirements verification at all stages, which provides an effective practical approach for the development of aero-engine.

Keywords: Model-based systems engineering · Turbine system · Harmony for Systems Engineering · System requirements · System model

1 Introduction

Aero-engine is a typical System of System complex engineering, composed of control and adjustment system, compressor system, combustor, turbine system, oil system and other systems. The turbine system working under high temperature and high pressure, driving the compressor and accessory system, is critical to aero-engine performance. The development process of a complex system similar to aero-engine produces a lot of design information, which is stored, transmitted and updated in the form of documents in the traditional way. However, with the increasing scale of the system, designers of different functions have different understanding of documents, and it is difficult to ensure the reusability of design information, the traceability of requirements and the top-level logic verification in traditional development way.

© Springer Nature Singapore Pte Ltd. 2019
X. Zhang (Ed.): APISAT 2018, LNEE 459, pp. 1730–1744, 2019.
https://doi.org/10.1007/978-981-13-3305-7_139

In response to these challenges, the International Systems Engineering Organization (INCOSE) proposed model-based systems engineering (MBSE) in 2007 and defined it as "the formalized application of modeling to support system requirements, design, analysis, verification, and validation activities beginning in the conceptual design phase and continuing throughout development and later life cycle phases" [1]. In the field of aerospace, MBSE has become a hotspot in research and application. In the development of spacesuit, Cordova use MBSE to realize the relationship between requirement and model and support integration of all products [2]. In the research on the interface Management of flight project, Vipavetz use MBSE based seven-step approach to identify, define, control, and verify the interfaces [3]. In Spangelo et al.'s study, they apply MBSE method to standard CubeSat and applying that model to an actual CubeSat mission [4, 5]. Parthasarathi has a system-level design for aero-engines subsystem: Full Authority Digital Electronic Controller and verifying system performance through simulation [6].

In summary, for aerospace products and missions, the MBSE method has been extensively studied and applied. However, in the field of aero-engine, the application to the core machine, and the layer-by-layer integration from components to the system is lack of research or rarely involved. Few attention has been paid to the definition and traceability of system requirements, which leads to the lack of in-depth discussion on the relationship between system model and requirements. In this paper, aiming at aero-engine turbine system, using System Modeling Language [7] and Harmony for system engineering [8], the requirement analysis, function analysis and architecture analysis are carried out, and the system requirements are refined, traced and logically verified based on the model.

2   Requirement Analysis

System requirements are the basis of system definition, the basis of system architecture and system verification, and requirements need to be verified and iterated at each stage of the systems engineering process, so the definition and analysis of turbine system requirements is the key of system development.

2.1   Requirements Definition and Classification

The requirements of the turbine system can be obtained by transforming the desired capacity of the turbine system from stakeholders and users to solution meeting the needs of the user's operation with standardized language. Then the requirements of the turbine system are classified according to function, port and interface, State and mode, physical, five principles, and corresponding-level ID is assigned to each category for more detailed division. The category of requirements is shown in Fig. 1.

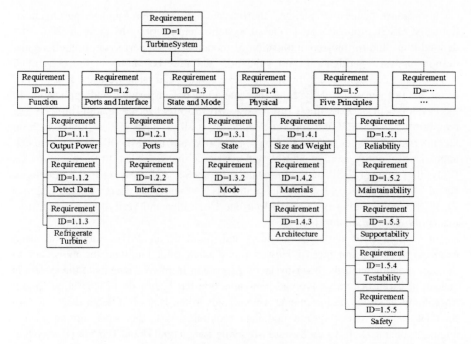

Fig. 1. Requirement classification

2.2   Use Case

Use cases represent the primary functions of turbine system. The use case model of turbine system consists of a system boundary that defines the scope of system's function, use cases that describe the service provided by the system, and an interaction relationship between the use cases and the participant. The turbine system use case model is shown in Fig. 2.

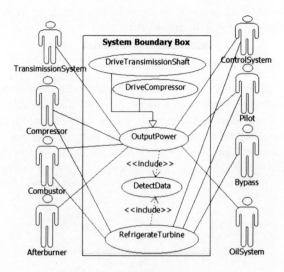

Fig. 2. Use cases of turbine system

Eight external participants (pilot, compressors, combustor, bypass, afterburner, oil systems, control systems and transmission system) are identified by defining the system boundaries and analyzing the external elements interacting with the turbine system. The main use case of the turbine system is Output Power, and drive compressor and drive transmission shaft are two generalization use cases. Refrigerate Turbine is the second main use case due to the limited material properties of the turbine system which working in a high temperature and high pressure environment. The different speed and temperature of the turbine will cause the system to be in different working states and refrigerate modes, so the use case of Detect Data is included in the two main use cases.

3   Function Analysis

Function analysis is the functional process analyzing each system use case at the system level, which can identify the interaction between the turbine system and the outside, and fully describe the state and behavior of the turbine system.

3.1   Process Analysis

Process analysis is used to transform the high-level requirements in the requirements analysis process into specific events by activity diagrams which focus on the internal system function. The turbine system's workflow is divided into four stages of testing, starting, running and shutdown, followed by a detailed description of each stage. The stage of starting the turbine with detailed description is as shown in Fig. 3.

The activity diagram focuses on the control flow of the turbine system from one activity to another and describes the interaction of the turbine with the external participants through ActorPin. At the beginning of starting the engine, the drive system is driven by the starter and drives high-pressure turbine to rotate through a bevel gear, high-pressure turbine drives high-pressure compressor, air flows from combustor into the turbine, refrigerating turbine and detecting data in parallel. The low-pressure turbine then starts to rotate and drive the low-pressure compressor. When the rotor rotational speed reaches the ignition speed, the engine ignites. The turbine receives gas from the combustor, while driving the high and low pressure compressor. The speed continues to increase until the starter is disconnected and the starting of the turbine system is completed.

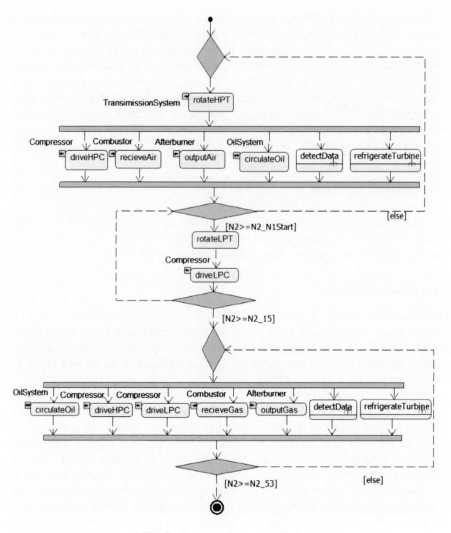

Fig. 3. Process of starting turbine

3.2 Scene Analysis

The purpose of scene analysis is to identify the interface between the system and the outside using sequence diagram by analyzing the interaction of the turbine system in different use scenarios. The focus of a sequence diagram is the interaction between the system and the environment. As shown in Fig. 4, the sequence diagram of the running turbine scene.

The different lifelines in the diagram represent the turbine system and the external participants. The turbine system interacts with the oil system to recycle oil, receive gas from the combustor, transmit torque to the high and low pressure compressor, transport the gas to the afterburner, drive the transmission shaft, refrigerate the turbine and detect data at the same time. And the transfer of material, data and instruction between the different participants and turbines is represented by data flow and events between lifelines.

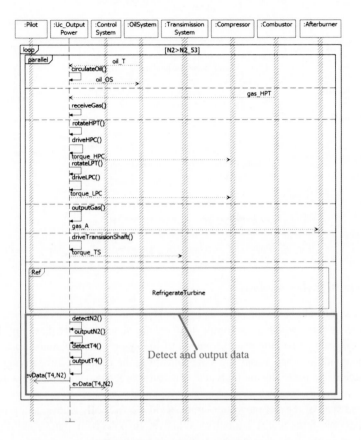

Fig. 4. Scene of running turbine

3.3 Ports and Interfaces

Based on process analysis and scenario analysis, it can be determined that the turbine system interacts with the specific external participants in different use cases to define the corresponding ports and interfaces shown in internal block diagram (IBD). As shown in Fig. 5, the IBD for the "output power" use case.

It is shown in the diagram that how the turbine system connects to external participants in the "Output power" use case to express the events, energy and data types that flow between connections and the services that are provided and requested through connections: For pilot and control system, the standard port and interface configured is used to describe the interaction of data information and instruction with the turbine system. For the transmission system, compressor, oil system, combustor and afterburner, the flow port with direction and media is used to describe the energy and material interaction with the turbine system.

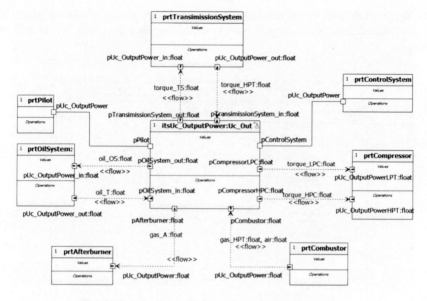

Fig. 5. Ports and interfaces of "output power" use case

3.4   State Behavior Analysis

The purpose of state behavior analysis is to analyze the complete dynamic behavior of the turbine system, which is accomplished by the state machine diagram. The state machine diagram shows the state sequence of the system, the behavior under specific states, the condition that causes the state transition, the action accompanying the state transfer, the response to the anomaly, etc. The turbine System state machine diagram is as shown in Fig. 6.

The top state of the turbine system consists of a power off, testing, and power on of three states, the three states mutually exclusive. In power on state, running turbine, detecting and outputting data are carried out in parallel with the refrigerating turbine. There are corresponding triggers when entering or exiting a state, for example, the acceptance of the refrigerate order and corresponding speed or temperature is required to transfer the "cold" mode into the "hot" mode when refrigerating turbine.

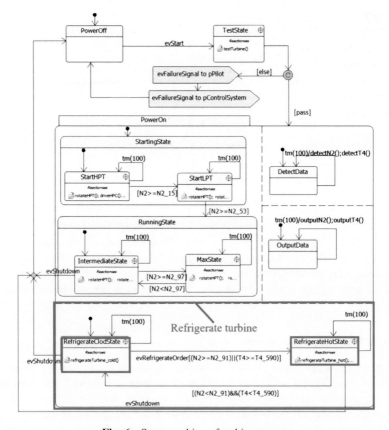

Fig. 6. State machine of turbine system

4 Design Synthesis

Design synthesis is the analysis and design of turbine system based on considering all system functions, also called as white-box analysis.

4.1 Architecture Design

Based on the composition and function of the system, the architecture is analyzed and designed and the system architecture is shown in the block definition diagram.

The turbine system block definition diagram is shown on the right side of Fig. 7. According to the main functions (drive high, low-voltage compressors and shafts, cyclic lubricating oil, cooling turbines, monitoring data) obtained from function analysis, combined with the existing turbine system architecture, the turbine system is divided into five subsystems (components): High-pressure turbine, low-pressure turbine, bearing block, refrigerate system and sensor.

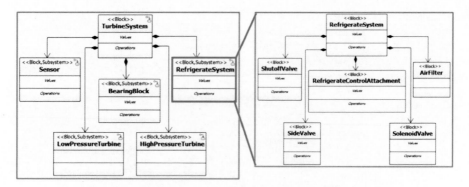

Fig. 7. Architecture of turbine system

The system is top-down and progressive architecture analyzed until the appropriate granularity. And the architecture of subsystems (components) can be obtained. The refrigerate system architecture is shown on the left side of Fig. 7, consisting of a shut-off valve, a refrigerate control attachment, an air filter, a side valve and a solenoid valve.

4.2   Use Case Implementation

The activities and operations of each subsystem are defined and function is allocated from the view inside the system by the white box activity diagram. Each swim lane in the activity diagram represents a subsystem or component. And the actions are assigned to the corresponding subsystems and components. For example, in the activity diagram of running turbine shown in Fig. 8, receive the gas from the combustor, drive the high pressure compressor and the transmission shaft are allocated to the high-pressure turbines.

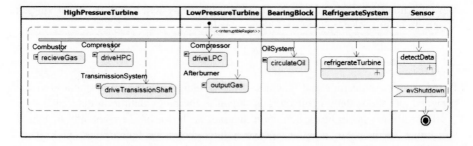

Fig. 8. Process of running turbine in white-box activity diagram

The scene analysis of white box analysis is still completed by the sequence diagram. Lifelines in sequence diagrams represent subsystems and external actors. After modeling, designers can get the actions that each subsystem performs and the

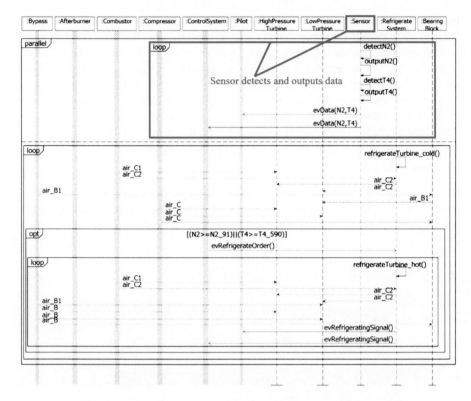

Fig. 9. Scene of refrigerate turbine in white-box sequence diagram

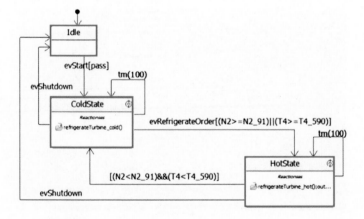

Fig. 10. State machine of refrigerate system

interaction between the external participant and the subsystem, and how the material and data are transmitted between subsystems. For example, in the sequence diagram of the refrigerate turbine shown in Fig. 9, sensor of turbine system detects data and send it to pilot and control system, the "air_B1" from the bypass refrigerates the bearing block first, then flows to the low-pressure compressor.

The main purpose of the state behavior analysis in the design synthesis phase is to analyze the dynamic behavior of each subsystem in its lifecycle. The three modes and their transition conditions of refrigerate system is shown in Fig. 10.

4.3   Use Case Merging

When all the use case analysis is completed, all system use cases should be merged. Combining the swim lane partitioning and the subsystem lifelines, the previously defined events, operations, attributes ports and interfaces are divided into different subsystems.

The system model of subsystem level is obtained through use case merging. The internal block diagram of the system is shown in Fig. 11.

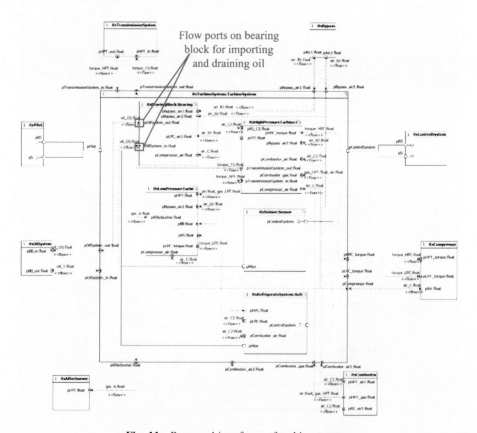

Fig. 11. Ports and interfaces of turbine system

The proxy port assigns the data and instructions that turbine interacts with the control system and the pilot to the standard ports on each subsystem. The proxy flow port assigns the flow and energy that turbine interacts with compressor, oil system, combustor, etc. to the corresponding flow ports on each subsystem. It is noteworthy that the standard port is usually implemented by the integration module, and the flow port is implemented by the flow channel, coupling and other structures.

After the requirement model, function model and architecture model of turbine system are established, the turbine system model can be delivered according to the subsystem, which makes the engineer of subsystem to understand the requirements and functions of turbine system accurately and comprehensively.

5   Requirements Management

Based on the system model, manage the system requirements by refining, tracing and verifying the requirements.

5.1   Refine Requirements

By "reading" the system model, designers can get detailed requirements description in a certain order that can be classified into corresponding requirement categories. Take the marks in each diagram as examples:

Function requirements: according to Fig. 4, "Turbine system should be able to send data to pilots and control systems". According to Fig. 9, the requirement can be further refined as "the sensor of the turbine system should be able to send data to the pilot and control system".

Port and interface requirements: according to Fig. 5, "the turbine system should have flow ports to introduce and drain oil from the oil system". According to Fig. 11, the requirement can be further refined as "the bearing block of turbine system should have flow ports to introduce and drain oil from the oil system".

State and mode requirements: according to Fig. 6, "the turbine system should have 'cold' and 'hot' two refrigerate modes." According to Fig. 10, it can be further refined that the refrigerate system of turbine system should have three modes: standby, cold and hot.

The process of requirement refinement and classification is the representation of integrity and reusability of MBSE method, at the same time, because of the unambiguous expression of the model, the communication efficiency of the designers has been improved.

According to the above, with the establishment of the model, the requirements are continuously identified and refined, the designers of different functions can get the precise requirements expression by reading the model of the turbine system, and the requirements of the turbine system can be traced back to the system model elements.

5.2   Trace Requirements

In order to trace the requirements, requirement diagrams are set up. Figure 12 is the "turbine refrigeration" requirement diagram. "Turbine refrigeration" is included in the requirements for turbine system, and the requirements for turbine system can be traced to aero-engine requirements. Logically, the requirement is verified by the "refrigerate turbine" state machine. "Import cooling air" and "adjusting flow" two requirements derive from "turbine refrigeration" requirement with the specific performance of the "refrigerate turbine" use case. The requirement "adjusting flow" is satisfied by the components in the system - the shut-off valve.

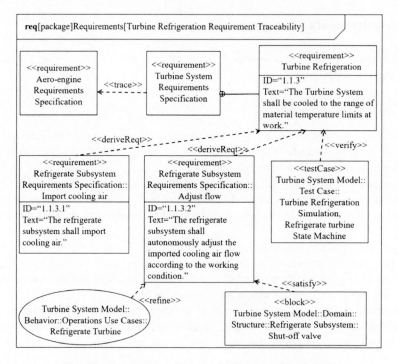

Fig. 12. Requirement diagram of "Turbine refrigeration"

By establishing the traceability of the model and requirements of the turbine system, the requirements can be traced to the elements of the system model which reflecting the characteristics of the traceability of MBSE. Further, designers can change the model to meet the requirements of system requirements change and design improvement, which is more accurate and efficient than the way to modify the document.

5.3   Verify Requirements

The function requirement of turbine system can be verified by compiling and executing the state machine of turbine and its subsystem. Compiling can verify whether the requirements are conflicting, overlapping or missing, and executing can verify whether the requirements are properly understood and whether the logic of the model conforms to the requirements of the system. The results of executing the state machine of refrigerate system and turbine system is as shown in Fig. 13.

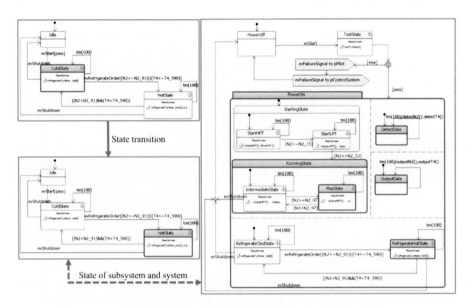

Fig. 13. Results of executing state machine

Shown in the upper left side of Fig. 13, when the high pressure speed is 80%, the refrigerate system is working in the "cold state" mode. The lower left corner: When the high pressure speed rises to 100% and refrigerate system receives the refrigerate order from the control system, the refrigerate system enters the "hot state" mode as shown in the lower left side of Fig. 13. At this time the entire turbine system state as shown on the right side of the diagram, the turbine working in the maximum state, while detecting and outputting data. The execution results of the state machine indicate that the behavior and state transition of the turbine and its subsystems conform to the system design. And the simulation verification of function requirement is realized.

It can be known that the system model is developing continuously from the process of refining, tracing and verifying the requirements. The requirement model, the function model and the architecture model are reciprocal feedback, association and iteration, and a comprehensive and reusable, traceable system model which supports the early system verification and lays the foundation for the subsequent development of turbine systems that gradually forms.

6   Conclusion

In this paper, the MBSE method is applied in the top-down design process of aero-engine turbine system. By defining and classifying system requirements, defining system use cases, analyzing system behavior and state, identifying ports and interfaces, designing system architectures, assigning functions to system subsystems, merging system use cases, the requirements model, function model and architecture model of turbine system are established. It provides a reference for further application of MBSE method in aero-engine development.

Based on the system model, the verification of the top-level logic requirements of the turbine system is realized to support the development process of the turbine system forward. The requirements of turbine system is refined, and the traceability of requirements is established, which supports the verification of requirement in every stage of system engineering process.

Through the practice of the MBSE method, it is proved that the model-based systems engineering is more intuitive to the system, and is helpful to improve the communication efficiency of the designers. Compared with the traditional development method, the design elements of model-based systems engineering with better reusability, traceability and completeness support the top logic verification in the early stage of system development.

References

1. Engineering ICOS (2015) INCOSE Systems Engineering Handbook: A Guide for System Life Cycle Processes and Activities
2. Cordova L, Kovich C, Sargusingh M (2012) An MBSE Approach to Space Suit Development
3. Vipavetz K, Shull TA, Infeld S, Price J (2016) Interface management for a NASA flight project using model-based systems engineering (MBSE). Incose Int Symp 26(1):1129–1144
4. Spangelo SC, Kaslow D, Delp C, Cole B (eds) (2012) Applying Model Based Systems Engineering (MBSE) to a standard CubeSat. In: Aerospace Conference
5. Spangelo SC, Cutler J, Anderson L, Fosse E, Cheng L, Yntema R et al (eds) (2013) Model based systems engineering (MBSE) applied to Radio Aurora Explorer (RAX) CubeSat mission operational scenarios. In: Aerospace Conference
6. Parthasarathi H, Ramachandra S, Srinivasamurthy PN (2016) Model-based systems engineering for aero gas turbine engine subsystems. Incose Int Symp 26(s1):70–82
7. Delligatti L (2013) SysML Distilled: A Brief Guide to the Systems Modeling Language. Addison-Wesley Professional
8. Hans-PeterHoffmann (2014) 基于模型的系统工程最佳实践: 航空工业出版社;

Design and Simulation of Turbofan Engine Digital Electronic Nozzle Control System

Huafeng Yu[✉], Yingqing Guo, and Jiawei Guo

Northwestern Polytechnical University, Chang'an Campus, Xi'an 710129, China
hfyu@mail.nwpu.edu.cn

Abstract. Fun surge often occurred for an augmented turbofan engine, which is related to the mechanical hydraulic nozzle control system (NCS). In order to fully realize the potential of the engine, it is now hoped that the mechanical hydraulic controller will be changed to a digital electronic controller. First of all, the control law of the engine is introduced. Secondly, modification scheme of a digital electronic NCS is proposed and a digital electronic controller is designed. Then the model of engine and NCS was established in AMESim and Simulink, and the closed-loop simulation of the NCS was realized. Finally, it is verified that the Digital Electronic NCS can ensure the normal operation of the engine in different inlet conditions and different dynamic processes, and its behavior is better than the mechanical hydraulic NCS. The digital electronic controller helps to reduce the occurrence of faults.

Keywords: Aeroengine · Nozzle control system ·
Digital electronic controller · Simulation

1 Engine Nozzle Control Introduction

The role of the engine control system is to ensure that the engine works well and its potential is fully utilized. Compared with the traditional mechanical hydraulic system, the full authority digital electronic control system (FADEC) has certain performance advantages in this respect. Due to its wide application prospects, FADEC has been a hotspot in the research of engine control systems, and related research involves many aspects. AECC Commercial Aircraft Engine Co., Ltd (AECC CAE) and AECC Aero Engine Control System Institute have conducted research on FADEC software development standards based on ClearCase [1] or SCADE [14]. AECC CAE [8] and Nanjing University of Aeronautics and Astronautics [7] have conducted research on FADEC maintenance strategies. Northwestern Polytechnical University (NWPU) [11], AECC CAE [13], and AECC Aero Engine Control System Institute [4] have built all-digital or hardware in loop simulation platforms for FADEC. Gou Linfeng of NWPU studied the optimal fault diagnosis method of FADEC system [3]. The digital electronic control system of the engine is constantly being researched and applied. This paper proposes a modification scheme for the NCS of an augmented turbofan engine, and proves that the modification is more beneficial to the performance of the engine through model simulation.

X. Zhang (Ed.): APISAT 2018, LNEE 459, pp. 1745–1756, 2019.
https://doi.org/10.1007/978-981-13-3305-7_140

NCS is an essential part of control system for modern aeroengine. At present, Laval nozzles are widely used. With a wide flight range, the available pressure drop of the nozzle changes greatly. This requires that the nozzle outlet area/throat area (A9/A8) can be changed. For an augmented engine A8 must be adjustable.

When the turbofan engine is in afterburner condition, in addition to providing additional thrust, it is also necessary to ensure that the core engine is operating in the maximum state and that the state of the afterburner cannot affect the operation of the core engine [2]. When the engine is in afterburner condition, the control law of that engine is shown in Eq. 1.

$$\begin{cases} q_m \rightarrow n_H = const \\ q_{mf} \rightarrow T_7 = const \\ A_8 \rightarrow \pi_T = f(P3/P2) \end{cases} \tag{1}$$

It can be described that the main fuel ensures that the high pressure rotor rotates at its maximum speed, allowing the core engine to operate at the maximum mechanical load. Afterburner fuel flow ensures that the T7 changes with the PLA, providing the appropriate thrust. The throat area of the nozzle is used to ensure that the turbine expansion ratio varies with the compressor pressure ratio. This function can ensure that the engine's bypass ratio B remains unchanged and the fan doesn't surge.

When the engine is not afterburning, the control law of an engine is shown in Eq. 2. The control law can be described that the main fuel controls the high pressure rotor speed as a function of PLA so as to provide the corresponding thrust. The A8 is not used as an engine control variable when it is not afterburning, but only performs open-loop control with the rotational speed.

$$\begin{cases} q_m \rightarrow n_H = f(PLA) \\ A_8 = f(n_H) \end{cases} \tag{2}$$

2 Design of Digital Electronic NCS

The nozzle control system for an engine is shown in Fig. 1. Mainly divided into four parts: pressure ration control unit, nozzle valve, nozzle lube pump and nozzle actuator. When the engine is afterburning, pressure ration control unit completes the pressure signal input, expecting turbine expansion ratio calculation, control variable calculation and output. When the engine is not afterburning, the nozzle valve completes the input of the rotational speed signal, expecting A8 calculation and control quantity calculation and output. These two mechanisms are equivalent to sensors and controllers. The nozzle lube pump and nozzle actuator are actuators that control the A8 as desired by the mechanical hydraulic controller. X6 in the Fig. 1 is a displacement, which corresponds to the swash plate angle of nozzle lube pump [9].

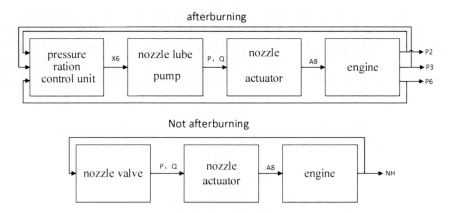

Fig. 1. Mechanical hydraulic nozzle control system

The guiding principle of the improvement is changing less and gaining more. If the controller is changed, pressure ration control unit and nozzle valve can be removed. The nozzle lube pump and nozzle actuator can be retained as actuators. At the same time, in order to change electrical signals to oil pressure, an oil pump control attachment needs to be added. In addition, sensors need to be installed to turn the speed and pressure into electrical signals.

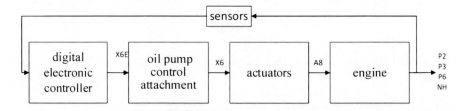

Fig. 2. Digital electronic nozzle control system

The modified NCS is shown in Fig. 2. The actuator refers to the nozzle lube pump and the nozzle actuator. Mechanical hydraulic logic switching and calculation structures are replaced with the electronic controller, reducing the possibility of complex mechanical hydraulic structures failing [10]. At the same time, the same control structure is applied whether afterburning or not, thereby reducing the hydraulic structures of the control system. In the end, the complicated mechanical hydraulic structure in the original control system is simplified, the probability of failure of the system is reduced, and the weight of the control system is also reduced.

The design of control circuit for digital electronic controller is shown in Fig. 3. The digital electronic controller is in the dotted line box. The control circuit is divided into three loops. The innermost loop is the closed loop feedback of X6. It is used to ensure the accuracy of the oil pump control attachment and allow the oil pump to reach the desired swash plate angle. The middle part of this circuit is A8 closed-loop feedback. This loop is added to ensure the accuracy of A8 control and resist the aerodynamic force of the nozzle. Also, it can make the response of system faster when the engine is afterburning. The outermost loop is the turbine expansion ratio and compressor pressure ratio control loop. It ensure that the engine achieves closed-loop control law in afterburner conditions. When engine isn't afterburning, A8 is not a control variable and the control system gives the expected A8 directly based on the NH. So, the outermost loop doesn't work.

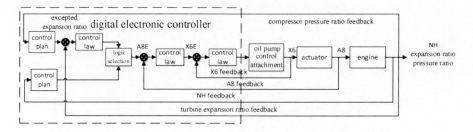

Fig. 3. Control circuit

The specific NCS control plan has two points: If the engine isn't afterburning, when the high-pressure rotor speed is less than 80%, the nozzle is at the pre-open area; after the high-pressure rotor speed is greater than 80%, the nozzle is at the minimum area; when the high-pressure rotor speed reaches 100%, the nozzle is at the pre-open area; the nozzle throat area does not change in any other conditions. If the engine begins to afterburn, the function between P2/P2 and P3/P6 is used to control within 0.05 s.

3 Establishment of Simulation Model for Digital Electronic NCS

The nozzle control of the engine consists of open-loop control part and closed-loop control part. In order to verify the digital electronic controller in different situations, it is necessary to establish the model of the engine and the control system at the same time.

Firstly, it is the model of the actuator in the control system that need to be established. An object-oriented model is built in AMESim. AMESim is widely used for modeling of mechanical hydraulic systems. As long as its actual parameters are known, an accurate model can be established easily. In order to make the digital electronic NCS model comparable to the real mechanical hydraulic control system, the actuator model should be the same as the real system.

The oil pump control attachment needs to convert the electrical signal into a hydraulic signal. Therefore, a small actuator controlling by an electro-hydraulic servo valve was built. In order to ensure that the actuator displacement (actually, X6 displacement) is accurate, it is controlled in a closed-loop manner. This part is the innermost loop of the control circuit. The actuator parameters are the same as the mechanism parameters used to push the swashplate to make the new attachment easier to match the original system in terms of power and flow rate. The nozzle lube pump was built from a universal variable displacement pump model, calibrated by experimental data. Referring to the actual structure, the leaks between the two chambers for cooling is taken into account when the model of nozzle actuator is built. The final model of executive mechanism is shown in Fig. 4.

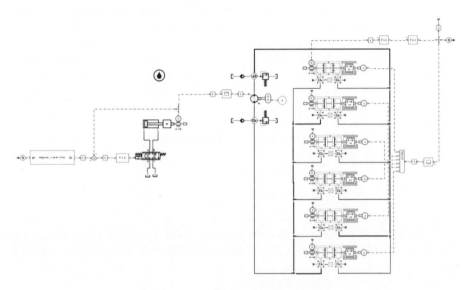

Fig. 4. Model of the actuator in AMESim

Secondly, the digital electronic controller mode is established. In order to achieve closed-loop simulation with the engine model, it is built in Simulink. The digital electronic controller mainly includes three parts: the control plan, the logic selection module, and the control law.

The control plan is divided into two parts. Without afterburning, the control plan is very simple, which is discussed before. However, the function of P3/P2 and P3/P6 are more complex and need to be derived from the original mechanical hydraulic control system.

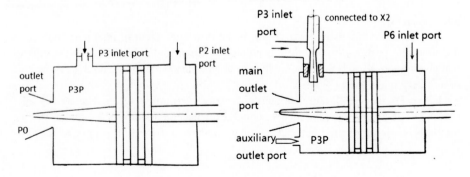

Fig. 5. X2 air piston and X6 air piston

The control plan of the mechanical hydraulic NCS is given by two air pistons in the pressure ration control unit, as shown in Fig. 5. The left piston is called an X2 air piston. The pressure in the two chambers is a partial pressure of P3 (P3P) and P2. The value of P3P is related to the value of P3 and the area of outlet port. And the area of the outlet port is related to the piston displacement. When the pressure in the two chambers is not equal, the piston moves under pressure differential until P3P = P2. Then, the value of P3/P2 is change into a piston displacement. The right-side piston is called the X6 air piston. The needle displacement at the P3 inlet port is in one-to-one correspondence with X2 displacement, so the P3 inlet area of X6 air piston depends on the value of X2 displacement. This P3P is the expected pressure after turbine P6E and the calculation of excepted P6/P3 is completed.

When the engine is at steady-state, the two pistons don't move. Therefore, the inlet flow should be equal to the outlet flow in a certain chamber of the piston. For X2 air piston, there is Eq. 3

$$K_q \frac{P_3}{\sqrt{T_3}} u_{x2i} A_{x2i} q(\lambda_{X2i}) = K_q \frac{P_{3p}}{\sqrt{T_{3p}}} u_{x20} A_{x20} q(\lambda_{X20}) \quad (3)$$

u_{x2i} is the flow coefficient of P3 inlet port; A_{x2i} is the area of P3 inlet port; u_{x20} the flow coefficient of P3P outlet port; A_{x20} is the area of P3P outlet port; q is flow function. Since the piston's inlet and outlet are all at supercritical condition, q is constant [12]. The Eq. 3 can be transformed into Eq. 4.

$$\frac{P3}{P2} = \frac{u_{x20}}{u_{x2i}} \cdot \frac{f_1(X2)}{A_{x2i}} \quad (4)$$

P3P = P2 and T3 = T3P is used to simplify Eq. 3. A_{x20} is a single-valued function of X2 displacement, so it is written as $f_1(X2)$. The ratio of the flow coefficient in the formula can be calculated from the experimental data [9]. Then the function between P3/P2 and X2 is found.

$$m_{in} = m_{out} \quad (5)$$

Where, $m_{in} = K_q \frac{P_3}{\sqrt{T_3}} u_{x6i} A_{x6i} q(\lambda_{X6i})$, $m_{out} = K_q \frac{P_{3p}}{\sqrt{T_{3p}}} u_{x60} A_{x60} q(\lambda_{X60}) + K_q \frac{P_{3p}}{\sqrt{T_{3p}}}$ $u_{x6f0} A_{x6fo} q(\lambda_{X6i})$.

For X6 air piston, there is Eq. 5. u_{x6i} is the flow coefficient of inlet port and A_{x6i} is the area inlet port. u_{x60} is the flow coefficient of main outlet port, A_{x60} is the area of main outlet port. u_{x6f0} is the flow coefficient of auxiliary outlet port, and A_{x6fo} is the area of auxiliary outlet port.

P3P = P6, T3 = T3P and q = constant is used to simplify Eq. 5. The area of inlet port is in one-to-one correspondence with X2 displacement, so it is written as $f_2(X2)$. Similarly, the area of the main outlet port is written as $f_3(X6)$. The ratio of the flow coefficient is also calculated with experimental data. The flow coefficient and area of the auxiliary outlet port are constant.

$$\frac{P3}{P6} = \frac{u_{x60}f_3(X6) + u_{X6fo}A_{X6fo}}{u_{x6i}f_2(X2)} \tag{6}$$

It can be seen from the Eq. 4 that knowing a P3/P2 can find an X2 displacement. And from Eq. 6 we know that an X2 displacement can be used to calculate an excepted P3/P6, as long as the X6 displacement is known. The X6 displacement in any conditions can be obtained by using the engine performance calculation program. Therefore, the calculation method of the close-loop control plan of P3/P2 and P3/P6 is established. The Fig. 6 shows the comparison between the calculated control plan and the actual engine operating point. It can be seen that the two are consistent. The calculation method of control plan is reasonable.

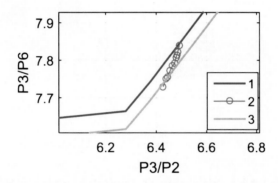

Fig. 6. Calculated control plan and the actual engine operating point (1 - calculated control plan with X6 at one operating point, 2 - actual engine operating point, 3 - calculated control plan with X6 at one operating point)

The function of logic selection module is to find whether the engine is in afterburner condition. This can be easily achieved through the value of PLA. The period of engine control is 0.02 s, so the controller can theoretically follow the closed-loop control plan in 0.02 s when the engine begins to afterburn, which is less than the required 0.05 s.

Traditional PID controller is used to design the control law. X6 and A8 in different conditions are made into two interpolation table as steady-state point data. When the state of the engine changes, this is equivalent to a feed-forward control, which can improve the steady and dynamic performance of the control system.

The simulation model of NCS was established. However, in order to verify the engine in afterburner conditions, it is necessary to establish an engine model. And the established engine model should cover multiple inlet conditions and different engine operating points to make sure the NCS can keep engine in safe states throughout the flight envelope.

The engine model is a second-order state space model established by linearization [5, 6]. The inputs of model are PLA and A8, the states are two speeds and outputs are P2, P3, P6 and some other gas path parameters. That model is only accurate near a certain stable point. Therefore, multiple linear models are established in different inlet conditions and different engine states. Then, a piecewise linearized model was used to simulate the nonlinear engine working process. The simulation model of Digital Electronic NCS is shown in Fig. 7.

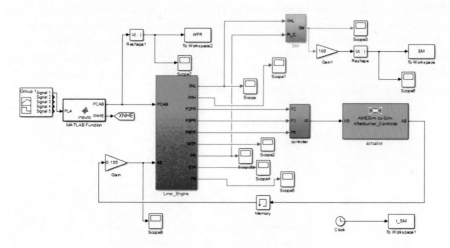

Fig. 7. Simulation model of Digital Electronic NCS

4 Simulation and Verification of Digital Electronic NCS

A wide range of afterburner ratio (afterburner fuel flow/maximum afterburner fuel flow) changing is one of the most significant changes in engine states and it is necessary to simulate. During the simulation, the afterburner ratio changes from 10% to 95% within 0.9 s. The simulation results are shown in Fig. 8, which is about the comparison between the mechanical hydraulic NCS and the digital electronic NCS with three groups of PID parameters. In the simulation, the entrance height is 0 and the Mach number is 0.

It can be seen that the digital electronic controller has a wider adjustment range than the mechanical hydraulic controller. Under the former two groups of PID parameters, the digital electronic control system can restore the surge margin faster and the settling time of A8 regulation is also shorter. The A8 adjustment with the third group of PID parameters is slightly slower than the mechanical hydraulic NCS.

The Fig. 8 also shows that the digital electronic controller starts to regulate the A8 earlier than the mechanical hydraulic controller. Because of the feed forward control, the digital electronic controller directly change the excepted A8 from PLA. As to the mechanical hydraulic controller, A8 will not change until the increase of the afterburner ratio causes change in the compressor pressure ratio, which causes great pressure and surge margin fluctuations. Also, this delay will cause the increase of A8 overshoot. The digital electronic controller has a better behavior in some conditions.

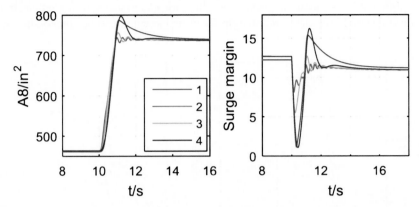

Fig. 8. Result of afterburner ratio changing from 10% to 95% (1 - mechanical hydraulic controller, 2 - digital electronic controller with first group of PID parameters, 3 - digital electronic controller with second group of PID parameters, 4 - digital electronic controller with third group of PID parameters)

Simulations for other engine inlet conditions are shown in Fig. 9. The second group of parameters is used for simulation. It can be seen that the A8 area has been controlled correctly in 7 different conditions, and the surge margin fluctuation has not changed much. Therefore digital electronic jet control system has a good adaptability.

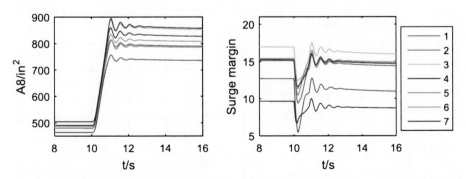

Fig. 9. Result of different import conditions (1 - 0 foot, Mach number 0; 2 - 20000 foot, Mach number 1.2; 3 - 20000 foot, Mach number 1.5; 4 - 20000 foot, Mach number 1.75; 5 - 40000 foot, Mach number 1.75; 6 - 40000 foot, Mach number 1.85; 7 - 40000 foot, Mach number 2)

Figure 10 shows the simulation of the afterburner ratio rising from 10% to 30% in 0.9 s. When the change in the afterburner ratio is small, it can be seen that the A8 overshoot of the digital electronic controller is slightly larger, but the fluctuation of the surge margin is significantly smaller than the mechanical hydraulic controller. There-fore, surge is less likely to occur controlled by the digital electronic NCS. Similar conclusions can be obtained by performing simulation of other afterburner changes under different inlet conditions. When the afterburner ratio changes, digital electronic NCS can generally recover surge margin faster and keep engine in safe conditions.

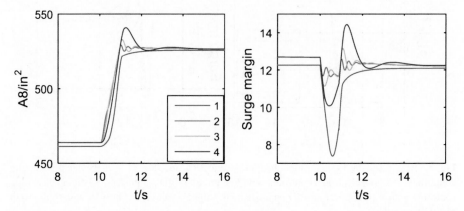

Fig. 10. Result of afterburner ratio rising from 10% to 30% (1 - mechanical hydraulic controller, 2 - digital electronic controller with first group of PID parameters, 3 - digital electronic controller with second group of PID parameters, 4 - digital electronic controller with third group of PID parameters)

Finally, open-loop control of A8 is simulated. The left graph shows that the NH exceeds 80% and A8 changes from the pre-open area to the minimum area. The right picture shows the NH achieve 100% and A8 changes from the minimum area to the pre-open area. The controller implements the control plan, and the settling time is short (Fig. 11).

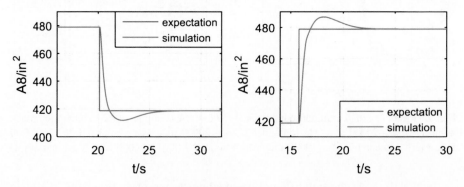

Fig. 11. Result of A8 open-loop control

5   Conclusion

This paper proposes a modification scheme of digital electronic NCS based on the existing mechanical hydraulic NCS, and gives its control circuit and control plan. The digital electronic NCS and engine model were respectively established to form a co-simulation system. This model can simulate the dynamic process of the engine under various import situations to verify the digital electronic control system. Finally, the simulation verification of digital electronic NCS in various import conditions and dynamic process is carried out. It shows that the digital electronic NCS can ensure the normal operation of the engine and, in most cases, can achieve better performance than mechanical hydraulic NCS. Especially the fluctuation of fun surge margin is small with digital electronic NCS, which helps to reduce the fun surge. So the change of controller do help to reduce the occurrence of faults and realize the potential of the engine.

The advantages and disadvantages of digital electronic NCS are as follows: (1) The mechanical hydraulic actuators is simplified, which helps to reduce the weight of the engine control system and improve engine performance. (2) The complex mechanical hydraulic logic switching and calculation structures is removed. So, the possibility of a fault is getting smaller. (3) Using more complicated control laws, the digital electronic NCS can fully realize the potential of the engine and reduce the occurrence of faults. (4) Sensors, cables and digital electronic controller are required in this scheme, so the reliability of the new control system need to be considered. This paper only provide a preliminary modification scheme and further study is necessary for the actual transformation plan.

References

1. Bao D, Liu F, Zhang H (2018) Application of ClearCase in collaborative development for FADEC software of aircraft engine. Process Autom Instrum 39(8):96–99
2. Fan S (2008) Aero engine control. Northwestern Polytechnical University Press
3. Gou L, Niu R, Shen Q (2013) Design approach of model-based optimal fault diagnosis. Comput Simul 30(5):90–93
4. Gu M, Zhang W, Gao Z (2017) Research and implementation on full digital simulation of a certain engine's control software. Aeronaut Comput Tech 5:123–126
5. Li J (2007) Dynamic model establishment of turbofan engine and design of digital electronic afterburner crotrol system. Northwestern Polytechnical University
6. Lu J, Guo Y, Chen X (2011) Establishment of aero-engine state variable model based on linear fitting method. J Aerosp Power 26(5):1172–1177
7. Lu Z, Rong X, Zhou J, Chen K (2015) TLD analysis method of dispatch with multiple faults based on monte carlo simulation for FADEC system. Acta Aeronautica ET Astronautica Sinica 36(12):3970–3979
8. Sun Y, Yang K, Hou N, She Y, Zhao X (2017) TLD and maintainability analysis of FADEC system. Syst Eng 6:152–158
9. Wang J (2017) Fault analysis and improvement of the augmented nozzle control system for a turbofan engine. Northwestern Polytechnical University
10. Wang L, Wang J, Yu H, Guo Y (2017) Simulation and analysis of cause for augmented engine surge. Aeronaut Comput Tech 47(2):72–75

11. Wu T, Fan D, Yang F, Peng K (2012) Design and simulation of FADEC system of one certain type aero-engine. Sci Technol Eng 12(1):106–111
12. Yuan X, Guo Y (2006) Calculation of flow coefficients in the flow equations of hydro-mechanical systems. Aeronaut Comput Tech 36(5):90–91
13. Zhang S, Lv X, Yin K, Zhang T, Hu Z (2013) Design of hardware-in-the-loop test system for FADEC. Measur Control Tech 32(9):81–84
14. Zhou Z, Huang H, Fang W, Zhu L (2018) Development of FADEC software for aero-engine based on SCADE. Measur Control Tech 38(1):110–115

A Reduction LPV Model Based on the Gas Dynamic Similarity for Turbofan Engine Dynamic Behavior

Zhanyue Zhao[✉] and Yingqing Guo

School of Power and Energy, Northwestern Polytechnical University,
Xi'an 710229, Shaanxi, People's Republic of China
zhaozhanyue@mail.nwpu.edu.cn

Abstract. In this paper, the reduction method using the gas dynamic similarity is analyzed, and an integrated reduction model using the gas dynamic similarity theory and the linear parameter-varying (LPV) strategy is proposed, which can describe the JT9D turbofan engine dynamic behavior at different powers around the full defined flight envelope. Compared with the general reduction method using the gas dynamic similarity, the proposed model is more accurate in the full envelope. To study the selection of scheduling parameters, several reduction LPV models are built by different parameters. The comparisons between these models and the reference model indicate that the model, which is constructed by the parameters consisting of polynomials of the corrected shaft speeds and the total temperature at the engine input, performs the similar accuracy with the models using more complicated parameters. This set of parameters take advantages considering both accuracy and complexity during the reduction LPV model constructing.

Keywords: Turbofan engine · LPV model · Gas dynamic similarity

1 Introduction

As a typical nonlinear system, the dynamic characteristics of aero engine are diversified around the large scale in flight envelope. Due to the wide use of linear models in control systems in the design stage, it is important to get more accurate integrated linear engine model at different powers around the full flight envelope. Simultaneously, all these models compromise the model accuracy with the model complexity, computational efforts and storage costs. The piecewise-linear model and the linear parameter varying (LPV) model are two effective ways using linear interpolation or fitting methods. The LPV concept is a major simplification with wide ranging implications for system identification, monitoring, and control [2, 5].

In order to model complex processes, there are some studies [6–8] about the identification of LPV models with multiple scheduling variables, different weighting functions or modified identification methods. Considering of the characteristics of the aero engines, different methods based on linearized state space model have been proposed, including partitioning the flight envelope [9], scheduling scheme with multi-

© Springer Nature Singapore Pte Ltd. 2019
X. Zhang (Ed.): APISAT 2018, LNEE 459, pp. 1757–1766, 2019.
https://doi.org/10.1007/978-981-13-3305-7_141

layers [10], polynomial LPV model [11–14]. However, these methods use multiple LPV models or piecewise-linear models around the flight envelope rather than an integrated model. This makes it difficult for controller design ensuring stability and performances over the whole operating range.

According to the original modeling method of aero engine, corrected parameters can be used to build the linear model which covers the entire operating envelop [1, 15]. But in some subareas of the flight envelope, the reduction relations between the corrected and physical parameters are true only qualitatively. It exists unneglectable error in models constructed by this method.

In this paper, to describe the JT9D turbofan engine dynamic behavior at different powers around the full defined flight envelope, an integrated reduction model is present using the gas dynamic similarity theory and the linear parameter-varying (LPV) strategy. This paper is organized as follows. Section 2 introduces a turbofan engine model and linearized at different powers in the defined envelope. Then Sect. 3 presents the reduction theory and the reduction LPV model of the engine using the gas dynamic similarity. In Sect. 4, the reduction LPV models are constructed using different parameters. Then comparison are performed to demonstrate the effectiveness of the proposed model method. Section 5 presents conclusions.

2   JT9D Model Description

A publicly available dual spool high bypass turbofan engine JT9D is studied in this paper. This model is provided by the T-MATS package [3, 4], a Simulink-based tool for thermodynamic system simulation. The JT9D turbofan engine model represents the dynamics of shaft speeds, pressures, and flows in various components of the engine. And it is developed and verified based on data from the numerical propulsion system simulation.

In this paper, an operational envelope encompasses the environmental conditions in which the engine is expected to operate, as shown in Fig. 1. The design point takes the cruise condition of 34,000 ft and Mach 0.8. The temperatures and pressures in the envelope take from the standard atmospheric environment.

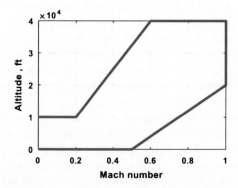

Fig. 1. JT9D operational envelope in this paper.

At conditions in the envelope, given a constant control input *uss*, the turbofan engine will reach a steady-state operating point (*uss*, *xss*, *yss*), where the dynamic performance is governed by the following space state model:

$$\delta \dot{x} = A\delta x + B\delta u$$
$$\delta y = C\delta x + D\delta u \tag{1}$$

$$\delta x = \begin{bmatrix} \delta N_{LP} \\ \delta N_{HP} \end{bmatrix}, \delta u = \delta W_f, \delta y = \begin{bmatrix} \delta N_{LP} \\ \delta N_{HP} \\ \delta P_{t,3} \\ \delta T_{t,45} \end{bmatrix} \tag{2}$$

where δx is the state, δu and δy are the input and output deviations from the operating point, Here, δN_{LP} and δN_{HP} are the low-pressure and high-pressure shaft speeds deviation, δW_f is the control input fuel flow deviation, $\delta P_{t,3}$ is the total pressure deviation at the outlet of the high pressure compressor, $\delta T_{t,45}$ is the total temperature deviation at the outlet of the high pressure turbine. The linearization block in T-MATS package is used to develop linear models at operating points.

3 Derivation of Reduction LPV Model

3.1 Reduction Model

The flight envelope can be divided into three subareas, different in terms of gas dynamic similarity, as shown in Fig. 2. In area I, the engine is described by a single line model in reduced coordinates. In subareas II and III, the reduction relations are true only qualitatively.

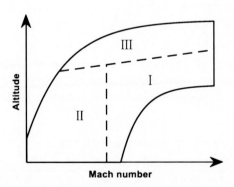

Fig. 2. Difference in terms of gas dynamic similarity around full envelope

Physical parameters can be converted to reduced coordinates by the following formulae:

$$N_{cor} = \sqrt{\frac{T_{t,std}}{T_{t,in}}}N, \qquad \dot{N}_{cor} = \frac{P_{t,std}}{P_{t,in}}\dot{N},$$
$$W_{f,cor} = \frac{P_{t,std}}{P_{t,in}}\sqrt{\frac{T_{t,std}}{T_{t,in}}}W_f, \qquad T_{cor} = \frac{T_{t,std}}{T_{t,in}}T, \qquad P_{cor} = \frac{P_{t,std}}{P_{t,in}}P \tag{3}$$

where N_{cor} is the corrected speed, N is mechanical shaft speed, $T_{t,in}$ and $P_{t,in}$ are the total temperature and pressure at the engine input, $T_{t,std}$ and $P_{t,std}$ are the total temperature and pressure at the engine input at the design condition. In this paper, the total parameters at the engine input use the corresponding parameters at the inlet of the fan.

Using the reduction formulae, the space state model in (1) can be defined in the reduced coordinates in the following way:

$$\delta\dot{x}_{cor} = A_{cor}\delta x_{cor} + B_{cor}\delta u_{cor}$$
$$\delta y_{cor} = C_{cor}x_{cor} + D_{cor}\delta u_{cor} \tag{4}$$

$$A_{cor} = P_{\dot{x}}AP_x^{-1} \quad B_{cor} = P_{\dot{x}}BP_u^{-1} \quad C_{cor} = P_yCP_x^{-1} \quad D_{cor} = P_yDP_u^{-1}$$
$$\delta\dot{x} = P_{\dot{x}}\delta\dot{x}_{cor} \quad \delta x = P_x\delta x_{cor} \quad \delta u = P_u\delta u_{cor} \quad \delta y = P_y\delta y_{cor} \tag{5}$$

The diagonal matrices $P_{\dot{x}}, P_x, P_u$ and P_y connect physical and reduced parameters:

$$P_{\dot{x}} = \begin{bmatrix} \frac{P_{t,in}}{P_{t,std}} & 0 \\ 0 & \frac{P_{t,in}}{P_{t,std}} \end{bmatrix}, \qquad P_x = \begin{bmatrix} \sqrt{\frac{T_{t,in}}{T_{t,std}}} & 0 \\ 0 & \sqrt{\frac{T_{t,in}}{T_{t,std}}} \end{bmatrix},$$

$$P_y = \begin{bmatrix} \sqrt{\frac{T_{t,in}}{T_{t,std}}} & 0 & 0 & 0 \\ 0 & \sqrt{\frac{T_{t,in}}{T_{t,std}}} & 0 & 0 \\ 0 & 0 & \frac{P_{t,in}}{P_{t,std}} & 0 \\ 0 & 0 & 0 & \frac{T_{t,in}}{T_{t,std}} \end{bmatrix}, \qquad P_u = \frac{P_{t,in}}{P_{t,std}}\sqrt{\frac{T_{t,in}}{T_{t,std}}} \tag{6}$$

For the JT9D engine, most area of the flight envelope is distributed in subarea II. It is necessary to analyze the steady and dynamic characteristics of the engine for different powers and flight conditions based on the reduction model described.

Table 1. Max steady relative error of y_{cor} at different powers around full envelope

Parameter	NL	NH	Pt3	Tt45
Max error %	6.4145	0.8851	3.4961	1.4696

When the $W_{f,cor}$ changes from 42% to 100% in the full flight envelope, it exists deviations for y_{cor} given the same input $W_{f,cor}$. The max steady relative error of y_{cor} is shown in Table 1. Given the 42% and 100% $W_{f,cor}$, the $N_{LP,cor}$ and $N_{HP,cor}$ in different conditions are shown in Fig. 3. There is no same one-to-one correspondence between $W_{f,cor}$ and $N_{LP,cor}$, $N_{HP,cor}$ around the flight envelope. The distribution of $N_{LP,cor}$ at 100% and 42% $W_{f,cor}$ shown in Fig. 3(a, b) illustrate that the deviations at different powers are also distinct, and larger deviations occur at low power. Around the same Mach number, the $N_{LP,cor}$ and $N_{HP,cor}$ have the similar values. The distribution characteristics of steady deviation also cause different in dynamic. The changes of elements and eigenvalues of A_{cor} is shown in Fig. 4, where the lines are the parameters at design point and the other colored areas are the distributions of corresponding parameters in the full flight envelope. As the Mach number decreases, the lines at design point move toward the other end of the colored area. Changes of the altitude can also influence the distribution of the lines in the colored areas.

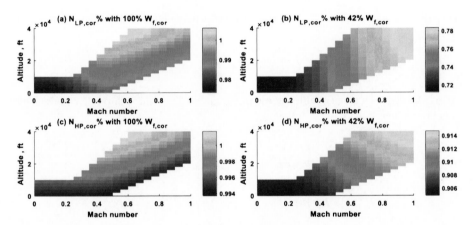

Fig. 3. $N_{LP,cor}$ and $N_{HP,cor}$ at 40% and 100% $W_{f,cor}$ around full envelope

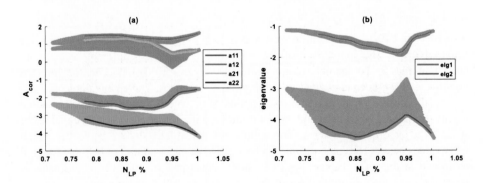

Fig. 4. Changes of the elements (a) and eigenvalues (b) of A_{cor} around full envelope

3.2   Reduction LPV Model

To get a linear engine model at different powers around the full flight envelope, the piecewise line model and the linear parameter varying (LPV) model are two effective ways, using linear interpolation or fitting methods. In fact, all these models compromise the model accuracy with the model complexity, computational efforts and storage costs, and it is difficult to get an integrated model around the large scale in flight envelope.

Consider the general reduction model in (4), an LPV description of the uncertain state space model has the form:

$$
\begin{aligned}
\delta \dot{x}_{cor} &= A_{cor}(p)\delta x_{cor} + B_{cor}(p)\delta u_{cor} \\
\delta y_{cor} &= C_{cor}(p)x_{cor} + D_{cor}(p)\delta u_{cor}
\end{aligned}
\tag{7}
$$

where $p = \begin{bmatrix} p_1 & p_2 & p_3 & \cdots & p_s \end{bmatrix}$ is a vector of s parameters. The system matrices are given by

$$
A_{cor}(p(t)) = A_0 + \sum_{i=1}^{s} p_i(t)A_i
\tag{8}
$$

$A_0, A_1 \ldots A_s$ is a set of coefficient matrices, can be determined through a generalized system of linear equations as following:

$$
\begin{bmatrix}
I & p_1(1)I & p_2(1)I & \cdots & p_s(1)I \\
I & p_1(2)I & p_2(2)I & \cdots & p_s(2)I \\
\vdots & & & & \vdots \\
I & p_1(r)I & p_2(r)I & \cdots & p_s(r)I
\end{bmatrix}
\begin{bmatrix}
A_0 \\ A_1 \\ \vdots \\ A_s
\end{bmatrix}
=
\begin{bmatrix}
A(p_1) \\ A(p_2) \\ \vdots \\ A(p_r)
\end{bmatrix}
\tag{9}
$$

The vector p can be given by N_{LP}, N_{HP}, altitude, Mach number or other parameters and their polynomials.

Considering the correspondence between $N_{cor,LP}$ and $N_{cor,HP}$ as mentioned above, use the vector $p1 = \begin{bmatrix} N_{LP} N_{LP}^2 N_{HP} N_{HP}^2 \end{bmatrix}$ and $p2 = \begin{bmatrix} N_{cor,LP} N_{cor,LP}^2 N_{cor,HP} N_{cor,HP}^2 \end{bmatrix}$ to construct the LPV decomposition of A and A_{cor} in (1) and (4) around the large scale flight envelope. The comparison of decomposition results using different vector p are shown in Fig. 5. To facilitate comparison, the eigenvalues of A_{cor} in the reduced coordinates have been converted to the original coordinates. The eigenvalues of A and A_{cor} at the design condition is shown in Fig. 5(a), where the circles are the actual eigenvalues, the dashed lines and the solid lines are the fitting curves of eigenvalues of A and A_{cor} at different powers. The solid lines constructed by $p2$ agree the actual eigenvalues better. The same results can be verified at the condition of 0 ft and Mach 0 as shown in Fig. 5(b). Hence, the PLV model constructed in reduced coordinates is more accuracy than in original coordinates around the full flight envelope.

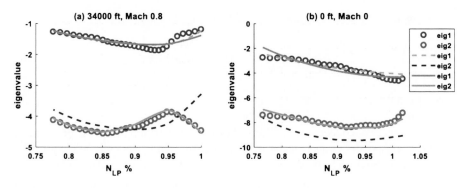

Fig. 5. Comparison of decomposition results using parameters in reduced and original coordinates.

4 Model Analysis

In order to verify the accuracy of the reduction LPV model and study the selection of parameters of vector p, several LPV decompositions are built and compared with the engine model linearized at different conditions and the general reduction model using gas dynamic similarity. The deviations of $A_{cor}, B_{cor}, C_{cor}, D_{cor}$ and eigenvalues of A_{cor} are shown in Table 2. The five models are constructed as follows:

I. general reduction model in (4),

II. reduction LPV model, $p = \left[N_{cor,LP}, N^2_{cor,LP}, N_{cor,HP}, N^2_{cor,HP}\right]$,

III. reduction LPV model, $p = \left[N_{cor,LP}, N^2_{cor,LP}, N_{cor,HP}, N^2_{cor,HP}, T_{t,in}\right]$,

IV. reduction LPV model, $p = \left[N_{cor,LP}, N^2_{cor,LP}, N_{cor,HP}, N^2_{cor,HP}, T_{t,in}, P_{t,in}\right]$,

V. reduction LPV model, $p = \left[N_{cor,LP}, N^2_{cor,LP}, N_{cor,HP}, N^2_{cor,HP}, T_{t,in}, P_{t,in}, W_{f,cor}\right]$.

Table 2. Deviations of $A_{cor}, B_{cor}, C_{cor}, D_{cor}$ and eigenvalues of A_{cor} in different model

		I	II	III	IV	V
A/%	Max	81.78	30.35	30.31	30.07	26.01
	Mean	20.55	9.74	9.58	8.88	8.61
B/%	Max	40.52	35.63	35.72	34.87	34.76
	Mean	14.13	10.07	8.35	7.51	7.26
C/%	Max	75.59	23.85	20.85	22.83	18.32
	Mean	18.83	6.82	6.20	6.15	5.69
D/%	Max	10.48	9.63	3.88	3.88	3.89
	Mean	2.64	3.11	0.57	0.57	0.57
Eig1	Max	0.51	0.34	0.33	0.32	0.31
	Mean	0.08	0.09	0.09	0.08	0.08
Eig2	Max	1.35	0.19	0.21	0.23	0.28
	Mean	0.32	0.06	0.05	0.05	0.04

The relative error is determined by the following formula:

$$err_A = \frac{\left\| \hat{A} - A \right\|}{\left\| A_{std} \right\|}, \left\| X \right\| = \sqrt{tr(X^T X)} \tag{10}$$

where $\hat{A}$ is a fitting matrix of A, elements of A_{std} are the variation ranges of elements of A at different powers in design condition.

For the general reduction model I, most of the mean error and max error is greater than the corresponding value of the other model in Table 2. For the reduction LPV models, more parameters in vector p lead better decomposition results. The set of $p = \left[N_{cor,LP}, N_{cor,LP}^2, N_{cor,HP}, N_{cor,HP}^2, T_{t,in} \right]$ in the model III gets the similar error compared with the model IV and V. This means the other parameters in later models are redundant practically.

The fitting error of model I and III is shown in Fig. 6, where 255 conditions in flight envelope are simulated with the $W_{f,cor}$ in the range of 42–100% and the Mach number increases along the x-axis. A general trend as shown in Fig. 6(a) is the mean error and max error increases as the Mach number decreases. This indicates the model I cannot agree the turbofan engine reliably in full area envelope. In contrast, the model III has closer and smaller fitting error in different conditions as shown in Fig. 6(b).

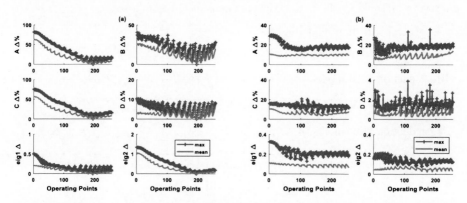

Fig. 6. General reduction model I(a) and reduction LPV model III (b) comparisons to the engine model linearized at 255 different conditions in flight envelope, each condition with the $W_{f,cor}$ in the range of 42–100%.

The frequency responses of model I and III compared with the reference engine model linearized at different conditions is shown in Figs. 7 and 8. The solid lines of two colors show the frequency responses of the model I and model III, and the dashed lines show the frequency responses of reference model at the same condition. For the condition of 0 ft and Mach 0 with 94% $W_{f,cor}$ shown in Fig. 7, the differences of frequency responses between model III and the reference model are smaller. For the condition of 40000 ft and Mach 1 with 42% $W_{f,cor}$ shown in Fig. 8, there are little differences between the frequency responses of three models.

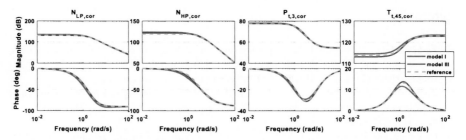

Fig. 7. Frequency response of model I and III compared with the engine model linearized at condition of 0 ft and Mach 0 with 94% $W_{f,cor}$.

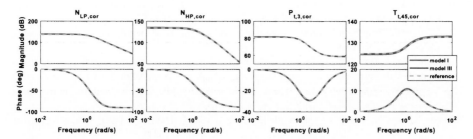

Fig. 8. Frequency response of model I and III compared with the engine model linearized at condition of 40000 ft and Mach 1 with 42% $W_{f,cor}$.

5 Conclusion

In this paper, an integrated reduction model using the gas dynamic similarity theory and the linear parameter-varying (LPV) strategy is proposed, which can describe the turbofan engine dynamic behavior at different powers around the full defined flight envelope. The proposed models constructed by different parameters are compared with the reference model and the general reduction model using the gas dynamic similarity. The results show that:

(1) It exists unneglectable error in the original reduction model using the gas dynamic similarity. And the LPV strategy is an effective method to correct the error. The proposed model performs more accurate than the original reduction model.

(2) The reduction LPV model proposed in this paper is more appropriate to describe the dynamic behavior of aero engine in the full envelope than the general LPV model. Compared with the model in reduced coordinates, the parameters of the model in original coordinates changes irregularly around the envelope.

(3) To describe the turbofan engine dynamic behavior at different powers in the full envelope, the parameters consisting of $N_{cor,LP}, N^2_{cor,LP}, N_{cor,HP}, N^2_{cor,HP}, T_{t,in}$ perform well. More complicated parameters selection will not improve the accuracy clearly.

References

1. Kulikov GG, Thompson HA (2004) Dynamic modelling of gas turbines: identification, simulation, condition monitoring and optimal control. Springer, London
2. Richter H (2012) Advanced control of turbofan engines. Springer, New York
3. Zinnecker A, Chapman JW, Lavelle TM, Litt JS (2014) Development of a twin-spool turbofan engine simulation using the toolbox for modeling and analysis of thermodynamic systems (T-MATS)
4. Chapman JW, Lavelle TM, Litt JS, Guo T (2014) A process for the creation of T-MATS propulsion system models from NPSS data
5. Gilbert W, Henrion D, Bernussou J, Boyer D (2010) Polynomial LPV synthesis applied to turbofan engines. Control Eng Pract 18(9):1077–1083
6. Huang J, Ji G, Zhu Y, van den Bosch P (2012) Identification of multi-model LPV models with two scheduling variables. J Process Control 22(7):1198–1208
7. Verdult V, Verhaegen M (2005) Kernel methods for subspace identification of multivariable LPV and bilinear systems. Automatica 41(9):1557–1565
8. You J, Lu J, Zhu Y, Yang Q, Zhu J, Huang J et al (2013) Identification of multimodel LPV models with asymmetric Gaussian weighting function. J Appl Math 2013:1–12
9. Yang S, Wang X, Long Y, Li Z, Hu Z, Yin K et al (2016) Partition method for improvement of piecewise linear model with full envelope covered
10. Du X, Sun X, Wang Z, Dai A (2017) A scheduling scheme of linear model predictive controllers for turbofan engines. IEEE Access 5:24533–24541
11. Lu F, Qian J, Huang J, Qiu X (2017) In-flight adaptive modeling using polynomial LPV approach for turbofan engine dynamic behavior. Aerosp Sci Techn 64:223–236
12. Wu B, Huang J (2016) Self-scheduled control method for aero-engine based on sum of squares programming of polynomial. J Aerosp Power (06):1460–1468
13. Li S, Zhang S, Hu W (2011) Equilibrium-manifold based linear parameter varying modeling for aeroengine. J Propul Tech (01):21–25
14. Liu X, Yuan Y, Shi J, Zhao L (2013) Adaptive modeling of aircraft engine performance degradation model based on the equilibrium manifold and expansion form. Proc Inst Mech Eng Part G: J Aerosp Eng 228(8):1246–1272
15. Wang B, Wang X, Shi Y, Wang H (2014) A real-time piecewise linear dynamic model of aeroengine. J Aerosp Power (03):696–701

A Frequency Domain Identification with Maximum Likelihood Method for Aircraft Engine

Nan Liu[(✉)]

Shanghai Aircraft Design and Research Institute, Commercial Aircraft
Corporation of China, Ltd., No. 5188, Jinke Road, Pudong New Area,
Shanghai 201210, China
liunan0530@163.com

Abstract. Due to the complicated non-linearity of aircraft engine, it is very
hard to design the control system directly. Now, linear model which could take
place of real engine is necessary during the design process of aircraft engine
control system. Normally, partial derivative method [1] and fitting technology
[2, 3] usually are used to the linear model establishment of aircraft engine.
Partial derivative method could be used for the steady state point identification,
and fitting could extend the state variable model to transient process. The
accuracy of fitting generally better than partial derivative method. However,
when the noise exists, the accuracy of previous two method is not satisfactory. In
this paper, the statistical characteristic of input, output and noise signal is
analyzed, and a maximum likelihood identification method in frequency domain
is proposed. This method establishes an optimized parameter estimation criterion function in frequency domain. By designing a multi-sinusoidal excitation
signal, the Levenberg-Marquardt algorithm is applied to solve the criterion
function to establish the control system model. The proposed method is applied
to identify the shaft speed system of aircraft engine and the linear models at
different points are established. From the simulation results, it can be seen that
the aircraft engine models established have high accuracy.

Keywords: Frequency domain · Maximum likelihood · Linear model ·
Aircraft engine

1 Frequency Domain Model of SISO System

For a single input single output control system, its frequency model is shown as Fig. 1.
$U(j\omega)$ is input, $Y(j\omega)$ is output, $N_x(j\omega)$ is noise of input end, $N_y(j\omega)$ is the noise of
output end, $U_r(j\omega)$ is the measurement of input, $Y_r(j\omega)$ is the measurement of output.

The frequency response function of the system is the ratio of output and input
Fourier transform.

$$H(j\omega) = \frac{Y(j\omega)}{U(j\omega)} \tag{1}$$

where, $Y(j\omega) = \int_0^\infty y(t)e^{-j\omega t}dt$, $U(j\omega) = \int_0^\infty u(t)e^{-j\omega t}dt$.

© Springer Nature Singapore Pte Ltd. 2019
X. Zhang (Ed.): APISAT 2018, LNEE 459, pp. 1767–1774, 2019.
https://doi.org/10.1007/978-981-13-3305-7_142

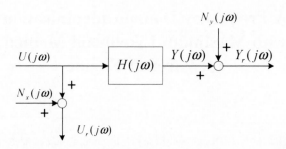

Fig. 1. Frequency domain model of single variable system

For $Y(j\omega)$ and $U(j\omega)$, their discrete form can be used for approximate calculation.

$$Y(j\omega_k) = T \sum_{l=0}^{N-1} y(lT)e^{-j\omega_k lT} \quad k = 1, 2, \cdots, m \tag{2}$$

$$U(j\omega_k) = T \sum_{l=0}^{N-1} u(lT)e^{-j\omega_k lT} \quad k = 1, 2, \cdots, m \tag{3}$$

where, T is sampling period, $m = N/2$, $\omega_k = 2\pi k/(NT)$.

The noise signal included in input and output measurement of the system will show approximate Gauss distribution with the increase of the sampling number [4]. When $N > 512$, the mean value in a period is approximate zero, and the impact of noise could be reduced by averaging the input and output signals.

$$\hat{H}_{EV}(j\omega_k) = \frac{\dfrac{1}{M}\sum_{m=1}^{M} Y_m(j\omega_k)}{\dfrac{1}{M}\sum_{m=1}^{M} U_m(j\omega_k)} = \frac{\bar{Y}(j\omega_k)}{\bar{U}(j\omega_k)} \tag{4}$$

where, $U_m(j\omega_k)$ and $Y_m(j\omega_k)$ are the spectrum of input and output signals in m period under the frequency ω_k respectively, M is the number of period. If input and output measurements are performed at the same time, the estimation is unbiased and efficient [5].

Assume the transfer function is

$$H(s) = \frac{b_0 + b_1 s + \ldots + b_m s^m}{a_0 + a_1 s + \ldots + a_n s^n} = \frac{N(s)}{D(s)} \tag{5}$$

where, m is the number of zero, n is the number of pole.

2 Maximum Likelihood Criterion Function

Define the parameter vector

$$p = [a_0, a_1, \cdots, a_n, b_0, b_1, \cdots b_m] \tag{6}$$

The implementation of maximum likelihood estimation can be based on the following criterion function [6].

$$K(p) = \frac{1}{2} \sum_{k=1}^{F} \frac{|N(j\omega_k, p)\bar{U}(j\omega_k) - D(j\omega_k, p)\bar{Y}(j\omega_k)|^2}{\sigma_y^2(\omega_k)|D(j\omega_k, p)|^2 + \sigma_u^2|N(j\omega_k, p)|^2 - 2\text{Re}\{\sigma_{xy}(j\omega_k)N^*(j\omega_k, p)D(j\omega_k, p)\}} \tag{7}$$

where, $\bar{U}(j\omega_k)$ and $\bar{Y}(j\omega_k)$ are the average of Fourier coefficients of input and output. F is the number of excitation frequency. $\sigma_u(\omega_k)$ 和 $\sigma_y(\omega_k)$ are the variance of the real and imaginary parts of the input and output. $\sigma_{uy}(j\omega_k)$ is the co-variance of the input and output noise. "*" means conjugated transform.

For $\sigma_u(\omega_k)$ and $\sigma_y(\omega_k)$, they could be got from measurement.

$$\sigma_u^2(\omega_k) = \frac{1}{M-1} \sum_{m=1}^{M} (U_m(j\omega_k) - \bar{U}_m(j\omega_k))(U_m(j\omega_k) - \bar{U}_m(j\omega_k))^* \tag{8}$$

$$\sigma_y^2(\omega_k) = \frac{1}{M-1} \sum_{m=1}^{M} (Y_m(j\omega_k) - \bar{Y}_m(j\omega_k))(Y_m(j\omega_k) - \bar{Y}_m(j\omega_k))^* \tag{9}$$

For $\sigma_{uy}(j\omega_k)$, it could be calculated as

$$\sigma_{uy}(\omega_k) = \frac{1}{M-1} \sum_{m=1}^{M} (U_m(j\omega_k) - \bar{U}_m(j\omega_k))(Y_m(j\omega_k) - \bar{Y}_m(j\omega_k)) \tag{10}$$

In this paper, Levenberg-Marquardt (L-M) iterative algorithm is studied to solve criterion function.

$$p_{i+1} = p_i + \Delta p_i \tag{11}$$

$$\Delta p_i = -[J(p_i)^T J(p_i) + \mu I]^{-1} J(p_i)e(p_i) \tag{12}$$

where, $J(p)$ is Jacobian matrix, μ is a constant coefficient, $\mu > 0$, $e(p_i) = [e_1(p_i) \quad \cdots \quad e_F(p_i)]$, $e_k(p_i)$ is

$$e_k(p_i) = \frac{1}{2} \frac{|N(j\omega_k, p_i)\bar{U}(j\omega_k) - D(j\omega_k, p_i)\bar{Y}(j\omega_k)|^2}{\sigma_y^2(\omega_k)|D(j\omega_k, p_i)|^2 + \sigma_u^2|N(j\omega_k, p_i)|^2 - 2\text{Re}\{\sigma_{xy}(j\omega_k)N^*(j\omega_k, p_i)D(j\omega_k, p_i)\}} \tag{13}$$

3 Multi-sinusoidal Excitation Signal

In this paper, a multi-sinusoidal signal is designed as following

$$u(t) = \sum_{k=1}^{F} a(k) \cos(i(k) \cdot 2\pi f_0 t + \phi(k)) \tag{14}$$

where, $\boldsymbol{a} = [a(1), \cdots, a(k), \cdots, a(F)]$ is amplitude vector, $\boldsymbol{i} = [i(1), \cdots, i(k), \cdots, i(F)]$ is harmonic vector, $\boldsymbol{\phi} = [\phi(1), \cdots, \phi(k), \cdots, \phi(F)]$ is phase vector, f_0 is base frequency.

Amplitude vector is set through Schroeder method, $a(k) = \left(\dfrac{p_k}{2}\right)^{\frac{1}{2}} \times P$. $\sum_{k=1}^{F} p_k = 1$,

$p_k = \dfrac{1}{F}$, $a(k) = \left(\dfrac{1}{2F}\right)^{\frac{1}{2}} \times P$.

Harmonic vector could be set based on the frequency characteristic of fuel loop and nozzle loop, and it is $\boldsymbol{i} = [1, 5, 9, \cdots, 157]$. Obviously, $F = 40$.

In order to obtain the phases vector $\boldsymbol{\phi}$, the crest factor (CF) $CF = \dfrac{\max\{|u(t)|\}}{\mathrm{rms}\{u(t)\}}$ should be introduced, where, rms is root mean square. Here, $L\infty$ method is employed to optimize CF [7]. Figures 2 and 3 are time domain graph of multi-sinusoidal excitation and its frequency domain spectrum for $H = 0$ m, Mach number $Ma = 0$.

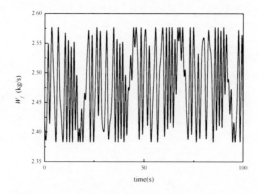

Fig. 2. Time domain graph of multi-sinusoidal

4 Dynamic Characteristic Identification of Aircraft Engine

In this paper, the structure of the linear model of aircraft engine is determined firstly, then the parameters are identified.

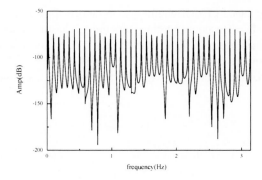

Fig. 3. Spectrum of multi-sinusoidal

4.1 Model Structure

For the low pressure rotor speed system, there are 3 kinds of potential structures [8], (1) zero Z = 0, pole P = 1, (2) zero Z = 1, pole P = 2, (3) zero Z = 2, pole P = 3.

For the condition of height $H = 0$ m, Mach number $Ma = 0$, the identification results obtained by the proposed maximum likelihood estimation in frequency domain are compared with the frequency characteristic of the nonlinear model, whose accuracy

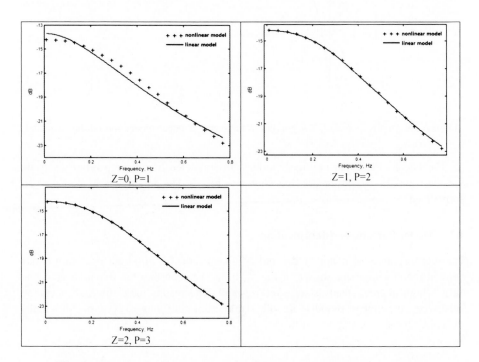

Fig. 4. Frequency characteristic of identification results and nonlinear model

is too high to take place of real engine, as shown by Fig. 4. It can be found that structure (1) Z = 0, P = 1 is poor in fitting, structure (3) has the poles in the right half S plane, which indicates unstable poles of the linear model. So these two structures are eliminated, and structure (2) Z = 1, P = 2 is the best choice.

Finally, low pressure rotor speed system transfer function can be determined as

$$G_{N_L}(s) = \frac{K^*(s-z)}{(s-p_1)(s-p_2)} \tag{15}$$

For height $H = 0$ m, Mach number $Ma = 0$, maximum likelihood identification method in frequency domain is applied to low pressure rotor speed system, the transfer function can be solved as

$$G_{N_L}(s) = \frac{0.35099(s-(-3.537))}{(s-(-2.207+1.237i))(s-(-2.207-1.237i))} \tag{16}$$

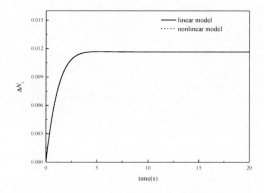

Fig. 5. The step response of low pressure rotor speed of linear model and nonlinear model

The linear model and nonlinear model are simulated by 2% step input verify the accuracy, and the system response is shown as Fig. 5. The parameters in the figure are normalized.

4.2   Model Parameter Identification

The step responses of linear model and nonlinear model in different operation conditions in the envelope are shown in Fig. 6. It can be seen that the dynamic processes have a high degree, steady-state error is small. The results have the high accuracy. Otherwise, the linear models are all stable minimum phase system. The multi-sinusoidal responses are shown in Fig. 7.

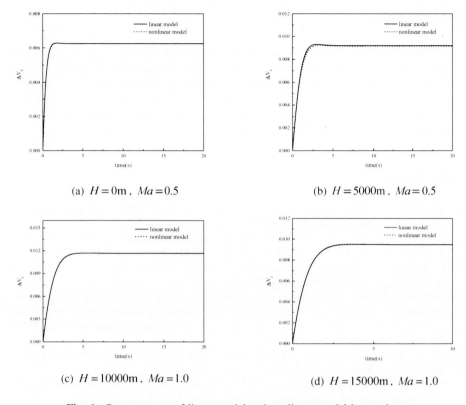

(a) $H = 0$m, $Ma = 0.5$

(b) $H = 5000$m, $Ma = 0.5$

(c) $H = 10000$m, $Ma = 1.0$

(d) $H = 15000$m, $Ma = 1.0$

Fig. 6. Step response of linear model and nonlinear model in envelope

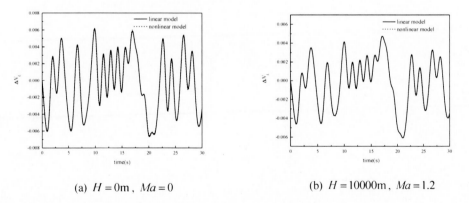

(a) $H = 0$m, $Ma = 0$

(b) $H = 10000$m, $Ma = 1.2$

Fig. 7. Multi-sinusoidal response of nonlinear and nonlinear model in envelope

5   Conclusion

A frequency domain maximum likelihood identification method is proposed in this paper. In this method, a criterion function with statistical characteristic is defined, and solved by L-M algorithm through a multi-sinusoidal excitation signal. The method is applied to rotor speed system of aircraft engine, and the simulation results shows that established linear model has high accuracy compared to nonlinear model in the envelope.

References

1. Sugiyama N (1992) Derivation of ABCD system matrices from nonlinear dynamic simulation of jet engines. Technical report, AIAA 92-3319
2. Lu J, Guo YQ, Chen XL (2011) Establishment of aero-engine state variable model based on linear fit-ting method. J Aerosp Power 26(5):1172–1177
3. Lu F, Huang JQ, She YF (2011) State space modeling based on QPSO hybrid method for aero-engines. J Propul Technol 32(5):722–727
4. Brillinger DR (1974) Fourier analysis of stationary processes. Proc IEEE 62(12):1628–1643
5. Guillaume P, Pintelon R, Schoukens J (1995) Nonparametric frequency response function estimators based on nonlinear averaging techniques. IEEE Trans Instrum Meas 41(6):739–746
6. Pintelon R, Guillaume P, Rolain Y et al (1992) Identification of linear system captured in a feedback loop. IEEE Trans IM-41(6):747–754
7. Kim WJ, Cho KJ, Stapleton SP et al (2007) An efficient crest factor reduction technique for wideband applications. Analog Integr Circ Sig Process 51(1):19–26
8. Evans C, Rees D, Hill D (1998) Frequency-domain identification of gas turbine dynamics. IEEE Trans Control Syst Technol 6(5):651–662

Extended Kalman Filter Infusion Algorithm Design and Application Characteristics Analysis to Stochastic Closed Loop Fan Speed Control of the Nonlinear Turbo-Fan Engine

Xiaowu Lv and Yuansuo Zhang[⊠]

AECC Commercial Aircraft Engine Co., Ltd., Shanghai, China
zhangyuansuo1983@163.com

Abstract. An Extended Kalman Filter (EKF) was proposed for fan speed signal infusion of the turbo fan engine. Firstly, a nonlinear discrete time analytical engine model was identified using a general nonlinear engine mathematical model based on least square method. Afterwards, fan speed signal infusion EKF algorithm was designed based on optimal filtering theory for nonlinear multistage dynamic process. Then, Kalman gains of EKF algorithm were offline tuned, and fan speed signal was synthesized by using 6 different combinations of 4 sensed parameters, T_{25}, T_3, P_{s3} and EGT as input of the infusion EKF algorithm, and the infused fan speed signal under different infusion combinations were compared with the actual fan speed signal sensed directly from nonlinear engine model using small step test cases. After that, fan speed closed loop control simulations were conducted using the infused fan speed signal as feedback, and algorithms with 6 different infusion combinations were analyzed with respects to control performance and stability using small step command test case under ideal environment. Finally, closed loop control simulation was conducted with the fan speed EKF infusion signal from selected 3 parameter infusion algorithm as feedback using both small and big step command test case under stochastic environment. The results show that, under both ideal (without consideration of noise) and stochastic conditions, the proposed EKF fan speed signal infusion algorithm has a good closed loop control performance

Keywords: Turbo-fan engine · Extend Kaman Filter (EKF) ·
Nonlinear analytic model · Control performance, signal infusion

1 Introduction

Aero gas turbine engine has become the main driving force of commercial aircraft since about 1950s worldwide. And its main performance characteristics (such as thrust, fuel consumption, noise and so on) have reached relatively mature levels. Nowadays, airlines and passengers are more concerned about flight safety, maintenance and operating costs. To find new potential way to improve the safety of the turbofan engine has become a hot topic in the field of aviation engine control field [1].

© Springer Nature Singapore Pte Ltd. 2019
X. Zhang (Ed.): APISAT 2018, LNEE 459, pp. 1775–1794, 2019.
https://doi.org/10.1007/978-981-13-3305-7_143

As an advanced power plant, the working environment of the aero gas turbine engine (noise, disturbance, etc.) is much better than some other application, such as rockets, ships and unmanned aerial vehicles. The parameters are used as input for closed loop control of the engine can be easily and reliably measured and processed. However, safety requirements derived from aircraft to aero engine have always been increasing, so new ways to assure the closed loop control still provide a potential choice for safety consideration. In this paper, the fan speed infusion algorithm is proposed as an alternate way for engine closed loop control. It is based on the Kalman Filter principle of the optimal estimation and optimal control process theory. Based on the thermodynamic parameters of the turbofan engine (such as temperature and pressure) the speed signal is infused, and then it is used as a feedback to implement the closed-loop control [2–6].

In the field of aero engine, the Kalman filter algorithm is mainly focused on the engine health diagnosis and fault detection. However, using Kalman filter in closed control engine control is rare.

In ref [7], one observer was designed and model verified for bias compensation with a truck engine application. In ref [8] a new method of target tracking by EKF using bearing and elevation measurements for underwater environment was investigated. In ref [9], A method for on-line map adaption was developed and EKF was utilized as a parameter estimator. In ref [10], a computationally efficient Kalman filter based estimator for updating lookup table applied to NOx estimation in diesel engines was verified. In ref [11], one EKF based quad-rotor state estimator based on asynchronous multi-sensor data was investigated. In ref [12], cycle-by-cycle Based In-cylinder temperature was estimated for diesel engines based on estimator. In ref [13], cylinder air charge was estimated for spark-ignition engines based on designed observer. In ref [14], one nonlinear recursive estimator was designed and analyzed for physical and semi physical Engine.

TAN [15] had compared the unknown input observer (UIO) and Kalman filter in his paper. Lu [16] had predicted the engine health parameters with high accuracy and robustness under the constraint of limited parameters based on EKF principle in the whole flight envelope. In ref [17], the authors had analyzed the different effects of tracking filters, extended Kalman Filters and predictive controllers on the dynamic characteristics of performance degradation engines.

From the engineering perspective, there are usually two important ways to estimate the main controlled parameters of aero engine [18, 19] (etc. core engine rotor speed, turbine pressure ratio, etc.).

The first is the equivalent table and formula method. The relationship between different parameters of aero engine can be interpolated as table or curve through engine test data, and based on the measurable parameters (such as temperature, pressure, etc.) as input, we could easily get the estimation of main controlled parameter with great credibility. But the drawback of this approach is the loss of control accuracy and dynamic performance.

The second way is the control law (in case of an fan speed signal failure, the use of more conservative or alternative controls such as engine core speed signal controls instead). Likewise, this approach also significantly reduces the performance of the engine control. Besides, shutting down of failed engine is always an acceptable to the certification authority.

In this paper, a complete fan speed EKF infusion algorithm design process was presented. Under different infusion parameter configurations, ideal/noise conditions and steady/transient cases were tested. It was shown that three parameters (T_{25}, T_3, P_{s3}) infusion algorithm was the best among the other infusion algorithms in terms of control performance and control stability.

2   EKF Infusion Algorithm Design

2.1   Structure Selection for Nonlinear Discrete Engine Analytic Model

Universal turbo fan engine model can be described as following format [20, 21]:

$$\dot{X} = f(X, U, t) \tag{1.1}$$

$$Y = g(X, U, t) \tag{1.2}$$

For the dynamic system, X is n-dimension state *variable* vector of the dynamic system, Y is m-dimension output variable vector, U is r-dimension input vector, f is the state vector function that describes dynamic characteristics between the input vector and state vector of the dynamic system, g is the output state vector function that describes dynamic characteristics between the state vector and output vector of the dynamic system.

Turbo fan engine is a typical dynamic system with deep coupling among combustion, heat transfer and dynamics. The Newton Law can be used to describe the key dynamics of the engine that unusually include two spool kinematics equations. And nonlinear terms in the equations are the key factors that can invoke the dynamic and static characteristics, not only in different state but of different environment working conditions. The variations of gas composition and thermodynamic parameters (specific heat, enthalpy, entropy, etc.) as constraints of the two base dynamic equations. Meanwhile, it is inevitably to consider the quality and energy conservation characteristics of the dynamic system as the constraints equations.

As shown in reference [18], there are 6 equations needed to describe a nonlinear engine dynamic system. Two of which are differentials equations and the others are algebraic equations.

$$\begin{cases} \dfrac{\pi}{30} \cdot J_2 \cdot \dot{N}_2 = \Delta MT_2 \\ \dfrac{\pi}{30} \cdot J_1 \cdot \dot{N}_1 = \Delta MT_1 \end{cases} \tag{1.3}$$

$$g_i(\pi_{CL}, \pi_{CH}, \pi_{TL}, \pi_{TH}) = 0, \; i = 1, 2, 3, 4 \tag{1.4}$$

As shown above, ΔMT_2, ΔMT_2 residual torques of the are low pressure spool and high-pressure spool respectively that accelerate or decelerate the rotor spools. J_2 and J_1

are the moment of inertia for the low and high spools. π_{CL} and π_{CH} are the pressure ratios of the low and high compressor of the engine, π_{TL}, π_{TH} are the pressure ratios of the low and high turbine of the engine.

As the system can be described by nonlinear Eqs. (1.3, 1.4), and there is no complete theory that can be used as the foundation from the design of the filter, controller or compensator to the integration of a whole control system. If the aero engine model can be expressed as the analytical Eq. (1.1, 1.2), the optimal control and estimation theory [2–6] and the nonlinear control theory [20, 21] can provide the solid foundation for the solution.

As it is often difficult to capture the analytical solutions of nonlinear Eq. (1.3, 1.4), so numerical method [22, 23] is commonly used such as Newton-Raphson iteration method. The nonlinear discrete dynamic equation of the turbofan engine is adopted as follows:

$$N_1(k+1) = AA(N_{2Rr}(k)) \cdot N_1(k) + BB(N_{2Rr}(k)) \cdot W_f(k) + EE(N_{2Rr}(k)) \quad (1.5)$$

$$Y(k) = CC(N_{2Rr}(k)) \cdot N_1(k) + DD(N_{2Rr}(k)) \cdot W_f(k) \quad (1.6)$$

In the Eqs. (1.5, 1.6), k is the sampling time instants with the units of sampling time interval, N_1 is fan physical speed, N2Rr is core relative corrected speed. Y is the column vector that consists of engine output signal $Y = [T_{25}, T_3, P_{s3}, EGT]^T$, CC is state output transfer vector $CC = [CCT25, CCT3, CCPs3, CCEGT]^T$. DD is output coefficient.

2.2   Nonlinear Discrete Engine Analytic Model Design

The CMAPSS model [24] is used as nonlinear object, it assist the design of EKF and verification of N1 infusion EKF algorithm. The model is based on the differential equations and algebraic equations, and the numerical solutions of the model (steady state and transient solution) are based on the solution of nonlinear equations. And it has DLL(Dynamic Link library) format that can be called in the MATLAB simulation environment. the numerical model (DLL format) is used as the controlled object. The time-domain method is used to identify the discrete-time nonlinear Eqs. (1.5, 1.6).

At a series of steady-state points of turbofan engine, fuel flow rate is Wf, and its increment is DWf, and taking $Wf + DWf$ as input for the numerical engine model. The time step of the non-linear engine model is set the same as the sampling period (such as 5–20 ms), and the numerical sequences of the engine are obtained at the sampling instants $\{Wf(k)\}, \{N_1(k)\}, \{N_2(k)\}, \{T_{25}(k)\}, \{EGT(k)\}, \{P_{s3}(k)\}$.

$$\frac{N_1(k+1) - N_1(1)}{N_1(1)} = a \cdot \left(\frac{N_1(k) - N_1(1)}{N_1(1)}\right) + b \cdot \left(\frac{W_f(k+1) - W_f(1)}{W_f(1)}\right) \quad (1.7)$$

$$\frac{T_{25}(k) - T_{25}(1)}{T_{25}(1)} = CCT25 \cdot \left(\frac{N_1(k) - N_1(1)}{N_1(1)}\right) \quad (1.8)$$

$$\frac{T_3(k) - T_3(1)}{T_3(1)} = CCT3 \cdot \left(\frac{N_1(k) - N_1(1)}{N_1(1)} \right) \tag{1.9}$$

$$\frac{P_{s3}(k) - P_{s3}(1)}{P_{s3}(1)} = CCPs3 \cdot \left(\frac{N_1(k) - N_1(1)}{N_1(1)} \right) \tag{1.10}$$

$$\frac{EGT(k) - EGT(1)}{EGT(1)} = CCEGT \cdot \left(\frac{N_1(k) - N_1(1)}{N_1(1)} \right) \tag{1.11}$$

The small step change of Wf is considered as the input of CMAPSS model, and the model's output such as T_{25}, T_3, P_{s3}, EGT are captured. The relationship between the Wf and output of engine are described using the linear Eqs. (1.7, 1.8, 1.9, 1.10, 1.11), and scalars a, b, $CCT25$, $CCT3$, $CCPs3$, $CCEGT$ are the parameters to be identified [25, 26].

Constructing the following vectors:

m-dimension vector N:

$$N = \left[\left\{ \frac{N_1(1) - N_1(1)}{N_1(1)} \right\}, \ldots \left\{ \frac{N_1(m) - N_1(1)}{N_1(1)} \right\} \right]^T;$$

m-dimension vector M:

$$M = \left[\left\{ \frac{N_1(2) - N_1(1)}{N_1(1)} \right\}, \ldots \left\{ \frac{N_1(m+1) - N_1(1)}{N_1(1)} \right\} \right]^T;$$

m-dimension vector M1:

$$M1 = \left[\left\{ \frac{T_{25}(1) - T_{25}(1)}{T_{25}(1)} \right\}, \ldots \left\{ \frac{T_{25}(m) - T_{25}(1)}{T_{25}(1)} \right\} \right]^T;$$

m-dimension vector M2:

$$M2 = \left[\left\{ \frac{T_3(1) - T_3(1)}{T_3(1)} \right\}, \ldots \left\{ \frac{T_3(m) - T_3(1)}{T_3(1)} \right\} \right]^T;$$

m-dimension vector M3:

$$M3 = \left[\left\{ \frac{P_{s3}(1) - P_{s3}(1)}{P_{s3}(1)} \right\}, \ldots \left\{ \frac{P_{s3}(m) - P_{s3}(1)}{P_{s3}(1)} \right\} \right]^T;$$

m-dimension vector M4:

$$M4 = \left[\left\{ \frac{EGT(1) - EGT(1)}{EGT(1)} \right\}, \ldots \left\{ \frac{EGT(m) - EGT(1)}{EGT(1)} \right\} \right]^T;$$

mx2-dimension vector N:

$$N = \left[\left\{ \frac{N_1(1) - N_1(1)}{N_1(1)} \right\}, \left(\frac{W_f(1) - W_f(1)}{W_f(1)} \right); \left\{ \frac{N_1(2) - N_1(1)}{N_1(1)} \right\}, \left(\frac{W_f(2) - W_f(1)}{W_f(1)} \right); \dots; \right.$$
$$\left. \left\{ \frac{N_1(m) - N_1(1)}{N_1(1)} \right\}, \left(\frac{W_f(j) - W_f(1)}{W_f(1)} \right) \right];$$

The linear matrix equations are constructed, and Least Squared Method could be easily used to capture the optimizing parameters $a, b, CCT25, CCT3, CCPs3, CCEGT$ as for Eqs. (1.12, 1.13, 1.14, 1.15, 1.16).

$$M = N \cdot [a, b]^T \tag{1.12}$$

$$M1 = CCT25 \cdot N \tag{1.13}$$

$$M2 = CCT25 \cdot N \tag{1.14}$$

$$M3 = CCT25 \cdot N \tag{1.15}$$

$$M4 = CCT25 \cdot N \tag{1.16}$$

It can be derived from Eqs. (1.7, 1.8, 1.9, 1.10, 1.11):

$$N_1(k+1) = a \cdot N_1(k) + \frac{b \cdot N_1(1)}{W_f(1)} \cdot W_f(k) + (1 - a - b) \cdot N_1(1) \tag{1.17}$$

$$T_{25}(k) = \left(\frac{T_{25}(1) \cdot CT25}{N_1(1)} \right) \cdot N_1(k) + (1 - CCT25) \cdot T_{25}(1) \tag{1.18}$$

$$T_3(k) = \left(\frac{T_3(1) \cdot CT3}{N_1(1)} \right) \cdot N_1(k) + (1 - CCT3) \cdot T_3(1) \tag{1.19}$$

$$P_{s3}(k) = \left(\frac{P_{s3}(1) \cdot CCPs3}{N_1(1)} \right) \cdot N_1(k) + (1 - CCT3) \cdot T_3(1) \tag{1.20}$$

$$EGT(k) = \left(\frac{EGT(1) \cdot CCEGT}{N_1(1)} \right) \cdot N_1(k) + (1 - CCEGT) \cdot EGT(1) \tag{1.21}$$

Comparing between (1.17, 1.18, 1.19, 1.20, 1.21) and (1.5, 1.6), we could get the following relationships:

$$AA = a \tag{1.22}$$

$$BB = \frac{b \cdot N_1(1)}{W_f(1)} \tag{1.23}$$

$$EE = (1 - a - b) \cdot N_1(1) \tag{1.24}$$

$$CCT25 = \frac{T_{25}(1) \cdot CT25}{N_1(1)} \tag{1.25}$$

$$CCT3 = \frac{T_3(1) \cdot CT3}{N_1(1)} \tag{1.26}$$

$$CCPs3 = \frac{P_{s3}(1) \cdot CPs3}{N_1(1)} \tag{1.27}$$

$$CCEGT = \frac{EGT(1) \cdot CEGT}{N_1(1)} \tag{1.28}$$

2.3   Structure Design of EKF Fan Speed Algorithm

Based on the optimal filtering principle of nonlinear dynamic process [6], a nonlinear multistage process filter suitable for a turbofan engine is designed. The filter is an Extended Kalman Filter, which is used to infuse the fan speed signal (N1) of the turbofan engine from the other parameters of the engine.

Considering a non-linear multi-stage dynamic process of a turbofan engine with additional pure random noise:

$$N_1(k + 1) = f(N_1(k)) + \Gamma(k)W_f(k) \tag{1.29}$$

$$Y(k) = h(N_1(k)) + v(k) \tag{1.30}$$

In the above Eqs. (1.29, 1.30), $\Gamma(k) = BB(N_{2Rr}(k))$, and $f(N_1(k)) = AA(N_{2Rr}(k)) \cdot N_1(k) + EE(N_{2Rr}(k))$.

The EKF algorithm for dynamic processes is represented by the following iterative calculation Eqs. (1.31, 1.32).

Prediction formula:

$$\hat{N}_1(k) = \bar{N}_1(k) + K(k)(Y(k) - h(\bar{X}(k))) \tag{1.31}$$

Optimized estimation formula:

$$\bar{N}_1(k + 1) = f(\hat{N}_1(k)) + \Gamma(k)W_f(k) \tag{1.32}$$

The formulation process of the Kalman gain is described using the Eqs. (1.33, 1.34, 1.35):

$$K(k) = P(k) \left(\frac{\partial h(k)}{\partial x(k)} \right)^T R(k)^{-1} \qquad (1.33)$$

$$P(k) = M(k) - M(k) \left(\frac{\partial h(k)}{\partial x(k)} \right)^T \left[\left(\frac{\partial h(k)}{\partial x(k)} \right) M(k) \left(\frac{\partial h(k)}{\partial x(k)} \right)^T + R(k) \right]^{-1} \left(\frac{\partial h(k)}{\partial x(k)} \right) M(k)$$

$$(1.34)$$

$$M(k+1) = \left(\frac{\partial f(k)}{\partial x(k)} \right) P(k) \left(\frac{\partial f(k)}{\partial x(k)} \right)^T + \Gamma(k) Q(k) \Gamma(k)^T \qquad (1.35)$$

2.4 Parameter Design of Fan Speed Infusion EKF Algorithm

The calculation principle of the parameters used in the EKK algorithm is shown as below.

$$\frac{\partial h(k)}{\partial x(k)} = [\text{CCT25}, \text{CCT3}, \text{CCPs3}, \text{CCEGT}]^T \qquad (1.36)$$

$$\frac{\partial f(k)}{\partial x(k)} = AA \qquad (1.37)$$

$$Q(k) = E(W_f(k) - \bar{W}_f(k))(W_f(k) - \bar{W}_f(k))^T = 25 \left(kg \cdot h^{-1} \right)^2 \qquad (1.38)$$

$$P(1) = E(N_1(k) - \bar{N}_1(k))(N_1(k) - \bar{N}_1(k))^T = 64(rpm)^2 \qquad (1.39)$$

$$R_{T25} = E(v_{T25}(k) - \bar{v}_{T25}(k))(v_{T25}(k) - \bar{v}_{T25}(k))^T = 25(K)^2 \qquad (1.40)$$

$$R_{T3} = E(v_{T3}(k) - \bar{v}_{T3}(k))(v_{T3}(k) - \bar{v}_{T3}(k))^T = 25(K)^2 \qquad (1.41)$$

$$R_{Ps3} = E(v_{Ps3}(k) - \bar{v}_{Ps3}(k))(v_{Ps3}(k) - \bar{v}_{Ps3}(k))^T = 100(kPa)^2 \qquad (1.42)$$

$$R_{EGT} = E(v_{EGT}(k) - \bar{v}_{EGT}(k))(v_{EGT}(k) - \bar{v}_{EGT}(k))^T = 25(K)^2 \qquad (1.43)$$

Based on Eqs. (1.5, 1.6) and Eqs. (1.22, 1.23, 1.24, 1.25, 1.26, 1.27, 1.28). The Kalman gain of EKF is obtained by using the Eqs. (1.38–1.43) as input conditions and the identified coefficient from linearized Eqs. (1.22–1.28) as engine model information [27].

In the Fig. 1, the curves of Kalman gains for the design EKF infusion are shown as for three parameters (T_{25}, T_3, P_{s3}) configuration. And KT25, KT3, KPs3 are the corresponding elements of $K(k) = [KT25, KT3, KPs3]$.

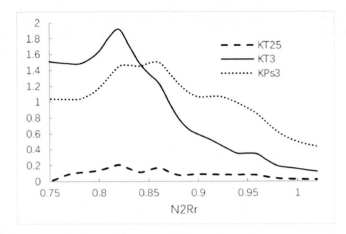

Fig. 1. Kalman gains variations of the designed EKF Algorithm based on triple (T_{25}, T_3, P_{s3}) configuration

3   Closed Control Simulation of EFK Infusion Algorithm

Based on the designed EKF speed fusion algorithm, six combinations (single, two, three and four parameters) are verified in the closed loop simulation environment.

3.1   Closed Loop Integration Principle of EKF Infusion Algorithm

The simulation flow diagram of N_1 closed-loop control simulations is shown in the Fig. 2. The simulation schematics mainly include N_1 control expectation demand, N_1 speed controller (proportional integral controller), fuel metering servo control actuator (servo controller, nonlinear fuel metering device model), nonlinear real-time universal turbofan engine model (CMPSS-40 K) [24].

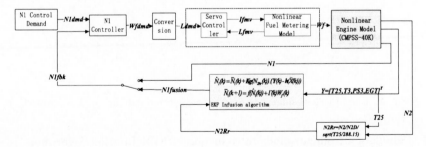

Fig. 2. Closed control simulation schematic diagram based on EKF fan speed infusion algorithm

The fan speed control desired (normalized) command signal is shown in the Fig. 3, the vertical coordinate N1Rr = N_1/N_{1D}, N_{1D} is the design point of fan physical speed.

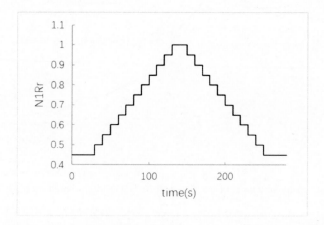

Fig. 3. Closed loop simulation fan speed control expectation demand (Normalized)

3.2   Closed Loop Simulation Diagram of EKF Infusion Algorithm

The simulation model is established in the MATLAB environment in the Fig. 4, an EKF infusion algorithm that is established in the MATLAB/Simulink environment in the Fig. 5.

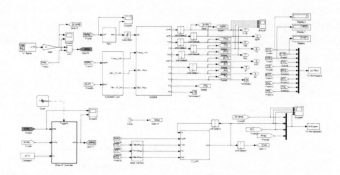

Fig. 4. Closed control simulation diagram of EKF infused fan speed as feedback

3.3   Simulation Configurations Introduction of EFK Infusion Algorithm

For a two-spool turbofan engine, at least two signals are needed for the accurate synthesis the fan speed signal [18, 19]. Therefore, the EKF fusion algorithm for synthesizing N_1 in this paper focuses on the control effect of two parameters, three parameters and four parameters combinations, and the *EGT* single parameter infusion case is taken as base simulation.

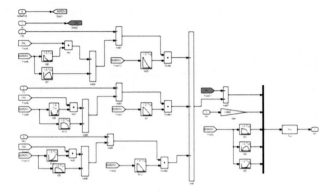

Fig. 5. Model diagram of EKF fan speed signal infusion algorithm

The parameters considered as input of the algorithm in the paper are the four typical state parameters of the typical two-spool turbo fan gas turbine engine, which include T_{25}, T_3, P_{s3} and EGT of the turbofan engine. The detailed combinations of parameters are shown in the following Table 1.

Table 1. Input parameter configuration of the designed EKF infusion algorithm

Single parameter	Double parameters	Triple parameters	Quadruple parameters
EGT	T_{25}, T_3	T_{25}, T_3, P_{s3}	T_{25}, T_3, P_{s3}, EGT
	T_{25}, P_{s3}		
	T_3, P_{s3}		

4 Characteristic Analysis of Closed Loop Control Based on EKF Algorithm

Based on the EKF infusion algorithm, the synthesized N1 signal with different parameters is used as the feedback of closed loop control of the engine actual fan speed. The main simulations results are shown in the following ranging from Figs. 6, 7, 8, 9, 10 and 11. For each pair of diagrams in these figures, the right one is the tracking effect of the infused N1 signal as the closed-loop feedback signal to the speed command. The left diagram is the response of the actual N1 signal (not involved in closed loop control) of the nonlinear engine model.

As shown in Fig. 6, the control performance of the single parameter infusion algorithm is the worst; the synthesized N_1 speed signal produces large amplitude high frequency chattering, but the actual engine fan speed N_1 still tracks the target N_1 command.

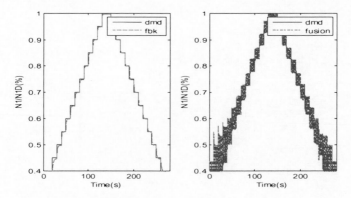

Fig. 6. Time response of closed loop control based on single parameter (EGT)

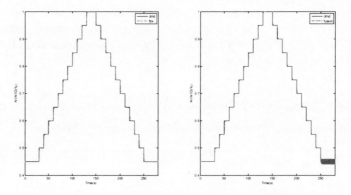

Fig. 7. Time response of closed loop control based on two parameters (T_{25}, T_3)

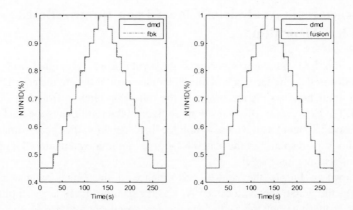

Fig. 8. Time response of closed loop control based on two parameters (T_3, P_{s3})

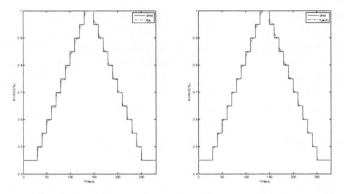

Fig. 9. Time response of closed loop control based on two parameters (T_{25}, P_{s3})

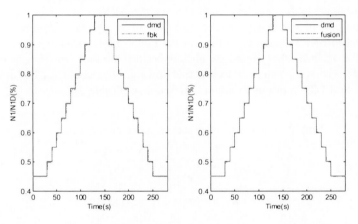

Fig. 10. Time response of closed loop control based on three parameters (T_{25}, T_3, P_{s3})

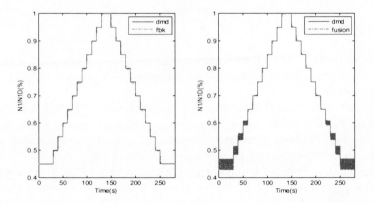

Fig. 11. Time response of closed loop control based on four parameters (T_{25}, T_3, P_{s3}, EGT)

The engine fan speed N_1 signal that is synthesized by T_{25} and T_3 is shown in Fig. 7. The N_1 closed loop dynamic response at low thrust state has a high frequency oscillation, meanwhile the engine actual fan speed approaches the command target as well. The other two parameters infusion configurations shown in the Figs. 8 and 9 indicate that good closed loop performance is achieved as for T_3P_{s3} and $T_{25}P_{s3}$ combination.

The closed-loop control performance of 3 parameter infusion configurations is shown in the Fig. 10 and the infused N_1 signal tracks N_1 control expectation demand very well.

Compared with the 3 parameters (T_{25}, T_3, P_{s3}) infusion, the closed loop control performance of 4 parameters configuration infusion algorithm actually drops slightly shown in Fig. 11, and the closed loop control process of N_1 parameter has a high frequency oscillation especially at low thrust state. It can be seen that the change of EGT has different characteristics from T_{25}, T_3 and P_{s3} for the infusion of fan speed, as the response of EGT to fuel flow rates much faster than that of T_{25}, T_3 and P_{s3} as to nonlinear engine both from theory and practice perspective.

The purpose here is to describe the rationality of the design method and the idea of the fan speed EKF infusion algorithm. In the future the infusion algorithm would be extended to include other sensed parameters besides the four typical selected parameters, and the control parameters such as fuel flow rate (Wf) would also be considered to capture a best EKF infusion algorithm configuration (Table 2).

Table 2. Coordinate index corresponding to each steady state

1	2	3	4	5	6
45–50	50–55	55–60	60–65	65–70	70–75
7	8	9	10	11	12
75–80	80–85	85–90	90–95	95–100	100–95
13	14	15	16	16	18
95–90	90–85	85–80	80–75	75–70	70–65
19	20	21	22		
65–60	60–55	55–50	50–45		

As shown in Figs. 12, 13, the variations of steady state accuracy and settling time of two parameters and three parameter are compared with respects to small step N1 command test case.

Figure 12 shows that steady accuracy of 3 parameters (T_{25}, T_3, P_{s3}) is better than 2 parameters (T_{25}, T_3), and the average steady state accuracy is improved from 0.28% to 0.18%. The Fig. 12 also shows that 3 parameters (T_{25}, T_3, P_{s3}) configuration infusion algorithm meet the engineering performance requirement [20].

Figure 13 shows that the settling time of three parameters (T_{25}, T_3, P_{s3}) is slightly longer than 2 parameters (T_{25}, T_3),and the average settling time is reduced from 1.04 s (T_{25}, T_3) to 0.92 s (T_{25}, T_3, P_{s3}). As shown in Fig. 13, both two parameters (T_{25}, T_3) and 3 parameters (T_{25}, T_3, P_{s3}) infusion EKF algorithm meet the engineering 2 s performance requirement [20].

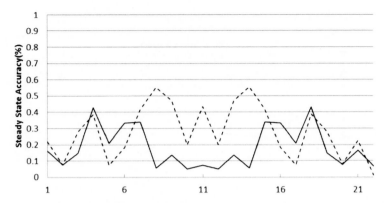

Fig. 12. Steady Accuracy variations on each small step steady state occasion between dashed line two parameters (T_{25}, T_3) and solid line three parameters (T_{25}, T_3, P_{s3})

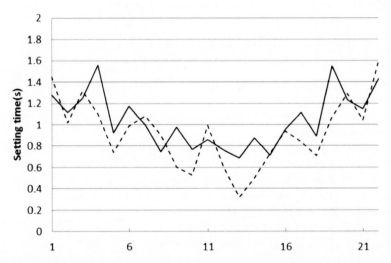

Fig. 13. Settling time index variations on each small step steady state for dashed line two parameters (T_{25}, T_3) and solid line three parameters (T_{25}, T_3, P_{s3}) configuration

As shown in Fig. 14, average steady state accuracy and average settling time of all six combination infusion algorithms are around 2 s and steady state accuracy are around 0.2%, which is decent compared with engineering requirement. The horizontal coordinate is explained in the Table 3. For steady state accuracy, 2 parameters (T_{25}, T_3) infusion algorithm is the worst and is beyond 0.2%, and for settling time, 2 parameters (T_3, P_{s3}) algorithm is the worst and is beyond 2 s.

Both 3 parameters (T_{25}, T_3, P_{s3}) and 4 parameters (T_{25}, T_3, P_{s3}, EGT) infusion algorithm have better weighted control performance over the other combinations. The steady state accuracy of 4 parameters is slightly better than 3 parameters, and the settling time of 3 parameters is slightly better than 4 parameters as shown in the

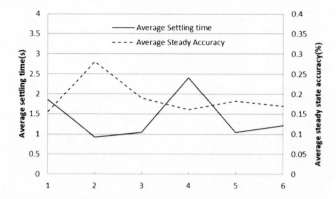

Fig. 14. Average steady accuracy (dashed line) and average settling time (solid line) variations curve under different parameters configurations

Table 3. Horizontal coordinate index corresponding to performance index variation curve

1	2	3	4	5	6
EGT	T_{25}, T_3	T_{25}, P_{s3}	T_3, P_{s3}	T_{25}, T_3, P_{s3}	T_{25}, T_3, P_{s3}, EGT

Fig. 14. Besides, Fig. 11 shows high frequency chattering happens especially when the engine is at low thrust.

The relative ideal fan speed infusion algorithm is 3 parameters (T_{25}, T_3, P_{s3}) among six parameter combinations.

As shown in Fig. 15, fan speed response of closed loop control simulation is plotted using 3 parameters (T_{25}, T_3, P_{s3}) infusion algorithm under stochastic noise condition. And Fig. 16 is the simulation diagram considering the stochastic noise characteristics. The relevant simulation parameters are set as follows:

Covariance of T_{25} signal: T25_Cov = 25 K^2

Covariance of T_3 signal: T3_Cov = 25 K^2

Covariance of P_{s3} signal: Ps3_Cov = 100 kPa^2

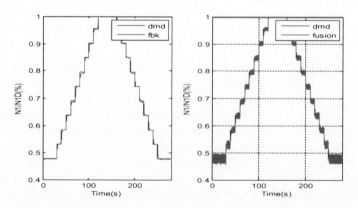

Fig. 15. Time response of closed loop control under stochastic noise based on triple parameters configuration (T_{25}, T_3, P_{s3}, EGT)

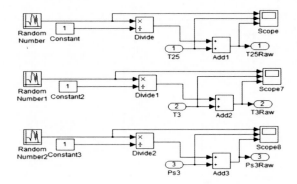

Fig. 16. Diagram for triple parameters infusion algorithm with stochastic noise

The control performance variations of 3 parameters infusion algorithm at every steady point is plotted in the Fig. 17. It can be seen that the 3 parameters fan speed signal infusion algorithm can produce better control performance under random noise environment.

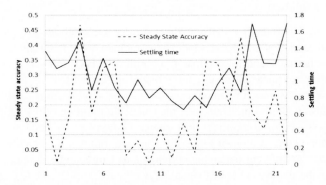

Fig. 17. Settling time and steady state accuracy variations on each small step steady state occasion for three parameters with stochastic noise

N_I command signal big step (snap acceleration or deceleration is between max power and idle) closed loop control simulation is also conducted based on three parameters (T_{25}, T_3, P_{s3}) infusion algorithm. Figure 18 shows the simulation results: (1) Acceleration time to 95% max power is about 2 s; (2) Deceleration time to idle is about 3 s; (3) Steady accuracy of max power state is about 0.8%; (4) Steady accuracy of idle state is about 0.3%. The snap acceleration and deceleration simulation show that the proposed 3 parameters (T_{25}, T_3, P_{s3}) EKF infusion algorithm has also behaved a relative acceptable performance.

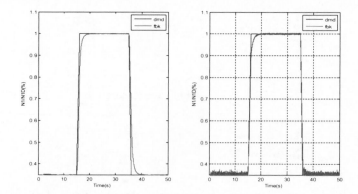

Fig. 18. Time response of closed loop control with stochastic noise based on three parameters (T_{25}, T_3, P_{s3}) infusion algorithm

5 Conclusion

The design process and simulation analysis of EKF infusion algorithm in the paper are summarized as follows:

(1) A method for obtaining discrete time nonlinear model of turbofan engine is proposed, and this nonlinear model is combined with the Kalman Filter with optimal filtering theory. An infusion algorithm of EKF fan speed N1, which is suitable for nonlinear turbofan engine model is designed.

(2) Based on the discrete-time nonlinear model, the Kalman gain is calculated for the EKF infusion algorithm. Therefore, the calculation burden of the Kalman gain in the process of closed loop real-time onboard control is avoided.

(3) Variable structure is applied to the designed algorithm, which can be easily used to analyze the different parameter combinations (from single to multiple) as input to infuse the engine fan speed signal (N_1).

(4) Three parameters (T_{25}, T_3, P_{s3}) infusion algorithm of N1 fan speed achieves good performance comparing with actual N_1 sensors as feedback in closed control. At the same time, the stability of the control (high frequency chattering of synthetic parameters) is better than that of other configurations (single, two and four parameters).

(5) The proposed three parameters (T_{25}, T_3, P_{s3}) infusion algorithm in the stochastic noise environment maintains excellent control performance, meanwhile, it still shows an acceptable performance during the transient control from idle to max power or reversely. It is an good design parameters combination as for the designed N1 EKF infusion algorithm.

In general, the EKF fan speed infusion algorithm has many advantages with respects to steady state accuracy and settling time of the synthesized parameters. It also overcomes high frequency control charttering. And the stability characteristics of the closed loop control would be further studied based on more rigorous theory. In addition, the author plan to introduce the design idea of the fan speed infusion algorithm into the fault tolerant logic design, which is the key of engineering engine safety application [28], and further to improve the performance and robustness of the actual engine control system under closed loop stability premise. The proposed algorithm should be validated and verified by manufacturer and certification authorities, before using in the commercial engine product.

Acknowledgments. The author wishes to thank dear senior engine expert Zheng Jianhong and colleagues from Controls Department of AECC CAE Company in China, for their support in the study of this paper.

Nomenclature

EGT	: Engines exhaust total temperature [K].
J	: Moment of inertia [kg.m2]
K	: Kalman filter state feedback gain.
MT	: Torsional moment [kg.m/s].
N_c	: Engine core rotor speed [rpm].
$N_1(N_f)$	: Engine fan speed [rpm].
Ndot	: Rate of engine core rotor or fan speed [rpm].
P_{s3}	: Engine high pressure compressor static pressure [kPa].
T_3	: Engine high pressure compressor total temperature [K].
T_{25}	: Engine high pressure compressor total temperature [K]
W_f	: Mass flow rate [kg/h]

References

1. Jaw LC, Mattingly JD (2009) Aircraft engine controls design, System Analysis, and Health Monitoring. AIAA Education Series, Virginia, USA
2. Wang X (2013) Foundation of inertial navigation. Northwestern Polytechnical University Press, Xi'an, China
3. Jing X (1973) Application foundation of Kalman filter. National Defense Industry Press, BEI Jing(China)
4. Kalman RE (1963) New methods in Wiener Filtering theory. In: Proceedings of the First Symposium on Engineering Applications of Random function theory and probability
5. Kalman RE (1960) A New Approach to linear Filtering and prediction problems. Trans ASME, Series D J Basic Eng 82:35–45
6. Bryson AE, Yu-Chi Ho (1975) Applied Optimal Control, 2nd edn. CRC Press
7. Hockerdal E, Frisk E, Eriksson L (2009) Observer design and model augmentation for bias compensation with a truck engine application. Control Eng Pract 17:408–417
8. Modalavalasa N, Rao GSB, Prasad KS, Ganesh L (2015) A new method of target tracking by EKF using bearing and elevation measurements for underwater environment. Robot Auton Syst 74:221–228
9. Hockerdal E, Frisk E, Eriksson L (2010) Model based engine map adaption using EKF. In: AAC 2010, Munich, Germany, 12–14 July 2010

10. Guardiola C, Pla B, Blanco-Rodriguez D, Eriksson L (2013) A computationally efficient Kalman filter based estimator for updating lookup table applied to NOx estimation in diesel engines. Control Eng Pract 21:1455–1468

11. Sarim M, Nemati A, Kumar M, Cohen K (2015). Extended Kalman Filter based quadrotor state estimation based on asynchronous multi-sensor data. In: ASME 2015 Dynamic Systems and Control Conference. DSCC 2015-9925

12. Chen S, Yan F (2013) Cycle-by-cycle Based In-cylinder temperature estimation for diesel engines. ASME 2013 Dyn Syst Control Conf. DSCC 2013-4005

13. Wang Z (2017) Observer-based cylinder air charge estimation for spark-Ignition engines. J Eng Gas Turbines Power from IC Engine Div of ASME 2017

14. Souflas I (2015) Nonlinear recursive estimation with estimability analysis for physical and semi physical engine. J Dyn Syst Measur Control Dyn Syst Div of ASME 2015

15. Tan D, He A, Kong X, Wang G (2011) UIO based on sensor fault diagnosis for aero engine control system. J Aerosp Power 26(6):1396–1404

16. Feng L (2016) An improved extended Kalman Filter with inequality constraints for gas turbine engine health monitoring. Aerosp Sci Technol 58:36–47

17. Viassolo D (2007) Advanced controls for fuel consumption and time-on-wing optimization in commercial aircraft engines. ASME Turbo Expo 2007: GT2007-27214

18. Fan S (2008) Aero Engine Control. Northwestern Polytechnical University Press, Xi'an, China

19. Walch PP, Fletch P (2004) Gas Turbine Performance, 2nd edn. American Society of Mechanical

20. Yao H (2014) Full authority digital electronic control system. Aviation Industry Press, Beijing, China)

21. Isdori (1995) Nonlinear Control systems, 3rd edn. Springer, London

22. Li Q (2001) Numerical Analysis, 4th edn. Tsinghua University Press, Beijing, China

23. Kincaid D (2009) Numerical Analysis. American Mathematical Society 2009

24. Frederick D, DeCastro J, Litt J (2007) User's Guide for the Commercial Modular Aero-Propulsion System Simulation (C-MAPSS). NASA TM-2007-215026, Glenn Research Center, Cleveland, Ohio

25. Ljung L (1999) System Identification: Theory for the user, 2nd edn. Prentice Hall Press, USA

26. Jun LU, Yingqing GUO, Xiaolei CHEN (2011) Establishment of aero-engine state variable model based on linear fitting method. J Aerosp Power 26(5):1172–1177

27. Scardua LA, da Cruz JJ (2017) Complete offline tuning of the unscented Kalman filter. Automatica 80:54–61

28. Zhang Y, Lu X, Tao Jin-wei, Xin-chen M (2017). Design and analysis of a sliding mode parameter limit regulating system for turbo fan engine. ASME Turbo Expo 2017, GT2017-64510

The Investigation of Fuel Effects on Industrial Gas Turbine Combustor Using OpenFOAM

Yinli Xiao[✉], Zhibo Cao, Changwu Wang, and Wenyan Song

School of Engine and Energy, Northwestern Polytechnical University,
Xi'an 710072, Shaanxi, People's Republic of China
xiaoyinli@nwpu.edu.cn

Abstract. To investigate the effect of fuel type on gas turbine combustor performance, a reacting solver using flamelet combustion model is developed on open source CFD code OpenFOAM. The flow field of an industrial gas turbine combustor under three different power setting conditions including 20%, 58% and 100% is simulated with the new solver. The major features of the flow field with two different fuels are analyzed. Furthermore, the main performance parameters of the combustor are discussed. The results indicate that the major features of velocity and temperature distribution using nature gas remains unchanged, in comparison with using the diesel oil. The natural gas fueled case produces more water in the whole combustion zone than the diesel oil case at the same power setting. However the combustion efficiency and total pressure recovery coefficient are both improved. Thus, the outlet temperature distributes more uniformly using the natural gas.

Keywords: Fuel effects · Steady laminar flamelet model · OpenFOAM ·
Gas turbine combustor

1 Introduction

Gas turbine has been used widely for power generation, natural gas booster station, mechanically driven apparatus, marine propulsion plant and adjusting peak power plant for many years. A considerable number of gas turbines in industrial fields are derived from gas turbines used for aircraft or warships. The liquid fuel such as diesel and kerosene is often used for these applications. The natural gas is more preferred for industrial application due to its low cost and easy accessibility. The effect on combustion of fuel type is of paramount importance due to different physical and chemical properties. In addition, the impact of fuel change varies with the operating power setting and combustor type. In this paper, simulations have been conducted and performance parameters of a gas turbine combustor are investigated to gain fundamental insights for impact of fuel type.

In the past, numerical simulation studies on gas turbine combustors are mostly performed with commercial CFD software such as Ansys Fluent, CFX, etc. due to friendly interfaces and powerful capabilities. With the rapid development of turbulent combustion solvers based on the open source CFD platform, OpenFOAM has been increasingly used to simulate the complex combustion flow at realistic gas turbine

© Springer Nature Singapore Pte Ltd. 2019
X. Zhang (Ed.): APISAT 2018, LNEE 459, pp. 1795–1805, 2019.
https://doi.org/10.1007/978-981-13-3305-7_144

combustors. Bulat (2015) investigated a commercial industrial gas turbine combustor using a reacting solver based on OpenFOAM. Four different global and skeletal reaction mechanisms are used for PaSR finite rate chemistry in this research. The results are compared with detailed experimental measurements and indicate that the species and temperature field is insensitive to the chemical mechanisms. Aligoodarz (2016) analyzed SGT600 gas turbine combustion chamber using turbulent flame speed model based on OpenFOAM. The outlet average temperature is only slightly 1.5 K less than the measured value. The effect of fuel nozzle orifice diameter on performance parameter is also investigated.

Many researchers conducted combustion simulation with OpenFOAM platform to predict the pollutant emissions. Benim (2016) numerically simulated a model gas turbine combustor applying laminar flamelet combustion model and different turbulent models based on OpenFOAM platform. Two distinct fuel including natural gas and syngas are considered in this research. The predictive capability of pollution emissions with the turbulent combustion model is assessed. Over prediction of NO was observed due to the higher temperature results. Koo (2016) integrated soot formation models in OpenFOAM platform to simulate the soot distribution. The configuration of simulation is a model gas turbine combustor using ethylene. The new developed solver reasonably predicted the soot tendencies. The intermittent phenomenon of soot generation is observed.

In order to investigate the dynamic features of combustion, many researchers conducted transient combustion simulation with OpenFOAM platform. Yunoki (2016) developed a large eddy simulation solver with hybrid turbulent combustion models based on the OpenFOAM platform. This solver is used to investigate an individual can of a multiple-injection low NOx combustor. The predictive NO concentrations tendency is in good agreement with experiment data. Xu (2016) developed a large eddy simulation approach and applied it to combustion flow field simulation of a realistic gas turbine combustor based on OpenFOAM. It is demonstrated that OpenFOAM is able to investigate the combustion dynamics and capture the main flow features such as the precessing vortex core (PVC) phenomena.

The above advances in reacting solvers demonstrate that OpenFOAM is reliable enough to be applied to the practical industrial engineering. In the current research, the reacting flow field of an industrial gas turbine combustor is investigated using steady laminar flamelet model with OpenFOAM. The first objective is to validate the combustion solver in OpenFOAM and understand its predictive capabilities and limitations. The second objective of this work is to investigate the influence of fuel type on flow field and performance of industrial gas turbine burner.

2   Geometric Configuration

The combustor considered in this study is derived from an industrial gas turbine with a nominal power output of up to 29 megawatt and total efficiency 36% Ren (2008). The industrial gas turbine comprises two-spool gas generation. The low pressure axial flow compressor consists of 9 stages and is driven by one stage axial low pressure turbine.

The high pressure axial flow compressor also consists of 9 stages and driven by the high pressure turbine. The free power turbine consist 4 stages and drives the power output shaft.

2.1   Combustor Features

The gas turbine employed reverse flow can-annular combustion chamber. It consists of 16 flame tubes fitted within an inner and outer air casing. In the present investigation, a full scale single flame tube is chosen to simulate and it consists of its own dome, axial flow swirler, fuel nozzle, a number of secondary and dilution air holes. Air enters the flame tube through the swirler and holes on the surface of the dome and flame tube.

Figure 1 illustrates the three dimensional model of one flame tube. The commercial software Solidworks is used to create the required physical model. The computational domain reproduces only 1/16 sector of the combustion chamber. For the two lateral sides, cyclic boundary condition is applied. The thickness of the vanes is neglected to simplify the overall geometry. The detailed dimensions of the combustor can be found in the references of Zheng (2013) and Ren (2008).

2.2   Grid

The computational model is meshed with the Gridgen software, which is capable of generating mesh for complex geometry and can directly deal with the iges file generated by solidworks. The total number of cells of the grids is 1.0×10^6.

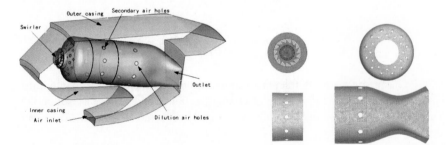

Fig. 1.   The 3D model of GT combustor   **Fig. 2.**   The grid of different parts

Figure 2 shows the detailed computational grids of different parts of the model. Due to the complexity of the geometry configuration, the combustor model is simplified to a certain extent. The film cooling air holes on the flame tube surfaces are replaced with equivalent area cooling slots. In addition, the thickness of the flame tube walls is neglected to facilitate the meshing. The grid clusters in the dome and the vicinity of the flame tube walls but coarse grid is used in remained parts. Tetrahedral mesh is employed for all the parts.

3   Numerical Model

In this study the open source CFD library OpenFoam is used as the computational platform. It is a C++ toolbox based on object oriented programming and gives a flexible framework which combines all the required tools for solving all kinds of CFD problem. It consists of enormous groups of libraries for different mathematical, numerical and physical models.

3.1   Mathematics Methods

The conservation equations for turbulent combustion flow include the continuity equation, momentum equations, and sensible enthalpy equation. The steady laminar flamelet model (SLFM) is used as turbulent combustion model. The SLFM views turbulent flame as an ensemble of stretched laminar flames. It is suitable for simulation of practical engineering problems with complex configuration because a large number of species transport equations are avoided to be resolved. As a result, the mixture fraction and its variance as the following are solved based on OpenFOAM platform.

$$\frac{\partial \bar{\rho}\tilde{Z}}{\partial t} + \frac{\partial \bar{\rho}\tilde{u}_i \tilde{Z}}{\partial x_i} = \frac{\partial}{\partial x_i}\left((\mu_l + \mu_t)\frac{\partial \tilde{Z}}{\partial x_i}\right)$$

$$\frac{\partial \bar{\rho}\widetilde{Z''^2}}{\partial t} + \frac{\partial \bar{\rho}\tilde{u}_i \widetilde{Z''^2}}{\partial x_i} = \frac{\partial}{\partial x_i}\left((\mu_l + \mu_t)\frac{\partial \widetilde{Z''^2}}{\partial x_i}\right) + 2(\mu_l + \mu_t)\left(\frac{\partial \tilde{Z}}{\partial x_i}\right)^2 - \bar{\rho}\tilde{\chi}$$

And the mean species mass fractions and temperature in the turbulent flame brush can be determined from the joint PDF of mixing fraction Z and scalar dissipation rate χ_{st}. Turbulence has been modeled with the realizable k-ε model with standard wall functions.

In the current research, the reacting solver is based on steady SIMPLE algorithm with two pressure correctors and two momentum correctors per time step. Bounded Gauss limited linear schemes are used for the convective terms. The Gauss linear corrected difference schemes are used for diffusion terms and mass fluxes at face centers from cell values. The preconditioner of diagonal incomplete LU and bi-conjugate gradient method (PBICG) by Jacobs is used for solving momentum equations with a local accuracy of 10-8 at each step. The flow reached the steady state conditions after approximately 4000 steps. All results are reported after 10000 steps to ensure that the reacting flow has totally reached the steady state conditions. Convergence was obtained when the residuals for all the variables were less than $1.0e^{-5}$.

In this study, the open-source chemistry software Cantera is used to pre-process the chemical mechanisms and generate the flamelets profiles. Then a look-up table (PDF table) for the scalar as a function of mean mixture fraction and mixing fraction variance is generated. All the required kinetic schemes, thermodynamic and transport properties file are in standard Chemkin format.

The porotype industrial GT uses diesel oil as the fuel, which heating value is 42700 kJ/kg. According to Zheng (2013), in the numerical simulation the normal heptane is used and its heating value (43000 kJ/kg) is close to diesel oil. The stoichiometric ratio of heptane is 0.066. The air and fuel temperature are defined as 760 K and 300 K at 100% power setting, respectively. Only the following species are considered, CH4, CO, CO2, C(S), H, H2, H2O, O2, O, OH, N, NO, N2. Then the PDF table is generated from equilibrium to extinction conditions. The mean temperature curve with scaled variance and mixing fraction is shown in Fig. 3a. The peak temperature is 2588 K corresponding to the stoichiometric ratio and zero scaled variance.

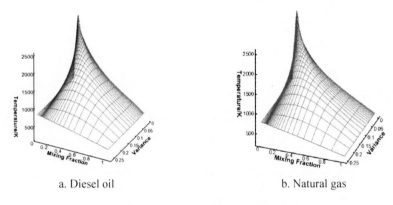

a. Diesel oil   b. Natural gas

Fig. 3. The generated PDF table

The alternative fuel is natural gas, in which the methane is the primary component and its mole fraction is 95–98%. Thus CH4 is used as the chemical formula of natural gas and its heating value is 46692 kJ/kg. The considered species and reactant temperatures are the same as the diesel oil. As shown in Fig. 3b, the peak temperature is 2526 K and about lower 60 K than the diesel oil at the stoichiometric ratio and zero scaled variance.

3.2   Boundary Conditions

Three different power setting conditions 100%, 58% and 20% are considered in this study. Based on the 300 h durability gas turbine experimental monitoring data, the incoming air conditions are as shown in Table 1. Uniform air values are given for all conditions. The turbulent parameters are set according to the turbulent intensity and hydraulic diameter.

For the fuel inlet, the injection pressure, temperature, mixture fraction and mass flow rate is given according to the air fuel ratio. The temperatures are both 300 K and mixture fraction is 1.0 for both diesel and natural gas. Tables 2 and 3 gives the

Table 1. Boundary condition of air inlet

Power setting	100%	58%	20%
Pressure (MPa)	1.98	1.50	0.90
Temperature (K)	761.5	703.6	609.5
Mass flow rate (kg/s)	4.40	3.57	2.45

Table 2. Boundary condition of diesel

Power setting		Pressure (MPa)	Mass flow rate (kg/s)
100%	Primary	4.88	0.102
	Secondary	3.49	
58%	Primary	3.66	0.071
	Secondary	2.11	
20%	Primary	2.40	0.036
	Secondary	0.98	

Table 3. Boundary condition of natural gas

	100%	58%	20%
Mass flow rate (kg/s)	0.0933	0.0649	0.0329
Pressure (MPa)	2.07	1.54	0.88

boundary conditions of diesel and natural gas, respectively. A dual orifice atomizer is used for diesel oil.

At the outlet, all flow quantities are extrapolated from the flow in the interior. No slip and adiabatic conditions are used on all the walls. Periodic boundary conditions are used for two lateral boundaries.

4   Results and Discussion

The flow features in the flame tube and the performance parameters for two fuels at three power setting conditions are discussed in the following.

4.1   Analysis of Flow Field

The results at 20% power setting conditions are used to analyze the flow field. Figures 4 and 5 gives the contours of velocity magnitude, total temperature and mole fraction of water at the center plane for two fuels. The velocity magnitude contour and streamlines illustrate the flow pattern and air distribution along the flame tube. The center circulation zone forms in the primary zone of the flame tube because the swirl assembly in the dome imparts a strong swirl to the air. The streamlines in the annular passages and rear parts of the flame tube are similar. However the streamlines are different and asymmetry especially in the dome due to the flow unsteady nature. It is observed that a small circulation zone forms behind the secondary air stream on the upper side but not on the lower side.

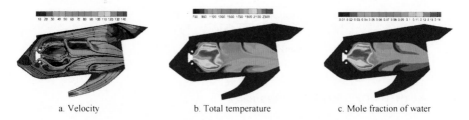

a. Velocity   b. Total temperature   c. Mole fraction of water

Fig. 4. Results of 20% power setting case with diesel oil

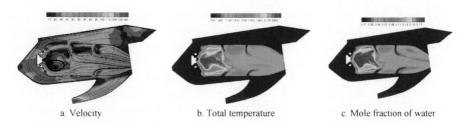

a. Velocity   b. Total temperature   c. Mole fraction of water

Fig. 5. Results of 20% power setting case with natural gas

As can be seen from the total temperature contour, the high temperature zone concentrates on central circulation zone in the primary zone for both fuels. The peak temperature reaches up to 2300 K, which is essential for maintaining stable and efficient combustion. Behind the secondary air streams, the temperature is still up to 1700 K. This demonstrates that incomplete fuel achieves further combustion in this zone. The dilution air streams significantly lower the gas temperature in the dilution zone before it leaves the combustion chamber. The distribution trend for diesel oil and natural gas are in good agreement except that the area of high temperature of natural gas is large. This is because the natural gas is better mixed with the incoming air, in comparison with diesel oil.

As can be seen from the water mole fraction contour, in the whole region of flame tube especially in the primary zone more water exists for the natural gas. The peak value of water mole fraction is 0.17 for natural gas while it is only 0.14 for diesel oil. This is because the major component of natural gas is methane, which has the least C/H ratio.

Table 4 gives the mean velocity at the outlet at three power settings for diesel oil and natural gas. It can be shown that the mean velocity remains almost constant for the same power setting and for the natural gas the mean velocity is only slightly higher than using the diesel oil. As the power setting increase, the mean velocity increases because the incoming air mass flow rate increases gradually and the incoming temperature increases.

Table 4. Mean velocity at the outlet (m/s)

Power setting	20%	58%	100%
Diesel oil	75.9	82.1	84.0
Natural gas	77.4	82.6	85.0

4.2   Analysis of Performance Parameter

(1) Outlet temperature

Figure 6 gives the minimum, average and maximum temperature at combustor outlet for both fuels at three power setting conditions and the average value is based on the mass flow weighted. From here we can see that the temperature only slightly changed at each power setting using natural gas, in comparison with the diesel oil. The most remarked difference of the average temperature is only 8 K which occurs at 58% power setting. However, the natural gas fueled cases have the smaller maximum temperature and larger minimum value. That means the outlet temperature distribution becomes more uniform. At the 58% power setting, the maximum temperature is 100 K lower than the diesel oil case and the minimum is higher 45 K. The experimental turbine inlet temperature (TIT) for the diesel oil at the 100% power setting is 1518 K. The TIT is 1564 in the current research and the relative error is only 3%. This confirms that the accuracy of the reacting solver is acceptable.

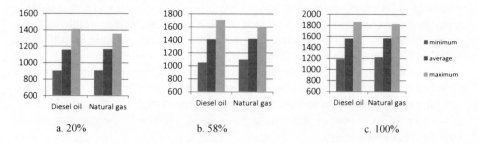

a. 20%   b. 58%   c. 100%

Fig. 6. Outlet temperature at three different power setting (unit: K)

This also can be seen in the overall temperature distribution factor (OTDF) of the combustor which is defined as the following:

$$\text{OTDF} = \frac{T_{4,max} - T_{4,aver}}{T_{4,aver} - T_{3,aver}}$$

Here, $T_{4,max}$, $T_{4,aver}$, $T_{3,aver}$ are the maximum outlet temperature, average outlet temperature and inlet average temperature.

Table 5. Overall temperature distribution factor (OTDF) at outlet

Power setting	20%	58%	100%
Diesel oil	0.4636	0.4195	0.3731
Natural gas	0.3453	0.2640	0.3177

From Table 5, it can be seen that the outlet average temperature remains constant, which is in agreement with the design requirements of the natural gas injector. The outlet temperature distributes more uniformly using natural gas for all conditions, which is helpful for prolonging the lifespan of the subsequent nozzle guide vanes. It should be noted that all the OTDF results in this table maybe a little bit higher than its normal value. It is because all the film cooling air holes are simplified to cooling slot, thus reducing the cooling air penetration depth and cooling effect.

(2) Combustion efficiency

The combustion efficiency is one of the most important performance parameter of the combustion chamber. In the current research, the temperature rise method is used to define the combustion efficiency. It equals the ratio of actual temperature rise and the theoretical temperature rise across the combustion chamber.

$$\eta = \frac{T_{4,aver} - T_3}{T_{4,th} - T_3} \times 100\%$$

Here, $T_{4,th}$, $T_{4,aver}$, T_3 are the theoretical temperature, actual average temperature at the outlet and the inlet temperature, respectively. Table 6 gives the theoretical outlet temperature from Chemkin for all the cases. As the power setting increases, the theoretical outlet temperature increases.

Table 6. Theoretical outlet temperature (K)

Power setting	20%	58%	100%
Diesel oil	1186.91	1435.23	1569.22
Natural gas	1188.38	1436.17	1569.79

Table 7 gives the combustion efficiency for all the cases. For the same power setting, the combustion efficiency using natural gas is higher than diesel oil. The combustion efficiency increases gradually as the power setting increases for both fuels.

At the maximum power setting, the combustion efficiency is more than 99% and satisfies the design need of the replacing fuel.

Table 7. Combustion efficiency

Power setting	20%	58%	100%
Diesel oil	95.17%	96.69%	99.35%
Natural gas	96.03%	97.68%	99.90%

(3) **Total pressure recovery coefficient**

Table 8 gives the total pressure recovery coefficient for all the cases. It is shown that for the same power setting the total pressure recovery coefficient increases about 1% when the natural gas is used. There are two possible reasons. Due to the lower stoichiometric ratio the natural gas flow rate is lower than the diesel for the same power setting. Even for the same mass flow, the natural gas flow causes less interference due to less molecular weight of methane.

For each fuel, as the power setting increases the total pressure recovery coefficient decreases. It is because the higher incoming air mass flow increases the overall velocity and results in more flow resistance and high fuel mass flow. It is worth noting that the total pressure recovery coefficient is usually about 93%. The result in the current research is a little bit higher than the normal value because the geometry configuration is simplified a lot to alleviate the computational cost and time, thus reduced the flow resistance.

Table 8. Total pressure coefficient

Power setting	20%	58%	100%
Diesel oil	96.0%	95.7%	95.3%
Natural gas	97.8%	96.8%	96.1%

5   Conclusion

Flow field in a gas turbine combustor is simulated using flamelet model with Open-FOAM and the prediction capability of new developed solver is validated. An acceptable level of accuracy is attained and the main conclusions can be summarized as follows:

(1) The simulation results reveal that the basic flow features using natural gas do not differ significantly from using the diesel oil in the porotype gas turbine. The velocity contour within the combustor is fairly self-similar and the air distribution of the flame tube remains constant using different fuels under three operating conditions.

(2) As the power setting increases, the zone of high temperature and the magnitude of velocity at the exit of combustor increases for both fuels. At the same time the total pressure recovery coefficient decreases while combustion efficiency increases.

(3) In comparison with diesel oil, the combustion efficiency and total pressure recovery efficiency both improved with natural gas. At the same time, the average temperature at the outlet is slightly higher (about 10 K) and temperature distribute more uniformly. Therefore, it is in favor of the increase of the thermal efficiency of the working cycle. It is also shown that the wall temperature of the flame tube especially at the dome is higher than the one using diesel oil, which is harmful for its durability.

Acknowledgement. This work has been funded by the National Natural Science Foundation of China under grant 51576164.

References

Bulat G, Fedina E, Fureby C, Meier W, Stopper U (2015) Reacting flow in an industrial gas turbine combustor: LES and experimental analysis. Proc Combust Inst 35(2015):3175–3183

Aligoodarz M, Soleimanitehrani M, Karrabi H, Ehsaniderakhshan F (2016) Numerical simulation of SGT-600 gas turbine combustor, flow characteristics analysis, and sensitivity measurement with respect to the main fuel holes diameter. Proc IMechE Part G: J Aerosp Eng 230 (13):2379–2391

Benim AC, Iqbal S, Joos F, Wiedermann A (2016) Numerical analysis of turbulent combustion in a model swirl gas turbine combustor. J Combust. Article ID 2572035

Koo H, Hassanaly M, Raman V (2016) Large eddy simulation of soot formation in a model gas turbine combustor. In: Proceedings of ASME turbo expo 2016: turbomachinery technical conference and exposition. GT2016-57952

Yunoki K, Murota T, Asai T, Okazaki T (2016) Large eddy simulation of a multiple-injection dry low nox combustor for hydrogen-rich syngas fuel at high pressure. In: Proceedings of ASME turbo expo 2016: turbomachinery technical conference and exposition 13–17 June 2016, Seoul, South Korea. GT2016-58119

Xu B, Liu Y, Xie R (2016) Large eddy simulation of a realistic gas turbine combustor. In: Proceedings of ASME turbo expo 2016: turbomachinery technical conference and exposition GT2016 13–17 June 2016, Seoul, South Korea. GT2016-57512

Maric T, Hopken J, Mooney K (2014) The OpenFOAM technology primer. Published by sourceflux UG

Zheng H, Pan G, Chen X, Hu, X (2013) Effect of dual fuel nozzle structures on combustion flow field in CRGT combustor. Math Probl Eng. Hindawi Publishing Corporation, Article ID 913837

Ren J (2008) Numerical analysis of the flow in the reverse flow type combustor. Dalian University of Technology Master Degree Thesis, China

Numerical Study on Combustion and Heat Transfer of a GOX/GCH4 Pintle Injector

Yibing Chang[1,2], Jianjun Zou[1,2(✉)], Qinglian Li[1,2], Peng Cheng[1,2], and Kang Zhou[1,2]

[1] College of Aerospace Science and Engineering, National University of Defense Technology, Changsha 410073, Hunan, People's Republic of China
zjj_xj@sina.com
[2] Science and Technology on Scramjet Laboratory, National University of Defense Technology, Changsha 410073, Hunan, People's Republic of China

Abstract. Aimed at providing a good thermal protection for a pintle injector from being overheated, a three-dimensional combustion and heat transfer coupled simulation is conducted based on a 500 N GOX/GCH4 pintle engine model. Further for such a pintle injector with double rows of injection holes, the effects of the 'skip distance' (Ls) and the interval between two rows of injection holes (Li) on the injector's characteristics of combustion and heat transfer are studied. The results indicate that given the diameter of pintle injector body (Dp), the maximum temperature on the outer surface of the pintle injector decreases from 1756 K to 1404 K when Ls/Dp increases from 0.64 to 1.5, meanwhile the combustion efficiency has a change less than 0.2% with an average value of 0.9510. But when it changes Li, the lowest maximum temperature on the pintle surface of 1326 K and the biggest combustion efficiency of 0.9525 occurs at a middle value of Li/Dp = 0.18 when Li/Dp increases from 0 to 0.38. This paper tries to explain the influences of Ls and Li from the view of the changes of flowfield structures, which show an intensive sensitivity with different pintle injector configurations. The conclusions this paper gets would be greatly helpful for the engineering design of a pintle injector with double rows of injection holes.

Keywords: GOX/GCH4 rocket engine ·
Pintle injector with double rows of injection holes ·
Characteristics of combustion and heat transfer · Numerical study

Nomenclature

Ls	Skip distance, mm
Li	Interval between two rows of injection holes along x direction, mm
Lc	Length of combustion chamber including contraction section, mm
Le	Length of divergent section of thrust chamber, mm
Dc	Diameter of combustion chamber cylinder section, mm
Dp	Diameter of pintle injector cylinder section, mm
Dt	Diameter of throat cross section, mm
De	Diameter of thrust chamber exit cross section, mm
Df	Diameter of first row of pintle injection holes, mm

© Springer Nature Singapore Pte Ltd. 2019
X. Zhang (Ed.): APISAT 2018, LNEE 459, pp. 1806–1825, 2019.
https://doi.org/10.1007/978-981-13-3305-7_145

Ds   Diameter of second row of pintle injection holes, mm

Gap   Width of circular seam between pintle injector and combustion chamber head, mm

O/F   Mixture ratio of oxidizer to fuel

ρ Density of combustion gas, kg/m^3

$\boldsymbol{u}$ Velocity vector, m/s

u_i Velocity component, m/s

p Pressure, Pa

$\bar{\tau}$ Stress tensor, Pa

E Total energy, J/kg

λ Thermal conductivity, W/(m•K)

T Temperature, K

h Enthalpy of formation, J/kg

J Diffusion flux of species

S_h Source term

k   Turbulent kinetic energy, m^2/s^2

μ Dynamic viscosity, Pa•s

ε Dissipation rate of turbulent kinetic energy, 1/s

Y Mass fraction of species

R Net rate of production of species

n Normal direction

Subscripts

eff   Effective parameters

i   Component of coordinates

j   Species index

fluid Fluid domain

solid Solid domain

1   Introduction

Out of the demand of decreasing cost for launching rockets, reusable launch vehicle (RLV) has been a developing trend of launch rocket today [1]. Featured of deep throttling capability, inherent combustion stability and simple mechanic structure [2], the pintle engine becomes a good choice for RLV and shows great potential in the applications such as planet landing, rocket recovery and orbit maneuvering. One of the most successful example toady is the Merlin series engine family served for Falcon 9 launch rocket developed by SpaceX Inc. [3], who had accomplished 23 times successful 1st stage recoveries after Falcon Heavy launched [4], fully demonstrating that the pintle injectors have excellent adaptability and reliability in an occasion of large thrust varying.

Although coming out from the Jet Propulsion Laboratory in 1957 [5], the designs of pintle injector are mostly based on thumb-rules nowadays and fundamental researches on pintle injectors are still scarce. Especially the pintle injector faces a rigorous thermal environment and is easy to be burnt in the combustion chamber. Casiano [6] regards it as a deficiency that hot gas would make pintle injector overheat, but it could be avoided if hot gas temperature can be controlled below permitted temperature of the injector material, as pointed out by Betts [7]. We want to make a reasonable geometrical design to form a good thermal protection for a pintle injector from high temperature gas. In the meantime, it should not cause combustion efficiency descent. Actually, only by combining the design of the pintle injector and the combustion chamber can we get a pintle engine with high efficiency and reliability, thought professor Heister [8]. Figure 1 shows typical flowfield structures inside a pintle engine thrust chamber, comprising two recirculation zones symmetrically, which play a key role on keeping combustion stable and cooling the pintle tip [5].

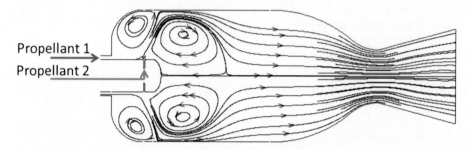

Fig. 1. Typical flowfield inside a pintle engine thrust chamber

Most scholars and institutes pay their main attention to how to increase the combustion efficiency by changing pintle injector's structure. However, few ones care whether pintle injectors could sustain the corresponding severe thermal environment or not. The simulation and experiment work of Son [9, 10], Fang [11], Sakaki [12, 13], mainly analyze with which configuration of a pintle injector the pintle engine will get a best combustion efficiency, but lack of discussion on the thermal protection of pintle injectors with the combustion efficiency increasing. Particularly, Bedard et al. [14, 15] from Purdue University, give a picture showing that pintle injector tip suffers obvious ablation after a hot test in a LOX/methane thrust chamber, which is much unfavorable for a pintle injector requiring multi-times work in a single task.

On the other hand, there are also some measures adopted for thermal protection for pintle injectors in the commercial rocket engines. First, active cooling, namely extracting a few proportion of central propellant (propellant 2 in Fig. 1) and injecting from pintle tip to impact and then form a liquid sheet protection, sees the patent held by Mueller [16], the designer of Merlin engines. Alternative example of active cooling is Advanced Columbium Liquid Apogee Engine (AC-LAE) [17] developed by TRW (Thompson-Ramo-Wooldridge Inc.), which lets the oxidizer flow through a regeneratively cooling channel to form a thermal isolation between pintle injector body and hot combustion chamber. Second, using heat-resistant and high conductivity materials to manufacture pintle injectors, sees TRW's 40 klbf Low Cost Pintle Engine (LCPE), whose pintle tip is made from nickel-beryllium copper alloy [18]. Third, optimizing head structure of pintle injectors including the shape of pintle tip, shape and arrangement of injection holes etc., sees Northrop Grumman's program TR202 [19], in which 15 injector configurations were tested to evaluate the injector heat transfer, but unfortunately, the test results are unavailable in the public reference.

This paper is trying to explore along the third way mentioned above, aimed at showing detailed flowfield structures around the pintle injector by a three-dimensional numerical investigation and interpreting effects of different configurations of pintle injectors on the characteristics of combustion and heat transfer. Precisely, for a pintle injector with two rows of injection holes, the effects of the skip distance and the interval between two rows of injection holes are studied based on a 500 N GOX/GCH4 pintle engine model used for the orbit maneuvering.

2 Computational Methodology

2.1 Physical Model

Figure 2 gives a schematic view of the 500 N GOX/GCH4 pintle engine model, in which the pintle injector contains two rows of stagger arranged injection holes, each row with 12 circular holes as shown in Fig. 3. The radial GCH4 jets from injection holes impact the axial GOX film from a circular gap bringing sufficient mixing and combusting.

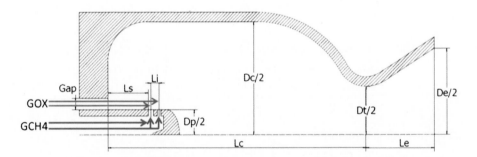

Fig. 2. Schematic view of 500 N GOX/GCH4 pintle engine model

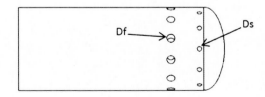

Fig. 3. Arrangement of injection holes of the pintle injector

There are several geometrical dimensionless parameters to express the pintle injector design, including BF, Ls/Dp, Li/Dp, and Dc/Dp etc. BF, blockage factor, as expressed in Eq. (1):

$$BF = \frac{N(Df + Ds)}{\pi Dp} \tag{1}$$

indicates the proportion of two propellants mixing, where N denotes the number of injection holes each row. Ls, the skip distance, is defined as a distance that axial film from circular gap passes away before it hits against radial jets from injection holes. Li is the axial interval between two rows of injection holes. Only the effects of Ls and Li are studied as variants listed in Table 1, whereas other parameters are constant listed in Table 2. From Eq. (1) and Table 2, it could be got that BF = 0.59. The case1 is set as a reference case.

Table 1. Dimensions of the variants studied

Case No.	Ls/mm	Ls/Dp	Li/mm	Li/Dp	Note
Case1	7	0.64	4.15	0.38	Reference
Case2	11	1	4.15	0.38	varying Ls
Case3	16.5	1.5	4.15	0.38	
Case4	7	0.64	0	0	varying Li
Case5	7	0.64	2	0.18	

Table 2. Dimensions of major geometrical parameters of the pintle engine

Parameter	Dp	Dc	Dt	De	Lc	Le	Gap	Df	Ds
Value/mm	11	45	19.1	57.3	93.5	30	1.3	1.1	0.6

2.2   Governing Equations

2.2.1   Combustion Gas Flow

For gas propellants and combustion gas, the Reynolds Averaged Navier-Stokes (RANS) equations govern the compressible steady three-dimensional flow, comprising mass, momentum and energy equation:

$$\nabla \cdot (\rho \boldsymbol{u}) = 0 \tag{2}$$

$$\nabla \cdot (\rho u_i \boldsymbol{u}) = -\nabla p + \nabla \cdot \bar{\tau} \tag{3}$$

$$\nabla \cdot [\boldsymbol{u}(\rho E + p)] = \nabla \cdot \left[\lambda_{eff} \nabla T - \sum_j h_j \bar{J}_j + (\bar{\tau}_{eff} \cdot \boldsymbol{u}) \right] + S_h \tag{4}$$

Reynolds-average introduces an unknown Reynolds stress, representing the effects of turbulent pulsation. Therefore, to close the Eqs. (3) and (4), the standard k-ε turbulence model is adopted by solving the transport equation of the turbulent kinetic energy k, and its rate of dissipation ε:

$$\frac{\partial}{\partial x_i}(\rho k u_i) = \frac{\partial}{\partial x_i}\left[\left(\mu + \frac{\mu_t}{\sigma_k}\right)\frac{\partial k}{\partial x_i}\right] + G_k + G_b - \rho\varepsilon - Y_M + S_k \tag{5}$$

$$\frac{\partial}{\partial x_i}(\rho\varepsilon u_j) = \frac{\partial}{\partial x_i}\left[\left(\mu + \frac{\mu_t}{\sigma_\varepsilon}\right)\frac{\partial\varepsilon}{\partial x_i}\right] + C_{1\varepsilon}\frac{\varepsilon}{k}(G_k + C_{3\varepsilon}G_b) - C_{2\varepsilon}\rho\frac{\varepsilon^2}{k} + S_\varepsilon \tag{6}$$

the constants appeared in the Eqs. (5) and (6) see the work of Launder and Spalding [20].

The diffusion flux $\bar{J}_j$ of species j and source item S_h are solved by equations of species transportation:

$$\nabla \cdot (\rho \boldsymbol{u} Y_j) = -\nabla \cdot \boldsymbol{J}_j + R_j \tag{7}$$

Considering computational cost and precision, a reduced Jones-Lindstedt 6-step mechanism (JL6) is used to describe the combustion reactions of methane and oxygen. The concrete forms of JL6 see the work of Frassoldati [21]. When dealing with interaction between chemical reactions and turbulence, an eddy-dissipation concept (EDC) model is employed, which could take detailed chemical mechanisms into consideration in a turbulent flow. Then the net producing rate of species j is calculated by the following equation:

$$R_j = \frac{\rho(\xi^*)^2}{\tau^*\left[1 - (\xi^*)^3\right]}\left(Y_j^* - Y_j\right) \tag{8}$$

The detailed definition of ξ^*, the length fraction of fine scales and corresponding time scale τ^*, sees the reference of Magnussen [22]. Further, S_h, the source term owing to chemical reactions, could be got

$$S_{h,rxn} = -\sum_j \frac{h_j^0}{M_j} R_j \tag{9}$$

In the calculation, the combustion gas is regarded as ideal gas, subjected to the Clapeyron equation:

$$p = \rho R T \tag{10}$$

2.2.2   Thermal Conduction

The Fourier equation governs the thermal conduction inside the pintle injector:

$$\nabla \cdot (\lambda \nabla T) = 0 \qquad (11)$$

2.3   Grid Generation and Boundary Conditions

Note that the pintle engine model is an axis-symmetric revolutionary body and the pintle injector has two rows of stagger arranged injection holes, each row with 12 holes. Thus a half of 1/12 of the engine model, namely 15 degrees along the circumferential direction, is selected as the computational domain. A structured grid is generated as shown in Fig. 4, which contains fluid domain and solid domain including 226290 and 13088 hexahedron elements respectively. Two domain's grids are generated separately, so bilateral interface meshes are non-matched, where data exchange is realized via interpolation to transfer heat.

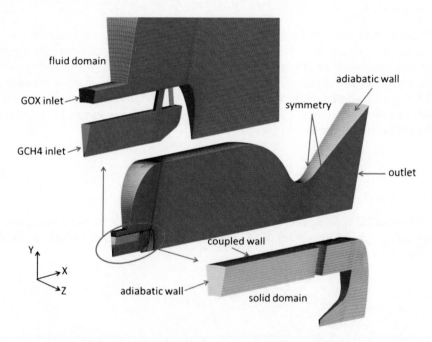

Fig. 4. Numerical grid for the computational domain

The boundary layer is added near the coupled walls in the fluid domain side, and the y^+ at the pintle tip wall-adjacent cell is set to be less than 5 to get a better resolution of calculation of heat transfer. The boundary conditions are listed in Table 3, where the interface walls denote the walls that stand between fluid domain and solid domain, while other thrust chamber walls are taken as an adiabatic wall. Both circumferential boundaries of the computation domain are set as symmetry. Additionally, the solid domain material is stainless steel 304 with a density of 8030 kg/m^3 and a thermal conductivity of 18.3 W/(m•K). All of cases studied share the same boundary conditions.

At the interface walls, the temperature and its 1st order derivative, heat flux, must keep continuous to ensure the energy balance:

$$T|_{fluid} = T|_{solid} \tag{12}$$

$$\lambda_{fluid}\frac{\partial T}{\partial n}\bigg|_{fluid} = \lambda_{solid}\frac{\partial T}{\partial n}\bigg|_{solid} \tag{13}$$

Therefore a simulation coupled of heat transfer and combustion is established, which is believed to give a more accurate result than the way computed by empirical formulas such as Bartz equation [23–25].

Table 3. Boundary conditions

Boundary	Type	Value
GOX inlet	Mass flow inlet	0.004825 kg/s, 300 K
GCH4 inlet	Mass flow inlet	0.001058 kg/s, 300 K
Outlet	Pressure outlet	10132.5 Pa
Interface walls	Coupled wall	
Non interface walls	Adiabatic wall	

2.4   Numerical Treatment

The second order upwind scheme is used for the spatial discretization of variants in the convective terms of the governing equations, and the pressure-velocity coupling is realized by the SIMPLE algorithm.

2.5   Grid Independence Validation

To make a quantitative contrast of results with different grid densities, temperature data from a sample line is extracted. The sample line (red line) and two symmetries (symmetry A with blue boundary and symmetry B with green boundary) for latter analysis are shown in Fig. 5.

There are three grids with different grid densities to be validated, named coarse, moderate and fine. As listed in Table 4, only the density of fluid domain has been adjusted while the density of solid domain keeps invariable because the calculation precision of heat transfer between hot gas and injector solid is mainly depends on the resolution of fluid domain grid.

Setting location of circular gap as the origin, Fig. 6 gives the temperature contrast result that moderate grid has a good conformance with the fine grid on the sample line's temperature distribution. The maximum temperature difference is −511 K (29.1%), 0 K (0%) and 64 K (3.64%) for coarse, moderate and fine respectively (take moderate as comparison reference). Thus out of the consideration of accuracy and computational cost, the moderate grid is selected for further study.

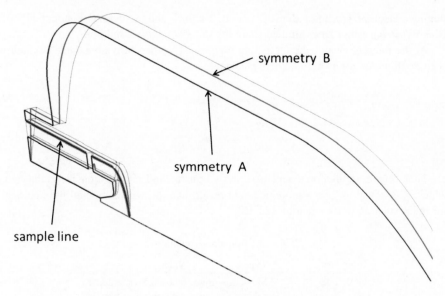

Fig. 5. Sample line and symmetries for analysis

Table 4. Grid density for independence validation

Grid	Fluid domain	Solid domain
Coarse	107484	13088
Moderate	226290	13088
Fine	487080	13088

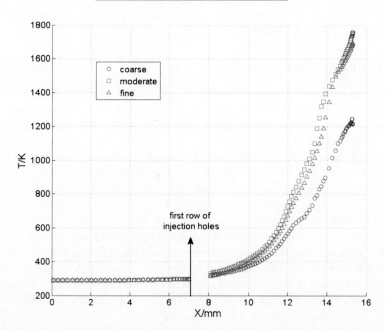

Fig. 6. Temperature distribution on the sample line for grid independence validation

3   Results and Discussion

This part will firstly study the flowfield structures and heat transfer on the pintle tip surface upon the reference case, and then inspect the effects of different Ls and Li on the thermal environment around the pintle injector. Last, the influences of Ls and Li on performance of the whole of pintle engine are also studied.

3.1   Flowfield Structures of Reference Case

Figure 7 shows the flowfield structures of reference case. Along the axial direction (x coordinate), there are two main recirculation zones as expected, one in the head-dome zone and another in the center of combustion chamber. In addition, a small recirculation zone appears under the pintle tip. In the various slices perpendicular to the axial direction, counter-rotating vertex pairs (CVP) are observed. Noticeably, the CVP structures and the center recirculation zone have the same end position along the axial direction, which implies some connectivity of two structures development. The maximum temperature inside the thrust chamber is 3427.64 K, 1% less than the theoretical value in the design point, 3459 K. Further, the methane jet from first row of injection holes has impacted the chamber wall due to relatively large momentum, as shown by the iso-surface of stoichiometric ratio of O/F = 4, which generally reflects main reaction zones and the location of flames [26].

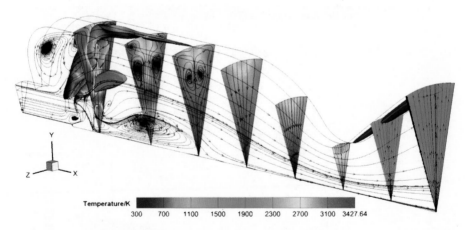

Fig. 7. Flowfield structures of reference case

From Figs. 8 and 9, it could be displayed more clearly where the methane and oxygen go. First, using the concept of BF, the methane jet from the first row of injection holes alone intercepts nominally about 38.2% of oxygen film. The majority of this stream of methane flows downstream close to the chamber wall, rendering zones neighbored to the wall fuel-rich. At the meantime, because of hitting against the chamber wall directly, the methane jet makes head-dome zone A relatively closed and blocks heat release from zone A, where consequently the temperature keeps high because the combustion continues due to ceaseless methane and oxygen being involved by head recirculation zone.

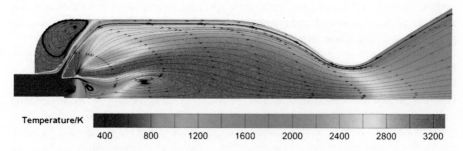

Fig. 8. Temperature contour and streamlines in the symmetry A

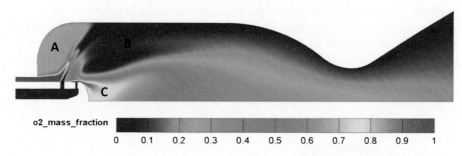

Fig. 9. Contour of oxygen mass fraction in the symmetry A

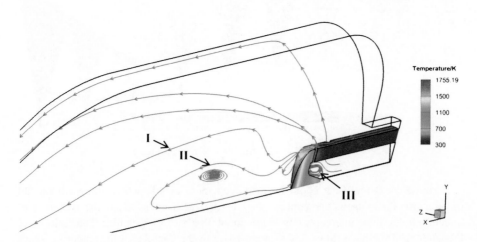

Fig. 10. Flowfield structures around the pintle tip of reference case

Second, about 20.1% of oxygen film is intercepted alone by the methane jet from the second row of injection holes. This stream of gas goes into the inner zone B and combusts efficiently.

Third, the remainder 41.7% of oxygen film escapes from the interspace between two rows of injection holes and mainly flows into the recirculation zones C under the pintle tip, which is consequently getting oxygen-rich and forming a relatively low temperature zone.

Further, to interpret the heat transfer from hot gas to the pintle tip, a three dimensional streamlines figure is made as shown in Fig. 10. The maximum temperature reaches to 1755.19 K on the pintle tip, higher than the pintle injector material's melting point, 1727 K. Therefore, the current design of reference case cannot ensure the safe work of the pintle injector. Figure 10 discovers the flowfield structures around the pintle tip and indicates that the oxygen film from the interspace of two rows of injection holes flows along two paths, forming two recirculation zones. One stream, I, recirculates just under the pintle tip and gets back to combustion chamber, which is almost low temperature and high concentration oxygen (mass fraction of oxygen: 0.7 to 0.9) and plays a role of cooling the pintle tip when it passes through the pintle tip. Another stream, II, recirculates in the big recirculation zone in the center of the chamber and flows back to the pintle tip to join the jet from second row of injection holes, which is high temperature and middle concentration oxygen (mass fraction of oxygen: 0.4 to 0.6) and plays a role of heating the pintle tip when it passes through the pintle tip.

Last but not least, the recirculation zone, III, formed by cool methane inside the pintle injector also plays a key role of cooling pintle tip.

Actually, main changes of flowfield structures due to different pintle injector configurations happen in the circumference of the pintle injector, which in turn influence the characteristics of heat transfer on the injector surface. Thus, the following study will focus on the changes of flowfield structures around the pintle injector.

3.2   Effects of Ls on Heat Transfer of Pintle Injector

It is observed clearly that temperature on surface of pintle injector has an apparent reduction with Ls increasing, as shown in Fig. 11, where the first row of injection holes' upper edge is set as the origin for the convenience of comparison. The maximum temperature on the sample line is 1756 K, 1510 K and 1404 K for Ls/Dp = 0.64, Ls/Dp = 1 and Ls/Dp = 1.5 respectively. The amplitude reduction up to 352 K is much considerable.

While in the Fig. 12, it could be found that the flowfield structures also have a huge transformation with Ls increasing. Firstly, in the case2 and case3, the recirculation zone in the center of combustion chamber is gone and the zone under the pintle tip is changed into a totally relatively low temperature (under 2400 K) zone extended downstream compared to case1, which greatly improves the thermal environment for service of the pintle injector. On the other hand, the temperature in the head-dome zone also declines with Ls increasing especially for case3.

In fact, the present flow of oxygen film could make an analogy with the backward-facing step flow. They both have a flow separation when enter an/a elliptical/sudden-expansion of cross sectional zone due to the pintle tip or step and then form some

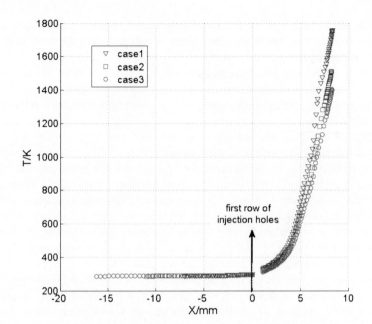

Fig. 11. Temperature distributions of cases with different Ls on the sample line

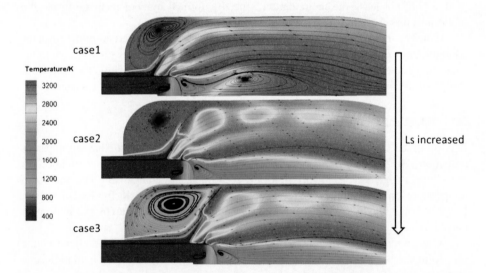

Fig. 12. Streamlines and temperature contours of cases with different Ls in the symmetry B

recirculation zones, in which the free shear flow plays a dominant role. The quantity and scale of the recirculation zones is closely related to the conditions of incoming flow, the geometrical structure of the pintle injector or step, and other perturbation factors [27]. In the present study, two methane jets from injection holes are an intensive perturbation. However, case1, case2 and case3 all have the same pintle tip and the same

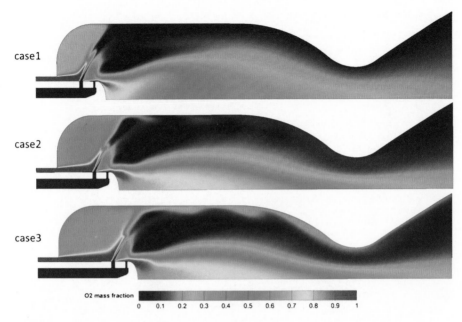

Fig. 13. Mass fraction contours of oxygen of cases with different Ls in the symmetry A

methane injection conditions, therefore, the only difference is different coming flow conditions induced by different Ls. Bigger Ls brings more complete development of oxygen flow along the axial direction (x direction) and stronger capability to resist the flow separation. But this explanation needs further deep research to sustain. To discover the underlying mechanism of the flowfield structure changes happened in the combustion chamber of the pintle engine, a dynamic analyzing method may be adopted, which is employed widely in the study of backward-facing step flow control [28, 29].

Apart from the significant temperature reduction in the pintle tip, temperature decline in the head-dome zone also should be noted. The temperature change generally originates from the species concentration change. As Figs. 12 and 13 show, with the Ls increasing, the scale of head recirculation zone increases, the more oxygen diffuses in the head-dome zone, making the zone oxygen-rich gradually and the temperature decline consequently.

3.3   Effects of Li on Heat Transfer of Pintle Injector

Unlike the gradual temperature reduction in the pintle tip with Ls increasing, as shown in Fig. 14, the lowest maximum temperature of 1326 K occurs at a middle value of Li/Dp = 0.18 (case5) when Li changes from 0 to 0.38, 430 K lower than Li/Dp = 0.38 (case1). While for Li/Dp = 0 (case4), namely when two rows injection holes transform into one row, the pintle tip suffers a very high temperature up to 2605 K, implying this configuration considerably deteriorates the thermal service environment of the pintle injector.

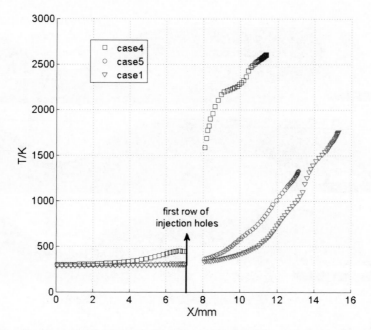

Fig. 14. Temperature distributions of cases with different Li on the sample line

From Fig. 15, it could be found that case4 do not have a recirculation zone under the pintle tip, and the whole of combustion chamber center under the pintle tip is involved in an approximately 3000 K high temperature zone. Because two rows injection holes are in the same axial position, the holes' arrangement gets much dense along the circumferential direction, which makes methane jets form something like a 'wall' blocking the channel that oxygen flows downstream. Therefore, hardly oxygen escapes from the interspace between injection holes to cool the pintle tip. Intensive impact between methane jets and oxygen film lifts the streamlines and enlarges the scale of the recirculation zone in the combustion chamber center.

For case5 and case1, their flowfield structures show a similarity except in the region enclosed by a black dashed rectangle, as shown in Fig. 15. There exists a hot gas stream with its temperature over 3000 K in the left side of methane jet from second row of injection holes for case1, which means a part of oxygen has fully combusted before it gets out the rectangle region, while case5 doesn't have the corresponding hot stream.

Therefore, to cool the pintle tip sufficiently, Li/Dp should be positioned at a proper value. If Li/Dp is too large, a part of oxygen will be dissipated in the combustion reaction between two methane jets along the axial direction, and too small Li/Dp will block the channel oxygen flows downstream.

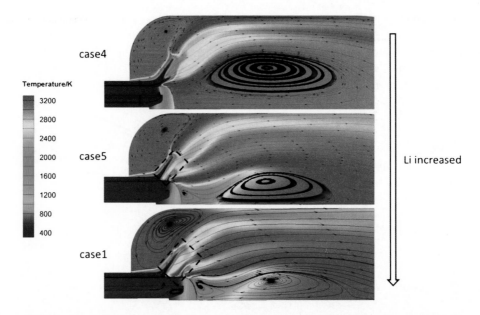

Fig. 15. Streamlines and temperature contours of cases with different Li in the symmetry B

3.4   Effects of Ls and Li on Characteristics of Combustion

In general, for a liquid rocket engine, it uses the combustion efficiency to express characteristics of combustion, which is defined as:

$$\text{combustion efficiency} = \frac{C^*}{C^*_{th}} \tag{14}$$

where C^* denotes the characteristics velocity, and the subscript th means its theoretical value. C^* is calculated by following equation:

$$C^* = \frac{P^*_c \cdot A_t}{\dot{m}_{o2} + \dot{m}_{CH4}} \tag{15}$$

where P^*_c is the stagnant pressure at the throat section and A_t is the cross sectional area at throat. C^*_{th} is solved by thermodynamic calculation with given propellants and mixture ratio. The results are shown in Figs. 16 and 17.

The combustion efficiency is 0.9516, 0.9497, 0.9515, 0.9486 and 0.9525 for case1 to case5 respectively. For the first three cases, the disparity on the combustion efficiency is less than 0.2%. It indicates that for a pintle injector with double rows of injection holes, Ls' variance exerts little influence on the combustion characteristics when Ls/Dp changes from 0.64 to 1.5 at Li/Dp = 0.38. However, generally speaking, the bigger Ls, the more space occupied by the injector, the smaller characteristic length, the lower combustion efficiency. Figure 18 may interpret this paradox. As aforementioned, the stoichiometric iso-surface reflects the main reaction zones and the position

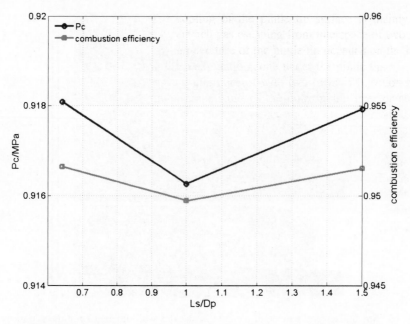

Fig. 16. Combustion chamber pressure and efficiency for different Ls/Dp

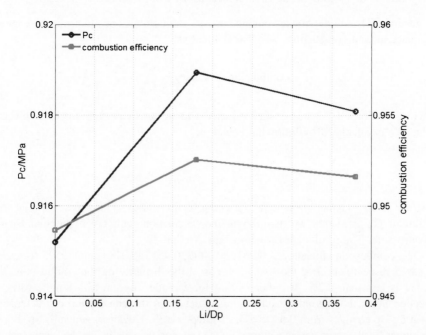

Fig. 17. Combustion chamber pressure and efficiency for different Li/Dp

of flames. When Ls increased, the iso-surface has not hit against the chamber wall like case1 and the iso-surface from second row of injection holes extends downstream gradually, which makes the reaction more complete and imbalances the effect of characteristic length decline.

For case1, case4 and case5, a relatively considerable difference up to 0.4% appears in the combustion efficiency. Figure 17 shows that Li/Dp = 0.18 (case5) has the highest combustion efficiency, 0.9525, and meanwhile the lowest temperature, 1326 K, on the surface of pintle injector within all cases studied. Form the view of stoichiometric iso-surfaces, proper Li/Dp makes both two rows of injection holes fully participate mixing and combusting compared to case4, further, merges the iso-surfaces of two rows of injection holes bringing incline of combustion efficiency due to additive effects compared to case1.

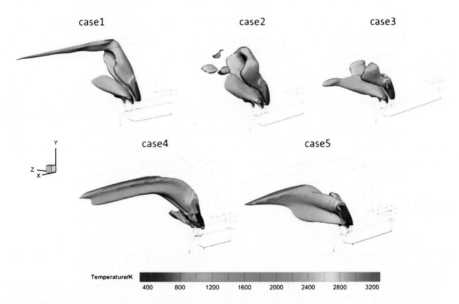

Fig. 18. Stoichiometric iso-surface of case1 to case5 colored by static temperature

4   Conclusions

This paper has proposed a three-dimensional numerical investigation on the characteristics of combustion and heat transfer of one GOX/CH4 pintle injector with two rows of injection holes. Five configurations of the injector were studied. The main conclusions could be drawn as following:

1. The basic flowfield structures contains a head recirculation zone in the head-dome zone, a center recirculation zone in the combustion chamber center and a small recirculation zone under the pintle tip. The scale and shape of the recirculation zones will change with different configurations of the pintle injector, sometimes significantly.

2. The cooling of the pintle tip surface is mainly realized by the recirculation of oxygen-rich gas escaping from interspace of two rows injection holes. The high or low temperature of the pintle tip depends on the high or low oxygen concentration in the recirculation zone under the pintle tip.
3. For a pintle injector with two rows of injection holes, at $Li/Dp = 0.38$, the maximum temperature of the outer pintle injector surface decreases over 20%, from 1756 K to 1404 K when Ls/Dp changes from 0.64 to 1.5, while meantime combustion efficiency has a small variance lower than 0.2% with an average value of 0.9510.
4. For a pintle injector with two rows of injection holes, at $Ls/Dp = 0.64$, temperature of the outer pintle injector surface significantly decreases over 24.48% to 1326 K at $Li/Dp = 0.18$ compared with the temperature of 1756 K at $Li/Dp = 0.38$, meanwhile a maximum combustion efficiency is gained of 0.9525 at $Li/Dp = 0.18$ too.

The results this paper gets would be greatly helpful for the engineering design of a pintle injector with two rows of injection holes. But there still requires further work to explain the fundamental mechanism that the flowfield structures change with different configurations of pintle injectors.

Acknowledgement. The authors would pay gratitude to the support of the Chinese Program for New Century Excellent Talents in University (Grant No. NCET-13-0156), the National Natural Science Foundation of China (No. 11472303 and 11402298) and the National Basic Research Program of China (Grant No. 613239). Additionally, a particular thank you to seniors from the research team of atomization and combustion for their suggestions and supports.

References

1. Powell RW, Lockwood MK, Cook SA (1998) The road from the NASA access to space study to a reusable launch vehicle. In: 49th international astronautical congress, Melbourne, Australia
2. Sackheim RL, Gavitt K (1999) An ultra low cost commercial launch vehicle, that you can truly afford to throw away. In: 35th AIAA/ASME/SAE/ASEE joint propulsion conference & exhibit
3. Vozoff M, Couluris J (2008) SpaceX products – advancing the use of space. In: AIAA SPACE 2008 conference & exposition
4. Davis LA (2018) Falcon Heavy. Engineering
5. Dressier GA, Bauer JM (2000) TRW pintle engine heritage and performance characteristics. In: 36th AiAA/ASME/SAE/ASEE joint propulsion conference and exhibit
6. Casiano MJ, Hulka JR, Yang V (2010) Liquid-propellant rocket engine throttling: a comprehensive review. J Propul Power 26(5):897–923
7. Betts EM, Frederick Jr RA (2010) A Historical systems study of liquid rocket engine throttling capabilities. In: 46th AIAA/ASME/SAE/ASEE joint propulsion conference & exhibit
8. Heister SD (2011) Pintle Injectors. Handbook of Atomization and Sprays. Springer, Heidelberg, pp 647–655
9. Radhakrishnan K, son M, Yu K, Ko J (2015) Numerical study on combustion characteristics of a pintle injector for liquid rocket engines. In: KSPE fall conference

10. Son M, Radhakrishnan K, Yoon Y, Koo J (2017) Numerical study on the combustion characteristics of a fuel-centered pintle injector for methane rocket engines. Acta Astronaut 135:139–149
11. X-X Fang, C-B Shen (2017) Study on atomization and combustion characteristics of LOX/methane pintle injectors. Acta Astronaut 136:369–379
12. Sakaki K, Kakudo H, Nakaya S, Tsue M (2015) Optical measurements of ethanol/liquid oxygen rocket engine combustor with planar pintle injector. In: 51st AIAA/SAE/ASEE joint propulsion conference, Orlando, FL
13. Sakaki K, Kakudo H, Nakaya S, Tsue M, Kanai R, Suzuki K, Inagawa T, Hiraiwa T (2016) Performance evaluation of rocket engine combustors using ethanol/liquid oxygen pintle injector. In: AIAA propulsion and energy forum 52nd AIAA/SAE/ASEE joint propulsion conference
14. Bedard MJ, Feldman TW, Rettenmaier A, Anderson W (2012) Student design/build/test of a throttleable LOX-LCH4 thrust chamber. In: 48th AIAA/ASME/SAE/ASEE joint propulsion conference & exhibit
15. Bedard MJ, Meier EJ, Anderson WE (2014) Student design/build/test of a throttleable LOX/LCH4 thrust chamber. In: 65th International Astronautical Congress, Toronto, Canada
16. Mueller TJ (2009) Space exploration technolohies, assignee. Pintle Injector Tip With Active Cooling. USA patent US 7,503,511 B2. Inventor
17. Ono DK, Dressler GA, Kruse WD, Solbes A (1998) The design, development, and qulification of an advanced columbium liquid apogee engiNE (AC-LAE). In: AIAA
18. Mueller T, Dressier G (2000) TRW 40 KLBf LOX/RP-1 low cost pintle enigine test results. In: 36th AIAA/ASWIE/SAE/ASEE joint propulsion conference and exhibit
19. Gromski JM, Majamaki AN, Chianese SG, Weinstock VD (2010) Northrop Grumman TR202 LOX/LH2 deep throttling engine technology project status. In: 46th AIAA/ASME/SAE/ASEE joint propulsion conference & exhibit
20. Launder BE, Spalding DB (1974) The numerical computation of turbulent flows. Comput Methods Appl Mech Eng 3:269–289
21. Frassoldati A, Cuoci A, Faravelli T, Ranzi E, Candusso C, Tolazzi D (2009) Simplified kinetic schemes for oxy-fuel combustion. In: 1st international conference on sustainable fuels for future energy - S4FE
22. Magnussen BF (1981) On the structure of turbulence and a generalized eddy dissipation concept for chemical reaction in turbulent flow. In: 19th AIAA meeting, St. Louis
23. Negishi H, Daimon Y, Kawashima H, Yamanishi N (2013) Conjugated combustion and heat transfer modeling for full-scale regeneratively cooled thrust chambers. In: 49th AIAA/ASME/SAE/ASEE joint propulsion conference, San Jose, CA
24. Song J, Sun B (2016) Coupled numerical simulation of combustion and regenerative cooling in LOX/Methane rocket engines. Appl Thermal Eng 106:762–773
25. Song J, Sun B (2017) Coupled heat transfer analysis of thrust chambers with recessed shear coaxial injectors. Acta Astronaut 132:150–160
26. Eiringhaus D, Rahn D, Riedmann H, Knab O, Haidn O (2017) Numerical investigation of a 7-element GOX/GCH4 subscale combustion chamber. In: 7TH european conference for aeronautics and aerospace scIENCES (EUCASS)
27. Hu R, Wang L, FU S (2015) Review of backward-facing step flow and separation reduction (in Chinese). Sci Sin-Phys Mech Astron 45:124704
28. Guoding C (2012) Research on the backward-facing step flow control Nanjing university of aeronautics and astronautics
29. Li B, Yao Y, Jiang Y, Yong H (2016) Experimental research of active flow control of turbulent separated flow on backward-facing step using synthetic jet perturbation. Acta Aeronaut et Astronaut Sinica. 37(2):545–554

Experimental Research on Air/Ethanol Mono-injector Gas Generator

Fang Zhao[1(✉)], Ze-bin Ren[1,2], Xian-feng Li[1], Long-de Guo[1],
Yu Tao[1], Yu Shi[1], and Zhi-feng Luo[3]

[1] Facility Design and Instrumentation Institute, China Aerodynamics
Research and Development Center, Mianyang 621000, China
1968683223@qq.com
[2] State Key Laboratory of Aerodynamics, China Aerodynamics
Research and Development Center, Mianyang 621000, China
[3] Aecc Sichuan Gas Turbine Research Establishment, Mianyang 621000, China

Abstract. A dual orifice gas generator project based on aero-engine combustor
was proposed. A set of mono-injector gas generator was designed and manufactured. Hot firing tests were conducted under different conditions to verify the
performance of the gas generator, and researches on the working pattern of
ethanol supply system, lean blowout limits tests and system sequence optimization were carried out. The test results show that the technical design scheme
of the generator is feasible, the mass flow rate, temperature and pressure all meet
the design target. The smooth ignition curve indicates a reliable ignition of the
generator and steady performance of combustion, and the gas generator is
characterized by advantages of step-start; through optimizing the ethanol supply
sequence, the problem about impact effect of supply pipeline caused by valve
work sequence is resolved, and combing with air pressure regulation optimization, the start-up time of gas generator is greatly shortened; the gas generator has a large lean blowout limits, and is able to work steadily under a wide
working range.

Keywords: Aero-engine · Combustor · Gas generator · Mono-injector ·
Blowout limits

1 Introduction

Driving gas source has an important impact on the ejecting efficiency, system scale,
operation safety and economy of ejecting system. Existing driving gas sources mainly
include compressed air, nitrogen, high temperature gas and so on. Compressed air and
nitrogen are easy to obtain and easy to use. It is widely used in wind tunnel test [1] and
ejector research [2–5], but the ejecting efficiency is low. The advantage of high temperature gas ejection lies in its higher efficiency and smaller scale. It is especially
suitable for the use of the ejector system which needs repeated start and short time
operation.

Gas generator is a kind of equipment to obtain high temperature gas through
combustion propellant. Current gas generator types mainly include liquid oxygen &

© Springer Nature Singapore Pte Ltd. 2019
X. Zhang (Ed.): APISAT 2018, LNEE 459, pp. 1826–1834, 2019.
https://doi.org/10.1007/978-981-13-3305-7_146

ethanol/kerosene/methane gas generator based on rocket engine combustor structure [6–8], other propellant combined gas generator [9–11], and high concentration hydrogen peroxide catalytic combustion gas generator [12] etc.

Based on the comprehensive analysis of the advantages and disadvantages of various schemes, the air/ethanol gas generator based on the structure of aero-engine single tube combustor is proposed as a new gas source for ejecting system. Compared with the gas generator based on the structure of liquid rocket engine combustor, the scheme uses gas film cooling instead of water cooling, greatly reduce the system scale. Compared to the high concentration hydrogen peroxide, liquid oxygen and other propellants, the application of air/ethanol obviously improves the system operation safety. In addition, the two propellants are safe and non-toxic, easy to transport and store, cheap and easy to obtain.

In the present study, gas generator test parts were developed, and a matching test system was set up to carry out the hot test, which verified the feasibility of the design scheme, meanwhile, obtained the preliminary research results.

2   Technology Proposal

In view of the characteristics and requirements of the overall scheme of an ejecting system, a gas generator based on the structure of aero-engine single tube combustor was designed. The main characteristic parameters are shown in Table 1.

Table 1. Design parameters of gas generator

No.	Parameter	Value
1	Air excess coefficient, α	3.0
2	Ethanol flow rate/(kg s^{-1})	0.924
3	Air flow rate/(kg s^{-1})	25
4	Gas total temperature/K	1100

2.1   Gas Generator Body

Gas generator body, as shown in Fig. 1, is mainly composed of combustion chamber casing (including flange before and after), swirler, injectors, flame tube, and the ignition system etc.

According to the starting work of the gas generator, injectors are designed to be dual oil circuit injectors (as shown in Fig. 2). The main oil circuit is used for ignition, and the secondary is used to meet the demand temperature.

The swirler shown in Fig. 3 uses two-stage axial swirler. The first stage swirler is mainly used for fuel atomization, and the swirling coefficient is 1.4. The second stage swirler with coefficient of 0.8, is mainly used to form a moderate reflux zone on the head of combustor.

The design of the flame tube adopts the "impact + gas film" cooling method, which greatly simplifies the auxiliary system.

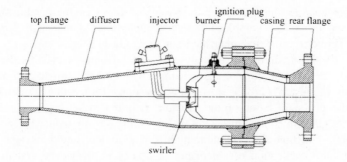

Fig. 1. Structure diagram of gas generator

2.2   Ignition Device

Common ignition methods include indirect ignition (pre combustion chamber) [13–15] and direct ignition (spark plug) [16, 17]. The ignition mode of the pre combustion chamber is mainly used in the early aero-engine combustion chamber and most of the industrial gas turbine combustor. Its advantages lies in its larger ignition energy and higher ignition reliability; and the shortcoming is that the ignition includes two stage, namely pre combustion chamber ignition and main combustion chamber ignition, thus the total ignition time is longer. Direct ignition is widely used in modern high performance aero-engine combustor with the continuous improvement of spark plug energy. Its advantages are light weight and fast ignition speed, and the disadvantage lies in its lower ignition reliability comparing with pre combustion chamber ignition. Considering the requirements of the ejector system for the start-up time of gas generator, the dual spark plug direct ignition method is used in this study, and the installation position is shown in Fig. 1.

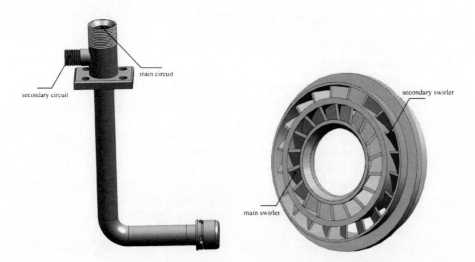

Fig. 2. Outside view of pressure-swirl injector   **Fig. 3.** Structure diagram of swirler

3   Hot Test System

The gas generator hot test system mainly includes air supply system, ethanol supply system, measurement and control system and so on, as shown in Fig. 4.

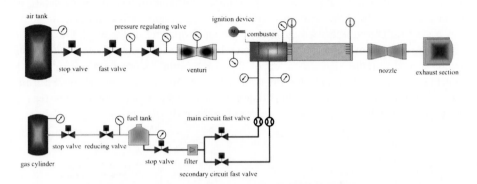

Fig. 4. Hot-test system schematic of gas generator

The air supply system consists of storage tanks, filters, valves, venturi pipes and pipelines, and the main air pipe are equipped with air filter, electric globe valve and air venturi tube in the downstream direction.

The ethanol supply system mainly consists of nitrogen path, ethanol storage tank, main/secondary oil supply pipeline, blow out gas pipe and so on. The ethanol is supplied by extrusion nitrogen, and the ethanol supply flow is regulated by controlling the pushing pressure of the storage tank.

The measurement and control system includes gas generator operation sequence control, safety interlocking, parameter measurement and data processing. The control object includes electric cut-off valve, pneumatic fast valve and blow off valve, and all the valves are controlled by open-loop except the air electric voltage regulating valve. In addition, the measurement and control system measures and records the pressure, temperature and mass flow data of each measuring point during the test. The pressure sensor adopts the piezoresistive sensor of Mike sensor company. The maximum response frequency is 30 kHz, and the measuring accuracy is 0.5% FS (full scale). The temperature is measured by temperature rake with precision of ±1 K. The US HOF-FER turbine flowmeter is used for flow measurement with precision of 1% FS.

4   Results and Analysis

For the sake of system safety, a number of small flow and single circuit ignition tests were carried.

4.1   Ignition Characteristic Analysis

Figure 5 shows the pressure curve with time during the hot test of gas generator under small flow and single oil circuit. The figure shows that ethanol is ignited immediately after entered into the gas generator, and the combustion chamber pressure is basically stable during the combustion process. In addition, the ignition test of gas generator is mainly divided into three processes: start up process, combustion process and shut-down process. During the starting process, the rapid opening of the ethanol fast valve causes the liquid flow impact, which is shown by the pressure peak of ethanol injection pressure in the figure. The first step of the combustion chamber pressure curve is generated by the cold air entering the combustor, and the second step is formed from ethanol combustion in the curve. In addition, the time interval between the two steps is about 0.1 s, which further indicates the rapid ignition of ethanol.

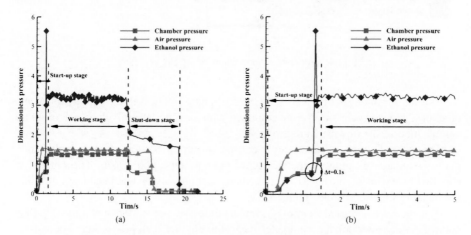

Fig. 5. Operation pressure curves of gas generator under single oil circuit ($\alpha \approx 2.9$)

Ignition tests under various air excess coefficient were carried out, aiming to explore the ignition limit of this gas generator. The correlation curves are shown in Figs. 6, 7 and 8.

It can be seen from the diagram that the gas generator has a large lean ignition limit and can work steadily at α 2.9 to 3.9. When the air excess coefficient increases to 4.1, the gas generator failed to ignite (as shown in Fig. 8). With the increase of air excess coefficient, the time interval between pressure producing of ethanol and the combustion chamber is prolonged, which causes the ignition delay of the gas generator. This is because the air excess coefficient increasing makes the area near the spark plug from the state of initial equivalent ratio of 1 to the oxygen-rich state, combined with the air velocity is larger, thus the ignition difficulty is bound to increase. In addition, the ethanol in the combustion chamber failed to ignite in time, and the accumulated ethanol is prone to deflagration. Meanwhile the intensity of the deflagration depends on the ignition delay time, resulting in greater impact damage to the structure.

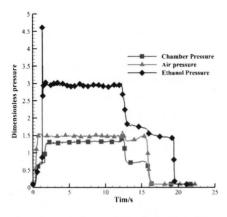

Fig. 6. Operation pressure curves of gas generator under single oil circuit ($\alpha \approx 3.2$)

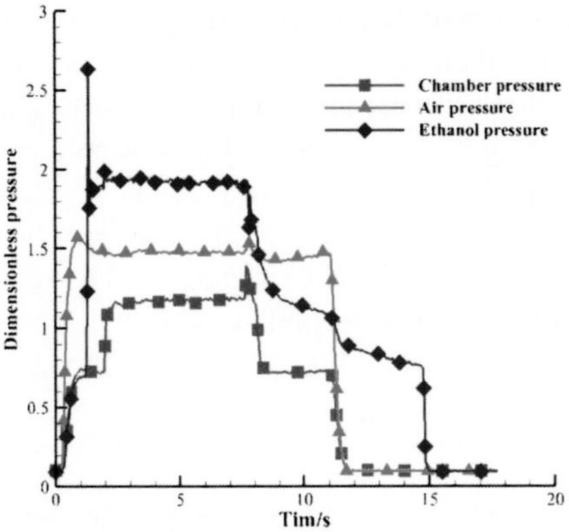

Fig. 7. Operation pressure curves of gas generator under single oil circuit ($\alpha \approx 3.9$)

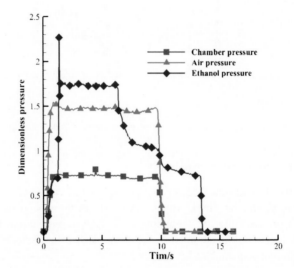

Fig. 8. Operation pressure curves of gas generator under single oil circuit ($\alpha \approx 4.1$)

After completing the small flow and single oil circuit ignition test, the dual oil circuit ignition test was carried out. At first, the main oil circuit and secondary oil circuit were opened simultaneously, while the test found that the secondary oil circuit suffered serious impact, causing the damage of solenoid valve in secondary oil circuit. The analysis shows that, as the flow velocity of the ethanol pipeline is fast, the main and secondary oil circuit is opened at the same time, and the mass flow rate of ethanol in the secondary oil circuit is greater than design value, which aggravates the flow velocity. Finally, it will bring considerable impact to the solenoid valve in secondary oil circuit. The time sequence of later adjustment of ethanol supply is in turn: opening of main oil circuit, opening of secondary oil circuit, closing of secondary and main oil circuit.

It is proved that the time sequence optimization solves the problem of impact. Typical dual oil circuit working pressure curve (after timing optimization), as shown in Fig. 9, is similar to the ignition process of single oil circuit, and divided into three processes: starting process, combustion process and shut-down process. The difference is that the ethanol is injected into the combustion chamber through two injections, forming three steps (marked by circle in the figure) of combustion chamber pressure, and the three steps represent cold air entering the combustion chamber, ignition of main oil circuit and ignition of secondary oil circuit, respectively.

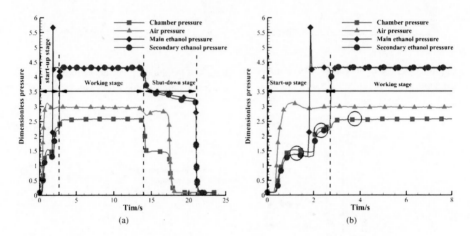

Fig. 9. Operation pressure curves of gas generator under dual circuit

4.2   Start Time Optimization Analysis

It is found that the starting time of gas generator is longer, reaching 3.1 s (combustion chamber pressure producing time), which can not meet the design index. The main reason is that air pressure after the regulating valve takes longer to reach the standard. Therefore, the relevant optimization work has been carried out, including air pressure regulation and timing sequence of ignition test.

Pressure curves before and after the air pressure optimization are shown in Fig. 10. They show the test curve under the same working condition (air pressure after regulating valve is about 3 MPa).

The criterion accuracy of air pressure reaching standard is controlled at ±0.1 MPa, and remains about 0.5 s stable time. By optimizing the pre-opening of pressure regulating valve and adjusting valve coefficients, time of air pressure reaches the standard obviously shortens from 1.7 s to 1 s after two times optimization. In addition, further optimization work is carried out for the ignition timing of gas generator, which mainly reduces the time difference between the main and secondary valve of ethanol pipeline.

Time sequence curves of the combustion chamber pressure producing before and after optimization as shown in Fig. 11, which is basically consistent with the time sequence of air pressure reaching the standard. After two times optimization, the start time of gas generator is shortened by 1.2 s, achieving the design requirement.

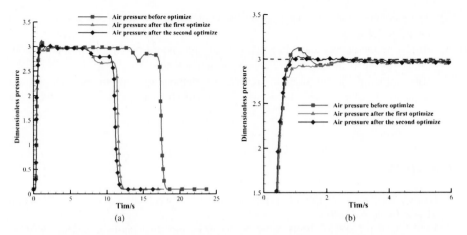

Fig. 10. Comparison of air pressure curves before and after optimization

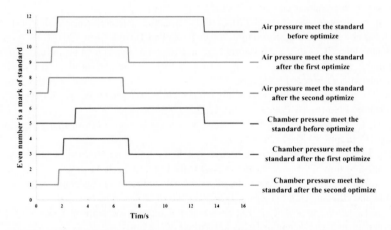

Fig. 11. Comparison of operation time sequence before and after optimization

5 Conclusion

An air/ethanol single tube gas generator based on the structure of aero-engine combustion chamber was developed for a high efficiency, large mass flow, miniaturized and fast start ejection gas source, and a series of hot tests were carried out for the ignition performance, the lean blowout limits and the timing optimization. The main conclusions are as follows:

(1) The technical scheme adopted is feasible, and the parameters such as flow rate, temperature and pressure of gas generator basically meet the design target;
(2) The gas generator is compact in structure, reliable in ignition and stable in operation performance;

(3) The gas generator has a larger lean blowout limits and can work stably in the a range of 2.9 to 3.9;
(4) Adjusting the supply sequence of main and secondary oil circuit, problem about impact of the If the secondary oil circuit get solved;
(5) By improving the air pressure regulating method and optimizing the ignition sequence of the system, the start-up time of the gas generator is effectively shortened.

References

1. Ling Q, Liao D, Tao Z (1994) Experimental research of wind tunnel injector. Aerodyn Exp Measur Control 8(2):10–17
2. Fabri J, Paulon J (1958) Theory and experiments on supersonic air-to-air ejectors. NACA-TM-1410
3. Xu W, Zou J, Wang Z et al (2005) Experimental investigation of the start performances of the supersonic annular ejector. J Rocket Propul 31(6):7–11
4. Chen J, Wang Z, Wu J et al (2012) Flow structure and performance of constant-area, supersonic-supersonic ejector. High Power Laser Part Beams 24(6):1301–1305
5. Connaughton JC (1977) Application of a hydrazine gas generator to vacuum ejector pumping of a chemical laser. In: AIAA-77-0892, presented at the AIAA/SAE 13th propulsion conference, 11–13 July 1977
6. Cao Z, Cai T, Tan Y (2005) Furnishing some theoretical basis for design of oxygen-rich gas generator. J Northwest Polytech Univ 23(5):557–561
7. Ma D, Lu G, Zhang X et al (2013) Research on hot tests of LOX/methane gas generator. J Rocket Propul 39(3):21–26
8. Feng J, Shen C, Zhao F (2012) Analysis of liquid and gaseous oxygen influence on the combustion flow field of air heater. J Natl Univ Defense Tech 34(4):43–48
9. Liu S, Hu X, Liu Q et al (2008) The experimental studies of gas generator based on nitrous oxide and ethanol. Ship Sci Tech 30(6):223–226
10. Li Q, Li Q, Wang Z (2010) Experimental research of the start-up process of gaseous oxygen/ethanol gas generator. J Rocket Propul 36(1):13–18
11. Liu J, Tan J, Yang T et al (2007) Research on low-frequency oscillation in hydrogen peroxide engine. J Natl Univ Defense Tech 29(5):23–26
12. Chen Z, Liao D, Liu Z et al (2007) Experimental studies of gas generator based on hydrogen peroxide and hypergolic ethanol. High Power Laser Part Beams 19(9):1409–1412
13. Yu N, Cai G, Zhang G et al (2006) Initial experiment research of oxidizer rich preburner. J Astronaut 27(5):834–838
14. Jin P, Yu N, Wu Z et al (2008) Experimental investigation of fuel rich preburner of hydrogen/ oxygen FFSC cycle engine. J Propul Tech 29(3):273–277
15. Li M, Gao Y, Chen Z et al (2011) Design and experiment for a GH2/ GO2 fuel rich preburner. J Aerosp Power 26(6):1426–1430
16. Kang Y, Lin Y, Huo W et al (2014) Effects of hardware geometry of dual-stage swirl cup on ignition performance. J Propul Tech 35(5):675–680
17. Jin S, Xu H, Ji L (2014) Technology research on air/jet fuel gas generator. J Rocket Propul 40(3):29–32

Analysis of Overall Performance of Multi-stage Combustor Scramjet Engine

Jinfeng Du$^{(\boxtimes)}$, Chun Guan$^{(\boxtimes)}$, Yuchun Chen$^{(\boxtimes)}$, Haomin Li$^{(\boxtimes)}$, and Zhihua Wang$^{(\boxtimes)}$

School of Power and Energy,
Northwestern Polytechnical University, Xi'an, China
nwpujin@126.com, guanchun10412@163.com,
chych888@nwpu.edu.cn, 1710274313@qq.com,
wzh0512@mail.nwpu.edu.cn

Abstract. In this paper, based on the scramjet, by the mode of characteristic calculation of one-dimensional scramjet engine, the effect of three and four-section combustion chamber geometry, fuel distribution ratio on scramjet performance is studied. The results of the comparison and evaluation of engine performance provide a certain reference for the geometric configuration of the combustion chamber and fuel injection parameters of the design point. It is mainly to determine the appropriate combustion chamber geometry and fuel injection method to obtain better combustion chamber performance and overall engine performance. The results show that increasing the thermal throat area of the combustion chamber can effectively increase the thrust of the aircraft during low-speed flight condition, and increase the fuel distribution ratio and the length of the expansion section in the front section of the combustion chamber without thermal choke and without affecting the inlet start, can improve the specific impulse of the engine at high speed.

Keywords: Scramjet · Overall performance · Combustion chamber · One-dimensional analysis

Nomenclature

Θ	Attack angle
Ma_0	Flight Mach number
Fn	Thrust
I_{SP}	Specific impulse
Q	Dynamic pressure head
H	Fight altitude (km)
$Width$	The width of the whole aircraft (m)

© Springer Nature Singapore Pte Ltd. 2019
X. Zhang (Ed.): APISAT 2018, LNEE 459, pp. 1835–1846, 2019.
https://doi.org/10.1007/978-981-13-3305-7_147

1 Introduction

At the beginning of the 21st century, the development of the world's military and war, the high speeds are the preconditions for achieving missile penetration and short-term strategic strikes. Therefore, the development of scramjet technology has become a research hotspot at present. Various countries consider scramjet technology as a key technology in the field of hypersonic research. At present, researchers have carried out a large number of theoretical studies [1–3], numerical simulation studies [2, 3], wind tunnel test studies and flight test studies [4–10].

The United States conducted two consecutive flight tests on the X-43A in 2004 and achieved flight targets of Mach 6.8 and Mach 9.7. On May 1, 2013, NASA completed the scheduled test target, the entire acceleration process lasted for 240 s, flight distance 426 km and no-power gliding (130 s); Japan, Germany, Australia [11–13], etc. are also carrying out their own hypersonic flight test plan. The choice of most propulsion systems is the scramjet engine. It can be seen that the scramjet engine technology has evolved from the theoretical research section of ground tests to the engineering application section.

Wu [14] and Wang [15] considered the design parameters of a three-section expansion type combustor, a single expansion ramp nozzle, and a three-section injection equivalence ratio and corresponding injection when studying scramjet engine combustion chambers. By changing the parameters of the above-mentioned design point, aiming at the maximum thrust and specific impulse, the optimization work is carried out and the test verification is carried out to provide a theoretical basis for the design of the combustion chamber and fuel distribution in the future.

For this reason, the performance of the multi-stage combustion chamber of the scramjet engine is studied in this paper, which provides a basis for future combustion chamber design and fuel distribution.

2 Scramjet Configuration

The working process of the scramjet engine is different from that of the traditional turbine engine. The components in the engine are also different. Therefore, the configuration of the scramjet engine needs to be explained.

2.1 Introduction to the Configuration of the Scramjet Engine

The scramjet engine consists of four parts: the inlet, the isolation, the combustion chamber, the nozzle. Figure 1 shows a schematic diagram of the components of the scramjet. Its cross-section definition and components are as follows (Table 1):

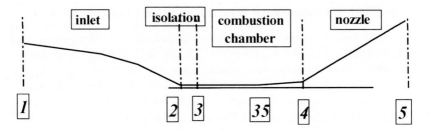

Fig. 1. Schematic diagram of scramjet components

Table 1. Section definition of scramjet

Section marking	Corresponding
Section 1	Undisturbed free flow section
Section 2	Outlet section of inlet/inlet section of isolation
Section 3	Outlet section of isolation/inlet section of first section of combustion chamber
Section 35	Outlet section of first section of combustion chamber/Inlet section of second section of combustion chamber
Section 4	Outlet section of second section of combustion chamber/inlet section of nozzle
Section 9	Outlet section of nozzle

2.2 Introduction to Design Point Operating Parameters of the Scramjet Engine

The engine uses kerosene ($C_{12}H_{23}$) as fuel and multi-point injection (Table 2).

Table 2. Scramjet engine design point parameters

Parameter	Value	Parameter	Value
Wa_0 (kg/s)	2.0	Width (m)	0.12
Q (kPa)	50.0	Ma_0	6.0
Θ (deg)	0	H (km)	26.5

3 Influence of Design Parameters of Three Sections of Combustion Chamber

The influence of the parameters of the three-section design of the combustion chamber is studied. Two variables are selected for study: A_{36}/A_3, Φ_2, and the parameters are similarly defined, in which A_{36} is the exit area of the second combustion chamber, and Φ_2 is the fuel supply ratio of the second section of the combustion chamber.

3.1   The Effect of Configuration of the Combustion Chamber

The engine reference configuration remains unchanged, and the expansion ratios of the first and the third sections of the combustion chamber are 1.0 and 2.0, respectively, and the expansion ratio of the combustion chamber in the middle section is $A_{36}/A_3 = 1.1$ to 1.5, and $\Delta A_{36}/A_3 = 0.1$. The specific profile configuration is shown in Fig. 2. The engine is operated at maximum fuel supply, and the overall performance of the engine at different flight speeds is calculated. The calculation results are shown in Fig. 3.

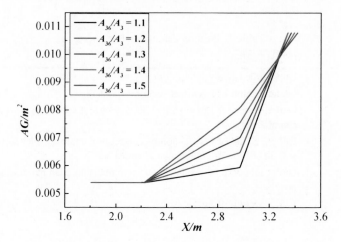

Fig. 2. The configuration of different second expansion ratio in three-section combustion chamber

From Fig. 3(a), we can see that for the same flight condition, when the maximum equivalent fuel-air ratio of the combustion chamber does not reach the equivalent fuel-air ratio $\Phi = 1.20$, the maximum equivalent fuel-air ratio increases with the increase of the second section expansion ratio A_{36}/A_3 of the combustion chamber. This is because the increase in the expansion ratio of the second section of the combustion chamber will lead to an increase in the throat area of the combustion chamber, which will allow more fuel to be added to the combustion chamber. Therefore, at the low Mach number, the maximum equivalent fuel-air ratio in the combustion chamber is large. When the expansion ratio A_{36}/A_3 of the second section of the combustion chamber is constant, with the increase of the flight Mach number Ma_0, the engine thrust appears to increase first and then decrease. The reason is that before the equivalent fuel-air ratio $\Phi = 1.20$, the increase in fuel-air ratio leads to an increase in thrust. The next reduction in thrust is due to the transition of the combustion mode of the engine to the direction of the purely supersonic mode of the boundary layer; for the same flight condition, the maximum equivalent fuel-air ratio in the combustion chamber reaches the equivalent of fuel-air ratio. Before $\Phi = 1.20$, the maximum state thrust increases with the increase of A_{36}/A_3, but when the equivalent fuel-air ratio $\Phi = 1.20$, the thrust decreases with the increase of A_{36}/A_3, but the effect is not significant. The engine specific thrust Fs and thrust Fn

change trend the same. As can be seen in Fig. 3(d), in general, the specific impulse increases with the increase of A_{36}/A_3 at low Mach number flight, while at high Mach numbers, the specific impulse decreases with the increase of A_{36}/A_3, and the impact of A_{36}/A_3 is getting smaller and smaller.

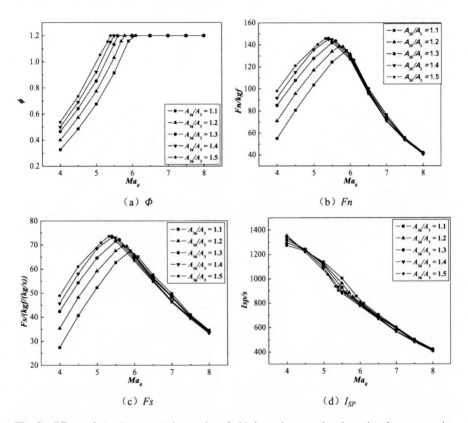

(a) Φ

(b) Fn

(c) Fs

(d) I_{SP}

Fig. 3. Effect of A_{36}/A_3 expansion ratio of third section combustion chamber on engine performance

3.2   The Proportion of Fuel Supply to the Combustion Chamber

The expansion ratios of the three sections of the combustion chamber are 1.0, 1.4 and 2.0, respectively. The other configuration parameters are unchanged, and the fuel supply rate of the three-section combustion chamber is analyzed and compared according to the following five conditions. The detailed fuel supply ratios in the five cases are shown in Table 3. The engine is operated at maximum fuel supply, and the overall performance of the engine at different flight speeds is calculated, as shown in Fig. 4.

Table 3. Distribution of fuel supply at each point of three-section combustion chamber

Num	Φ_1	Φ_2	Φ_3
1	0.20	0.25	0.55
2	0.20	0.30	0.50
3	0.20	0.35	0.45
4	0.20	0.40	0.40
5	0.20	0.45	0.35

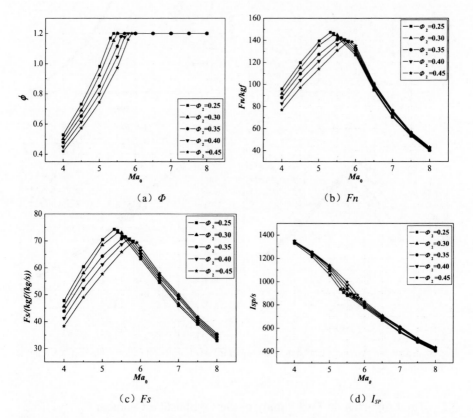

Fig. 4. The effect of the fuel supply ratio Φ_2 on the performance of the third section combustion chamber

From Fig. 4(a), it can be seen that for the same flight condition, when the maximum equivalent fuel-air ratio of the combustion chamber does not reach the equivalent fuel-air ratio $\Phi = 1.20$, the maximum equivalent fuel-air ratio of the engine is reduced with the decrease of the second-section fuel supply ratio Φ_2 of the combustion chamber. This is because the fuel supply ratio Φ_2 of the second section of the combustion chamber is reduced, that is, the heat absorption of the gas upstream of the combustion chamber is reduced, the overall Mach number of the combustion chamber is increased as a whole, and the thermal resistance loss during the heating process is increased. The Mach number downstream of the chamber decreases, but it is reflected in the geometric relation that the throat of the combustion chamber moves backwards, the area of the throat increases, and more fuel can be added in the combustion chamber. When the fuel supply ratio Φ_2 in the second section of the combustor is constant, the maximum engine thrust increases first and then decreases with the increase of the flying Mach number. The reason is that before the equivalent fuel-air ratio $\Phi = 1.20$, the engine equivalent fuel-air ratio, the increase in thrust leads to an increase in thrust; the reason for the next decrease in thrust is that the overall Mach number of the combustion chamber increases, the thermal resistance loss increases, and the combustion mode of the engine transitions to the purely scramjet mode direction in which the boundary layer is not separated; for the same in the flight condition, the maximum state thrust decreases with the increase of Φ_2 before the maximum equivalent fuel-air ratio of the combustion chamber reaches the equivalent fuel-air ratio $\Phi = 1.20$, but the maximum state thrust increases with the increase of Φ_2 when the equivalent fuel-air ratio $\Phi = 1.20$. However, the impact is not great. The engine specific thrust Fs and the thrust Fn have the same tendency (Fig. 4(c)). From Fig. 4(d) we can see that the specific impulse increases with the increase of the fuel supply ratio Φ_2 in the second section of the combustion chamber. The reason is that the larger the fuel injection quantity upstream of the combustion chamber, the higher the heat absorption in the upstream of the combustion chamber, resulting in an overall upper Mach number deviation low, the thermal resistance loss during heating decreases, and the total engine pressure loss decreases, so the engine specific impulse increases.

4   The Influence of Design Parameters of the Four Sections of the Combustion Chamber

To study the influence of the design parameters of the four-section combustion chamber, this section selects two variables for study: A_4/A_3, Φ_4. The parameter definition is similar to Fig. 1, in which A_4 is the fourth section of the combustion chamber exit area, and Φ_4 is the fuel supply rate of the fourth section combustion chamber.

4.1   The Geometry of the Combustion Chamber

The fuel distribution ratio of the four sections of the combustion chamber is kept constant at 0.20, 0.25, 0.25, and 0.30, respectively. The configuration of the following five cases with different expansion ratios is used, as shown in Fig. 5. Detailed configuration parameters are shown in Table 4. The engine is operated at maximum fuel supply, and the overall performance of the engine at different flight speeds is calculated, as shown in Fig. 6.

From Fig. 6(a), it can be seen that the expansion ratio of the fourth section of the combustion chamber has little effect on the maximum equivalent fuel-air ratio of the engine, because the configuration of the first three sections of the combustion chamber is certain, and the four-section fuel distribution ratio remains unchanged, making the combustion chamber thermal choke, the flow rate at the throat is constant, and the air

Table 4. Four-section combustion chamber expansion ratio

Num	A_{35}/A_3	A_{36}/A_3	A_{37}/A_3	A_4/A_3
1	1.0	1.2	1.3	2.0
2	1.0	1.2	1.3	2.2
3	1.0	1.2	1.3	2.4
4	1.0	1.2	1.3	2.6
5	1.0	1.2	1.3	2.8

flow already adheres to the fourth section of the combustion chamber, so the maximum equivalent fuel-air ratio remains unchanged. For the engine thrust, it can be seen from Fig. 6(b) that under the condition that the expansion ratio A_4/A_3 of the fourth section of the combustion chamber is not changed, the maximum state thrust of the engine increases first and then decreases with the increase of the flying Mach number $Ma_0 = 5.5$, the thrust of the engine reaches its maximum; for the same flight condition, it can be seen that the maximum engine thrust increases with the increase of A_4/A_3, because the air flow passes through the fourth section of the combustion chamber and is already an adherent flow. For the supersonic flow, in the expansion plane, the flow is accelerated, and the larger the expansion angle is, the faster the air flow is accelerated. The engine specific thrust Fs and the thrust Fn change in the same direction, as shown in Fig. 6(c). For the engine specific impulse, it can be seen from Fig. 6(d) that the effect of the A_4/A_3 is not significant at the low Mach number flight because the maximum amount of fuel that can be added to the engine is small when the aircraft is flying at a low Mach number. So that the combustion chamber exit temperature is low, fuel heat dissociation can be ignored; while flying at high Mach number, the impact of A_4/A_3 is increasing, and the larger the A_4/A_3, the higher the engine specific impulse.

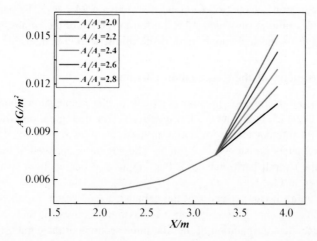

Fig. 5. Four-section combustion chamber with different fourth section expansion ratio A_4/A_3 configuration

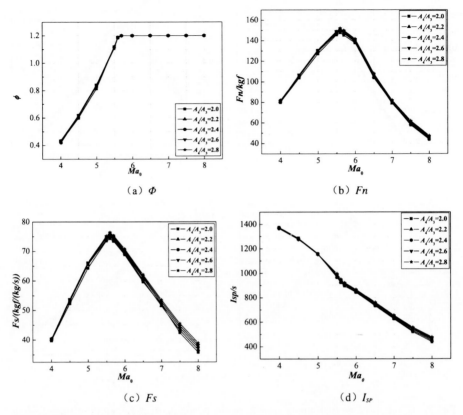

(a) Φ

(b) Fn

(c) Fs

(d) I_{SP}

Fig. 6. Effect of A_4/A_3 expansion ratio on the fourth section combustion chamber on engine performance

4.2 The Proportion of Fuel Supply to the Combustion Chamber

In this section, the expansion ratio of the four sections of the combustion chamber is 1.0, 1.2, 1.3, and 2.0, respectively, and the other configuration parameters remain unchanged. The fuel supply rate of the first three sections of the combustion chamber remains unchanged, and only the fourth section of the fuel supply parameters is changed. According to the following five conditions (Table 5) for analysis and comparison, the engine is operated at maximum fuel supply, and the overall performance of the engine at different flight speeds is calculated, as shown in Fig. 7.

Table 5. Distribution of fuel supply at each point of four- section combustion chamber

Num	Φ_1	Φ_2	Φ_3	Φ_4
1	0.20	0.25	0.25	0.30
2	0.20	0.25	0.25	0.35
3	0.20	0.25	0.25	0.40
4	0.20	0.25	0.25	0.45
5	0.20	0.25	0.25	0.50

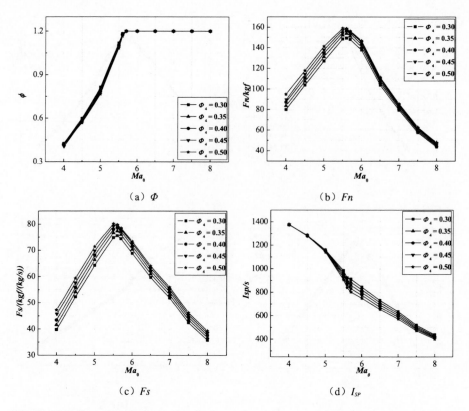

Fig. 7. The effect of the fuel supply ratio Φ_4 on the performance of the fourth section of a four-section combustion chamber

From Fig. 7(a), it can be seen that for the same flight condition, the fuel supply ratio Φ_4 in the fourth section of the combustion chamber has little effect on the maximum equivalent fuel-air ratio of the engine because the parameters of the inlet air flow are unchanged, and the configuration of the four-section combustion chamber is certain, and before the three-section fuel distribution ratio does not change, so the flow in the first three sections of the combustion chamber is almost the same, and the thermal throat of the combustion chamber often appears in the straight section of the combustion chamber and the first expansion section, so the geometric area and flow rate at the thermal throat are not change, so the maximum equivalent fuel-air ratio does not change. Under the condition that the fuel supply ratio of the fourth section of the combustion chamber is constant, the maximum equivalent fuel-air ratio of the engine appears to increase first and then remain unchanged with the increase of the flight Mach number; for the engine thrust, as shown in Fig. 7(b), When the fuel supply ratio Φ_4 in the fourth section of the combustion chamber is constant, the maximum engine thrust increases first and then decreases with the increase of the flight Mach number Ma_0, reaching the maximum at Ma = 5.5; for the same flight condition, we can see that the thrust of the engine increases with increasing Φ_4, and at low Mach numbers, Φ_4 has a greater influence on the engine thrust and has a smaller effect at high Mach numbers because the flow velocity of the combustion chamber is slower at low Mach numbers, although the fuel injected into the fourth section of the combustion chamber is more, but most of the fuels are burned completely and the combustion efficiency is high. Therefore, Φ_4 has a greater influence on the engine thrust; however, at high Mach numbers, although the combustion chamber adds more fuel, the airflow faster, the residence time of the fuel in the combustion chamber is shorter, and many fuels do not enter the nozzle directly without combustion, so the Φ_4 has little effect on engine thrust. As can be seen from Fig. 7(d), the specific impulse decreases as Φ_4 increases, and the influence at low Mach numbers is small.

5   Conclusion

With the increase of the flight Mach number, the engine maximum thrust first increases and then decreases. When $Ma_0 = 5.5$, the thrust of the engine reaches its maximum.

Through the study of different geometric configurations and fuel injection methods, it is found that increasing the thermal throat area of the combustion chamber can effectively increase the thrust of the aircraft during low-speed flight.

In the case where there is no thermal choking and no influence on the intake start, increasing the fuel distribution ratio in front of the combustion chamber and the length of the expansion section can improve the specific impulse at high engine speed.

References

1. White ME, Drummond JP, Kumar A (1986) Evolution and status of CFD techniques for Scramjet applications. In: Joint technology office of hypersonics. Roadmap for the hypersonic programs of the department of defense. Pub. L. No. 109-364, AIAA 1986, p 00160
2. Heiser WH, Pratt DT, Daley DH, Mehta UB (2002) Hypersonic Airbreathing Propulsion. AIAA education series
3. Cockrell CE, Auslender AH, Wayne Guy R (2002) Technology roadmap for dual-mode Scramjet propulsion to support space-access vision vehicle development. In: AIAA 2002, p 05188
4. Hirschel EJ, Erbland PJ Critical technologies for hypersonic vehicle development. RTO-EN-AVT-116
5. Pamadi BN, Brauckmann GJ (2001) Aerodynamic characteristics, database development, and flight simulation of the X-34 vehicle. In: AIAA 2001, p 03706
6. Faulkner RF, Pratt, Whitney (2003) The evolution of the HYSET hydrocarbon fueled Scramjet engine. In: AIAA 2003, p 07005
7. Waltrup PJ, White ME, Zarlingo F (2002) History of U.S. Navy Ramjet, Scramjet, and mixed-cycle propulsion development. In: AIAA 2002, p 05928
8. Prisell E (2005) The Scramjet - a solution for hypersonic aerodynamic propulsion. In: AIAA 2005, p 03550
9. Ferlemann SM, McClinton CR, Rock KE (2005) Hyper-X mach 7 Scramjet design, ground test and flight results. In: AIAA 2005, p 03322

1846   J. Du et al.

10. Hellman BM, Hartong AR (2007) Conceptual level off-design Scramjet performance modeling. In: AIAA 2007, p 05031
11. Mitani T, Tomioka S, Kanda T, Chinzei N, Kouchi T (2003) Scramjet performance achieved in engine tests from M4 to M8 flight conditions. In: AIAA 2003, p 7009
12. Heitmeir F, Lederer R, Herrmann O (1992) German hypersonic technology programme airbreathing propulsion activities. In: AIAA 1992, p 5057
13. Hass NE, Smart MK, Paull A (2005) Flight data analysis of HyShot2. In: AIAA 2005, p 3354
14. Wu X (2007) Research on design optimization of integrated flow passage for scramjet engine. Doctoral dissertation of National University of Defense Technology, p 10
15. Wang C (2011) Research on the overall design and optimization of Scramjet ramjet. National University of Defense Technology Ph.D. thesis, p 11

An Experimental and Computational Study of Freestream Condition in an Oxygen/Oil Gas-Jet Facility

Ling Zhao$^{(\boxtimes)}$, Xin Zhang, Bin Qi, and Yanghui Zou

Beijing Institute of Space Long March Vehicle,
Beijing 100076, People's Republic of China
zhaoling_zlg@163.com

Abstract. A numerical simulation and experimental measurements are performed to characterize the flowfield of oxygen/oil gas-jet facility. Chemical non-equilibrium Navier-Stokes equations have been solved to acquire the axisymmetric combustion-heated stream condition. The numerical results indicate that chemical reactions have an impact on flow oscillation, which alleviate the rate of decay of flow and make shock region move downstream. Flow diagnostics has been conducted to measure flow properties. Distributions of the pressure, temperature and Mach number along the centerline of the exit plane are experimentally obtained. The comparison has been presented between simulation and measurement. Static pressure distribution measurements show a good agreement with the numerical results. The computed temperatures are higher than measured ones, but the trends of oscillation and declension are nearly the same. Possible reasons of the difference are mainly because of uncertainly in chamber pressure and combustion efficiency.

Keywords: Gas jet facility · Chemical reaction · Experimental measurement · Jet flow

1 Introduction

The high temperature gas-jet facility is a combustion-heated hypersonic free-steam facility that provides high enthalpy simulation for thermal protection system components at test section. The test medium is the combustion products of oxygen and oil that are burned in a combustion chamber and through a conical nozzle to achieve the desired test condition. The flow field characteristic of freestream gas-jet facility is important to test and qualify thermal protection system material for aerospace vehicles. However, the freestream flow is not well characterized. Shock wave/boundary layer interactions form a complex jet flow wave system. Some residual oil may burn again with oxygen outside in the air. Those induce the pressure and temperature distribution changing dramatically. The non-equilibrium flow in and outside combustion nozzle is considerably more complicated than conventional gas dynamic.

Since the thermo-chemical state of the test gas has a large impact on flow fields around material samples or test models. Thermo-chemical models have been wide studied to get detail flow parameters for some ground-based facilities like arc-heated, or

© Springer Nature Singapore Pte Ltd. 2019
X. Zhang (Ed.): APISAT 2018, LNEE 459, pp. 1847–1856, 2019.
https://doi.org/10.1007/978-981-13-3305-7_148

impulse tunnels, whose test medium are air [1, 2]. For combustion-heated facilities, the freestream distribution, the flow field around test article and thermal response of test material are not accurately known. Therefore, it is necessary to better characterize the gas-jet freestream first.

In this study, a numerical simulation and experimental measurements are performed to study the jet flow of oxygen/oil gas-jet facility. Multi-reaction finite rate thermo-chemical models have been used to simulate the combustion-heated stream. Progress has been made to measure flow properties including static temperature, Mach number, static pressure and total pressure. Some comparisons between simulated and measured free-stream properties have been done, which offer an opportunity to test the current computational model.

2   High Temperature Combustion Gas-Jet Facility

A schematic of the oxygen/oil gas-jet facility is shown in Fig. 1. It consists of a combustion heater, control system, measuring system, watching system, fuel system and model support. The power of combustion heater is varying from 10 MW to 80 MW. In the combustion heater, the incoming test gas flows is speeding through the converging-diverging conical nozzle. Various exit diameter nozzles sections can be used from 64.8 mm to 166 mm, but the exit diameter of 64.8 mm is used for all the runs considered in this study.

Protection system material and models are conducted downstream of the nozzle exit on the nozzle centerline. Point pressure and heating are usually measured with a water cooled calibrated model to confirm test conditions. Finally, the flow exits test section through exhaust system. Compared with arc-heated facility, combustion-heated facility is easy to achieve high power condition which is situated for large scale models and higher heating environment.

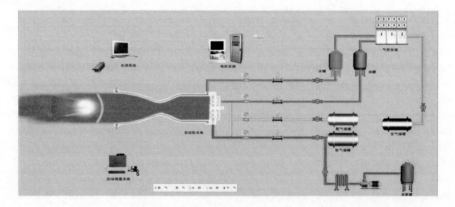

Fig. 1. Schematic representation of the oxygen/oil gas-jet facility

3   Experimental Approach

A ground experiment has been conducted on the gas-jet facility in order to get the characteristics of the freestream flow. Distributions of the static pressure, temperature and Mach number along the centerline of the exit plane are experimentally obtained. It is anticipated that the planned measurements will identify typical characteristics of the flowfield and provide validation data for simulation model.

3.1   Pressure Measurement

A single water-cooled probe is installed at the downstream centerline of the nozzle. The location of the probe is changed to measure static and total pressure at fix axial location. It is shown that how the probe and tube are installed in the flow field in Fig. 2. A support plate is mounted to provide the mechanical attachment of the probe. Linear motion is done by the adjustment of support platform. According to the measure environment, the measure time is controlled between 0.4 s and 0.8 s. The valid data of six measure points are obtained for static pressure and nineteen measure points are got for total pressure measurement.

Fig. 2.   Picture of pressure measurement

3.2   Temperature Measurement

Combustion measurements of temperature are conducted along the axis by IMPAC IPE140/45 infrared temperature sensors, which is a special pyrometer for the temperature measurement of hot CO_2 containing in combustion gas and its range is from 400 °C to 2000 °C. During the experiment, the pyrometer is placed on a removable optical platform and perpendicular to the jet flow. The gas emissivity is set to 1.0. It is shown that the curve of temperature data of 39 points vary with x coordinate in Fig. 3. Because of pyrometer range limiting, there are eight points exceeding the high limit and represented by blue square which means no valid data are obtained.

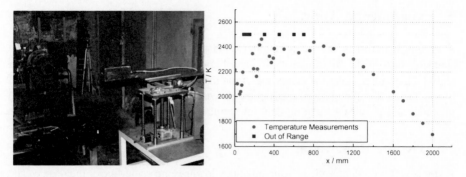

Fig. 3. Picture of temperature measurement and the result curve of T vs. x

3.3 Mach Measurement

Based on self luminescence characteristics of combustion gas, a water-cooled wedge model is placed into the flow to get Mach number. The image of flow around model is obtained by high speed photography, and then Mach number would be calculated from the shock angle and model parameters. The image of flow around wedge model is shown in Fig. 4.

Fig. 4. Picture of Mach measurement

4   Numerical Simulation Method

The combustion chamber and conical nozzle has L = 330 mm long. The nozzle is a typically conical nozzle, which consists of a converging section 85.4 mm long and a diverging section that has a halt angle of $20°$. The exit diameter of the nozzle is 65 mm.

Numerical simulation of flowfield from the combustion chamber to the exhaust flame is conducted. The grid depicted in Fig. 5 is used for the computations. The dimensions x and y are in millimeter with the nozzle exit located at x = 0. The computational domain includes the combustion chamber aft end, the nozzle and the external filed. Inside the nozzle, axial grid spacing is refined at throat to capture rapid flow expansion downstream and normal spacing is refined at the nozzle wall to resolve the boundary layer. Outside the nozzle, normal spacing is also refined at the mixing layer to consider gas mixing of exhaust gas with freestream air.

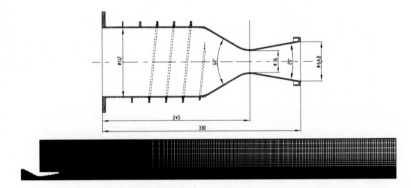

Fig. 5. Schematic of mesh grid

The equations of component transport and chemical reaction source terms are added to two-dimensional axisymmetric Navier-Stokes equations. The conservation equations for reaction flow are derived by a time-marching method and numerically solved using implicit second upwind scheme. The SST k-ω turbulence model and finite rate chemical kinetics method are taken to describe turbulence and chemical non-equilibrium reaction. The effect of turbulence on combustion is ignored in calculation.

$$\frac{\partial Q}{\partial t} + \frac{\partial E}{\partial x} + \frac{\partial F}{\partial y} + H = \frac{1}{\text{Re}}\left(\frac{\partial E_v}{\partial x} + \frac{\partial F_v}{\partial y} + H_v\right) + W \tag{1}$$

The flow of the combustion products in the nozzle and reaction with the ambient air are simulated. Nine main gaseous species (H_2O; CO_2; CO; H_2; OH; O_2; H; O and N_2) are considered in the computations. The eight species H_2, O_2, H_2O, H, O, OH, CO, CO_2 are treated as gaseous combustion products, and N_2 is treated as a chemically inert component. The gas phase fluid medium is assumed to be a thermally perfect gas mixture, with specific heats for all species given by polynomial functions of temperature.

The eleven reaction model [3–5] used for the finite-rate chemistry is given in Table 1. The first 1 to 8 reactions are accounted for the H, O and OH radicals in the H_2/O_2 system. The 9 to 11 reactions are employed for the CO/CO_2 system. The third body M is for N_2. This reaction mechanism represents a practical set for the after-burning environment. The direct reaction rate coefficients expressed in the generalized Arrhenius form $K_{f,b} = A_{f,b}T^{n_{f,b}}\exp\left[-E_{f,b}\left(R/_uT\right)\right]$ and are in cm^3/mole-sec units.

Table 1. The reaction model

Number	Reaction	Forward rate constant			Reverse rate constant		
		A	n	E/R	A	n	E/R
1	$H + O_2 = OH + O$	2.2×10^{14}	0	8455	1.5×10^{13}	0	0
2	$O + H_2 = OH + H$	7.5×10^{13}	0	5586	3.0×10^{13}	0	4429
3	$H_2 + OH = H + H_2O$	2.0×10^{13}	0	2600	8.4×10^{13}	0	10116
4	$OH + OH = O + H_2O$	5.3×10^{12}	0	503	5.8×10^{13}	0	9059
5	$H + H + M = H_2 + M$	1.8×10^{18}	−1.0	0	5.5×10^{18}	−1.0	51987
6	$H + OH + M = H_2O + M$	4.4×10^{20}	−1.5	0	5.2×10^{21}	−1.5	59386
7	$H + O + M = OH + M$	7.1×10^{18}	−1.0	0	8.5×10^{18}	−1.0	50830
8	$O + O + M = O_2 + M$	4.0×10^{17}	−1.0	0	7.2×10^{18}	−1.0	59340
9	$CO + OH = CO_2 + H$	1.7×10^{7}	1.3	−330	1.8×10^{9}	1.3	10550
10	$CO + O_2 = CO_2 + O$	2.5×10^{12}	0	24000	2.1×10^{13}	0	26890
11	$CO + O + M = CO_2 + M$	4.2×10^{-9}	0	2200	5.0×10^{17}	1.3	64670

The entrance flow conditions of the nozzle are specified as combustion gas at chamber temperature and pressure. Considering that a sufficiently detailed reaction model to predict the actual nozzle flow composition is not available, it is preferred to specify a simplified nozzle entrance condition. A uniform entrance flow is assumed with combustion efficiency equal to 1. In the computation uniform mass fraction of the considered chemical species is given (combustion products for nozzle and air for outside domain). Different temperature and velocity profiles are specified at the inlet to far field initially. The chamber pressure is given by the experiment, and chamber temperature and species mass fractions are provided by theory at the actual mixture ratio and chamber pressure. The exit boundary is specified as extrapolation condition. The water-called nozzle wall is assumed to be non-catalytic and an isothermal wall temperature of 300 K is used (Table 2).

Table 2. Initial data for inlet flow

Parameter name	Data		Name	Data	Name	Data
Chamber pressure, P_0 (MPa)	1.5	Mole fractions	H_2O	0.3053	OH	0.0548
Chamber temperature, T_0 (K)	3470		CO_2	0.1157	O_2	0.0166
Residual oxygen coefficient, α	0.7		CO	0.3353	H	0.0462
			H_2	0.1104	O	0.0156

5 Result and Discussion

5.1 Flow Parameter at Nozzle Exit

The parameter of nozzle exit is first studied numerically. Three computations are made: (1) in chemical non-equilibrium, (2) in chemical equilibrium, and (3) in a mixture model. The results of centerline Mach number, temperature and velocity at nozzle exit section are presented in Table 3. As shown in Table 3, the freezing of chemical energy can result in a decrease of the exit temperature. Temperatures predicted by chemical non-equilibrium model and chemical equilibrium model are obviously higher than mixture flow model. A variation of about 20% in the axial temperature between mixture flow and non-equilibrium flow is predicted for the reason of heat release by chemical reaction which emphasizes the importance of the chemical kinetics effects. The Mach number behaves contrarily trends.

Table 3. Parameters at nozzle exit

Parameter	Chemical non-equilibrium	Chemical equilibrium	Mixture
Mach number, Ma	2.38	2.3	2.4
Temperature (K)	2560	2676	2210
Velocity, U (m/s)	2394	2456	2390

5.2 Flow Structure of Freestream

In Fig. 6, the global structure of the flame calculated by numerical simulation is illustrated by temperature fields. At the nozzle exit, the jet flow are expanded and accelerated which corresponds to a temperature decrease. Temperature is still decreasing until an oblique shock is forming which result in a temperature disconti-nuity. Then the shock is reflected to the mixing layer and a newer expansion is propagating. Succession of expansions and compressions forms a string of Mach diamonds which causes the temperature decrease and increase along the axis till the pressure stabilize at the ambient pressure. As the axial distance increasing, the intensity of expansion and compression zones is weakened gradually. The amplitudes of tem-perature oscillation are also falling down due to the mixing of inner jet gas with outside air. Figure 7 is visualization of the exhaust flame taken by camera in which the shock wave system structure of the jet flow in the near field is acquired. By simple com-parison with simulation results, the structures of flow field and location of the shock wave are basically consistent.

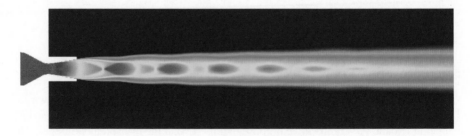

Fig. 6. Temperature field of computation

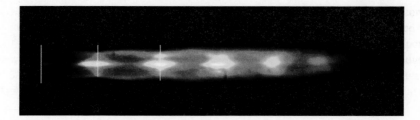

Fig. 7. Visualization of the exhaust flame

To investigate the chemical reaction effects in the external flow field, the variations of temperature from nozzle exit to outlet at the axial distance are compared between reacting and mixture models with the same nozzle exit conditions. In Fig. 8, the axial variations of temperature along the centerline with two models are displayed. Comparison of the data with reaction and mixture curves indicates that the shifts of the curves due to heat release are clearly evident which reveals the existence of the result of finite rate chemical kinetics method. The chemical reaction reduces the peak temperature and the slope of temperature profiles is less stiff than mixture flow. For reacting flow, the shock region is situated downstream the mixture case and the interaction between shocks are weakened. At far-field region, the decrease rate of temperature is alleviated due to chemical reactions between exhaust and atmospheric species.

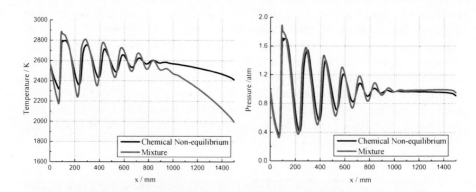

Fig. 8. Temperature and pressure distribution

5.3 Comparison with Experiment

Comparisons between measured and computed temperature axis profiles is shown in Fig. 9(a) in which the X axis is the distance from the nozzle exit and the Y axis is the temperature data. For comparative purpose, points beyond the measurement range with a blue square mark are specified as some virtual values to ensure the continuity of the measurement profile. The calculation results are higher than the measured values, but they are roughly consistent with the trend of complex wave system changes. The calculated successive and compression zones and the decrease rate of the temperature wave trends are similar to experiment. The first computed flow compression zone take place around 23 mm downstream compared to measure ones. The distance between two successive and compression zones is also displayed with a decreasing trend. All above indicate an overreaction effect. Considering there is an uncertainly in entrance parameter, the comparison in the way is expected for the computed temperature is particularly sensitive to modeling parameters. The difference can be explained by uncertain chamber pressure and idealized combustion efficiency which have an impact on the computation results.

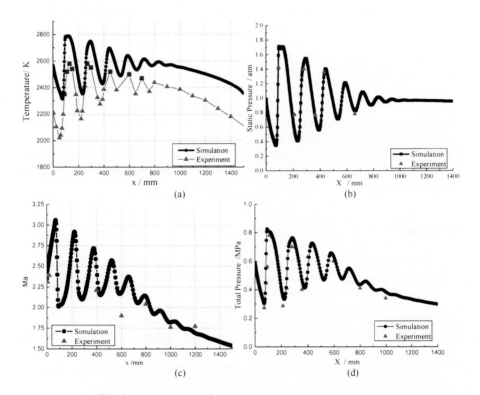

Fig. 9. Comparisons of measured and computed parameters

The measured and computed static pressure is shown in Fig. 9(b). Although the static pressure measurement points are less, the test data and the calculated data also maintain a good matching since the pressure is mainly determined by fluid dynamic effects and not chemical effects [1].

Figure 9(c) shows the measured and calculated Mach number. The computation shows a trend of increasing Mach number with increasing temperature, and Mach number is quickly decreased in far field. The measured values also show the same trend. The presence of shock waves in the jet region brings obvious interference for shock angle evaluation which may cause difference between computation and measurement.

The total pressure computations are presented in Fig. 9(d). The computed total pressure is derived based on the static pressure and Mach number calculations. The comparison between experiment and numerical results shows similar behaviors. The numerical results are roughly consistent with the actual test results and both of them better capture complex wave system changes in the gas jet flow field.

6   Conclusions

A numerical simulation and experiment have been performed to study the jet flow of oxygen/oil gas-jet facility. The numerical results indicate that chemical reactions had an impact on shock region which make shock region moves downstream and the inter-action between shocks are weakened. Meanwhile chemical reactions also alleviated the rate of flow decay due to the air/exhaust mixture chemical reaction process. Mea-surements have been performed along with temperature, pressure and Mach number. Comparisons between computed and measured flow properties reveal that the calcu-lation results are generally higher than the measured values, but they are roughly consistent with the trend of complex wave system changes. The results suggest that the chemical model used in this computation is basically satisfying the requirement. Sensitivity study should be conducted future to access the effect of chamber pressure and combustion efficiency uncertainties on the flow properties.

References

1. Joseph O, Douglas GF (2000) An experimental and computational study of the freestream conditions in an arc-jet facility. In: 31st AIAA plasmadynamics and lasers conference, AIAA 2000, p 2567
2. Scott C (1995) Surveys of measurements of flow properties in arcjets. J Thermophys Heat Transf 9:26–33
3. Baulch DL (1992) Evaluated kinetic data for combustion modelling. J Phys Chem Ref Data 21(3):411–734
4. Wang TS (2001) Thermophysics characterization of kerosene combustion. J Thermophys Heat Transf 15(2):140–147
5. Troyes J, Dubois I, Borie V et al (2006) Multi-phase reactive numerical simulations of a model solid rocket motor exhaust jet. In: 42nd AIAA/ASME/SAE/ASEE joint propulsion conference, July 2006, AIAA 2006, p 4414

Application of the Projective Method in the Numerical Simulation of Combustion

Yang Liu[1($\boxtimes$)] and Zheng Chen[2]

[1] Shanghai Aircraft Design and Research Institute, Shanghai 201210, China
liuyang6@comac.cc
[2] College of Engineering, Peking University, Beijing 100871, China
cz@pku.edu.cn

Abstract. The numerical simulation of combustion is one of the most important research tools for developing alternative fuels and high-performance combustion engines, which is widely used in the field of aerospace. However, the broad range of time scales and complex chemical kinetics bring great challenge to combustion simulation. With traditional implicit methods, most of the CPU time will be spent on solving the stiff ODEs caused by detailed chemistry mechanism. So, the object of this study is to use an accurate and efficient explicit method to solve the stiff ODEs, by which the computational efficiency can be greatly improved. Developed by Gear and co-workers, the projective method is utilized to solve stiff ODEs with a broad range of time scales, which is suitable for combustion problems. In this study, the homogeneous ignition process and spherical flame propagation process of methane are both investigated. Compared to the results from a traditional implicit method, VODE, projective method gives almost the same results, while the calculation speed is one-order faster than that of the VODE method. Due to its high accuracy and efficiency, the projective method will have great potentials in combustion simulation.

Keywords: Projective method · Combustion · Numerical simulation ·
Stiff equations

1 Introduction

Combustion and its relevant theories are widely used in propulsion technology in the field of aerospace. As a powerful tool, numerical simulation is helpful for understanding fundamental combustion phenomena and for revealing the physical-chemical mechanisms in combustion processes. Therefore, numerical simulation becomes one of the most important research tools for developing alternative fuels and high-performance combustion engines [4].

However, combustion involves multiple physical-chemical processes, broad range of time and spatial scales and complex chemical kinetics, bringing great challenge to numerical simulation [14]. For accurate simulation and quantitative prediction, detailed chemistry must be investigated, which will cause significant difficulty to simulation. On one hand, different components have different characteristic time, leading to a wide range of time scale. For example, the characteristic time of hydrogen and methane is

© Springer Nature Singapore Pte Ltd. 2019
X. Zhang (Ed.): APISAT 2018, LNEE 459, pp. 1857–1864, 2019.
https://doi.org/10.1007/978-981-13-3305-7_149

10^{-9} to 10^{-5} s, while for n-heptane, it is 10^{-13} to 10^{-5} s [1]. So, with the increase of complexity of molecule structure, the time scale of components in detailed chemistry will be more and more extensive, leading to very severe stiffness of relevant differential equations. One the other hand, with the increase of carbon number, the number of components and elementary reactions will increase exponentially, which also leads to a sharp increase of computation quantity [9]. For example, for fuels such as n-heptane and iso-octane, the number of components in detailed chemistry is about one thousand. As a result, nearly one thousand conservation equations concerning each component need to be solved, leading to a huge challenge to the simulation. During the simulation of propagation of methane/air premixed flame, solving stiff ODEs caused by detailed chemistry mechanisms will take most of the CPU time, if the traditional implicit method VODE is used. So, the key point to simulate the chemistry reaction flow efficiently is to deal with the wide range of time scales and complex chemistry kinetics mechanism.

To solve the stiff problem caused by the diverse time scale of different components, researchers proposed kinds of methods. For example, quasi-steady-state approximation (QSSA) and partial equilibrium approximation (PEA) [13] are classical methods to reduce stiffness. However, both of them need to experiential prediction and are difficult to adapt the constantly-changing conditions in the process of combustion. Recently, Lu and co-workers put forward a correctional method [10], which employed a different way to deal with QSS and PE components, according to the sparseness of complex coupling, to reduce the stiffness. But a shortage of this method is the difficulty to use explicit integration with big time step. Another way to solve stiff problems is to develop accurate numerical methods with high efficiency to solve stiff ODEs. Traditionally, implicit methods, such as VODE, are often employed to solve stiff ODEs, in which Jacobi Matrices are needed to be calculated. Thus the computation quantity is proportional to the square or even cube of the component number, which influence the computation efficiency very much. So, there are researchers who try to advance the explicit method or develop explicit-implicit-multi methods. For instance, Gou and co-workers recently develop a multi-time-scale method (MTS) [7, 8], by which the components are divided into different groups according to their characteristic time. In the fast group and slow group, different time step is used to conduct the Euler integration. As a result, the computation speed can be enhanced by one order. However, as the mass conservation of components cannot be ensured during the computation process, sometimes the error will accumulate to some extent and lead to a mistake.

In this paper, a totally explicit numerical algorithm, the projective method [5, 6], will be employed to simulate the homogeneous ignition process and spherical flame propagation process of methane, with the computation result and efficiency compared to that of implicit VODE method and MTS method, to reveal the advantage of the projective method in dealing with complex chemistry mechanism.

2 Principles of the Projective Method

Considering the following linear ordinary differential equations (ODEs):

$$dy/dt = Ay + g(t) \qquad (1)$$

where y is a vector function of m dimension. It is supposed that λ_j is eigenvalue of Matrix A and all of them are negative. The stiff ODEs mean that there are faster-change and slower-change parts in the solution, corresponding to bigger and smaller eigenvalues respectively. Faster-change parts only contribute to the solution in a very short time, and then the slower-change parts dominate the solution. Nevertheless, the stability requirement of numerical method is required by faster-change parts all the time. So, to get the numerical integration solution, time step is strictly constrained by the faster-change parts but the length of integration interval is decided by slower-change parts, leading to a huge number of integration steps, which is the main difficulty for solving the numerical solution of stiff ODEs.

Gear and co-workers put forward the so-called *Projective Method* (referred to as 'PM' in the following text) firstly in 2003 to solve stiff ODEs with a broad range of time scales. The basic principle of PM is as follows: Initially, integrate for some steps with a small time step, corresponding to time constant of faster-change parts, to decrease the disturbed error and faster-change parts of the solution, which could be called 'inner layer integration'. Next, the character of the solution is mainly depended on the slower-change parts, so the trend could be simulated on a longer interval by an extrapolation based on the last points of inner integration. For example, if the last two points of inner integration is used to conduct a linear extrapolation, the trend of slower-change parts is approximately represented by the slope of the line connected by that two points, which could be called Projective Forward Euler method [5]. Suppose the length of inner time step is h and numerical solution y_n at time $t = t_n$ has been obtained, then the calculate procedure in next $(k + 1 + M)$ time steps is listed as follows:

(1) from t_n to t_{n+k}, integrate with explicit method (e.g. Euler method) for k steps. Here h is rather smaller, so the high order accuracy is not necessary and basic requirement is just stability;

(2) from t_{n+k} to t_{n+k+1}, conduct inner integration for another step, and y_{n+k+1} will be obtained;

(3) base on y_{n+k} and y_{n+k+1}, extrapolate to an interval with length equaling to Mh, that is, $y_{n+k+1+M} = (M + 1) y_{n+k+1} - My_{n+k}$

The key point of above procedure is inner integration for $(k + 1)$ steps and an outer extrapolation on an interval of Mh, which could be called 'PkM' method. Figure 1 gives a sketch of the projective method [5] with k = 2.

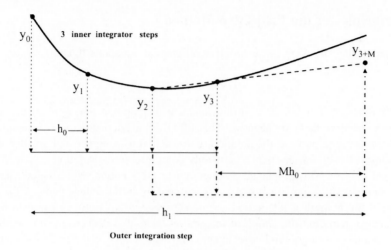

Fig. 1. Procedure of P2M method

According to the opinion of Gear etc., the procedure above can be 'iterated' [6], that is to say, a 'PkM' procedure can be packaged as a whole step with length equaling to $h_1 = (k + 1 + M)h$, which can be regarded a new inner layer. Based on the new layer, several step of integration and an extrapolation on Mh_1 interval can be conducted, where a longer step $h_2 = (k + 1 + M)h_1$ can be obtained. By this kind of procedure, multiple layers of integration step can be built, which is named 'Telescopic Projective Method' by Gear etc. [6]. Usually, a two-layer projective method could be named as P$k_1M_1-k_2M_2$.

3   Simulation for Homogeneous Ignition Process by Projective Method

In this section, a zero-dimension insulated homogeneous system with stable mass and pressure is studied, with pressure set to 1 atm and equivalent ration set to 1.0.

For projective method (PM), the integration step of innermost layer is set to $h = 10^{-9}$s, which is close to the maximum step that satisfies the stability of explicit Euler method. Based on this, PM of one, two or three layers could be built, and corresponding outer step is 10^{-8}s, 10^{-7}s and 10^{-6}s. The result of PM is compared to that of VODE and, to evaluate the calculation efficiency of different methods, the ratio of the CPU time cost of VODE against other methods with same t_{base} (for PM, t_{base} is the outer layer step) is defined as speed-up. The speed-up of VODE is defined as 1.

For methane/air ignition process, a detailed chemistry mechanism called GRI-Mech3.0 [11] is employed, which contains 53 kinds of components and 325 base reactions. Figure 2 illustrates the time history of temperature and mass percentage of CH_4 and OH, with initial temperate $T_0 = 1400$ K. Figures 2(a) and (b) are results of

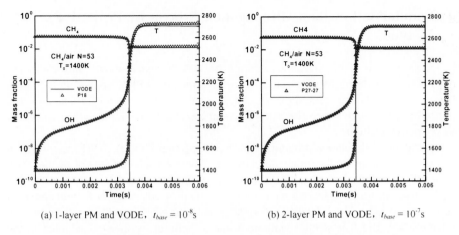

(a) 1-layer PM and VODE, $t_{base} = 10^{-8}$s

(b) 2-layer PM and VODE, $t_{base} = 10^{-7}$s

Fig. 2. Time history of temperature and mass percentage of CH_4 and OH

one and two layer PM compared with VODE. It is demonstrated that the calculation results from PM and VODE are almost the same.

As to the calculation efficiency, Fig. 3 show the speed-up of different methods. It reveals that the efficiency of PM rises with the decrease of number of layer. For one-layer PM, the calculation speed is 30 times faster than VODE. So, PM can be used to simulate the process of methane homogeneous ignition with good accuracy and high efficiency.

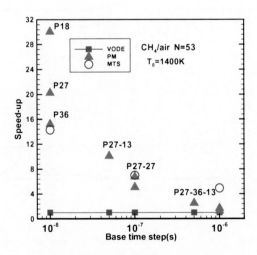

Fig. 3. Speed-up of different methods

4 Simulation for Flame Propagation Process by Projective Method

In the simulation of homogeneous ignition process discussed above, only chemistry reaction is considered. In this section, projective method will be applied in the simulation of flame propagation process, which contains both chemistry reaction and transport process like convection and diffusion, to validate the accuracy and efficiency of projective method.

An adaptive simulation program of unsteady reaction flow (A-SURF) is applied in this section to simulate the propagation of one dimension sphere flame of methane/air. A-SURF is widely applied in the simulation of sphere flame [2, 3], the governing equations of that are conservation equations of 1-D unsteady reaction flow in spherical coordinate system. More detailed information about A-SURF could be referred to the reference [3], and it is worth mentioning that a 'splitting' strategy is adopted in the program to separate the calculation process into two steps [12]: one is solving the partial differential equations (PDEs) only for convection and diffusion, the other is solving ODEs only describing the chemistry reaction, which is in fact the homogeneous ignition process discussed in last section.

Figure 4 shows the calculation results of the methane flame propagation, where Fig. 4(a) is the time history of flame radius and Fig. 4(b) is the spatial distribution of temperature, heat release rate and mass percentage of CH_4 and CO at time t = 3 ms. It is demonstrated that results obtained by PM and VODE are almost the same, which validates that PM can simulate the process of flame propagation accurately.

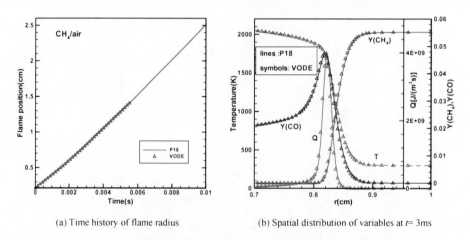

(a) Time history of flame radius   (b) Spatial distribution of variables at t = 3ms

Fig. 4. Results of spherical flame propagation of CH_4/air

Table 1 gives the CPU time cost of different methods. The physical time is from 0 to 1 ms while the length of time step is 10 ns, so there are totally 100,000 time steps. The 'speed-up' in the table is the ratio of the CPU time cost of VODE against PM, and this definition is the same with that in Sect. 2.

Table 1. Time cost for simulation of spherical flame propagation until t = 1 ms

CH₄/air	P18		VODE		Speed-up
	Time(s)	Proportion	Time(s)	Proportion	
Total	12400	100%	192760	100%	15.55
PDEs	6230.4	50.2%	6313.2	3.3%	1.01
ODEs	6132.8	49.5%	186409	96.7%	30.40
Diffusion	3909.2	31.5%	3972.6	2.1%	1.02
Viscosity	1435.6	11.6%	1428.9	0.7%	1.00
Others	885.6	7.1%	911.7	0.5%	1.03

Because of the splitting strategy, the two methods only have difference in the solution of stiff ODEs, so the computation time on other terms is theoretically the same, that is, the speed-up is one. A little error (within 3%) in Table 1 is caused by CPU disturbance during the calculation. Table 1 illustrates that if VODE method is employed, most CPU time is occupied by solving stiff ODEs. So the key point of enhancing the efficiency of combustion simulation is how to solve the stiff ODEs caused by detailed chemistry mechanism rapidly. Table 1 shows that if PM is employed, the time cost of ODEs is only 1/30 of that of VODE, and the total time cost is 1/15 of VODE. So, as to the simulation of flame propagation, PM can enhance the calculation speed by one order. Figure 5 illustrates the time cost percentage of each term in the equations, which obviously shows that time cost of solving stiff ODEs decreases significantly with PM replacing VODE method.

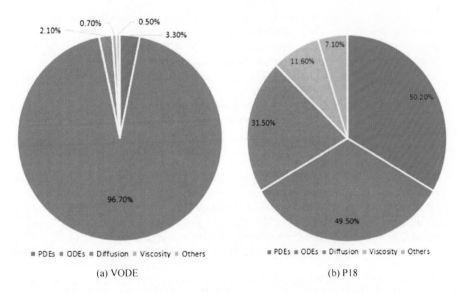

(a) VODE   (b) P18

Fig. 5. Time cost percentage for solving each item of equations

5   Conclusions

In this paper, the projective method is applied to the numerical simulation of methane combustion process. The research object includes both the zero-dimensional homogeneous ignition problem involving chemical reaction only, and the flame propagation problem involving both the chemical reaction and transport (convection and diffusion) process. Numerical test results show that:

(1) The projective method can guarantee the calculation accuracy. For methane ignition and flame propagation, it accurately predicts the change of temperature and main components over time and space, and the process of flame propagation.
(2) The projective method can greatly improve computational efficiency. Compared with the conventional VODE method, the projective method can enhance the computational speed by approximately one order of magnitude. Due to this, the projective method will have great potentials in combustion simulation.

References

1. Chen Z (2009) Multi-scale Simulation of Chemistry Reaction Flow. In: CCTAM2009, Zhengzhou, China
2. Chen Z, Burke MP, Ju Y (2009) Effects of compression and stretch on the determination of laminar flame speeds using propagating spherical flames. Combust Theor Model 13:343–364
3. Chen Z, Burke MP, Ju Y (2009) Effects of Lewis number and ignition energy on the determination of laminar flame speed using propagating spherical flames. Proc Combust Inst 32:1253–1260
4. Dec J (2009) Advanced compression-ignition engines-understanding the in-cylinder processes. Proc Combust Inst 32:2727–2742
5. Gear C, Kevrekidis IG (2003) Projective methods for stiff differential equations: problems with gaps in their eigenvalue spectrum. SIAM J Sci Comput 24(4):1091–1106
6. Gear C, Kevrekidis IG (2003) Telescopic projective methods for parabolic differential equations. J Comput Phys 187:95–109
7. Gou X, Sun W, Chen Z, Ju Y (2010) A dynamic multi-timescale method for combustion modeling with detailed and reduced chemical kinetic mechanisms. Combust Flame 157 (6):1111–1121
8. Gou X, Sun W, Chen Z, Ju Y (2010) Multi-timescales simulation for air ignition process. Science and technology of Combustion 16(5):452–455
9. Lu T, Law C (2009) Towards accommodating realistic fuel chemistry in large scale computations. Prog Energy Combust Sci 35:192–215
10. Lu T, Law C, Yoo C, Chen J (2009) Dynamic stiffness removal for direct numerical simulations. Combust Flame 156:1542–1551
11. Smith G et al GRI-MECH 3.0. http://www.me.berkeley.edu/grimech
12. Strang G (1968) On the construction and comparison of difference schemes. SIAM J Numer Anal 5:506–517
13. Turns SR (2009) An Introduction to Combustion: Concepts and Applications, 2nd edn
14. Westbrook C, Mizobuchi Y, Poinsot T, Smith P, Warnatz E (2005) Computational combustion. Proc Combust Inst 30:125–157

Experimental Study of High-Efficiency Loop Heat Pipe for High Power Avionics Cooling

Zhihu Xue[(✉)], Minghui Xie, Jiangfei Duan, and Wei Qu

China Academy of Aerospace Aerodynamics (CAAA),
Yungang West Road 17, Fengtai District, Beijing 100074, China
xuezhihu9@hotmail.com

Abstract. Avionics cooling is quickly becoming the limiting factor of aircraft/spacecraft performance and reliability, particularly with the rapidly increasing power density, and decreasing module size. This paper looks at the high-efficiency heat removal using loop heat pipe technology, as an advanced two-phase thermal control method by reducing the thermal resistance between the heat sources and heat sinks. Two high performance loop heat pipes (LHPs) are developed and experimented. The test results show that LHPs can well work at the heat load up to 663 W, with low thermal resistance of 0.042°C/W.

Keywords: Loop heat pipe · Thermal resistance ·
Two-phase technology · Avionics cooling

1 Introduction

The electronic device and flight control system on avionics is one of the most critical systems devised for aircraft and spacecraft during the whole mission. However, it is always beard to the high heat fluxes due to its continuously increasing power in chips for higher performance, decreasing in the physical dimensions of electronic packages for lighter weight, and serious dangerous surroundings which will worsen the performance of the electrical components. As a result, it is becoming more and more meaningful and challengeable for us to satisfy the need of high heat flux thermal control for electronics cooling [1–3]. Forced air-cooling systems are hard to control the operating temperature of chips or electronics below 100°C, limited by their outstanding constraints of small cooing capability and low cooling efficiency, while the power dissipation is over 300 W in a small area of 4×4 cm^2.

The heat pipe cooling technology, as an advanced two-phase thermal control method by reducing the thermal resistance between the heat sources and heat sinks, plays an important role in aircraft and spacecraft thermal management for its high efficiency, high stability, and good compatibility. Loop heat pipes (LHPs) are one kind of heat pipes which have a high heat transfer capability to meet the temperature control need of chips or electronics and an unique advantage or attraction with long distance heat transport compared with other heat pipes [4–6]. That is to say, the heat generated by chips or electronics can be removed outside of the electronic equipments faraway by LHPs, in consideration of that the space limits or secure safety constrains, such as liquid cooling.

© Springer Nature Singapore Pte Ltd. 2019
X. Zhang (Ed.): APISAT 2018, LNEE 459, pp. 1865–1871, 2019.
https://doi.org/10.1007/978-981-13-3305-7_150

This paper presents two loop heat pipes with flat evaporator for the direct cooling of the sources. The two loop heat pipes are coupled with different cooling methods, indicating that the condensers of LHPs are absolute varied. In order to study the heat transfer and temperature control levels to heater surface by LHPs, the thermal performance tests of two LHPs are experimented.

2   Experiment Design

Two aluminum LHPs with flat evaporator for the direct cooling of the chip are shown in Figs. 1 and 2. The two LHPs with an effective length of 300 mm and ID/OD line diameters of 4/6 mm correspondingly are designed and tested. The evaporator has a thickness of 13 mm, a length of 60 mm and a width of 40 mm. The working fluid of LHP is anhydrous ammonia with a purity of 99.995%. One LHP is called water cooling loop heat pipe, that is, its condenser is cooled by the external water circuit, shown as Fig. 1. Another LHP is called wind cooling loop heat pipe, that is, its condenser is cooled by the external wind, shown as Fig. 2. The volume of fins in air cooling LHP is $150 \times 80 \times 22 \ mm^3$, with 0.5 mm thickness for each fin.

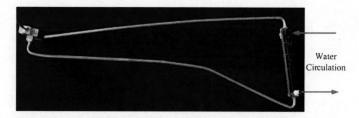

Fig. 1.   Photo of water cooling LHP

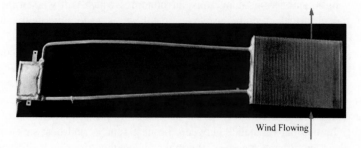

Fig. 2.   Photo of wind cooling LHP

The schematics of experiment system are shown in Figs. 3 and 4. The water cooling LHP is tested with the evaporator in horizontal orientation. A ceramic electrical heater with 40×40 mm^2 area is used as the input power device, which is analogous to the standard CPU chip. The heat power can be adjusted from 0 to 700 W with the accuracy of $\pm$ 5 W. Eight micro-scale thermocouples (K type) are attached onto the positions as shown Fig. 3 to measure the following temperatures: the temperature at the bottom surface of heater (101), the temperature at the upper surface of evaporator (102), the temperature at the exit of evaporator (103), the temperature at the entrance of condenser (109), the temperature at the inlet and outlet of water container (111) and (112) respectively, the temperature at the exit of condenser (114), and the temperature at the entrance of evaporator (118). The accuracy of all thermocouples is $\pm$ 0.25°C. The inlet temperature of cooling water is maintained at 23.5°C with $\pm$ 0.25°C accuracy by a cold bath appliance. All temperature data are recorded by the highly sensitive temperature logger (Agilent 34970A with resolution 0.1°C, accuracy $\pm$ 0.2°C) and connected to a PC for scanning the data every two second. There are a very thin layer of thermal grease (4.5 W/m.K) to fill in the gap between the top surface of heater and the bottom surface of evaporator. The whole experiment system including heater, LHP, water container or wind passage, is carefully covered insulation to reduce the heat loss to the environment. The environment temperature is 25°C for water cooling LHP test and air cooling LHP test.

The whole thermal resistance of LHP (R_{all}), from the chip to the condenser, is defined as:

$$R_{all} = \frac{T_{chip} - T_c}{Q} = \frac{T_{chip} - T_e}{Q} + \frac{T_e - T_c}{Q} = R_e + R_{e'} \tag{1}$$

Where T_{chip}, T_e, T_c, Q, R_e, $R_{e'}$ are the surface temperature of chip, the average temperature of evaporator, the average temperature of condenser, the heat load of heater, the thermal resistance of evaporator, the thermal resistance of LHP except evaporator.

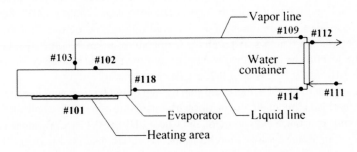

Fig. 3. Schematic of water cooling LHP experimental setup

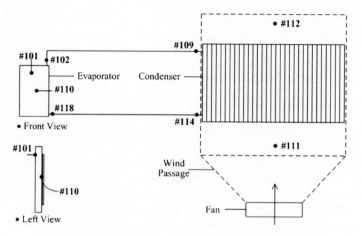

Fig. 4. Schematic of wind cooling LHP experimental setup

3   Results and Discussions

3.1   Water Cooling LHP

In order to investigate the heat transfer performance of LHP and temperature control level for heater, the power cycle tests for water cooling LHP are carried out. Figure 5 shows the temperature response results and operating characteristics of water cooling LHP at different heat loads. The heat loads are increased step by step, with maximum power of 663 W in this test. As shown in Fig. 5, the black curve indicated as num. 101 is on behalf of the surface operating temperature of heater, representing the temperature control ability of water cooling LHP.

When the heat load is 25 W, the temperature at all the measurement positions are increased fast simultaneously if no water is provided in the water container, indicating that the LHP could start quickly. After circulation water in water container is turn-on, the temperature curves will fall down and become steady-state in a short time, as seen in the Fig. 5. From the heat input to steady-state with water cooling LHP, there are no temperature oscillations or fluctuations occurred during this period, and the time of whole period is short than 5 s. After all the temperature curves become steady-state, the temperature level of heater surface is 33.5°C, and the thermal resistance of evaporator and whole LHP are 0.2°C/W and 0.225°C/W, respectively.

The heat input of heater is increased step by step through increasing its input voltage. When the heat load is about 90 W, the surface temperature of heater is controlled to 49°C, and the thermal resistance of evaporator and whole LHP are 0.116°C/W and 0.122°C/W, respectively. For water cooling loop heat pipe, when the heat load is over 200 W, the surface temperature of heater do not exceed 67°C, and the thermal resistance of evaporator and whole LHP are 0.073°C/W and 0.077°C/W, respectively. In test, the tested maximum heat load is 663 W. In this case, the surface temperature of heater is maintained at 95°C, lower than 100°C. And the thermal resistance of evaporator and whole LHP are 0.039°C/W and 0.042°C/W, respectively, with heat load of 663 W. From Fig. 5, it is apparent that all the temperature curves are kept smooth and of no fluctuations, meaning that water cooling LHP could work well at heat load of 663 W, and far away from its critical heat transfer performance.

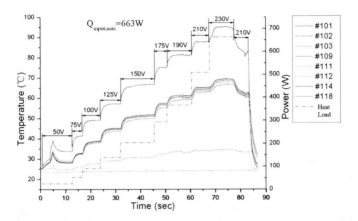

Fig. 5. Temperature response to varied heat load for water cooling LHP

3.2   Wind Cooling LHP

For wind cooling LHP, its condenser design is more crucial compared to water cooling LHP, with respect to the thermal resistance of condenser far higher than other parts in wind cooling LHP. The desired heat transport of wind cooling LHP is over 350 W and the thermal resistance of whole LHP could be as low as possible, to obtain good temperature control performance. Therefore, the condenser need enough outside surface area for heat transfer with air firstly. Then, in order to improve LHP the thermal performance, the whole condenser is conceived with an internal cavity structure coupling with some channels, instead of the traditional structure just as snake-bend pipe with covering fins.

Figure 6 shows the temperature response results and operating characteristics of wind cooling LHP at different heat loads. The heat loads are increased step by step, with maximum power of 513 W in this test. Before the heat is inputted, the air fan is turn-on in the wind passage. As shown in Fig. 6, the greenish blue curve indicated as num. 110 is on behalf of the surface operating temperature of heater. When the heat load is 29 W, the temperature at all the measurement positions are increased simultaneously with no sharp or peak temperature oscillations, presenting that the LHP could start in a smooth state. From the heat input to steady-state with wind cooling LHP, the time of whole period is short than 3 s. After all the temperature curves become steady-state, the temperature level of heater surface is 30.5°C, and the thermal resistance of evaporator and whole LHP are 0.069°C/W and 0.121°C/W, respectively. For wind cooling loop heat pipe, when the heat load is over 200 W, the surface temperature of heater do not exceed 55°C, and the thermal resistance of evaporator and whole LHP are 0.014°C/W and 0.086°C/W, respectively. In test, the tested maximum heat load is

513 W. In this case, the surface temperature of heater is maintained at 72.3°C, and the thermal resistance of evaporator and whole LHP are 0.011°C/W and 0.058°C/W, respectively.

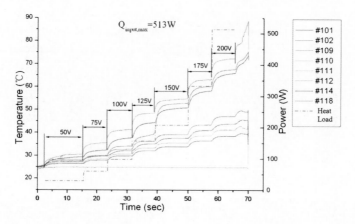

Fig. 6. Temperature response to varied heat load for wind cooling LHP

If the heat load continually increases forward 670 W after 513 W in test, the condenser structure of wind cooling LHP will be broken and could not work. As seen in Fig. 6, the temperature curves of evaporator will increase sharply, but the temperature curves of condenser will decrease on the contrary, with a few seconds at heat load of 670 W. At this instance, the inner saturated temperature of LHP fluid reaches to 72°C, corresponding to the inner pressure of fluid up to 3.7 MPa, which exceeds the safety operating pressure of cavity structure with restriction of weld methods.

4 Conclusions

The paper presents the thermal performance tests of two loop heat pipes with flat evaporator for the application of high power avionics cooling. The LHPs test results have shown that water cooling LHP can well work at the heat load up to 663 W, with low thermal resistance of 0.042°C/W and surface temperature of heater not exceeding than 95°C. Similarly, the wind cooling LHP results show that it can well work at the heat load up to 513 W, with low thermal resistance of 0.058°C/W and surface temperature of heater not exceeding than 72.3°C.

References

1. Sauciuc I, Prasher R, Chang J et al (2005) Thermal performance and key challenges for future CPU cooling technologies. In: Proceedings of IPACK2005, San Francisco, California
2. Wei J (2006) Challenges in package-level high density cooling. In: Proceedings of international symposium on transport phenomena, Toyama, Japan
3. Chen PH, Chang SW, Chiang KF, Li J (2008) High power electronic component: review. Recent Patents Eng 2(3):174–188
4. Maydanik YF, Vershini SV, Korukov MA et al (2005) Miniature loop heat pipe, a promising means for cooling electronics. IEEE Trans Comp Pack Teck 28(2):290–296
5. Pastukhov VG, Maydanik YF (2007) Low noise cooling of PC on the base of loop heat pipe. App Therm Eng 27:894–901
6. Singh R, Akbarzadeh A, Dixon C et al (2007) Miniature loop heat pipe with flat evaporator for cooling computer CPU. IEEE Trans Comp Pack Teck 30(1):42–49

Design and Experimental Study of Spherical Calorimeter in Arc-Heated Wind Tunnel

Jinlong Peng[✉], Jianqiang Tu, Guosheng Lin, and Dongbin Ou

Beijing Key Laboratory of Arc Plasma Application Equipment,
China Academy of Aerospace Aerodynamics, Beijing 100074, China
fzp_86@126.com

Abstract. In order to improve its performance and reduce the cost, hypersonic vehicle urgently needs to design the thermal structure with low redundancy. This requires the ground electric arc heating test to be more closer to the real aerodynamic thermal environment of the aircraft with better accuracy and repeatability. The surface heat flux density of the model is the important basis for determining the surface ablation state and total heat addition in the test parameters. In particular, the accuracy and repeatability of the heat flux density on the model surface directly affect the ablation and thermal insulation performance of the models. Therefore, it is necessary to accurately monitor and measure the heat flux density of model surface during long time ground assessment test in arc-heated wind tunnel. In this paper, based on the conventional water calorimeter measurement principle, combining with theoretical analysis, we design a new type spherical water calorimeter, which can measure the sum of spherical heat addition and stationary point heat flux density of the model. The internal cooling structure of spherical water calorimeter is optimized by means of three-dimensional heat transfer structure optimization, and the experimental verification is carried out. The results show that the quantity of the calorimeter with good accuracy and repeatability. This calorimeter can get the stationary point heat flux of the model during the long time test, and can diagnose the parameter fluctuation of the flow field in arc-heated wind tunnel.

Keywords: Heat flux · Arc-heated wind tunnel · Calorimeter · Parameter fluctuation

1 Introduction

In the development of a hypersonic vehicle, a large number of aerodynamic thermal simulation tests are required on the ground to examine the performance of thermal protection materials and the thermal structural performance of the aircraft [1–3]. From the ground simulation test assessment point of view, after the arc heating wind tunnel and the nozzle selected, the total airflow enthalpy, pressure and the heat flux of model surface become the most important flow field characterization parameters [4–6]. Because of the ablation and other factors of the arc heater electrodes, the air flow parameters will fluctuate during the long-term evaluation test, thus affecting the thermal environment of the model surface. In the long-term material assessment tests, the accuracy and repeatability of the heat flux on the surface of the model directly affect the

© Springer Nature Singapore Pte Ltd. 2019
X. Zhang (Ed.): APISAT 2018, LNEE 459, pp. 1872–1879, 2019.
https://doi.org/10.1007/978-981-13-3305-7_151

ablation and thermal insulation properties of the model materials [7, 8]. Therefore, the accurate monitoring and measurement of the heat flux on the surface of the model during the long-term ground assessment test is urgently necessary.

In the simulated thermal protection test parameters, the heat flux of the model surface is an important basis for determining the ablation state and total heating of the model surface. In the case of the low redundancy design of thermal protection model structures, the accuracy and repeatability of the heat flux measurement on the surface of the model directly affect the ablation and thermal insulation properties of the micro ablation material [9]. For the measurement of high heat flux, the transient calorimeter can not meet the requirements, and the steady state measurement is needed.

The traditional water calorimeter is assembled from various small parts. Installation and matching errors will have a great influence on the measurement accuracy. Traditional water calorimeters cannot directly measure the stagnation point temperature because of structural design limitations. Due to structural design and assembly factors, the traditional water card has a large three-dimensional heat transfer effect and causes a large error in measurement. Therefore, the existing water calorimeter needs to be improved and redesigned.

In this paper, a new type of spherical water calorimeter was designed based on theoretical analysis and numerical optimization technology. The influence of the fluctuation of airflow parameters on the surface heat flow of the ball head model in the arc wind tunnel was experimentally studied using the supersonic stagnation point test technology. Through data analysis, deviations of the stagnation heat flow and air flow parameters of the ball head model can be obtained.

2   Test Methods and Equipment

The arc wind tunnel ground ablation test is an effective method to evaluate the performance of thermal protection materials. The arc wind tunnel supersonic stagnation jet test technology is one of the commonly used technologies. This article uses the FD-04 high enthalpy plasma wind tunnel of China Academy of Aerospace Aerodynamics for verification tests. The schematic diagram of the test system is shown in Fig. 1. The simulation parameters mainly include the total air flow enthalpy, the heat flux at the stagnation point, and the pressure at the stagnation point.

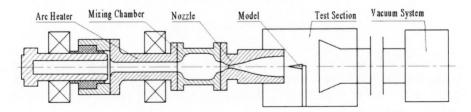

Fig. 1. Supersonic station free jet test schematic in arc wind tunnel.

The wind tunnel is a pressure-vacuum type supersonic arc wind tunnel, which is mainly composed of an arc heater, a mixing section, nozzle, test section, diffuser section, cooler and vacuum system.

The working principle of the wind tunnel is as follows: the high-pressure air is heated by an arc heater, and the high-temperature and high-pressure air is accelerated by the supersonic nozzle to form a high temperature and supersonic flow field in the test section, and the hypersonic aircraft aerodynamic heating environment can be simulated in a certain range. After the test, the gas is decelerated through the diffuser section, cooled down through the cooler, and discharged into the vacuum system.

3   Analysis of Sensitivity of Stationary Water Calorimeter

The structural schematic of the water calorimeter is shown in Fig. 2. According to the principle of heat balance, the heat flow from the high-temperature air flow to the end of the water card probe should be equal to the heat absorbed by the test cooling water. Therefore, the heat flux can be calculated by measuring the water flow rate and the water temperature rise:

$$q = \frac{m}{A} \cdot C_p \cdot \Delta T \qquad (1)$$

Where: $q(\text{kW/m}^2)$ is surface heat flux of calorimeter, $m(\text{kg/s})$ is the mass flow rate of cooling water, $C_p(\text{kJ/(kg.K)})$ is the specific heat capacity of water, $\Delta T(\text{K})$ is cooling water temperature rise, unit: K, $A(\text{m}^2)$ is the heated area of water-cooled calorimeter.

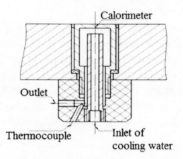

Fig. 2. The structural schematic of water calorimeter.

For the hemispherical model commonly used in the supersonic stagnation test, the stagnation heat flux is measured with a water calorimeter. The stagnation heat flux can be estimated by the formula of Fay-Reddell. The heat flux calculation formula of spherical model is:

$$\frac{q_{wb}}{q_{ws}} = \frac{2 \cdot \theta \cdot \sin\theta \left\{ \left[1 - \frac{1}{(\gamma_1 M_1^2)}\right] \cos^2\theta \right\}}{[D(\theta)]^{1/2}} \tag{2}$$

$$D(\theta) = \left[1 - \frac{1}{\gamma_1 M_1^2}\right](\theta^2 - \frac{\theta \cdot \sin 4\theta}{2} + \frac{1 - \cos 4\theta}{8}) \\ + \frac{4}{\gamma_1 M_1^2}(\theta^2 - \frac{\theta \cdot \sin 2\theta}{1} + \frac{1 - \cos 2\theta}{2}) \tag{3}$$

θ is the central angle, it can be seen from the formula that the spherical heat flux distribution is only related to the flow conditions and the geometric parameters of the model. For a certain ball nose radius $R = 35$ mm, the Mach number and adiabatic coefficient of the outlet gas flow in a given conical nozzle change very little (<0.1%) and can be ignored.

For the ball head, the hemisphere is divided into n equal parts along the central angle of the center of the circle, and the heating amount of each ring surface is added. That is, the heat amount taken by the water flow meter is taken away, which is:

Add the amount of heating to each ring surface, which is the amount of heat taken away by the cooling water in the water calorimeter.

$$G \cdot c_p \cdot (T_{out} - T_{in}) = \int q_\theta \cdot A_\theta d\theta \tag{4}$$

Therefore, for a calorimeter of the same size, if the cooling water flow is constant,

$$G \cdot c_p \cdot (T_{out} - T_{in} + dT) = (q_{ws} + dq) \int A(R, \theta) d\theta \tag{5}$$

$$\frac{T_{out} - T_{in} + dT}{T_{out} - T_{in}} = \frac{q_{ws} + dq}{q_{ws}} \tag{6}$$

$$\frac{dT}{T_{out} - T_{in}} = \frac{dq}{q_{ws}} \tag{7}$$

That is, the temperature response dT of the stagnation card calorimeter is proportional to the degree of deviation of the heat flow, and is proportional to the temperature rise of water in and out of the water card cooling water regardless of the size of the heated area of the water card.

It can be seen that the temperature response of the water calorimeter for measuring the total heat flow of the entire hemispherical sphere is consistent with the temperature response of the water calorimeter for measuring the hemisphere stagnation point area. Therefore, designing a spherical water calorimeter can avoid the assembly difficulties and deviations of the regional mini water calorimeter. It also can avoid the three-dimensional heat transfer effect, and the total heating amount of the entire sphere and the heat flux density of the model stagnation point can be obtained. It provides a new technical means for accurately monitoring and measuring the heat flux on the surface of the model during long-term ground assessment tests.

4   New Spherical Water Calorimeter Structure Design

The three-dimensional structure diagram of the integrated spherical water calorimeter designed is shown in Fig. 3. It includes the water card cooling water inlet, the cooling water outlet, the water cooling channel, the cooling water outlet temperature measuring channel and the ball head body. The heat quantity of the entire sphere can be calculated by measuring the water temperature rise and the flow rate at the inlet and outlet of the water cooling water in the test. According to the relationship between the heat flux density on the surface and the heat flux at the stagnation point, the heat flux at the stagnation point can be obtained through integral solution.

In this paper, an integral spherical water calorimeter with a ball radius $R = 35$ mm was designed according to the actual test requirements. The structure and dimensions of the waterway were finally determined by heat transfer coupling calculation and optimization. Figure 4 shows the temperature and pressure distribution of the optimized spherical water card.

Due to the axial symmetry of the internal structure of the spherical water card calorimeter, one quarter model is physically modeled. Among them, the inner diameter and length of each water flow channel and the wall thickness of the spherical water card outer circle are variables of optimization design. The heat flux density on the surface of the spherical calorimeter and the water flow rate at the inlet of the cooling channel are certain, and the spherical temperature distribution needs to be optimized. The spherical water card structure with uniform cooling water outlet temperature.

The numerical model is used to optimize the coupled heat transfer analysis of the model. The boundary condition for cooling heat transfer calculation is that the inlet of the fluid domain cooling water adopts the speed inlet condition, and the spherical heat flux density is set according to the heat flow density of the ball-point stagnation point to

Fig. 3. Spherical water calorimeter 3D structure diagram.

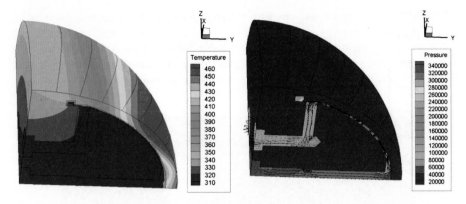

Fig. 4. Spherical water calorimeter temperature and pressure distribution cloud.

be measured. Distribution, additionally set the cooling water inlet flow rate is 10 g/s, the outlet uses pressure outlet conditions; the inner wall surface of the solid domain is the heating surface, given the boundary condition of the heat flow density; the interface between the cooling water and the solid domain is the coupling heat transfer condition, the interface The heat flux is equal, and other solid walls are insulated. Figure 4 shows the calculated temperature distribution of the spherical water card structure and the pressure distribution of the cooling water.

5   Test Results and Discussion

This paper uses three airflow states to verify the spherical water calorimeter under the same experiment. Table 1 shows the test condition parameters and temperature rise of the water calorimeter cooling water. The initial temperature of the cooling water in the test was 13.5 °C and the cooling water flow rate was 160 g/s. During the entire long test period, the surface of the spherical water calorimeter remained intact, and the test measurement data was intact without any abnormalities. After the test, all components of the spherical water calorimeter were intact and no water leakage occurred.

Table 1. The operating parameters of wind tunnel used in the current study

Condition	Mach number Ma	Total pressure P_0, kPa	Total enthalpy H_0, MJ/kg	Cooling water temperature rise DT, K
1	3.6	226	14	20.5
2	3.6	190	14	18
3	3.6	246	14	22.5

A photograph of the spherical water calorimeter during the test is shown in Fig. 5. The total air pressure during the test and the spherical water calorimeter cooling water outlet temperature curve are shown in Fig. 6. The water temperature rise response of the water card calorimeter changes with the change of the incoming heat parameters, and the response speed is within 1 s. It can be seen from the test results that the spherical water calorimeter designed in this paper has a fast response speed, and can operate stably for a long time, and can well monitor and measure the surface heat flux density of the stagnation model in long-term ground assessment tests.

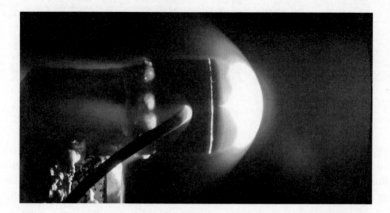

Fig. 5. Photograph of spherical water calorimeter in arc wind tunnel test.

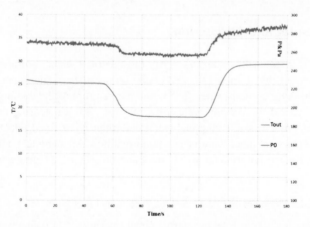

Fig. 6. Total pressure of air flow and cooling water outlet temperature curve during the test.

Table 2 shows the measured data of the spherical water card heat flow meter and the stagnation point heat flux data measured with the transient plug heat flow meter and their relative deviations. It can be seen from the results that the spherical water calorimeter designed in this paper has a relative deviation less than 3% compared with the transient heat flow calorimeter and can operate stably for a long time.

Table 2. Spherical water calorimeter test data

Condition	Total heat flow of spherical calorimeter, kW	Spherical water calorimeter stagnation point heat flux, kW/m^2	Stationary heat flux measured by plug calorimeter, kW/m^2	Relative deviation %
1	16.4	6050	5950	1.6
2	15.5	5600	5480	2.1
3	17.1	6435	6330	1.6

6   Conclusions

This article optimizes the internal flow channel structure of the spherical water calorimeter through three-dimensional heat transfer optimization analysis, and carries out experimental verification. The results show that the spherical water calorimeter designed in this paper has good accuracy and repeatability, and can obtain the value of the stagnation point heat flux of the model under long-term conditions. The calorimeter can be used for the arc wind tunnel flow field for a long time. Parameter fluctuations during operation are diagnosed.

References

1. Felderman EJ, Chapman R Jacocks JI et al (1994) Development of a high-pressure, high power arc heater: modeling requirements and status. AIAA Paper No. 94-2658, June 1994
2. Balter-Peterson A, Nichols F, Mifsud B, Love W (1992) Arc Jet Testing in NASA Ames Research Center Thermophysics Facilities. AIAA Paper 92-5041, December 1992
3. Bugel M, Reynier P, Smith A (2008) Review of European aerodynamics and aerothermo-dynamics capabilities for sample return missions. In: Proceedings of the 6th European symposium on aerodynamics for space vehicles
4. Hanson RK (2011) Applications of quantitative laser sensors to kinetics, propulsion and practical energy systems. Proc Combust Inst 33(1):1–40
5. Smith DM, Felderman EJ (2006) Aerothermal testing of space and missile materials development center arc jet facilities. In: 25th AIAA aerodynamic measurement technology and ground testing conference, San Francisco, 5–8 June 2006
6. Di Clemente M, Rufolo GC, Battista F An extrapolation from flight methodology for a re-entry vehicle wing leading edge test in a plasma wind tunnel facility. AIAA-2007-3895
7. Auweter-Kurtz M, Beck W, Bottin B, Boubert P, Bourdon A, Bultel A, Carbonaro M, Chazot O, Feigl M, Fletcher D (2000) Measurement Techniques for High Enthalpy and Plasma Flows
8. Kim S (2004) Development of tunable diode laser absorption sensors for a large-scale arc-heated-plasma wind tunnelD. Stanford University
9. Smith DM, Moody H, Wanstall C, Terrazas-Salinas I (2002) The design and use of calorimeters for characterization of high enthalpy flows in arc-heated test facilities. AIAA 2002-5236

Transient Simulation for the Gas Ingestion Through Turbine's Rim Seal

Jianping Hu, Zhenxia Liu[(⊠)], and Pengfei Zhu

School of Power and Energy, Northwestern Polytechnical University,
Xi'an, Shaanxi, China
zxliu@nwpu.edu.cn

Abstract. As the turbine front temperature of the modern aero-engine gets higher and higher, the problem of gas ingestion and the related rim sealing is becoming more and more serious. In this paper, the unsteady numerical simulation of the rim seal model with cavity is carried out, based on the gas ingestion experiment rig of Bath University. Firstly, the CFD results is compared with the experiments to valid the accuracy. The comparison for the pressure distribution shows that the simulation agrees well with the experiments. But the results for the sealing efficiency have some relatively errors. Secondly, the change process for some parameters about the ingestion is discussed. The simulation shows the pressure distribution of main flow is the decisive factor, and the interaction between the blades and the vanes affects the pressure distribution. Thirdly, the effect of the interaction between blade and vanes is provided a close analysis. Interestingly, the periodicity of the sealing efficiency is not consistent with that of the rotating blades. Also, the effect of the axial spacing is obtained through reducing the spacing by twenty percent. The decrease of axial spacing between vane and blade will cause the increase of the peak value of the pressure in the high pressure region.

Keywords: Rim seal · Gas ingestion · Unsteady simulation ·
Interaction between blades and vanes

1 Introduction

For a gas turbine, one of the functions of the air extracted from the compressor is to prevent the ingestion of hot mainstream gas into the turbine disc cavities. A significant amount of the seal air is required, otherwise hot gas probably ingresses into the wheelspace due to the pumping effect and uniform pressure distribution. Extracting air from the compressor means there is an associated performance penalty to the engine. In order to minimize the purge flow, the engineers try to quantify the minimum seal flow and the seal efficiency through experiments and simulations.

Early researches on ingestion neglected the effects of the main flow and showed that the minimum sealing flow rate has direct relationship with rotating speed of turbine disc and the clearance of the rim seal. Owen proposed the minimum seal flow is proportional to the axial seal clearance and the rotational Reynolds number [1]. Subsequently, a large number of studies had been carried out to consider the impact of

© Springer Nature Singapore Pte Ltd. 2019
X. Zhang (Ed.): APISAT 2018, LNEE 459, pp. 1880–1887, 2019.
https://doi.org/10.1007/978-981-13-3305-7_152

main flow on gas ingestion. By means of pressure measurement, flow visualization and tracer gas concentration monitoring, Phadke and Owen put forward the ingestion can be divided into two regions [2–4]. The first is the rotationally induced region, where Re_w/Re_ϕ is small. If Re_w/Re_ϕ , there is a proportional relation between the dimensionless minimum seal flow and turbine rotating speed. The second is the externally induced region, where Re_w/Re_φ is so large enough that the rotating speed of the rotating disc has rarely influence on the dimensionless minimum seal flow.

Owen [5–7] team had been studied theoretically and experimentally in great detail in the field of gas ingestion. Especially, Sangan [8, 9] and other scholars carried out a series of experimental to identify the difference between single and double seal structures. The results showed that the pressure distribution in the annulus is the dominant factor of gas ingestion and the correctness of the orifice model is verified, and the sealing efficiency of double seal structure is significantly higher than that of single seal structure. Bohn's team also built their own gas ingestion experiment rig, a series of experiments were carried out [10–13]. Two kinds of seal structures were measured by pressure and carbon dioxide gas concentration measurement and an empirical formula of the minimum seal flow under the influence of the main flow is given. The experiments show that the sealing efficiency reduces with the increase of Re_w, and the blades have influence on the pressure distribution of the trailing edge of vane. Roy used the high frequency pressure sensor to measure the unsteady pressure change, results show that the influence of blade on the asymmetrical pressure distribution of the main flow is as important as that of vane [14–16]. Zhou carried out the unsteady numerical calculation based on the gas ingestion experiment rig of Bath university [10].

This paper referred to the factors on pressure distribution near rim seal. By unsteady simulation, the pressure distributions at different moments are present. The factor of the interaction between blade and vane is taken account in this paper, and the axial spacing between the blades and vanes is a decisive effect on unsteady characteristic of rim seal. The work may be a part of compliment for the research of rim seal.

2   Computation Model

In this paper, the rim experiment platform for rim seal from the literature [7] is simplified to obtain the computational domain. The experiment platform contains 32 vanes and 41 blades. While, a stage of turbine, containing 32 vanes and 32 blades, have been simplified for numerical simulation. Hence, an 11.25° sector model is considered, including only one vane and one blade in the mainstream annulus. The simplified model and the type of boundary conditions are shown in Fig. 2. In the computational model, the cavity part and the rotor part are specified as rotating domain (Fig. 1).

Both of the mainstream inlet and the cooling inlet are specified as mass flow inlet, and mainstream outlet is defined as static pressure outlet. The mainstream Reynolds number $Re_w = 4.4 \times 10^5$ (mass flow = 0.016 kg/s), the rotational Reynolds number $Re_\varphi = 8.17 \times 10^5$. The commercial software package CFX 16.0 was employed for this study. Hexahedral structured grids were generated in the entire computational domain. The total grid number is 1.36 million, including contains 0.82 million elements for the rotating

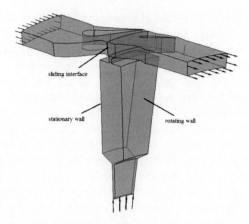

Fig. 1. Geometrical model and boundary conditions

domain and 0.54 million elements for the stationary domain. Mesh independence checks were carried out for the annulus pressure distribution downstream of the vane trailing edge. The y+ values were less than 5 in the cavity and mainstream. The computational meshes of mainstream and cavity and the mesh details are shown in Fig. 2.

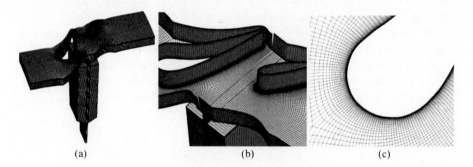

(a)   (b)   (c)

Fig. 2. Computational meshes for the model

The unsteady RANS equations with the SST turbulence model were solved in the computations. Upwind scheme was used for the advection terms and the second order backward Euler scheme was used for the transient calculations. The unsteady numerical calculation was carried out and the time step used is 1.5625×10^{-5}s, corresponding to 40 time steps for each blade passing period. The results are recorded at the moments of T/5, 2T/5, 3T/5, 4T/5 and T, respectively, where T is the time used for the blade passing a flow passage.

3   Validation Results

To measure the pressure distribution in the annulus, a row of pressure measuring holes are set at 2.5 mm behind the trailing edge of vane to measure the pressure distribution in the rig of Bath university [13]. In the following, the sealing efficiency is obtained by simulating the concentration distribution of the tracer gas, i.e. carbon dioxide. In the flow field, the concentration of carbon dioxide is between 0 and 1, and the sealing efficiency is expressed in the terms of the concentration of carbon dioxide as followed.

$$\varepsilon = \frac{c - c_a}{c_0 - c_a} \tag{1}$$

In the above expression, c is the concentration of CO_2, and the subscripts of a and o denote the annulus and coolant flow, respectively. The comparison of circumferential pressure coefficient (Cp) and sealing efficiency between the CFD results and the experimental results [13] is shown in Fig. 3. Circumferential pressure coefficient is defined as:

$$C_p = (p - \bar{p})/(0.5\rho\Omega^2 b) \tag{2}$$

where p is the absolute pressure near the hub and 2.5 mm behind the trailing edge of vane, $\bar{p}$ is the average value of absolute pressure in a flow passage, Ω is the angular speed.

Figure 3 shows that the numerical results of C_p and the experimental data versus the circumferential coordinates. As can be seen, the simulations are in good agreement with the experimental results, and the pressure is maximum in the region where locates at the trailing edge of the vane. That is because that boundary separation occurs at the trailing edge of vane when the air flows through the vane, resulting in local high pressure regions near the trailing edge of vane. These regions followed the ingesting

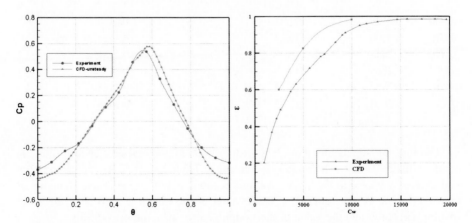

Fig. 3. Comparison of pressure coefficient Cp   **Fig. 4.** Comparison of seal efficiency

regions in the rim seal clearance, where the radial velocity V_r is below 0. Figure 4 shows the sealing efficiency versus the sealing flow rate. Relatively speaking, the sealing efficiency is not predicted correctly. The simulation for the seal efficiency is larger than the experiments.

4   Results and Discussion

4.1   Transient Performance of Gas Ingestion

For the three-dimensional unsteady rotating, the two types of high pressure regions induced by blades and vanes will interact with each other. The interaction leads to the change of the pressure distribution at the seal clearance with time, which further induces the change of the gas ingestion region size and extent of the gas ingestion. In order to evaluate the extent of gas ingestion, two parameters, i.e. the region ration A_i/A_c and the maximum ingestion velocity $V_{i,max}$, are applied to characterize the area of the gas ingestion region and the velocity of ingestion, respectively. Both of them are the critical parameters in the mass flow efficiency method. In the following, ε_m indicates the sealing efficiency of the disc cavity. If $\varepsilon_m = 1$, it means no hot gas ingress into the cavity.

Table 1 lists simulation results at different moments, including the region ratio, the maximum ingestion velocity and the mass flow sealing efficiency at different time. It is clearly that the distance mentioned above is smallest at the moment of t = 4T/10, and the area of gas ingestion region is the largest, the gas ingestion velocity is at a high level and the sealing efficiency is the least. The gas ingestion is the most serious at the moment of the 4T/10.

Table 1. Results for the change of the parameters with time

Time(t)	Area ratio (A_i/A_c)	The maximum ingestion velocity ($V_{i,max}$)	Sealing efficiency (ε_m)
2T/10	0.3529	11.41	0.2459
4T/10	0.5104	16.24	0.1521
6T/10	0.2897	16.28	0.2234
7T/10	0.2868	17.66	0.3237
8T/10	0.2882	16.24	0.4485
T	0.2618	12.92	0.5448

Through the simulation above, the distributions of the seal efficiency in the circumferential direction is shown in Fig. 5 from t = T/5 to t = 7T/5. As can be seen, the seal efficiency waved in the blade rotating direction, and the value is also changing at the every point. At the same time, the location of the trough is transfer in the same rotating direction, and the value of the trough is also changing periodically.

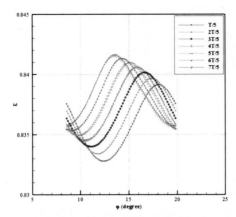

Fig. 5. The circumferential seal efficiency distribution (R = 28.6198 mm)

It is interesting that the period of the seal efficiency changing is rather longer than that of the blade rotating. This means that the seal efficiency is not only determined by the pressure distribution. This phenomenon has been particularly reported in the literature [17].

4.2 Influence of Interaction Between Blade and Vane

This section investigates the effect of interaction between blade and vane on pressure distribution of main flow. Along with the blade close to the vane, high pressure region near the seal clearance gradually expanded and the pressure level increased obviously, because this region forms due to the interaction between the high pressure region at the leading edge of blade and the high pressure region at the trailing edge of vane.

Figure 6 shows the change of the pressure distribution on the hub at the location x = 28.6198 mm at different times, where is the middle position of trailing edge of vane and leading edge of blade. It is clear that each curve has a peak value, and the absolute positions of the peaks are quite same. We can determine the position of the peak is at the location of the high pressure region of the trailing edge of the vane. From t = T/5 to t = 3T/5, the peak value increases gradually, and the peak value decreases gradually from t = 3T/5 to t = T, and the maximum value is reached at t = 3T/5. While, due to the vanes, there is a smaller peak value at the pressure curve corresponding to the t = T/5 moment. The smaller peak is caused by the high pressure region at the leading edge of blade. This peak shift along the circumferential direction not only leads to large increase in the peak value at the above t = 2T/5 and t = 3T/5 moments, but also leads to a small deviation of the absolute circumferential position of the larger peak.

Figure 7 shows the effect on the pressure distribution from the axial spacing between the vanes and the blades. Another case is carried out by shortening 20% off the axial spacing. As can be seen from the figure, there are two obvious high pressure regions at moment t = T/5. It means the effect of the interaction will be greater.

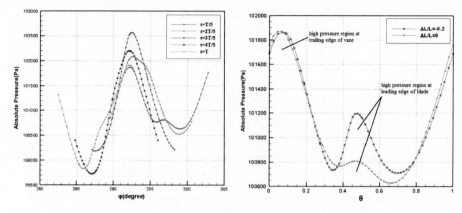

Fig. 6. The pressure distribution with time (x = 28.6198 mm)

Fig. 7. The effect of axial spacing on the pressure distribution (t = T/5)

5   Conclusions

(1)   The pressure distribution of main flow is the decisive factor of gas ingestion, and the high pressure region corresponds to the region of gas ingestion. The rotation of blade causes the variation of pressure distribution in the seal clearance and further leading to gas ingestion. The extent of gas ingestion varies with time.

(2)   In a period of the blades rotating, some parameters vary all the time. The period of the seal efficiency is longer than that of the blade passing a passage. It is probably cause by the rotating flow in the seal cavity, and the period of the rotating flow is longer than that of the rotating of blades.

(3)   The presence of the blade effects the pressure distribution of the main flow, which leads to the increase of the level of pressure peak and further leads to gas ingestion. Sometimes, the interaction causes the expansion of high pressure region outer the seal clearance. When the leading edge of the blade becomes close to the trailing edge of the vane, the high pressure region expands.

References

1. Bayley F, Owen JM (1970) The fluid dynamics of a shrouded disk system with a radial outflow of coolant. ASME J Eng Power 92:335–341
2. Phadke UP, Owen JM (1988) Aerodynamic aspects of the sealing of gas-turbine rotor-stator systems: part 1: the behavior of simple shrouded rotating-disk systems in a quiescent environment. Int J Heat Fluid Flow 9(2):98–105
3. Phadke UP, Owen JM (1988) Aerodynamic aspects of the sealing of gas-turbine rotor-stator systems: part 2: the performance of simple seals in a quasi-axisymmetric external flow. Int J Heat Fluid Flow 9(2):106–112

4. Phadke UP, Owen JM (1988) Aerodynamic aspects of the sealing of gas-turbine rotor-stator systems: part 3: the effect of nonaxisymmetric external flow on seal performance. Int J Heat Fluid Flow 9(2):113–117
5. Owen JM Prediction of ingestion through turbine rim seals part 1: rotationally-induced ingress. ASME GT2009-59121
6. Owen JM Prediction of ingestion through turbine rim seals part 2: externally-induced and combined ingress. ASME GT2009-59122
7. Owen JM, Zhou K, Wilson M Prediction of ingress through turbine rim seals part 1: externally-induced ingress. ASME GT2010-23346
8. Sangan CM, Zhou K, Owen JM Experimental measurements of ingestion through turbine rim seals. Part 1: externally-induced ingress. ASME GT2011-45310
9. Sangan CM, Pountney OJ, Zhou K, Wilson M, Owen JM (2011) Experimental Measurements of Ingress through Turbine Rim Seals. Part2: Rotationally-Induced Ingress. ASME GT2011-45313
10. Bhon D, Rudzinski B (1999) Influence of rim seal geometry on hot gas ingestion into the upstream cavity of an axial turbine stage. In: ASME international gas turbine and aeroengine congress and exposition. ASME GT1999-0248
11. Bhon D, Bernd R, Norbert S (2000) Experimental and numerical investigation of the influence of rotor blades on hot gas ingestion into the upstream cavity of an axial turbine stage. In: ASME GT2000-0284
12. Bhon D, Michael W (2003) Improved formulation to determine minimum sealing flow – C_w, M_{in} – for different sealing configurations. In: ASME GT2003-38465
13. Bhon D, Achim D, Ma H (2003) Influence of sealing air mass flow on the velocity distribution in and inside the rim seal of the upstream cavity of A 1.5-stage turbine. In: ASME GT2003-38459
14. Roy RP, Xu G, Feng J (2001) Pressure field and main-stream gas ingestion in a rotor-stator disk cavity. In: ASME GT2001-0564
15. Roy RP, Feng J, Narzary D, Paolollo RE (2005) Experiment on gas ingestion through axial-flow turbine rim seals. J Eng Gas Turbines Power 127:573–582
16. Zhou DW, Roy RP, Wang CZ et al (2009) Main gas ingestion in a turbine stage for three rim cavity configurations. In: ASME, GT2009-59851
17. Zhou K, Wilson M, Owen JM Computation of ingestion through gas turbine rim seals. ASME, GT2011-45314

Numerical Investigation on Intersecting-Grids Composite Cooling Structure with Internal Network Channels

Guanghua Zheng[✉], Yong Chen, Jialin Li, and Chengcheng Hui

School of Power and Energy,
Northwestern Polytechnical University, Xi'an, China
zhengguanghua@nwpu.edu.cn

Abstract. As the Mach number and range of aerial vehicles increases, the heat protection of the combustion chamber has been a more and more important issue. Besides developing heat-tolerant materials and new combustion chamber cooling techniques, the development of advanced cooling technology is also a significant way to the thermal protection of combustion chambers, such as film cooling, intersecting-grids cooling configuration with internal network channels, laminate cooling and the combination of thermal barrier and film cooling. The internal flow condition and cooling features of the intersecting-grids cooling structure that can be applied in the combustion chambers and turbine vanes and blades of modern gas turbine engines is investigated numerically in the paper, to get knowledge of the cooling features of such structure and discuss the feasibility of applying it in cooling systems of aero-engines.

Keywords: Numerical investigation ·
Intersecting-grids cooling configuration · Secondary flow

1 Introduction

As the Mach number and range of aerial vehicles increases, the heat protection of the combustion chamber has been a more and more important issue [1, 2]. Besides developing heat-tolerant materials and new combustion chamber cooling techniques, the development of advanced cooling technology is also a significant way to the thermal protection of combustion chambers.

Film cooling is the most conventionally and widely applied cooling method. The gas film from the holes of combustion chambers forms a piece of film on the internal surface of them, which separates the wall from the hot gas and meanwhile takes away the radiation of hot gas and flame. Thus cooling and heat insulation can be achieved.

Intersecting-grids cooling configuration with internal network channels is a new scheme of cooling configuration with wall cooling and film cooling combined. Such film cooling structure could strengthen the cooling effect and reduce the usage of coolant. If the mechanics of enhancement of heat transfer and the dominating factor for cooling features can be discovered, then disciplines of the application of such structures in the design process of wall cooling in combustion chambers are available. Highly efficient cooling

© Springer Nature Singapore Pte Ltd. 2019
X. Zhang (Ed.): APISAT 2018, LNEE 459, pp. 1888–1899, 2019.
https://doi.org/10.1007/978-981-13-3305-7_153

structure ensures higher combustion chamber temperatures and consequently increases the thrust and efficiency of aero engines. Usually the combination of full-coverage film cooling and impingement, convection and film cooling are employed in flame tube cooling. In recent years composite cooling structures are being investigated here and abroad.

Except conventional film cooling, new cooling techniques are being developed, for example the combination of thermal barrier and film cooling [3–5], taking advantage of each, could greatly enhance the cooling effect, which has good prospect in aero engines. However the effectiveness is related to the quality of coating craftwork. Laminate cooling also belongs to a new technology and USA and Russia have long before applied it to flame tube cooling. It is essentially a composite of impingement, convective and film cooling, which has good cooling performance but with complex manufacturability. In applications blocking appears very often and the weld spots crack easily.

In this cooling configuration the cooled wall is divided into units, in which the cooling gas flows intersected and cools down the wall and then forms the gas film through the holes on the external surface of the wall, which provides further heat protection. The intersecting gas flow inside the wall can at meantime result in impingement effect, which is more effective. Furthermore, the arrangement of the units can be adjusted due to the thermal load of the external surface. It shows that the manufacturing of this cooling structure is feasible and has pretty good prospect in applications. This technique is found at present only in patent and no experiments or numerical investigations are carried out, which provides large development potential.

In this paper, the internal flow condition and cooling features of the intersecting-grids cooling structure that can be applied in the combustion chambers and turbine vanes and blades of modern gas turbine engines is investigated numerically, to get knowledge of the cooling features of such structure and discuss the feasibility of applying it in cooling of combustion chambers and turbine vanes and blades in aero engines.

2 Physical Model

The configuration of internal channel networks is shown in Fig. 1. This structure is able to be applied to the cooling of either a turbine blade or the wall of combustion chamber. In this paragraph it will be demonstrated for a turbine blade.

Each set of internal network channels forms a single unit. And a certain number of unites locate in the solid body of the blade, protecting the entire entity. The coolant is supplied from the root into the blade cavity, some of which flows into the network channels transferring heat convectively, then out of the holes and forms a cooling film. Detailed structure is specified in Fig. 2.

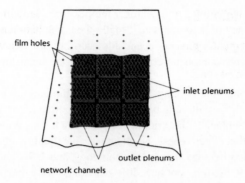

Fig. 1. Configuration of internal network channels within a turbine blade

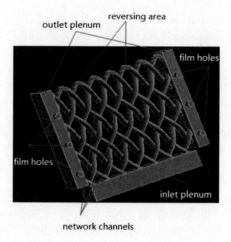

Fig. 2. Configuration of a single internal network channel unit

The coolant first enters the inlet plenum which is connected with the channels where the cooling gas exchanges heat with the solid body. The coolant flows out of the channel into the outlet plenum and forms film through the holes, thus the wall temperature is reduced. The channels wind in the wall, interfering each other.

The front view of the cooling unit is specified in Fig. 3. It can be observed that the reversing angle on the top of the channel is 60°, after which the coolant continues its way to the outlet plenum. Disturbing between gas flows exists at the interfering area of the channels, which increases the turbulence intensity, consequently the heat transfer of the coolant is enhanced. The A-A and B-B cross sections are shown in Figs. 4 and 5.

A clear view of the degree of the interfering is specified in stream wise in Fig. 4. The total length of the cooling unit is 27 mm, width 24 mm, thickness 3 mm. The outlet plenum (height 2 mm) on lateral sides is connected with film holes which is 0.5 mm long. In the middle are the network channels with a diameter of 1 mm. The distance between the centers of the interfering circles is 0.6 mm.

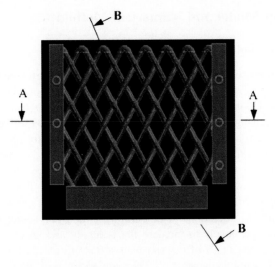

Fig. 3. Front view of internal network channel unit

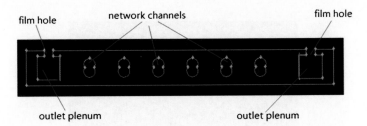

Fig. 4. View of A-A cross section

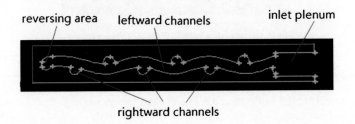

Fig. 5. View of B-B cross section

The disturbance of a flow in a single channel is illustrated in Fig. 5. The flow in the channel can be affected by the other 6 channels, and meanwhile influencing the others, which is of great benefit to the convective heat transfer.

3 Calculation Model and Numerical Method

3.1 Calculation Model of the Internal Network Channels

The target calculation zone in this paper is illustrated in Fig. 6. It consists of two zones, one is the fluid area, and the other is the metallic wall, inside of which lay the internal network channels. Coolant enters the network channels through the inlet plenum on lateral side, and exchanges heat with the wall. After that the cooling gas flows into the outlet plenums which are connected to the main stream via six film holes. The main stream enters through the inlet with a height of 12 mm and a width of 24 mm. The main stream heats the wall convectively and mixes with the secondary flow, after which the mixture flows out through the outlet. Conjugate flow and heat transfer calculation is employed, as the heat transfer between the two parts of the calculation model is taken into consideration. Tetrahedral mesh and polyhedral mesh are used to divide the model successively to increase the accuracy and speed of the simulation. Encrypt the surface layer grid of the end wall surface to ensure the y^+ that near wall turbulence model required near 1.0. The numerical grid consists totally about 2,600,000 points.

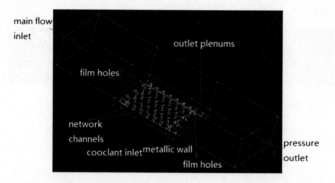

Fig. 6. Calculation zone of the network channels

3.2 Numerical Method

In the numerical model both the main stream and the secondary flow are set to mass flow inlet, normal to the inlet boundary. The outlet boundary is pressure outlet, which equals one standard atmosphere.

To investigate the cooling efficiency in real conditions, the calculation is made under a large temperature difference, with the temperature of main flow 1800 K and the secondary flow 800 K. In this condition the properties of air are handled as follows: density considered as ideal gas, heat conductivity, heat capacity, viscosity and other properties are derived by Lagrange interpolation.

The numerical simulation employs the separated implicit solver in Fluent 6.3. The method is Semi-Implicit Method for Pressure Linked Equations, i.e. the SIMPLE. A turbulence model of Realizable k-ε is applied and the near wall treatment is Enhanced Wall Function method. A double-preciseness upwind formula is employed for the segregation of each parameter. The pressure correction equation, mass conservation equation, momentum equation, k and ε equation are confronted with under-relaxation treatment, in which the under-relaxation factors are adjusted during the iteration process. The judging of convergence is determined by the comparative residual of lower than 1×10^{-5}, with no tend of further increase.

The calculation is made with a blow ratio of 1:1 and the Reynolds number of the main stream 40000.

4 Results and Analysis

The directions in the calculation model are defined as follows: the stream wise of the main flow is x axis, the inlet direction of secondary flow as y axis and the normal direction of the metallic wall side exposed to the main flow as z axis.

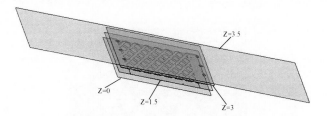

Fig. 7. Positions of the cross sections along the z axis

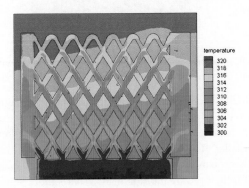

Fig. 8. Temperature distribution of z = 1.5 cross section

Positions of cross sections alongside the thickness of the metal are specified in Fig. 7. The thickness of the metallic wall is 3 mm, thus the three representative cross sections are z = 0 cross section as the bottom boundary, z = 3 cross section as the border of wall and main flow z = 1.5 as the middle cross section of the wall, containing the internal network channels and z = 3.5 as a cross section within the main flow near the wall.

Figure 8 illustrates the temperature distribution on the middle cross section of the wall, which tells that the temperature increases upward within the wall. The temperature at the beginning of the channels stays low then an obvious raise can be observed until the maximal magnitude at the reversing area. Due to the disturbance of gas flows at the interfering cross sections, the temperature there varies a lot, especially when the reverse flow confronts the previous secondary flow. This is to be observed in the upper half of the metallic wall. As a result of the high heat conductivity the temperature change within the metallic wall is not too sharp. The temperature gradient change broadwise is relatively small while the temperature gradient endwise varies gradually. The top area in Fig. 8 has the highest temperature on account of no internal channels. The large temperature gradient in outlet plenums relates to the temperature magnitudes of the gas flows into the outlet plenum. There's a total trend of rising of the temperature in y axis direction.

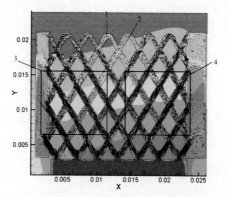

Fig. 9. Velocity vector distribution on z = 1.5 cross section

In Fig. 9 the velocity vectors are illustrated. The velocity magnitude near the inlet plenum is relatively high, while goes down as the gas flows toward the reversing area. The mass flow rate in channels without a reversing area is larger because the resistance is relatively much smaller. Two channels named 1 and 2 in the figure have extremely low mass flow, that's probably because the entrance of them locate at the two ends of the inlet plenum and influenced by the side wall. What's more significant, the channels linked to them leads directly to the outlet plenum without too much resistance, consequently gains more mass flow and reducing that in 1 and 2. Therefore the temperature near 1 and 2 is higher. Local vector distribution in zone 3 and 4 is specified in Fig. 10.

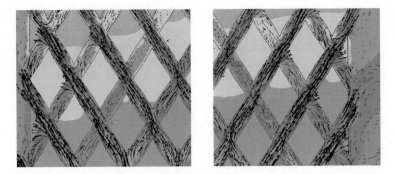

Fig. 10. Local velocity vectors on z = 1.5 cross section

The vectors upwards in Fig. 10 indicate the gas flow that has not jet encounters the reversing areas, while those downwise have confronted and head toward the outlet plenum. It is to be seen that some of the gas flow upward turns into the passage downwards and meantime some downwards goes up. So the direction change in the channels is a common phenomenon. What is certain is that the gas in all the channels flows from the inlet plenum towards the outlet plenum. In Fig. 9 it is found that the mass flow rate after reversing in channels increases as it passes each crossing till the arrival in outlet plenum.

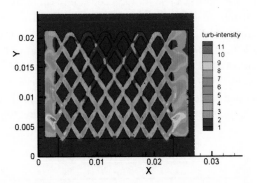

Fig. 11. Distribution of turbulence intensity on z = 1.5 cross section

Figure 11 illustrates the turbulent intensity of the cross section of z = 1.5. It's obviously seen that the turbulence intensity of the gas flow gains a sudden raise at the crossings and falls again in the channel, varying regularly inside the channels. The 1 and 2 area in Fig. 9 have relatively smaller turbulence intensity owing to the small mass flow rate. Besides, the turbulence intensity change at the crossings varies obviously. In case the directions of flow of two channels have a intersection angle of 60°, the maximal turbulence intensity at the next crossing is definitely smaller than that of the last crossing. However, when the flow directions at the crossings have a intersection angel of 120, the maximal turbulence intensity at the next crossing turns out to be

1896   G. Zheng et al.

greater than the last one. In conclusion, the turbulence intensity of the gas flow at the crossings from inlet plenum toward the reversing areas becomes smaller and smaller, and that of the opposite direction gets gradually larger and reaches a peak value at the outlet plenum.

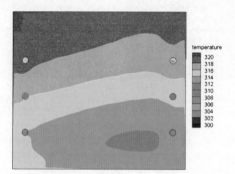

Fig. 12. Temperature distribution of z = 3 cross section

Figure 12 shows the temperature distribution of the intersection of metallic wall and main flow, which is generally in accordance of that in Fig. 8. The total tendency is the lower area at right hand has a lower temperature while the upper area at left hand has a higher temperature. This is a result of the configuration of the model investigated. A single unit is investigated in this paper and no others exist in upstream or downstream. The high temperature main flow from upstream hits the metallic wall directly thus creating a high temperature. In contrast, the right hand part of the metallic wall has a relatively lower temperature as it's protected by both the film holes at front and in the rear. Minimal temperature did not occur at the lower right, but a little upper left where the influence of inlet and outlet plenum both take effect. The high temperature in both the upper area and the main stream create the high temperature zone in the upper left.

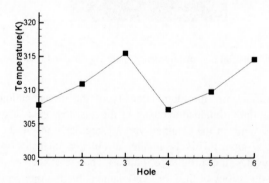

Fig. 13. Mean temperatures of each film holes

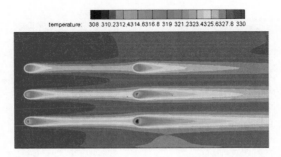

Fig. 14. Temperature distribution of z = 3.5 cross section

Mean temperatures of the gas in six film holes are specified in Fig. 13. It can be perceived that the corresponding holes in the main flow direction have little difference in temperature, while the temperature of gas in holes connected to the same outlet plenum varies much. The gas in upper holes has higher temperature and its potential developed relatively better.

Figure 14 illustrates the temperature distribution of the cross section in middle downstream of main flow, 0.5 mm from the surface of the metallic wall. The temperature difference of the gas from film holes can also be observed in this figure. The affection of the cooler gas from the holes below sustains farther yet that of the upper holes lasts relatively short, consequently provides less protection from the thermal load. The holes in later positions have better performance due to the absence of other units and the boundary condition of adiabatic wall, what' s more, the composition of the upstream and downstream coolant flow at the same time contributes to the longer effective distance in downstream. The well-bedded core area and dissipation area of the film hole protection further convinced the good convergence of the calculation.

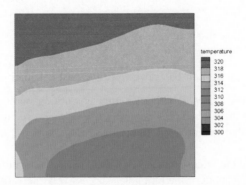

Fig. 15. Temperature distribution of z = 0 cross section

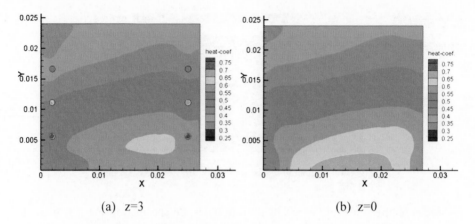

(a) z=3   (b) z=0

Fig. 16. Cooling effectiveness distribution on two surfaces of the metallic wall

Table 1. Cooling effectiveness of wall under standard working condition of the unit

Cross section	z = 0	z = 3	Average
Cooling effectiveness	0.504	0.482	0.493

The temperature distribution of the bottom of the metallic wall is shown in Fig. 15. The boundary condition of this cross section is adiabatic wall temperature. In regard of the good heat conductivity of metal, the temperature distribution is approximate to the other two cross sections. As the influence of the film holes is weakened, cooling effect of the secondary flow into the inlet plenum appears clearer. As a result the lowest temperature appears at the foot, meanwhile the upper temperature distribution stays the same (Fig. 16, Table 1).

5   Conclusions

The following conclusions can be drawn from the past calculation and analysis:

1. The secondary flow after passing the internal network channels in the wall performs more potential of the coolant. The cooling effect of such structure can be as much as three times of that with round film holes only, which has excellent cooling performance.
2. In the direction of the inlet secondary flow, the mean temperature and mean velocity magnitude of the gas out of the film holes successively increases. Meanwhile in the same direction the temperature of the metallic wall increases gradually.
3. Mixing appears at the crossings of the channels where change of directions of some flows take place. And the turbulence intensity augments at the interfering area, especially obvious when the intersection angel is an obtuse angel, which is beneficial to internal convective heat transfer.

The investigation in this paper indicates that the intersecting-grids composite cooling structure with internal network channels has pretty good cooling performance under calculation condition of normal temperature.

References

1. Li MK, He LM, Jiang YJ (2007) Numerical simulation of 2-D wall temperature of combustor's maze composite cooling structure. Mech Sci Technol 26(2):245–248
2. Zhao QJ, Li B (2001) Application and development of floating panel cooling structure in combustors. Gas Turbine Exp Res 14(1):10–13
3. Li B, Ji HH, Jiang YJ, Li F (2007) Numerical simulation for impingement-counterflow convection-film cooling of a combustor wall. J Aerosp Power 22(3):365–369
4. Jubran BA, Maiteh BY (1999) Film cooling and heat transfer from a combination of two rows of simple and/or compound argle holes in incline and/or staggered configuration. Heat Mass Transf 34(6): 495–502
5. Xu GQ, Xie Y, Ding ST, Sun JN, Tao Z (2009) Numerical simulation on cooling effectiveness of combined impingement and film cooling. J Therm Sci Technol 8(1):1–7

Quantitative Relationship Between Fluorescence Intensity and Equivalence Ratio of Kerosene

Yongsheng Zhao[1], Junfei Wu[1], Xiaohu Tian[1], Yuzhen Lin[2], and Wei Wei[1(✉)]

[1] China Academy of Aerospace Aerodynamics, Beijing 100074, China
zhaoyongshengying@163.com
[2] School of Energy and Power Engineering, Beihang University, Beijing 100091, China

Abstract. With planar laser-induced fluorescence (PLIF) technique and multi-points injection mode, the quantitative relationship between the equivalence ratio and fluorescence intensity of kerosene is studied. Using the research results, the quantitative distribution of local equivalence ratio of kerosene in the crossflow is analyzed. The operating conditions are as follows: pressure ranging from 0.15 to 0.25 MPa, temperature ranging from 580 to 700 K, and the equivalence ratio ranging from 0.3 to 0.8. The results show that the equivalence ratio and fluorescence intensity of kerosene satisfy linear relationship; and the local equivalence ratio of kerosene in the crossflow shows an obvious delamination phenomenon; the spray combustible area decreases gradually with the increase of pressure and equivalence ratio.

Keywords: Planar laser-induced fluorescence · Kerosene · Equivalence ratio · Quantitative relationship

1 Introduction

With the continuous development of lean pre-mixed pre-evaporation (LPP) combustion technology, direct injection of kerosene through a direct-injection nozzle into the channels of the premixed section to achieve rapid kerosene/air mixing is the simplest and most effective injection mode [1]. With this injection mode, the equivalence ratio distribution of kerosene in the crossflow is decisive of the sufficient and efficient combustion of kerosene [2], and it is always the research focus worldwide.

Planar laser induced fluorescence (PLIF) is one of the important technical means to study the equivalence ratio distribution of kerosene. The principle of PLIF is that the kerosene molecules are excited into an excited state under the laser energy input and produce fluorescence, and the concentration of kerosene is characterized by the fluorescence intensity [3–5]. Kojima et al. [6] study the determinants of the equivalence ratio distribution of the sprays with PLIF, it shows that the equivalence ratio distribution is determined by the injector orifice and is independent of the injection pressure; Muehlfriedel et al. [7] study the concentration distribution of the sprays during mass

© Springer Nature Singapore Pte Ltd. 2019
X. Zhang (Ed.): APISAT 2018, LNEE 459, pp. 1900–1909, 2019.
https://doi.org/10.1007/978-981-13-3305-7_154

transfer, and focus on the physical mechanism of the components transfer. Deshmukh et al. [8] study the distribution of the sprays and the liquid volume fraction in dense sprays by using PLIF and Mie's scattering theory, and the relative concentration distribution characteristics of sprays are obtained. Clayfield et al. [9] obtain the equivalence ratio distribution of sprays in the coaxial jet using PIV and PLIF. However, the results are only applied to the injection conditions with Reynolds number of 6500. Lei et al. [10] obtain the relative distribution characteristics of the two-dimensional concentration fields of kerosene in both liquid and gaseous phases. Liu et al. [11] study the atomization and breakup characteristics of kerosene in the crossflow using PLIF and high-speed shadowgraph.

The penetration depth of sprays can indirectly characterize the equivalence ratio distribution of kerosene. Therefore, the researches on the sprays in the crossflow using PLIF are mainly focused on the penetration depth. New et al. [12] study the flow fields and vortex development of the crossflow using PLIF, and point out that the injector orifice shape can affect the penetration depth. Xue et al. [13] study the penetration depth of kerosene injected into a high-temperature and high-pressure crossflow under subcritical and supercritical conditions after being heated, and fit a formula to the outer edge trajectory.

It should be pointed out that the quantitative relationship between the fluorescence intensity and the kerosene equivalence ratio has not been calibrated in the above research work. The qualitative analysis methods are mainly adopted for the study of the concentration distribution of sprays in the crossflow. However, the quantitative distribution of equivalence ratio has not been obtained, even it is crucial for the rational organization of the combustion zone and the design of the aero-engine.

The relationship between the fluorescence intensity and the equivalence ratio of kerosene is analysed with PLIF in this paper, a theoretical model is established, and a multi-points injection mode for kerosene mixing is designed. The theoretical model is verified with the uniform flow field delivered by the multi-points injection mode. In addition, the quantitative distribution of the local equivalence ratio of kerosene in the crossflow is preliminarily analysed using the calibration results.

2 Experiments

2.1 Experiment System

The experiment system is composed of a high-pressure air supply system, a flow reactor system, a kerosene supply system, and a PLIF test system, as shown in Fig. 1. High temperature and high pressure air (T $\leq$ 1200 K; P $\leq$ 3 MPa) is supplied by the high pressure air supply system using the heat accumulator to heat pure air [14].

The flow reactor consists of a measuring section, a uniform sprays section, a crossflow experiment section, and an exhaust section. (1) The measuring section monitors parameters such as the flow temperature, pressure, and flow-rate. (2) The multi-points injection method is used to generate the uniform sprays with a constant equivalence ratio. The nozzle of the multi-points injection system consists of 15 orifices (each with a diameter of 0.2 mm) and 15 venture channels, as shown in Fig. 2.

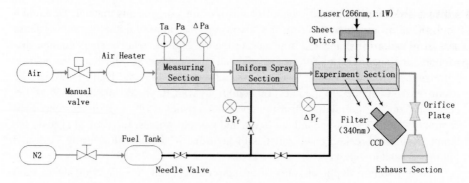

Fig. 1. Test system diagram

After injected into the venture, the kerosene is rapidly atomized into fine particles by the high-temperature air, which is accelerated by the throat. Then the fine particles mix with the air, and a uniform kerosene spray field is formed. Then the uniform field flows into the measuring area, and the quantitative relationship between equivalence ratio and fluorescence intensity can be studied [15]. (3) The kerosene is injected into the crossflow experiment section by a direct-injection nozzle (with a diameter of 0.5 mm and a length-to-diameter ratio of 4). (4) The corresponding pressure in the experiment section is guaranteed by the exhaust section orifice plate.

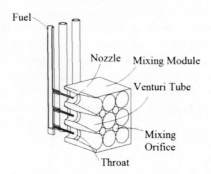

Fig. 2. Multi-points injection nozzles [15]

PLIF test system includes: an Nd: YAG laser, a light sheet generator, a CCD camera and an image intensifier. The Nd: YAG laser produces dot laser (with wavelength of 266 nm and energy of 1.1 W) which is converted into sheet laser by sheet laser light generator to excite kerosene. The thickness of the sheet laser is less than 0.5 mm. The resolution of the CCD camera is 1600×1200 pixels, with dynamic range of 14 bit and shooting frequency of 30 Hz. The fluorescence produced by kerosene is received by the CCD camera with a 340 ± 5 nm filter. In the experiment, the image intensifier has a delay time of 10 ns and a gate width of 180 ns–220 ns. The sampling frequency is 30 Hz, and the sampling area is 25×50 mm.

2.2 Experimental Principles and Operation Conditions

2.2.1 Calibration Experiments

It can be assumed that in a certain space, the generated energy is constant when each excited kerosene molecule jumps, and the induced fluorescence intensity (I) is linear with the number of molecules (n_f). Equation 1 can be obtained.

$$I = a + b \cdot n_f \tag{1}$$

Equation 2 represents the relationship between the equivalence ratio (Φ) of kerosene and the number of molecules, where n is the number of molecules, M is the relative molecular mass, subscript 'a' represents air, and subscript 'f' represents kerosene.

$$\varnothing = \frac{n_f}{n_a} \cdot \frac{M_f}{M_a} \cdot \frac{1}{0.068} \tag{2}$$

Therefore, if the space remains unchanged, then Eq. 3 can be obtained.

$$I = a + c \cdot \frac{V_m \cdot 1.97 \cdot \varnothing}{141.6 + 1.97 \cdot \varnothing} \tag{3}$$

Since the equivalence ratio range of the kerosene flammability limit is from 0.2 to 3.75, the 1.97 • Φ is much smaller than 141.6. Equation 5 can be obtained by combining Eq. 4 with Eq. 3.

$$V_m = \frac{RT}{P} \tag{4}$$

$$I = a + c \cdot \frac{1.97 \cdot \varnothing \cdot R \cdot T}{141.6P} \tag{5}$$

That is, in a certain space, the quantitative relationship between fluorescence intensity and equivalence ratio can be expressed as Eq. 6.

$$I = A + B \cdot \frac{\varnothing \cdot T}{P} \tag{6}$$

The calibration operation conditions are determined based on the above analysis, as shown in Table 1. Case 1 is used to construct the theoretical model, and Case 2 is used to verify the accuracy of the model.

Table 1. Operation conditions of calibration experiments

Case	T/K	P/MPa	Φ
1	663	0.15	0.3, 0.4, 0.5, 0.6, 0.7, 0.8
2	697	0.25	0.4, 0.5, 0.6

2.2.2   Measurement of the Local Equivalence Ratio in the Crossflow

The fluorescence intensity of kerosene in the crossflow is measured based on the same laser intensity and PLIF test system parameter settings. The quantitative distribution characteristics of the local equivalence ratio are studied using the calibration test results. The flow velocity is 30 m/s. The operation conditions are shown in Table 2.

Table 2. Operation conditions for the local equivalence ratio measurement

Case	T/K	P/MPa	Φ
3	589	0.15	0.3–0.8
4	589	0.25	0.3–0.8

3   Experimental Results and Analysis

3.1   Quantitative Relational Model

Under the same conditions, 500 fluorescence intensity pictures are taken using the PLIF system, and after each background is subtracted, the average picture is taken as the final result [13]. Figure 3(a) shows the fluorescence intensity of the uniform sprays field with the equivalence ratio of 0.3. From the results, it can be seen that the distribution of the fluorescence intensity is uniform and consistent. The results also prove that the uniformity of the sprays field formed by the calibration nozzle is satisfactory, and there is no local rich kerosene area or local lean kerosene area. The probability distribution density (PDF) of the measured fluorescence intensity is calculated using statistical methods, as shown in Fig. 3(b). It can be seen that the PDF is approximately Gaussian and the distribution has a bandwidth of 50. There are two reasons for this distribution. Firstly, the intensity of the sheet laser cannot be exactly the same; secondly, the transition process of the excited molecular is random. However, the overall

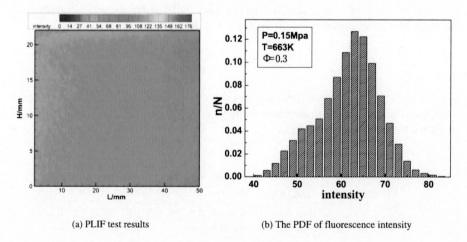

(a) PLIF test results   (b) The PDF of fluorescence intensity

Fig. 3. Fluorescence intensity distribution (Φ = 0.3)

fluorescence intensity of the PLIF test ranges from 0 to 2500, the measured fluorescence distribution is relatively concentrated compared to the 2500 range, so the equivalence ratio distribution can be considered as uniform.

With the same experimental method and data processing method, the fluorescence intensity values of different equivalence ratios under the operation conditions of Case 1 are obtained. The results are shown in Fig. 4. It can be seen from Fig. 4: (1) the fluorescence intensity distributions are all Gaussian distribution under different equivalence ratios and the distribution bandwidths are about 50. (2) With the increase of the equivalence ratio, the fluorescence intensity values at the maximum frequency gradually increase and the entire bands move to the right.

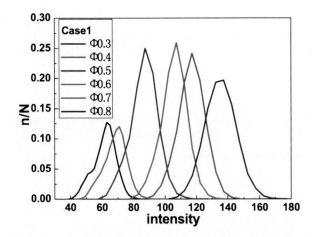

Fig. 4. Relationship between PDF of fluorescence intensity and equivalence ratio

The average fluorescence intensity and the fluorescence intensity at the maximum frequency are obtained, as shown in Fig. 5. It can be seen from the results that the average fluorescence intensity is the same as the fluorescence intensity at the maximum

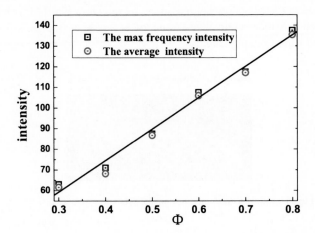

Fig. 5. Fluorescence intensity distributions with equivalence ratio

frequency. Two kinds of fluorescence intensity are both linear to the equivalence ratio, as shown in Eq. 7. The fitting factor is $R^2 = 0.98$.

$$I = 152\Phi + 13.73 \tag{7}$$

The quantitative relationship between kerosene equivalence ratio and fluorescence intensity can be obtained by combining Eq. 7 with operation conditions, as shown in Eq. 8.

$$I = 13.73 + 34389 \cdot \frac{\varnothing \cdot T}{P} \tag{8}$$

In order to verify the accuracy of the Eq. 8, the fluorescence intensity values with equivalence ratios in Case 2 are tested. The results are similar to those of Case 1. The fluorescence intensity values are also calculated by Eq. 8. The experimental results are compared with the analytical results, as shown in Table 3, and the equation fitting errors are less than 2.5%.

Table 3. Comparison of fluorescence intensity and experimental statistics

Φ	Experimental value		Analytical value	Error
	Average	Maximum frequency		
0.4	52.8	53	51.7	2.2%
0.5	62.6	61	61.3	2.0%
0.6	75.6	75	70.9	2.3%

3.2   The Local Equivalence Ratio in the Crossflow

With the same laser intensity and PLIF test system parameters, the fluorescence intensity values in the crossflow are measured, as shown in Fig. 6. It can be seen from the results that the local equivalence ratio of kerosene in the crossflow shows an obvious delamination phenomenon.

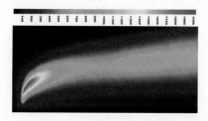

Fig. 6. Fluorescence intensity in the crossflow (Case 3, $\Phi = 0.7$)

Based on the theoretical analysis in Sect. 2.2.1, Eq. 8 can only be used for the calculation of gaseous kerosene equivalence ratio. It can be obtained by theoretical analysis that the maximum time required for liquid kerosene to evaporate completely is 1.5 ms under the operation conditions of Case 3–4 [16]. Hence, the kerosene has evaporated completely at 50 mm downstream of the nozzle, where the equivalence ratio distribution is studied. The quantitative distributions of the equivalence ratio on the central line are obtained by Eq. 8, as shown in Fig. 7. From the test results, it can be known that: (1) In the height direction, the local equivalence ratio is almost symmetrical, and as the injection equivalence ratio increases, the symmetry axis moves higher; (2) As the injection equivalence ratio increases, the equivalence ratios of the lower surfaces are basically the same, but there is a significant difference in the ones of the upper surfaces. (3) The flammable limit range decreases gradually as the equivalence ratio increases. (4) Comparing the flammable limit ranges of Case 3 with Case 4, it is found that the local combustible area of Case 3 is larger than that of Case 4, that is, when the other parameters are the same, the higher the pressure, the smaller the local combustible area.

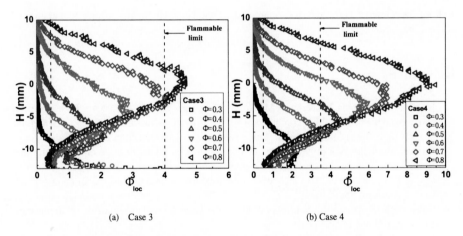

(a) Case 3   (b) Case 4

Fig. 7. Local equivalence ratio distributions in the crossflow

4 Conclusions

Based on the PLIF test technology, this article designs a multi-points injection mode to construct a constant and uniform kerosene spray field. Then the quantitative relationship between the kerosene fluorescence intensity and the equivalence ratio is calibrated, and the theoretical model is constructed. The accuracy of the model is verified by experiments. Using the calibration results, a preliminary quantitative analysis of the local equivalence ratio in the crossflow is performed. The conclusions are as follows:

(1) The quantitative relationship between the fluorescence intensity and the kerosene equivalence ratio under the operation conditions is shown below.

$$I = 13.73 + 34389 \cdot (\varnothing \cdot T)/P$$

(2) The prediction error of the model is less than 2.5%.

(3) The local equivalence ratio of kerosene in the crossflow shows an obvious delamination phenomenon.

(4) For the crossflow, in the height direction, the local equivalence ratio is almost symmetrical, and as the injection equivalence ratio increases, the symmetry axis moves higher.

(5) As the pressure and the injection equivalence ratio increase, the combustible area in the center of the sprays gradually decreases.

The fluorescence characteristics of gaseous kerosene are analyzed in this paper. The fluorescence characteristics of kerosene in gas-liquid mixed state still need further research.

Acknowledgement. This study was supported by the National Nature Science Foundation of China (Grant No. 11602263).

References

1. Imamura A, Yoshida M, Kawano M, Aruga N, Nagata Y, Kawagishi M (2001). Research and development of a LPP combustor with swirling flow for low NO(x). In: 37th joint propulsion conference and exhibit. American Institute of Aeronautics and Astronautics
2. Barnes JC, Mellor AM (1998) Effects of unmixedness in piloted-lean premixed gas-turbine combustors. J Propul Power A1(6):967–973
3. Lin Y, Li L, Zhang C (2014) Progress on the mixing of liquid jet injected into a crossflow. Chin J Aeronaut 35(1):46–57
4. Hanson RK, Seitzman JM, Paul PH (1990) Planar laser-fluorescence imaging of combustion gases. Appl Phys B 50:441–454
5. Lozano A, Yip B, Hanson RK (1992) Acetone: a tracer for concentration measurements in gaseous flows by planar laser-induced fluorescence. Exp Fluids 13:369–376
6. Kojima H, Kawanabe H (2010) PLIF measurement of kerosene concentration in a diesel spray. Trans Jpn Soc Mech Eng Part B 76(768):1326–1333
7. Muehlfriedel K, Baumann K-H (2000) Concentration measurements during mass transfer across liquid-phase boundaries using planar laser induced fluorescence (PLIF). Exp Fluids 28(3):279–281
8. Deshmukh D, Ravikrishna R (2013) A method for measurement of planar liquid volume fraction in dense sprays. Exp Therm Fluid Sci 46(4):254–258
9. Clayfield K, Kelso RM (2003) Simultaneous planar measurement of velocity and concentration using PIV and acetone PLIF. In: 33rd AIAA fluid dynamics conference and exhibit, Orlando, FL, 23–26 June
10. Lei W, Su W (2004) A PLIF test bench for fuel spray characteristics of HCCI diesel engines. Automot Eng 26(4):405–408
11. Liu L, Xu S, Zheng R (2010) Studies on kerosene injection and mixing with cavity in supersonic flow. J Propul Tech 31(6):721–729

12. New TH, Lim TT, Luo SC (2003) Elliptic jets in cross-flow. J Fluid Mech 494:119–140
13. Xue X, Gao W, Xu Q (2011) Injection of subcritical and supercritical kerosene into a high-temperature and high-pressure crossflow. In: ASME 2011 turbo expo: turbine technical conference and exposition, GT2011
14. Wang J, Lin Y (2014) Research of clean air storage heater for direct-connected test-bed. J Propul Tech 10:1392–1397
15. Li W (2015) Equivalence ratio pulsation mechanism of partially premixed flame. Master thesis, Beihang University. Beijing
16. Zhao Y, Zhang C, Lin Y (2016) Random behavior of kerosene spray autoignition in crossflow. In: Proceedings of ASME turbo expo 2016: turbomachinery technical conference and exposition, GT2016-57043

Measuring Method of Micro Cone Hole Based on Depth from Focus

Chengxing Bao$^{(\boxtimes)}$, JingLiang Liu, Xue Hao, and Dongwei Wang

Aviation Key Laboratory of Science and Technology on Precision
Manufacturing Technology, Beijing Precision Engineering Institute
for Aircraft Industry, Beijing, China
326187214@qq.com

Abstract. In the aero engine, the nozzle is one of the most important parts.
Nozzles are very small and difficult to be measured. The cone hole in the nozzle
is hard to be machined, the quality of the cone hole seriously affects the
atomization performance of the nozzle. Different from conventional measuring
methods, this paper uses image processing to measure the angle of the cone
hole. A method based on Depth From Focus was proposed, which makes use of
a series of images to reconstruct the 3D model of the nozzle. This paper uses
telecentric lenses and CCD to get images from different distance, then some
experiments were made to prove this method can goodly measure the angle of
the cone hole. Also this paper finds out that the smaller the moving interval is,
the more accuracy there will be, but the more time will be cost.

Keywords: Nozzle · Cone hole · Depth from focus · Image processing

1 Introduction

The fuel nozzle is a key part of the aero engine. The processing quality affects the
performance of the whole engine to a large extent [1]. The size of the nozzle cone hole
is very small, the length is generally only 1–10 mm, the angle is 60–120°. The shape
accuracy of the cone hole determines the atomization performance of the nozzle. The
traditional manual detection method can not meet the requirements of detection
accuracy and detection efficiency. Therefore, how to realize the rapid detection and
accurate detection of the nozzle cone hole becomes an important issue in the aviation
engine manufacturing industry (Fig. 1).

At present, there are two ways measuring the cone hole of the aero-engine nozzle.
One is called Double-Ball method and another is the conventional three-coordinate
measuring machine. The Double-Ball method (shown in Fig. 2) uses two steel balls of
different diameters and two inductive displacement sensors to form a contact-type
probe. The steel ball is mounted on the bottom of the displacement sensor, and the
probe moves to press cone hole's surface with the steel balls separately. By getting the
two centers of the two balls, the angle of the cone hole can be calculated. The shape
accuracy of the steel ball has a great influence on the measurement accuracy, but high
accuracy steel balls are difficult to machine. The coordinate measuring machine (shown
in Fig. 2) is a conventional equipment for the inspection industry. The coordinate

© Springer Nature Singapore Pte Ltd. 2019
X. Zhang (Ed.): APISAT 2018, LNEE 459, pp. 1910–1916, 2019.
https://doi.org/10.1007/978-981-13-3305-7_155

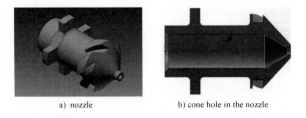

a) nozzle   b) cone hole in the nozzle

Fig. 1. Fuel nozzle in the aero engine

measuring machine includes four types, such as a moving bridge coordinate measuring machine, a fixed bridge coordinate measuring machine, a mobile working table coordinate measuring machine, a cantilever coordinate measuring machine, etc. At present, the technology of small or medium-sized coordinate measuring machines has become mature. The coordinate measuring machines of various precision grades such as production type, precision type and metering type, they can meet the needs of modern advanced manufacturing and scientific research. The measurement uncertainty of production type measuring machine is greater than 3 μm, the measurement accuracy can meet the requirements of nozzle cone hole measurement accuracy, but because the cone hole is too small, the measurement operation is inconvenient and resulting in low detection efficiency.

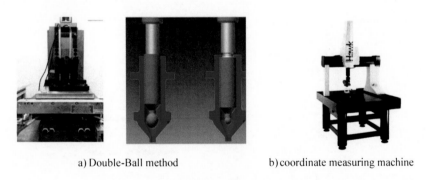

a) Double-Ball method   b) coordinate measuring machine

Fig. 2. Two ways of measuring the nozzle cone hole

With the development of CCD technology, the pixels of CCD are getting higher and higher, which can cope with the demand of high-precision detection [2]. With the development of visual measurement technology, various algorithms are becoming more and more mature, greatly improving the detection efficiency. The visual measurement technology has become one of the most famous non-contact methods. Guo [3] has studied the forming structure characteristics and edge optical characteristics of micro-cone holes, and proposed a method for effectively detecting micro-cone holes. Liang [4] analyzed the error of CCD microscopic detection system and compensated for various errors. Based on the above research results and the image processing technology of CCD, aiming at the characteristics of aero-engine nozzles, this paper studied the Depth-From-Focus method, and applied this detection method to measure the cone hole of the aero-engine nozzle. In the actual detection process, the detection accuracy can meet the requirements and greatly improve detection efficiency.

2   The Method of Depth from Focus

The method of Depth From Focus uses a series of images which taken from different distance between target and camera to reconstruct 3D model of the target, than the information of the surface can be extracted from the 3D model. This method is a more precise method than other methods, such as stereo method or laser triangulation [6]. Furthermore, this method only needs one camera, so it is affordable than other methods. In order to achieve a parallel projection, the method requires telecentric lenses or microscope lenses, so it maybe only suitable for small target.

The points on the target surface have different distance to the camera, and the visual system has a limited depth of field. Based on these knowledge, the method of Depth From Focus gets the most sharply parts from each image, and combines all these focused parts and the distances of each image to reconstruct the 3D surface of the target. Each point of the target is displayed sharply in one image, so the distance of each point to the camera can be calculated by determining in which image the point is focused. This principle is show in Fig. 3.

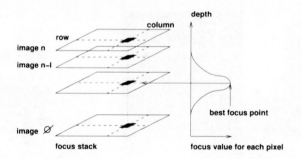

Fig. 3. The principle of depth from focus

Depth of field is a technical term of the visual system, it is a range of distance in which the image is mostly sharped, the points within the distance are best focused. The DOF depends on several factors such as the pixel size, the aperture, the focal length and the focusing distance. A high DOF means that the more part of the image or maybe the whole image is sharp, whereas a low DOF means only a small part of the image is sharp (see Fig. 4). For the method of Depth From Focus, the DOF determines the precision of the measurement, so the lower the DOF is, the more precise there will be.

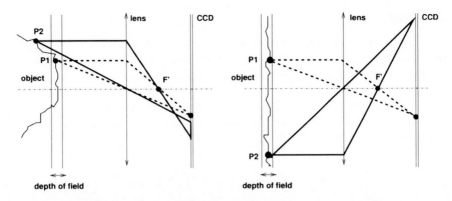

Fig. 4. The conception of depth of field

3   Measuring Method of the Cone Hole Based on DFF

3.1   Measurement Overall Plan

The cone hole detection process based on image Depth-From-Focus has 5 steps (see Fig. 5). First is to collect a set of nozzle cone images at different distances by a telecentric lens with a small depth of field. Second, extracts the sharpest points in each image to form a depth image. The third step is to generate a three-dimensional mode by using an image containing the X coordinate, an image containing the Y coordinate and the depth image, which is a three-dimensional model of the surface of the nozzle cone. The fourth step is to obtain a contour image of the cone hole by generating a plane that crosses the axis of the cone and intersecting with the three-dimensional model of the cone hole. In the contour image we can extract two straight line segments which represent the cone hole, and the angle between the two straight line segments can be calculated, which is equal to the angle of the nozzle cone hole.

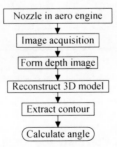

Fig. 5. Measure process

3.2   Image Capture

High-quality images are the premise of image processing, and good image quality can improve detection accuracy. When performing nozzle cone image acquisition, the industrial camera and lens is installed in the parallel direction of the cone axis, the camera moves in the direction of the cone axis, and collects images at every move interval. The group of images should cover the whole nozzle cone. Each image refers to a different distance, and focuses on the different positions of the nozzle cone, as shown in the following Fig. 6.

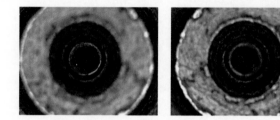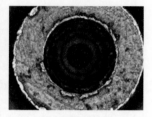

Fig. 6.   Different focused image

3.3   3D Model Reconstruction by Depth Image

The clearest point on each image represents the surface information of the part to be tested, and the clearest point is extracted from each image to form a depth image, as shown in the Fig. 7 below. Then using the depth image to generate the 3D model, the 3D model of the cone is obtained as shown in the figure, which is reconstructed based on the Depth Form Focus method.

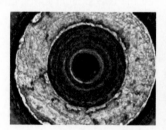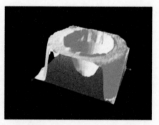

Fig. 7.   Depth image and reconstructed 3D model

3.4   Extract the Contour and Calculate the Angle

The parameter information of the nozzle cone can be obtained from the three-dimensional model. The angle of the nozzle cone refers to the angle between the two straight lines obtained by intersecting the plane of the cone axis with the cone. Therefore, first establish a plane according to the axis of the cone hole, as shown in the following figure, and then the plane intersects the nozzle cone model to obtain the contour curve of the nozzle cone, as shown in the figure. The two straight line segments representing the nozzle cone are extracted on the curve, and the angle between the straight segments is calculated, which is the angle of the nozzle cone (see Fig. 8).

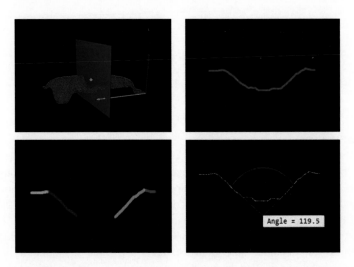

Fig. 8. The contour of the cone hole and calculate the angle

4 Measuring Experiments

Based on the method of image Depth Form Focus, a nozzle with a cone angle of 120° was tested to verify the measurement accuracy of different moving intervals and the repeat measurement accuracy at fixed intervals.

The first experiment uses five kinds of moving intervals for image acquisition, and then calculates the cone angle. The size of the moving interval determines the resolution of the axis direction of the cone. It can be seen from the experimental results that the smaller the moving interval is, the higher the measurement accuracy there will be, but the more time the single measurement will cost (Table 1).

Table 1. Experiment of different move intervals

Move interval (unit: mm)	Cone angle (unit: degree)	Single time (unit: s)
0.01	119.6	200
0.02	119.5	120
0.05	119.3	75
0.10	121.1	60
0.20	118.8	45

In the second experiment, the move interval of the image acquisition was 0.02 mm, and the experiment was repeated 10 times. The experimental results showed that the measurement repeatability is less than 1° (Table 2).

Table 2. Experiment of repeated measurements

No.	1	2	3	4	5	6	7	8	9	10	AVG
Value	119.6	119.6	119.5	119.9	119.5	120.1	119.5	119.6	120.2	119.9	119.8

(unit: degree)

5 Conclusions

The cone hole in the aero-engine nozzle parts are not easy to detect by conventional methods because of the small size. This paper realizes the detection of the nozzle cone angle based on the image Depth-From-Focus method. The method obtains the surface images of the nozzle cones at different heights by the industrial camera and telecentric lense, then reconstructs the three-dimensional model of the nozzle cone surface by the Depth-From-Focus method, and finally extracts the cone contours on the three-dimensional model to calculate the angle of the cone hole. The moving interval during image acquisition determines the resolution of the axial direction of the cone hole. The smaller the moving interval is, the higher the measurement accuracy is, but the more images will be captured by a single measurement, and the more time will be cost in the image processing, and this greatly affects the detection efficiency.

References

1. Gan X (2006) Aviation gas turbine fuel nozzle technology. Beijing National Defense Industry Press
2. Lei Y, Luo H, Ma J (2004) CCD Camera Error and its Calibration. Opt Optoelectron Technol 2(4):49–52
3. Guo Z, Fan J (2010) Micro-hole Optical Feature Recognition Technology Based on Image Processing. Tool Technol 44(9):102–105
4. Liang B (2008) Error analysis and compensation of CCD microscopic detection system. Harbin Institute of Technology, Haerbin
5. Jiang Z, Shi W (2004) Three-dimensional MicroScopy image system based on depth from focus. CT Theory Appl 13(4):9–15
6. Gong J, Xu X (2006) 3-D surface reconstruction of grinding wheel topography based on depth from focus. Diamond Abrasives Eng 21(5):23–28

Study on Non-contact Measurement Technology for Swirling Slot of Aero Engine Fuel Nozzle

Lei Wang[(✉)], Chenxing Bao, Wenming Lei, and Chunchun Tang

Aviation Key Laboratory of Science and Technology on Precision
Manufacturing Technology, Beijing Precision Engineering Institute for Aircraft
Industry, Beijing, China
wanglei1771@163.com

Abstract. Fuel nozzle is an important component of aerospace engine, with the typical features are swirling slots, micro-holes and micro-cones. The consistency in structure and size has a major impact on the performance of the engine's combustion chamber. Aiming at the requirement of the nozzle swirling slot measurement, a four-axis non-contact measuring system based on the laser probe with conoscopic holography is built. According to the spatial geometric distribution characteristics of the swirling slot, the inspection path planning of laser scanning based on the probe spot is designed to measure the spiral line, then the point on the spiral of the swirling slot can be obtained. The coordinate values of points are quickly picked up, and then the fitting analysis is performed to get the relevant characteristic parameter values of swirling slot. The experimental results show that the measurement method can obtain the structure size of the nozzle swirling slot quantitatively, meeting the measurement accuracy requirements, and realize the judgment of the consistency of the size and structure of the swirling slot.

Keywords: Nozzle · Swirling slot · Conoscopic holography ·
Non-contact measurement · Fitting analysis

1 Introduction

The aero engine is known as the "flower of industry", It is the heart of the aircraft, and is also an important symbol of the country's industrial level and comprehensive strength. It has extremely high economic value, military value and political value. If the aero engine is the crown of the highest manufacturing industry, then the aero engine fuel nozzle is the pearl on the crown. The fuel nozzle of the engine chamber is one of the important components of the aero engine. It is an important and small part. Its main function is to supply the engine chamber with a suitable amount of fuel with good oil atomization quality according to different working conditions of the engine [1]. The nozzle parts have the characteristics of precise structure, high technical requirements and complicated manufacturing process. The processing precision and quality have direct and significant influence on the working performance, combustion efficiency and reliability of the aero engine chamber. The study on the precise measurement

© Springer Nature Singapore Pte Ltd. 2019
X. Zhang (Ed.): APISAT 2018, LNEE 459, pp. 1917–1930, 2019.
https://doi.org/10.1007/978-981-13-3305-7_156

technology of the important features of the nozzle has become a key link to improve the overall performance of the aero engine.

The nozzle structure is complex and precise, with its typical features are micro-holes, micro-cones and swirling slots [2]. The consistency in structure and size has an important influence on the performance of the engine chamber. The nozzle swirling slot is similar to the thread structure, and is currently mainly processed by a multi-axis milling machine through a numerical control program [3]. When the engine is working, the fuel is rotated by the swirling slot under the action of oil pressure to realize centrifugal spraying [4]. The parameters such as the slot width, slot depth and spiral angle of the swirl slot have a direct influence on the atomization performance indexes such as nozzle flow rate and atomization cone angle. Therefore, the precise measurement technology of the nozzle swirling slot is needed to accurately obtain the parameters of the swirling slot, and the deterministic judgment of the processing quality of the swirling slot can be realized, thereby ensuring the consistency in structure and size of the nozzle swirling slot.

Currently, foreign countries have made many research results in the measurement of geometric features of small parts such as nozzles. National Jet Corporation of the United States is a manufacturer specializing in the production of aerospace nozzles. It uses an optical inspection system with a Nikon lens to achieve fast and accurate measurement of the tiny aperture of the nozzle [6]. Werth in Germany has developed a contact fiber optic sensor probe with a probe radius of only 10 μm. Its micro-contact force is less than 1 μN, enabling automatic scanning and measurement of the full-size and contour of small parts, especially for microscopic geometric size measurement of nozzles [8]. Most of the domestic aero-engine research and manufacturing institutes only use the universal gage and some simple gauges to qualitatively judge the nozzle structure size, and it is impossible to quantitatively obtain accurate dimensional information, and even some typical features such as the slot width, slot depth and the spiral angle of the micro slots cannot be measured. Therefore, there is a big gap in precision measurement of nozzles between the domestic and foreign countries. Only qualified judgments or repair tests can be used to screen out qualified nozzle parts, which reduces the quality and working performance of the nozzles. Aiming at the precise measurement requirements of the feature size of the nozzle swirling slot, this paper proposes a fast laser scanning method of spiral line boundary based on the four-axis laser measuring system to accurately measure the swirling slot, and obtain key characteristic information such as slot width and spiral angle to realize deterministic measurement of the nozzle swirling slot.

After a brief review and introduction, the rest of the paper is organized as follows. First of all, the structure features of the nozzle swirling slot and the measurement principle are described briefly in Sect. 2. And then, Sect. 3 introduces the four-axis non-contact measuring system based on the laser probe with conoscopic holography. Next, the method of swirling slot measurement based on laser scanning is presented in detail in Sect. 4. Afterwards, the inspecting experiments of the nozzle swirling slot are shown in Sect. 5, as well as the results. Finally, conclusions of this piece of work are summarized in Sect. 6.

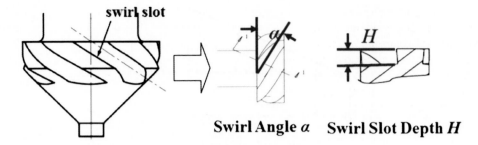

Swirl Angle α **Swirl Slot Depth** H

Fig. 1. Nozzle swirling slot and its characteristic parameters

2 Structure Characteristics of the Swirling Slot and Measurement Principle

The swirling slot is a typical structure of the aero engine nozzle. The aviation fuel generates high-speed rotation through the swirling slot under the action of pressure. The characteristic parameters such as the slot depth, the slot width and the spiral angle of the swirling slot have important influence on the fuel atomization performance. The swirling slot belongs to the structure of the rotating body, which is usually processed by multi-axis numerical control milling; its intersection with the outer cylindrical surface forms a spiral, and each swirling slot has two spirals. The measurement method designed in this paper is to measure the shape and relative position of two spirals (i.e., the sharp edges formed by the intersection of the swirling slot and the cylinder) to achieve the slot width, slot depth, spiral angle of the swirl slot (Fig. 1).

The processing method of the swirling slot is similar to the thread. The cylindrical coordinate equation of the spiral line is as follows:

$$\begin{cases} \rho = \dfrac{D}{2} \\ Z = Z_0 + k \cdot \rho \cdot \alpha \end{cases} \tag{1}$$

In the Eq. (1), D is the diameter of the spiral, α is the central angle, and the slope k is determined by $k = \tan(\theta)$. θ is the spiral angle.

After the spiral line is expanded, a right angled triangle is shown. As shown in Fig. 2 the spiral shape characteristics of the measured parts are mainly determined by the diameter of the cylinder D and the spiral angle θ. Assuming that the spiral angle is certain, the spiral can be determined at a given position O ($\alpha = 0$, $Z = Z_0$), as long as the two points of the spiral are obtained.

Therefore, to determine the shape of the sharp edge of a spiral line on the nozzle swirling slot, we need to confirm the diameter of the spiral D and the spiral angle θ. The diameter D can be directly obtained by measuring the cylindrical surface using laser scanning where the spiral line is on. The spiral angle can be obtained by measuring the two point coordinates at different heights in the direction of the rotating axis, and then combining the center angle with the spiral cylinder equation. So the key of this measurement method is to get two coordinates at different heights on the sharp edges of the spiral line.

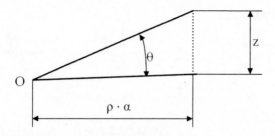

Fig. 2. Spiral expansion

3   Four-Axis Non-contact Measuring System Based on the Conoprobe

The structure of aero engine nozzle swirling slot is small and the measured surface is complicated. It is difficult and inefficient for the conventional three coordinate measuring machine (CMM) to measure the swirling slot. And most of the nozzle parts are revolving, the three axis measurement method is difficult to obtain the characteristics of the spiral line quickly and accurately. In view of obtaining the micro structure characteristics of the nozzle swirling slot and according to the above measurement principle, a four axis non-contact measuring system based on the laser probe using conoscopic holography is designed to measure the swirl slot.

3.1   Four-Axis Measuring Platform

The mechanical part of the nozzle swirl slot measuring system is designed in four axis, that is X, Y, Z three Cartesian coordinates and C axis of rotary shaft, as shown in Fig. 3.

Fig. 3. Four-axis measuring platform

The mechanical part of the measuring system is mainly composed of three parts: the base, the rotary shaft, and the X, Y and Z linear coordinates. Among them, three linear axes are integrated together, the rotation axis C is arranged separately, and the measured object is placed on the rotary shaft C through a special fixture, and the laser probe is fixed on the Z axis. When measuring the nozzle, the three linear axis drives the probe to carry out the spatial orientation movement and adjusts the distance between the probe and the nozzle. The rotating platform adjusts the attitude angle of the nozzle to make the direction of the measured surface match the beam vector of laser probe.

3.2   Laser Probe Using Conoscopic Holography

In order to achieve rapid measurement of nozzle swirl slot, the four axis non-contact measuring system designed in this paper is equipped with a laser probe based on conoscopic holography. Compared with the general holography, the advantages of the conoscopic holography are the large measurement range and the characteristics of the common optical path, especially this method can produce the hologram without the use of the coherent light source. Therefore, the dependence of the light source is reduced and the strong anti-interference energy makes the method have a strong environmental adaptability.

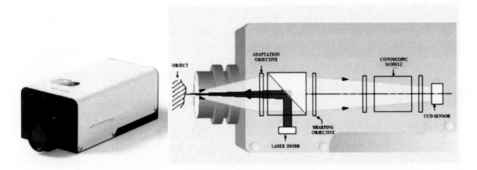

Fig. 4. Laser probe based on conoscopic holography and its principle

As Fig. 4 shows the internal structure of a conical polarizing holographic laser probe, the measurement principle is that the laser beam returned by the surface of the measured object becomes polarized by the polarizer, and then it is split into unusual and unusual light by the uniaxial crystal. The velocity of the unusual light depends on the incident angle of the beam. The analyzer is located at the rear of the uniaxial crystal and is perpendicular to the direction of the polarizer. Two beams of light interfere with each other from the Gabor observation mirror, and the hologram of interference is formed [9].

The intensity of interference fringes can be expressed as:

$$I(\rho) = I_0 * \left[1 + \cos(K \frac{r^2}{Z_c^2}) \right] \tag{2}$$

In the equation, K is the wave number, I_0 is the intensity of the incident light, Z_c is the distance between the measured point and the point of the coordinates, that is the relative height of the measured point. r is the distance between the one point ρ and the center of the mirror surface in the Gabor observation mirror. According to formula (2), the intensity of each point on the map is related to the measured distance and the radius of the circular fringe. Therefore, by calibrating the relationship between light intensity and distance Z_c, the height of the object in the optical axis can be measured.

Compared with the contact probe, the speed of measurement using conoprobe is high, and it can detect the deep hole, and the measurement error of the sharp angle or the tip edge is small. Compared with the laser triangulation probe, it has the characteristics of large measuring angle and high coaxiality (Figs. 5 and 6).

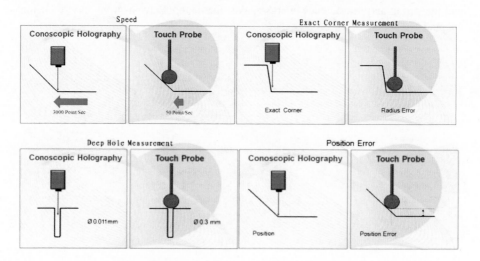

Fig. 5. Cono probe vs. contact probe

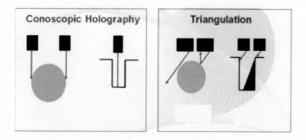

Fig. 6. Cono probe vs. triangulation probe

The diameter of the cono probe spout is 6 μm and the resolution is 0.1 μm. The measuring accuracy is 1 μm and the scanning frequency can reach 10 kHz. The sensitivity of the probe to the sharp edge is high and the size of the spot is small. By judging the noise ratio parameters of the probe signal, the identification of the characteristic points on the sharp edge of the nozzle swirl slot can be realized.

4 Measurement of Swirling Slot by Optical Laser Scanning

The key to the measurement of swirling slots in the nozzle is to get the spiral line on the cylinder. So long as the characteristic point coordinates of the spiral line are obtained, combined with the center angle, the spiral angle and the slot width can be calculated or fitted. The depth of the swirling slot can be achieved by measuring the radius of the bottom surface and the top surface of the slot.

4.1 The Acquisition of Spiral Line on Swirling Slot

In this paper, two way of laser scanning are compared to obtain the sharp edge of the spiral line. One way is that laser probe moves in one direction with a equal step until it finds the sharp edge. The second way is that laser probe moves alternately in two directions with a step which changes when the laser probe signal changes. The two ways are used to measure the length of gauge block to verify which is better to identify sharp edge of spiral line. The length of the standard block used in the experiment is 26.255 mm.

4.1.1 Laser Scanning in One Direction with a Equal Step

The four-axis non-contact measuring system is used to measure the position of the two sharp edges of the gauge block, and then the length of block can be calculated. As shown by Fig. 7, the sharp edge is measured by laser scanning using a given feed step distance in the specified direction, and the change of the light intensity in different planes is obtained which can be used to determine where the sharp edge is. The process of measuring the gauge block is that the gauge block is placed on the horizontal surface surf2, and the surf2 is parallel to the surf1 on the surface of the gauge block. The probe beam vector is perpendicular to the surface surf1, and the surf1 is within the measuring range of the conoprobe. Then the probe uses a certain feed distance d to scan along the length of the gauge block. Due to the distance measured by the probe is limited, the intensity information of laser probe in the two planes which located on both sides of the sharp edge of the gauge block will change sharply, thus the sharp edge of the block can be found, then the length of block can also be obtained.

In this paper, the four-axis non-contact measuring system moves the probe in the specified direction to find the sharp edge positions of the gauge block with the step distance is 0.01 mm. After repeated 10 measurements, the data results is shown by Table 1. According to the data results, the mean value of the 10 measurement is 26.2458 mm, the peak to peak is 0.013 mm. The measurement error is 0.0092 mm and the maximum error is 0.015 mm as compared with the standard length.

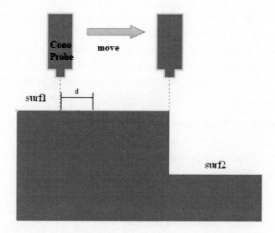

Fig. 7. The diagram of Laser scanning in one direction

Table 1. Measurements data using the way of scanning in one direction

Order	1	2	3	4	5	6	7	8	9	10
Data/mm	26.253	26.24	26.24	26.24	26.247	26.247	26.247	24.248	26.248	26.248

Conclusion:

(1) The precision of the scanning method in one direction depends on the size of the scanning step, and the smaller the step distance is, the higher the precision can reach. The scanning distance is restricted by the motion accuracy of the measurement system.

(2) The scanning step adopted by this method is given a certain value. When the step is smaller to move on both sides of the sharp edge, the moving speed is slow, the scanning time is long, and the efficiency is low.

4.1.2   Laser Scanning in Two Directions with a Changing Step

The way to locate the sharp edges is that laser probe moves alternately in two directions with a step which changes when the laser probe signal changes. The method is described in details in combination with the standard gauge block measurement experiment, as shown by Fig. 8.

(1) First, the gauge block is placed on the horizontal surface surf2, and the surf2 is parallel to the surf1 on the surface of the gauge block. The probe beam vector is perpendicular to the surface surf1, and the surf1 is within the measuring range of the conoprobe.

(2) Next, set the moving step distance is d_n, and its value can be bigger. The probe moves in the direction of " + ". When the laser spot moves from the surf1 surface of the gauge block to the surf2 surface, the light intensity information which have changed obviously obtained by the probe can be used to judge whether the probe have moved from one side of sharp edge to the other or not.

(3) The laser probe moves in the "−" direction according to the step d_n, so that the laser spot moves to the last point S_n on the surf1 surface of the gauge block. Then set the step distance $d_{n+1} = {d_n}/{2}$, and made the probe move in the " + " direction.

(4) Repeat steps 2 and 3, until the d_n value is less than the set threshold d_∞, it is determined that the laser spot has been positioned on the sharp edge of block.

The scanning method with self-variable step distance makes the laser spot scan forth and back on the sharp edge, and adaptively reduces the scanning step according to the probe signal changing result of each time, until the step distance is small enough to meet the requirement of measurement accuracy, then it is determined that the laser spot has been positioned to the edge of the sharp edge.

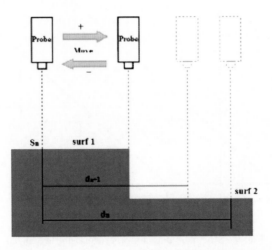

Fig. 8. Diagram of laser scanning with self-variable step

In this paper, the four-axis non-contact measuring system moves the probe alternately in two directions with a step which changes when the laser probe signal changes to find the sharp edge positions of the gauge block. After repeated 10 measurements, the data results is shown by Table 2. According to the data results, the mean value of the 10 measurement is 26.2585 mm, the peak to peak is 0.004 mm. The measurement error is 0.0035 mm and the maximum error is 0.005 mm as compared with the standard length.

Table 2. Measurements data using the way of scanning alternately in two directions

Order	1	2	3	4	5	6	7	8	9	10
Data/mm	26.258	26.258	26.259	26.259	26.26	26.258	26.259	24.259	26.256	26.259

Conclusion:

(1) Compared with the method of scanning in one direction with a certain step, the bidirectional scanning method with self-variable step distance has high precision and good repeatability. The initial step is bigger, when the sharp edge is nearly found, the step gets smaller, so the speed of its positioning to the edge is faster.

(2) The positioning accuracy of this method also depends on the value of the minimum feed threshold d_∞. The smaller the value d_∞ is, the higher the precision can reach.

4.2   The Measurement of Swirling Slot by Laser Scanning

In this paper, the four-axis non-contact measuring system is designed to measure the swirl slot of the nozzle using the scanning method with self-variable step distance, which can quickly and accurately identify the sharp edges of the spiral line, so that the related characteristic parameters of the swirling slot can be obtained. As shown in Fig. 9, the measurement and calculation process of the swirling slot is as follows:

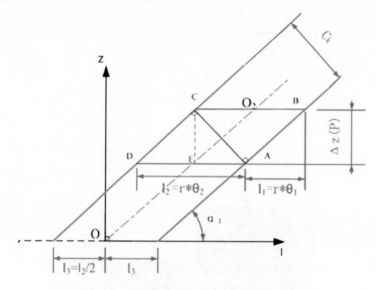

Fig. 9. Schematic diagram of swirling slots expanded

Fig. 10. Schematic diagram of swirling slot measurement

(1) The nozzle is fixed on the special fixture. Through the calibration and adjustment of the cylindrical area in the middle part of the nozzle, the nozzle axis is parallel to the direction of the Z axis of the measuring machine and perpendicular to the direction of X and Y. The laser probe is used to measure the cylindrical surface where the spiral line is, and the radius r_w of the outer cylinder is calculated (Fig. 10).

(2) The outer cylinder surface is located in the measuring range of the probe. Rotates the table, and the point A is obtained by the scanning method with self-variable step distance.

(3) Rotate the table to a certain position and record the rotation angle. Then the point B on the spiral line (AB) is found when the probe moves in the direction of the nozzle axis, and the moving distance $\Delta Z = Z_0$ is recorded. At this time, the spiral angle of the top spiral can be calculated $\alpha_1 = \arctan(k) = \arctan[z_0/(\theta_1 \times r_w)]$.

(4) Next, the point C can be found using the scanning method with self-variable step distance, meanwhile record the rotation angle $\theta = \theta_2$. The arc length of the BC section (BC $= AD = l_2 = \theta_2 \times r_w$) and the width of the slot ($C_k = l_2 \times \sin(\alpha_1)$) can be calculated.

(5) Set the rotation angle $\theta = \theta_2/2$ and make the table rotate reversely, the laser spot of the probe is located on the bottom cylinder of swirl slot. Gathering the measuring point on the cylinder, then the radius of the bottom cylinder r_{d1} can be fitted and calculated. Also, the depth of swirl slot can be calculated $d = r_w - r_{d1}$.

(6) The outer spiral line and the bottom spiral line of the swirl slot have the same helix lead. According to this principle, the spiral angle α_{d1} of the bottom spiral line can be calculated:

$$P = r_w \bullet \tan(\alpha_1) \bullet 2\pi = r_{d1} \bullet \tan(\alpha_{d1}) \bullet 2\pi$$
$$\Rightarrow \alpha_{d1} = \arctan[\frac{r_w}{r_{d1}} \bullet \tan(\alpha_1)] \tag{3}$$

5 Measuring Experiments

By using four-axis non-contact measuring system based on the cono probe, the experiment of measuring the nozzle swirling slot is carried out through utilizing the laser scanning method with self-variable step distance, thus the main parameters of the nozzle swirling slot such as the width of the slot, the depth of the slot and the spiral angle of the slot can be obtained (Fig. 11).

A single nozzle of a certain type contains 6 swirling slots, each of which is repeated 7 times in the experiment. The result is shown in the following table.

As can be seen in the Table 3, the measurement repeatability of the spiral angle of the single swirl slot can reach 30′, the measurement repeatability of the slot width is 0.05 mm, and the measurement repeatability of the slot depth is 0.013 mm; it can meet the measurement requirement of the nozzle slot.

Fig. 11. Schematic diagram of swirling slot measuring experiment

Table 3. The results of swirling slot measurements

Measurements	Angle/°	Width/mm	Depth/mm
Slot1			
1	30.662	1.707	1.221
2	31.036	1.727	1.229
3	31.107	1.732	1.223
4	31.138	1.733	1.224
5	30.648	1.706	1.226
6	31.097	1.731	1.219
7	30.628	1.717	1.222
Max	31.138	1.733	1.229
Min	30.628	1.706	1.219
Slot 2			
1	30.391	1.68	1.218
2	30.341	1.677	1.222
3	30.39	1.68	1.225
4	30.405	1.68	1.229
5	30.403	1.68	1.219
6	30.412	1.681	1.225
7	30.419	1.681	1.227
Max	30.419	1.681	1.229
Min	30.341	1.677	1.218
Slot 3			
1	29.927	1.664	1.226
2	30.737	1.71	1.222
3	29.98	1.653	1.224
4	30.528	1.687	1.226
5	30.571	1.69	1.228

(*continued*)

Table 3. (*continued*)

Measurements	Angle/°	Width/mm	Depth/mm
6	30.595	1.693	1.217
7	30.714	1.702	1.218
Max	30.737	1.71	1.228
Min	29.927	1.653	1.217
Slot 4			
1	30.901	1.719	1.224
2	30.882	1.717	1.228
3	30.86	1.721	1.227
4	30.864	1.722	1.229
5	30.905	1.725	1.219
6	30.891	1.724	1.224
7	30.9	1.724	1.222
Max	30.905	1.725	1.229
Min	30.86	1.717	1.219
Slot 5			
1	30.611	1.726	1.217
2	30.55	1.723	1.228
3	30.483	1.72	1.222
4	30.521	1.722	1.224
5	30.535	1.721	1.226
6	30.523	1.72	1.223
7	30.524	1.72	1.221
Max	30.611	1.726	1.228
Min	30.483	1.72	1.217
Slot 6			
1	30.114	1.668	1.216
2	30.035	1.663	1.229
3	30.041	1.663	1.219
4	30.071	1.665	1.222
5	30.356	1.680	1.223
6	30.346	1.680	1.225
7	30.347	1.680	1.227
Max	30.356	1.680	1.229
Min	30.035	1.663	1.216

6 Conclusions

The swirling slot is a typical structure of the aero engine nozzle, which has a direct effect on the fuel atomization. In this paper, a four-axis non-contact measurement system using the optical laser probe based on conoscopic holography is built to meet the requirements of measuring the slot width, the slot depth and the spiral angle. The system utilize the way of laser scanning with self-variable step distance in bidirectional direction, and then the points on the spiral line can be obtained accurately. The coordinate values of these points can be used to calculate or fit the characteristic parameters of swirling slot, realizing the deterministic measurement of the nozzle swirling slot. The repeatability of the spiral angle can basically be 30′, the repeatability of the slot width is 0.05 mm, and the repeatability of the slot depth is 0.013 mm. The factors that affect the measurement precision mainly include the accuracy of the four-axis non-contact measuring system, the spot size of the laser probe, the sensitivity of the probe to sharp edge, and whether spiral line of slot is sharp edge or not. The measurement method proposed in this paper can meet the requirement of the nozzle swirling slot, and it can be used to judge the consistency of the swirl slot size and structure. It can provide the data basis for further improvement of the engine performance.

Acknowledgements. This research was supported by the Technical innovation foundation of Aviation industry corporation of china (No. 2013F30327). The authors would like to thank all the editors and anonymous reviewers for their help in improving the paper.

References

1. Zheng X, Huang X, Chen Y (2014) Influencing factor and debugging technology of major performance parameter for fuel nozzle. Aeronaut Manuf Technol 466(22):155–157
2. Liu G, Pan G, Zheng H (2015) Numerical investigation for effect of structural parameters on nozzle performance. Aeroengine 41(5):28–32
3. Liu K, Zhang B (2012) Investigation of effect of machining quality on characteristics for aero-engine fuel injector. Aeroengine 38(1):55–59
4. Dang L, Yan Y (2013) Experimental study on fuel spray characteristics of pressure-swirl atomizer. J Nanjing Univ Aeronaut Astronaut 45(4):453–460
5. Di D, Liu Y (2017) Experimental study on characteristics of fuel manifold and injector. Aeroengine 43(2):67–74
6. Huang J, Xu B (2016) A machine vision system for position measurement of small holes on spherical surface. Nanotechnol Precis Eng 14(1):28–34
7. Han Y, Zhang Z, Dai M (2011) Monocular vision system for distance measurement based on feature points. Opt Precis Eng 19(5):1110–1118
8. Zhu X, Du H, Wang W (2010) Development and application of fiber probe. Measure Tech 2:12–15
9. Zhou A, Shao W, Wu Y (2017) A new method of aviation blade measurement via applications of conoscopic holography principle. China Mech Eng 28(12):1394–1399
10. Ren S, Yang Y, Duan H (2015) Design and uncertainty analysis of non-contact measurement system based on conoscopic holography. J Electron Measure Instrum 24(7):616–620
11. Xie Z, Xu Y, Zhang G (2002) A four-axis instrument for measuring free-form surface. Aviat Precis Manuf Technol 38(1):31–33
12. Wei Z, Yang L, Guo Y (2015) Research of distance measuring system based on the principle of conoscopic holography. J Optoelectron Laser 10:1967–1973

Computational Study on Two Dimensional Electrothermal Deicing Problem

Chunhua Xiao[✉], Kunlong Yu, Yubiao Jiang, Ming Li,
and Zhangsong Ni

State Key Laboratory of Aerodynamics, China Aerodynamics Research
and Development Center, Mianyang 621000, China
dragoner76@163.com

Abstract. For airplane deicing problem, two dimensional electrothermal deicing model is established. The numerical study on heat transfer characteristics during electrothermal deicing process is presented. The Enthalpy-Porous Medium method is applied to describe the phase change process for the numerical simulation, which is a kind of enthalpy model. The computational domain is treated as a porous medium including solid ice and water liquid and mushy zone. The structured mesh topology is used to distribute the computational domain. The finite volume method is adopted to discretize the governing equations. The temperature is obtained by iteration of the energy equation coupled with the liquid volume fraction formula. The properties such as thermal conductivity of the mushy zone can be obtained by linear interpolation. The heat transfer characteristics are studied systematically for deicer pad including phase change process. The numerical study emphasizes on the effects of heating mode, cooling time, heater power and heater gap on phase change heat transfer characteristics. It shows that periodic heating mode for high heater power is superior to continuous heating mode for low heater power if reasonable combination of cooling time and heater power are adopted to obtain better deicing performance and less energy consumption. The existence of heater gap will decrease energy consumption less and make the temperature distribution more reasonable. Therefore, the reasonable distribution of heater gap can largely improve the deicing efficiency which is beneficial for the design of electrothermal deicer.

Keywords: Computational study · Heat transfer · Electrothermal · Deicing · Phase change

1 Introduction

Electrothermal deicing usually uses a combination of continuous and periodic heating strategy [1]. The continuous heater pad is located near the stagnation line along the chordwise and spanwise direction. This arrangement can prevent ice from covering the leading edge, which can decrease the difficulty of deicing [2]. But the power requirements must follow the anti-icing design. Periodic heater pad is primarily distributed on the major deicing region, which can save the heat energy and cause the periodic ice shedding. The combination of these two heating strategies are comprehensive method

© Springer Nature Singapore Pte Ltd. 2019
X. Zhang (Ed.): APISAT 2018, LNEE 459, pp. 1931–1947, 2019.
https://doi.org/10.1007/978-981-13-3305-7_157

currently used in the airplane [3]. The relevant research about electrothermal deicing is concentrated on heat transfer characteristics of the multilayer electrothermal deicer pad.

From the 1970s to the early 1980s, the researchers focused on the modeling and computation of one-dimensional electrothermal deicing problems [4–7]. Each material layer is assumed to be infinite plane in one-dimensional problem. The thicknesswise temperature is varied and the lengthwise one is constant. The property of each material layer is constant. The contact thermal resistance between layers is not considered. Stallabrass [4] firstly modeled and calculated one-dimensional electrothermal deicing problem. The heat conduction equation was discretized by using the explicit difference scheme for time and space. The influence of thickness ratio between inner insulation and outer insulation on electrothermal deicing was studied. It was recommended that the thickness ratio of the inner and outer insulation should be greater than 2.

In order to consider the latent heat effect of the phase change process, Baliga [5] used the sensible heat capacity method to improve one-dimensional electrothermal deicing computational model. The Crank-Nicholson implicit difference scheme was used to discretize the equation. Gauss-Jordan method was used to solve the linear algebraic equations. And the phase change heat transfer characteristics of electrothermal deicing process were studied.

Marano [6] further developed one-dimensional model using the enthalpy method. Then he studied the phase change heat transfer characteristics during the electrothermal deicing process. Roelke [7] also developed a one-dimensional model in which the enthalpy and the temperature are both functions to be solved. A unified energy equation was established for the whole region (including the liquid, solid and interface) which could improve the computational efficiency. The enthalpy distribution was obtained numerically and the interface was determined by enthalpy value.

From the early 1980s to the early 1990s, the researchers focused on the modeling and computation of two-dimensional electrothermal deicing problems [8–15]. Chao [8] directly extended Marano's one-dimensional model to two-dimensional model. The finite difference method was used to discretize the equations. And the effect of heater gap on electrothermal deicing was studied. Leffel [9] carried out a transient thermal response experiment on the electrothermal deicer in a rotating helicopter blade. The result of Marano's computation was in good agreement with the experimental result when each layer of blade was thin enough and the curvature was close to zero. However, there was a large error between the computation and experiment at the region with large curvature or the region close to the heater. Masiulaniec [10] developed a method of coordinate transformation and established an electrothermal deicer pad model which could overcome the problem of curvature. The result of two-dimensional electrothermal deicer pad model was in good agreement with Leffel's experimental result.

Wright [11–13] extended the two-dimensional electrothermal deicer pad model to study the influence of various numerical methods (ADI, ADE, SIP, MSIP, etc.) on the speed of computation. ADI Method was found to be the most efficient method. Keith et al. [14] compared the superiority among three models (one-dimensional, two-dimensional plane, two-dimensional coordinate transformation). Wright [15] developed a two-dimensional model with icing, melting and ice-shedding effects considered. This model was successfully implanted into NASA/Lewis's famous icing code, LEWICE.

From the early 1990s, the researchers mainly studied three-dimensional electrothermal deicer pad modeling and computation [16–20]. Yaslik et al. [16] also used the assumed phase method (MOAS) to study three-dimensional electrothermal deicing problem using alternating Douglas finite difference scheme. The ice usually has very irregular shape, so the finite difference method is very difficult to simulate. Therefore, the study was just for two-dimensional and three-dimensional rectangular model.

In order to simulate the irregular shape of ice, Huang et al. [17–20] used a finite element method to simulate two-dimensional and three-dimensional electrothermal deicing problems coupled with the assumed phase method. The curvature, chordwise and spanwise distribution of heater were studied. The chordwise curvature effect and the spanwise swept effect was found to be more important in practical computation. The improved finite element method had low computational efficiency in three-dimensional electrothermal deicing simulation. And divergence existed if the heater element is very thin. Chang et al. [21, 22] studied the influence of periodic heating pulse on deicing surface temperature. The block-correction technique was used to solve the discrete equations. Xiao et al. [23] studied systematically the heat transfer characteristics of airplane electrothermal deicing process and analyzed the influence of different factors.

These studies above mentioned were mainly concentrated on the computational method research [8–15]. In fact, the study on the heat transfer characteristics of the electrothermal deicing process also is very important for the interface of ice-deicer melting, which can be divided into two cases (Fig. 1). One case is that the ice partially melts and some ice still bond to solid wall surface. Thus the ice shedding doesn't happen. This is a problem about phase change heat transfer characteristics of the electrothermal deicer pad [24]. This is the key study point in the present paper. The other case is that the ice bonding to the solid wall surface melts completely. Thus the ice shedding will happen and the effect of ice shedding on the heat transfer characteristics of electrothermal deicer pad should be considered [15].

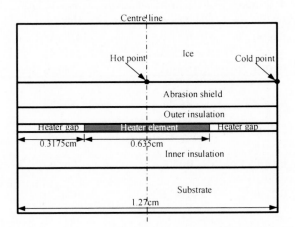

Fig. 1. Sketch of electrothermal deicer pad with heater gap

Numerical simulation method is used to study the electrothermal deicing process in the present study. The traditional phase change heat transfer model based on enthalpy method is applied to describe the electrothermal deicing process. The finite volume method and hybrid mesh technique are adopted to discretize the equations. The effects of heating mode, cooling time, heater power and heater gap on the heat transfer characteristics are studied systematically. The ice shedding is not considered in the present study. It can provide a theoretic basis for the optimization of heat transfer characteristics of electrothermal deicer.

2 Methodology

The heater element in the electrothermal deicer pad are periodic distributed. The multilayer structure is composed of substrate, inner insulation, heater element, outer insulation, abrasion shield and ice. One section of the electrothermal deicer is taken to study for simplifying problem. Figure 1 presents the electrothermal deicer pad with heater gap, which has length of 1.27 cm. The heater element with length of 0.635 cm is located in the middle, which is symmetrical along the centre line. There are two important points for heat transfer and deicing analysis. One is hot point which is the intersecting point between the symmetrical center line and the interface of ice-shield. The other is cold point intersected between side boundary and the interface of ice-shield. These two points represent the highest and lowest temperature on the interface of ice-shield which are representative for analysis. In Fig. 1, the heater gap length is 0.3175 cm. In fact, the deicer pad becomes one-dimensional model if the length of heater gap decreases to zero. Table 1 shows the physical properties and thickness of each layer of electrothermal deicer pad.

Table 1. Physical properties and thickness of each layer

Layer name	Thickness (cm)	Thermal conductivity [J/(m s K)]	Specific heat [J/(kg K)]	Density (kg/m^3)
Water	0.6350	0.5538	4186.55	999.55
Ice		2.2325	2110.02	919.46
Abrasion shield	0.0305	15.0566	489.80	7929.11
Outer insulation	0.0254	0.3807	962.90	1762.02
Heater element	0.0102	13.1529	418.66	882.13
Inner insulation	0.1270	0.3807	962.90	1762.02
Substrate	0.2210	115.0876	1088.50	2482.85

The following assumptions are made to simplify the problem.

a. property of each material layer is constant;
b. no contact thermal resistance between layers;
c. neglecting the density change caused by ice melt;
d. phase change occurs in the temperature range near the melting point ([273 K, 273.15 K]);

Based on the above assumptions, the Enthalpy-Porous Medium method [25–27] is adopted. The computational domain is treated as a porous medium composed of solid ice and liquid water and mushy zone. The governing equations are shown as follows:

$$\frac{\partial u}{\partial x} + \frac{\partial v}{\partial y} = 0 \tag{1.1}$$

$$\frac{\partial u}{\partial t} + u\frac{\partial u}{\partial x} + v\frac{\partial u}{\partial y} = -\frac{1}{\rho}\frac{\partial P}{\partial x} + \mu(\frac{\partial^2 u}{\partial x^2} + \frac{\partial^2 u}{\partial y^2}) + S_x \tag{1.2}$$

$$\frac{\partial v}{\partial t} + u\frac{\partial v}{\partial x} + v\frac{\partial v}{\partial y} = -\frac{1}{\rho}\frac{\partial P}{\partial y} + \mu(\frac{\partial^2 v}{\partial x^2} + \frac{\partial^2 v}{\partial y^2}) + S_y + S_b \tag{1.3}$$

$$\frac{\partial H}{\partial t} + \frac{\partial(uH)}{\partial x} + \frac{\partial(vH)}{\partial y} = \frac{k}{\rho}(\frac{\partial^2 T}{\partial x^2} + \frac{\partial^2 T}{\partial y^2}) + Q \tag{1.4}$$

$$\beta = \frac{T - T_S}{T_L - T_S} \quad T_S < T < T_L \tag{1.5}$$

When the liquid volume fraction is 1, the porous medium is water completely. When the liquid volume fraction is reduced, the flow velocity is reduced correspondingly. When the liquid volume fraction is zero, the flow is also stationary and the porous medium becomes ice. It means that the water is completely solidified within mushy zone. The relationship between the source term and the velocity is expressed as follows:

$$S_x = -\omega u, \quad S_y = -\omega v \tag{1.6}$$

ω is a large number in the solid phase and reduced to 0 in the liquid phase. For liquid phase, the flow velocity is solved through momentum equations when the additional source term is zero. For the ice-water mushy zone, the proportion of the additional source term increases. Thus the additional source term dominates the momentum equations so that the flow velocity can be forced to zero for the solid phase. The additional source terms unify the momentum equations of the liquid, mushy and solid phase. It makes the whole field solution possible. According to Carman-Koseny equation [26], ω is chosen according to the following formula.

$$\omega = -A_{mush}\frac{(1-\beta)^2}{(\beta^3 + \varepsilon)} \tag{1.7}$$

The enthalpy (H) of each material is the sum of the sensible and latent heats and is calculated by the following formula:

$$H = h + \Delta H \tag{1.8}$$

The relationship between sensible heat (h) is shown as follows:

$$h = h_{ref} + \int_{T_{ref}}^{T} C_P dT \tag{1.9}$$

The release of latent heat in the ice-water mushy zone is linear in the present work. So the latent heat of the mushy zone is regarded to be the liquid volume fraction times the latent heat, which is shown as follows.

$$\Delta H = \beta L \tag{1.10}$$

The enthalpy and temperature satisfy the following relationship:

$$H = \begin{cases} C_S(T - T_{ref}), & T < T_S \\ C_S(T_S - T_{ref}) + \frac{T-T_S}{T_L-T_S} \cdot L, & T_S \leq T \leq T_L \\ C_S(T_S - T_{ref}) + L + C_L(T - T_L), & T > T_L \end{cases} \tag{1.11}$$

The temperature is obtained by iteration of the energy equation coupled with the liquid volume fraction. To enhance the convergence, the liquid volume fraction is updated by the method proposed by Voller [27].

The properties such as thermal conductivity of the mushy zone can be obtained by linear interpolation, which has the following relationship (Fig. 2).

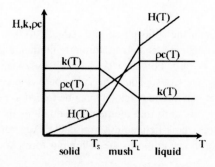

Fig. 2. Linear variation on physical properties of mushy zone

$$k = \begin{cases} k_S & T < T_S \\ k_S + (k_L - k_S)\frac{T-T_S}{T_L-T_S} & T_S < T < T_L \\ k_L & T > T_L \end{cases} \tag{1.12}$$

The equations from (1.1) to (1.5) are the basic forms of the Enthalpy-Porous Medium model, which are established on the condition that the phase change occurs in a certain temperature range and the ice-water mushy zone exists. The model can be used to simulate the phase change occurs at a certain temperature when the phase change temperature range is small enough.

The initial flow velocity is zero and the initial temperature is same as the ambient temperature. Forced convective heat transfer condition is applied to the outer ice surface adjacent to the external flow field due to the relative strong external flow in the present work [9, 28–33]. So the convective heat transfer coefficient is relatively large. Accordingly, the natural convection heat transfer condition is adopted on the inner surface of substrate connecting with the internal flow [9]. And the natural convective heat transfer coefficient is much smaller than that of forced convection. Thus the boundary adjacent to external and internal flow will satisfy the following formula:

$$-k\nabla T \cdot \vec{n}\big|_{interface} = h_{out}(T - T_{out}) \tag{1.13}$$

$$-k\nabla T \cdot \vec{n}\big|_{interface} = h_{in}(T - T_{in}) \tag{1.14}$$

Both the temperature and the heat flow are continuous at the interface of ith and $(i + 1)^{th}$ layer. Due to the symmetry of one section of electrothermal deicer pad model, the symmetry boundary condition is adopted on the left and right sides in the present work. It means that the normal gradient of temperature along the symmetry boundary is zero [34].

For a multilayer deicer pad with a heater element, the volumetric heat source term Q is larger than zero which is adopted by the following formula. $D = t_h/(t_w - t_h)$ is defined as heating/cooling time ratio for one period (Fig. 3).

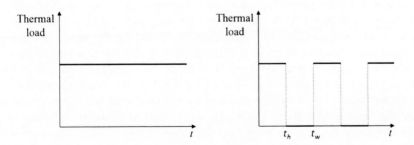

Fig. 3. Thermal load variation of continuous and periodic heating

$$Q = \begin{cases} const & 0 + i \cdot t_w \leq t \leq t_h + i \cdot t_w \\ 0 & t_h + i \cdot t_w < t \leq i + i \cdot t_w \end{cases} \qquad (1.15)$$

The finite volume method is used to discretize the governing equations. The structured mesh topology is used to distribute the computational domain. The energy equation and the momentum equations are discretized by second-order scheme for simulation accuracy. The separation method is adopted and the sub-relaxation technique is implemented during numerical simulation. The SIMPLE algorithm is used for pressure-velocity coupling.

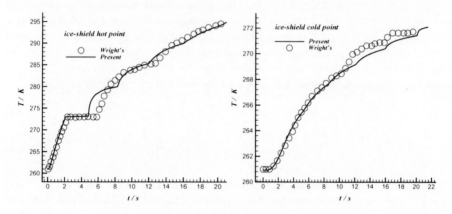

Fig. 4. Interfacial temperature history at hot and cold point of ice-shield with heater gap

In order to further verify the reliability of the present method, the phase change heat transfer of electrothermal deicer pad is also validated numerically. Figure 1 presents sketch of the computational model and Table 1 gives the computational parameters. Figure 4 shows the temperature history at hot point and cold point of ice-shield interface with heater gap respectively. All the present results are in good agreement with the results in Ref. [11]. It validates that the computational method is reliable and can be used for next further study. In addition, it is found that the natural convection in the liquid phase has little effect on the heat transfer of electrothermal deicing process.

The grid independence validation has been implemented. Four sets of grids including coarse mesh (48 × 41), medium mesh (95 × 41), dense mesh (189 × 41) and refined mesh (377 × 41) are designed for the numerical simulation on the electrothermal deicer pad without heater gap. The interfacial temperature history of ice-shield is obtained for 30 s continuous heating shown in Fig. 5. The temperature history is much better agreement with the reference results when the grid is medium level. Therefore, the formal computation is implemented by the medium mesh (95 × 41).

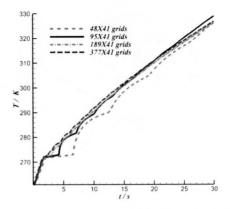

Fig. 5. Comparison of results for different meshes

3   Results and Discussion

The phase change of ice is the most characteristics and important phenomenon in the electrothermal deicing process. And the heating mode, cooling time, heater power and heater gap are four important factors that can influence the phase change and heat transfer characteristics.

3.1   Effect of Heating Mode

Electrothermal deicing usually uses a combination of continuous heating and periodic heating strategy. The heater element for continuous heating is installed on the stagnation line to prevent the ice from covering the leading edge region in spanwise and chordwise direction. On the other hand, the heater element for periodic heating is installed on the deicing region and to split ice which is easier for shedding [35]. Therefore, the influence of continuous and periodic heating mode on the heat transfer characteristics is deserved to be studied and necessary for electrothermal deicing strategy optimization.

3.1.1   Continuous Heating Mode

Figure 6 shows the interfacial temperature history of deicer pad without heater gap for 30 s continuous heating. The interfacial temperature history of ice-shield appears an apparent platform, which reflects the ice melting process. Excluding the temperature platform, all the other interfacial temperature histories are on the rise. And the temperature of the region closer to heater element rises relatively faster than others. The interfacial temperature of insulation-substrate rises slowest. Compared with outer insulation, the thickness of inner insulation is 5 times more and the thermal resistance is much larger accordingly. This kind of structure can reduce heat loss to the substrate and assure more heat energy being transferred to deicing region.

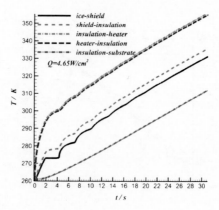

Fig. 6. Temperature history of different interface of deicer pad without heater gap (30 s continuous heating)

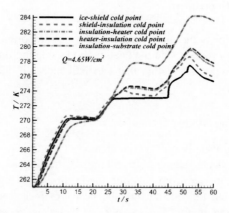

Fig. 7. Temperature history at cold point of different interface of deicer pad (3 periods, $t_w = 20$ s, $D = 10$ s/10 s)

3.1.2 Periodic Heating Mode

Figure 7 presents the temperature history at cold point of different interface of deicer pad with heater gap for 3 periods ($t_w = 20$ s, $D = 10$ s/10 s). This phenomena is much different from that of Fig. 6. The phase change platform is much longer, which even lasts within the second and third deicing period. And it is mainly the contribution of heater gap which will influence the phase change heat transfer characteristics seriously.

3.1.3 Comparison of Two Heating Mode

Figure 8 presents the comparison of interfacial temperature history at cold point of ice-shield of deicer pad with heater gap for two heating modes for 60 s heating. For deicing process, the study on the temperature history at cold point is more important than that at hot point, which decides the ice shedding moment. Due to the existence of heater gap, the phase change time becomes larger. So it can be found that the phase change will

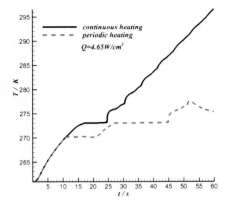

Fig. 8. Comparison interfacial temperature history at cold point of ice-shield of deicer pad for two heating modes (3 periods, t_w = 20 s, D = 10 s/10 s)

last from the second period to the third period. It costs 25 s of phase change for continuous heating. Actually, the continuous heating is just a special case without cooling time for periodic heating. If the ice melt must be finished within one period, the minimum continuous heating time within one period should be more than 25 s for the periodic heating mode.

3.2   Effect of Cooling Time

The heating/cooling time ratio is an important parameter for the periodic heating mode. The function of cooling time is to save energy and prevent the ice ridge formation downstream of the deicing region. The study on the cooling time is very important for optimizing the heat transfer characteristics of deicer pad and improving the efficiency of periodic deicing. The interfacial temperature history of cold point and hot point at the ice-shield of deicer pad with heater gap for three periods of 60 s heating under different cooling time respectively are shown in Fig. 9. Due to the existence of heater gap and cooling time, the phase change platform becomes much longer compared with the deicer pad without heater gap and cooling time. When t_w is fixed to 20 s, the temperature rise rate becomes larger when the cooling time is decreased no matter for the hot point and cold point. The cooling time can't be too large, because the melting water on hot point will freeze again when the ice on the cold point is not still melt. This phenomena should be avoided during the design of electrothermal deicer pad.

Figure 10 shows the comparison of deicing time and heating time at the cold point of ice-shield under different cooling time. It shows that the deicing time increases monotonously along with the increasing of cooling time. But it is clear that the heating time doesn't follow this regulation and doesn't change monotonously along with the cooling time. There is a positive peak at the cooling time of 6 s. And a negative peak appears at the cooling time of 8 s where the heating time is the least. So the corresponding energy consumption is also the least. The reason is that deicing behavior need

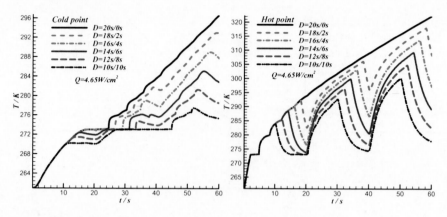

Fig. 9. Interfacial temperature history of cold and hot point at ice-shield of deicer pad under cooling time (3 periods, t_w = 20 s)

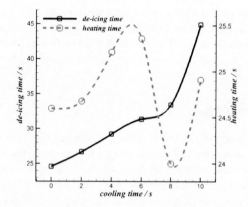

Fig. 10. Interfacial comparison of deicing and heating time of cold point of ice-shield

to take full advantage of the residual heat energy by setting the reasonable value of the cooling time. It is an important method to reduce the deicing energy consumption.

Figure 11 shows that temperature history of the cold point and hot point at ice-shield of deicer pad with heater gap for two heating modes (6 periods, t_w = 20 s) respectively. The heater power is 1.55 W/cm² for the continuous heating mode and 4.65 W/cm² for periodic heating mode. The time of one period is 20 s and the heating/cooling time ratio is 10 s/10 s. It can be found that the ice at cold point melt completely at the time of 74.6 s for the continuous heating, whose energy consumption is 7322.8 J. But the corresponding time is 44.8 s for periodic heating at same cold point, which is just 60% of that for continuous heating. But the energy consumption is 7342.5 J for the continuous heating, which is a little higher than that for periodic heating. Therefore, the combination of the cooling time and heater power should be adjusted reasonably which can make the high-power periodic heating mode more superior to the low-power continuous heating mode.

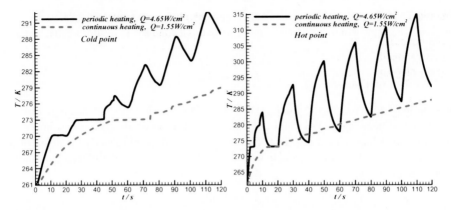

Fig. 11. Temperature history of cold and hot point at ice-shield of deicer pad for two heating modes (6 periods, $t_w = 20$ s)

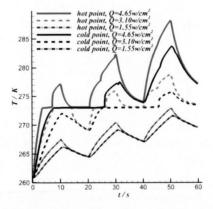

Fig. 12. Temperature history of cold and hot point at ice-shield of deicer pad (3 periods, $t_w = 20$ s, D = 10 s/10 s)

3.3   Effect of Heater Power

The heater power is an important influential factor for the deicing process. There is specific proportion of energy for deicing operation in airplane. Thus it is impossible to supply too large heater power for deicing. And if the power is too small, it is difficult to finish deicing operation timely. Figure 12 shows the temperature history of cold point and hot point at ice-shield of deicer pad with heater gap for 3 periods ($t_w = 20$ s, D = 10 s/10 s). The heater power are 4.65 W/cm^2, 3.10 W/cm^2, 1.55 W/cm^2 respectively. It can be seen that the ice starts to melt earlier for higher heater power, which also increases the temperature rise rate at both cold point and hot point. In this case, the melting time is equal to deicing time. So the deicing time will be less for higher heater power. If the heater power is too small (1.55 W/cm^2), the heat energy will be not enough to melt ice during three periods. When the medium heater power (3.10 W/cm^2) is used, the melt water will freeze again to form secondary icing during the cooling period.

3.4   Effect of Heater Gap

The heater gap will greatly reduce the heat energy consumption of deicing. But the existence of heater gap will yield cold point and hot point which are two important locations to influence deicing effect. Therefore, it is significant to study the distribution of heater gap within deicer pad. It will be helpful to improve the efficiency of electrothermal deicing.

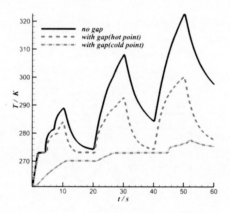

Fig. 13. Interfacial temperature history of ice-shield of deicer pad with/without heater gap (60 s periodic heating)

Figure 13 shows the comparison of interfacial temperature history of ice-shield of deicer pad with/without heater gap for 60 s periodic heating. The existence of heater gap greatly affects the phase change and heat transfer characteristics of the deicer pad. For deicer pad with heater gap, the ice at cold point starts to melt at time of 16 s, which seriously lags behind that at hot point. The temperature rise rate of cold point is much lower than that of hot point due to the existence of heater gap. The heater gap is the most important factor for the uniform melting of ice. The smaller the heater gap is, the more uniform the ice melt is. In practice, it is impossible to eliminate heater gap which increases the heater element number and the heat energy consumption tremendously. Therefore, reasonable heater gap is the key point for deicer pad design.

Figure 14 shows the comparison of interfacial temperature history of cold point at ice-shield of deicer pad varied with heater gap power for 30 s continuous heating. If an adjustable power is added on the heater gap, whose maximum power (Q) is equal to the value of heater power. And the heater gap power is changed from 0 to Q. The function is same as changing the length of heater gap. And the case with power Q is the case without heater gap.

Computational Study on Two Dimensional Electrothermal Deicing Problem 1945

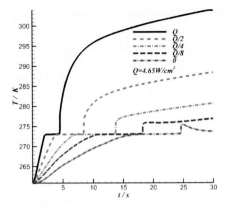

Fig. 14. Interfacial temperature history of cold point at ice-shield of deicer pad varied with heater gap power (30 s continuous heating)

Thus the influence of heater gap on the temperature history can be studied in this special view. All of the temperature histories show a gradual increasing trend with different slope, which is larger for higher power. For higher power, the temperature history has a trend of rapid rise firstly and then gradual increasing later after ice shedding. When the heater gap power decreases to $Q/8$, the temperature history tendency is appropriate flat. If the heater gap power is reduced to zero, the temperature appear a trend of decline. The influence of heater gap on the temperature history can be found clearly.

4 Conclusions

The heat transfer characteristics of electrothermal deicing process are studied by numerical computation method. The influences of ice melt and ice shedding on the heat transfer characteristics are mainly considered. The conclusions are made as follows.

(1) Deicing operation needs fewer time under same heating mode and more heater power. Increasing the heating/cooling time ratio, the deicing time can be decreased when one period time is fixed. The cooling time and heater power combined reasonably, the periodic heating mode with higher power is more superior to the continuous heating mode with low power for less deicing time and less heat energy consumption.

(2) If the cooling time is too small, excessive water will be melt to runback downstream and form ice ridge beyond anti-icing region. On the other hand, if the cooling time is too large, the water on hot point will easily freeze again but the ice on cold point is still frozen. The balance of heating time and cooling time should be balanced for deicing strategy.

(3) The mushy zone is equal to heat sink which absorbs heat from the surrounding. So the temperature history decreases more rapidly at region closer to mushy zone. The thicker the inner insulation is, the more heat energy it prevents from being lost.

For periodic heating, the temperature rise and drop are both more rapid at region closer to heater element.

(4) The existence of heater gap will yield cold and hot point which both are important for deicing effect. The heater gap will make the slope of temperature history smaller, which has much less effect on deicer pad without heater gap. The reasonable distribution of heater gap can largely improve the deicing efficiency and effect.

Acknowledgments. This work is supported by China National Science Funding (No: 11572338) and The National Basic Research Program (No: 2015CB755804).

References

1. Lee S, Kim HS, Bragg MB (2013) Investigation of factors that influence iced-airfoil aerodynamics. AIAA-2000-0099
2. Cebeci T, Kafyeke F (2003) Aircraft icing. Annu Rev Fluid Mech 35:11–21
3. (2000) Aircraft icing handbook (version I), civil aviation authority, New Zealand
4. Stallabrass JR (1972) Thermal aspects of de-icer design. In: The international helicopter icing conference, Ottawa, Canada
5. Baliga G (1980) Numerical simulation of one-dimensional heat transfer in composite bodies with phase change. M.Sc. thesis. The University of Toledo, Toledo, Ohio
6. Marano JJ (1982) Numerical simulation of an electrothermal de-icer pad. M.Sc. thesis. The University of Toledo, Toledo, Ohio
7. Roelke RJ, Keith TG, Dewitt KJ, Wright WB (1988) Efficient numerical simulation of a one-dimensional electrothermal deicer pad. J Aircr 25(12):1097–1101
8. Chao DF (1983) Numerical simulation of two-dimensional heat transfer in composite bodies with application to de-icing of aircraft components. Ph.D. Dissertation. University of Toledo, Toledo, Ohio
9. Leffel KL (1986) A numerical and experimental investigation of electrothermal aircraft deicing. M.S. thesis. University of Toledo
10. Masiulaniec KC (1987) A numerical simulation of the full two-dimensional electrothermal de-icer pad. Ph.D. thesis. University of Toledo, Toledo, Ohio
11. Wright WB (1987) A comparison of numerical methods for the prediction of two-dimensional heat transfer in an electrothermal deicer pad. M.Sc. thesis. University of Toledo, Toledo, Ohio
12. Wright WB (1988) A comparison of numerical methods for the prediction of two-dimensional heat transfer in an electrothermal deicer pad. NASA-CR-4202
13. Wright WB, Keith TG, DeWitt KJ (1988) Transient two-dimensional heat transfer through a composite body with application to deicing of aircraft components. AIAA-88-0358
14. Keith TG, DeWitt KJ, Wright WB (1988) Overview of numerical codes developed for predicted electrothermal de-icing of aircraft blades. AIAA-88-0288
15. Wright WB, Keith TG, DeWitt KJ (1992) Numerical analysis of a thermal deicer. AIAA-92-0527
16. Yaslik AD, DeWitt KJ, Keith TG (1992) Further developments in three-dimensional numerical simulation of electrothermal deicing systems. AIAA-92-0528
17. Huang JR, Keith TG, DeWitt KJ (1992) An efficient finite element method for aircraft de-icing problems. AIAA-92-0532

18. Huang JR (1993) Numerical simulation of an electrothermally de-iced aircraft surface using the finite element method. Ph.D. thesis. University of Toledo, Toledo, Ohio
19. Huang JR, Keith TG, DeWitt KJ (1993) Investigation of an electrothermal de-icer pad using a three-dimensional finite element simulation. AIAA-93-0397
20. Huang JR, Keith TG, Dewitt KJ (1995) Effect of curvature in the numerical simulation of an electrothermal de-icer pad. J Aircr 32(1):84–91
21. 常士楠, 侯雅琴, 袁修干 (2007) 周期电加热控制律对除冰表面温度的影响. 航空动力学报, 22(8):1247–1251. Chang S, Hou Y, Yuan X (2007) Influence of periodic electro-heating pulse on deicing surface temperature. J Aerosp Power 22(8):1247–1251 (in Chinese)
22. 常士楠, 艾素霄, 霍西恒, 袁修干 (2008) 改进的电热除冰系统仿真. 航空动力学报, 23 (10):1753–1758. Chang S, Ai S, Huo X, Yuan X (2008) Improved simulation of electrothermal de-icing system. J Aerosp Power 23(10):1753–1758 (in Chinese)
23. 肖春华 (2010) 飞机电热除冰过程的传热特性及其影响研究. 绵阳:中国空气动力研究与发展中心. Xiao C (2010) Study on heat transfer characteristics and effects of electrothermal aircraft deicing. Mianyang: China Aerodynamics Research and Development Center (in Chinese)
24. Thomas S, Cassoni R, MacArthur C (1996) Aircraft anti-icing and deicing techniques and modeling. AIAA-96-0390
25. Voller VR, Cross M (1981) Accurate solutions of moving boundary problems using the enthalpy method. Int J Heat Mass Transf 24(3):545–556
26. Voller VR, Prakash C (1987) A fixed grid numerical modeling methodology for convection/diffusion mushy phase change problems. Int J Heat Mass Transf 30(9):1709–1719
27. Voller VR, Swaminathan CR (1991) Generalized source-based method for solidification phase change. Numer Heat Transf B 19(2):175–189
28. Kirby MS, Hansman RJ (1986) Experimental measurement of heat transfer from an iced surface during artificial and natural cloud icing conditions. AIAA-86-1352
29. Arimilli RV, Keshock EG (1984) Measurements of local convective heat transfer coefficients on ice accretion shapes. AIAA-84-0018
30. Bragg MB, Cummings MJ, Lee S, Henze CM (1996) Boundary-layer and heat-transfer measurements on an airfoil with simulated ice roughness. AIAA-96-0866
31. Dukhan N, James GVF, Masiulaniec KC, DeWitt KJ (1996) Convective heat transfer coefficients from various types of ice roughened surfaces in parallel and accelerated flow. AIAA-96-0867
32. Sherif SA, Pasumarthi N (1995) Local heat transfer and ice accretion in high speed subsonic flow over an airfoil. AIAA-95-2104
33. Robert C, Henry, Didier G, Sndre B, Gerard G (1999) Heat transfer coefficient measurement on iced airfoil in a small icing wind tunnel. AIAA-99-0372
34. Incropera FP, Dewitt DP, Bergman TL et al (2006) Fundamentals of heat and mass transfer. Wiley, Hoboken
35. Wright WB, DeWitt KJ, Keith TGJ (1991) Numerical Simulation of Icing, Deicing, and Shedding. AIAA-91-0655

The Effects of Swirl on Low Power Arcjet Thruster Flowfield and Heat Transfer Characteristics

Xin-ai Zhang[1($\boxtimes$)] and Hai-bin Tang[2]

[1] Shanghai Aircraft Design and Research Institute,
No. 5188, Road Jinke, Shanghai, China
buaa_zhangshenai@sina.com
[2] Beijing University of Aeronautics and Astronautics,
No. 37, Road Xueyuan, Beijing, China

Abstract. A steady, three-dimensional numerical model is developed for the low power arcjet thrusters. Numerical investigation was carried out to determine the effects of swirl on the flowfield and heat transfer characteristics inside the arcjet nozzle. Two different three-dimensional arcjet nozzle configurations with axial injectors and radial injectors are used in this numerical study in order to reveal the effects of swirl on the flowfield and heat transfer characteristics inside the arcjet nozzle. Comparisons of the flowfield and heat transfer characteristics in the two different three-dimensional configurations, i.e., with axial injectors and radial injectors respectively, are performed to reveal the effects of the swirl flow. The distributions of pressure, temperature, axial velocity and tangential velocity in the two different three-dimensional configurations are presented. The modeling results show that there exist evident three-dimensional effects of the flowfield in both the configuration with axial injectors and the nozzle with radial injectors. The swirl in the configuration with axial injectors is dissipated with small magnitude while it is inerratic for the case with radial injectors. The tangential velocity magnitude in the nozzle with radial injectors is big enough to persist downstream and is enhanced at the constrictor entrance. It is found that the swirl flow has evident influences on the flowfield, heat transfer characteristics inside the arcjet nozzle and little effect on electric field and thrust of the thruster when the arcjet thruster operated under the operation condition used in the study.

Keywords: Low power arcjet · Plasma flow and heat transfer · Swirl flow · Numerical study

1 Introduction

Thermal arcjet propulsion systems with high specific impulse and moderate thrust levels are playing an increasingly important role in the satellite applications, and the improvement of performance will further extend the application fields of the arcjet propulsion systems in the space propulsion. The low power hydrazine arcjets now are applied for North South Stationkeeping (NSSK) and East West Stationkeeping (EWSK) of commercial geostationary satellites [1–10]. Future applications of medium and high power arcjet propulsion systems may include low earth orbit - geosynchronous

© Springer Nature Singapore Pte Ltd. 2019
X. Zhang (Ed.): APISAT 2018, LNEE 459, pp. 1948–1968, 2019.
https://doi.org/10.1007/978-981-13-3305-7_158

(LEO - GEO) orbit transfer, repositioning, maneuvering, and drag compensation of large satellites and space stations [11]. In those applications, arcjet propulsion systems could provide significant propellant savings, launch mass reduction and longer service life in comparison with the conventional chemical propulsion systems.

From energy conservation, the I_{sp} of a propulsion system scales as: $I_{sp} \sim (T_o/M)^{1/2}$, where I_{sp}, T_o and M are the specific impulse, chamber temperature and propellant molecular weight respectively. It suggests that higher specific impulse can be achieved through increasing the chamber temperature for a certain propellant. As an electrothermal device, the arcjet thruster uses electric arc, instead of a chemical reaction, to heat the gaseous propellant. The electric arc emanates from the cathode, passes through the constrictor, and attaches to the anode along the divergent side in a diffuse mode. The gaseous propellant is injected upstream of the cathode tip, and merges with the electric arc. The gaseous propellant is heated by the arc to a temperature as high as about 20000 K, which is much higher than the chamber temperature of the conventional chemical propulsion system. Subsequently, the hot gas (fully ionized and weakly ionized plasma) is expanded through a Laval nozzle, which also forms the anode. The internal energy is partially converted in to kinetic energy of the exhausting supersonic jet and produces a thrust force. The higher chamber temperature in the arcjet thruster conduces a resultant specific impulse higher than that of the conventional chemical propulsion system.

Unfortunately, the electric arc, which is used to heat the propellant in the arcjet thruster, is inherently unstable due to the "Pinch effect". For example, if any "sausage" constriction occurs in the arc, the magnetic pinching force at the small-radius portions will enhance the constriction until the arc column becomes severed. In the same way, when a slight bend in the arc occurs, the excessive magnetic forces on the concave sides, produced by self-magnetic field, will also amplify the bend, and distort the arc toward the constrictor wall. If the constriction or the kink is not opposed by a certain stabilization mechanism, the electric arc will interrupt or shunt to the constrictor wall, affecting performance of the arcjet thruster [12, 14, 15]. As a result, it is necessary to invoke some external stabilization mechanisms to confine the intrinsic instabilities of the electric arc. From the view of space propulsion, one stabilization technique is to let electric arc pass through a "tube" which is designed to be relatively long and narrow. Such constricted arcs are found to be stable against radial kinking, and also embody sufficiently steep temperature gradients to protect the constrictor wall from melting. As a second stabilization technique, a swirling vortex of the propellant gas is used to sustain the arc. It was found by Schonherr in 1909 that the rotational motion of the injected propellant gas has multiple benefits of constraining the hot gas discharge column to the axis of the vortex, and of cooling the electrodes [12]. The two arc stabilization techniques mentioned above are widely used in the arcjet thrusters now successfully used on the satellites, such as the MR-510 hydrazine arcjet system. Although the feasibilities of the two arc stabilization techniques are approved by previous researches, the understanding on the mechanisms of the two arc stabilization techniques and the effects of swirling flow on flowfield and heat transfer characteristics are still incomplete and almost empirical.

The need for a complete description of the arc stabilization mechanisms of the swirling gas flow and the effects of swirling flow on flowfield and heat transfer charac-teristics has resulted in a variety of numerical efforts. The early modeling researches by

Neuberger and Shaeffer involved an axisymmetric model in a constant area geometry with viscous, swirling flow [13–15]. An axisymmetric steady flow model of viscous, compressible plasma in local thermodynamic equilibrium (LTE) state was developed to investigate the influence of an azimuthal velocity component on the radial and axial distributions of pressure, enthalpy, density, velocity components and electric field strength in a tube with constant cross section. The modeling results indicated that viscous effects and swirl decrease the mass flow rate and hence the thrust in a converging-diverging nozzle with the same inlet stagnation pressure and stagnation temperature. Meanwhile, it was found that the effect of swirl was small in geometries with large constrictor length/diameter ratios [13]. An axisymmetric two-dimensional model for swirling, turbulent, radiative arc heater flowfields was developed to simulate the plasma processes in a high-pressure air arc heater. The balance between the destabilizing arc column kink forces and the swirl stabilizing pressure forces was predicted [14, 15]. However, in both Neuberger and Shaeffer's researches, it was assumed that the radial velocity were small compared to the axial and swirl velocities, and the axial gradients were negligible in comparison to the radial gradients, which is termed the weak swirl approximation. As a result of the approximation, a great deal of the physics is neglected in the assumption of weakly swirling flow [16]. An axisymmetric 2-D model in local thermodynamic equilibrium (LTE) state incorporating the effects of swirl and viscosity for low power arcjet thrusters was developed by Pawlas. Compared to unswirled flows, the effects of swirl flows on the arcjet thruster performance were determined. It was found that swirl included in the model caused a decrease in the mass flow rate, and a resultant decrease in the thrust of the arcjet thruster with the same inlet pressure. Meanwhile, the effects of swirl were most strongly felt in the subsonic portion of the nozzle. However, it was also found that the assumed inlet swirl profile and maximum swirl angle greatly affects the resulting flowfield. It was necessary to measure the accurate inlet swirl profile experimentally to obtain a precise representation of the flowfield in the arcjet [16–18]. An axisymmetric 2-D model was devised to model the plasma processes in arcjet thrusters. The effects of injected swirl on both the cold flow and the plasma flow inside a nominal 30 kW arcjet were investigated using this model. In the case of cold flow (mass flow rates on the order of mg/s), it is found that the injected swirl persists downstream and is enhanced at the constrictor entrance, near the cathode tip. The results also show that mass flow rate and inlet plenum length are key design parameters in a swirl stabilized arcjet thruster [19]. More recently, several models were developed to study the flowfield and to predict the performance of the arcjet thrusters. Although the azimuthal velocity component is incorporated in some models to account for the swirl, few discuss about the effects of the swirling gas flows on the flowfield and heat transfer characteristics inside the arcjet nozzle were mentioned [20–22]. Meanwhile, previous simulation researches all treat the models as two-dimensional and axisymmetric. The inlet swirl velocity profiles specified along the inlet are artificial, which could not represent the real flow conditions at the inlet, and the assumed inlet swirl profiles and maximum swirl angle greatly affects the resulting flowfield. To this point, a real three-dimensional model is needed to clarify the influence of the vortex injection. Fortunately, the developments of the computer and commercial software make it available to model the plasma processes in a real 3D configuration.

The objective of this paper is to determine the effects of the swirling gas flow on the flowfield and heat transfer characteristics within the arcjet thruster nozzle.

A three-dimensional numerical model incorporating the effects of viscous dissipation, ohmic heating, heat conduction, radiation loss and pressure work is developed. In order to reveal the effects of the swirl injection, comparisons of the flowfield and heat transfer characteristics in two different three-dimensional configurations, i.e., with axial injectors and radial injectors respectively, are carried out. The distributions of pressure, temperature, axial velocity and tangential velocity in the two different three-dimensional configurations are presented.

2   Arcjet Physical Model

The Fluent is a software designed for modeling fluid flow and heat transfer. Meanwhile, the user-defined function (UDF) and user-defined scalar (UDS) provided in the Fluent allow one user to customize the Fluent to enhance the standard features of the code [23]. In this modeling study, a steady, three-dimensional numerical model incorporating the effects of viscous dissipation, Lorentz force, ohmic heating, heat conduction, radiation loss and pressure work is developed for the low power arcjet thrusters. The source terms, i.e. the Lorentz force, ohmic heating, net radiation loss etc., were added to the momentum and energy conservation equations by using the user-defined function (UDF) to model the basic plasma physics processes taking place inside the arcjet nozzle. Also, the user-defined function (UDF) is used to specify the boundary conditions and to allow the thermodynamic and transport properties of the plasma to be functions of the local gas temperature and pressure. The Navier-Stokes equations and electromagnetic equations are solved by the density-based coupled solution algorithm. The coupled solver in Fluent solves the governing equations of continuity, momentum, and energy simultaneously as a set of equations, and the governing equations for additional scalars are solved sequentially [23]. The validation of the model presented in this study have been approved by other previous modeling studies on the free burning arc configuration and the plasma torch [24–27].

2.1   Assumptions

The following assumptions are employed to simulate the plasma processes in the low power arcjet thruster:

(1)   The flow inside the arcjet nozzle is three-dimensional, steady, laminar, continuum, viscous and compressible.
(2)   The plasma is assumed to be in the local thermodynamic equilibrium (LTE) state and thus the thermodynamic and transport properties of nitrogen-hydrogen plasma are the function of local gas temperature and pressure.
(3)   The plasma is optically thin.
(4)   The Lorentz force components are included in the momentum equations, and the Joule heating, radiation loss, viscous dissipation and pressure work are included in the energy equation.
(5)   The induced electric field is negligible in comparison with the electric field.
(6)   The effects of gravity are neglected.

2.2   Governing Equations

Based on the foregoing assumptions as mentioned above, in the 3-D Cartesian coordinate system (x, y, z), the governing equations in the finite volume method to simulate the 3-D plasma arc flow inside the arcjet nozzle can be written as the general form:

$$\frac{\partial}{\partial x}(\rho v_x \Phi) + \frac{\partial}{\partial y}(\rho v_y \Phi) + \frac{\partial}{\partial z}(\rho v_z \Phi) = \frac{\partial}{\partial x}\left(\Gamma_\Phi \frac{\partial \Phi}{\partial x}\right) + \frac{\partial}{\partial y}\left(\Gamma_\Phi \frac{\partial \Phi}{\partial y}\right) + \frac{\partial}{\partial z}\left(\Gamma_\Phi \frac{\partial \Phi}{\partial z}\right) + S_\Phi \quad (1)$$

Table 1. Equations in three dimensions

Conservation equations	Φ	Γ_Φ	S_Φ
Mass (2)	1	0	0
x momentum (3)	v_x	μ	$-\frac{\partial p}{\partial x} + \frac{\partial}{\partial y}\left(\mu\frac{\partial v_y}{\partial x}\right) + \frac{\partial}{\partial z}\left(\mu\frac{\partial v_z}{\partial x}\right) + \frac{1}{3}\frac{\partial}{\partial x}\left(\mu\frac{\partial v_x}{\partial x}\right)$ $-\frac{2}{3}\frac{\partial}{\partial x}\left(\mu\left(\frac{\partial v_y}{\partial y} + \frac{\partial v_z}{\partial z}\right)\right) + j_y B_z - j_z B_y$
y momentum (4)	v_y	μ	$-\frac{\partial p}{\partial y} + \frac{\partial}{\partial z}\left(\mu\frac{\partial v_z}{\partial y}\right) + \frac{\partial}{\partial x}\left(\mu\frac{\partial v_x}{\partial y}\right) + \frac{1}{3}\frac{\partial}{\partial y}\left(\mu\frac{\partial v_y}{\partial y}\right)$ $-\frac{2}{3}\frac{\partial}{\partial y}\left(\mu\left(\frac{\partial v_x}{\partial x} + \frac{\partial v_z}{\partial z}\right)\right) + j_z B_x - j_x B_z$
z momentum (5)	v_z	μ	$-\frac{\partial p}{\partial z} + \frac{\partial}{\partial x}\left(\mu\frac{\partial v_x}{\partial z}\right) + \frac{\partial}{\partial y}\left(\mu\frac{\partial v_y}{\partial z}\right) + \frac{1}{3}\frac{\partial}{\partial z}\left(\mu\frac{\partial v_z}{\partial z}\right)$ $-\frac{2}{3}\frac{\partial}{\partial z}\left(\mu\left(\frac{\partial v_x}{\partial x} + \frac{\partial v_y}{\partial y}\right)\right) + j_x B_y - j_y B_x$
Energy (6)	T	λ	$v_x\frac{\partial p}{\partial x} + v_y\frac{\partial p}{\partial y} + v_z\frac{\partial p}{\partial z} + \mu\left\{2\left[\left(\frac{\partial v_x}{\partial x}\right)^2 + \left(\frac{\partial v_y}{\partial y}\right)^2 + \left(\frac{\partial v_z}{\partial z}\right)^2\right]\right.$ $\left[\frac{\partial v_y}{\partial x} + \frac{\partial v_x}{\partial y}\right]^2 + \left[\frac{\partial v_z}{\partial y} + \frac{\partial v_y}{\partial z}\right]^2 + \left[\frac{\partial v_x}{\partial z} + \frac{\partial v_z}{\partial x}\right]^2 - \frac{2}{3}\left[\frac{\partial v_x}{\partial x} + \frac{\partial v_y}{\partial y} + \frac{\partial v_z}{\partial z}\right]^2\right\}$ $+ \frac{j_x^2 + j_y^2 + j_z^2}{\sigma} + \frac{5}{2}\frac{k_B}{e}\left(j_x\frac{\partial T}{\partial x} + j_y\frac{\partial T}{\partial y} + j_z\frac{\partial T}{\partial z}\right) - U_r$
Potential (7)	V	σ	0
x potential vector (8)	A_x	1	$\mu_0 j_x$
y potential vector (9)	A_y	1	$\mu_0 j_y$
z potential vector (10)	A_z	1	$\mu_0 j_z$

In Table 1, ρ is the flow mass density, v_x, v_y and v_z are the x, y, and z components of the velocity. Φ presents the scalar variable that must be solved in the various conservation equations, while Γ_Φ and S_Φ are the diffusion coefficient and the source term, respectively. For each equation, the scalar variable (Φ), diffusion coefficient (Γ_Φ) and source term (S_Φ) are tabulated in Table 1. p, T and V are the gas static pressure, the temperature and the electric potential, A_x, A_y and A_z are, x, y, and z components of the potential vector, B_x, B_y and B_z are, x, y, and z components of the magnetic field, j_x, j_y and j_z are, x, y, and z current density vector components, respectively. μ, λ, σ, and U_r

are the viscosity, the thermal conductivity, the electric conductivity and the volumetric radiation power loss, which are the function of the local gas temperature and pressure. The material parameters that used in the equations were tabulated as functions of the local gas temperature and pressure. The thermodynamic and transport properties of the simulated hydrazine (N_2+2H_2) plasma at a temperature range of 300–29900 K (with 200 K temperature intervals) at eleven different pressure levels are used in this modeling study, which are calculated by Murphy A. B. [28]. k_B, μ_o and e are the Boltzmann constant, magnetic permeability and the elementary charge.

The electric potential equation is solved to determine the electric field, while the potential vectors equations are solved to estimate the magnetic field. In Eqs. (7)–(10), the convective terms are set to be zero. The current density vector components are deduced from the electric potential according to the Ohm's law:

$$\vec{j} = \sigma(-\vec{\nabla}V) \tag{11}$$

$$j_x = -\sigma\frac{\partial V}{\partial x}, \ j_y = -\sigma\frac{\partial V}{\partial y} \text{ and } j_z = -\sigma\frac{\partial V}{\partial z} \tag{12}$$

The magnetic induction vector components are deduced from the potential vectors and can be written as:

$$\nabla^2\vec{A} = -\mu_o\vec{j} \tag{13}$$

$$\vec{B} = \vec{\nabla} \times \vec{A} \tag{14}$$

$$B_x = \frac{\partial A_z}{\partial y} - \frac{\partial A_y}{\partial z}, B_y = \frac{\partial A_x}{\partial z} - \frac{\partial A_z}{\partial x} \text{ and } B_z = \frac{\partial A_y}{\partial x} - \frac{\partial A_x}{\partial y} \tag{15}$$

The plasma in DC electric arc is assumed to be in the local thermodynamic equilibrium (LTE) state in many previous modeling studies. Simultaneity, the thermodynamic and transport properties become the function of the gas temperature and pressure [24–27]. The gas temperature near the anode wall is relatively low. As a result, it underpredicts the plasma electrical conductivity and further affects current distribution and the local electrical power dissipation. In previous study, Bulter and Rhodes found that reasonable modeling results could be obtained by specifying a artificial electrical conductivity [29, 30]. In Reference 29, a minimum value of the electrical conductivity specified according to different propellant was used to compensate the underprediction of the electrical conductivity. In Reference 30, the electrical conductivity was defined as (in the near anode region when the gas temperature is lower than 10000 K):

$$\sigma = f\sigma_f + (1-f)\sigma_{eq} \tag{16}$$

Where σ is the electrical conductivity used in the calculation, f is a wall pressure dependent weighting factor, σ_f is a 'frozen' electrical conductivity, σ_{eq} is the local electrical conductivity and determined by the local gas temperature and pressure. Equation 17 is used to calculate the wall pressure dependent weighting factor f, where the p_c is constrictor pressure, and p_n is the exit plane pressure. σ_f is calculated by Eq. 18, where σ_{10000} is the gas electrical conductivity at 10000 K for the LTE plasma. 10 S/m is chosen as the minimum electrical conductivity.

$$f = (p_c - p)/(p_c - p_n) \tag{17}$$

$$\sigma_f = \frac{(\sigma_{10000} - 10)}{10000} + 10 \tag{18}$$

The approach mentioned in Reference 30 is employed in this model. In the present modeling of a 1 kW arcjet, the predicted arc voltage is smaller than the measured arc voltage data if the critical temperature for electrical conductivity modification is set to be 10000 K. In this study, the critical temperature for electrical conductivity modification is set to be 6500 K. As a result, the predicted arc voltage could agree with the arc voltage value measured in the corresponding operation condition test.

2.3   Calculation Domain and Boundary Conditions

The nozzle geometry is identical to the geometry of the 1 kW hydrazine Engineering Design Model (EDM) arcjet thruster [31, 32]. The convergent side of the nozzle is conical and has a half angle of 30°, while the divergent or the downstream side of the nozzle is also conical and but has a 20° half angle. The constrictor is 0.7 mm in diameter and 0.5 mm in length. The cathode is made from 2% thoriated tungsten rod of 3 mm diameter with a tip of 30° half angle. In the computational domain, a plane with a diameter of 0.1 mm is located on the cathode tip in order to apply the electric field boundary condition. The insulators are made of boron nitride. The arc gap is set to be 0.4 mm. In order to determine the effects of swirl on the flowfield, heat transfer characteristics inside the arcjet nozzle, two different three-dimensional configurations are used in this numerical study. The schematic of the arcjet thrusters different injectors are shown in Fig. 1. As shown in Fig. 1(a), there are four radial injectors located at the head of the nozzle which is designed to provide the flow with an azimuthal component of velocity. The axis of the radial injectors is tangent with the anode wall and normal to the axis of the nozzle. In the arcjet thruster with axial propellant injection configuration, there are four axial injectors located at the head of the nozzle as shown in Fig. 1(b). The lengths and diameters of the radial and axial injectors are identical, i.e. 0.2 mm in diameter and 1.5 mm in length. In order to avoid the affects of the assumed inlet velocity profile on modeling results, no assumption of the inlet swirl profile and swirl angle is used at the inlet of the injectors. The velocity vector is set to be normal to the inlet surface of the injector. It is believed that the treatment of the inlet velocity profile would make it available to present the actual flowfield properly. With the consideration of the large computer memory and computational effort required in three-dimensional arcjet modeling, one quarter of the two different arcjet nozzle configurations are chosen as the computational domains. Hybrid grids composed of 37822 cells and 25168 nodes and 36953 cells and 23101 nodes are used to discretize the computational domains. The mesh at the injector, constrictor and near the anode wall, whose dimensions are smaller than the other dimensions of the arcjet nozzle and where steep temperature and pressure gradients exist, are refined.

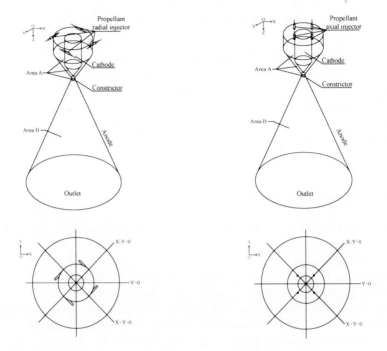

(a) Arcjet thruster with four radial injectors   (b) Arcjet thruster with four axial injectors

Fig. 1. Schematic of the arcjet thruster different injectors

The boundary conditions used in the modeling of the two configurations are listed in Table 2. In order to compare the flowfield and heat transfer characteristics inside the two configurations, the propellant mass flow rates at the radial and axial injector inlet surfaces are set to be the same value. The propellant gas enters the computational domain at a uniform temperature of 600 K and the velocity vectors are both set to be normal to the injector inlet surfaces. The temperature distribution of the convergent and constrictor sector of the nozzle wall (Area A in Fig. 1) follows the Eq. 19, and the temperature distribution of the divergent nozzle wall (i.e. Area B in Fig. 1) follows the Eq. 20, i.e. the anode temperature is held at 1000 K at the head of the nozzle and ramped linearly to 1600 K until the exit of the constrictor, and then decreased linearly to 1000 K at the nozzle exit plane. The cathode wall temperature is set to be 3500 K. The current density imposed on cathode tip plane is calculated according to a given arc current. The zero potential condition is imposed on divergent section of the nozzle, while it is assumed the current density is zero at other section of the anode. The temperature and velocity at the nozzle exit plane are calculated by extrapolating the interior flowfield.

$$T(z) = 1000 + (600 \times z)/0.00896\,(K) \tag{19}$$

$$T(z) = 1600 - 600 \times (z - 0.00896)/0.02239\,(K) \tag{20}$$

Table 2. Boundary conditions for the arcjet model

	v	T	V	A
Inlet	Mass flow rate (mg/s)	600 K	$\partial V/\partial n = 0$	$\partial A/\partial n = 0$
Outlet	$\partial v/\partial n = 0$	$\partial T/\partial z = 0$	$\partial V/\partial n = 0$	$\partial A/\partial n = 0$
Cathode	0	3500 K	$\partial V/\partial n = 0$	$\partial A/\partial n = 0$
Cathode tip plane	0	3500 K	$\sigma(\partial V/\partial n) = -j$	$\partial A/\partial n = 0$
Area A	0	Equation (17)	$\partial V/\partial n = 0$	$\partial A/\partial n = 0$
Area B	0	Equation (18)	0	$\partial A/\partial n = 0$

3   Results and Discussion

The numerical modeling of the two different arcjet configurations, i.e. with axial injectors and radial injectors respectively, at the input power level of 1 kW class were conducted. The simulated hydrazine (N_2+2H_2) was chosen as the propellant. The same propellant mass flow rate and arc current i.e. simulated hydrazine of 24 mg/s and arc current of 10 A, were imposed in two arcjet configurations so that the effects of swirling flow on the flowfield and heat transfer characteristics inside the two configurations could be revealed and compared under the same operation condition.

Characteristics of the flowfield and the heat transfer inside the arcjet nozzle with four axial injectors are presented in Fig. 2. Figure 2(a) shows the temperature distribution inside the thruster nozzle. The gaseous propellant enters the thruster nozzle through the injectors at a temperature of 600 K and then the temperature is increased in response of the heating by the electrodes. The cathode temperature is set to be 3500 K which is higher than the anode temperature. As a result, the gas temperature close to the cathode is higher than that of the gas near the anode wall. The gas temperature is increased under the effect of the arc heating in the constrictor. The maximum temperature of 17843 K occurs at the center of the constrictor near the cathode tip. The anode wall temperatures are on the order of about 1000 K which is much lower than the temperature in the core flow region. Consequently, the temperature distribution inside the nozzle exhibits evident temperature gradients in the radial direction. Especially, the dimension of the constrictor is so small that the temperature gradient in radial direction in the constrictor is larger. The steep temperature gradients in radial direction would protect the constrictor wall from melting effectively. In the axial direction, the temperature gradually falls with downstream distance owing to the conversion of gas internal energy into kinetic energy in the divergent nozzle. The pressure, velocity magnitude and density contours are shown in Fig. 2(b), (c) and (d). The gas pressure decreases from 260900 to about 218000 Pa when the gaseous propellant passes through the narrow axial injectors. As shown in Fig. 2(c), the propellant gas is rapidly accelerated when the gas passes through the constrictor which results in steep pressure fall in the constrictor. In the divergent part of the nozzle, the propellant gas is expanded through the divergent nozzle to be supersonic. Correspondingly, the gas pressure falls continuously in the axial direction due to the conversion of the pressure energy into kinetic energy. The maximum velocity of about 9481 m/s appears at the centerline

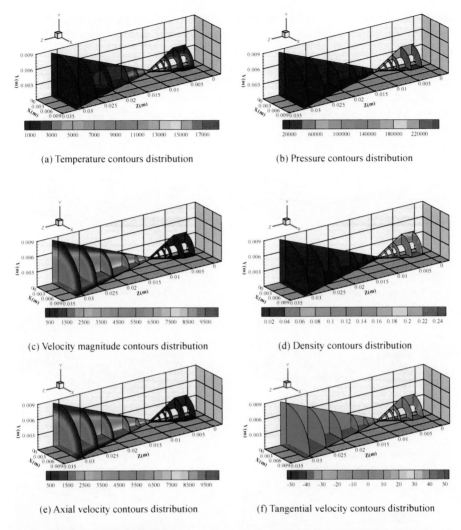

(a) Temperature contours distribution

(b) Pressure contours distribution

(c) Velocity magnitude contours distribution

(d) Density contours distribution

(e) Axial velocity contours distribution

(f) Tangential velocity contours distribution

Fig. 2. Typical modeling results of the arcjet with four axial injectors operated with 24 mg/s simulated hydrazine and $I = 10$ A

downstream constrictor exit instead of at the nozzle exit. It is believed that the decrease of the velocity at the nozzle centerline is contributed to the effects of viscosity. The density contours follows the pressure and temperature distributions. At the injector section, the gas pressure is high and the temperature is low relatively, and the gas density is high. According to the temperature distribution in the constrictor and divergent section of the nozzle, the gas density in the core flow is much lower than that near the anode wall region for a given axial location. The axial and tangential velocity contours inside the nozzle are illustrated in Fig. 2(e) and (f). As seen in Fig. 2(e), the axial velocity distribution is similar to the velocity magnitude distribution. Evident three-dimensional effects of the computed contours exist in the subsonic section of the nozzle due to four axial injectors' locations. The pressure, density and axial velocity at the injector ports are relatively high, while the pressure at the interspaces between the injectors is relatively low. Backflow of the gas occurs inside the interspaces at the section in front of the constrictor due to the interesting pressure distribution. Correspondingly, the axial velocity vector towards the inlet and tangential velocity appear at the subsonic section of the nozzle, which locates in front of the constrictor. But the magnitudes of the backflow velocity and tangential velocity are relatively low.

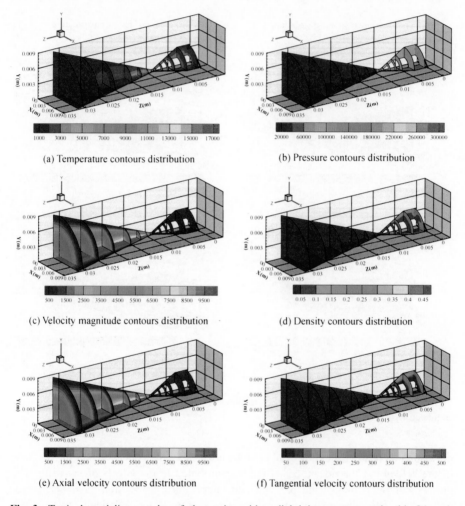

(a) Temperature contours distribution

(b) Pressure contours distribution

(c) Velocity magnitude contours distribution

(d) Density contours distribution

(e) Axial velocity contours distribution

(f) Tangential velocity contours distribution

Fig. 3. Typical modeling results of the arcjet with radial injectors operated with 24 mg/s simulated hydrazine and $I = 10$ A

The computed flowfield contours inside the arcjet nozzle configuration with four radial injectors operated with 24 mg/s simulated hydrazine when the arc current is set to be 10 A are presented in Fig. 3. Figure 3(a) shows the temperature contours inside the nozzle configuration with four radial injectors. The gas temperature is increased from 600 to about 1000 K when the propellant passes through the radial injector. There exists an annular relatively low temperature zone around the cathode root due to the frame of the four radial injectors. The temperature decreases from the cathode temperature to the anode temperature on account of different temperature imposed on the cathode and anode wall. Meanwhile, the use of radial injector makes the temperature distribution inside the nozzle (especially in the subsonic section of the nozzle) differ from that for the case with axial injectors. The gas temperature increases to the maximum value of 17938 K in response of the arc heating, which is about 100 K higher than the maximum temperature for the case with axial injector. Similarly, the temperature gradually decreases in the axial direction as a result of the conversion of gas internal energy into kinetic energy in the divergent nozzle. Figure 3(b), (c) and (d) illustrate the pressure, velocity magnitude and density distribution inside the arcjet nozzle with four radial injectors. It should be noticed that the maximum pressure at the radial injector inlet is about 300600 Pa which is higher than that for the case with axial injectors. Meanwhile, the gas pressure decreases from 300600 to about 221070 Pa when the gaseous propellant passes through the narrow radial injector. It is found in previous studies that the swirl would decrease the mass flow rate with the same inlet pressure and temperature [13, 16]. In another word, the decreased mass flow rate can be offset by increasing the inlet pressure. Compared with the arcjet nozzle configuration with axial injectors, a higher inlet pressure is needed in the configuration with radial injectors to maintain the identical propellant mass flow rate imposed at the injector inlet. It can be seen in Fig. 3(c) that the velocity magnitude contours is similar with that for the case with axial injectors. But the maximum velocity is 9502 m/s which is a little higher than that for the case with axial injectors. It can be seen in Fig. 3(d) that the gas density is high at the injector section where the gas pressure is high and the temperature is low relatively. As mentioned above, the propellant is heated by the electric arc in the constrictor. The hot gas flow is expanded in the divergent nozzle. As a result, the axial velocity increases when the internal energy is converted into kinetic energy as seen in Fig. 3(e). Figure 3(f) shows the tangential velocity contours inside the arcjet nozzle with radial injectors. The axis of the radial injector is tangent with the anode wall and normal to the axis of the nozzle which allows the gas propellant to enter the nozzle with tangential velocity. Meanwhile, the swirl distribution and the swirl magnitude are only depend on the structure of the actual configuration, and can accurately present the real flowfield in the arcjet nozzle with radial injectors, which is helpful to avoid the affects of artificial assumed inlet swirl profile on the modeling results. The maximum tangential velocity of about 500 m/s appears the outlet of the injector. It also can be seen in Fig. 3(f) that the swirl are most strongly felt in the subsonic portion of the nozzle, and are enhanced somewhat at the constrictor entrance.

Figures 4 and 5 show the axial velocity and tangential velocity contours at different axial locations inside the two different nozzle configurations in order to visualize the axial and tangential velocity distributions throughout the nozzles. In the figures, the head surface of the nozzle chamber is set to be the starting point of z axis. The axial velocity profiles on several cross-sections at different axial locations in the nozzle configurations with both axial injectors and radial injectors are shown in Fig. 4. The maximum axial velocity on each cross-section is also given in the figures.

Three-dimensional effect due to the frame of the injectors is seen both in the nozzle with four axial injectors and the nozzle with four radial injectors. In the arcjet nozzle configuration with four axial injectors, the axial velocity is relatively high at the injection port as mentioned above and can persists downstream for a distance. In the interspaces between the injection ports where the gas pressure is somewhat lower, and resultant backflow exists as seen in Fig. 4. The axial velocity magnitude becomes uniform with the increased axial distance in response of the diffusion of momentum. As a result, the maximum axial velocity for a certain cross-section decreases with the increase of the axial distance in the annular channel in front of the constrictor. The gas propellant is accelerated when passes through the constrictor and the maximum axial velocity of the whole flowfield is achieved at the centerline downstream constrictor exit instead of at the nozzle exit. In the configuration with four radial injectors, the gas propellant is injected out of the radial injectors with relatively high tangential but low axial velocity. The axial velocity in the nozzle with radial injectors is smaller than that in the nozzle with axial injectors for a given axial location in the subsonic section of the nozzle. It should be noticed in the constrictor and divergent section of the nozzle that the axial velocity at the centerline of the nozzle in the configuration with radial injectors is appreciably higher than that in the in the configuration with axial injectors. This trend is also found in previous numerical studies [16, 33].

Figure 5 shows the tangential velocity contours and velocity vector on different cross-sections with various axial locations. Swirl exists in the subsonic section of the nozzle in both the configuration with axial injectors and the configuration with radial injectors. It can be seen that the swirl is dissipated in the configuration with axial injectors while it is inerratic for the case with radial injectors. The swirl is most strongly felt in the subsonic portion of the nozzle. From energy conservation, the enhancement or decay of swirl can be explained from simple considerations of conservation of angular momentum. Thus the efflux of angular momentum will equal the total influx minus the total torque due to viscous forces for any location downstream in the nozzle. The magnitude of the tangential velocity in the nozzle with axial injectors is much lower than that with radial injectors. The intensity of swirl in the nozzle with axial injectors is so small that the effect of the swirl can hardly be felt in the constrictor and divergent section of the nozzle. In contrast, the gas propellant is injected out of the injectors with tangential velocity of 500 m/s in the configuration with radial injectors. The swirl is big enough to persist till the constrictor exit (i.e. $z = 0.00896$ m) even though there exists decay of the angular momentum. Meanwhile, the tangential velocity at a cross-section of a nozzle varies inversely to area of the cross-section. Although the tangential velocity decays with the increase of the axial distance, the swirl is enhanced appreciably in the constrictor whose radius is the smallest, i.e. the area of the constrictor across-section is the smallest. With the consideration of viscous dissipation, the enhanced tangential velocity at the constrictor is still smaller than the maximum tangential velocity at the head of the nozzle plenum, as seen in Figs. 3(f) and 5. The tangential velocity component of gas flow can provide a stabilizing force by establishing a radial pressure gradient, which will oppose any kinking occurring in the arc column [12, 16]. The relatively higher tangential velocity at the constrictor section in the configuration with radial injectors will confine the intrinsic instabilities of the electric arc.

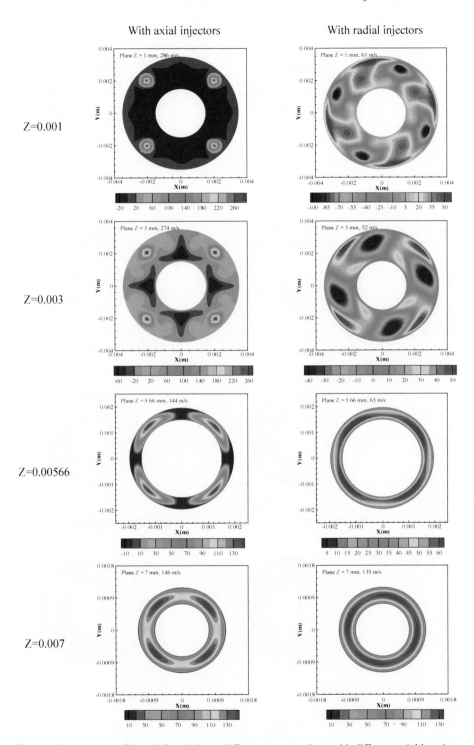

Fig. 4. Comparisons of the axial velocity at different cross-sections with different axial locations in the arcjet nozzle with different injectors operated with 24 mg/s simulated hydrazine and $I = 10$ A

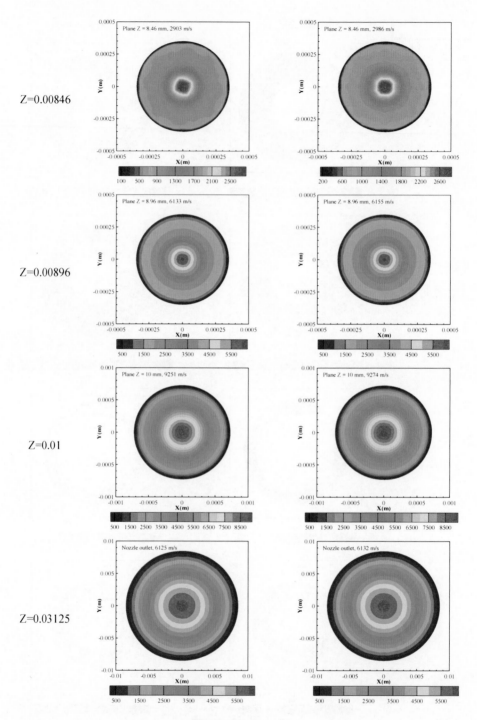

Fig. 4. (*continued*)

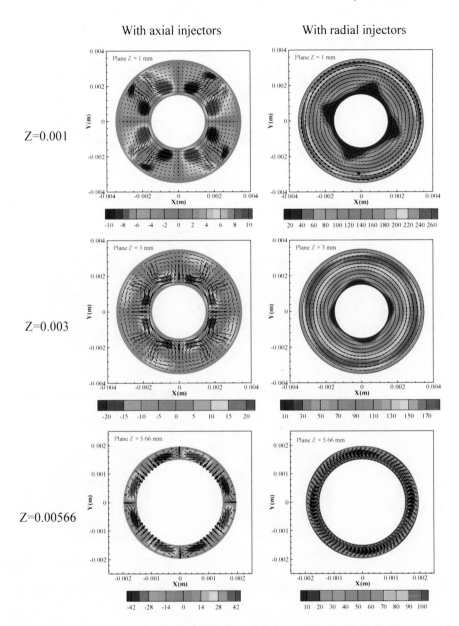

Fig. 5. Comparisons of the tangential velocity at different cross-sections with different axial locations in the arcjet nozzle with different injectors operated with 24 mg/s simulated hydrazine and $I = 10$ A

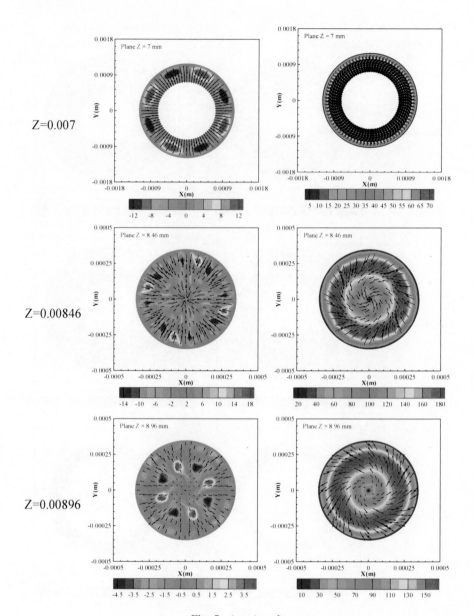

Fig. 5. (*continued*)

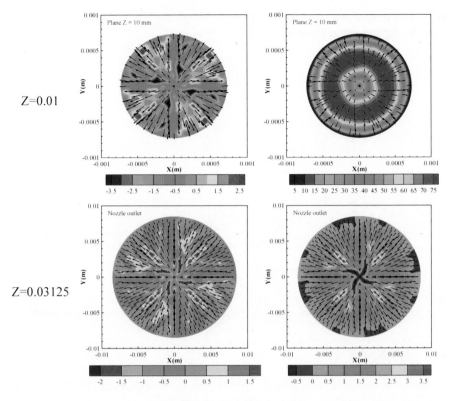

Z=0.01

Z=0.03125

Fig. 5. (*continued*)

The swirl number is introduced in order to quantify the intense of the swirl, which is defined as [14, 16]:

$$S = \frac{\int_{rc}^{ra} \rho v_x w r^2 dr}{\bar{R} \int_{rc}^{ra} \rho v_x^2 r dr} \tag{21}$$

Where S is the swirl number, r_a is the radius of anode inner wall, r_c is the radius of cathode, the $\bar{R}$ is the hydraulic radius of the cross-section, and w is the tangential velocity. The swirl numbers at cross-sections with different axial locations for the arcjet nozzle with both axial injectors and radial injectors are plotted in Fig. 6. The swirl numbers at different cross-sections for the case with axial injectors are much smaller than those for the case with radial injectors, which is contributed the small tangential velocity magnitude in the nozzle configuration with axial injectors. The swirl number profiles in the two cases show the similar trend that the swirl number appreciably increases until the gas flow enters the convergent portion of the nozzle, and decreases in the constrictor and divergent sections of the nozzle. The gas propellant is continuously accelerated in the convergent, constrictor and divergent portions of the nozzle. Meanwhile, the magnitude of the tangential velocity decreased with the increase of the radius of the nozzle as mentioned above. Consequently, the swirl numbers at cross-sections in the divergent portion of the nozzle is small.

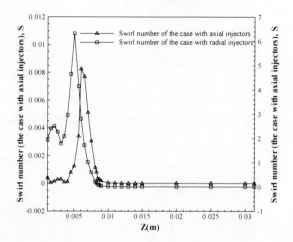

Fig. 6. Variations of the swirl number inside the arcjet nozzles with different injectors

4 Conclusions

Numerical investigation was conducted to determine the effects of swirl on the flowfield and heat transfer characteristics inside the arcjet nozzle. A steady, three-dimensional numerical model incorporating the effects of viscous dissipation, Lorentz force, ohmic heating, heat conduction, radiation loss and pressure work is developed for the low power arcjet thrusters. Two different three-dimensional arcjet nozzle configurations with axial injectors and axial injectors are used in this numerical study so that the effects of swirl on the flowfield, heat transfer characteristics inside the arcjet nozzle could be revealed. Typical modeling results of both the arcjet with axial injectors and the arcjet with radial injectors are presented.

Evident three-dimensional effects of the flowfield exist in both the configuration with axial injectors and the nozzle with radial injectors, especially in the subsonic section of nozzle. Although swirl exists in two configurations, the swirl is dissipated in the configuration with axial injectors while it is inerratic for the case with radial injectors. The tangential velocity magnitude in the nozzle with radial injectors is higher than that in the nozzle with axial injectors. Consequently, the injected swirl in the nozzle with radial injectors can persist downstream and is enhanced at the constrictor entrance, while the dissipated swirl in the in the nozzle with axial injectors can hardly be felt in the constrictor. The relatively higher tangential velocity at the constrictor section in the configuration with radial injectors is helpful to confine the intrinsic instabilities of the electric arc. In the divergent section of the nozzle, the tangential velocity magnitude is smaller in comparison with the axial velocity which results in the relatively low swirl intense.

The modeling results show that the swirl flow has evident influence on the flow-field, heat transfer characteristics inside the arcjet nozzle. The relatively more intensive swirl in the nozzle with radial injectors need a higher inlet pressure to maintain the identical propellant mass flow rate imposed at the injector inlet. The inlet pressure in the nozzle with radial injectors is 15% higher than that for the case with axial injectors. In the constrictor and divergent section of the nozzle, the axial velocity at the centerline of the nozzle in the configuration with radial injectors is appreciably higher than that in the in the configuration with axial injectors. The maximum temperature for the case with radial injectors is 17938 K which is about 100 K higher than the maximum temperature for the case with axial injector. It is found that different swirl intense in two nozzle configurations has little effect on electric field and thrust of the thruster when the arcjet thruster operated under the operation condition used in the study. The potential for the case with radial injectors is 95.80 V while the potential for the case with axial injectors is 95.84 V.

References

1. Oleson SR, Myers RM, Kluever CA, Riehl JP, Curran FM (1997) Advance propulsion for geostationary orbit insertion and north-south station keeping. J Spacecr Rocket 34(1):22–28
2. Curran FM, Hagg TW (1992) Extended life and performance test of a low-power arcjet. J Spacecr Rocket 29(4):444–452
3. Lichon PG, Sankovic JM (1996) Development and demonstration of a 600-second mission-average Isp arcjet. J Propuls Power 12(6):1018–1025
4. Smith RD, Yano SE, Armbruster K, Roberts CR, Lichtin DA, Beck JW (1993) Flight qualification of a 1.8 kW hydrazine arcjet system. In: International electric propulsion conference. IEPC-93-007
5. Smith RD, Roberts CR, Aadland RS, Lichtin DA, Davies K.(1997) Flight qualification of the 1.8 kW MR-509 hydrazine arcjet system. In: International electric propulsion conference. IEPC-97-081
6. Smith RD, Aadland RS, Roberts CR, Lichtin DA (1997) Flight qualification of the 2.2 kW MR-510 hydrazine arcjet system. In: International electric propulsion conference. IEPC-97-082
7. Lichtin DA, Chilelli NV, Henderson JB, Rauscher RA, Young KJ, Mckinnon DV, Bailey JA, Roberts CR, Zube DM, Fisher JR (2009) AMC-1 (GE-1) arcjet at 12-plus years on-orbit. AIAA-2009-5364
8. Auweter-Kurtz M, Glocker B, Golz T, Kurtz HL, Messerschmid W, Riehle M, Zube DM (1996) Arcjet thruster development. J Propuls Power 12(6):1077–1083
9. Messerschmid EW, Zube DM, Meinzer K, Kurtz HL (1996) Arcjet development for amateur radio satellite. J Spacecr Rocket 33(1):86–91
10. Bock D, Auweter-Kurtz M, Kurtz HL (2005) 1 kW ammonia arcjet system development for a science mission to the moon. In: International electric propulsion conference. IEPC-2005-075
11. Auweter-Kurtz M, Golz T, Habiger H, Hammer F (1997) 100 kW class hydrogen arcjet thruster. In: International electric propulsion conference. IEPC-97-007
12. Robert GJ (1968) Physics of electric propulsion. McGRAW-HILL, New York
13. Neuberger AW (1975) Thermo-gasdynamical and electrical behavior of the wall and vortex stabilized arc. Translation of DLR-FB-75-38, European Space-Agency, ESA-TF-220

14. Schaeffer JF (1978) Swirl arc, a model for swirling, turbulent, radiative arc heater flowfields. AIAA-78-68
15. Shaeffer JF (1978) Swirl arc: a model for swirling, turbulent, radiative arc heater flowfields. AIAA J 16(10):1068–1075
16. Pawlas GE (1991) Numerical modeling of a vortex stabilized arcjet. Ph.D. thesis, University of Toledo
17. Pawlas GE, Keith TG (1989) Numerical modeling of a vortex stabilized arcjet thruster. AIAA-89-2724
18. Pawlas GE, Keith TG (1991) Analysis of a swirl stabilized arcjet thruster. AIAA-1991-463
19. Babu V, Aithal S, Subramaniam VV (1993) On the effects of swirl in arcjet thruster flows. IEPC 93-183
20. Miller SA, Martinez-Sanchez M (1996) Two-fluid nonequilibrium simulation of hydrogen arcjet thrusters. J Propuls Power 12(1):112–119
21. Megli TW, Lu J, Krier H, Burton RL (1998) Modeling plasma processes in 1-kilowatt hydrazine arcjet thrusters. J Propuls Power 14(1):29–36
22. Fujita K, Arakawa Y (1999) Performance computation of a low-power hydrogen arcjet. J Propuls Power 15(1):144–150
23. ANSYS Inc. (2009) ANSYS Fluent User's Manual
24. Freton P, Gonzalez JJ, Gleizes A (2000) Comparison between a two- and a three-dimensional arc plasma. J Phys D Appl Phys 2000(33):2442–2452
25. Lago F, Gonzalez JJ, Freton P et al (2004) A numerical modelling of an electric arc and its interaction with the anode: part I. the two-dimensional model. J Phys D Appl Phys 37 (2004):883–897
26. Gonzalez JJ, Lago F, Freton PM et al (2005) Numerical modelling of an electric arc and its interaction with the anode: part II. The three-dimensional model influence of external forces on the arc column. J Phys D Appl Phys 2005(38):306–318
27. Lago F, Gonzalez JJ, Freton P et al (2006) A numerical modelling of an electric arc and its interaction with the anode: part III. Application to the interaction of a lightning strike and an aircraft in flight. J Phys D Appl Phys 2006(39):2294–2310
28. Murphy AB, Arundell CJ (1994) Transport coefficients of argon, nitrogen, oxygen, argon-nitrogen, and argon-oxygen plasmas. Plasma Chem Plasma Process 14(4):451–490
29. Bulter GW, King DQ (1992) Single and two fluid simulations of arcjet performance. AIAA Paper 92-3104
30. Rhodes RP, Keefer D (1990) Numerical modeling of an arcjet thruster. AIAA Paper 90-2614
31. Tang HB, Zhang XA, Liu Y, Wang HX, Shi CB, Cai B Experimental Study of startup characteristics and performance of a low-power arcjet. J Propuls Power. https://doi.org/10.2514/1.47380
32. Tang HB, Zhang XA, Liu Y, Wang HX, Shen Y Performance and preliminary life test of a low power hydrazine engineering design model arcjet. Aerosp Sci Technol https://doi.org/10.1016/j.ast.2010.12.001
33. Dutton JC (1987) Swirling supersonic nozzle flow. J Propuls Power 3(4):342–349

Investigation of Influence of Magnet Thickness on Performance of Cusped Field Thruster via Multi-objective Design Optimization

Suk H. Yeo[(✉)] and Hideaki Ogawa

School of Engineering, RMIT University, Melbourne, VIC 3001, Australia
suk.hyun.yeo@student.rmit.edu.au

Abstract. The cusped field thruster (CFT) is a class of advanced electric propulsion (EP) technology for satellite and space missions, offering advantages over other types of EP including enhanced electron confinement owing to the magnetic mirror and reduced particle loss effects at the dielectric wall. The increasing demand for downscaling for micro-satellite class platforms while keeping performance at similar level has led to considerable efforts dedicated to physical modeling and performance characterization of downsized CFT. Multi-objective design optimization is conducted in this study by employing performance parameters of downscaled CFT, namely, thrust, total efficiency, and specific impulse as the objective functions to maximize and design parameters including anode voltage and current, mass flow rate, and inner and outer magnet radii as the decision variables. Two geometric configurations are considered, *i.e.*, those comprising three magnets with fixed thickness and four magnets with variable thickness to gain insights into the influence of magnet thickness on the performance. Considerable effects of magnet thickness on the performance have been found, including thrust increase of up to approximately 20% and increase in specific impulse by up to approximately 10%, as compared to the configuration with fixed thickness magnets.

Keywords: Electric propulsion · Cusped field thruster ·
Multi-objective design optimization

Nomenclature

B	= magnetic field [T]
B_0	= low field region [T]
B_m	= high field region [T]
q	= elementary charge [C]
I_a	= anode current [A]
I_{sp}	= specific impulse [s]
M	= particle mass [kg]
m_a	= anode mass flow [sccm]
N	= population size
P	= power [W]
S_i	= first-order sensitivity index
S_{Ti}	= total-effect sensitivity index

© Springer Nature Singapore Pte Ltd. 2019
X. Zhang (Ed.): APISAT 2018, LNEE 459, pp. 1969–1989, 2019.
https://doi.org/10.1007/978-981-13-3305-7_159

T = thrust [mN]
v = velocity [m/s]
μ = magnetic moment
η_t = total efficiency
η_u = mass utilization efficiency
η_b = beam efficiency
U_a = anode potential [V]

1   Introduction

In-space electric propulsion (EP) technologies have been studied and developed over many years for spacecraft propulsion for scientific missions and satellites station keeping in orbit due to the performance benefits of electric propulsion such as a longer operational lifetime, better fuel efficiency and less weight [10]. Generally, EP increases the propellant exhaust velocity to achieve thrust with high ΔV. Most common and well known types of EP are grid ion thruster (GIT) and Hall-effect thruster (HET) offering a longer lifetime over 10,000 h, and higher specific impulse I_{sp} of 1600–6000 s, but relatively lower thruster values of 30–230 mN, as compared to chemical propulsions [2, 25].

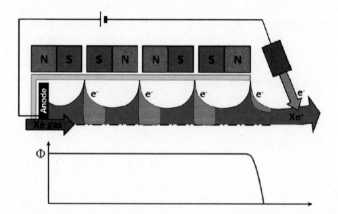

Fig. 1. Ionization region (orange) and potential profile of the thruster [16].

The High-Efficiency Multistage Plasma Thruster (HEMP-T) was firstly developed by Thales Electron Devices (TED) in 1999, and a similar thruster design known as cusped field thruster (CFT) was developed by Harbin Institute of Technology [15, 18]. These technologies use a series of ring-shaped permanent periodic magnets (PPM) to confine the plasma beam through the magnetic mirror effect. These magnets are aligned co-axially along the chamber with reversing polarity to produce the magnetic mirror effect. Cusped regions are formed at between the states of PPM where the magnetic field runs radially from the chamber wall, so the electrons oscillate on Larmor radii to increase the interaction length and to reduce wall erosion and to yield high ion beam efficiencies in the range of 80–90% [17]. This effectively confines the electrons to the center of the engine that in turn helps to electro-statically confine the ions, upon which the magnetic field has little effect because the mean free path of the ion is significantly smaller than its gyro-radius and is not considered magnetized [6]. The electrons emitted from the cathode form the cloud and they are confined at the exit cusp that acts as a virtual acceleration grid, like the GIT, resulting in high efficiencies and ions close to the potential of the anode so the thruster features a steep potential drop occurring after the exit cusp while the plasma potential is constant throughout the engine, as schematically presented in Fig. 1 [9, 16, 17, 23].

A miniature concept of CFT would bring further advantages over current full-scaled EP thrusters in terms of weight reduction and fuel consumption but a preceding experimental investigation of down-scaled EP revealed significantly low performance due to the complexity of CFT system [14]. The main problem arises in the complex interaction of plasma beam with the magnetic field, anode current, anode power, mass flow rate and geometric considerations. A state-of-the-art framework has been developed by coupling an analytical global-variable based steady-state power and plasma model solver and multi-objective design optimization (MDO) capability based on evolutionary algorithms, aiming to achieve such optimal design requirements for down-scaled EP [7]. This methodology allows robust and efficient population-based global search in the design space, assisted by surrogate modeling that reduces the computational cost effectively by approximating the model calculation in lieu of expensive computational evaluation [7]. This would subsequently result in saving on launch cost for satellites applications [7]. The effect of the thickness of magnets is not well understood, as little research has been conducted to examine its influence, except for a few studies such as an investigation conducted on the effect of the thickness ratio of magnets on the performance of downscaled CFT by keeping the maximum possible thickness in the middle magnet [18].

This paper presents the results and insights obtained from an MDO study that has been conducted to simultaneously maximize three objectives, *i.e.*, thrust T, total efficiency η_t, and specific impulse I_{sp} of CFT with two different magnetic configurations; (1) three magnets with fixed thickness; and (2) three magnets with variable thickness. Representative and selected magnet geometries are examined to investigate the key design factors and underlying physics. Surrogate models are trained using the archive of solutions evaluated in the course of optimization. A large-scale optimization solely employing the surrogate prediction has been performed additionally to better capture the effects of the design parameters on the CFT performance.

2 Methodology

2.1 Performance Characterization

The basic principle of EP is accelerating mass and ejecting it from the vehicle at higher
exhaust velocity than the other propulsion systems. Due to operational similarities of
HET and CFT, the basic relation of performance parametric such as thrust T, specific
impulse I_{sp}, power P_a and ideal power to thrust ratio $PTTR$ can be defined as follows
[10, 11]:

$$T = v\dot{m}_p = qE \tag{1}$$

$$I_{sp} = \frac{T}{\dot{m}_p g} \tag{2}$$

$$P_a = \frac{1}{2}\dot{m}_p v^2 = U_a I_a \tag{3}$$

$$PTTE = \frac{P}{T} = \frac{U_a I_a}{I_a\sqrt{2\frac{M}{q}U_a}} = \sqrt{\frac{qU_a}{2M}} \tag{4}$$

where E is the potential difference in an electric field, U_a is the anode voltage, I_a is the
anode current, M is the propellant molecular mass (2.18×10^{-25} kg for Xe), and q is
the elementary charge (1.602×10^{-19}C).

The anode efficiency η_a can be derived and accurately calculated using voltage
efficiency, beam efficiency (η_b), utilization efficiency (η_u) the coefficient of plume
divergence and effective divergence angle (θ_{eff}) and the beam current (I_b) can be
determined by the mass flow rate [10].

$$I_b = \frac{q\dot{m}_a}{M} \tag{5}$$

$$\eta_v = \frac{U_b}{U_a} \tag{6}$$

$$\eta_b = \frac{P_b}{P_a} = \frac{I_b}{I_a} \tag{7}$$

$$\eta_u = \frac{I_b}{\dot{m}_p q/M} \tag{8}$$

$$\eta_a = \frac{T^2}{2\dot{m}_p I_b U_a} = \frac{I_b}{I_a} \tag{9}$$

$$\eta_a = \frac{1}{Q^2} \eta_v \eta_b \eta_u \cos^2 \theta_{eff} \tag{10}$$

where the subscript b indicates the ion beam, Q is the average ionic charge, I_b is the total ion current in the beam, U_b is the total ion voltage in the beam, and $\cos^2 \theta_{eff}$ accounts for plume divergence where ions are not accelerated parallel to the engine axis. The propellant is assumed to be Xenon, which is typically used for CFT. Equation (10) is a basic equation that assumes 100% ionization and singly charged ions for thruster, while it would not fully describe low ionization at low voltages and multiply charged ions at high voltages, which could occur in practical applications.

The mass utilization efficiency, described in part of Eq. (11), is sensitive to the beam current and mass flow. As a result, this study assumes that the ion beam current can be sufficiently approximated by Eq. (5). The mass utilization efficiency correction factor to consider the effect of multiply charged ions is given by α_m in Eq. (12) [10]

$$\eta_m = \frac{\dot{m}_a}{\dot{m}_p} = \frac{I_b}{q} \frac{M}{\dot{m}_p} \tag{11}$$

$$\eta_m = \alpha_m \frac{I_b}{q} \frac{M}{\dot{m}_p} \tag{12}$$

$$\alpha_m = \frac{1 + \frac{1 I^{++}}{2 I^+}}{1 + \frac{I^{++}}{I^+}} \tag{13}$$

The present study does not consider the effects of multiple ion species on the correction factor. However, the effect of 20% doubly charged ions in the mass utilization efficiency ($\alpha_m = 0.9$) and other assumptions made regarding the acceleration, divergence and utilization efficiencies are taken into account in the post-processing phase. This is because while these factors have uniform effects on the objective parameters, they are difficult to be determined accurately with the present methodology [7].

A simplified power balance description of HEMP-T based on plasma fluid theory yields a one-dimensional set of equations (28 in total), which can be solved simultaneously to allow for reasonable estimate of the thruster performance [17]. As the only known values in the equation set are the probabilities to reach the channel wall at the cusp locations and these probabilities are based on the magnetic field strength, the thruster performance can be estimated through only a few parameters. These are anode potential, anode current and the ratio of magnetic field strength from the axially aligned region where the fields radially cross the discharge channel walls, i.e., the magnetic mirror strength. It is also important to note that the ratios of power transferred to excitation, ionization, and thermalization are only estimations, and a full description of the power model can be found in Ref. [17].

CFT operation is mainly characterized by the PPMs that are used to create magnetic mirror effect to reduce electrons losses due to impingement on the walls, as prescribed by the Lorentz force equation below in Eq. (14). It follows that the magnetic field does not have impact on the particle directly, but it exerts longitudinal axial force F_z when its strength increases in the opposite direction of its motion with constant kinetic energy K [8–10, 12].

$$F_z = \frac{mv_\perp^2}{2B} \nabla_\parallel B \tag{14}$$

where $v_\perp$ is the cyclotron motion of the particle in terms of a magnetic moment.

$$F_k = -\mu \nabla_k B \tag{15}$$

The magnetic moment is constant by equating and balancing the magnetic moment at the high and low field regions when the particle is moving through a magnetic field of increasing strength:

$$\frac{v_{\perp 0}^2}{B_0} = \frac{v_{\perp m}^2}{B_m} \tag{16}$$

where subscripts 0 and m refer to low and high field regions, respectively.

The equation of velocity of the particle can be solved by conservation of the particle kinetic energy:

$$K = \frac{1}{2}m\left(v_\parallel^2 + v_\perp^2\right) \tag{17}$$

Therefore,

$$v_\parallel = \left[\frac{2}{m}(K - \mu B)\right]^{\frac{1}{2}} \tag{18}$$

To ensure magnet mirror effect, the vector of velocity is required to be within acceptable angle, and the equation can be solved using the conservation of kinetic energy and derived as follows [12]:

$$\frac{B_m}{B_0} = \frac{v_{\parallel 0}^2 + v_{\perp 0}^2}{v_{\perp 0}^2} = \frac{1}{\sin^2 \theta_m} \tag{19}$$

Therefore,

$$\theta_m \leq \sin^{-1}\left(\frac{B_0}{B_m}\right) \tag{20}$$

Arrival probabilities of the electrons at the cusp region can be determined as [17]

$$P_c = \frac{2\pi \int_0^{a_c} \sin\theta d\theta}{4\pi} \tag{21}$$

The cusp arrival probabilities are directly related to the accuracy of the simulated magnetic topologies of the thruster, which is calculated using two-dimensional electromagnetic field analysis.

Variable magnet length cusped field thruster (VML-CFT) from Ma et al. [18] and another CFT design from Hu et al. [13] are used to verify this approach as it considered the physical dimensions and material properties for the thruster and the subsequent analysis of the results provides a robust range of observation points that can be modeled and compared (readers are referred to Ref. [7] for verification).

2.2 Performance Characterization

MDO is performed in a chain process consisting of sequential phases, as displayed in Fig. 2 [7]; (1) The decision variables are examined in the pre-processing phase to assure the geometry is physically viable prior to simulation; (2) ANSYS Maxwell [1] constructs a model for the given geometry and calculates the magnetic field by means of magnetostatic analysis, and then electrostatic analysis is performed to calculate the potential at the thruster exit and in the plume. The data is then extracted from the magnetic topology to compute the cusp arrival probability for each location throughout the thruster; (3) These conditions are subsequently passed to be used in the power balance calculation; (4) The resultant solutions from this are post-processed to deliver the objectives and assess if they lie within the set of physical constraints; (5) They are then submitted to the MDO algorithms for evaluation. The evolutionary algorithm, in particular, the elitist non-dominated sorting genetic algorithm assisted by surrogate modeling, is employed for a population-based optimization approach, where the members in the population pool evolve over generations [5]. This iterative cycle continues, yielding more designs to be evaluated according to the set criteria.

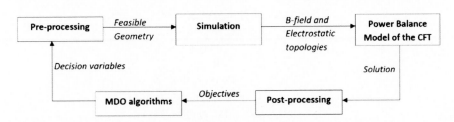

Fig. 2. Multi-objective design optimization (MDO) process chain [7]

A population size of $N = 64$ is used in this study to be evolved over 50 generations for the initial MDO study and that of $N = 100$ over 100 generations in the MDO run performed solely based on surrogate prediction. These values have been chosen to sufficiently explore the design space for a three-objective design problem with five decision variables within reasonable computational effort. Recombination operators are applied to the previous generation's decision variable values to create offspring. A simulated binary crossover and polynomial mutation are used as recombination operators at a given probability (1.0 and 0.1, respectively, in this study) with a specified distribution index (10 and 20, respectively) [21]. The use of a strongly elitist non-dominated sorting genetic algorithm always retains the best solutions across generations.

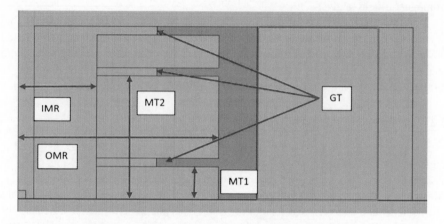

Fig. 3. Magnet configurations of simplified thruster model

Three objective functions are considered and employed to evaluate design performance, namely thrust T, total efficiency η_t, and specific impulse I_{sp}. Total efficiency η_t is comprised of the measures of efficiencies within the thruster model, that is, the beam efficiency η_b, the mass utilization efficiency η_m, and the voltage efficiency η_v (Table 1).

Table 1. Optimization problem

Maximize:	T, η_t, I_{sp}
Subject to:	$0 \leq U_a \text{ (V)} \leq 1000$
	$0 \leq I_a \text{ (A)} \leq 10$
	$0.2 \leq m_a(\text{sccm}) \leq 50$
	$2 \leq \text{IMR(mm)} \leq 15$
	$2 \leq \text{OMR(mm)} \leq 28$
	$2 \leq \text{MT1(mm)} \leq 15$
	$2 \leq \text{MT2(mm)} \leq 15$
	$0.5 \leq \text{GT(mm)} \leq 2$

The schematic diagram of magnet configuration of the simplified thruster model presented in Fig. 3. The decision variables chosen to represent the main design factors investigated in this study are U_a (V), I_a (A), $\dot{m}_a$ (sccm), inner magnetic radius IMR (mm) and outer magnetic radius OMR (mm), first magnet thickness MT1 (mm), second magnet thickness MT2 (mm) and gap spacing, GT (mm). The negative sign for the objective functions denotes a maximization problem (converted from a minimization problem). The decision variables U_a (V), I_a (A) and $\dot{m}_a$ (sccm) relate to the objective functions through Eqs. (1)–(10) and the initial conditions of the CFT at the anode. The magnetic radii, IMR and OMR are related to the objective functions through the one-dimensional simplified power balance model.

Variance-based global sensitivity analysis is performed to examine the influence of each decision variable, x_i as input (i.e., design parameters) on the objective function y as output (i.e., performance parameters). A numerical procedure is employed to derive the sensitivity indices, facilitated by surrogate modeling [20]. Input matrices $\mathbf{X}$ of a base sample quantity of 10,000 and multiple columns for the decision variables are built by using quasi-random numbers within the range for each variable [24]. Output vectors $\mathbf{Y}$ are obtained by forwarding the input matrices to the surrogate model that is of the greatest prediction accuracy. The first-order indices S_i and total-effect indices S_{Ti} in Eqs. (22) and (23) are calculated by the method described in Ref. [22].

$$S_i = V[E(Y|X_i)]/V(Y) \tag{22}$$

$$S_{Ti} = 1 - V[E(Y|X_{-i})]/V(Y) \tag{23}$$

A simplified two-dimensional CFT model is used for the calculations using ANSYS Maxwell to mitigate the computational load, which would otherwise be expensive due to the nature of the population-based MDO process and to facilitate the identification of the relationships between the output objectives and the decision variables. This model assuming axis symmetry at the engine axis consists of a consistently straight chamber made of BN Ceramic, three Samarium-Cobalt (SmCo) 27 megagauss-oersteds (MGOe) magnets with spacers made of pure iron (due to high conductivity and to improve commonality with other CFT design) and the thruster housing made of Al 6061T6 [4, 15, 16, 26]. Geometric constraints are applied to restrict the scope and output of the design space, including geometries (inner magnetic radius *IMR* and outer magnetic radius *OMR*. Designs that overlap are deemed infeasible solutions in MDO. The upper limit is assumed within the limits of the standard CubeSat design [7].

3   Results

Figures 4a and b present the results obtained from the MDO runs for the configurations with fixed and variable thickness of magnets, respectively. The green points indicate the feasible geometries, the blue points represent non-dominated solutions and the red points indicate the infeasible solutions that have not satisfied all feasibility criteria imposed by constraint functions. Discrete Pareto optimal fronts can be seen, indicative of the convergence achieved as a result of evolutionary optimization. Presented in Figs. 5a and b are parallel coordinate plots that visualize the trends and relations among the decision variables and objective and constraint functions for the non-dominated solutions identified by the MDO.

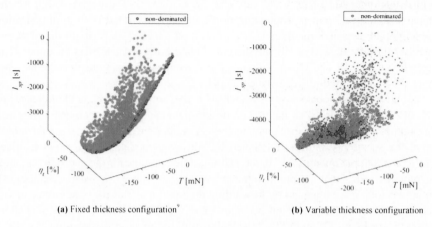

(a) Fixed thickness configuration[9]   **(b)** Variable thickness configuration

Fig. 4. Optimization results at the 100th generation from MDO with surrogate prediction

Table 2 shows the representative solutions, primarily non-dominated hence optimal design points that have been selected from the Pareto optimal front (Figs. 4a and b) that resulted from the MDO performed for design configurations comprising magnets of fixed and variable thickness, respectively, except S_4, which is a suboptimal solution selected for comparison, as described below.

The first selected solutions S_1 feature the largest thrust T among all designs for both configurations. They are characterized by decision variables of mass flow and anode potential at their upper limits. The second selected points S_2 attain the highest measured total efficiency η_t of the evaluated feasible designs for both configurations, while they produce the lowest specific impulse (albeit similar in values), as compared to their non-dominated peers, $i.e.$, S_1 and S_3. The third selected solutions S_3 have the highest specific impulse I_{sp} of all evaluated designs with the lowest mass flow rate. They achieve appreciably high I_{sp} as a trade-off with thrust and efficiency. In comparison to their non-dominated peers, they incur approximately 90mN reduction in thrust for the fixed magnet thickness configuration, whereas no thrust penalty is incurred for the variable magnet thickness configuration. Further, reduction of about 40%–50% in efficiency is commonly observed for S_3 for both configurations. A suboptimal solution has been selected as the fourth design point for the fixed thickness configuration, as it represents similar influence of design as S_2 of the fixed thickness configuration. For the variable thickness configuration, an optimal solution O_1 has been selected additionally as the fourth design point, which possesses similar characteristics in design variables to all selected points, particularly S_2, but offers balanced performance in all criteria.

Tables 3 and 4 display the first-order indices S_i and the total-effect indices S_{Ti} identified by the covariance-based sensitivity analysis performed for both magnetic configurations. They quantitatively indicate the main and overall effects of the input parameters ($i.e.$, decision variables) namely U_a, I_a, $\dot{m}_a$, IMR, OMR, MT1, MT2 and GT on the output parameters ($i.e.$, objective functions) namely thrust T, total efficiency η_t, and specific impulse I_{sp}. The difference between the total-effect index S_{Ti} and the first-order index S_i is indicative of the degree of the influence of the decision variable in combination with other decision variables (*i.e.*, interactions) on the objective functions [24]. If the decision variables are characterized by the sum of both first-order and total-effect indices being near unity (*i.e.*, $\sum S_i \approx 1$ and $\sum S_{Ti} \approx 1$), it follows that the effects of individual decision variables are linearly additive. Tables 3 and 4 indicate the influence of design parameters on the design objectives for both design configurations. The most influential design variable on specific impulse I_{sp} for the fixed thickness configuration is anode current I_a, while the anode potential U_a exerts the dominant influence on thrust T and specific impulse I_{sp} for the variable thickness configuration.

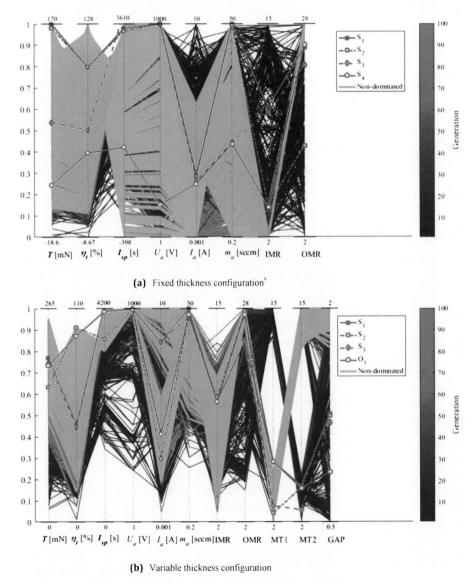

(a) Fixed thickness configuration[9]

(b) Variable thickness configuration

Fig. 5. Parallel coordinate plots at the 100th generation from MDO with surrogate prediction

The considerably large difference between the summations of S_i and S_{Ti} is indicative of highly nonlinear behavior between the design variables and output parameters of the fixed magnet configuration.

Table 2. Selected design configurations of three magnets

Fixed thickness solutions [7]	T (mN)	η_t (%)	I_{sp} (s)	U_a (V)	I_a (A)	$\dot{m}_a$ (sccm)	IMR (mm)	OMR (mm)	MT1 (mm)	MT2 (mm)	GT (mm)
S_1	**169.9**	99.6	3526	999.9	2.94	49.98	9.91	25.10	4.00	10.0	1.00
S_2	166.2	**99.9**	3469	998.4	2.83	49.69	3.91	13.13	4.00	10.0	1.00
S_3	82.9	59.8	**3825**	997.8	2.61	22.51	9.59	22.97	4.00	10.0	1.00
S_4	27.4	45.2	1291	153.3	2.49	21.98	3.83	25.45	4.00	10.0	1.00
Variable thickness solutions	T (mN)	η_t (%)	I_{sp} (s)	U_a (V)	I_a (A)	$\dot{m}_a$ (sccm)	IMR (mm)	OMR (mm)	MT1 (mm)	MT2 (mm)	GT (mm)
S_1	**203.9**	50.4	4233	997.2	8.41	49.98	9.30	24.05	5.65	4.20	1.26
S_2	167.0	**99.9**	3604	996.8	2.96	48.07	9.61	27.69	2.53	4.10	1.25
S_3	196.3	48.6	**4240**	997.2	8.41	48.02	9.30	23.95	2.98	2.72	1.20
O_1	173.3	99.9	3596	997.8	3.06	49.98	9.30	24.12	5.28	4.18	1.35

Table 3. First-order sensitivity indices of influence of design parameters on performance parameters.

Fixed thickness solutions [7]	U_a	I_a	$\dot{m}_a$	IMR	OMR	MT1	MT2	GT	SUM
T	0.015	0.084	0.050	0.018	0.072	–	–	–	0.239
η_t	0.007	0.051	0.026	0.042	0.153	–	–	–	0.279
I_{sp}	0.009	0.097	0.034	0.036	0.167	–	–	–	0.343
Variable thickness solutions	U_a	I_a	$\dot{m}_a$	IMR	OMR	MT1	MT2	GT	SUM
T	0.912	0.004	0.053	0.002	0.003	0.003	0.002	0.003	0.982
η_t	0.001	0.520	0.425	0.013	0.003	0.001	0.003	0.001	0.967
I_{sp}	0.818	0.028	0.009	0.026	0.012	0.010	0.056	0.006	0.965

Table 4. Total-effect sensitivity indices of influence of design parameters on performance parameters.

Fixed thickness solutions [7]	U_a	I_a	$\dot{m}_a$	IMR	OMR	MT1	MT2	GT	SUM
T	0.111	0.597	0.430	0.433	0.565	–	–	–	2.136
η_t	0.165	0.549	0.328	0.363	0.702	–	–	–	2.107
I_{sp}	0.089	0.644	0.308	0.409	0.600	–	–	–	2.050
Variable thickness solutions	U_a	I_a	$\dot{m}_a$	IMR	OMR	MT1	MT2	GT	SUM
T	0.943	0.008	0.061	0.008	0.013	0.012	0.007	0.011	1.063
η_t	0.005	0.546	0.447	0.020	0.006	0.003	0.014	0.003	1.044
I_{sp}	0.825	0.051	0.008	0.028	0.033	0.010	0.063	0.002	1.020

Examination of the decision variables in relation to the output parameters in Figs. 6, 7, 8 and 9 reveals several interesting relationships. The results from both MDO studies indicate a trend toward maximizing the anode potential (1000 V) in Fig. 6 and mass flow rate (50sccm) in Fig. 8. Expansion of the upper limit of the anode potential and mass flow rate is required to explore an optimal values of anode potential and mass flow rate and identify the starting point of adverse effects on the objective functions. The anode current presented in Fig. 7a converges towards 2.5–3.0A for fixed thickness configuration while anode current exerts large influence on efficiency performance of the variable thickness configuration. 2.5–4.0A in Fig. 7b offers a higher efficiency similar to that for the fixed configuration but this is at the expense of thrust performance. Figure 9 presents the optimal ranges for IMR being 3–4 mm and 9–10 mm for the fixed thickness configuration, and 9–11 mm for the variable thickness configuration. These IMR ranges would suggest the optimal ratio of magnetic strength at the cusp regions and in the far field as lower *IMR* would result in higher magnetic field strength due to calculation of cusp arrival probabilities as per Eq. (21). OMR has similar influence as IMR on the performance, as expected from the sensitivity analysis (Tables 3 and 4), The solution ranges of anode current I_a, inner magnet radius *IMR*, and outer magnet radius *OMR* for high-performance solutions were found to be approximately 2–3A (Fig. 7), 9–10 mm (Fig. 9), and 22 mm–25 mm, respectively, for the variable thickness configuration.

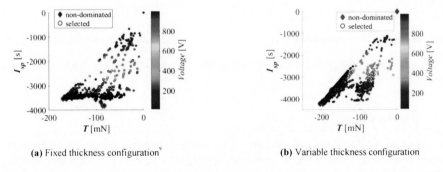

(a) Fixed thickness configuration[9]   **(b)** Variable thickness configuration

Fig. 6. Anode potential color map of the feasible solutions with respect to thrust T and specific impulse I_{sp}

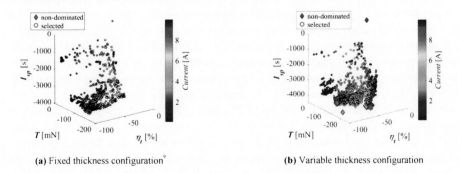

(a) Fixed thickness configuration[9]   **(b)** Variable thickness configuration

Fig. 7. Anode current color map of the feasible solutions with respect to efficiency η_t, thrust T and specific impulse I_{sp}

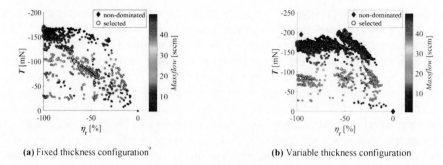

(a) Fixed thickness configuration[9]

(b) Variable thickness configuration

Fig. 8. Mass flow rate color map of the feasible solutions with respect to efficiency η_t and thrust T

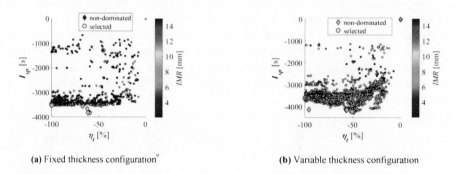

(a) Fixed thickness configuration[9]

(b) Variable thickness configuration

Fig. 9. Inner Magnet Radii color map of the feasible solutions with respect to efficiency η_t and specific impulse I_{sp}

3.1   Verification

To verify the MDO results and validate the approach in the present study, a model of the experimental setup of the CFT thruster from the available literature is examined by using ANSYS Maxwell [1]. The selected experimental model for the fixed thickness configuration is the VML-CFT from Ma et al. [18], and the CFT design from Hu et al. [13] is used for the variable thickness configuration. These models are considered because of similarities in the design parameters and their experimental results, which will serve as useful guidelines to validate the results based on simulation. Firstly, the magnetic field strength is calculated using the magnetostatic solver to determine the cusps region and the magnetic mirror ratio at each cusp. This is then used to calculate the cusp arrival probabilities at each magnetic cusp, while their accuracy is difficult to assess due to the lack of available reference data. From literature [17], it is expected that the cusp probabilities can differ significantly in experimental probing of thrusters. The second analysis performed is electrostatic simulation, which is employed to calculate the potential drop in the plume region. Using the electrostatic analysis results, the difference between the cusp potentials at the plume cell and the thruster exit can be predicted more accurately in line with observations made in the literature. The simplified HEMP-T model by Kornfeld et al. [17]

tended to over-predict the potential in the plume region, which was assumed to be responsible for the inaccuracy in the calculation of the grid efficiency and acceleration efficiency as well as the over-prediction of the beam current I_b. Likewise, the potential at the exit of the engine is over-predicted in the present methodology, where particle losses are not taken into account. In this verification study, therefore, hypothetical assumptions are made for the efficiencies so as to account for particle losses, mass utilization, acceleration and plume divergence efficiencies, based on the experimental studies by Keller et al. [15], Ma et al. [18] and Matlock [19], in conjunction with Eqs. (5)–(13); a correction factor of $\alpha_m = 0.9$ is used for the mass utilization efficiency as per Eq. (13), based on an assumption of 20% doubly charged ions [7] and; the divergence angle is assumed to be 60° to calculate the plume divergence efficiency [15], which presents a combined acceleration and divergence efficiency of 40.7% (i.e., $\eta = \eta_{acc} \cdot \eta_{div} = 0.407$), based on the acceleration efficiency extrapolated from the Faraday measurement values [15]. The loss in acceleration efficiency is not accounted for in this study, as it requires significant CFT measurement to determine and thus beyond the scope of this study, but it consequently yields more conservative estimation for the performance parameters.

Table 5 shows the comparison between the results from Ma et al. [18] using the magnetic cell length ratios of 2:9:0.5 and the analytical study for the three magnet configuration with fixed thickness, employing the hypothetical assumptions described above. Table 6 compares the result from Hu et al. [13] using the magnetic cell length ratios of 1:2:7 (except for specific impulse, which was not presented) and that from the present analytical study using the three magnet configuration with variable thickness. These specific experimental models represented the peak performance characteristics of the thruster in terms of thrust, specific impulse and efficiency. The performance parameters are a little lower, but comparable to the values measured in the experimental studies. The thrust is measured by means of two different methods from Kornfeld et al. [17] and Keller et al. [15] and calculated using Eq. (10). These comparisons indicate reasonable agreement and consistent tendencies, verifying the validity and effectiveness of the present methodology used for optimization studies.

Table 5. Comparison of the results for fixed thickness configuration with experiment at mass flow rate of 30sccm

Anode potential (V)	Ma et al. [18]			Fixed thickness			Thrust error (%)
	Total efficiency (%)	Specific impulse (s)	Thrust (mN)	Total efficiency (%)	Specific impulse (s)	Thrust (mN)	
100	12	486	14.2	15.0	525	15.2	7.1
200	25	1116	32.1	19.4	958	27.7	13.7
300	30	1565	45.2	23.5	1367	39.5	12.5
400	32	1909	55.1	24.7	1657	47.9	13.0
500	35	2290	66.3	25.5	1921	55.6	16.1

Table 6. Comparison of the results for variable thickness configuration with experiment at mass flow rate of 10 sccm

Anode potential (V)	Hu et al. [13]			Variable thickness			Thrust error (%)
	Total efficiency (%)	Specific impulse (s)	Thrust (mN)	Total efficiency (%)	Specific impulse (s)	Thrust (mN)	
300	25	N/A	9.6	24	956	9.2	4.0
400	28	N/A	11.6	24	1115	10.7	7.3
500	29	N/A	13.4	23	1251	12.0	10.0
600	29	N/A	14.8	25	1439	13.2	6.3
700	30	N/A	16.3	24	1513	14.3	10.5
800	30	N/A	17.7	24	1633	15.5	10.8

Table 7. Adjusted performance of selected design configurations taking acceleration, divergence efficiencies and multiply charged ions into account

Fixed thickness [7]	Solution	T (mN)	η_t (%)	I_{sp} (s)
	S_1	**102.7**	36.5	2131
	S_2	100.5	**36.6**	2098
	S_3	50.2	21.6	**2313**
	S_4	16.5	16.5	778
Variable thickness	Solution	T (mN)	η_t (%)	I_{sp} (s)
	S_1	**123.4**	18.5	2562
	S_2	101.1	36.6	2181
	S_3	118.8	17.8	**2566**
	O_1	104.9	36.6	2176

Table 7 presents adjusted performance for the selected points displayed in Table 2 by applying the correction factors based on the aforementioned assumptions. The upper limit of the maximum thrust, T and specific impulse I_{sp} have been improved by approximately 20% in S_1 and 10% in S_3 with the variable thickness configuration. Approximately 4% performance improvement has also been observed in thrust and specific impulse with the variable configurations.

3.2   Discussion

Magnetic field strength topologies calculated by magnetostatic analysis of ANSYS Maxwell is presented in Figs. 10, 11 and 12. The magnetic field strengths of the fixed thickness configuration are very similar on the whole as they commonly have a larger cusped region at the mid-magnet location. On the other hand, the largest cusped regions are observed at the exit of the thruster for the variable thickness configuration. The higher cusped field at the exit is advantageous to keep the plasma at the center of the thruster at the exit and the plume region, which has led to performance improvements of variable thickness configuration in thrust T and specific impulse I_{sp}. Figure 11 compares the magnetic field strength of S_2 solutions, which have attained the highest efficiency. This indicates the importance of the strength of the third magnet in terms of cusped field for ion confinement, because higher performance of thrust T and specific impulse I_{sp} has been observed for variable thickness configuration (Fig. 11b) while the fixed magnet config- uration has formed the strongest cusped fields near the chamfer of the thruster at mid- magnet to confine the plasma in the middle of the thruster (Fig. 11a).

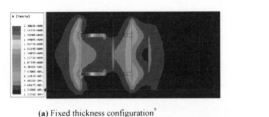
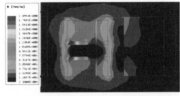

(a) Fixed thickness configuration[9]   (b) Variable thickness configuration

Fig. 10. Magnetic field strength topology of non-dominated solution S_1

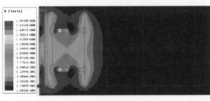
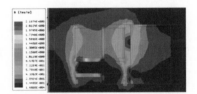

(a) Fixed thickness configuration[9]   (b) Variable thickness configuration

Fig. 11. Magnetic field strength topology of non-dominated solution S_2

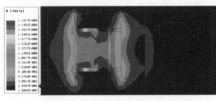

(a) Fixed thickness configuration[9]   (b) Variable thickness configuration

Fig. 12. Magnetic field strength topology of non-dominated solution S_3

The sensitivity analysis has identified the key design parameters to be anode power and mass flow rate. The MDO results presented in Table 2 show some trends to maximize U_a and $\dot{m}_a$ for higher performance. Optimum designs have been found to be commonly characterized by IMR lying between 9.0 and 10.0 mm and OMR between 23.0 and 27.0 mm for both configurations.

Table 8 compares the design parameters and performance of the selected points of both configurations from the present study and those from experimental studies reported in available literature. These experimental data are used as the baseline models for simulation validation. The optimal thrust for both design configurations has been determined to be approximately 103mN and 104mN, respectively, which are considerably larger than all data from the experimental studies. Due to the lack of studies of the variable thickness thruster model, the performance objectives and decision variables were not well defined in the previous studies [17]. The estimated efficiency of the

Table 8. Comparison of performance and design parameters

	Ma et al. [18]	Young et al. [26]	Courtney et al. [3]	Keller et al. [15]	Kornfeld et al. [17]	Keller et al. [14]	Hu et al. [13]	S_1 [7] (Fixed)	S_2 (Variable)
Performance parameters									
Efficiency η_t (%)	35.4	21.7	44.5	40.7	45	–	30	36.5	36.6
Anode Potential U_a (V)	500	300	550	1100	1000	700	800	1000	998
Anode Current I_a (A)	4.1	0.37	0.44	–	1.5	–	–	2.94	3.06
Mass flow rate $\dot{m}_a$ (sccm)	30	8.2	8.5	0.48	17.5	0.59	10	50	50
Thrust T (mN)	66	4.9	13.4	0.36	50	0.19	17.7	103	104.9
Specific impulse I_{sp} (s)	2287	1239	1640	860	3000	360	1836	2131	2176
Design parameters									
Number of magnets	3	3	3	3	3	4	3	3	3
IMR (mm)	20	8.75	–	2.2	–	2.2	10.5	9.91	9.30
OMR (mm)	32	23.9	32	15	–	15	–	25.1	24.12
MT1 (mm)	16	–	–	–	–	–	8	4	5.28
MT2 (mm)	72	–	–	–	–	–	16	10	4.18
MT3 (mm)	4	–	–	–	–	–	56	4	7.49
GT (mm)	–	–	–	2.5	–	2.5	–	1	1.35
Chamber length (mm)	96	39.7	40	–	–	–	–	50	50
Magnet material	2Sm17Co	SmCo	3212SmCo	SmCo	SmCo	SmCo	–	2Sm17Co	2Sm17Co
Propellant	Xe	Kr	Xe	Xe	Xe	Xe	Xe	Xe	Xe
Chamber wall material	BN	BN	BN	BN	BN	5BN	–	BN	BN

S_1 design for the fixed thickness configuration and that of the S_2 design from variable thickness configuration are reasonably aligned with the experimental results of Ma et al. [18]. The anode potential U_a, correlates to the experimental setups of Keller et al. [15] and Kornfeld et al. [17]. The anode current I_a, of the selected solution is between 1.5A [17] and 4.1A [18]. The design variables are comparable to the experimental study of Young et al. [26] while the objective performance comparable to the exper-imental study of Ma et al. [18] and Kornfeld et al. [17], consequently leading to reduction in thruster weight. The comparative size of the thruster magnets can be optimized to produce high performance for the CFT by applying an MDO algorithm.

4   Conclusion

The influence of the magnet thickness on CFT performance has been investigated by performing multi-objective design optimization for the CFT configuration with variable thickness of magnets in comparison with that with fixed magnet thickness. They have yielded insights into the key design factors and underlying physics that are responsible for effective downscaling of CFT models while maintaining high performance. Com-parison of the results has identified the third magnet to be one of the key elements that crucially determine the performance of the thruster. The maximum possible thickness of the third magnet has been found to produce higher performance, as compared to the performance with the maximum thickness for the second magnet. This is directly related to the location of the largest cusped region for plasma confinement. The selected points S_1 for the fixed thickness model and O_1 for the variable thickness model (shown in Table 8) have been identified as the candidates for the optimal thruster design to achieve the most balanced, beneficial performance metrics. Performance improvements have been observed in thrust T and specific impulse I_{sp} by up to approximately 20% and 10%, respectively, as compared to the fixed thickness magnet configuration, albeit reduction in the other performance parameters. The balanced optimal solution O_1 of the configuration with variable thickness of magnets has achieved thrust and specific impulse improvements of approximately 4% on average, compared to the fixed thickness case. The influence of the gap spacing between magnets is a subject for further exploration, and physical uncertainties such as ion or electron density, wall erosion, sheath effects are also to be considered in future work in order to enhance fidelity and accuracy of the model and analysis for practical design.

Acknowledgment. The authors are grateful to Angus Muffatti and Thomas Fahey for the original development of the methodology and process chain used in this study. They are also thankful to Tapabrata Ray for the original MDO framework developed in the group.

References

1. ANSYS Inc. (2016) ANSYS Electronics Desktop Suite ver. 17.2 Users' Guide
2. Boyd ID (2011) Simulation of electric propulsion thrusters. Technical report, Michigan Univ Ann Arbor

3. Courtney D, Lozano P, Martinez-Sanchez M (2008) Continued investigation of diverging cusped field thruster. In: 44th AIAA/ASME/SAE/ASEE joint propulsion conference & exhibit, Hartford, CT, p 4631. https://doi.org/10.2514/6.2008-4631

4. Courtney DG (2008) Development and characterization of a diverging cusped field thruster and a lanthanum hexaboride hollow cathode. Ph.D. thesis, Massachusetts Institute of Technology

5. Deb K, Pratap A, Agarwal S, Meyarivan T (2002) A fast and elitist multiobjective genetic algorithm: NSGA-II. IEEE Trans Evol Comput 6(2):182–197

6. Eichmeier JA, Thumm M (2008) Vacuum electronics: components and devices. Springer Science & Business Media

7. Fahey T, Muffatti A, Ogawa H (2017) High fidelity multi-objective design optimization of a downscaled cusped field thruster. Aerospace 4(4):55

8. Gallimore A (2008) The physics of spacecraft hall-effect thrusters. In: APS division of fluid dynamics meeting abstracts

9. Genovese A, Lazurenko A, Koch N, Weis S, Schirra M, van Reijen B, Haderspeck J, Holtmann P (2011) Endurance testing of hempt-based ion propulsion modules for smallgeo. In: 32nd international electric propulsion conference, Wiesbaden

10. Goebel DM, Katz I (2008) Fundamentals of electric propulsion: ion and Hall thrusters, vol 1. Wiley

11. Hofer R, Gallimore A (2004) Efficiency analysis of a high-specific impulse hall thruster. In: 40th AIAA/ASME/SAE/ASEE joint propulsion conference and exhibit, Fort Lauderdale, FL, p 3602. https://doi.org/10.2514/6.2004-3602

12. Howard J (2002) Introduction to plasma physics c17 lecture notes. http://people.physics.anu. edu.au/~jnh112/AIIM/c17/chap04.pdf. Accessed 15 Apr 2018

13. Hu P, Liu H, Gao Y, Mao W, Yu D (2016) An experimental study of the effect of magnet length on the performance of a multi-cusped field thruster. J Phys D: Appl Phys 49 (28):285201. http://stacks.iop.org/0022-3727/49/i=28/a=285201

14. Keller A, Kohler P, Feili D, Berger M, Braxmaier C, Weise D, Johann U (2014) Feasibility of a down-scaled HEMP Thruster. Verlag Dr. Hut

15. Keller A, Khler P, Hey FG, Berger M, Braxmaier C, Feili D, Weise D, Johann U (2015) Parametric study of hemp-thruster downscaling to μn thrust levels. IEEE Trans Plasma Sci 43(1):45–53. https://doi.org/10.1109/TPS.2014.2321095

16. Koch N, Schirra M, Weis S, Lazurenko A, van Reijen B, Haderspeck J, Genovese A, Holtmann P, Schneider R, Matyash K et al (2011) The hempt concept-a survey on theoretical considerations and experimental evidences. In: 32nd international electric propulsion conference, Wiesbaden

17. Kornfeld G, Koch N, Harmann HP (2007) Physics and evolution of hemp-thrusters. In: Proceedings of the 30th international electric propulsion conference, Florence, pp 17–20

18. Ma C, Liu H, Hu Y, Yu D, Chen P, Sun G, Zhao Y (2015) Experimental study on a variable magnet length cusped field thruster. Vacuum 115:101–107

19. Matlock TS (2012) An exploration of prominent cusped-field thruster phenomena: the hollow conical plume and anode current bifurcation. Ph.D. thesis, Massachusetts Institute of Technology

20. Queipo NV, Haftka RT, Shyy W, Goel T, Vaidyanathan R, Tucker PK (2005) Surrogate-based analysis and optimization. Progress Aerosp Sci 41(1):1–28

21. Ray T, Isaacs A, Smith W (2008) Multi-objective optimization using surrogate assisted evolutionary algorithm. In: Introduction (GP Rangaiah), multi-objective optimization: techniques and applications in chemical engineering, pp 131–151

22. Saltelli A, Ratto M, Andres T, Campolongo F, Cariboni J, Gatelli D, Saisana M, Tarantola S (2008) Global sensitivity analysis: the primer. Wiley

23. Schneider R, Matyash K, Kalentev O, Taccogna F, Koch N, Schirra M (2009) Particle-in-cell simulations for ion thrusters. Contrib Plasma Phys 49(9):655–661

24. Soboĺ I (1976) Uniformly distributed sequences with additional uniformity properties. USSR Comput Math Math Phys 16:1332–1337

25. Van Noord J (2007) Lifetime assessment of the next ion thruster. In: 43rd AIAA/ASME/SAE/ASEE joint propulsion conference & exhibit, Cincinnati, OH, p 5274. https://doi.org/10.2514/6.2007-5274

26. Young CV, Smith AW, Cappelli MA (2009) Preliminary characterization of a diverging cusped field (DCF) thruster. In: Proceeding of 31st international electric propulsion conference, Ann Arbor, MI

Performance Evaluation of Magnetic Nozzle by Using Thermal Plasma

Tatsumasa Hagiwara[1]([✉]), Yoshihiro Kajimura[2]([✉]), Yuya Oshio[3]([✉]),
Ikkoh Funaki[4]([✉]), and Hiroshi Yamakawa[5]([✉])

[1] Kyoto University Research Institute of Sustainable Humanosphere,
Uji, Kyoto 6110011, Japan
tatsumasa_hagiwara@rish.kyoto-u.ac.jp
[2] National Institute of Technology, Akashi College,
Akashi, Hyogo 6478501, Japan
kajimura@akashi.ac.jp
[3] Tokyo University of Agriculture and Technology,
Koganei, Tokyo 1848588, Japan
y-oshio@cc.tuat.ac.jp
[4] Japan Aerospace Exploration Agency, Sagamihara, Kanagawa 2525210, Japan
funaki.ikkoh@jaxa.jp
[5] Japan Aerospace Exploration Agency, Chiyoda, Tokyo 1018008, Japan
yamakawa.hiroshi@jaxa.jp

Abstract. The development of new propulsion system for exploring the moon and other planets in the solar system or deep space is necessary to achieve short mission term and large payload ratio. In recent years, magnetic nozzle is focused on as the candidate system to attain above objectives. Magnetic nozzle is the system which generates the thrust by converting the thermal energy of the plasma injected in the nozzle magnetic field formed by radial magnetic field into the directed kinetic energy. The objective of the present study is to clarify the performance of the new proposed plasma source which consists of LaB6 cathode. Furthermore, it is also the objective to measure the thrust and evaluate the performance of magnetic nozzle with LaB6 plasma source. The performance evaluation of LaB6 plasma source is conducted by measuring the plasma temperature and density by applying the double probe method. The thrust evaluation is conducted in the vacuum chamber. In the present experiment, we succeeded in measuring the thrust of thermal plasma and magnetic nozzle. As a result, the thrust of thermal plasma is 4.7 mN, and the thrust of magnetic nozzle is 16 mN. Therefore, the thrust of magnetic nozzle is 3.3 times larger than that of thermal plasma. Furthermore, thrust-power-ratio is 8.99 mN/kW, specific impulse is 4437.8 s and propulsive efficiency is 19.5%. Every performance of magnetic nozzle is also improved than those of thermal plasma.

Keywords: Magnetic nozzle · Thermal plasma · LaB6 cathode

X. Zhang (Ed.): APISAT 2018, LNEE 459, pp. 1990–1998, 2019.
https://doi.org/10.1007/978-981-13-3305-7_160

1   Introduction

The development of new propulsion system for exploring the moon and other planets in the solar system or deep space is necessary to achieve short mission term and large payload ratio. In recent years, magnetic nozzle is focused on as the candidate system to attain above objectives. The schematic illustration of magnetic nozzle is shown in Fig. 1.

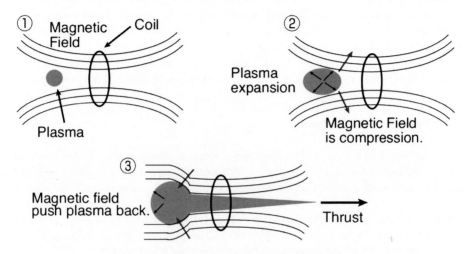

Fig. 1. The schematic illustration of magnetic nozzle

Magnetic nozzle is the propulsion system that generates the thrust by converting the thermal energy of the injected plasma in the nozzle magnetic field into the directed kinetic energy. First of all, as shown in Fig. 1, the thermal plasma which have thermal velocity is injected into the radial magnetic field. The injected plasma is expanded by the thermal energy, and then magnetic field is compressed. Magnetic field push plasma back, so that the thrust is generated. The feature of magnetic nozzle is not only the reduction of energy loss but also to prevent the damage of the emission parts due to the interaction between the plasma and the magnetic wall. Therefore, magnetic nozzle could achieve the high specific impulse and specific power simultaneously comparing with other electric propulsion systems. Magnetic nozzle system is developed by many agencies and universities in the world that VASIMR [2] in NASA, HITOP [1] in Tohoku university and LFR [5] in Kyusyu university.

However, it needs strong magnetic field to obtain the sufficiently thrust by only magnetic nozzle, so that this system is required high electric power. Accordingly, the electric power scale is too large to use the spacecraft which could escape the solar system within 10 years.

In the past study, we conducted the thrust measurement of magnetic nozzle by using MPD(Magneto Plasma Dynamic) Arcjet [3, 4] as the plasma injection source. The plasma is produced by arc discharge in MPD arcjet. Arc discharge is required high electric power, MW order, it is impossible to generate such high electric power in the space.

In the present research, we focused on LaB6 cathode which is widely utilized for thermionic emission source. LaB6 cathode is easy to emit the thermal electrons with low energy due to low work function of LaB6. We considered that it is possible to realize the low electric power plasma source by using the LaB6 cathode as a thermionic emission source. Consequently, we proposed and developed the new plasma source with LaB6 cathode.

The objectives in the present research are not only to measure the surface temperature of LaB6 cathode and temperature and density distribution of produced plasma to evaluate the performance of LaB6 plasma source but also to clarify how plasma temperature and plasma density change by the effect of magnetic nozzle. Moreover, it is also the objectives to evaluate the performance of magnetic nozzle by using LaB6 plasma source and to compare the performance difference between MPD arcjet and LaB6 plasma source by carrying out the measurement of the thrust of magnetic nozzle.

2 Experimental Method

The schematic illustration of the experimental setup is shown in Fig. 2. As shown in Fig. 2, the experimental devices consist of a vacuum chamber, a solenoid coil, the LaB6 plasma source, power and measurement devices. A vacuum chamber was to simulate the space environment, a solenoid coil was to generate magnetic field for magnetic nozzle, LaB6 plasma source was to emit the thermal plasma at the center of the solenoid coil. The schematic illustration of LaB6 plasma source is shown in Fig. 3 and the photograph of the experimental devices is shown in Fig. 4.

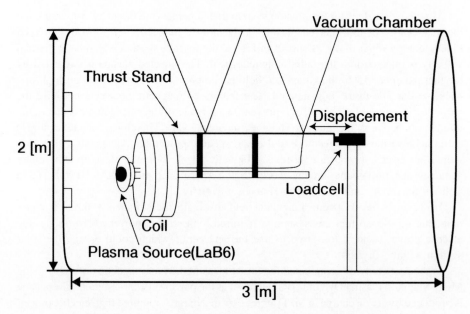

Fig. 2. The schematic illustration of the experimental setup

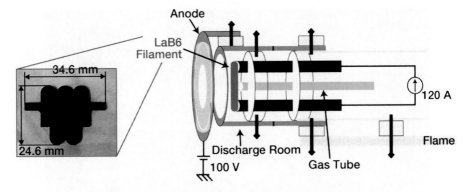

Fig. 3. The schematic illustration of LaB6 plasma source

The principle of plasma production by LaB6 plasma source is, first of all, LaB6 cathode heats by Joule heat, and then thermionic emission is occurred. Secondly, the thermo electron excited by thermionic emission effect is emitted from the LaB6 cathode. The plasma is produced so that the electrons collide with the neutral gas. Produced plasma is emitted by being attracted to the anode applied the positive voltage.

The measurement of surface temperature of LaB6 cathode is utilized for a two-color pyrometer. Tow-color pyrometer measure as opposed to radiation pyrometers in two spectral ranges simultaneously and determine the temperature by calculating the radiation ratio from the object. Wave-length in this experiment are 950 μm and 990 μm. The measurement of plasma temperature and plasma density is utilized for double probe method. Double probe method is the method which measure the plasma parameter with two probes which have electrode. Tow probes which consist of Langmuir probe are placed closely and their shape and size are same. This method could measure the plasma temperature and plasma density by the current-voltage characteristic which is obtained by applying the voltage to the probes.

In order to measure the thrust of magnetic nozzle, the coil and LaB6 plasma source operated simultaneously. In the present study, the coil and LaB6 plasma source have a quasi-steady operating period during 5 s. the displacement caused by the thrust is detected by the loadcell. A loadcell is a transducer that measure force, and outputs this force as an electrical signal. A loadcell uses a strain gauge to detect the force. The strain gauge is deformed by the force, and the deformation is measured as change in electrical signal. A loadcell is PCB Piezotronics Inc., and model No. is 208C01, the relation between output voltage and force is 0.112 V/N. Parameter of experimental devices is shown in Table 1.

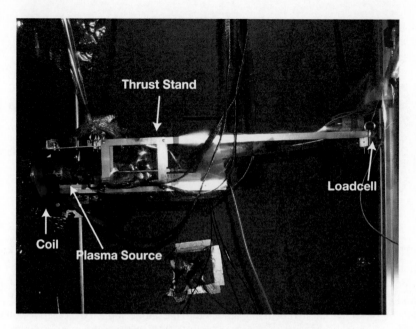

Fig. 4. The photograph of experimental devices

Table 1. Parameter of experimental devices

LaB6 plasma source	
Gas flow rate [sccm]	8, 10, 12
Type of gas	Argon
LaB6 current [A]	120
Coil	
Radius [m]	0.06
Turn	20
Current [A]	100
Magnetic flux density at the center [T]	0.02

3 Experimental Results

The surface temperature distribution of LaB6 cathode is shown in Fig. 5. As shown in Fig. 5, the surface temperature of LaB6 cathode is heated more than 1500 K. Comparing the Fig. 6, which shows the characteristic of surface temperature and thermionic current of LaB6 [6], the thermo electron is sufficiently emitted for producing the plasma.

Fig. 5. The surface temperature distribution of LaB6 cathode

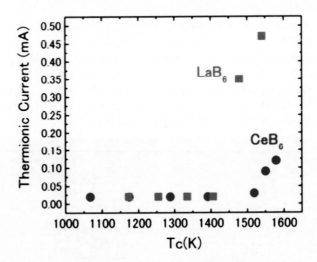

Fig. 6. Characteristic of surface temperature and thermionic current of LaB6 [6]

The result of produced plasma temperature and density distribution is shown in Fig. 7. As shown in Fig. 7, at the 7.5 cm from the center of the coil, the emitted plasma temperature is 7.16 eV in the case that thermal plasma operation, and that is 5.96 eV in the case that magnetic nozzle operation. Also, at the 10 cm from the center of the coil, the emitted plasma temperature is 7.96 eV in the case that thermal plasma operation, and that is 5.70 eV in the case that magnetic nozzle operation. Therefore, the emitted plasma temperature in the case that magnetic nozzle operation is lower than that of thermal plasma operation. This phenomenon shows that approximately 6 eV plasma is only trapped by the magnetic field and convert the thrust. Higher plasma than 6 eV is

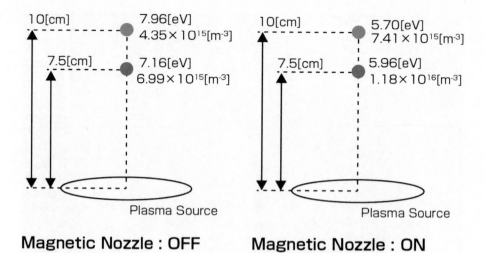

Magnetic Nozzle : OFF   Magnetic Nozzle : ON
                                      (Coil + Plasma)

Fig. 7. The result of produced plasma temperature and density distribution

not trapped by magnetic field. The photograph of thermal plasma operation is shown in Fig. 8, and the photograph of magnetic nozzle operation is shown in Fig. 9. As shown in Fig. 8, the thermal plasma is emitted isotropically from the plasma source. Comparing Fig. 8 with Fig. 9, the emitted plasma of magnetic nozzle mode which operate the coil and the plasma source simultaneously is concentrated at the center of the coil due to the effect of magnetic nozzle formed by magnetic field of the coil. Also, the emission intensity corresponds to the plasma density. For the above reason, from the view of the emission intensity, we could confirm that the emitted plasma is collected at the center of the coil due to the effect of magnetic nozzle.

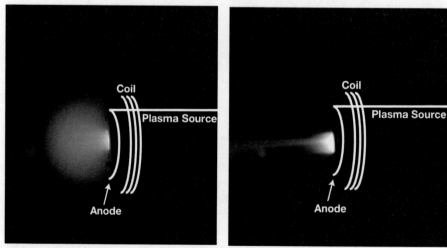

Fig. 8. The photograph of thermal plasma operation

Fig. 9. The photograph magnetic nozzle operation

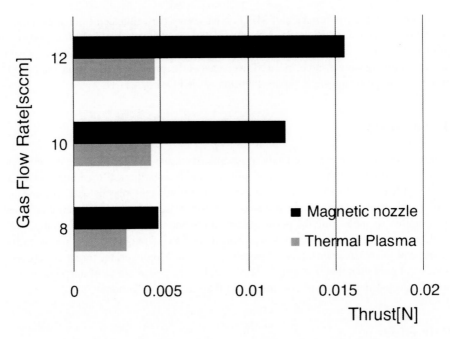

Fig. 10. The thrust result of thermal plasma operation and magnetic nozzle operation

The thrust result of thermal plasma operation and magnetic nozzle operation is shown in Fig. 10. As shown in Fig. 10, in the case that gas flow rate is 8 sccm, the thrust of magnetic nozzle is 1.6 times larger than that of thermal plasma. Moreover, in the case that gas flow rate is 10 sccm, that of magnetic nozzle is 2.7 times larger than that of thermal plasma, and in the case that gas flow rate is 12 sccm, that of magnetic nozzle is 3.3 times larger than that of thermal plasma. We could confirm the increase of thrust quantitatively. As the gas flow rate increases, the thrust of thermal plasma is getting larger. However, comparing 10 and 12 sccm, the thrust of thermal plasma is not increase due to the saturation of the produced plasma. The produced plasma doesn't change because the number of the emitted electron from LaB6 cathode is same, so if the gas flow rate is increased, the number of the produced plasma doesn't change. The performance comparison of magnetic nozzle with MPD Arcjet and LaB6 plasma source is shown in Table 2.

Table 2. Performance comparison of magnetic nozzle mode with MPD Arcjet and LaB6 plasma source

	MPD arcjet	LaB6 cathode
Thrust-power-ratio [mN/kW]	3.03	8.99
Specific impulse [s]	859.8	4437.8
Propulsive efficiency [%]	1.28	19.5

Comparing the performance of MPD arcjet, every performance of magnetic nozzle is improved by using the LaB6 cathode. MPD arcjet needs high electric power to produce the plasma, so the performance is very low. However, the proposed plasma source, LaB6 plasma source, needs lower energy to produce the plasma than MPD arcjet, so the performance is higher than that of MPD arcjet.

4 Summary

In this paper, we have evaluated experimentally the thrust and performance of magnetic nozzle. We proposed the new plasma source, which uses the LaB6 cathode which is widely utilized for thermionic emission source to produce the plasma with low energy. We also have evaluated the performance of LaB6 plasma source, and then confirmed the usability by measuring the plasma temperature and density. According to the result of density measurement, we could confirm the effect of magnetic nozzle due to the increase of the plasma density at the center of the coil. Moreover, the thrust of magnetic nozzle is larger than that of thermal plasma, and every performance of magnetic nozzle with LaB6 plasma source is also improved than that of MPD arcjet.

References

1. Ando A, Inutake M, Hattori K, Shibata M, Kasashima Y (2007) ICRF heating and plasma acceleration with an open magnetic field for the advanced space thruster. Trans Fusion Sci Technol 51(2T):72–74
2. Chang Diaz FR (2000) The VASIMR rocket. Sci Am 283(5):90
3. Hagiwara T, Kajimura Y, Oshio Y, Funaki I (2015) Thrust measurement of magneto plasma sail with magnetic nozzle by using thermal plasma injection. In: Joint conference of 30th international symposium on space technology and science, IEPC-2015-461/ISTS-2015-b-461
4. Kajimura Y, Hagiwara T, Oshio Y, Funaki I, Yamakawa H (2015) Thrust performance of magneto plasma sail with a magnetic nozzle. In: Joint conference of 30th international symposium on space technology and science, IEPC-2015-329/ISTS-2015-b-329
5. Konstantin V, Nakashima H, Ichikawa F, Zakharov Y (2004) Optimization of thrust efficiency in laser fusion rocket by using three-dimensional hybrid particle-in-cell code. Vacuum 73(3–4):427–432
6. Morita K, Zen H, Masuda K, Torgasin K, Suphakul S, Katsurayama T, Yamashita H, Kii T, Nagasaki K, Ohgaki H (2016) Dependence of LaB6 and CeB6 photocathodes on temperature and incident laser wavelength. In: Proceedings of the 13th annual meeting of particle accelerator society of Japan, PASF2016 WEOM03, pp 204–207

Printed by Books on Demand, Germany